GW01158477

Carbohydrate-Active Enzymes

Carbohydrate-Active Enzymes

Structure, Activity and Reaction Products

Special Issue Editor

Stefano Benini

MDPI • Basel • Beijing • Wuhan • Barcelona • Belgrade • Manchester • Tokyo • Cluj • Tianjin

Special Issue Editor
Stefano Benini
Free University of Bozen-Bolzano
Italy

Editorial Office
MDPI
St. Alban-Anlage 66
4052 Basel, Switzerland

This is a reprint of articles from the Special Issue published online in the open access journal *International Journal of Molecular Sciences* (ISSN 1422-0067) (available at: https://www.mdpi.com/journal/ijms/special_issues/carbohydrate-active_enzymes).

For citation purposes, cite each article independently as indicated on the article page online and as indicated below:

LastName, A.A.; LastName, B.B.; LastName, C.C. Article Title. *Journal Name* **Year**, *Article Number*, Page Range.

ISBN 978-3-03936-090-1 (Pbk)
ISBN 978-3-03936-091-8 (PDF)

Contents

About the Special Issue Editor

Stefano Benini, Ph.D. (Assistant Professor). Employment: 2009–present, Bioorganic Chemistry and Bio-Crystallography Laboratory, Faculty of Science and Technology, Libera Università di Bolzano; 2007–2009, AstraZeneca (UK); 2002–2007, York Structural Biology Laboratory (YSBL); 2000–2002, International Centre for Genetic Engineering and Biotechnology (ICGEB)/Elettra Synchrotron Trieste, Italy; 1996–2000, European Molecular Biology Laboratory (EMBL), Hamburg, Germany; 1992–1996, Institute of Agricultural Chemistry, University of Bologna. Education: 2001, Ph.D. in Chemistry, EMBL Hamburg (DE)/University of York (UK); 1991, Master in Agricultural Science, University of Bologna (Italy).

International Journal of
Molecular Sciences

Editorial

Carbohydrate-Active Enzymes: Structure, Activity, and Reaction Products

Stefano Benini

Bioorganic Chemistry and Bio-Crystallography laboratory (B2Cl), Libera Università di Bolzano, 39100 Bolzano, Italy; Stefano.Benini@unibz.it

Received: 7 April 2020; Accepted: 11 April 2020; Published: 15 April 2020

Carbohydrate-active enzymes are responsible for both the biosynthesis and breakdown of carbohydrates and glycoconjugates. They are involved in many metabolic pathways, in the biosynthesis and degradation of various biomolecules, such as bacterial exopolysaccharides, starch, cellulose, and lignin, and in the glycosylation of proteins and lipids. Their interesting reactions have attracted the attention of researchers belonging to different scientific fields ranging from basic research to biotechnology. Interest in carbohydrate-active enzymes is due not only to their ability to build and degrade biopolymers, which is highly relevant in biotechnology, but also because they are involved in bacterial biofilm formation, and in the glycosylation of proteins and lipids, which has important health implications.

This Special Issue features a collection of research papers and reviews representing an up-to-date state of the art in this growing field of research to broaden our understanding about carbohydrate-active enzymes, their mutants, and their reaction products at the molecular level.

Tindara Venuto et al. wrote a paper about alpha2,8-sialyltransferases from ray-finned fish and showed that, in tetrapods, duplicated *st8sia* genes like *st8sia7*, *st8sia8*, and *st8sia9* have disappeared while orthologues are maintained in teleosts [1]. They reconstructed the evolutionary history of *st8sia* genes in fish genomes and by bioinformatics analysis showed changes in the conserved polysialyltransferase domain of the fish sequences possibly accounting for variable enzymatic activities [1].

PhNah20A, a β-N-acetylhexosaminidase from the marine bacterium *Paraglaciecola hydrolytica* S66T, was identified and characterized. By phylogenetic analysis, the authors discovered that PhNah20A is located outside clusters of other studied β-N-acetylhexosaminidases, in a unique position between bacterial and eukaryotic enzymes. PhNah20A is the first characterized member of a distinct subgroup within GH20 β-N-acetylhexosaminidases. The recombinant PhNah20A showed optimum hydrolytic activity on GlcNAc β-1,4 and β-1,3 linkages in chitobiose (GlcNAc)2 and GlcNAc-1,3-β-Gal-1,4-β-Glc (LNT2), the core structure of a human milk oligosaccharide, at pH 6.0 and 50 °C. Interestingly PhNah20A catalyzed the formation of LNT2, the non-reducing trisaccharide β-Gal-1,4-β-Glc-1,1-β-GlcNAc, and in low amounts the β-1,2- or β-1,3-linked trisaccharide β-Gal-1,4(β-GlcNAc)-1,x-Glc by a transglycosylation of lactose using 2-methyl-(1,2-dideoxy-α-d-glucopyrano)-oxazoline (NAG-oxazoline) as the donor [2].

The production of biofuels from lignocellulosic biomasses is hampered by the saccharification step of the biomasses requiring the concerted action of lignocellulose cellulases and hemicellulases.

Zhang et al. characterized the secretome of *Trichoderma harzianum* EM0925 under induction of lignocellulose cellulases and hemicellulases [3]. Compared with the commercial enzyme preparations, the *T. harzianum* EM0925 enzyme cocktail presented significantly higher lignocellulolytic enzyme activities and hydrolysis efficiency against lignocellulosic biomasses. The enzyme mixture used on ultrafine grinding and alkali pretreated corn stover yielded 100% glucose and xylose. The authors suggest that natural cellulases mixed with the hemicellulases enzyme cocktail from *T. harzianum* EM0925 used for complete conversion of lignocellulose biomasses have a great potential for industrial applications [3].

BaAG2 from the early-diverged yeast *Blastobotrys (Arxula) adeninivorans* is an α-glucosidase of the GH13 family. The enzyme was overexpressed in *Escherichia coli*, purified, and characterized. BaAG2's preferred substrates were maltose, other maltose-like molecules and malto-oligosaccharides with degrees of polymerization (DP) up to 4-7, and sucrose, whereas isomaltose and isomaltose-like substrates were not hydrolyzed, confirming that BaAG2 is a maltase. At high maltose concentrations, BaAG2 exhibited transglycosylating ability, producing di- and trisaccharides. BaAG2 was inhibited by acarbose, Tris, and also by isomaltose-like sugars and glucose [4].

Levansucrases (LSCs, EC 2.4.1.10) are enzymes studied for their potential use in biotechnological application for the production of fructooligosaccharides and the transfructosylation of a wide range of acceptors. While LSCs from Gram-positive bacteria can produce long-chain levan (e.g., *Bacillus megaterium* SacB), LSCs from Gram-negative bacteria such as *Erwinia tasmaniensis* (EtLSC) are interesting biocatalysts for the production of fructooligosaccharides (FOSs). The crystal structure of EtLSC in complex with levanbiose (LBS) was obtained by soaking the preformed crystals in a solution containing a high concentration of sucrose followed by crystal storage in liquid N_2. LBS is a di-fructose FOS intermediate formed during the synthesis of longer-chain FOSs. The binding pocket discovered in this crystal structure represents a starting point for planning mutagenesis studies in order to understand its biological relevance and role in FOS chain elongation [5].

Two major lipoglycans, lipomannan (LM) and lipoarabinomannan (LAM), are produced by *Mycobacteria*. Their structural characterization is very difficult due to their heterogeneity in terms of internal and terminal covalent modifications and branching patterns. To allow identification of the reducing end of these molecules, enzymes such as endo-α-(1→6)-D-mannanase from *Bacillus circulans* proved useful in cleaving the mannan backbone of LM and LAM. Angala et al. report the production and purification of the glycosyl hydrolase domain of the endo-α-(1→6)-D-mannanase gene from *B. circulans* TN-31 and studied its substrate specificity [6].

A GH3 β-glucosidase from the thermophilic fungus *Chaetomium thermophilum* was structurally characterized by Mohsin et al [7]. The structure provides useful information for determining a better use of enzymes in biotechnological applications for cellulose degradation in the production of biofuels, as thermophilic fungi are a promising source of enzymes with improved stability. The enzyme displays moderate glycosylation compared to other GH3 family β-glucosidases with similar structure, and a comparison of several thermostability parameters suggests that glycosylation and electrostatic interactions between charged residues may contribute to the enzyme's stability at high temperatures. The structure provides insights into the GH3 enzyme family for further improvements of the β-glucosidases used in biotechnological cellulose degradation [7].

Two homologous uridine-5'-diphosphate (UDP)-glucose pyrophosphorylase enzymes produced by *Rhodococcus opacus*, 1CP—RoGalU1 and RoGalU2, their genes were cloned and the corresponding proteins characterized by Kumpf et al [8]. UDP-glucose is a versatile building block in both prokaryotes and eukaryotes. UDP-glucose is synthesized by the enzyme UDP-glucose pyrophosphorylase (GalU). The two enzymes are encoded by most *Rhodococcus* strains, known to synthesize natural products (i.e., trehalose-containing biosurfactants). Like other GalUs, RoGalU2 seems to be rather specific for the substrates UTP and glucose 1-phosphate, while dTTP and galactose 1-phoshate are substrates with merely 2% of residual activity. In comparison to other bacterial GalU, RoGalU2 activity was greater even at high temperatures [8].

The peptidoglycan of bacterial cell walls is hydrolyzed by muramidases/lysozymes, a lysozyme glycoside hydrolase (GH) family 22 C-type, from the upper gastrointestinal tract of the folivorous bird *Opisthocomus hoazin* in the native form and various mutants were expressed in *Aspergillus oryzae*. All mutants were enzymatically active; four of them with improved thermostability at pH 4.7, compared to the wild type. The X-ray structure of the enzyme was determined in the apo form and in complex, with chitin oligomers providing valuable information on substrate binding. Bioinformatic analysis of avian GH22 amino acid sequences showed that they split into three distinct subgroups, and the enzyme from *O. hoazin* is found in the "other birds" group. Taylor et al. propose that this represents a new cluster of avian upper-gut enzymes [9].

The use of deep eutectic solvents (DES) for the synthesis of alkyl glycosides catalyzed by the thermostable α-amylase from *Thermotoga maritima* Amy A was investigated by Miranda-Molina et al [10]. While DES containing alcohols, sugars, and amides as hydrogen-bond donors (HBD) as cosolvents performed best, pure DES almost completely deactivated the enzyme. Circular dichroism measurements of Amy A showed that large conformational changes were observed above 60 °C, changes not observed in aqueous medium. The authors claim that this is the first report on the effect of DES and temperature on an enzyme, to establish the temperature limit for a thermostable enzyme in DES [10].

The 15-O-glycosylation of ganoderic acid A (GAA) into GAA-15-O-β-glucoside is carried out by the enzyme BtGT_16345 from *Bacillus thuringiensis* GA A07. The enzyme was identified by whole genome sequencing of strain GA A07. A phylogenomic analysis revealed the species of the GA A07 strain to be *B. thuringiensis*. The enzyme is also regioselective toward triterpenoid substrates. BtGT_16345 showed glycosylation activity toward seven flavonoids (apigenin, quercetin, naringenin, resveratrol, genistein, daidzein, and 8-hydroxydaidzein) and two triterpenoids (GAA and antcin K) [11].

Roth et al. describe novel fungal amylases from *Thamnidium elegans* and *Cordyceps farinosa* with potential industrial applications, as they feature activity and high stability in a broad pH spectrum extending to pH 8 [12]. These enzymes have the typical GH13 α-amylase fold but feature shortened loops flanking the substrate-binding cleft, creating a large crevice. The inhibitor acarbose, in form of a transglycosylation product, was bound in the active site in the amylases from *T. elegans* and *C. farinosa*. Moreover, a potential novel binding site in the C-terminal domain of the *Cordyceps* enzyme was identified, which might be part of a starch interaction site [12].

A directed evolution approach was used by Yoav et al. to enhance the thermostability of *Clostridium thermocellum* β-Glucosidase [13]. In the process of cellulose utilization, β-glucosidases are key enzymes that convert cellobiose, produced by cellulose hydrolysis, to glucose. A thermostable mutant with kcat and *Km*, similar to those of the wild-type enzyme, was produced. The addition of the thermostable double mutant (A17S and K268N) to *C. thermocellum* secretome in order to carry out the hydrolysis of microcrystalline cellulose at 70 °C drastically increased the soluble glucose yield compared to the activity of the secretome supplemented with the wild-type enzyme [13].

A series of crystal structures of the wild type of the cold-adapted β-d-galactosidase from *Arthrobacter* sp. 32cB, as well as its mutant E441Q, in complex with various ligands were obtained in order to describe the reaction mechanism. Comparative analysis with mesophilic homologs revealed striking differences in the enzymatic reaction. Access of the substrate to the active site is facilitated by the fact that in ArthβDG a 10-aa loop is moved outward, while in mesophilic GH2 βDGs the same loop hampers access to the active site. The authors suggest that the exposure of the active site entrance improves the turnover rate of the cold-adapted enzyme [14].

Franceus et al. report the characterization of a sucrose 6F-phosphate phosphorylases member (SPP) of the family GH13 subfamily 18 from *Ilumatobacter coccineus* and its comparison with the enzyme from *Thermoanaerobacterium thermosaccharolyticum* [15]. Crystal structures of both SPPs were determined to provide insight into their similarities and differences. These enzymes are interesting because of their reversible phosphorolysis reactions in the synthesis of various glycosidic compounds. *I. coccineus* has stricter specificity than the promiscuous SPP from *T. thermosaccharolyticum*. Considerable differences between the two enzymes were found in the residues responsible for binding the fructose 6-phosphate group in subsite +1. Mutants that provided a higher degree of substrate promiscuity in the SPP from *I. coccineus* were probed, thus paving the way to rational enzyme engineering in biotechnology [15].

The degradation of pectins carried out by pectate lyases are of great interest in food and textile industrial production. An alkaline pectate lyase gene (*pppel9a*) from *Paenibacillus polymyxa* KF-1 belonging to the polysaccharide lyase family 9 (PL9) was cloned and the corresponding enzyme was expressed and characterized by Yuan et al [16]. The enzyme substrates are homogalacturonan-type (HG) pectins vis-à-vis rhamnogalacturonan-I (RG-I)-type pectins. The lyase activity was found to be optimal at pH 10.0 and 40 °C. The molecular mass of citrus pectin (~230.2 kDa) was reduced by PpPel9a action to ~24 kDa. PpPel9a was shown to be an endo-pectate lyase, acting primarily on the HG domain of citrus pectin. The enzyme was proposed to have potential in the preparation of pharmacologically active pectin products [16].

A xylanase belonging to family GH10 from the halo-thermophilic bacterium *Roseithermus sacchariphilus* strain RA (XynRA2) was characterized by Teo et al. [17]. XynRA2 is composed of a family 4_9 carbohydrate-binding module (CBM4_9), a family 10 glycoside hydrolase catalytic domain (GH10), and a C-terminal domain (CTD) for a type IX secretion system (T9SS). This study reports the identification and biochemical characterization of XynRA2 and its CBM-truncated variant (XynRA2ΔCBM). The mutant XynRA2ΔCBM showed a lower activity compared to the wild type, underlying the importance of XynRA2 carbohydrate-binding module in determining enzyme performance [17].

The review by Roman et al. reports an up-to-date analysis of the structural and functional diversity of β-xylosidases and discusses their inhibition by monosaccharides. β-xylosidases are crucial enzymes in biotechnology because they hydrolyze the glycosidic bonds of cellulosic biomasses with a high potential for degradation into reducing sugars, which can be used in the subsequent fermentation into bioethanol [18].

The review by Mestrom discusses the use of Leloir glycosyltransferases, as they offer excellent control over the reactivity and selectivity of glycosylation reactions. One-pot multi-enzyme glycosylation cascades have been achieved by the development of nucleotide-recycling cascades for the production and reuse of nucleotide sugar donors, and complex stereochemistry glycans and glycoconjugates can be constructed. Therefore, Leloir glycosyltransferases, on account of their reactivity and selectivity of glycosylation reactions with unprotected carbohydrates, are good candidates for biotechnological industrial applications in multi-enzyme glycosylation cascades [19].

In a review about the ketal-pyruvylation of monosaccharides, Hager et al. illustrate different classes of pyruvylated glycoconjugates and their associated functions, as well as the pyruvyltransferases responsible of the modification. Ketal-pyruvylation is present in diverse classes of glycoconjugates, produced in bacteria, algae, and yeast, but is absent in humans. The authors collected up-to-date information on the prevalent ketal-pyruvylation of monosaccharides, including specificity and sequence space of pyruvyltransferases, and provided insights into pyruvate analytics [20].

Conflicts of Interest: The authors declare no conflict of interest.

References

1. Venuto, M.T.; Decloquement, M.; Martorell Ribera, J.; Noel, M.; Rebl, A.; Cogez, V.; Petit, D.; Galuska, S.P.; Harduin-Lepers, A. Vertebrate Alpha2,8-Sialyltransferases (ST8Sia): A Teleost Perspective. *Int. J. Mol. Sci.* **2020**, *21*, 513. [CrossRef] [PubMed]
2. Visnapuu, T.; Teze, D.; Kjeldsen, C.; Lie, A.; Duus, J.Ø.; André-Miral, C.; Pedersen, L.H.; Stougaard, P.; Svensson, B. Identification and Characterization of a β-*N*-Acetylhexosaminidase with a Biosynthetic Activity from the Marine Bacterium *Paraglaciecola hydrolytica* S66^{T}. *Int. J. Mol. Sci.* **2020**, *21*, 417. [CrossRef] [PubMed]
3. Zhang, Y.; Yang, J.; Luo, L.; Wang, E.; Wang, R.; Liu, L.; Liu, J.; Yuan, H. Low-Cost Cellulase-Hemicellulase Mixture Secreted by *Trichoderma harzianum* EM0925 with Complete Saccharification Efficacy of Lignocellulose. *Int. J. Mol. Sci.* **2020**, *21*, 371. [CrossRef] [PubMed]
4. Visnapuu, T.; Meldre, A.; Põšnograjeva, K.; Viigand, K.; Ernits, K.; Alamäe, T. Characterization of a Maltase from an Early-Diverged Non-Conventional Yeast *Blastobotrys adeninivorans*. *Int. J. Mol. Sci.* **2020**, *21*, 297. [CrossRef] [PubMed]
5. Polsinelli, I.; Caliandro, R.; Demitri, N.; Benini, S. The Structure of Sucrose-Soaked Levansucrase Crystals from *Erwinia tasmaniensis* reveals a Binding Pocket for Levanbiose. *Int. J. Mol. Sci.* **2020**, *21*, 83. [CrossRef] [PubMed]

6. Angala, S.; Li, W.; Palčeková, Z.; Zou, L.; Lowary, T.L.; McNeil, M.R.; Jackson, M. Cloning and Partial Characterization of an Endo-α-(1→6)-D-Mannanase Gene from *Bacillus circulans*. *Int. J. Mol. Sci.* **2019**, *20*, 6244. [CrossRef] [PubMed]
7. Mohsin, I.; Poudel, N.; Li, D.-C.; Papageorgiou, A.C. Crystal Structure of a GH3 β-Glucosidase from the Thermophilic Fungus *Chaetomium thermophilum*. *Int. J. Mol. Sci.* **2019**, *20*, 5962. [CrossRef] [PubMed]
8. Kumpf, A.; Partzsch, A.; Pollender, A.; Bento, I.; Tischler, D. Two Homologous Enzymes of the GalU Family in *Rhodococcus opacus* 1CP—*Ro*GalU1 and *Ro*GalU2. *Int. J. Mol. Sci.* **2019**, *20*, 5809. [CrossRef] [PubMed]
9. Taylor, E.J.; Skjøt, M.; Skov, L.K.; Klausen, M.; De Maria, L.; Gippert, G.P.; Turkenburg, J.P.; Davies, G.J.; Wilson, K.S. The C-Type Lysozyme from the upper Gastrointestinal Tract of *Opisthocomus hoatzin*, the Stinkbird. *Int. J. Mol. Sci.* **2019**, *20*, 5531. [CrossRef] [PubMed]
10. Miranda-Molina, A.; Xolalpa, W.; Strompen, S.; Arreola-Barroso, R.; Olvera, L.; López-Munguía, A.; Castillo, E.; Saab-Rincon, G. Deep Eutectic Solvents as New Reaction Media to Produce Alkyl-Glycosides Using Alpha-Amylase from *Thermotoga maritima*. *Int. J. Mol. Sci.* **2019**, *20*, 5439. [CrossRef] [PubMed]
11. Chang, T.-S.; Wang, T.-Y.; Hsueh, T.-Y.; Lee, Y.-W.; Chuang, H.-M.; Cai, W.-X.; Wu, J.-Y.; Chiang, C.-M.; Wu, Y.-W. A Genome-Centric Approach Reveals a Novel Glycosyltransferase from the GA A07 Strain of *Bacillus thuringiensis* Responsible for Catalyzing 15-*O*-Glycosylation of Ganoderic Acid A. *Int. J. Mol. Sci.* **2019**, *20*, 5192. [CrossRef] [PubMed]
12. Roth, C.; Moroz, O.V.; Turkenburg, J.P.; Blagova, E.; Waterman, J.; Ariza, A.; Ming, L.; Tianqi, S.; Andersen, C.; Davies, G.J.; et al. Structural and Functional Characterization of Three Novel Fungal Amylases with Enhanced Stability and pH Tolerance. *Int. J. Mol. Sci.* **2019**, *20*, 4902. [CrossRef] [PubMed]
13. Yoav, S.; Stern, J.; Salama-Alber, O.; Frolow, F.; Anbar, M.; Karpol, A.; Hadar, Y.; Morag, E.; Bayer, E.A. Directed Evolution of *Clostridium thermocellum* β-Glucosidase A Towards Enhanced Thermostability. *Int. J. Mol. Sci.* **2019**, *20*, 4701. [CrossRef] [PubMed]
14. Rutkiewicz, M.; Bujacz, A.; Wanarska, M.; Wierzbicka-Wos, A.; Cieslinski, H. Active Site Architecture and Reaction Mechanism Determination of Cold Adapted β-D-galactosidase from *Arthrobacter* sp. 32cB. *Int. J. Mol. Sci.* **2019**, *20*, 4301. [CrossRef] [PubMed]
15. Franceus, J.; Capra, N.; Desmet, T.; Thunnissen, A.-M.W. Structural Comparison of a Promiscuous and a Highly Specific Sucrose 6^F-Phosphate Phosphorylase. *Int. J. Mol. Sci.* **2019**, *20*, 3906. [CrossRef] [PubMed]
16. Yuan, Y.; Zhang, X.-Y.; Zhao, Y.; Zhang, H.; Zhou, Y.-F.; Gao, J. A Novel PL9 Pectate Lyase from *Paenibacillus polymyxa* KF-1: Cloning, Expression, and Its Application in Pectin Degradation. *Int. J. Mol. Sci.* **2019**, *20*, 3060. [CrossRef] [PubMed]
17. Teo, S.C.; Liew, K.J.; Shamsir, M.S.; Chong, C.S.; Bruce, N.C.; Chan, K.-G.; Goh, K.M. Characterizing a Halo-Tolerant GH10 Xylanase from *Roseithermus sacchariphilus* Strain RA and Its CBM-Truncated Variant. *Int. J. Mol. Sci.* **2019**, *20*, 2284. [CrossRef] [PubMed]
18. Rohman, A.; Dijkstra, B.W.; Puspaningsih, N.N.T. β-Xylosidases: Structural Diversity, Catalytic Mechanism, and Inhibition by Monosaccharides. *Int. J. Mol. Sci.* **2019**, *20*, 5524. [CrossRef] [PubMed]
19. Mestrom, L.; Przypis, M.; Kowalczykiewicz, D.; Pollender, A.; Kumpf, A.; Marsden, S.R.; Bento, I.; Jarzębski, A.B.; Szymańska, K.; Chruściel, A.; et al. Leloir Glycosyltransferases in Applied Biocatalysis: A Multidisciplinary Approach. *Int. J. Mol. Sci.* **2019**, *20*, 5263. [CrossRef] [PubMed]
20. Hager, F.F.; Sützl, L.; Stefanović, C.; Blaukopf, M.; Schäffer, C. Pyruvate Substitutions on Glycoconjugates. *Int. J. Mol. Sci.* **2019**, *20*, 4929. [CrossRef] [PubMed]

International Journal of
Molecular Sciences

Article

Vertebrate Alpha2,8-Sialyltransferases (ST8Sia): A Teleost Perspective

Marzia Tindara Venuto [1], Mathieu Decloquement [2], Joan Martorell Ribera [3], Maxence Noel [2], Alexander Rebl [3], Virginie Cogez [2], Daniel Petit [4], Sebastian Peter Galuska [1] and Anne Harduin-Lepers [2,*]

1. Institute of Reproductive Biology, Leibniz Institute for Farm Animal Biology (FBN), Wilhelm-Stahl-Allee 2, 18196 Dummerstorf, Germany; venuto@fbn-dummerstorf.de (M.T.V.); galuska.Sebastian@fbn-dummerstorf.de (S.P.G.)
2. Université de Lille, CNRS, UMR 8576-UGSF-Unité de Glycobiologie Structurale et Fonctionnelle, F-59000 Lille, France; mathieu.decloquement.etu@univ-lille.fr (M.D.); maxence.noel@univ-lille.fr (M.N.); virginie.cogez@univ-lille.fr (V.C.)
3. Institute of Genome Biology Leibniz Institute for Farm Animal Biology (FBN), Wilhelm-Stahl-Allee 2, 18196 Dummerstorf, Germany; martorell-ribera@fbn-dummerstorf.de (J.M.R.); rebl@fbn-dummerstorf.de (A.R.)
4. Glycosylation et Différenciation Cellulaire, EA 7500, Laboratoire PEIRENE, Université de Limoges, 123 Avenue Albert Thomas, 87060 Limoges CEDEX, France; daniel.petit@unilim.fr

* Correspondence: anne.harduin-lepers@univ-lille.fr; Tel.: +33-320-33-62-46

Received: 29 November 2019; Accepted: 10 January 2020; Published: 14 January 2020

Abstract: We identified and analyzed α2,8-sialyltransferases sequences among 71 ray-finned fish species to provide the first comprehensive view of the Teleost ST8Sia repertoire. This repertoire expanded over the course of Vertebrate evolution and was primarily shaped by the whole genome events R1 and R2, but not by the Teleost-specific R3. We showed that duplicated *st8sia* genes like *st8sia7*, *st8sia8*, and *st8sia9* have disappeared from Tetrapods, whereas their orthologues were maintained in Teleosts. Furthermore, several fish species specific genome duplications account for the presence of multiple poly-α2,8-sialyltransferases in the Salmonidae (ST8Sia II-r1 and ST8Sia II-r2) and in *Cyprinus carpio* (ST8Sia IV-r1 and ST8Sia IV-r2). Paralogy and synteny analyses provided more relevant and solid information that enabled us to reconstruct the evolutionary history of *st8sia* genes in fish genomes. Our data also indicated that, while the mammalian ST8Sia family is comprised of six subfamilies forming di-, oligo-, or polymers of α2,8-linked sialic acids, the fish ST8Sia family, amounting to a total of 10 genes in fish, appears to be much more diverse and shows a patchy distribution among fish species. A focus on Salmonidae showed that (i) the two copies of *st8sia2* genes have overall contrasted tissue-specific expressions, with noticeable changes when compared with human co-orthologue, and that (ii) *st8sia4* is weakly expressed. Multiple sequence alignments enabled us to detect changes in the conserved polysialyltransferase domain (PSTD) of the fish sequences that could account for variable enzymatic activities. These data provide the bases for further functional studies using recombinant enzymes.

Keywords: molecular phylogeny; α2,8-sialyltransferases; polySia motifs; evolution; ST8Sia; functional genomics

1. Introduction

Glycoproteins and glycolipids can be modified with numerous different glycans during their transit to the cell surface. Here, these glycoconjugates form a dense meshwork, the glycocalyx, influencing several essential processes, such as adhesion and migration mechanisms in addition to cell signaling. Intriguingly, all living cells are surrounded by such a sugar-coat, which demonstrates the importance of glycans for all living organisms [1]. However, glycoconjugates are not only found on the cellular membranes, but also on released extracellular vesicles and soluble glycoconjugates; likewise, various physiological and pathological can be targeted by their released forms. Several different monosaccharides are utilized for the formation of glycans. Nevertheless, a very special position among the building blocks of glycans takes the family of sialic acids [2,3]. These α-keto acids consist of a nine-carbon backbone with a carboxylic acid group at C1 and a ketone group at C2 [4]. Remarkably, more than 50 derivatives are known in nature. Besides N-acetylneuraminic acid (Neu5Ac), N-glycolylneuraminic acid (Neu5Gc) is the most common sialic acid and the hydroxyl groups of both can be additionally substituted, for example, by acetylation. The same applies for a further common sialic acid, which is mainly used in lower vertebrates, deaminated neuraminic acid (KDN, 2-keto-3-deoxy-D-glycero-D-galacto-nononic acid) [5]. All three of these sialic acids are frequently added by α2,3- and α2,6-sialyltransferases (ST3Gal, ST6Gal and ST6GalNAc) to nascent glycans. However, in contrast to other commonly utilized monosaccharides of glycans, an attached sialic acid residue can only be used to add another sialic acid residue, which explains their outermost position on sialylated glycans. The elongation at position C8 of α2,3- or α2,6-linked sialic acid residues is catalyzed by sialyltransferases belonging to the group of α2,8-sialyltransferases (ST8Sia) and long polymers of sialic acids can be enzymatically synthesized in this way [6–8].

All those animal sialyltransferases (α2,3-, α2,6- α2,8-sialyltransferases) belong to the CAZy glycosyltransferase family GT29, which indicates their common modular organization (GT-A-like fold) and their common ancestral origin [8,9]. These protein sequences are characterized by the presence of four consensus motifs called sialylmotifs (L (Large), S (Small), III, and VS (Very Small)) involved in 3D structure maintenance, substrate binding, and catalysis [10,11]. The sialylmotifs are very useful for in silico identification of sialyltransferases-related sequences [12]. On the basis of their sugar acceptor specificity and glycosidic linkage formed, GT29 is subdivided into four families ST3Gal, ST6Gal, ST6GalNAc, and ST8Sia in vertebrates [7,13], each of which is characterized by family motifs likely involved in linkage specificity [14–16]. The biosynthesis of α2,8-sialylated molecules is an ancient pathway achieved by the ST8Sia, a group of enzymes that emerge in the first eukaryotes [8] and expanded very early in animal evolution [14]. Up to now, the ST8Sia enzymes have been studied and characterized in mammalian tissues and primarily in the adult brain. The human and mouse genomes show six ST8Sia subfamilies: ST8Sia I, ST8Sia V, and ST8Sia VI are mono-α2,8-sialyltransferases and constitute a first group of ST8Sia enzymes involved in di-sialylation of glycoconjugates, while ST8Sia III in addition to ST8Sia II and ST8Sia IV form a second group of oligo- and poly-α2,8-sialyltransferases implicated in the polysialylation of glycoproteins [15].

Interestingly, our recent studies pointed to the fact that the *st8sia* gene family appears to be much larger in teleost fish genomes [14,17]. The emergence of several novel vertebrate mono-α2,8-sialyltransferases subfamilies like ST8Sia VII and ST8Sia VIII was described in this first group of ST8Sia and their enzymatic specificities remain to be determined. These mono-α2,8-sialyltransferase genes have arisen as a consequence of whole genome duplications (WGDs, R1 and R2) at the base of vertebrates and were maintained in fish, whereas some others such as *st8sia6*, maintained in Tetrapods, have disappeared in fish [17,18]. In the second ST8Sia group, the enzymes responsible for the biosynthesis of sialic acid polymers, the poly-α2,8-sialyltransferases ST8Sia II and ST8Sia IV and the oligo- α2,8-sialyltransferase ST8Sia III, have been cloned and characterized from mammalian tissues, essentially the brain, where they act on α2,3-sialylated N-glycans of the neural cell adhesion molecule (NCAM), leading to an increased neuronal plasticity in embryos [19–26]. From a structural point of view, the poly-α2,8-sialyltransferases share a high degree of similarity in their sequence and structure [27–29] and are characterized by two additional sequence motifs, termed the polysialyltransferase domain (PSTD), of 32 amino acids located upstream of the sialylmotif S, and the polybasic region (PBR), of 35 amino acids located in the stem region of the enzymes involved in protein-specific polysialylation [30,31]. The oligo-α2,8-sialyltransferases ST8Sia III also show additional broadly conserved motifs with respect to ST8Sia II and ST8Sia IV (motifs III-1 and III-2) [14] with potential implication in the oligosialylation activity [32]. Their fish orthologues have been identified, cloned, and characterized in zebrafish (*Danio rerio*) in addition to rainbow trout (*Oncorhynchus mykiss*) [18,33,34].

Our previous phylogenetic studies also identified novel α2,8-sialyltransferases-related sequences like the ST8Sia III-related (ST8Sia III-r) found in a few fish orders like Perciformes, Tetraodontiformes, and Beloniformes, whereas the ST8Sia IV disappeared from the Neognathi fish [14]. It has long been appreciated that gene-, segmental-, and genome duplication, as well as gene loss events, have played important role in evolution, providing new genetic materials, which may facilitate new adaptation for the organism [35,36].

In this study, we used a BLAST strategy to identify over 700 ST8Sia-related sequences from ray-finned fish genomes and performed phylogenetic analyses and sequences alignments to reevaluate their evolutionary relationships and fate, focusing on those responsible for polysialic acid (polySia) biosynthesis with implications for the evolution of nervous system, immunological system, and cell–cell interactions. Our findings point to a particular distribution of ST8Sia in fish, revealing novel *st8sia* gene members and further suggesting their functional divergence in vertebrates.

2. Results and Discussion

2.1. In Silico Identification and Phylogenetic Reconstruction of ST8Sia Sequences

To investigate *st8sia* genes' expansion and distribution in vertebrates, we performed public database screenings in the National Center for Biotechnology Information (NCBI), ENSEMBL, and Phylofish databases [37] using a BLAST strategy [38]. The obtained results led to the identification of more than 700 ST8Sia-related sequences (Supplemental Data 1) in chordate genomes, including 71 ray-finned fish genomes (68 Teleosts genomes). Putative ST8Sia sequences with significant similarity to the known human ST8Sia based on the presence of the sialylmotifs L, S, III and VS found in all GT29 sialyltransferases, and of family motifs characteristic for the ST8Sia family were selected, and multiple sequence alignments were performed to select the complete open reading frame. The orthologues of ST8Sia I and ST8Sia V involved in gangliosides biosynthesis are identified in all the investigated genomes, suggesting a high conservation of the gangliosides biosynthetic pathways in vertebrates (Supplemental Table S1). Similarly, the ST8Sia III and the recently described fish ST8Sia VIII [17] could be found in all the Actinopterygii (ray-finned fishes) genomes (Figure 1; Supplemental Table S1). Intriguingly, multiple copies of *st8sia*-related gene sequences were identified in Teleost genomes and their number varied considerably from one fish order or species to another. For example, there are 6 *st8sia*-related genes in the medaka (*Oryzias latipes*), 8 in the clownfish (*Amphiprion oscellaris*) and the common carp (*C. carpio*), and up to 10 in the rainbow trout. Indeed, multiple copies of ST8Sia VIII (>3) were found in Perciformes, Cichliiformes, and Cyprinodontiformes; that is, two copies of the ST8Sia VII in Cypriniformes, two copies of the ST8Sia II in Salmoniformes, and two copies of the ST8Sia IV were found in the Cypriniforme *C. carpio* (Supplemental Table S1). In addition, some other *st8sia* genes could not be found like ST8Sia VI in Teleosts and Chondrostei [17]; ST8Sia IV in Neoteleostei genomes [14]; or ST8Sia II in Esociformes, Siluriformes, or Gymnotiformes (except *Electrophorus electricus*) genomes (Table 1). This resulted in a particular distribution of ST8Sia observed in the Actinopterygii compared with the Sarcopterygii (lobbed-finned fishes and Tetrapods) and Chondrichtyes (sharks) (Figure 1; Supplemental Table S1), which might have facilitated the acquisition of evolutionary innovations during vertebrate evolution [35]. These observations prompted us to re-examine the genetic events, which have shaped α2,8-sialylation in Teleosts.

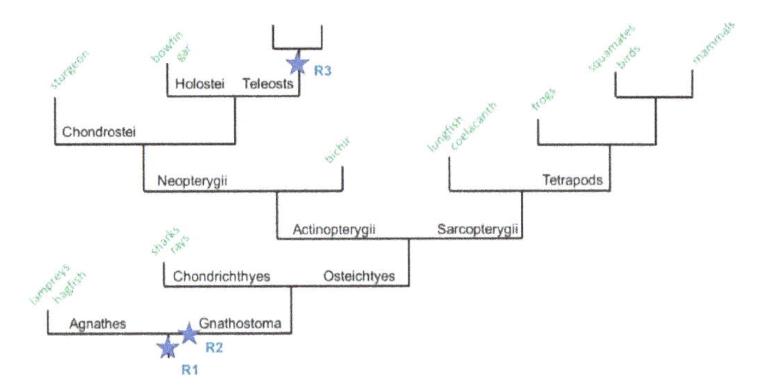

Figure 1. A schematic phylogenetic tree of vertebrate evolution. A simplified phylogenetic tree depicting the evolution of the jawed vertebrates Gnathostomes after the two rounds of whole genome duplication (WGD, R1 and R2). It is hypothesized here that WGD-R2 occurred after the Gnathostomes-Agnathes (jawless vertebrates) split. The Gnathostomes branch is divided into two categories: the cartilaginous fish Chondrychtyes (sharks and rays) and the bony fish Osteichthyes. The Osteichthyes are split into the lobe-finned fish Sarcopterygii that contain Tetrapods, and the ray-finned fish Actinopterygii that contain Neopterygii (Chondrostei, Holostei, and Teleosts).

Table 1. Fish orders that have lost *st8sia* genes.

Missing Sialyltransferase	Fish Order
st8sia2	Siluriformes, Gymnotiformes, Esociformes
st8sia4	Perciformes, Tetraodontiformes, Beloniformes, Cichliiformes, Cyprinodontiformes, Gadiformes, Gobiiformes, Pleuronectiformes, Anabantiformes, Syngnathiforme, Synbranchiformes
st8sia9	Cypriniformes, Siluriformes, Clupeiformes, Gymnotiformes, Characiformes, Lepisosteiformes, Amiiformes
st8sia7	Osteoglossiformes, Cichliiformes, Clupeiformes, Cyprinodontiformes, Gadiformes, Gobiiformes, Pleuronectiformes, Siluriformes

To determine whether the expansion of *st8sia* genes observed in Actinopterygii could be associated to WGD or smaller scale duplication events, we took advantage of the improved genome sequencing of several critical species for basal Vertebrates as Agnathans (Lampreys and Hagfish) and for Actinopterygii as Chondrostei (Sturgeons) and Holostei (Gars and Bowfin) (Figure 1). A simplified dataset was constructed including sequences of Agnathans (*Lethenteron camtschaticum*, *Petromyzon marinus*, *Eptatretus burgii*), Chondrichthyans (*Callorhinchus milii*, *Squalus acanthias*, and *Heterodontus zebra*), basal Actinopterygians (*Acipenser sinensis*, *Amia calva*, and *Lepisosteus oculatus*) and basal Teleosteans such as the Elopomorphs *Anguilla anguilla* and *Mastacembelus armatus*, in addition to two Teleosts, the Beloniforme *O. latipes* (medaka) and the Characiforme *Astyanax mexicanus* (cave fish). The potential orthology of the selected sequences was assessed through the construction of phylogenetic trees (Figure 2). The topology of these trees indicated two major phylogenetic groups of mono-α2,8-sialyltransferases on one hand, and oligo- and poly-α2,8-sialyltransferases on the other, as previously described [7,14].

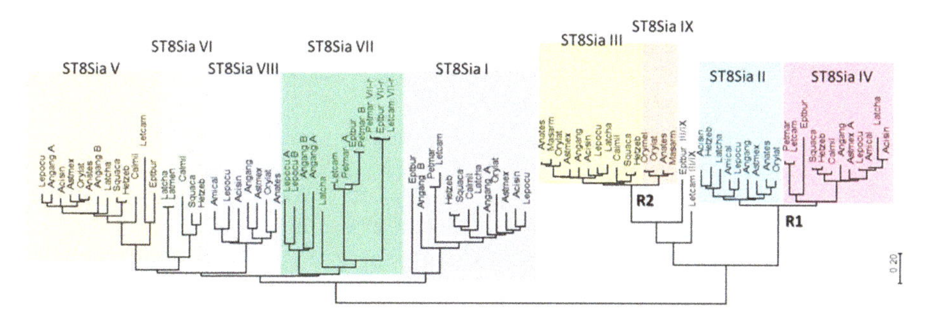

Figure 2. Minimum evolution phylogenetic tree of 89 chordates ST8Sia. The evolutionary history of 89 ST8Sia (see names and sequences in Supplemental Data 1) was inferred using the minimum evolution (ME) method. The optimal tree drawn to scale with the sum of branch length = 16.02931149 is shown. The evolutionary distances were computed using the JTT (Jones-Taylor-Thornton) matrix-based method and the rate variation among sites was modeled with a gamma distribution (shape parameter = 5). The ME tree was searched using the close-neighbor-interchange (CNI) algorithm at a search level of 1. The neighbor-joining algorithm [39] was used to generate the initial tree. The analysis involved 89 amino acid sequences and all positions with less than 95% site coverage were eliminated. A total of 226 positions were in the final dataset (see multiple sequence alignments in Supplemental Data 2). Evolutionary analyses were conducted in MEGA7.0 [40]. The nine Vertebrate subfamilies of ST8Sia (ST8Sia I to ST8Sia IX) are indicated by various colors.

In the mono-α2,8-sialyltransferases group, a series of Agnathan sequences are found at the base of each of ST8Sia I and ST8Sia V. The results corroborate previous findings suggesting the emergence of these two subfamilies around 596 and 563 million years ago (MYA), well before vertebrates emergence and prior WGD R1 and R2 [14]. Consistent with our previous data [17], we identified *st8sia7* genes in the jawless vertebrates *Lethenteron camtschaticum*, *Petromyzon marinum*, and *Eptatretus burgeri* genomes. Thus, these genes might have arisen from the ancestral *st8sia6/7/8* gene after the first WGD R1 event (~552 MYA), although timing of these events with respect to the divergence of agnathans is still a matter of debate [41,42]. Interestingly, Agnathans possess two copies of this later enzyme, named ST8Sia VII and ST8Sia VII-r in Figure 2, likely resulting from species specific large-scale gene duplication events. Similarly, in Teleosts, the eel *A. anguilla* (Elopomorphes, see the work of [43]) also harbors two copies of ST8Sia VII, ST8Sia I, and ST8Sia V enzymes (Figure 2). This observation is in favor of a large-scale genome duplication event different from the Teleost specific third round of WGD R3 (TGD) [44,45], which may have taken place in a common ancestor of freshwater eels sometime after the split of Elopomorpha and Osteoglossomorpha [46]. The ST8Sia VI and ST8Sia VIII subfamilies likely have arisen from the second WGD at the base of Vertebrates; the first one was maintained in Sarcopterygii and disappeared in Actinopterygii, and vice versa for ST8Sia VIII [17]. The many gene copies of *st8sia7* and *st8sia8* identified in Teleosts genomes (Supplemental Table S1) are likely the result of single gene duplication events because they were identified on the same piece of chromosome (data not shown), and were thus noted with -A, -B, or -C extension. However, it is difficult to infer the origin of these segmental duplications as they have occurred in many, but not all terminal branches of clades.

The second branch encompasses both oligo- and poly-α2,8-sialyltransferases. Regarding poly-α2,8-sialyltransferases, the Agnathan sequences were attributable only to ST8Sia IV, indicative of a divergence between ST8Sia II and ST8Sia IV dating back to WGD-R1 (Figure 2) [14] followed by *st8sia2* gene loss in Agnathans. In contrast, the Agnathan sequences of oligo-α2,8-sialyltransferases are at the base of the ST8Sia III and ST8Sia III-r subfamilies, while there are orthologues to the ST8Sia III from sharks to Tetrapod lineages, suggesting a genome duplication event linked to WGD-R2 consistent with previous dating around 474 MYA [14]. Despite the fact that the ST8Sia III-r sequences appear to be restricted to Teleosteans, including Elopomorphes, and are lost in Chondrichthyans and Tetrapods lineages, they were not issued from the Teleost specific WGD, and thus were renamed ST8Sia IX according to the previously described nomenclature [12].

2.2. Identification and Phylogenetic Analysis of the Fish St8sia Genes (st8sia2, st8sia4, st8sia3, and st8sia9)

Interestingly, in the oligo- and poly-α2,8-sialyltransferases group, the ST8Sia II and ST8Sia IV appeared to be duplicated or lost in several Teleost lineages after divergence of Actinopterygii from Sarcopterygii [47,48], whereas the ST8Sia III was found in all the Actinopterygii. In the basal Elopomorphes and Osteoglossiformes branches, the four *st8sia* genes (*st8sia2*, *st8sia3*, *st8sia4*, and *st8sia9*) could be identified. The results indicate that these genes already existed in the common ancestor of the 68 Teleost fishes examined. All Otocephalan lineages lack the *st8sia9* gene and the Siluriformes lack both the *st8sia9* and *st8sia2* genes. Consequently, the *st8sia9* gene was lost shortly after Otocephala emergence around 176.2 MYA and the *st8sia2* gene was lost more recently (~82.6 MYA) during siluriformes evolution [49]. As previously observed, all Neoteleostei fish lack the *st8sia4* gene [14], which was lost at the basis of Neoteleostei lineage. Finally, the Esociformes lack the *st8sia2* gene only (Table 1). Furthermore, two ST8Sia II-related sequences were identified in all the investigated Salmoniformes (*Oncorhynchus*, *Coregonus*, *Salmo*, *Salvelinus*, and *Thymallus*) and two ST8Sia IV-related sequences were identified only in the Cypriniformes *C. carpio* and *Sinocyclocheilus anhuiensis* (Supplemental Table S1). We took advantage of the improved genome and transcriptome sequencing of several fish [37,50], selected several representative Salmoniformes and Cypriniformes ST8Sia sequences, and constructed phylogenetic trees (Supplemental Figure S1). The topology of these trees indicated that the later duplications of *st8sia* genes were not associated to the Teleost specific genome duplication (TGD, WGD R3), but rather to more recent lineage-specific genome duplication events described in Salmonidae (SGD) lineage [51] and in *C. carpio* species [52].

2.3. Synteny and Paralogy Analyses of the st8sia2, st8sia4, st8sia3, and st8sia9 Gene Loci

To explain the gain or loss of ST8Sia subfamilies, we further analyzed the evolutionary relationships between these *st8sia* genes. The kind of event that created duplication was characterized by analyzing the conserved synteny between ST8Sia paralogues. It was expected that the *st8sia* genes created by a WGD would be far apart on different chromosomes in one genome, but surrounded by similar genes in each of the duplicated regions (i.e., paralogons). Significant Tetrapod paralogons containing *st8sia2* and *st8sia4* genes were found and a well conserved synteny could be established for *st8sia2* and *st8sia4* gene loci in Tetrapods (i.e., human, mouse, chicken, and xenopus) genomes (Figure 3A) as previously described [14]. However, in the fish genomes, as the *st8sia2* gene was absent in Esociformes and Siluriformes, we considered the neighboring *furin*, *fes*, *sv2b*, *fam147b*, *mctp2*, and *chd2* genes around *st8sia2* on the medaka chromosome 6 to retrieve the synteny on *Esox lucius* LG19 and on *Ictularus punctatus* chromosome 4. Similarly, *ppip5k2*, *pam*, *chd1 erap1a*, and *syk* genes conserved around the *st8sia4* locus were used to retrieve the synteny on *O. latipes* chromosome 12, *Gasterosteus oculatus* chromosome XIV, and *Xiphophorus maculatus* chromosome 8. Interestingly, paralogues of these genes could be identified on other chromosomes in the various fish genomes indicative of an ancient Teleost specific WGD (TGD) followed by intense gene rearrangements. This further suggests that the *st8sia* genes have undergone the TGD and the duplicated *st8sia* genes were rapidly lost during Teleost evolution. In the Salmoniformes, a highly conserved synteny was found around the two *st8sia2-r* gene loci corresponding to one ohnologous region in the spotted gar (*L. oculatus*), likely resulting from the fourth round of WGD (SGD) that took place more recently in the Salmoniforme genomes [51]. The two *st8sia4-r* genes were localized on two distinct chromosomes in *C. carpio* genome, supporting the hypothesis of a more recent species-specific genome duplication event in *C. carpio* [52] in spite of a weak synteny conservation (Figure 3A).

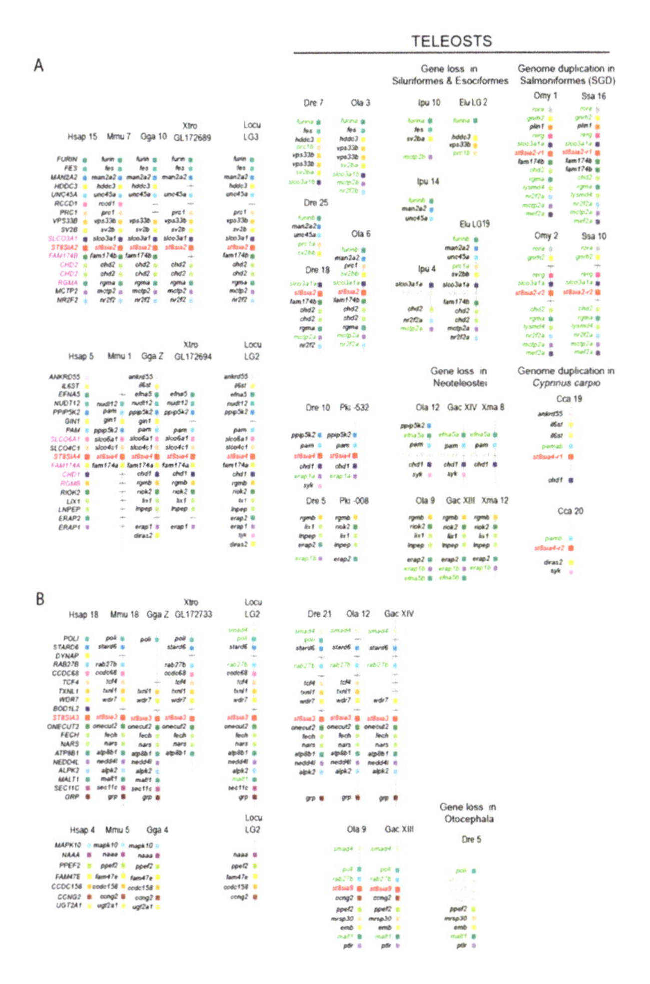

Figure 3. Syntenic relationships of the oligo- and poly-α2,8-sialyltransferases gene loci in vertebrates. Chromosomal locations of the *st8sia* genes and neighboring gene loci were determined in the human (*Homo sapiens*, Hsa), the mouse (*Mus musculus* (Mmu), the chicken (*Gallus gallus*, Gga), the spotted gar (*L. oculatus*, Locu), the western clawed frog (*Xenopus tropicalis*, Xtro), the zebrafish (*D. rerio*, Dre), the Japanese medaka (*O. latipes*, Ola), the channel catfish (*I. punctatus*, Ipu), the northern pike (*E. lucius*, Elu), the rainbow trout (*O. mykiss*, Omy), the Atlantic salmon (*Salmo salar*, Ssa), the African weakly electric fish (*Paramormyrops kingsleyae*, Pki), the three-spined stickleback (*G. aculeatus*, Gac), the southern platyfish (*X. maculatus*, Xma), and the European carp (*C. carpio*, Cca). Information from the National Center for Biotechnology Information (NCBI) and ENSEMBL release 97 was used to identify putative orthologues, which were visualized using the Genomicus 97.01 [53]. Paralogous genes in the fish genomes are indicated in green and in purple in the human genome. The *st8sia* genes are indicated in red or in grey when lost. (**A**) Syntenic relationships of the *st8sia2* and *st8sia4* gene loci in vertebrates. (**B**) Syntenic relationships of the *st8sia3* and *st8sia9* gene loci in vertebrates.

The synteny around the *st8sia3* gene locus including *wdr7*, *onecut2*, and *fech* genes is highly conserved in vertebrate lineages from fish to mammals (Figure 3B). Synteny around *st8sia9* locus is less conserved and is limited to a smaller syntenic block with *ccng2* and *ppef2* genes, which is reminiscent of ancient WGD followed by intrachromosomal rearrangement in the ancestral fish genome.

Altogether, our phylogenetic analyses enabled us to refine the evolutionary history of the fish ST8Sia and to propose a model of their evolution illustrated in Figure 4, which agrees with the fish phylogenetic tree of life [54]. It is interesting to note that, while Braasch and Postlethwait (2012) determined duplicated gene retention rates of 12–24% after the TGD 320 MYA [55], we observed no remaining *st8sia* gene copy from this event and no modification on the fish ST8Sia repertoire. However, more recent polyploidization events were recorded in several families (Salmonidae, 80 MYA), genera (Anguilla) or species (*C. carpio*, 8 MYA), which impacted the overall poly-α2,8-sialyltransferases repertoire. In Salmonidae, we described only two remaining *st8sia2* duplicates after the Ss4R among the eight ancestral *st8sia* genes (12% duplicate retention), while Lien et al. (2016) revealed a global retention rate around 55% [56]. In the carp *C. carpio*, two *st8sia4* genes were retained as duplicates among the seven *st8sia* genes (14% duplicate retention), while Li et al. (2015) calculated a global value of 92% [57]. Furthermore, these studies highlighted the fact that the retained genes after tetraploidization were specifically involved in signal transduction, protein complex formation, and immune system, which prompted us to focus on the functional divergence of these poly-α2,8-sialyltransferase duplicated genes (neofunctionalization) and on their expression divergence (subfunctionalization).

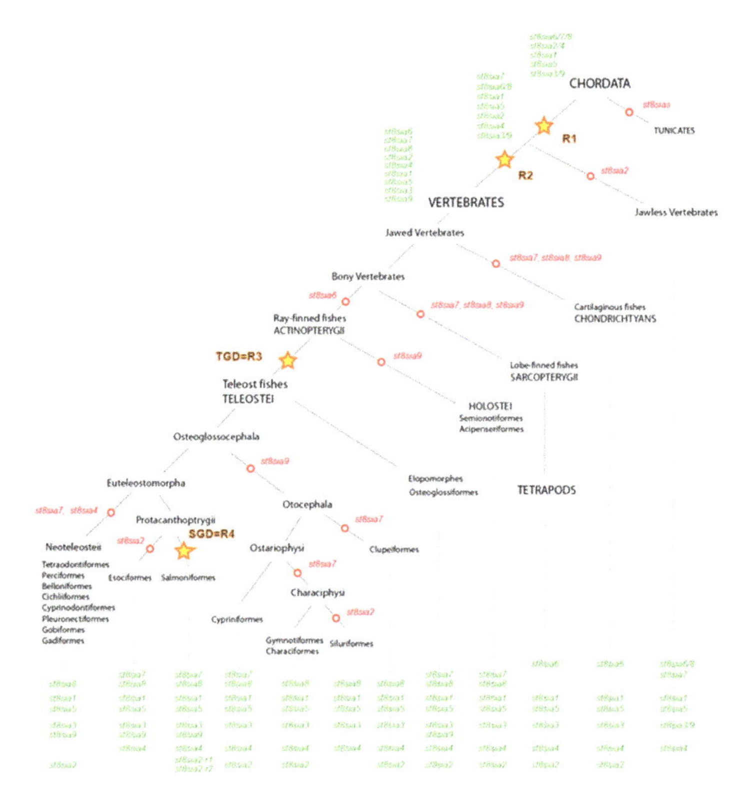

Figure 4. Schematic representation of the ST8Sia family evolution in the ray-finned fishes. This model for the evolution of *st8sia* genes is based on the evidence from protein sequence phylogeny, conserved synteny of genomic *st8sia* loci in vertebrate species and their paralogous relationships in fish genomes. The model takes into account the evolution of five ancestral groups of ST8Sia (*st8sia6/7/8*, *st8sia2/4*, *st8sia1*, *st8sia5*, and *st8sia3/9*) indicated in green and present in the ancestor of Chordates that predate the WGD R1 and WGD R2. Open red circles depict gene losses on the phylogenetic tree and yellow stars correspond to the WGDs R1, R2, R3 (teleost specific duplication, TGD), and R4 (salmonids specific duplication, SGD).

2.4. Molecular Evolution of the Poly-α2,8-Sialyltransferases

A remarkable difference between α2,8-linked polySia chains found in mammals and salmonid fish seems to be the structural diversity of polySia in fish [58–60]. Whereas in mammals, homopolymers of Neu5Ac residues are typically formed [61], in rainbow trout eggs, polymers can consist of Neu5Ac, Neu5Gc, and KDN in addition to their O-acetylated forms [62]. One explanation could be a better accessibility to different sialic acids in fish, because, in transgenic mice—showing a Neu5Gc overexpression in brain—besides Neu5Ac, Neu5Gc also seems to be utilized to build polySia [63].

Another explanation might be the occurrence of structural changes of the protein backbone during the evolution of the polysialyltransferases. We thus investigated the potential consequences of specific-lineages' *st8sia* gene loss and duplication on the functional fate of duplicates, an issue that is still poorly understood [64,65]. Substitution rate analysis of the duplicated *st8sia2* genes maintained in Salmoniformes genome after the SDG event indicated four amino acid substitutions in the ST8Sia II-r2 coding sequences compared with ST8Sia II-r1 and the rest of Teleost ST8Sia II sequences, while there were only two substitutions in the ST8Sia II-r1 sequence. Of particular note, among the four substitutions found in ST8Sia II-r2, the H → Y is recorded in sialylmotif L, and the R → Q between sialylmotifs S and III, whereas the two substitutions in the ST8Sia II-r1 sequence are located nearby the PSTD motif (Figure 5A). In addition, two convergent substitutions leading to the same amino acid were identified near the end of sialylmotif L (i.e., acquisition of a G from a Q) and beyond the sialylmotif III (i.e., acquisition of an H from an S), respectively. These drastic modifications in amino acid properties in functionally important locations in the catalytic domain of these salmonid ST8Sia II let us suggest profound changes in both ST8Sia II functions (i.e., neofunctionalization). Likewise, we examined the impact of *st8sia4* loss on the remaining *st8sia2* gene in Neoteleostei using parsimony analysis. We found two substitutions, A → S and Q → S, located in the sialylmotif L and between the sialylmotifs III and VS that of Neosteleostei ST8Sia II, respectively (Figure 5B). Interestingly, we also found a convergent T → K substitution located between the sialylmotifs III and VS that of Neosteleostei ST8Sia II that restores the K amino acid characteristic of all the ST8Sia IV sequences (Figure 5B), further suggesting changes in ST8Sia II functions in Neoteleotei. No substitution could be detected in ST8Sia IV sequences after the loss of *st8sia2* gene in Esociformes and Osmeriformes. Finally, we recorded the substitutions on the ancestral sequence of ST8Sia III after ST8Sia IX loss in Otocephala. We observed three substitutions in ST8Sia III sequence: V → T near the sialylmotif L, A → T in the sialylmotif VS, and Y → F beyond (Figure 5C).

The most striking domain of both polysialyltransferases—ST8Sia II and ST8Sia IV—is PSTD, which is essential for the polysialylation of NCAM [31,66]. This motif contains a high number of basic amino acids and is important for substrate binding and the catalytic activity. Troy and co-workers exchanged several of these amino acids to determine their distinct impact on the enzymatic activity of human ST8Sia IV [31]. Doubled substituent mutants with an exchange of the first basic residues (declared as K2 and K4 in Figure 6) by neutral amino acids retained approximately 80% of the enzyme activity and comparable values were determined, when only K6 was replaced. Stronger effects were observed in single substituted mutants where R8, H18, K28, K32, or R33 was replaced by a neutral amino acid. All these changes reduced activity by more than 50%. Their experiments demonstrated that, in addition to the neutral amino acid I31 (mutants retained only 6% of their activity), especially the basic amino acids of PSTD were key elements for polysialylation. Most of these important amino acids of the human ST8Sia IV are also highly conserved in the fish enzyme. Changes occurred sporadically at K2, K4, K6, and R8 in individual fish species (Figure 6). On the basis of the work of Troy and co-workers [31], the R8 change may have the highest impact on the general enzyme activity, as a replacement of this amino acid reduced the activity to less than 25%. However, we observed an exchange of R8 only in three fish species including *I. punctatus*. Nevertheless, as mentioned above, other substituted amino acids may also influence the interaction with the nascent sialic acid chain, depending on the composition (Neu5Ac, Neu5Gc, KDN, and O-acetylated variations) of the polySia chain.

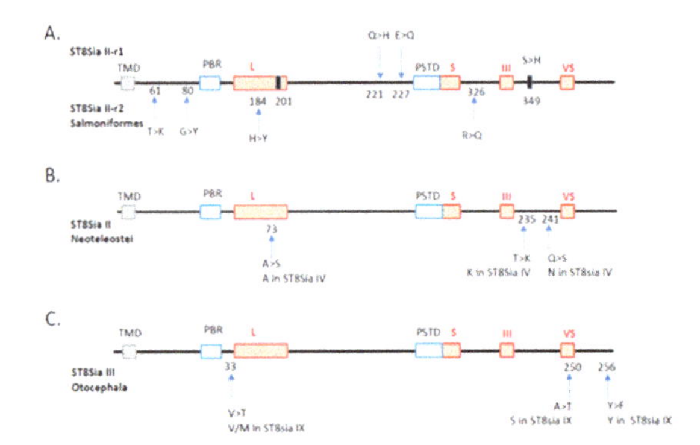

Figure 5. Substitution rate analysis of the impact of *st8sia* gene duplications and losses. The sialylmotifs are indicated by red boxes and the transmembrane domain by a grey box. (**A**) Duplication of *st8sia2* genes in Salmoniformes. The substitutions observed in ST8Sia II-r1 and ST8Sia II-r2 are indicated by an arrow above and below, respectively. The position of the substitutions corresponds to the alignment in Supplemental Data 2. The black rectangles correspond to convergent mutations retrieved in both sequences. In T > K, for example, T is the ancestral state and K is the derived one. (**B**) Impact of *st8sia4* gene loss in Neoteleostei on the remaining fish ST8Sia II sequences. The code for substitution is the same as in A. The corresponding amino acid present in the paralogue ST8Sia IV sequence is given below. (**C**) Impact of *st8sia9* gene loss in Otocephala on the remaining fish ST8Sia III sequences (same abbreviations as in B).

More consistent variations were observed when ST8Sia II sequences were compared. In addition to the mentioned K2 and H4 (K instead of H in ST8Sia IV), an exchange of a basic amino acid occurs more frequently and is often highly conserved within one family. For instance, in Salmoniformes, lysine residues at position 2 and 28 are changed with apolar amino acids and the strongly basic R8 residue is exchanged with histidine, which is only partly positively charged at neutral pH. On the basis of the studies of Nakata et al. using human ST8Sia IV, we can also assume remarkable changes in the enzymatic activity of ST8Sia II [31]. For instance, ST8Sia IV mutants with a neutral amino acid at position K28 retained less than 25% of their enzymatic activity. This is in line with studies by Kitajima and co-workers demonstrating that rainbow trout ST8Sia II isoforms show only low enzymatic activity in vitro [33]. Intriguingly, in Neoteleostei, the very important lysine at position 28 was also exchanged with a neutral amino acid. Notably, in contrast to Salmoniformes, in Neoteleostei, ST8Sia II is the only polysialyltransferase because ST8Sia IV is absent. The presence of only one polysialyltransferase in Neoteleostei, which additionally includes such a striking mutation, suggests that polysialylation significantly changed in Neoteleostei in comparison with other vertebrates.

In addition to sequence alignments, we simulated the PSTD 3D structure of fish ST8Sia II and ST8Sia IV, based on the determined 3D structure of human ST8Sia IV PSTD (PDB 6AHZ) (Figure 7), which were published by Peng and colleagues [66]. Volkers et al. described that PSTD acts as a basic furrow, leading the nascent sialic acid chain to the active site of the polysialyltransferase [32]. The 3D simulation of the human ST8Sia IV PSTD shows that only significant differences between the electrostatic potential surfaces are detectable at the N-terminal region. Especially the orientation of the basic areas changed between the species. In contrast, the central and C-terminal area exhibited only minor changes. In the case of ST8Sia II, the most prominent alterations also occurred at the N-terminal domain (Figure 8). However, exchanging the N6 with aspartate, an exposed acidic segment is formed in Salmoniformes and Neoteleostei, which may influence the interaction between PSTD and the negatively charged sialic acid polymers. However, regarding the 3D simulation of PSTD, it has to be noted that a simulation is only a simulation and crystal structures of PSTD in addition to the whole enzymes are necessary for the generation of unambiguous 3D models.

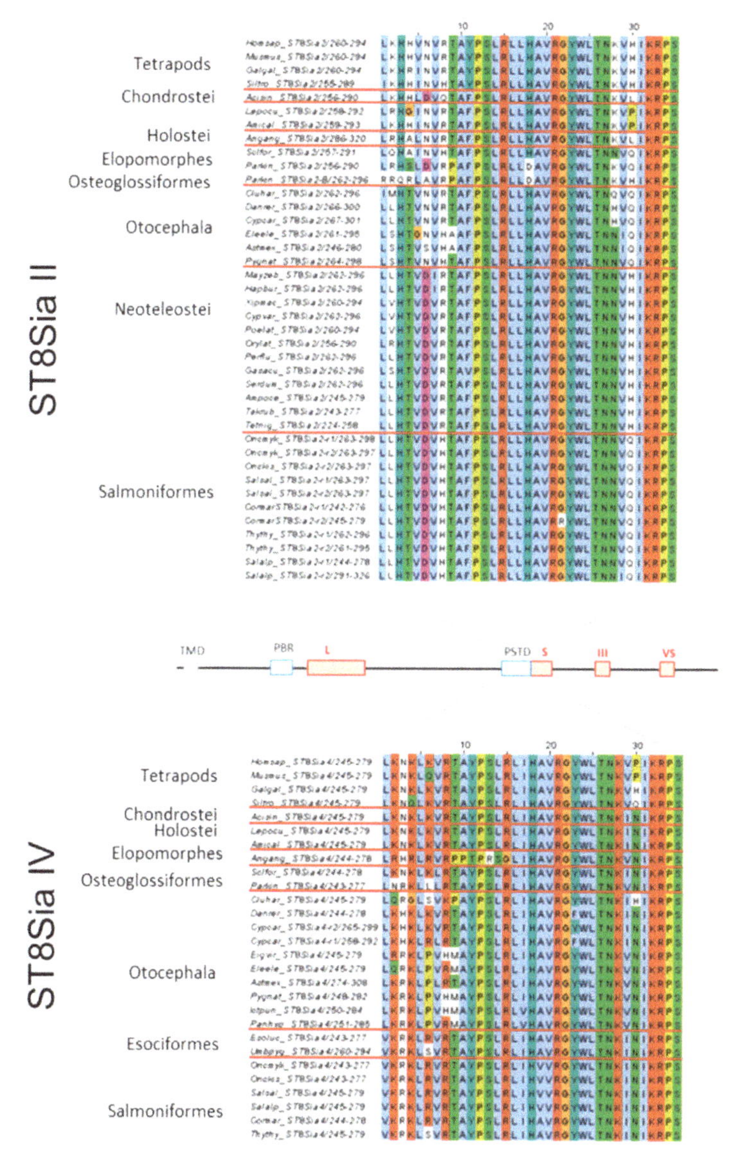

Figure 6. Sequence-based analysis of the polysialyltransferase domain (PSTD) in fish ST8Sia II and ST8Sia IV. Multiple sequence alignment of PSTD were performed with CLUSTAL OMEGA of EMBL-EBI by MUSCLE (3.8) edited and annotated in Jawa Alignment Jalview [67]. The used protein entries from different species are listed in Supplemental Table S1. The different colors from Clustal X scheme codes indicate the following characteristics: hydrophobic (blue), positive charge (red), negative charge (magenta), polar (green), cysteine (pink), glycine (orange), proline (yellow), aromatic (cyan), and gap (white). It should be noted that one additional amino acid was added to the N-terminus and two additional amino acids to the C-terminus of PSTD.

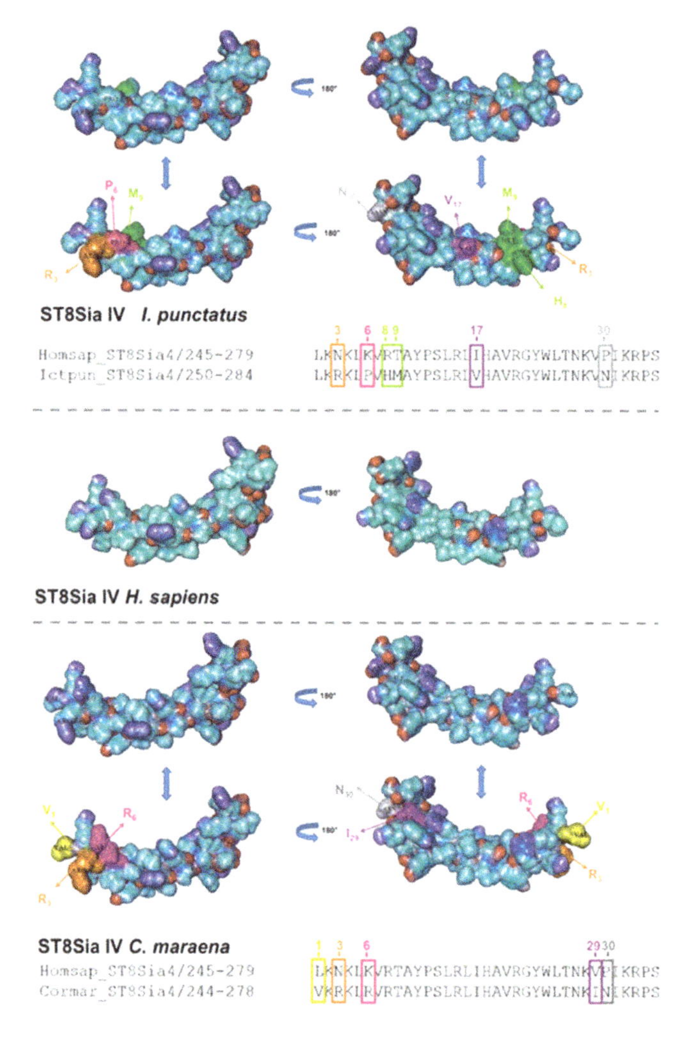

Figure 7. Three-dimensional (3D) structure of PSTD motifs in fish ST8Sia IV. The 3D model of human ST8Sia IV PSTD (Protein Data Bank entry 6AHZ)—electrostatic potential surfaces—is displayed in addition to the simulated structure of PSTD from I. punctatus and *C. maraena* using YASARA. The exchanged amino acids are colored in an additional version of the 3D structure to highlight the position of the exchange: N3 → R3 (orange), K6 → P6 (magenta), R8 → H8 (green), T9 → M9 (green), I17 → V17 (violet), and P30 → N30 (grey) for *I. punctatus* and L1 → V1 (yellow), N3 → R3 (orange), K6 → R6 (magenta), V29 → I29, and P30 → N30 (grey) for *C. maraena*. It should be noted that, for the determination of the 3D structure of human ST8Sia IV PSTD, a peptide was used with one additional amino acid on the N-terminus and two additional amino acids on the C-terminus of PSTD [66].

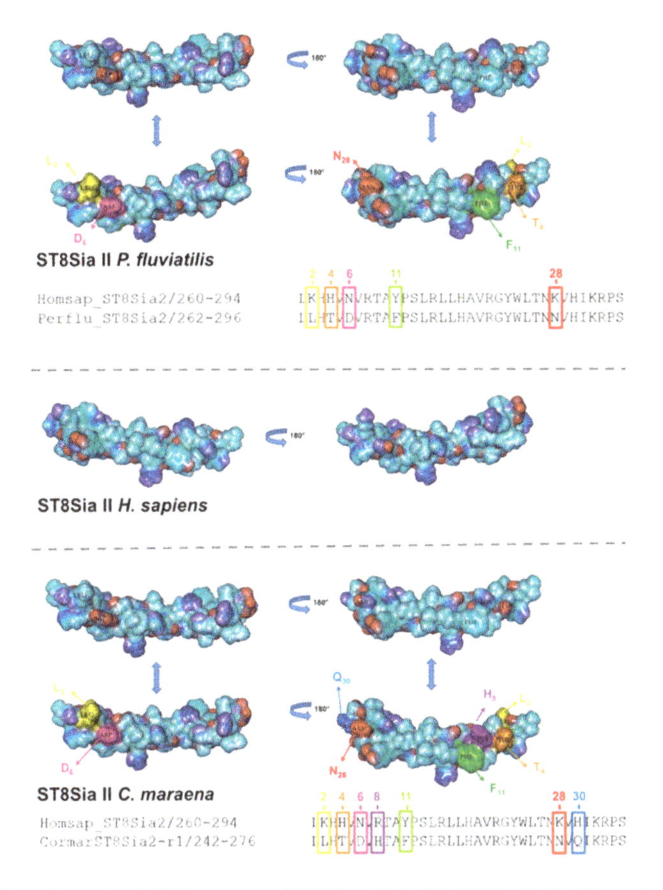

Figure 8. Three-dimensional (3D) structure of PSTD motifs in fish ST8Sia II. The 3D model of human ST8Sia II PSTD in addition to PSTD from *P. fluviatilis* and *C. maraena* was simulated, based on the 3D model of human ST8Sia IV PSTD (Protein Data Bank entry 6AHZ) using YASARA. The electrostatic potential surfaces are displayed. The exchanged amino acids are colored in an additional version of the 3D structure to highlight the position of the exchange: K2 → L2 (yellow), H4 → T4 (orange), N6 → D6 (magenta), Y11 → F11 (green), and K28 → N28 (red) for *P. fluviatilis* and K2 → L2 (yellow), H4 → T4 (orange), N6 → D6 (magenta), R8 → H8 (violet), Y11 → F11 (green), K28 → N28 (red), and H30 → Q30 (light blue) for *C. maraena*. For the determination of the 3D structure of human ST8Sia IV PSTD, a peptide with one additional amino acid on the N-terminus and two additional amino acids on the C-terminus of PSTD were used [66].

Taken together, our sequence alignments and 3D simulations demonstrate that, in fish, characteristic alterations of the amino acid sequences occurred within PSTD and that several of these replaced amino acids are important for the enzymatic activity in the case of human ST8Sia IV, as demonstrated by Troy and co-workers [31]. These variations might also influence the ability of PSTD to interact with sialic acid chains consisting of other sialic acids than Neu5Ac, such as Neu5Gc and KDN, as well as their O-acetylated forms. However, to definitively proof this hypothesis of neofunctionalization of fish polysialyltransferases, their enzymatic activity has to be characterized in more detail.

2.5. Expression of Polyα2,8-Sialyltransferase Genes in C. Maraena Tissues

Having characterized the chromosomal localization, evolutionary history, and structure of the poly-α2,8-sialyltransferases ST8Sia II and ST8Sia IV encoded by the *st8sia2* and *st8sia4* genes, respectively, we eventually profiled their expression in ten organs and tissues of *C. maraena* as a representative of the Salmoniformes (Figure 9A,B). As *st8sia2* is duplicated in salmonid fishes, we investigated whether the expression of both genes is tissue-specific, and thus possibly function-specific. To this end, discriminating primer pairs for *st8sia2-r1* and *st8sia2-r2* as well as for *st8sia4* transcripts were designed. The RT-qPCR analysis revealed that *st8sia2-r1* transcripts were on low levels in liver, heart, spleen, head kidney, gills, hypothalamus, and hind brain (>300 copies/ng RNA), and almost absent in muscle (>10 copies/ng RNA) (Figure 9A). In stark contrast, the copy numbers of *st8sia2-r1* were at a high level in gonads (~1700 copies/ng RNA) and telencephalon (~ 2140 copies/ng RNA) (Figure 9B). The transcript levels of the gene copy *st8sia2-r2* were generally higher compared with its paralogue, ranging from a 1.5-fold difference in gonads to a 233-fold difference in spleen (Figure 9A). While the expression of *st8sia2-r2* was not detectable in hind brain and telencephalon, it exceeded the expression level of *st8sia2-r1* by 4622-fold in the hypothalamus.

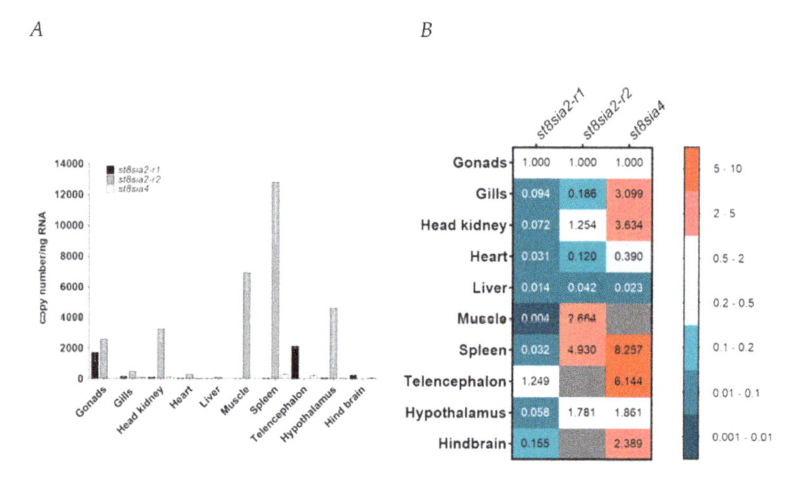

Figure 9. Expression profiling of poly-α2,8-sialyltransferase-encoding genes in maraena whitefish. (**A**) Transcript levels of *st8sia2-r1* (black bars), *st8sia2-r2* (gray), and *st8sia4* (blank) were determined in ten tissues from maraena whitefish ($n = 4$), as indicated on the abscissa. Bars represent the averaged copy numbers normalized against three reference genes; error bars represent the standard deviation. (**B**) A heat map represents the same copy numbers per target gene as shown in (**A**) relative to the expression in gonads (set as 1.0). These relative expression values are colored according to the code given at the right. Non-detectable transcript numbers are indicated by gray fields.

The expression level of *st8sia4* was at a similarly low or even significantly lower level compared with that of *st8sia2-r1* with the highest copy numbers in spleen (~330 copies/ng RNA). No or only very few *st8sia4* transcripts were detectable in liver, muscle, and heart (Figure 9B). The results are partially different in comparison with the determined mRNA levels in rainbow trout using Northern blot analysis and semi-quantitative PCR [33]. For instance, spleen samples were negative for *st8sia2* transcripts, which might not only be the result of differences in the applied methods, but also in general differences between these two Salmoniformes.

Taken together, profiling the expression of the poly-α2,8-sialyltransferase genes revealed a tissue-specific expression pattern of *st8sia2* genes in *C. maraena* tissues indicative of their subfunctionalization. Probably one of the most striking differences between the expression profiles in maraena whitefish and humans is the presence of *st8sia2* and *st8sia4* transcripts in the reproductive tract. Whereas in humans, only a weak signal for *st8sia2* mRNA and no signal for *st8sia4* mRNA could be detected by Northern blotting [68], in *C. maraena*, the gonads belongs to the tissues with the highest expression levels of polysialyltransferases. This was already described by Kitajima and colleagues using rainbow trout ovaries [33]. Besides the gonads, remarkable differences were also observed in spleen. Contrary to humans, where no *st8sia2* mRNA was detectable [68], *st8sia2-r2* expression was extremely high in the spleen of *C. maraena*, indicating that ST8Sia II-r2 might play a role during immunologic reactions in maraena whitefish. Altogether, these results let us suggest that, in addition to the number of active polysialyltransferases, as well as their enzymatic activity, the physiological roles of these polysialyltransferases may have changed during the evolution of vertebrates.

3. Materials and Methods

3.1. Materials and Animals

Maraena whitefish were provided by the Institute for Fisheries of the State Research Centre for Agriculture and Fishery Mecklenburg-Western Pomerania (Born, Germany), and BiMES, Binnenfischerei GmbH (Friedrichsruhe, Germany). Fish were held in fresh-water recirculation systems with a 12:12 day-and-night cycle at 18 °C. Water quality was maintained by automated purification and disinfection (bio-filter and UV light). In addition, the concentrations of selected chemical and physical water parameters were constantly determined.

Sampling of ten organs or tissues (gills; gonads; head kidney; heart; liver; muscle; spleen; and the brain regions hypothalamus, hind brain, and telencephalon) from four maraena whitefish followed the standards described in the German Animal Welfare and was approved by the Landesamt für Landwirtschaft, Lebensmittelsicherheit und Fischerei, Mecklenburg-Vorpommern, Germany (LALLF M-V/TSD/7221.3-1-069/18) in November 2018. The tissues were sampled rapidly and immediately frozen in liquid nitrogen to be kept at −80 °C until RNA extraction.

3.2. In Silico Identification and Phylogenetic Analysis of ST8Sia Sequences

A local alignment BLAST approach was used to retrieve the vertebrate *st8sia* nucleotide sequences with significant homology to the mammalian sequences from the genomic and Transcriptome Shotgun Assembly (TSA) divisions of the GenBank/EBI databases at the National Center for Biotechnology Information (NCBI) (last accessed on 27 September 2019), ENSEMBL (release 97) and from the PhyloFish database [7,14,37]. The protein sequence analysis was performed using the Expert Protein Analysis System (ExPASy; Swiss Institute of Bioinformatics, Switzerland; website (https://www.expasy.org/)). Sequence alignments were performed using the clustalW (PRABI; https://npsa-prabi.ibcp.fr/cgi-bin/npsa_automat.pl?page=/NPSA/npsa_clustalw.html). Phylogeny was determined aligning the known vertebrate ST8Sia sequences with MUSCLE in MEGA7.0 [40]. The multiple sequence alignments of the selected vertebrate ST8Sia sequences were conducted using MUSCLE and Clustal Omega algorithms in MEGA7.0 and manually refined (see Supplementary Data 1 and 2). Phylogenetic trees were produced by the neighbor-joining (NJ), maximum likelihood, and minimum evolution method in MEGA 7.0 [40,69].

To determine the consequences of duplication or loss of genes of a given order of Actinopterygian, we considered what happened on its closest paralogue. The amino acid substitutions that occurred at its base were deduced using the parsimony method implemented in Protpars program (PHYLIP package vers. 3.69) [70].

3.3. Synteny Analysis, Paralogon Detection, and Ancestral Genome Reconstruction

Synteny between the *st8sia* gene loci and neighbour genes in vertebrate genomes was assessed by manual chromosome walking and reciprocal BLAST. Detection of paralogous blocks was visualized with Genomicus (version 97.01) http://www.genomicus.biologie.ens.fr/genomicus-92.01/cgi-bin/search.pl, last accessed August 2019 [53]. When the *st8sia* gene of interest was not found in a genome, physically close genes were used as a seed to identify syntenic segments.

3.4. Sequence Alignments, Motifs Analysis, and 3D Simulation of PSTD

Multiple sequence alignments were performed with MUSCLE of EMBL-EBI (version 3.8; https://www.ebi.ac.uk/Tools/msa/muscle/) from selected species using published sequences (accession numbers in Supplemental Data 1). The sequences were conducted, edited, and annotated in Jawa Alignment Viewer Jalview 2.11.0, and manually refined [67]. The 3D structure of the human PSTD of ST8Sia II as well as ST8Sia IV was generated in YASARA (Version 19.9.17) using the following Protein Data Bank (PDB) entries: ST8Sia IV (code: 6AHZ) (PDB, https://www.rcsb.org/pdb/home/sitemap.do https://www.wwpdb.org/pdb?id=pdb_00006ahz). The ST8Sia II was generated in YASARA changing the amino acid in the positions N3, K4, L5, and K6, corresponding to H3, H4, V5, and N6, respectively. The human amino acid sequences of ST8Sia II and ST8Sia IV were modified at the positions with the following amino acids according to the different fish PSTD sequences: *C. maraena* ST8Sia II, L2, T4, D6, H8, F11, N28, and Q30; *P. fluviatilis* ST8Sia II, L2, T4, D6, F11, and N28; I. punctatus ST8Sia IV, R3, P6, H8, M9, V17, and N30; and *C. maraena* ST8Sia IV, V1, R3, R6, I29, and N30. We designed only one PSTD motif of *C. maraena* because there is only one difference at position R22 between ST8Sia II-r1 and ST8Sia II- r2.

3.5. RNA Extraction, cDNA Synthesis, Primer Design, and RT-qPCR

Total RNA was isolated from the individually homogenized organs and tissues using TRIzol (Invitrogen/Thermo Fisher Scientific, Darmstadt Germany), followed by an additional purification step (RNeasy Mini Kit, Qiagen). The quantity and integrity of the isolated RNA were determined using the NanoDrop 2000 photometer (Thermo Fisher Scientific) and agarose gel electrophoresis. Subsequently, we reverse-transcribed the total RNA using the SuperScript II Reverse Transcriptase (Thermo Fisher Scientific) and a mixture of oligo-d(T) and random hexanucleotides. This reaction was carried out at 42 °C (50 min), followed by an inactivation step (70 °C, 15 min). The resulting cDNA was diluted in 100 µL distilled water.

Real-time fluorescence-based quantitative RT-PCR (RT-qPCR) was used to determine the mRNA abundance of the two *st8sia2* gene variants in the above ten organs and tissues of maraena whitefish ($n = 4$). To this end, we identified discriminating sequence motifs to derive the oligonucleotides for *st8sia2-r1* (sense, 5′-AGCCTCATCAGGAAGAACATCC-3′; antisense, 5′-TTCCCTACGATGGCACAGCGT-3′) and *st8sia2-r2* (sense, 5′-CGTTCAACAGGAGCCTCTCTAA-3′; antisense, 5′-TTCCCTACGATGGCACAGCGC-3′). Moreover, we designed a *st8sia4*-specific primer pair (sense, 5′-ATGATAAGGAAGGACGTGCTGC-3′; antisense, 5′-TGTTGAGCGTTCGGCGTCTGT-3′). These RT-qPCR primers were designed (Pyrosequencing Assay Design software v.1.0.6; Biotage, Uppsala, Sweden) to synthesize amplicons between 121 bp and 226 bp. *eef1a1a2* (encoding eukaryotic translation elongation factor, variant a2), *rpl9*, and *rpl32* (ribosomal proteins L9 and L32) were selected as reference genes [71]. The RT-qPCR analyses were conducted with the LightCycler 96 System (Roche, Mannheim, Germany) using the SensiFAST SYBR No-ROX Kit (Bioline, Luckenwalde, Germany). We only considered crossing point (CP) values >35 for The expression analysis of the *st8sia2-r1* *st8sia2-r2*, and *st8sia4*. The calculation of their copy numbers was based on standard curves having been generated on 10-fold dilutions of the respective PCR-generated fragments (1×10^3 to 1×10^6 copies). Melting-curve analyses validated the amplification of the distinct products. Amplicons were visualized on 3% agarose gels in order to assess product size and quality.

3.6. Data Availability

To identify the maraena whitefish ST8Sia II sequences, the orthologous sequences from rainbow trout and Atlantic salmon were aligned with the software Bowtie2 (v 2.2.4) to our RNA-seq read collection from maraena whitefish [72]. The alignments were then indexed and sorted with the software package Samtools (v.16) and the final consensus sequences were obtained with the Ugene software (v 1.29).

4. Conclusions

In this study, we highlighted an expansion and particular distribution of the ray-finned fish ST8Sia repertoire owing to several duplications and loss events of *st8sia* genes, and we refined their evolutionary history. Our analyses of the molecular evolution in ST8Sia sequences and in key functional motifs (i.e., motif L and PSTD) let us suggest that the polysialyltransferases might evolved new enzymatic activities and/or specificities in the course of Vertebrate evolution. Their expression profiles in Salmonid tissues differ from those observed in mammals and further point to a subfunctionalization of these poly-α2,8-sialylatransferases. Altogether, we have laid the foundation for further studies towards understanding of the remarkable differences between α2,8-linked polySia chains found in mammals and fish.

Supplementary Materials: Supplementary materials can be found at http://www.mdpi.com/1422-0067/21/2/513/s1.

Author Contributions: Conceptualization, data curation, and writing—review and editing: S.P.G., A.R., D.P., and A.H.-L.; Funding acquisition: S.P.G., A.R., and A.H.-L.; Investigation and methodology: M.T.V., M.D., J.M.R., M.N., V.C., S.P.G., A.R., D.P., and A.H.-L.; Supervision: S.P.G., A.R., and A.H.-L.; Writing—original draft: M.T.V., S.P.G., A.R., D.P., and A.H.-L. All authors have read and agreed to the published version of the manuscript.

Funding: The authors acknowledge the financial support of the CNRS, the University of Lille (FST), the program PHC Procope 2019 (project 42533RC), and the German Academic Exchange Service (DAAD) for financial support (PN:57446225).

Acknowledgments: The authors are very grateful to Brigitte Schöpel, Christian Plinski, Torsten Viergutz for the Laboratory support, and to Olga Plechakova for her assistance with the GT-database. We thank Ralf Bochert (Landesforschungsanstalt Mecklenburg-Vorpommern, Germany) for providing maraena whitefish. The authors acknowledge the Research Federation FRABio (Univ. Lille, CNRS, FR3688, Biochimie Structurale et Fonctionnelle des assemblages Biomoléculaires) for providing the scientific and technical environment conducive to achieving this work, the Ministère de l'enseignement supérieur, de la Recherche et de l'innovation, and the Région Hauts de France for providing research fellowships to M.N. and M.D., and the contribution of the COST Action CA18103-INNOGLY supported by the European Cooperation in Science and Technology (COST). This paper is dedicated to the memory of Roland Schauer.

Conflicts of Interest: The authors declare no conflict of interest.

Abbreviations

DP	degree of polymerization
Neu5Ac	N-acetylneuraminic acid
Neu5Gc	N-glycolylneuraminic
oligoSia	oligosialic acid
polySia or PSA	polysialic acid
PDB	Protein Data Bank
PSGP	salmonid egg polysialoglycoprotein
ST8Sia	α2,8-sialyltransferase
WGD	whole genome duplication
KDN	2-keto-3-deoxynononic acid
MSA	multiple sequence alignment

References

1. Varki, A.; Freeze, H.H.; Gagneux, P. Evolution of glycan diversity. In *Essentials of Glycobiology*; Varki, A., Cummings, R.D., Esko, J.D., Freeze, H.H., Stanley, P., Bertozzi, C.R., Hart, G.W., Etzler, M.E., Eds.; Cold Spring Harbor: New York, NY, USA, 2009.
2. Schauer, R. Sialic acids: Fascinating sugars in higher animals and man. *Zoology* **2004**, *107*, 49–64. [CrossRef] [PubMed]

3. Traving, C.; Schauer, R. Structure, function and metabolism of sialic acids. *Cell. Mol. Life Sci.* **1998**, *54*, 1330–1349. [CrossRef] [PubMed]
4. Angata, T.; Varki, A. Chemical diversity in the sialic acids and related αketo acids: An evolutionary perspective. *Chem. Rev.* **2002**, *102*, 439–469. [CrossRef] [PubMed]
5. Inoue, S.; Kitajima, K. Kdn (deaminated neuraminic acid): Dreamful past and exciting future of the newest member of the sialic acid family. *Glycoconj. J.* **2006**, *23*, 277–290. [CrossRef]
6. Harduin-Lepers, A.; Krzewinski-Recchi, M.A.; Hebbar, M.; Samyn-Petit, B.; Vallejo-Ruiz, V.; Julien, S.; Peyrat, J.P.; Delannoy, P. Sialyltransferases and breast cancer. *Recent Res. Dev. Cancer* **2001**, *3*, 111–126.
7. Harduin-Lepers, A.; Mollicone, R.; Delannoy, P.; Oriol, R. The animal sialyltransferases and sialyltransferase-related genes: A phylogenetic approach. *Glycobiology* **2005**, *15*, 805–817. [CrossRef]
8. Petit, D.; Teppa, E.; Cenci, U.; Ball, S.; Harduin-Lepers, A. Reconstruction of the sialylation pathway in the ancestor of eukaryotes. *Sci. Rep.* **2018**, *8*, 2946. [CrossRef]
9. Lombard, V.; Golaconda Ramulu, H.; Drula, E.; Coutinho, P.M.; Henrissat, B. The carbohydrate-active enzymes database (cazy) in 2013. *Nucleic Acids Res.* **2014**, *42*, D490–D495. [CrossRef]
10. Audry, M.; Jeanneau, C.; Imberty, A.; Harduin-Lepers, A.; Delannoy, P.; Breton, C. Current trends in the structure-activity relationships of sialyltransferases. *Glycobiology* **2011**, *21*, 716–726. [CrossRef]
11. Datta, A.K. Comparative sequence analysis in the sialyltransferase protein family: Analysis of motifs. *Curr. Drug Targets* **2009**, *10*, 483–498. [CrossRef]
12. Petit, D.; Teppa, R.E.; Petit, J.M.; Harduin-Lepers, A. A practical approach to reconstruct evolutionary history of animal sialyltransferases and gain insights into the sequence-function relationships of golgi-glycosyltransferases. *Methods Mol. Biol.* **2013**, *1022*, 73–97.
13. Harduin-Lepers, A. Vertebrate sialyltransferases. In *Sialobiology: Structure, Biosynthesis and Function. Sialic Acid Glycoconjugates in Health and Diseases*; Tiralongo, J., Martinez-Duncker, I., Eds.; Bentham Science: Schipol, The Netherlands, 2013; Volume 5, pp. 139–187.
14. Harduin-Lepers, A.; Petit, D.; Mollicone, R.; Delannoy, P.; Petit, J.M.; Oriol, R. Evolutionary history of the alpha2,8-sialyltransferase (st8sia) gene family: Tandem duplications in early deuterostomes explain most of the diversity found in the vertebrate st8sia genes. *BMC Evol. Biol.* **2008**, *8*, 258. [CrossRef] [PubMed]
15. Harduin-Lepers, A. Comprehensive analysis of sialyltransferases in vertebrate genomes. *Glycobiol. Insights* **2010**, *2*, 29–61. [CrossRef]
16. Patel, R.Y.; Balaji, P.V. Identification of linkage-specific sequence motifs in sialyltransferases. *Glycobiology* **2006**, *16*, 108–116. [CrossRef]
17. Chang, L.Y.; Teppa, E.; Noel, M.; Gilormini, P.A.; Decloquement, M.; Lion, C.; Biot, C.; Mir, A.M.; Cogez, V.; Delannoy, P.; et al. Novel zebrafish mono-alpha2,8-sialyltransferase (st8sia viii): An evolutionary perspective of alpha2,8-sialylation. *Int. J. Mol. Sci.* **2019**, *20*, 622. [CrossRef]
18. Chang, L.Y.; Mir, A.M.; Thisse, C.; Guerardel, Y.; Delannoy, P.; Thisse, B.; Harduin-Lepers, A. Molecular cloning and characterization of the expression pattern of the zebrafish alpha2, 8-sialyltransferases (st8sia) in the developing nervous system. *Glycoconj. J.* **2009**, *26*, 263–275. [CrossRef] [PubMed]
19. Eckhardt, M.; Muhlenhoff, M.; Bethe, A.; Koopman, J.; Frosch, M.; Gerardy-Schahn, R. Molecular characterization of eukaryotic polysialyltransferase-1. *Nature* **1995**, *373*, 715–718. [CrossRef]
20. Kojima, N.; Yoshida, Y.; Tsuji, S. A developmentally regulated member of the sialyltransferase family (st8sia ii, stx) is a polysialic acid synthase. *FEBS Lett.* **1995**, *373*, 119–122. [CrossRef]
21. Nakayama, J.; Fukuda, M.N.; Fredette, B.; Ranscht, B.; Fukuda, M. Expression cloning of a human polysialyltransferase that forms the polysialylated neural cell adhesion molecule present in embryonic brain. *Proc. Natl. Acad. Sci. USA* **1995**, *92*, 7031–7035. [CrossRef]
22. Scheidegger, E.P.; Sternberg, L.R.; Roth, J.; Lowe, J.B. A human stx cdna confers polysialic acid expression in mammalian cells. *J. Biol. Chem.* **1995**, *270*, 22685–22688. [CrossRef]
23. Yoshida, Y.; Kojima, N.; Tsuji, S. Molecular cloning and characterization of a third type of n-glycan alpha 2,8-sialyltransferase from mouse lung. *J. Biochem.* **1995**, *118*, 658–664. [CrossRef] [PubMed]
24. Lee, Y.C.; Kim, Y.J.; Lee, K.Y.; Kim, K.S.; Kim, B.U.; Kim, H.N.; Kim, C.H.; Do, S.I. Cloning and expression of cdna for a human sia alpha 2,3gal beta 1, 4glcna:Alpha 2,8-sialyltransferase (hst8sia iii). *Arch. Biochem. Biophys.* **1998**, *360*, 41–46. [CrossRef] [PubMed]
25. Yoshida, Y.; Kojima, N.; Kurosawa, N.; Hamamoto, T.; Tsuji, S. Molecular cloning of sia α2,3galβ1,4glcnac α2,8-sialyltransferase from mouse brain. *J. Biol. Chem.* **1995**, *270*, 14628–14633. [CrossRef] [PubMed]

26. Angata, K.; Suzuki, M.; McAuliffe, J.; Ding, Y.; Hindsgaul, O.; Fukuda, M. Differential biosynthesis of polysialic acid on neural cell adhesion molecule (ncam) and oligosaccharide acceptors by three distinct α2,8-sialyltransferases, st8sia iv (pst), st8sia ii (stx), and st8sia iii. *J. Biol. Chem.* **2000**, *275*, 18594–18601. [CrossRef] [PubMed]
27. Bhide, G.P.; Fernandes, N.R.; Colley, K.J. Sequence requirements for neuropilin-2 recognition by st8siaiv and polysialylation of its o-glycans. *J. Biol. Chem.* **2016**, *291*, 9444–9457. [CrossRef] [PubMed]
28. Huang, R.B.; Cheng, D.; Liao, S.M.; Lu, B.; Wang, Q.Y.; Xie, N.Z.; Troy Ii, F.A.; Zhou, G.P. The intrinsic relationship between structure and function of the sialyltransferase st8sia family members. *Curr. Top. Med. Chem.* **2017**, *17*, 2359–2369. [CrossRef]
29. Zhou, G.P.; Huang, R.B.; Troy, F.A., 2nd. 3d structural conformation and functional domains of polysialyltransferase st8sia iv required for polysialylation of neural cell adhesion molecules. *Protein Pept. Lett.* **2015**, *22*, 137–148. [CrossRef]
30. Foley, D.A.; Swartzentruber, K.G.; Colley, K.J. Identification of sequences in the polysialyltransferases st8sia ii and st8sia iv that are required for the protein-specific polysialylation of the neural cell adhesion molecule, ncam. *J. Biol. Chem.* **2009**, *284*, 15505–15516. [CrossRef]
31. Nakata, D.; Zhang, L.; Troy, F.A., 2nd. Molecular basis for polysialylation: A novel polybasic polysialyltransferase domain (pstd) of 32 amino acids unique to the α2,8-polysialyltransferases is essential for polysialylation. *Glycoconj. J.* **2006**, *23*, 423–436. [CrossRef]
32. Volkers, G.; Worrall, L.J.; Kwan, D.H.; Yu, C.C.; Baumann, L.; Lameignere, E.; Wasney, G.A.; Scott, N.E.; Wakarchuk, W.; Foster, L.J.; et al. Structure of human st8siaiii sialyltransferase provides insight into cell-surface polysialylation. *Nat. Struct. Mol. Biol.* **2015**, *22*, 627–635. [CrossRef]
33. Asahina, S.; Sato, C.; Matsuno, M.; Matsuda, T.; Colley, K.; Kitajima, K. Involvement of the α2,8-polysialyltransferases ii/stx and iv/pst in the biosynthesis of polysialic acid chains on the o-linked glycoproteins in rainbow trout ovary. *J. Biochem.* **2006**, *140*, 687–701. [CrossRef] [PubMed]
34. Bentrop, J.; Marx, M.; Schattschneider, S.; Rivera-Milla, E.; Bastmeyer, M. Molecular evolution and expression of zebrafish st8siaiii, an alpha-2,8-sialyltransferase involved in myotome development. *Dev. Dyn.* **2008**, *237*, 808–818. [CrossRef] [PubMed]
35. Canestro, C.; Albalat, R.; Irimia, M.; Garcia-Fernandez, J. Impact of gene gains, losses and duplication modes on the origin and diversification of vertebrates. *Semin. Cell Dev. Biol.* **2013**, *24*, 83–94. [CrossRef] [PubMed]
36. Onho, S. Gene duplication and the uniqueness of vertebrate genomes circa 1970–1999. *Semin. Cell Dev. Biol.* **1999**, *10*, 517–522.
37. Pasquier, J.; Cabau, C.; Nguyen, T.; Jouanno, E.; Severac, D.; Braasch, I.; Journot, L.; Pontarotti, P.; Klopp, C.; Postlethwait, J.H.; et al. Gene evolution and gene expression after whole genome duplication in fish: The phylofish database. *BMC Genom.* **2016**, *17*, 368. [CrossRef]
38. Altschul, S.F.; Madden, T.L.; Schaffer, A.A.; Zhang, J.; Zhang, Z.; Miller, W.; Lipman, D.J. Gapped blast and psi-blast: A new generation of protein database search programs. *Nucleic Acids Res.* **1997**, *25*, 3389–3402. [CrossRef] [PubMed]
39. Saitou, N.; Nei, M. The neighbor-joining method: A new method for reconstructing phylogenetic trees. *Mol. Biol. Evol.* **1987**, *4*, 406–425.
40. Kumar, S.; Stecher, G.; Tamura, K. Mega7: Molecular evolutionary genetics analysis version 7.0 for bigger datasets. *Mol. Biol. Evol.* **2016**, *33*, 1870–1874. [CrossRef]
41. Smith, J.J.; Keinath, M.C. The sea lamprey meiotic map improves resolution of ancient vertebrate genome duplications. *Genome Res.* **2015**, *25*, 1081–1090. [CrossRef]
42. Holland, L.Z.; Ocampo Daza, D. A new look at an old question: When did the second whole genome duplication occur in vertebrate evolution? *Genome Biol.* **2018**, *19*, 209. [CrossRef]
43. Inoue, J.G.; Miya, M.; Tsukamoto, K.; Nishida, M. Mitogenomic evidence for the monophyly of elopomorph fishes (teleostei) and the evolutionary origin of the leptocephalus larva. *Mol. Phylogenet. Evol.* **2004**, *32*, 274–286. [CrossRef] [PubMed]
44. Christoffels, A.; Koh, E.G.; Chia, J.M.; Brenner, S.; Aparicio, S.; Venkatesh, B. Fugu genome analysis provides evidence for a whole-genome duplication early during the evolution of ray-finned fishes. *Mol. Biol. Evol.* **2004**, *21*, 1146–1151. [CrossRef] [PubMed]

45. Vandepoele, K.; De Vos, W.; Taylor, J.S.; Meyer, A.; Van de Peer, Y. Major events in the genome evolution of vertebrates: Paranome age and size differ considerably between ray-finned fishes and land vertebrates. *Proc. Natl. Acad. Sci. USA* **2004**, *101*, 1638–1643. [CrossRef] [PubMed]
46. Rozenfeld, C.; Blanca, J.; Gallego, V.; Garcia-Carpintero, V.; Herranz-Jusdado, J.G.; Perez, L.; Asturiano, J.F.; Canizares, J.; Penaranda, D.S. De novo european eel transcriptome provides insights into the evolutionary history of duplicated genes in teleost lineages. *PLoS ONE* **2019**, *14*, e0218085. [CrossRef] [PubMed]
47. Amores, A.; Force, A.; Yan, Y.L.; Joly, L.; Amemiya, C.; Fritz, A.; Ho, R.K.; Langeland, J.; Prince, V.; Wang, Y.L.; et al. Zebrafish hox clusters and vertebrate genome evolution. *Science* **1998**, *282*, 1711–1714. [CrossRef] [PubMed]
48. Robinson-Rechavi, M.; Laudet, V. Evolutionary rates of duplicate genes in fish and mammals. *Mol. Biol. Evol.* **2001**, *18*, 681–683. [CrossRef]
49. Dai, W.; Zou, M.; Yang, L.; Du, K.; Chen, W.; Shen, Y.; Mayden, R.L.; He, S. Phylogenomic perspective on the relationships and evolutionary history of the major otocephalan lineages. *Sci. Rep.* **2018**, *8*, 205. [CrossRef]
50. Hughes, L.C.; Orti, G.; Huang, Y.; Sun, Y.; Baldwin, C.C.; Thompson, A.W.; Arcila, D.; Betancur, R.R.; Li, C.; Becker, L.; et al. Comprehensive phylogeny of ray-finned fishes (actinopterygii) based on transcriptomic and genomic data. *Proc. Natl. Acad. Sci. USA* **2018**, *115*, 6249–6254. [CrossRef]
51. Macqueen, D.J.; Johnston, I.A. A well-constrained estimate for the timing of the salmonid whole genome duplication reveals major decoupling from species diversification. *Proc. Biol. Sci.* **2014**, *281*, 20132881. [CrossRef] [PubMed]
52. Xu, P.; Zhang, X.; Wang, X.; Li, J.; Liu, G.; Kuang, Y.; Xu, J.; Zheng, X.; Ren, L.; Wang, G.; et al. Genome sequence and genetic diversity of the common carp, cyprinus carpio. *Nat. Genet.* **2014**, *46*, 1212–1219. [CrossRef] [PubMed]
53. Nguyen, N.T.T.; Vincens, P.; Roest Crollius, H.; Louis, A. Genomicus 2018: Karyotype evolutionary trees and on-the-fly synteny computing. *Nucleic Acids Res.* **2018**, *46*, D816–D822. [CrossRef] [PubMed]
54. Rabosky, D.L.; Chang, J.; Title, P.O.; Cowman, P.F.; Sallan, L.; Friedman, M.; Kaschner, K.; Garilao, C.; Near, T.J.; Coll, M.; et al. An inverse latitudinal gradient in speciation rate for marine fishes. *Nature* **2018**, *559*, 392–395. [CrossRef] [PubMed]
55. Braasch, I.; Postlethwait, J.H. Polyploidy in fish and the teleost genome duplication. In *Polyploidy and Genome Evolution*; Soltis, P.S., Soltis, D.E., Eds.; Springer: Berlin/Heidelberg, Germany, 2012; pp. 341–383.
56. Lien, S.; Koop, B.F.; Sandve, S.R.; Miller, J.R.; Kent, M.P.; Nome, T.; Hvidsten, T.R.; Leong, J.S.; Minkley, D.R.; Zimin, A.; et al. The atlantic salmon genome provides insights into rediploidization. *Nature* **2016**, *533*, 200–205. [CrossRef] [PubMed]
57. Li, J.T.; Hou, G.Y.; Kong, X.F.; Li, C.Y.; Zeng, J.M.; Li, H.D.; Xiao, G.B.; Li, X.M.; Sun, X.W. The fate of recent duplicated genes following a fourth-round whole genome duplication in a tetraploid fish, common carp (cyprinus carpio). *Sci. Rep.* **2015**, *5*, 8199. [CrossRef] [PubMed]
58. Colley, K.J.; Kitajima, K.; Sato, C. Polysialic acid: Biosynthesis, novel functions and applications. *Crit. Rev. Biochem. Mol. Biol.* **2014**, *49*, 498–532. [CrossRef] [PubMed]
59. Sato, C. Chain length diversity of sialic acids and its biological significance. *Trends Glycosci. Glycotechnol.* **2004**, *16*, 331–344. [CrossRef]
60. Sato, C.; Kitajima, K. Disialic, oligosialic and polysialic acids: Distribution, functions and related disease. *J. Biochem.* **2013**, *154*, 115–136. [CrossRef]
61. Guo, X.; Elkashef, S.M.; Loadman, P.M.; Patterson, L.H.; Falconer, R.A. Recent advances in the analysis of polysialic acid from complex biological systems. *Carbohydr. Polym.* **2019**, *224*, 115145. [CrossRef]
62. Sato, C.; Kitajima, K.; Tazawa, I.; Inoue, Y.; Inoue, S.; Troy, F.A., 2nd. Structural diversity in the alpha 2–>8-linked polysialic acid chains in salmonid fish egg glycoproteins. Occurrence of poly(neu5ac), poly(neu5gc), poly(neu5ac, neu5gc), poly(kdn), and their partially acetylated forms. *J. Biol. Chem.* **1993**, *268*, 23675–23684.
63. Naito-Matsui, Y.; Davies, L.R.; Takematsu, H.; Chou, H.H.; Tangvoranuntakul, P.; Carlin, A.F.; Verhagen, A.; Heyser, C.J.; Yoo, S.W.; Choudhury, B.; et al. Physiological exploration of the long term evolutionary selection against expression of n-glycolylneuraminic acid in the brain. *J. Biol. Chem.* **2017**, *292*, 2557–2570. [CrossRef]
64. Krylov, D.M.; Wolf, Y.I.; Rogozin, I.B.; Koonin, E.V. Gene loss, protein sequence divergence, gene dispensability, expression level, and interactivity are correlated in eukaryotic evolution. *Genome Res.* **2003**, *13*, 2229–2235. [CrossRef] [PubMed]

65. Petit, D.; Teppa, E.; Mir, A.M.; Vicogne, D.; Thisse, C.; Thisse, B.; Filloux, C.; Harduin-Lepers, A. Integrative view of alpha2,3-sialyltransferases (st3gal) molecular and functional evolution in deuterostomes: Significance of lineage-specific losses. *Mol. Biol. Evol.* **2015**, *32*, 906–927. [CrossRef] [PubMed]
66. Peng, L.X.; Liu, X.H.; Lu, B.; Liao, S.M.; Zhou, F.; Huang, J.M.; Chen, D.; Troy, F.A., II; Zhou, G.P.; Huang, R.B. The inhibition of polysialyltranseferase st8siaiv through heparin binding to polysialyltransferase domain (pstd). *Med. Chem.* **2019**, *15*, 486–495. [CrossRef] [PubMed]
67. Waterhouse, A.M.; Procter, J.B.; Martin, D.M.A.; Clamp, M.; Barton, G.J. Jalview version 2—A multiple sequence alignment editor and analysis workbench. *Bioinformatics* **2009**, *25*, 1189–1191. [CrossRef]
68. Angata, K.; Nakayama, J.; Fredette, B.; Chong, K.; Ranscht, B.; Fukuda, M. Human stx polysialyltransferase forms the embryonic form of the neural cell adhesion molecule. Tissue-specific expression, neurite outgrowth, and chromosomal localization in comparison with another polysialyltransferase, PST. *J. Biol. Chem.* **1997**, *272*, 7182–7190. [CrossRef]
69. Hall, B.G. Building phylogenetic trees from molecular data with mega. *Mol. Biol. Evol.* **2013**, *30*, 1229–1235. [CrossRef]
70. Felsenstein, J. Confidence limits on phylogenies: An approach using the bootstrap. *Evolution* **1985**, *39*, 783–791. [CrossRef]
71. Altmann, S.; Rebl, A.; Kuhn, C.; Goldammer, T. Identification and de novo sequencing of housekeeping genes appropriate for gene expression analyses in farmed maraena whitefish (coregonus maraena) during crowding stress. *Fish Physiol. Biochem.* **2015**, *41*, 397–412. [CrossRef]
72. Brietzke, A.; Borchel, A.; Altmann, S.; Nipkow, M.; Rebl, A.; Brunner, R.M.; Goldammer, T. Transcriptome sequencing of maraena whitefish (coregonus maraena). *Mar. Genom.* **2016**, *29*, 27–29. [CrossRef]

International Journal of
Molecular Sciences

Article

Identification and Characterization of a β-*N*-Acetylhexosaminidase with a Biosynthetic Activity from the Marine Bacterium *Paraglaciecola hydrolytica* S66^{T}

Triinu Visnapuu [1,2,*], David Teze [1], Christian Kjeldsen [3], Aleksander Lie [4,†], Jens Øllgaard Duus [3], Corinne André-Miral [5], Lars Haastrup Pedersen [4], Peter Stougaard [6,‡] and Birte Svensson [1,*]

1. Department of Biotechnology and Biomedicine, Technical University of Denmark, Søltofts Plads, Building 224, DK-2800 Kgs. Lyngby, Denmark; david.teze@gmail.com
2. Institute of Molecular and Cell Biology, University of Tartu, Riia 23, 51010 Tartu, Estonia
3. Department of Chemistry, Technical University of Denmark, Kemitorvet, Building 207, DK-2800 Kgs. Lyngby, Denmark; chkje@kemi.dtu.dk (C.K.); jduus@kemi.dtu.dk (J.Ø.D.)
4. Department of Chemistry and Bioscience, Aalborg University, Fredrik Bajers Vej 7H, DK-9220 Aalborg, Denmark; al@kebony.com (A.L.); lhp@bio.aau.dk (L.H.P.)
5. Unité de Fonctionnalité et Ingénierie des Protéines (UFIP), UMR CNRS 6286, Université de Nantes, F-44000 Nantes, France; corinne.miral@univ-nantes.fr
6. Department of Plant and Environmental Sciences, University of Copenhagen, Thorvaldsensvej 40, DK-1871 Frederiksberg C, Denmark; pst@envs.au.dk

* Correspondence: triinu.visnapuu@ut.ee (T.V.); bis@bio.dtu.dk (B.S.); Tel.: +372-737-5013 (T.V.); +45-4525-2740 (B.S.)

† Present affiliation: Kebony Norge AS, Havneveien 35, NO-3739 Skien, Norway.

‡ Present affiliation: Department of Environmental Science—Environmental Microbiology and Circular Resource Flow, Aarhus University, Frederiksborgvej 399, Building 7411, B2.12, DK-4000 Roskilde, Denmark.

Received: 1 December 2019; Accepted: 7 January 2020; Published: 9 January 2020

Abstract: β-*N*-Acetylhexosaminidases are glycoside hydrolases (GHs) acting on *N*-acetylated carbohydrates and glycoproteins with the release of *N*-acetylhexosamines. Members of the family GH20 have been reported to catalyze the transfer of *N*-acetylglucosamine (GlcNAc) to an acceptor, i.e., the reverse of hydrolysis, thus representing an alternative to chemical oligosaccharide synthesis. Two putative GH20 β-*N*-acetylhexosaminidases, *Ph*Nah20A and *Ph*Nah20B, encoded by the marine bacterium *Paraglaciecola hydrolytica* S66^{T}, are distantly related to previously characterized enzymes. Remarkably, *Ph*Nah20A was located by phylogenetic analysis outside clusters of other studied β-*N*-acetylhexosaminidases, in a unique position between bacterial and eukaryotic enzymes. We successfully produced recombinant *Ph*Nah20A showing optimum activity at pH 6.0 and 50 °C, hydrolysis of GlcNAc β-1,4 and β-1,3 linkages in chitobiose $(GlcNAc)_2$ and GlcNAc-1,3-β-Gal-1,4-β-Glc (LNT2), a human milk oligosaccharide core structure. The kinetic parameters of *Ph*Nah20A for *p*-nitrophenyl-GlcNAc and *p*-nitrophenyl-GalNAc were highly similar: k_{cat}/K_M being 341 and 344 $mM^{-1}\cdot s^{-1}$, respectively. *Ph*Nah20A was unstable in dilute solution, but retained full activity in the presence of 0.5% bovine serum albumin (BSA). *Ph*Nah20A catalyzed the formation of LNT2, the non-reducing trisaccharide β-Gal-1,4-β-Glc-1,1-β-GlcNAc, and in low amounts the β-1,2- or β-1,3-linked trisaccharide β-Gal-1,4(β-GlcNAc)-1,*x*-Glc by a transglycosylation of lactose using 2-methyl-(1,2-dideoxy-α-D-glucopyrano)-oxazoline (NAG-oxazoline) as the donor. *Ph*Nah20A is the first characterized member of a distinct subgroup within GH20 β-*N*-acetylhexosaminidases.

Keywords: *N*-acetylhexosamine specificity; glycoside hydrolase; GH20; phylogenetic analysis; transglycosylation; NAG-oxazoline; acceptor diversity; lacto-*N*-triose II; human milk oligosaccharides; NMR

---

1. Introduction

A new marine bacterial species *Paraglaciecola hydrolytica* S66^T of the family *Alteromonadaceae* isolated from eelgrass (*Zostera* sp.) was shown by genome-sequencing [1] to encode 270 protein modules potentially acting on carbohydrates, 188 of which belong to enzyme families involved in degradation of carbohydrates [2,3]. The algal polysaccharides agar, agarose, alginate, porphyran or laminarin, but not carrageenans, fucoidan and ulvan, sustained the growth of *P. hydrolytica* as a sole carbon source, and the bacterium also grew on the plant polysaccharides: starch, amylopectin, amylose, xylan and pectin [2]. Overall, the large number of encoded carbohydrate-active enzymes (CAZymes) [4] and the flexibility with regard to carbon source indicates a very promising potential of the genome of *P. hydrolytica* for the discovery of enzymes with rare or not yet described activities.

Enzymes hydrolyzing glycosidic bonds with the release of *N*-acetylglucosamine (GlcNAc) are in focus since these carbohydrate residues occur in vital complex glycans, such as milk oligosaccharides and glycosphingolipids, for which there is a great demand [5]. Human milk oligosaccharides (HMOs) in particular are considered beneficial and needed for research and clinical trials within nutrition and as ingredients in functional foods and infant formulas [6–8]. HMOs are also regarded as emerging prebiotics or novel foods with positive health effects [9,10]. However, the chemical and enzymatic production of HMOs and their precursors or purification from natural sources are problematical [6,11,12], which creates bottlenecks for assessing the functional roles and applications of HMOs [13–15].

Lacto-*N*-triose II (LNT2, β-GlcNAc-1,3-β-Gal-1,4-Glc) is an HMO core structure in which *N*-acetylglucosamine is β-1,3-linked to lactose [6,16,17]. A few β-*N*-acetylhexosaminidases (β-NAHAs; EC 3.2.1.52) of the glycoside hydrolase family 20 (GH20) from bacteria, fungi and plants are reported to produce HMO-type GlcNAc-containing oligosaccharides with 1,3 linkages [15,18,19], as well as chitooligosaccharides and their analogs in transglycosylation reactions with the formation of 1,6 rather than 1,4 linkages [18,20–22]. In Nature, β-NAHAs from GH3, 20, 84, 109 and 116 [5,23,24] categorized in the CAZy database (www.cazy.org) [4] degrade *N*-acetylhexosamine-containing compounds by releasing GlcNAc and GalNAc from the non-reducing ends of *N*-acetylglucosides, *N*-acetylgalactosides, glycosphingolipids and glycoproteins [5,25–27]. Interestingly, these families display a variety of mechanisms, either retaining via a substrate-assisted mechanism (GH20 and GH84) [28,29] or a glycosyl-enzyme intermediate (GH3 and GH116) [30], or inverting via an oxidized form of nicotinamide adenine dinucleotide (NAD^+)-depending mechanism (GH109) [31]. While being represented in five distinct GH families, the large majority of β-NAHAs belong to GH20.

N-acetylated oligo- and polysaccharides, e.g., chitooligosaccharides and chitin, are prevalent in marine organisms, thus crustaceans represent an abundant source of GlcNAc in marine environments. The National Center for Biotechnology Information (NCBI) database (https://www.ncbi.nlm.nih.gov/) currently has more than 112,000 predicted β-NAHAs, but out of the more than 200 characterized EC 3.2.1.52 enzymes (www.brenda-enzymes.org) [32], only a small number are of marine origin [21,26,33–36]. Accordingly, only six out of the 133 characterized GH20 β-NAHAs are from a marine organism (from www.cazy.org, 21st of November, 2019) even though a large number of sequences, also of marine origin, are annotated in genomes and metagenomes. These six characterized marine GH20 enzymes comprise Hex99 and Hex86 from *Pseudoalteromonas piscicida* (previously *Alteromonas* sp.) [21,35], Nag20A [36] and NagB [34] from the widespread *Aeromonas hydrophila*, chitobiase of *Vibrio harveyi* [37] and ExoI from *Vibrio furnissii* [33]. However, of these, only Hex99 from *P. piscicida* was examined for its ability to catalyze transglycosylation reactions [21].

Biochemical characteristics such as pH optimum (between pH 6.0-7.0) and temperature optimum (37-50 °C) of the six enzymes are rather similar. Moreover, based on K_M and V_{max} values, most of the enzymes have higher specific activity towards *p*-nitrophenyl-GlcNAc (*p*NPGlcNAc) compared to *p*-nitrophenyl-GalNAc (*p*NPGalNAc) [25,33,34,36].

It has not been possible to clearly distinguish between GH20 β-NAHAs from water-living and terrestrial organisms or from bacterial and eukaryotic organisms based solely on different functional features of the enzymes. The biochemical characteristics of GH20 β-NAHAs vary considerably, as it has been reviewed recently by Zhang et al. [25]. For example, pH optima of GH20 enzymes range from pH 3.0 for Hex of *Streptomyces plicatus* to pH 8.0 for Hex1 (from a metagenomic library) [18,25,38]. Affinities as given by K_M values for *p*NPGlcNAc range from 53 μM for *Cf*Hex20 from *Cellulomonas fimi* [39] to 120 mM for BbhI of *Bifidobacterium bifidum* [40]. Murine cytosolic β-NAHA shows K_M = 0.25 mM on *p*NPGalNAc, which it preferred over *p*NPGlcNAc [41]. Similarly, human plasma and pig brain β-NAHAs have a lower K_M for *p*NPGalNAc of 0.17 mM and 0.2 mM, respectively [42,43]. Interestingly, salt-tolerant HJ5Nag from *Microbacterium* sp. has a high V_{max} towards *p*NPGlcNAc of 3097 $\mu mol \cdot mg^{-1} \cdot min^{-1}$ [44]. One of the highest k_{cat} and catalytic efficiency values reported towards *p*NPGlcNAc are for *Cf*Hex20 of *C. fimi* reaching 480 s^{-1} and 9000 $mM^{-1} \cdot s^{-1}$, respectively [39]. Crystal structures are available for several terrestrial GH20 β-NAHAs, e.g., Hex1T from *Paenibacillus* sp. TS12 [45], StrH from *Streptococcus pneumoniae* [46], HexA from *Streptomyces coelicolor* [47], Hex from *S. plicatus* [38] and Am2301 from *Akkermansia muciniphila* [48], but not for any aquatic GH20 enzymes.

Here, the genome of *P. hydrolytica* S66^T encoding 113 predicted glycoside hydrolases [1,3] was mined for β-NAHAs potentially acting on GlcNAc-containing compounds, e.g., chitooligosaccharides, which are abundant in the marine environment. Two putative GH20 encoding genes were identified in the genome, and one of the corresponding enzymes, *Ph*Nah20A, was produced recombinantly, characterized biochemically and moreover shown to catalyze transglycosylation using the GH20 reaction intermediate NAG-oxazoline [2-methyl-(1,2-dideoxy-α-D-glucopyrano)-oxazoline] as donor and lactose as well as a series of monosaccharides as acceptors.

2. Results and Discussion

2.1. Identification of Putative β-NAHAs in P. hydrolytica and Organization of Vicinal Genomic Regions

The marine bacterium *P. hydrolytica* degrades effectively many different polysaccharides [2] and its genome exhibits potential for the degradation of chitin and chitooligosaccharides. *P. hydrolytica* was grown in marine mineral medium supplemented with a mixture of chitooligosaccharides $(GlcNAc)_{1-6}$ as the sole carbon source, which were hydrolyzed to GlcNAc (Supplementary Information 1, Figure S1A,B). *P. hydrolytica*, however, did not hydrolyze α-chitin from crab shells used to supplement the marine mineral medium, as neither GlcNAc nor chitooligosaccharides appeared during the incubation (Figure S1C). β-NAHA activity from *P. hydrolytica* was detected by a hydrolysis of the chromogenic 5-bromo-4-chloro-3-indolyl *N*-acetyl-β-D-glucosaminide (X-GlcNAc) on a complex marine agar medium (Figure S1D). These results indicated that the bacterium produced at least one β-NAHA which was active towards chitooligosaccharides.

The draft genome sequence of *P. hydrolytica* [1], deposited on the RAST server (http://rast.nmpdr.org/), encodes two putative GH20 β-NAHAs (EC 3.2.1.52) based on automatic annotation. Both genes were found in contig 11 of the *P. hydrolytica* whole genome shotgun sequence (NCBI accession: NZ_LSNE01000003.1). The protein sequence identity between full-length *Ph*Nah20A (WP_068373836.1) and *Ph*Nah20B (WP_082768773.1) was 23%.

Top hits of protein BLAST, showing up to 54% to *Ph*Nah20A and up to 49% sequence identity to *Ph*Nah20B, were GH20 β-NAHAs or chitobiases from phylogenetically closely related marine and soil bacteria belonging mostly to the same order as *P. hydrolytica*—*Alteromonadales* (Table S1). None of these proteins, encoded by genes from *Paraglaciecola* or related bacteria (Table S1), had been recombinantly produced or characterized.

The closest relatives of *Ph*Nah20A are GH20 β-NAHAs from *Bowmanella denitrificans* and *Lacimicrobium alkaliphilum* with 53%-54% sequence identity (Table S1). *Ph*Nah20A contains two domains, the GH20 catalytic $(\beta/\alpha)_8$-barrel domain (Pfam: PF00728) and the N-terminal GH20b domain (also referred to as GH20 domain 2; Pfam: PF02838) of a predicted zincin-like fold similar to zinc-dependent metalloproteases [49] consisting of four antiparallel β-strands and an α-helix [27,50]. These two domains are typical for GH20 enzymes [50], and importantly they constitute an active and stable minimum functional unit of GH20 enzymes, thus requiring both a catalytic GH20 and a GH20b domain [50]. *Ph*Nah20A has no predicted signal peptide sequence and most probably is not secreted, whereas a 28 residues N-terminal signal peptide was predicted for the hypothetical *Ph*Nah20B (Figure 1A). Therefore, during the growth of *P. hydrolytica* on chitooligosaccharides, *Ph*Nah20B probably performs the initial degradation of these substrates. *Ph*Nah20B, in addition to the GH20b and GH20 domains, contains a putative carbohydrate binding domain of the CHB_HEX superfamily (Pfam: PF03173) having a predicted β-sandwich structure similar to cellulose binding domains in cellulases [51], and a C-terminal CHB_HEX_C domain (Pfam: PF03174) of unknown function resembling an immunoglobulin-like fold [50,51]. A similar four-domain architecture was seen in the crystal structure of a chitobiase from *S. marcescens* [51], and has only been reported for bacterial GH20 enzymes [50,51]. Based on its protein sequence identity and domain architecture, *Ph*Nah20B resembles a biochemically uncharacterized GH20 chitobiase from *Aliiglaciecola lipolytica* and β-NAHAs from other phylogenetically close marine bacteria (Table S1). It can be concluded that one of the reasons for low sequence identity, i.e., 23%, between two putative GH20 enzymes of *P. hydrolytica*, was the different domain architecture of *Ph*Nah20A and *Ph*Nah20B (Figure 1A), as *Ph*Nah20B has two additional domains besides the GH20 catalytic domain and an N-terminal GH20b domain. The identity between the two proteins remained low when only the predicted GH20b and GH20 domain sequences were compared, as some regions are not aligning between proteins (Supplementary Information 2).

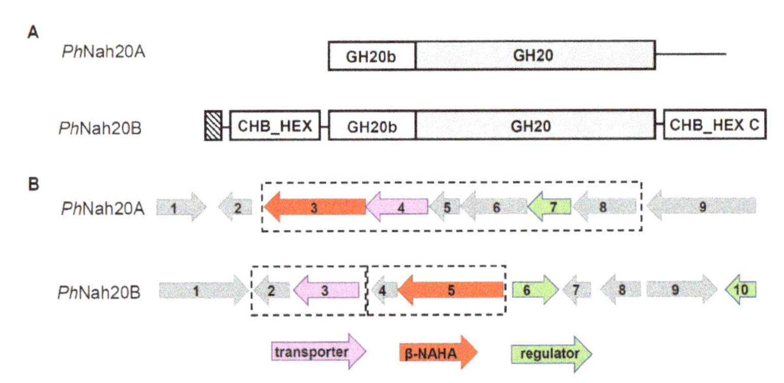

Figure 1. Schematic domain architecture of *P. hydrolytica Ph*Nah20A and *Ph*Nah20B (**A**) and of genomic regions flanking the two putative β-*N*-acetylhexosaminidases (β-NAHAs) (red arrows) (**B**). (**A**) GH20 catalytic domains are gray and the *N*-terminal signal peptide is striped. (**B**) Predicted protein functions are color coded. The information was retrieved from the National Center for Biotechnology Information (NCBI) database (NZ_LSNE01000003.1), Uniprot and Pfam databases. The regions flanking *Ph*Nah20A (3): 1, LemA family protein; 2, hypothetical protein; 4, sodium:solute symporter, putative SLC5sbd family protein; 5, RidA (reactive intermediate/imine deaminase A) family protein; 6, D-aminoacylase; 7, MurR/RpiR family transcriptional regulator; 8, amino acid deaminase; 9, sodium/proton-translocating pyrophosphatase. The regions flanking *Ph*Nah20B (5): 1, TonB-dependent receptor; 2, DUF1624 domain-containing protein, putative acyltransferase; 3, glucose/galactose MFS transporter; 4, hypothetical protein, putative BadF-type ATPase; 6, LacI family DNA-binding transcriptional regulator; 7, dCTP deaminase; 8, iron–sulfur cluster carrier protein ApbC; 9, methionine-tRNA ligase; 10, TetR/AcrR family transcriptional regulator. Predicted operons are in dashed frames.

Genomic regions flanking the two annotated *P. hydrolytica* β-NAHAs, *Ph*Nah20A and *Ph*Nah20B, were examined for the presence of operons (Figure 1B), but were found not to be organized similarly to the operon responsible for chitobiose-utilization in *Escherichia coli* [52]. Surrounding putative genes, however, encoded proteins potentially participating in the modification of acetylated compounds, the transporter function and transcription regulation (Figure 1B; Table S2). Notably, a predicted operon of six genes that harbors *Ph*Nah20A (Figure 1B) included a putative amino acid deaminase, D-aminoacylase and the RidA (reactive intermediate/imine deaminase A) family protein, possibly associated with the processing of acylated compounds or amino acids [53]. A two-gene operon was predicted to harbor *Ph*Nah20B and a putative ATPase (Figure 1B, Table S2). Thus, GH20 β-NAHAs genes of *P. hydrolytica* were not situated adjacent to genes encoding proteins directly coupled to β-NAHA activity, but flanking genes may be important for regulation or substrate transport.

2.2. Phylogenetic Analysis of PhNah20A and PhNah20B

Sequences of *Ph*Nah20A, *Ph*Nah20B and 41 characterized GH20 enzymes were aligned (Supplementary Information 2). *Ph*Nah20A and *Ph*Nah20B shared a low sequence identity with the other GH20 enzymes (up to 34.1% for *Ph*Nah20A and 37.9% for *Ph*Nah20B) and only a few highly conserved regions were identified among these GH20 members (Supplementary Information 2). The closest homologs of *Ph*Nah20A were Hex2 of an uncultured *Bacteroidetes* (34.1% identity) and ExoI of the marine bacterium *V. furnissii* (33.1% identity). Remarkably, GH20 sequences from eukaryotes (human and mouse) were 31.3% and 30.9% identical and more similar to *Ph*Nah20A than most other included bacterial sequences. The *Ph*Nah20B sequence was most similar to chitobiases from *S. marcescens* (37.9% identity) and *V. harveyi* (36.4% identity). The evolutionary relationship illustrated by a radial phylogenetic tree (Figure 2; for bootstrap values see Figure S2) showed that bacterial GH20s segregate into three groups.

*Ph*Nah20B clustered with β-NAHAs from water-living bacteria from the phylogenetically close species such as *V. harveyi*, *P. piscicida* and *A. hydrophila*. However, *Ph*Nah20A did not cluster with characterized bacterial β-NAHAs but seems to represent a new distinct group of GH20 enzymes situated between predominantly water-living bacteria and the eukaryotes (Figure 2).

NagA of the slime mold *Dictyostelium discoideum* which clusters not far from *Ph*Nah20A (Figure 2), is a lysosomal enzyme that maintains the size of pseudoplasmodia [54], and shares 28.5% sequence identity with *Ph*Nah20A. According to the BLAST analysis, *Ph*Nah20A has higher sequence identity to biochemically uncharacterized β-NAHAs from phylogenetically close marine bacteria (Table S1). Additionally, protein sequences with 44–47% identity to *Ph*Nah20A were found in compost, hydrothermal vent and marine sediment metagenomes (Table S1) highlighting unexplored resources harbouring a new group of β-NAHAs.

According to the literature, substrate specificities and biochemical features (e.g., pH and temperature optima) are reported for 41 β-NAHAs of GH20 [4,32] mostly from terrestrial organisms. The few enzymes being from marine bacteria comprise ExoI and chitobiase from *Vibrio* sp. [33,37], Hex99 and Hex86 from *P. piscicida* [21,35] and Nag20A and NagB from *A. hydrophila* [34,36]. The limited knowledge on GH20 from marine organisms motivated the present characterisation of β-NAHA from *P. hydrolytica* S66^{T}.

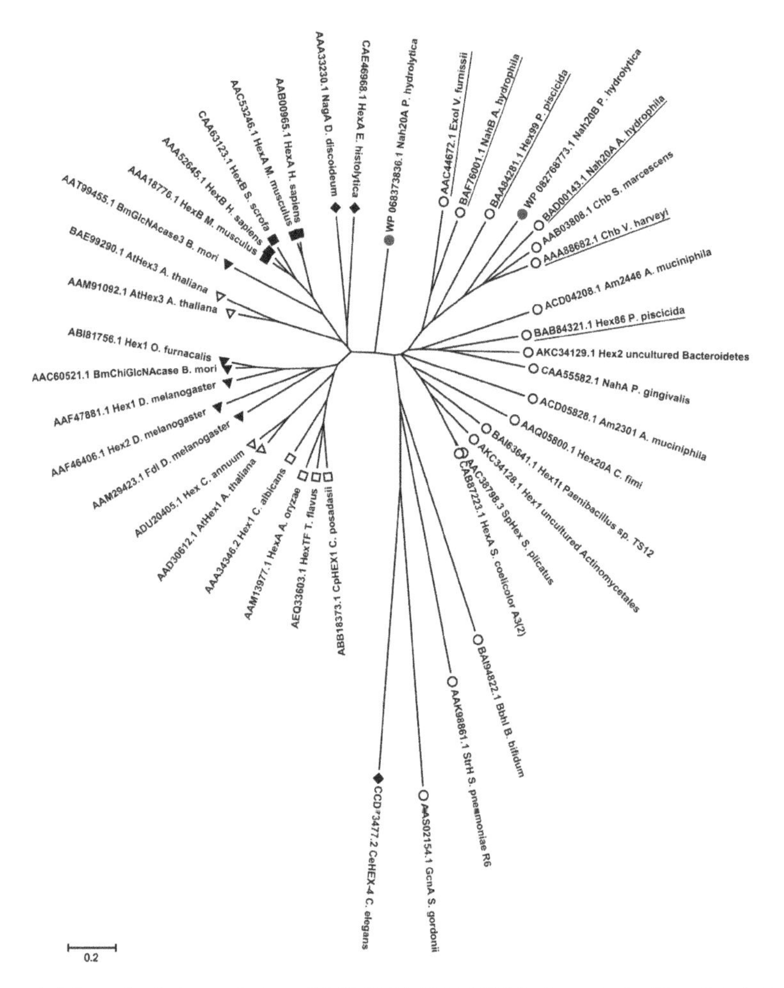

Figure 2. Schematic phylogenetic tree of *Ph*Nah20A, *Ph*Nah20B (both marked with red circles) and 41 biochemically characterized GH20 (EC 3.2.1.52) enzymes. Evolutionary analyzes were conducted, and the tree was composed and visualized using MEGA v 7.0.26 [55]. Protein sequences were aligned with Clustal Omega and the BLOSUM62 protein weight matrix was used. Evolutionary relationships were calculated using the Neighbor-Joining method. Evolutionary distances were computed using the Poisson correction method. All positions containing gaps and missing data were eliminated, and there was in total 292 positions in the final dataset. The tree is in scale with branch lengths in the same units as those of the evolutionary distances used to infer the phylogenetic tree. Bacterial (○), fungal (□), plant (Δ), insect (▲) and mammal (■) sequences. Amoebae and *C. elegans* sequences are marked with a filled diamond (♦). Characterized GH20 enzymes from marine organisms are underlined.

2.3. Cloning and Production of β-NAHA

From the two candidate β-NAHA genes (Figure 1A), only recombinant *Ph*Nah20A was successfully produced in *E. coli* (Figure 3). *Ph*Nah20B cloned without the N-terminal signal peptide (Figure S3) was not obtained despite expression attempts in three *E. coli* strains [BL21(DE3), BL21(DE3)ΔlacZ and Rosetta], using different induction methods: isopropyl thio-β-D-galactoside (IPTG)-induction in lysogeny broth (LB) or auto-induction. *Ph*Nah20B was not found in the insoluble fraction by analyzing whole cells from IPTG-induced cultures (Figure S4). The yield of *Ph*Nah20A was modest, probably due to a low expression level. Using different strains and induction strategies resulted in the highest β-NAHA activity of 6 μmol *p*-nitrophenol released per min and per mg protein in the *E. coli* cell lysate for IPTG-induced BL21(DE3) transformants in LB medium (Figure 3).

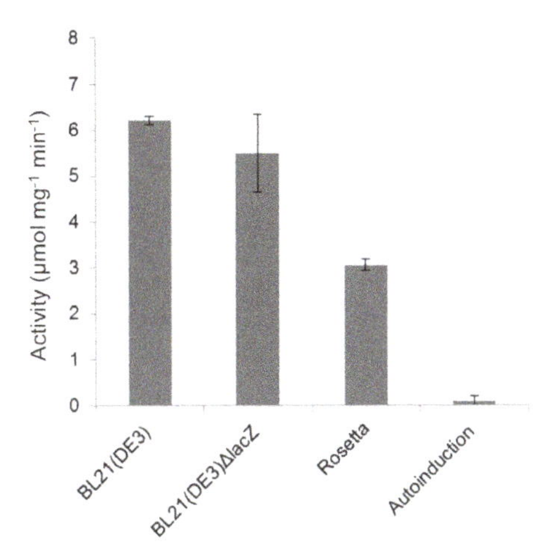

Figure 3. β-NAHA activity in μmol per min per mg of total protein in the lysates of three different *Escherichia coli* (*E. coli*) strains harboring *Ph*Nah20A grown in lysogeny broth (LB) induced by isopropyl thio-β-D-galactoside (IPTG) or in the auto-induction medium ZYM-5052 [56] (30 h). *E. coli* BL21(DE3) transformants carrying the full-length *Ph*Nah20A gene in an pURI3TEV expression vector were used in the auto-induction experiment. Values are given as the average of three independent experiments and the standard deviation (SD) is shown.

Previously, an increased expression of GH20 β-NAHAs from a metagenome [18] was achieved in *E. coli* strains BL21(DE3), Turner, C41 or C43 grown in an auto-induction medium ZYM-5052 [56], but this medium gave a very low yield of *Ph*Nah20A (Figure 3) and failed to lead to *Ph*Nah20B production.

Up to 2 mg of *Ph*Nah20A was purified in two chromatographic steps from one liter of *E. coli* BL21(DE3) culture (see Section 3.4). Expression of truncated *Ph*Nah20A and *Ph*Nah20B, containing only the catalytic and not the GH20b domain (see Figure S3), did not result in protein production which is in agreement with previous findings that GH20b is essential for enzyme production and activity [50]. Attempts to produce *Ph*Nah20B without the CHB_HEX domains (Figure S3) also gave no detected protein or β-NAHA activity.

2.4. *Characterization of* Ph*Nah20A*

2.4.1. Enzyme Stability

The activity of *Ph*Nah20A decreased immediately after dilution to the low concentration of 5 μg·mL^{-1}, even when kept on ice (Figure 4). By contrast, 1 mg·mL^{-1} *Ph*Nah20A retained activity at least four months at 4 °C in 50 mM sodium phosphate pH 7.0, 0.3 M NaCl and 0.02% NaN_3. The presence of 0.5% BSA or 0.5% Triton X-100 efficiently stabilized *Ph*Nah20A at 5 μg·mL^{-1} and pH 6.0 (see Figure 5A), whereas 0.5 and 2 M NaCl had no effect (Figure 4). This behavior and the absence of a signal peptide suggest *Ph*Nah20A is an intracellular enzyme. Without a stabilizing agent, 5 μg·mL^{-1} *Ph*Nah20A was completely inactivated within 5 min at 50 °C, while 50% and 3% activity were retained after 20 min and 4 h, respectively, in 0.5% BSA (Figure S5), and activity was fully retained after 4 d at 37 °C. β-NAHAs from *E. coli* [57], *Prunus serotina* [58], *Bos taurus* [59], *Hordeum vulgare* [60] and *Streptomyces plicatus* [61] were similarly found to lose activity by dilution. BSA has been identified as an activating compound to some β-NAHAs, e.g., from *Mus musculus* [41] and human plasma [42].

Notably, Hex, the commercial *S. plicatus* β-NAHA, is produced as a fusion with maltose-binding protein to secure stability and the Hex reaction mixture contained 0.3% of BSA to maintain activity [38].

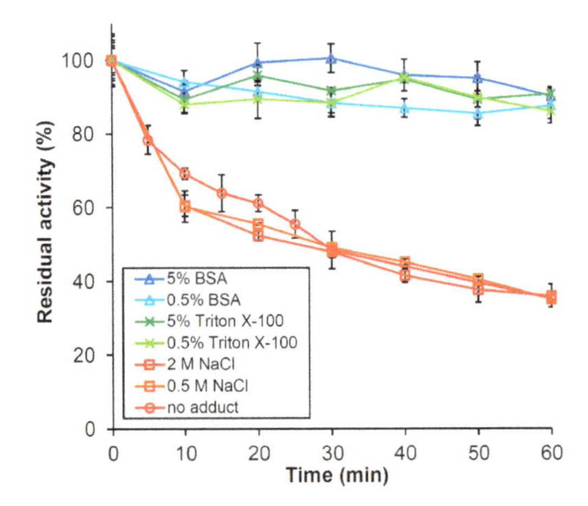

Figure 4. Effect of bovine serum albumin (BSA), Triton X-100, and NaCl on the stability of 5 μg·mL^{-1} *Ph*Nah20A on ice. The retained activity was measured at 37 °C in McIlvaine buffer, pH 6.0, using 2 mM *p*NPGlcNAc as the substrate. *Ph*Nah20A, BSA, NaCl and Triton X-100 were further 10 times diluted in the activity assay done at 0.5 μg·mL^{-1} *Ph*Nah20A, 0.05% or 0.5% BSA, 0.05% or 0.5% Triton X-100, 0.05 or 0.2 M NaCl.

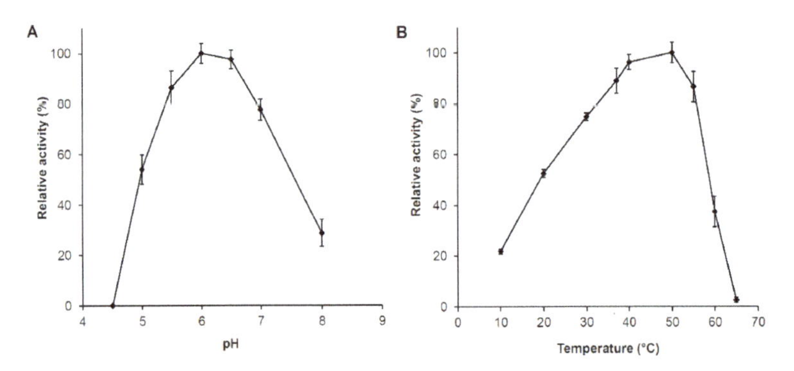

Figure 5. *Ph*Nah20A pH and temperature optima. The effect of pH (**A**) at 37 °C and the temperature at pH 6.0 (**B**) on initial rates of the hydrolysis of 2 mM *p*NPGlcNAc are both expressed as relative activity (%) from optimal activity values.

2.4.2. pH and Temperature Optima

*Ph*Nah20A was most active at pH 5.0–7.5 with a maximum around pH 6.0 (Figure 5A) and a temperature optimum at 50 °C (Figure 5B). The pH optimum of *Ph*Nah20A is highly similar to numerous characterized GH20 β-NAHAs [25], e.g., from *Microbacterium* sp. [44], *Paenibacillus* sp. [45] and *A. hydrophila* [34]. Some fungal β-NAHAs have more acidic pH optima (pH 4–5) [62–64]. Similarly high temperature optima as for *Ph*Nah20A were found for chitinases from *Salinivibrio costicola* [65], β-NAHAs from *Serratia marcescens* [66], *A. hydrophila* [36] as well as *Penicillium oxalicum* [27]. *Ph*Nah20A, however, when diluted in buffer lost activity completely within 5 min at 50 °C in the absence of stabilizers (Figure S5), emphasizing the importance of an environment with high protein concentration for the stability of *Ph*Nah20A. The optimal growth temperature of *P. hydrolytica* was 20–25 °C [2], but the temperature optimum for *Ph*Nah20A activity was much higher, which is a common phenomenon reported for other bacterial GH20 enzymes [27,66].

2.4.3. Substrate Specificity and Kinetic Parameters of *Ph*Nah20A

*Ph*Nah20A hydrolyzed *N,N′*-diacetylchitobiose [chitobiose, $(GlcNAc)_2$, β-GlcNAc-1,4-GlcNAc] and lacto-*N*-triose II (β-GlcNAc-1,3-β-Gal-1,4-Glc, LNT2) with the release of GlcNAc. Chitobiose was a poor substrate and 1 $U{\cdot}mL^{-1}$ (11.6 $\mu g{\cdot}mL^{-1}$) *Ph*Nah20A converted only 25% of 200 mM chitobiose in 20 h at pH 6.0 as analyzed by high-performance anion exchange chromatography with pulsed amperometric detector (HPAEC-PAD). Similarly, only trace amounts of GlcNAc were released from chitobiose by the GH20 BbhI from *B. bifidum* [40]. The action on LNT2 motivated assaying for transglycosylation activity (see Section 2.5), i.e., the ability to catalyze the reverse reaction of hydrolysis and in particular to produce HMOs, as described for the BbhI from *B. bifidum* [15,40].

Kinetic parameters for *Ph*Nah20A hydrolyzing *p*NPGlcNAc and *p*NPGalNAc (Table 1) were very similar, k_{cat} being slightly higher on *p*NPGalNAc. This identified *Ph*Nah20A as an *N*-acetylhexosaminidase rather than either an *N*-acetylglucosaminidase or an *N*-acetylgalactosaminidase. Most β-NAHAs, especially bacterial GH20 enzymes, prefer *p*NPGlcNAc (Table 1) and are referred to as *N*-acetylglucosaminidases. For instance, *S. marcescens* β-NAHA showed only 28.1% activity on *p*NPGalNAc compared to *p*NPGlcNAc [66]. Similarly, HexA from the ameba *E. histolytica* had 38% activity on *p*NPGalNAc compared to *p*NPGlcNAc [67]. Nag20A from *A. hydrophila* had very similar K_M for *p*NPGlcNAc and *p*NPGalNAc, but V_{max} for *p*NPGalNAc was only 13% of V_{max} for *p*NPGlcNAc [36]. Nag20B, also from *A. hydrophila*, showed about 20 times higher K_M for *p*NPGlcNAc and *p*NPGalNAc [34] than *Ph*Nah20A. Other differences include *V. furnissii* ExoI showing 3.6 times lower K_M towards *p*NPGlcNAc than *p*NPGalNAc [33]. Similarly, Hex1 and Hex2 from a metagenomic library showed very poor activity for *p*NPGalNAc [18]. On the other hand, β-NAHAs of human and mouse prefer *p*NPGalNAc as a substrate over *p*NPGlcNAc and have a high affinity towards it (K_M of 0.17 and 0.25 mM, respectively) [41,42]. Another eukaryotic β-NAHA from *D. discoideum* showed equal affinity (K_M of 1.5 mM) for both substrates [68], thus resembling more *Ph*Nah20A and some other bacterial enzymes (Table 1). Interestingly, BbhI had very high K_M of 120 mM for *p*NPGlcNAc (Table 1), but much lower K_M of 0.36 mM for LNT2 [40].

Table 1. Kinetic parameters of *Ph*Nah20A and β-NAHAs from the literature on *p*NPGlcNAc and *p*NPGalNAc. *Sm*—*S. marcescens*; *Ah*—*A. hydrophila*; *Bb*—*B. bifidum*; *Cf*—*C. fimi*; *Vf*—*V. furnissii*; *Eh*—*E. histolytica*; *Tr*—*Trichoderma reesei*.

Enzyme	Substrate	K_M (mM)	V_{max} (μmoL·mg^{-1}·min^{-1})	k_{cat} (s^{-1})	k_{cat}/K_M (mM^{-1}·s^{-1})
*Ph*Nah20A	*p*NPGlcNAc	0.43 ± 0.07	93.7 ± 5.0	146.8	341
	*p*NPGalNAc	0.56 ± 0.11	123.0 ± 7.0	192.7	344
*Sm*Chb [1]	*p*NPGlcNAc	56.7 ± 4.3	NI	111.0	1.95
*Ah*Nag20A [2]	*p*NPGlcNAc	0.52	115	NI	NI
	*p*NPGalNAc	0.5	7.6	NI	NI
*Ah*NagB [2]	*p*NPGlcNAc	8.57	25	NI	NI
	*p*NPGalNAc	11.1	11	NI	NI
BbhI of Bb [3]	*p*NPGlcNAc	120.0 ± 0.2	NI	213	178
	*p*NPGalNAc	NA	NA	NA	NA
*Cf*Hex20 [4]	*p*NPGlcNAc	0.053	NI	482.3	9090
	*p*NPGalNAc	0.066	NI	129.1	1950
*Vf*ExoI [5]	*p*NPGlcNAc	0.09	270	NI	NI
	*p*NPGalNAc	0.33	130	NI	NI
Hex2 [6]	*p*NPGlcNAc	0.48	NI	60.0 ± 1.7	NI
*Eh*HexA [7]	*p*NPGlcNAc	0.1	3.8	NI	NI
*Tr*Nag1	*p*NPGlcNAc	69.4 ± 4.0	NI	NI	1023 ± 23

Data from [1] [69], [2] [36] and [34], [3] [40], [4] [39], [5] [33], [6] [18], [7] [67]. NI—not indicated; NA—no detected activity.

2.5. *Transglycosylation by* PhNah20A

There are a few reports on LNT2 formation by GH20 catalyzed transglycosylation with $(GlcNAc)_2$ or *p*NPGlcNAc as donors and lactose as the acceptor [15,18,64] (see Figure 6). Hydrolysis of LNT2 by *Ph*Nah20A, an HMO core structure [70], warranted the investigation of the transglycosylation with $(GlcNAc)_2$ and the GH20 reaction intermediate NAG-oxazoline [2-methyl-(1,2-dideoxy-α-D-glucopyrano)-oxazoline] [71] as a donor and lactose as an acceptor (Figure 7A). We here also demonstrated transglycosylation by the commercial GH20 *N*-acetylglucosaminidase from *S. plicatus* (*Sp*Hex) [38] (see Figure S6), which has not been previously reported. Notably, the protein sequence identity between *Sp*Hex and a bacterial transglycosylating enzyme Hex1 isolated from a metagenome [18] was as high as 53.6%.

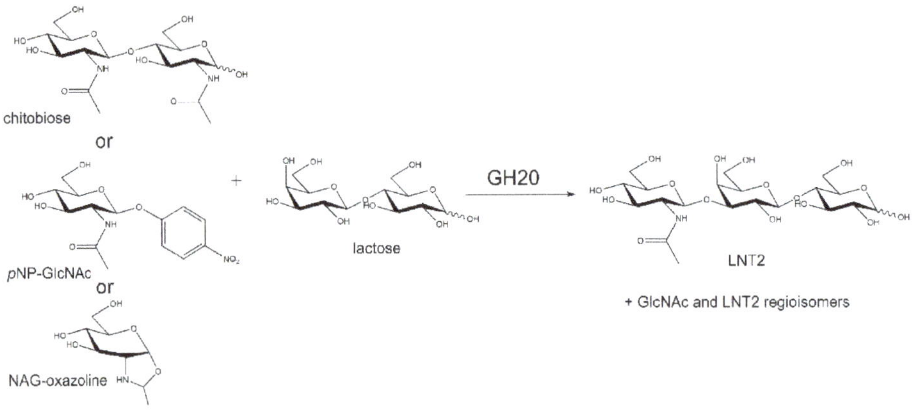

chitobiose
or
pNP-GlcNAc
or
NAG-oxazoline
+
lactose
GH20
LNT2
+ GlcNAc and LNT2 regioisomers

Figure 6. Scheme of transglycosylation reactions catalyzed by GH20 enzymes showing three different possible donors and lactose as an acceptor.

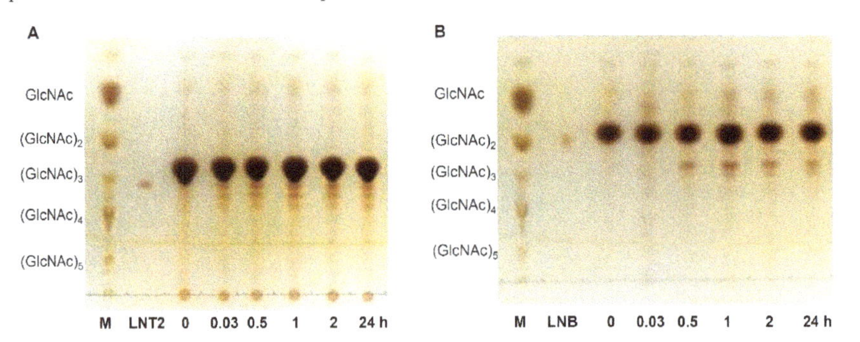

Figure 7. Time course of transglycosylation by *Ph*Nah20A (10 $U \cdot mL^{-1}$) with 100 mM NAG-oxazoline as donor and either 200 mM lactose (**A**) or D-galactose (**B**) as acceptor (see Section 3.8 for details). Chitooligosaccharides (M), lacto-*N*-triose II (LNT2) and lacto-*N*-biose (LNB) were used as references.

Transglycosylation by β-NAHAs has been rarely investigated, and in one case there is a report on a bacterial GH20 enzyme for which no transglycosylation was detected [72], indicating that not all GH20 enzymes have the ability to transglycosylate. A GH20 chitobiase Hex99 from the *Alteromonas* sp. strain O-7 (currently classified as *P. piscicida*) of the order *Alteromonadales* formed β-GlcNAc-1,6-GlcNAc from $(GlcNAc)_2$ by transglycosylation. It is to date the only marine GH20 enzyme reported to produce GlcNAc-containing oligosaccharides [21]. Notably, *P. piscicida* belongs to the same bacterial order as *P. hydrolytica*. Hex99 has a unique substrate specificity, as it hydrolyzed only chitobiose and *p*NP(GlcNAc)$_2$, but neither other chitooligosaccharides nor *p*NPGlcNAc.

*Ph*Nah20A transglycosylated lactose with NAG-oxazoline as the donor (Figure 7A), resulting in three trisaccharides (Figure 8). **2**, purified by gel permeation chromatography (GPC) (Figure S7) migrated similarly to LNT2 in thin-layer chromatography (TLC), and nuclear magnetic resonance (NMR) spectroscopy confirmed the product structure (Figure S9). **1** was determined to be a non-reducing trisaccharide, β-Gal-1,4-β-Glc-1,1-β-GlcNAc (Figure 8 and Figure S8; Tables S3 and S4), once reported as a transglycosylation product of a β-NAHA from *Aspergillus flavofurcatis* CCF 3061 [73]. For full NMR assignment as well as all measurable $^3J_{H,H}$ coupling constants of **1**, see Tables S3 and S4. The 1,1-linkage was supported by heteronuclear multiple-bond correlation spectroscopy (HMBC) and rotating frame nuclear Overhauser effect spectroscopy (ROESY) correlations between the two anomeric positions as well as by lack of a reducing end. Lastly, the β-configuration was determined of the anomeric positions using the $^3J_{H,H}$ coupling constants between the anomeric proton and the neighboring proton (Table S4). A third trisaccharide (**3**) was detected, but not fully characterized due to low abundance. Based on chemical shifts of **3** (Figure S10), however, it seemed unlikely that the galactose moiety in lactose acted as an acceptor, as none of the corresponding chemical shifts were affected. Consequently, most probably the glucose moiety was the acceptor. As O6 was determined to be unsubstituted and glucose was the reducing end residue, therefore either β-Gal-(β-GlcNAc)-1,2-Glc or β-Gal-(β-GlcNAc)-1,3-Glc was produced (Figure S10).

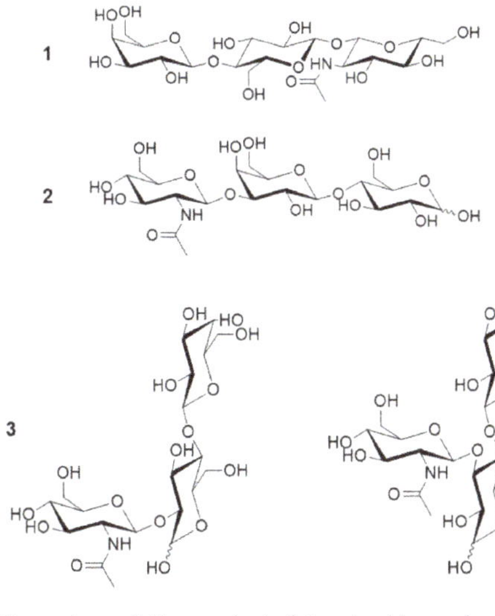

Figure 8. Structures of transglycosylation products determined by nuclear magnetic resonance (NMR). The three detected regioisomers are β-Gal-1,4-β-Glc-1,1-β-GlcNAc (**1**), β-GlcNAc-1,3-β-Gal-1,4-Glc, LNT2 (**2**) and β-Gal-1,4-(β-GlcNAc)-1,*x*-Glc (**3**), (*x* = 2, 3). Both possible structures of **3** are shown.

Several examples exist in Nature of β-Gal-(β-GlcNAc)-1,2-Glc and β-GlcNAc-1,3-Glc being part of polysaccharide backbones, such as the *O*-antigens (*O*-polysaccharides) of lipopolysaccharides from Gram-negative bacteria, i.e., *Proteus* sp., *Hafnia alvei* and *Citrobacter werkmanii* [74–76].

The overall transglycosylation yield for trisaccharides was estimated from the high-performance anion exchange chromatography with pulsed amperometric detector (HPAEC-PAD) chromatogram to 3.8% obtained with 200 mM acceptor and 100 mM donor. Since other trisaccharides were formed, no further optimization of transglycosylation conditions were pursued, even though LNT2 was the major product. Notably, the three trisaccharides were not completely separated by gel permeation chromatography (GPC) (Figure S7), but thin-layer chromatography (TLC) and HPAEC-PAD analysis (Figures S7 and S11) showed products consistent with trisaccharides **1** and **3** (Figure 8).

The acceptor specificity of *Ph*Nah20A was explored using D-galactose, D-glucose, 2-deoxy-D-glucose or L-fucose as an acceptor and NAG-oxazoline as a donor. These monosaccharides all proved to be transglycosylated (Figure 7B and Figure S12) with the similar velocity and transglycosylation products being detected in the most cases already after 0.03 h (2 min) incubation. Therefore, 2 h incubation was sufficient to assess the transglycosylation ability of *Ph*Nah20A (Figure 7 and Figure S12). *Ph*Nah20A thus showed unusual promiscuity towards acceptor molecules, but due to the low yields and formation of several products as seen by TLC (Figure 7, Figure S12), purification and NMR analysis were not pursued. Remarkably, however, the ability to transglycosylate a wide range of acceptors has very rarely been reported for GH20 enzymes [18] and perhaps is associated with the marine origin and the unique phylogenetic relation of *Ph*Nah20A. *S. marcescens* Chb (see Section 2.2) is able to transglycosylate several alcohols, albeit sugar alcohols were not effective acceptors [66]. Some bacterial and fungal β-NAHAs can use lactose as their acceptor [15,18,64], and two Hex enzymes from uncultured bacteria were reported to transfer GlcNAc to D-glucose, D-galactose, sucrose and maltose [18].

3. Materials and Methods

3.1. Materials

LNT2 was purchased from Elicityl Oligotech (Crolles, France). Lactose, *p*NPGalNAc and 5-bromo-4-chloro-3-indolyl *N*-acetyl-β-D-glucosaminide (X-GlcNAc) were from Carbosynth (San Diego, CA, USA), *N,N'*-diacetylchitobiose [$(GlcNAc)_2$] from Omicron Biochemicals (South Bend, IN, USA), and *N,N',N''*-triacetylchitotriose [$(GlcNAc)_3$] and *p*NPGlcNAc from Megazyme (Bray, Co. Wicklow, Ireland). A mixture of chitooligosaccharides, $(GlcNAc)_{1–6}$, was from Koyo Chemicals (Osaka, Japan). All other chemicals were purchased from Sigma-Aldrich (Merck, Darmstadt, Germany) and used without further purification. *S. plicatus* β-NAHA in fusion with maltose-binding protein was purchased from New England Biolabs (Ipswich, MA, USA).

3.2. Bacterial Strains and Media

Paraglaciecola hydrolytica (type strain S66^T) [1,2] was grown at 23 °C in Difco Marine Broth 2216 (BD, Franklin Lakes, NJ, USA) or on Marine Broth supplemented with 15 g·L^{-1} agar. X-GlcNAc was added to the marine agar medium to 20 mg·L^{-1}. Hydrolytic activity was assessed in 5 mL marine mineral medium [77] supplemented with chitooligosaccharides (5 g·L^{-1}) or α-chitin (2 g·L^{-1}) at 23 °C. *E. coli* DH5α was used for molecular cloning, *E. coli* BL21(DE3), BL21(DE3)ΔlacZ [78] and Rosetta (Novagen, Merck, Darmstadt, Germany) for gene expression and *E. coli* BL21(DE3) for recombinant protein production. *E. coli* was grown in Lysogeny Broth (LB; MoBio Laboratories, Carlsbad, CA, USA) or on LB agar plates at 37 °C. Media were supplemented with 100 mg·L^{-1} ampicillin for selection. Auto-induction medium ZYM-5052 was prepared as described [56]. Liquid cultures were aerated on a shaker (160 rpm).

3.3. Molecular Cloning and Plasmids

P. hydrolytica genomic DNA was purified using the Gentra Puregene Yeast/Bact kit B (Qiagen, Venlo, The Netherlands) and plasmid DNA was isolated using the GeneJET Plasmid Miniprep kit (Thermo Fisher Scientific, Waltham, MA, USA). DNA content was determined on NanoDrop Lite (Thermo Fisher Scientific, Waltham, MA, USA). Two putative *P. hydrolytica* β-NAHA-encoding genes were amplified from genomic DNA by Phusion high-fidelity polymerase (Thermo Fisher Scientific, Waltham, MA, USA) using specific primers (Table S5). Genes were cloned as full-length or truncated variants (see Figure S3) into the pURI3TEV vector by PCR cloning [79].

DNA sequencing (Eurofins Genomics, Ebersberg, Germany) verified that cloned sequences matched the sequences in the *P. hydrolytica* genome. Plasmids were transformed into *E. coli* DH5α or BL21(DE3) by electroporation.

3.4. Recombinant Protein Production and Purification

For initial expression analysis *E. coli* BL21(DE3) harboring *Ph*Nah20A in pURI3TEV grown in 20 mL LB medium at 37 °C until $OD_{600nm} \approx 0.5$ was induced by 0.5 mM isopropyl thio-β-D-galactoside (IPTG), and incubated at 22 °C. Aliquots (10 μL) were mixed at 0, 4, 20 h with 4 μL SDS-PAGE sample buffer, heated (10 min, 80 °C) to lyse cells and denature proteins, centrifuged (12,000× *g*, 1 min, RT) and analyzed on pre-cast SDS-polyacrylamide gels according to the manufacturers' instructions (NuPAGE, Thermo Fisher Scientific, Waltham, MA, USA) in an XCell SureLock mini-cell electrophoresis system (Thermo Fisher Scientific, USA). Gels were stained by Coomassie Brilliant Blue G-250. Cell lysates were prepared from cell pellets after IPTG-induction by suspension in 0.4 mL 50 mM sodium phosphate pH 7.0, added 0.4 mL BugBuster protein extraction reagent (Merck, Darmstadt, Germany), approx. 100 U Benzonase nuclease (Merck, Darmstadt, Germany), and centrifuged (12,000× *g*, 20 min, 4 °C).

For enzyme preparation *E. coli* BL21(DE3) harboring *Ph*Nah20A in pURI3TEV was grown in 1 L LB medium at 37 °C to $OD_{600nm} \approx 0.5$, induced by 0.5 mM IPTG, and incubated (20 h, 22 °C). Cells collected by centrifugation (10,000× *g*, 15 min, 4 °C) were resuspended in 50 mL lysis buffer (50 mM sodium phosphate, pH 7.0, 0.3 M NaCl, 20 mM imidazole containing 250 U Benzonase nuclease), disrupted (Cell Pressure Homogenizer, Stansted, UK) and centrifuged to remove debris (25,000× *g*, 20 min, 4 °C). The supernatant was filtered (0.45 μm sterile polyvinylidene fluoride (PVDF) membrane filter, Millex-HV, Merck, Darmstadt, Germany) and *Ph*Nah20A purified by Ni^{2+}-affinity chromatography (HisTrapHP, GE Healthcare, Uppsala, Sweden) followed by size-exclusion chromatography (HiLoad 16/60 Superdex 200 pg; ÄKTA Avant chromatography system, GE Healthcare, Uppsala, Sweden) in 50 mM sodium phosphate, pH 7.0, 0.3 M NaCl at a flow rate of 2 $mL{\cdot}min^{-1}$. Eluate was analyzed by SDS-PAGE and fractions containing *Ph*Nah20A were pooled, concentrated (Amicon ultra-15 30K centrifugal filter device, Merck, Darmstadt, Germany), and had added to them 0.02% Na-azide, and then were stored in the above-mentioned buffer at 4 °C. Protein concentration was determined by the Pierce Coomassie (Bradford) Protein Assay Kit (Thermo Fisher Scientific, Waltham, MA, USA) for cell lysates and NanoDrop Lite (Themo Fisher Scientific, Waltham, MA, USA) for purified protein using the calculated ε_{280} = 136,835 $M^{-1}{\cdot}cm^{-1}$ (ExPasy server; https://web.expasy.org/protparam/). After spectrophotometric determination of the concentration of *Ph*Nah20A, bovine serum albumin (BSA) was added to 0.5% of final concentration for storage.

3.5. Activity Assays

*Ph*Nah20A activity was routinely determined on 2 mM *p*NPGlcNAc at 37 °C in two-fold diluted McIlvaine buffer pH 6.0 (63 mM Na_2HPO_4; 18 mM citric acid), containing 0.05% BSA. The reaction (total volume 500 μL) was performed in McIlvaine buffer, pH 6.0 (250 μl), 100 μl milliQ water and 100 μL of substrate was added. The reaction was initiated by adding 50 μL of *Ph*Nah20A (prepared immediately before use in McIlvaine buffer, pH 6.0, 0.5% BSA, and kept on ice) to the reaction mixture yielding a final concentration of 0.3–5 $\mu g{\cdot}mL^{-1}$. The reaction was stopped typically after 2–5 min by 250 μL 1 M Na_2CO_3 and the product was measured spectrophotometrically at 400 nm (Ultrospec 3100 pro UV/Visible spectrophotometer, GE Healthcare, Uppsala, Sweden) using *p*NP (ε_{400} = 18,000 $M^{-1}{\cdot}cm^{-1}$) as the standard. One U of activity was defined as the amount of enzyme releasing 1 μmol *p*NP per min from 2 mM *p*NPGlcNAc. pH activity optimum was determined for *Ph*Nah20A in McIlvaine buffers (pH 4.0–8.0) at 37 °C towards 2 mM *p*NPGlcNAc and the temperature optimum was determined from the initial rates of *p*NP release at temperatures in the range 10–65 °C at pH 6.0.

To determine the hydrolysis by *Ph*Nah20A 200 mM $(GlcNAc)_2$ was incubated with 1 $U{\cdot}mL^{-1}$ (11.6 $\mu g{\cdot}mL^{-1}$) or 5 mM LNT2 with 10 $U{\cdot}mL^{-1}$ (116 $\mu g{\cdot}mL^{-1}$) in 50 mM sodium phosphate, pH 6.0, 0.5% BSA, at 37 °C for 20 h. The release of GlcNAc was monitored by high-performance anion exchange chromatography with pulsed amperometric detector (HPAEC-PAD) for $(GlcNAc)_2$ and by thin-layer chromatography (TLC) for LNT2 (see Section 3.9).

3.6. Kinetics

*Ph*Nah20A (final concentration 0.3–1.2 μg·mL^{-1}) was added to initiate the hydrolysis of 0.05–2 mM *p*NPGlcNAc (six concentrations) and 0.1–2 mM *p*NPGalNAc (five concentrations) in 500 μL two-fold diluted McIlvaine buffer pH 6.0, 0.05% BSA at 37 °C. The reaction was stopped at suitable time points by the addition of 250 μL 1 M Na_2CO_3 and quantified spectrophotometrically as above. Initial rates calculated from *p*NP formation versus time were plotted against substrate concentration and fitted to the Michaelis-Menten equation using OriginPro 2015 (OriginLab, Northampton, MA, USA) to obtain k_{cat} and K_M. The k_{cat}/K_M values were either calculated or determined from rates of hydrolysis at low substrate concentration.

3.7. Synthesis of NAG-Oxazoline

NAG-oxazoline [2-methyl-(1,2-dideoxy-α-D-glucopyrano)-oxazoline] was synthesized and purified as described previously [71]. Briefly, 2 g GlcNAc (9 mmol) was dissolved in 20 mL acetic anhydride, then we added 10 mL pyridine and stirred overnight at room temperature (RT). After extraction by dichloromethane (DCM) and successive washings (Na_2CO_3, H_2O, H_2SO_4, H_2O) the organic layer was dried and evaporated. Trimethylsilyl trifluoromethanesulfonate (0.8 mL) was added to 1.5 g peracetylated glucosamine dissolved in 1,2-dichloroethane and stirred at 50 °C until completion of the reaction (about 4 h). Trimethylamine was added (2 mL) followed by 50 mL DCM, washed with cold water, dried and evaporated. The product was purified by flash chromatography (cyclohexane: 1% triethylamine in ethyl acetate 100:0 to 40:60). Peracetylated oxazole (300 mg) in 10 mL anhydrous methanol at 0 °C was added 15 μL 5.3 M sodium methanolate in methanol and stirred at RT until the reaction was completed (about 3 h). The resulting NAG-oxazoline was dried and used without further purification.

3.8. Transglycosylation

Reaction mixtures for transglycosylation contained either 100 mM NAG-oxazoline (from 1 M stock in 50 mM sodium borate, pH 9.3) or 100 mM $(GlcNAc)_2$ donor, and as acceptor 200 mM lactose, D-galactose, D-glucose, 2-deoxy-D-glucose or L-fucose; 1 or 10 U·mL^{-1} (11.6 or 116 μg·mL^{-1}) *Ph*Nah20A or 10 U·mL^{-1} *S. plicatus* β-NAHA in 50 mM sodium phosphate, pH 8.0, 0.5% BSA, at 37 °C. The reaction volume was typically 20 μL for TLC analysis and 250 μL for product yield and structure determination. Slightly basic conditions were required as NAG-oxazoline is not stable at neutral or acidic pH [71]. Reactions were stopped at various time points by heating (5 min, 90 °C), cooled to RT and centrifuged (12,000× *g*, 1 min, 4 °C). Samples were diluted four- and 150-fold in milliQ water for TLC and HPAEC-PAD (see Section 3.9), respectively. For the reaction mixtures for the analysis of transglycosylation products after gel permeation chromatography (GPC), containing 10 U·mL^{-1} *Ph*Nah20A, 100 mM NAG-oxazoline and 200 mM lactose in 50 mM sodium phosphate pH 8.0, 0.5% BSA were incubated 2 h at 37 °C followed by heating (5 min, 90 °C). To the sample was added three volumes of sterile milliQ water, and the enzyme was removed (Amicon Ultra 0.5 mL centrifugal device, Mw cut-off 30 kDa; Merck, Darmstadt, Germany) followed by filtration (0.45 μm filters; Millex-HV, Merck, Darmstadt, Germany) prior to GPC.

3.9. Chromatographic Methods

Reaction mixtures containing 15–30 μg carbohydrate were spotted onto TLC plates (Silica Gel 60 F254 plates; Merck, Darmstadt, Germany) developed twice in chloroform:acetic acid:water (6:7:1; *v:v:v*) [80,81] or n-butanol: ethanol: water (5:3:2; *v:v:v*) [82]. Carbohydrates were visualized with orcinol (0.5% 5-methyl resorcinol and 10% H_2SO_4 in ethanol) or aniline dye (1.2% aniline hydrochloride and 1.2% diphenylamine in acidic methanol).

Oligosaccharides were also separated by HPAEC-PAD at 22 °C (Dionex CarboPac PA1 column, 250 × 4 mm with 50 × 4 mm Guard, Thermo Fisher Scientific, Waltham, MA, USA) using an ICS-5000 system (Thermo Fisher Scientific, Waltham, MA, USA) equipped with AS autosampler and pulsed amperometric detector (carbohydrate four-potential waveform, sampling rate 2 Hz) with a gold electrode (Au) and an Ag/AgCl reference electrode.

The elution was done with (A) water; (B) 1 M NaOH; (C) 200 mM NaOH + 800 mM NaOAc isocratically using 7.5% B in A (25 min) followed by 100% C (1 min) and column re-equilibration (9 min) at 7.5% B in A at 1.0 mL·min^{-1}. Oligosaccharides in water (10 μL) containing 9 μM L-fucose as standard were injected by autosampler kept at 5 °C. LNT2, glucose, galactose, lactose, GlcNAc, $(GlcNAc)_2$ and chitooligosaccharides were used as standards for calibration. Reaction mixtures (0.5 mL) containing approximately 10 mg oligosaccharides were separated by GPC (Bio-Gel P-2, Bio-Rad Laboratories, Hercules, CA, USA; 16 × 900 mm XK16/100 mounted on an ÄKTAprime plus chromatography system, GE Healthcare, Sweden), eluted by degassed milliQ water at flow rate of 0.1 mL·min^{-1} at RT and pressure limit set to 0.3 MPa. Reducing sugar in collected fractions (2 mL) were quantified by the Nelson-Somogyi method [83] using glucose and GlcNAc as standards. Fractions containing trisaccharides were dried (SpeedVac, Thermo Fisher Scientific, Waltham, MA, USA) at 50 °C, dissolved in 50 μL milliQ water and subjected to TLC for the preliminary identification of transglycosylation products. For NMR analysis, identical trisaccharide-containing fractions from two GPC runs were pooled, dried (SpeedVac) and dissolved in 0.5 mL D_2O (Sigma-Aldrich, USA). Each fraction contained a major component and trace amounts of one or two other products.

3.10. Nuclear Magnetic Resonance (NMR)

All NMR spectra were recorded on an 800 MHz Bruker Avance III (799.75 MHz for ^{1}H and 201.10 MHz for ^{13}C) equipped with a 5 mm TCI cryoprobe. Acetone was used as internal reference (2.22 ppm and 30.89 ppm for ^{1}H and ^{13}C, respectively). The following experiments were used for the structure elucidation: ^{1}H with presaturation, double quantum filtered correlation spectroscopy (DQF-COSY), rotating frame nuclear Overhauser effect spectroscopy (ROESY), heteronuclear single-quantum correlation spectroscopy (HSQC), heteronuclear single-quantum correlation spectroscopy-total correlation spectroscopy (HSQC-TOCSY) and heteronuclear multiple-bond correlation spectroscopy (HMBC) all performed using standard Bruker pulse sequences. LNT2 and lactose were used as reference compounds. Structural elucidation was carried out by first identifying all ^{1}H and corresponding ^{13}C chemical shifts using ^{1}H with presaturation and HSQC. Subsequently, the different signals belonging to each position in each monosaccharide were determined, primarily using DQF-COSY and HSQC-TOCSY, and finally the connections between the monosaccharides were determined using HMBC and ROESY, as well as comparing chemical shifts to reference compounds.

3.11. In Silico Methods

The draft genome sequence of *P. hydrolytica* S66^{T} [1] annotated on the RAST server (http://rast.nmpdr.org/) [84] was mined on 20 March 2016, to identify putative β-NAHAs. Visualization of the RAST-annotated proteins was done on the SEED Viewer v 2.0 (www.theSEED.org).

Protein sequences of characterized β-NAHAs were retrieved from UniprotKB (https://www.uniprot.org/) on 10 February 2019. Nucleotide BLAST and protein BLAST tools (https://blast.ncbi.nlm.nih.gov/Blast.cgi) were used in 10 February 2019 and 3 January 2020 for identity analysis of nucleotide and protein sequences, respectively. Multiple sequence alignments were carried out with Clustal Omega v 2.1 (https://www.ebi.ac.uk/Tools/msa/clustalo/) and visualized by BioEdit v 7.0.5.3 (https://www.softpedia.com/get/Science-CAD/BioEdit.shtml).

The phylogenetic tree was constructed and visualized using MEGA v 7.0.26 (https://megasoftware.net/) [55].

N-terminal signal peptide prediction was done by SignalP v 4.1 with sensitive default cut-off values (http://www.cbs.dtu.dk/services/SignalP/) [85]. Promoter locations were predicted by SoftBerry tool BPROM (http://www.softberry.com/berry.phtml?topic=bprom&group=programs&subgroup=gfindb).

4. Conclusions

The genome of the marine bacterium *P. hydrolytica* S66^T encodes two putative GH20 β-*N*-acetylhexosaminidase (EC 3.2.1.52) having protein sequences that differed remarkably from earlier characterized β-NAHAs (≤30% identity). *Ph*Nah20A was positioned on a phylogenetic tree between β-NAHAs of water-associated bacteria, i.e., *Vibrio furnissii* and *Aeromonas hydrophila*, and unicellular eukaryotes (amobae). *Ph*Nah20A, produced in *E. coli*, was unstable if diluted, but was stabilized by BSA or Triton X-100. *Ph*Nah20A is a genuine β-NAHA with essentially the same catalytic efficiency for *p*NPGlcNAc and *p*NPGalNAc, and thus differs from most of the previously studied bacterial β-NAHAs, which prefer *p*NPGlcNAc as a substrate while some eukaryotic GH20 prefer *p*NPGalNAc. *Ph*Nah20A also hydrolyzed LNT2, a core structure of human milk oligosaccharides, and showed biosynthetic activity (transglycosylation) which is a poorly studied aspect of GH20 β-NAHAs, especially from eukaryotes and water-living prokaryotes. *Ph*Nah20A was able to form LTN2 by transglycosylation using NAG-oxazoline as a donor and lactose as an acceptor, LNT2, β-Gal-1,4-β-Glc-1,1-β-GlcNAc and β-Gal-1,4-(β-GlcNAc)-1,2/3-Glc being identified by NMR as main transglycosylation products. Several monosaccharides were also recognized as acceptors by *Ph*Nah20A. To date, based on pH and temperature optima, kinetic parameters or stability characteristics alone, no clear distinction can be made between eukaryotic versus prokaryotic or terrestrial versus aquatic GH20 β-NAHAs. However, this may be due to the very limited number of characterized β-NAHAs of salt or fresh water origin. *Ph*Nah20A is the first characterized member of a distinct group of GH20 β-NAHAs located phylogenetically between eukaryotic and prokaryotic enzymes.

Supplementary Materials: The following are available online at http://www.mdpi.com/1422-0067/21/2/417/s1. The following materials are available online: Supplementary Information 1 containing Figure S1. TLC analysis of growth media of *P. hydrolytica* (A, B, C) and growth phenotype on marine agar medium with X-GlcNAc (5-bromo-4-chloro-3-indolyl *N*-acetyl-β-D-glucosaminide) (D); Figure S2. Phylogenetic tree with bootstrap test (1000 replicates) of *Ph*Nah20A, *Ph*Nah20B (both marked with red circles) and 41 biochemically characterised GH20 (EC 3.2.1.52) enzymes; Figure S3. Schematic domain architecture of full-length and truncated variants of *Ph*Nah20A and *Ph*Nah20B; Figure S4. IPTG-induced *E. coli* transformants growing in LB analysed by SDS-PAGE; Figure S5. Inactivation of 5 μg·mL^{-1} *Ph*Nah20A at 50 °C and pH 6.0 in the presence of 0.5% BSA; Figure S6. Time course of transglycosylation by *S. plicatus* Hex (10 U·mL^{-1}) with 100 mM NAG-oxazoline as donor and 200 mM lactose as acceptor; Figure S7. TLC of trisaccharide-containing fractions of the *Ph*Nah20A reaction (2 h; Figure 7A) separated by gel-permeation chromatography; Figure S8. HSCQ spectrum of the chromatographic fraction 50 (see Figure S7) containing over 80% of compound **1**; Figure S9. HSCQ spectrum of the chromatographic fraction 51 (see Figure S7) containing over 75% of 2 (LNT2); Figure S10. HSCQ spectrum of the chromatographic fraction 53 (see Figure S7) containing primarily 3; Figure S11. Extraction of HPAEC-PAD analysis of transglycosylation products by *Ph*Nah20A (10 U·mL^{-1}) reacting 2 h with 100 mM NAG-oxazoline as donor and 200 mM lactose as acceptor (blue line); Figure S12. Time course of transglycosylation by *Ph*Nah20A (10 U·mL^{-1}) with 100 mM NAG-oxazoline as donor and 200 mM D-glucose (A), 2-deoxy-D-glucose (B) or L-fucose (C) as acceptor; Table S1. BLAST analysis of putative β-NAHAs (EC 3.2.1.52) from *P. hydrolytica*. Table S2. Information on proteins flanking identified β-NAHAs (presented in Figure 1B); Table S3. NMR assignment of 1. The methyl of the GlcNAc acetyl group was at 2.090 ppm for ^{1}H and 22.81 ppm for ^{13}C and the carbonyl of the acetyl was at 176.06 ppm ^{13}C; Table S4. ^{3}H-H coupling constants for 1 measured through DQF-COSY; Table S5. PCR primers to isolate full-length β-NAHA encoding genes and indicated truncated variants. Underlined sequences are priming with pURI3-TEV expression vector; Supplementary Information 2 containing multiple sequence alignment.

Author Contributions: Conceptualization, P.S., T.V. and B.S.; methodology, T.V., C.K., A.L. and L.H.P.; validation, T.V., D.T., C.K. and L.H.P.; formal analysis, T.V., C.K., A.L. and D.T.; investigation, T.V., C.K., D.T., A.L. and L.H.P.; resources, B.S., J.Ø.D., L.H.P., C.A.-M., D.T. and P.S.; data curation, T.V., B.S.; writing—original draft preparation, T.V., D.T., C.K., P.S. and B.S.; writing—review and editing, T.V., D.T., C.K., P.S. and B.S.; visualization, T.V. and D.T.; supervision, P.S., J.Ø.D., L.H.P. and B.S.; project administration, P.S., T.V., B.S.; funding acquisition, P.S., D.T., T.V. and B.S. All authors have read and agreed to the published version of the manuscript.

Funding: This research was funded by Innovation Fund Denmark to the project "OliGram. Design and gram scale enzymatic synthesis of human milk oligosaccharides", grant number 1308-00014B having P.S. as PI. D.T. is grateful to the Novo Nordisk Foundation for a postdoctoral fellowship (grant NNF17OC0025660). The APC was funded by University of Tartu Feasibility Fund grant PLTMRARENG13 to T.V. and the NNF17OC0025660 grant to D.T.

Acknowledgments: Karina Jansen (Technical University of Denmark) is thanked for general technical assistance, Pernille K. Bech and Mikkel Schultz-Johansen (University of Copenhagen) for providing the *P. hydrolytica* strain, Corinna Schiano di Cola for preparing autoinduction medium and Tiina Alamäe (University of Tartu) for fruitful discussions.

Conflicts of Interest: The authors declare no conflict of interest. The funders had no role in the design of the study; in the collection, analyzes, or interpretation of data; in the writing of the manuscript, or in the decision to publish the results.

Abbreviations

BSA	bovine serum albumin
DCM	dichloromethane
GH	glycoside hydrolase
GlcNAc	*N*-acetylglucosamine
$(GlcNAc)_2$	*N,N′*-diacetylchitobiose, chitobiose
GPC	gel permeation chromatography
HMOs	human milk oligosaccharides
HPAEC-PAD	high-performance anion exchange chromatography with pulsed amperometric detector
IPTG	isopropyl thio-β-D-galactoside
LNT2	lacto-*N*-triose II
NAG-oxazoline	2-methyl-(1,2-dideoxy-α-D-glucopyrano)-oxazoline
β-NAHA	β-*N*-acetylhexosaminidases
NCBI	National Center for Biotechnology Information
NMR	nuclear magnetic resonance
*p*NPGlcNAc	*p*-nitrophenyl-GlcNAc
*p*NPGalNAc	*p*-nitrophenyl-GalNAc
TLC	thin layer chromatography
X-GlcNAc	5-bromo-4-chloro-3-indolyl *N*-acetyl-β-D-glucosaminide

References

1. Schultz-Johansen, M.; Glaring, M.A.; Bech, P.K.; Stougaard, P. Draft genome sequence of a novel marine bacterium, *Paraglaciecola* sp. strain S66, with hydrolytic activity against seaweed. *Genome Announc.* **2016**, *4*, e00304-16. [CrossRef]
2. Bech, P.K.; Schultz-Johansen, M.; Glaring, M.A.; Barbeyron, T.; Czjzek, M.; Stougaard, P. *Paraglaciecola hydrolytica* sp. nov., a bacterium with hydrolytic activity against multiple seaweed-derived polysaccharides. *Int. J. Syst. Evol. Microbiol.* **2017**, *67*, 2242–2247. [CrossRef]
3. Schultz-Johansen, M.; Bech, P.; Hennessy, R.; Glaring, M.; Barbeyron, T.; Czjzek, M.; Stougaard, P. A novel enzyme portfolio for red algal polysaccharide degradation in the marine bacterium *Paraglaciecola hydrolytica* S66[T] encoded in a sizeable polysaccharide. *Front. Microbiol.* **2018**, *9*, 839. [CrossRef]
4. Lombard, V.; Golaconda Ramulu, H.; Drula, E.; Coutinho, P.M.; Henrissat, B. The carbohydrate-active enzymes database (CAZy) in 2013. *Nucleic Acids Res.* **2014**, *42*, 490–495. [CrossRef]
5. Slámová, K.; Bojarová, P. Engineered *N*-acetylhexosamine-active enzymes in glycoscience. *Biochim. Biophys. Acta* **2017**, *1861*, 2070–2087. [CrossRef]
6. Bode, L. The functional biology of human milk oligosaccharides. *Early Hum. Dev.* **2015**, *91*, 619–622. [CrossRef]
7. Kunz, C.; Kuntz, S.; Rudloff, S. Bioactivity of human milk oligosaccharides. In *Food Oligosaccharides: Production, Analysis and Bioactivity*; Moreno, J.F., Sanz, M.L., Eds.; John Wiley & Sons Ltd.: Hoboken, NJ, USA, 2014; pp. 1–20.

8. Barile, D.; Rastall, R.A. Human milk and related oligosaccharides as prebiotics. *Curr. Opin. Biotechnol.* **2013**, *24*, 214–219. [CrossRef]
9. Verspreet, J.; Damen, B.; Broekaert, W.F.; Verbeke, K.; Delcour, J.A.; Courtin, C.M. A critical look at prebiotics within the dietary fiber concept. *Annu. Rev. Food Sci. Technol.* **2016**, *7*, 167–190. [CrossRef]
10. Gibson, G.R.; Hutkins, R.; Sanders, M.E.; Prescott, S.L.; Reimer, R.A.; Salminen, S.J.; Scott, K.; Stanton, C.; Swanson, K.S.; Cani, P.D.; et al. Expert consensus document: The International Scientific Association for Probiotics and Prebiotics (ISAPP) consensus statement on the definition and scope of prebiotics. *Nat. Rev. Gastroenterol. Hepatol.* **2017**, *14*, 491–502. [CrossRef]
11. Boltje, T.J.; Buskas, T.; Boons, G.J. Opportunities and challenges in synthetic oligosaccharide and glycoconjugate research. *Nat. Chem.* **2009**, *1*, 611–622. [CrossRef]
12. Zeuner, B.; Teze, D.; Muschiol, J.; Meyer, A.S. Synthesis of human milk oligosaccharides: Protein engineering strategies for improved enzymatic transglycosylation. *Molecules* **2019**, *24*, 2033. [CrossRef] [PubMed]
13. Zeuner, B.; Jers, C.; Mikkelsen, J.D.; Meyer, A.S. Methods for improving enzymatic trans-glycosylation for synthesis of human milk oligosaccharide biomimetics. *J. Agric. Food Chem.* **2014**, *62*, 9615–9631. [CrossRef] [PubMed]
14. Jamek, S.B.; Nyffenegger, C.; Muschiol, J.; Holck, J.; Meyer, A.S.; Mikkelsen, J.D. Characterization of two novel bacterial type A exo-chitobiose hydrolases having C-terminal 5/12-type carbohydrate-binding modules. *Appl. Microbiol. Biotechnol.* **2017**, *101*, 4533–4546. [CrossRef] [PubMed]
15. Chen, X.; Xu, L.; Jin, L.; Sun, B.; Gu, G.; Lu, L.; Xiao, M. Efficient and regioselective synthesis of β-GalNAc/GlcNAc-lactose by a bifunctional transglycosylating β-*N*-acetylhexosaminidase from *Bifidobacterium bifidum*. *Appl. Environ. Microbiol.* **2016**, *82*, 5642–5652. [CrossRef]
16. Baumgärtner, F.; Conrad, J.; Sprenger, G.A.; Albermann, C. Synthesis of the human milk oligosaccharide lacto-*N*-tetraose in metabolically engineered, plasmid-free *E. coli*. *ChemBioChem* **2014**, *15*, 1896–1900. [CrossRef]
17. Wang, M.; Li, M.; Wu, S.; Lebrilla, C.B.; Chapkin, R.S.; Ivanov, I.; Donovan, S.M. Fecal microbiota composition of breast-fed infants is correlated with human milk oligosaccharides consumed. *J. Pediatr. Gastroenterol. Nutr.* **2015**, *60*, 825–833. [CrossRef]
18. Nyffenegger, C.; Nordvang, R.T.; Zeuner, B.; Łężyk, M.; Difilippo, E.; Logtenberg, M.J.; Schols, H.A.; Meyer, A.S.; Mikkelsen, J.D. Backbone structures in human milk oligosaccharides: Trans-glycosylation by metagenomic β-*N*-acetylhexosaminidases. *Appl. Microbiol. Biotechnol.* **2015**, *99*, 7997–8009. [CrossRef]
19. Murata, T.; Tashiro, A.; Itoh, T.; Usui, T. Enzymic synthesis of 3′-*O*- and 6′-*O*-*N*-acetylglucosaminyl-*N*-acetyllactosaminide glycosides catalyzed by β-*N*-acetyl-D-hexosaminidase from *Nocardia orientalis*. *Biochim. Biophys. Acta* **1997**, *1335*, 326–334. [CrossRef]
20. Slámová, K.; Gažák, R.; Bojarová, P.; Kulik, N.; Ettrich, R.; Pelantová, H.; Sedmera, P.; Křen, V. 4-deoxy-substrates for β-*N*-acetylhexosaminidases: How to make use of their loose specificity. *Glycobiology* **2010**, *20*, 1002–1009. [CrossRef]
21. Tsujibo, H.; Kondo, N.; Tanaka, K.; Miyamoto, K.; Baba, N.; Inamori, Y. Molecular analysis of the gene encoding a novel transglycosylative enzyme from *Alteromonas* sp. strain O-7 and its physiological role in the chitinolytic system. *J. Bacteriol.* **1999**, *181*, 5461–5466. [CrossRef]
22. Lakshmanan, T.; Loganathan, D. Enzymatic synthesis of *N*-glycoprotein linkage region disaccharide mimetics using β-*N*-acetylhexosaminidases from *Aspergillus oryzae* and *Vigna radiata*. *Tetrahedron Asymmetry* **2005**, *16*, 255–260. [CrossRef]
23. Beier, S.; Bertilsson, S. Bacterial chitin degradation-mechanisms and ecophysiological strategies. *Front. Microbiol.* **2013**, *4*, 149. [CrossRef]
24. Slámová, K.; Bojarová, P.; Petrásková, L.; Křen, V. β-*N*-Acetylhexosaminidase: What's in a name...? *Biotechnol. Adv.* **2010**, *28*, 682–693. [CrossRef]
25. Zhang, R.; Zhou, J.; Song, Z.; Huang, Z. Enzymatic properties of beta-*N*-acetylglucosaminidases. *Appl. Microbiol. Biotechnol.* **2018**, *102*, 93–103. [CrossRef]

26. Choi, K.H.; Seo, J.Y.; Park, K.M.; Park, C.S.; Cha, J. Characterization of glycosyl hydrolase family 3β-*N*-acetylglucosaminidases from *Thermotoga maritima* and *Thermotoga neapolitana*. *J. Biosci. Bioeng.* **2009**, *108*, 455–459. [CrossRef]
27. Ryšlavá, H.; Kalendová, A.; Doubnerová, V.; Skočdopol, P.; Kumar, V.; Kukačka, Z.; Pompach, P.; Vaněk, O.; Slámová, K.; Bojarová, P.; et al. Enzymatic characterization and molecular modeling of an evolutionarily interesting fungal β-*N*-acetylhexosaminidase. *FEBS J.* **2011**, *278*, 2469–2484. [CrossRef]
28. Piszkiewicz, D.; Bruice, T.C.; Glycoside, H., III. Intramolecular acetamido group participation in the specific acid catalyzed hydrolysis of methyl 2-acetamido-2-deoxy-β-D-glucopyranoside. *J. Am. Chem. Soc.* **1968**, *378*, 5844–5848. [CrossRef]
29. Knapp, S.; Vocadlo, D.; Gao, Z.; Kirk, B.; Lou, J.; Withers, S.G. NAG-thiazoline, an *N*-acetyl-β-hexosaminidase inhibitor that implicates acetamido participation. *J. Am. Chem. Soc.* **1996**, *118*, 6804–6805. [CrossRef]
30. Ferrara, M.C.; Cobucci-Ponzano, B.; Carpentieri, A.; Henrissat, B.; Rossi, M.; Amoresano, A.; Moracci, M. The identification and molecular characterization of the first archaeal bifunctional exo-β-glucosidase/*N*-acetyl-β-glucosaminidase demonstrate that family GH116 is made of three functionally distinct subfamilies. *Biochim. Biophys. Acta* **2014**, *1840*, 367–377. [CrossRef]
31. Teze, D.; Shuoker, B.; Chaberski, E.K.; Kunstmann, R.S.; Fredslund, F.; Peters, G.H.J.; Karlsson, E.N.; Welner, D.H.; Abou Hachem, M. The catalytic acid-base in GH109 resides in a conserved GGHGG loop and allows for comparable α-retaining and β-inverting activity in an *N*-acetylgalactosaminidase from *Akkermansia muciniphila*. *ChemRxiv* **2019**, preprint.
32. Placzek, S.; Schomburg, I.; Chang, A.; Jeske, L.; Ulbrich, M.; Tillack, J.; Schomburg, D. BRENDA in 2017: New perspectives and new tools in BRENDA. *Nucleic Acids Res.* **2017**, *45*, D380–D388. [CrossRef] [PubMed]
33. Keyhani, N.O.; Roseman, S. The chitin catabolic cascade in the marine bacterium *Vibrio furnissii*. *J. Biol. Chem.* **1996**, *271*, 33425–33432. [CrossRef] [PubMed]
34. Lan, X.; Zhang, X.; Kodaira, R.; Zhou, Z.; Shimosaka, M. Gene cloning, expression, and characterization of a second β-*N*-acetylglucosaminidase from the chitinolytic bacterium *Aeromonas hydrophila* strain SUWA-9. *Biosci. Biotechnol. Biochem.* **2008**, *72*, 492–498. [CrossRef] [PubMed]
35. Tsujibo, H.; Miyamoto, K.; Yoshimura, M.; Takata, M.; Miyamoto, J.; Inamori, Y. Molecular cloning of the gene encoding a novel β-*N*-acetylhexosaminidase from a marine bacterium, *Alteromonas* sp. strain O-7, and characterization of the cloned enzyme. *Biosci. Biotechnol. Biochem.* **2002**, *66*, 471–475. [CrossRef] [PubMed]
36. Lan, X.; Ozawa, N.; Nishiwaki, N.; Kodaira, R.; Okazaki, M.; Shimosaka, M. Purification, cloning, and sequence analysis of β-*N*-acetylglucosaminidase from the chitinolytic bacterium *Aeromonas hydrophila* strain SUWA-9. *Biosci. Biotechnol. Biochem.* **2004**, *68*, 1082–1090. [CrossRef]
37. Soto-Gil, R.W.; Zyskind, J.W. *N,N'*-Diacetylchitobiase of *Vibrio harveyi*. Primary structure, processing, and evolutionary relationships. *J. Biol. Chem.* **1989**, *264*, 14778–14783.
38. Mark, B.L.; Wasney, G.A.; Salo, T.J.; Khan, A.R.; Cao, Z.; Robbins, P.W.; James, M.N.; Triggs-Raine, B.L. Structural and functional characterization of *Streptomyces plicatus* beta-*N*-acetylhexosaminidase by comparative molecular modeling and site-directed mutagenesis. *J. Biol. Chem.* **1998**, *273*, 19618–19624. [CrossRef]
39. Mayer, C.; Vocadlo, D.J.; Mah, M.; Rupitz, K.; Stoll, D.; Warren, R.A.J.; Withers, S.G. Characterization of a β-*N*-acetylhexosaminidase and a β-*N*-acetylglucosaminidase/β-glucosidase from *Cellulomonas fimi*. *FEBS J.* **2006**, *273*, 2929–2941. [CrossRef]
40. Miwa, M.; Horimoto, T.; Kiyohara, M.; Katayama, T.; Kitaoka, M.; Ashida, H.; Yamamoto, K. Cooperation of β-galactosidase and β-*N*-acetylhexosaminidase from bifidobacteria in assimilation of human milk oligosaccharides with type 2 structure. *Glycobiology* **2010**, *20*, 1402–1409. [CrossRef]
41. Gutternigg, M.; Rendić, D.; Voglauer, R.; Iskratsch, T.; Wilson, I.B.H. Mammalian cells contain a second nucleocytoplasmic hexosaminidase. *Biochem. J.* **2009**, *419*, 83–90. [CrossRef]
42. Verpoorte, J.A. Isolation and characterization of the major beta-*N*-acetyl-D-glucosaminidase from human plasma. *Biochemistry* **1974**, *13*, 793–799. [CrossRef] [PubMed]
43. Garcia-Alonso, J.; Reglero, A.; Cabezas, J.A. Purification and properties of β-*N*-acetylhexosaminidase a from pig brain. *Int. J. Biochem.* **1990**, *22*, 645–651. [CrossRef]

44. Zhou, J.; Song, Z.; Zhang, R.; Liu, R.; Wu, Q.; Li, J.; Tang, X.; Xu, B.; Ding, J.; Han, N.; et al. Distinctive molecular and biochemical characteristics of a glycoside hydrolase family 20 β-*N*-acetylglucosaminidase and salt tolerance. *BMC Biotechnol.* **2017**, *17*, 37. [CrossRef] [PubMed]
45. Sumida, T.; Ishii, R.; Yanagisawa, T.; Yokoyama, S.; Ito, M. Molecular coning and crystal structural analysis of a novel β-*N*-acetylhexosaminidase from *Paenibacillus* sp. TS12 capable of degrading glycosphingolipids. *J. Mol. Biol.* **2009**, *392*, 87–99. [CrossRef] [PubMed]
46. Jiang, Y.-L.; Yu, W.-L.; Zhang, J.-W.; Frolet, C.; Di Guilmi, A.-M.; Zhou, C.-Z.; Vernet, T.; Chen, Y. Structural basis for the substrate specificity of a novel β-*N*-acetylhexosaminidase StrH protein from *Streptococcus pneumoniae* R6. *J. Biol. Chem.* **2011**, *286*, 43004–43012. [CrossRef] [PubMed]
47. Thi, N.N.; Offen, W.A.; Shareck, F.; Davies, G.J.; Doucet, N. Structure and activity of the *Streptomyces coelicolor* A3(2) β-*N*-acetylhexosaminidase provides further insight into GH20 family catalysis and inhibition. *Biochemistry* **2014**, *53*, 1789–1800. [CrossRef] [PubMed]
48. Chen, X.; Wang, J.; Liu, M.; Yang, W.; Wang, Y.; Tang, R.; Zhang, M. Crystallographic evidence for substrate-assisted catalysis of β-*N*-acetylhexosaminidas from *Akkermansia muciniphila*. *Biochem. Biophys. Res. Commun.* **2019**, *511*, 833–839. [CrossRef] [PubMed]
49. Lenart, A.; Dudkiewicz, M.; Grynberg, M.; Pawłowski, K. CLCAs—A family of metalloproteases of intriguing phylogenetic distribution and with cases of substituted catalytic sites. *PLoS ONE* **2013**, *8*, e62272. [CrossRef]
50. Val-Cid, C.; Biarnés, X.; Faijes, M.; Planas, A. Structural-functional analysis reveals a specific domain organization in family GH20 hexosaminidases. *PLoS ONE* **2015**, *10*, e0128075. [CrossRef]
51. Tews, I.; Perrakis, A.; Wilson, K.S.; Oppenheim, A.; Dauter, Z.; Vorgias, C.E.; Tews, I. Bacterial chitobiase structure provides insight into catalytic mechanism and the basis of Tay–Sachs disease. *Nat. Struct. Biol.* **1996**, *3*, 638–648. [CrossRef]
52. Verma, S.C.; Mahadevan, S. The ChbG gene of the chitobiose (*chb*) operon of *Escherichia coli* encodes a chitooligosaccharide deacetylase. *J. Bacteriol.* **2012**, *194*, 4959–4971. [CrossRef]
53. Wang, W.; Xi, H.; Bi, Q.; Hu, Y.; Zhang, Y.; Ni, M. Cloning, expression and characterization of D-aminoacylase from *Achromobacter xylosoxidans* subsp. denitrificans ATCC 15173. *Microbiol. Res.* **2013**, *168*, 360–366. [CrossRef]
54. Graham, T.R.; Zassenhaus, H.P.; Kaplan, A. Molecular cloning of the cDNA which encodes β-*N*-acetylhexosaminidase A from *Dictyostelium discoideum*. Complete amino acid sequence and homology with the human sequence. *J. Biol. Chem.* **1988**, *263*, 16823–16829. [PubMed]
55. Kumar, S.; Stecher, G.; Tamura, K. MEGA7: Molecular Evolutionary Genetics Analysis version 7.0 for bigger datasets. *Mol. Biol. Evol.* **2016**, *33*, 1870–1874. [CrossRef] [PubMed]
56. Studier, F.W. Protein production by auto-induction in high-density shaking cultures. *Protein Expr. Purif.* **2005**, *41*, 207–234. [CrossRef] [PubMed]
57. Yem, D.W.; Wu, H.C. Purification and properties of beta-*N*-acetylglucosaminidase from *Escherichia coli*. *J. Bacteriol.* **1976**, *125*, 324–331. [CrossRef] [PubMed]
58. Poulton, J.E.; Thomas, M.A.; Ottwell, K.K.; McCormick, S.J. Partial purification and characterization of a β-*N*-acetylhexosaminidase from black cherry (*Prunus serotina* EHRH.) seeds. *Plant Sci.* **1985**, *42*, 107–114. [CrossRef]
59. Vehpoorte, J.A. Purification of two β-*N*-acetyl-D-glucosaminidases from beef spleen. *J. Biol. Chem.* **1972**, *247*, 4787–4793.
60. Mitchell, E.D.; Houston, W.C.; Latimer, S.B. Purification and properties of a β-*N*-acetylaminoglucohydrolase from malted barley. *Phytochemistry* **1976**, *15*, 1869–1871. [CrossRef]
61. Trimble, R.B.; Evans, G.; Maley, F. Purification and properties of endo-beta-*N*-acetylglucosaminidase L from *Streptomyces plicatus*. *J. Biol. Chem.* **1979**, *254*, 9708–9713.
62. Lisboa De Marco, J.; Valadares-Inglis, M.C.; Felix, C.R. Purification and characterization of an *N*-acetylglucosaminidase produced by a *Trichoderma harzianum* strain which controls *Crinipellis perniciosa*. *Appl. Microbiol. Biotechnol.* **2004**, *64*, 70–75. [CrossRef] [PubMed]
63. Chen, F.; Chen, X.-Z.; Qin, L.-N.; Tao, Y.; Dong, Z.-Y. Characterization and homologous overexpression of an *N*-acetylglucosaminidase Nag1 from *Trichoderma reesei*. *Biochem. Biophys. Res. Commun.* **2015**, *459*, 184–188. [CrossRef] [PubMed]

64. Matsuo, I.; Kim, S.; Yamamoto, Y.; Ajisaka, K.; Maruyama, J.; Nakajima, H.; Kitamoto, K. Cloning and overexpression of beta-*N*-acetylglucosaminidase encoding gene *nagA* from *Aspergillus oryzae* and enzyme-catalyzed synthesis of human milk oligosaccharide. *Biosci. Biotechnol. Biochem.* **2003**, *67*, 646–650. [CrossRef] [PubMed]
65. Aunpad, R.; Rice, D.W.; Sedelnikova, S.; Panbangred, W. Biochemical characterisation of two forms of halo- and thermo-tolerant chitinase C of *Salinivibrio costicola* expressed in *Escherichia coli*. *Ann. Microbiol.* **2007**, *57*, 249–257. [CrossRef]
66. Kurakake, M.; Goto, T.; Ashiki, K.; Suenaga, Y.; Komaki, T. Synthesis of new glycosides by transglycosylation of *N*-acetylhexosaminidase from *Serratia marcescens* YS-1. *J. Agric. Food Chem.* **2003**, *51*, 1701–1705. [CrossRef]
67. Flockenhaus, B.; Kieß, M.; Müller, M.C.M.; Leippe, M.; Scholze, H.; Riekenberg, S.; Vahrmann, A. The β-*N*-acetylhexosaminidase of *Entamoeba histolytica* is composed of two homologous chains and has been localized to cytoplasmic granules. *Mol. Biochem. Parasitol.* **2004**, *138*, 217–225.
68. Dimond, R.L.; Loomis, W.F. Vegetative isozyme of *N*-acetylglucosaminidase in *Dictyostelium discoideum*. *J. Biol. Chem.* **1974**, *249*, 5628–5632.
69. Drouillard, S.; Armand, S.; Davies, J.G.; Vorgias, E.C.; Henrissat, B. *Serratia marcescens* chitobiase is a retaining glycosidase utilizing substrate acetamido group participation. *Biochem. J.* **1997**, *328*, 945–949. [CrossRef]
70. Kobata, A. Structures and application of oligosaccharides in human milk. *Proc. Jpn. Acad. Ser. B Phys. Biol. Sci.* **2010**, *86*, 731–747. [CrossRef]
71. André-Miral, C.; Koné, F.M.; Solleux, C.; Grandjean, C.; Dion, M.; Tran, V.; Tellier, C. *De novo* design of a trans-β-*N*-acetylglucosaminidase activity from a GH1 β-glycosidase by mechanism engineering. *Glycobiology* **2015**, *25*, 394–402. [CrossRef]
72. Nguyen, H.A.; Nguyen, T.; K Křen, V.; Eijsink, V.G.H.; Haltrich, D.; Peterbauer, C.K. Heterologous expression and characterization of an *N*-acetyl-β-D-hexosaminidase from *Lactococcus lactis* ssp. lactis IL1403. *J. Agric. Food Chem.* **2012**, *60*, 3275–3281. [CrossRef] [PubMed]
73. Rauvolfová, J.; Kuzma, M.; Weignerová, L.; Fialová, P.; Přikrylová, V.; Pišvejcová, A.; Macková, M.; Křen, V. β-*N*-Acetylhexosaminidase-catalysed synthesis of non-reducing oligosaccharides. *J. Mol. Catal. B Enzym.* **2004**, *29*, 233–239. [CrossRef]
74. Sidorczyk, Z.; Senchenkova, S.N.; Perepelov, A.V.; Kondakova, A.N.; Kaca, W.; Rozalski, A.; Knirel, Y.A. Structure and serology of *O*-antigens as the basis for classification of *Proteus* strains. *Innate Immun.* **2010**, *17*, 70–96.
75. Katzenellenbogen, E.; Kocharova, N.A.; Korzeniowska-Kowal, A.; Bogulska, M.; Rybka, J.; Gamian, A.; Kachala, V.V.; Shashkov, A.S.; Knirel, Y.A. Structure of the glycerol phosphate-containing *O*-specific polysaccharide and serological studies on the lipopolysaccharides of *Citrobacter werkmanii* PCM 1548 and PCM 1549 (serogroup O14). *FEMS Immunol. Med. Microbiol.* **2008**, *54*, 255–262. [CrossRef] [PubMed]
76. Katzenellenbogen, E.; Kocharova, N.A.; Zatonsky, G.V.; Shashkov, A.S.; Bogulska, M.; Knirel, Y.A. Structures of the biological repeating units in the *O*-chain polysaccharides of *Hafnia alvei* strains having a typical lipopolysaccharide outer core region. *FEMS Immunol. Med. Microbiol.* **2005**, *45*, 269–278. [CrossRef] [PubMed]
77. Thomas, F.; Barbeyron, T.; Michel, G. Evaluation of reference genes for real-time quantitative PCR in the marine flavobacterium *Zobellia galactanivorans*. *J. Microbiol. Methods* **2011**, *84*, 61–66. [CrossRef]
78. Ashida, H.; Miyake, A.; Kiyohara, M.; Yoshida, E.; Kumagai, H.; Yamamoto, K. Two distinct α-L -fucosidases from *Bifidobacterium bifidum* are essential for the utilization of fucosylated milk oligosaccharides and glycoconjugates. *Glycobiology* **2009**, *19*, 1010–1017. [CrossRef]
79. Curiel, J.A.; De Las Rivas, B.; Mancheño, J.M.; Muñoz, R. The pURI family of expression vectors: A versatile set of ligation independent cloning plasmids for producing recombinant His-fusion proteins. *Protein Expr. Purif.* **2011**, *76*, 44–53. [CrossRef]
80. Stingele, F.; Newell, J.W.; Neeser, J.R. Unraveling the function of glycosyltransferases in *Streptococcus thermophilus* Sfi6. *J. Bacteriol.* **1999**, *181*, 6354–6360. [CrossRef]
81. Viigand, K.; Visnapuu, T.; Mardo, K.; Aasamets, A.; Alamäe, T. Maltase protein of *Ogataea* (*Hansenula*) polymorpha is a counterpart to the resurrected ancestor protein ancMALS of yeast maltases and isomaltases. *Yeast* **2016**, *33*, 415–432. [CrossRef]
82. Reiffová, K.; Nemcová, R. Thin-layer chromatography analysis of fructooligosaccharides in biological samples. *J. Chromatogr. A* **2006**, *1110*, 214–221. [CrossRef] [PubMed]

83. McCleary, B.V.; McGeough, P. A Comparison of polysaccharide substrates and reducing sugar methods for the measurement of endo-1,4-β-xylanase. *Appl. Biochem. Biotechnol.* **2015**, *177*, 1152–1163. [CrossRef] [PubMed]
84. Overbeek, R.; Olson, R.; Pusch, G.D.; Olsen, G.J.; Davis, J.J.; Disz, T.; Edwards, R.A.; Gerdes, S.; Parrello, B.; Shukla, M.; et al. The SEED and the Rapid Annotation of microbial genomes using Subsystems Technology (RAST). *Nucleic Acids Res.* **2014**, *42*, 206–214. [CrossRef] [PubMed]
85. Petersen, T.N.; Brunak, S.; Von Heijne, G.; Nielsen, H. SignalP 4.0: Discriminating signal peptides from transmembrane regions. *Nat. Methods* **2011**, *8*, 785–786. [CrossRef]

International Journal of
Molecular Sciences

Article

Low-Cost Cellulase-Hemicellulase Mixture Secreted by *Trichoderma harzianum* EM0925 with Complete Saccharification Efficacy of Lignocellulose

Yu Zhang [1], Jinshui Yang [1], Lijin Luo [2], Entao Wang [3], Ruonan Wang [1], Liang Liu [1], Jiawen Liu [1] and Hongli Yuan [1,*]

1 State Key Laboratory of Agrobiotechnology and Key Laboratory of Soil Microbiology, Ministry of Agriculture, College of Biological Sciences, China Agricultural University, Beijing 100193, China; B20173020090@cau.edu.cn (Y.Z.); yangjsh1999@cau.edu.cn (J.Y.); rnwang@cau.edu.cn (R.W.); B20163020078@cau.edu.cn (L.L.); SZ20143020044@cau.edu.cn (J.L.)

2 Fujian Institute of Microbiology, Fuzhou 350007, China; luolijin@sina.com

3 Departamento de Microbiología, Escuela Nacional de Ciencias Biológicas, Instituto Politécnico Nacional, Mexico City 11340, Mexico; entaowang@yahoo.com.mx

* Correspondence: hlyuan@cau.edu.cn

Received: 9 November 2019; Accepted: 2 January 2020; Published: 7 January 2020

Abstract: Fermentable sugars are important intermediate products in the conversion of lignocellulosic biomass to biofuels and other value-added bio-products. The main bottlenecks limiting the production of fermentable sugars from lignocellulosic biomass are the high cost and the low saccharification efficiency of degradation enzymes. Herein, we report the secretome of *Trichoderma harzianum* EM0925 under induction of lignocellulose. Numerously and quantitatively balanced cellulases and hemicellulases, especially high levels of glycosidases, could be secreted by *T. harzianum* EM0925. Compared with the commercial enzyme preparations, the *T. harzianum* EM0925 enzyme cocktail presented significantly higher lignocellulolytic enzyme activities and hydrolysis efficiency against lignocellulosic biomass. Moreover, 100% yields of glucose and xylose were obtained simultaneously from ultrafine grinding and alkali pretreated corn stover. These findings demonstrate a natural cellulases and hemicellulases mixture for complete conversion of biomass polysaccharide, suggesting *T. harzianum* EM0925 enzymes have great potential for industrial applications.

Keywords: glycoside hydrolyase; *Trichoderma harzianum*; complete saccharification; lignocellulose

---

1. Introduction

Recently, bio-based industries have developed rapidly, and potential bio-based products including ethanol, aldehydes, organic acids, polyhydric alcohols, and other bio-chemicals and biomaterials have attracted a lot of attention. As the most important intermediate compounds of biological and chemical transformation from biomass, fermentable sugars (mainly glucose and xylose) are crucial for the production of downstream products [1,2]. Due to the structural complexity of lignocellulose, complete enzymatic deconstruction requires the synergistic action of cellulases, hemicellulases, and ligninases [3]. In order to obtain high yield of glucose from cellulose, endoglucanase, cellobiohydrolase, and β-glucosidase are required to work together. Endoglucanases randomly cleave the internal *O*-glycosidic bonds and produce glucan chains with different lengths, cellobiohydrolases attack the reducing and non-reducing ends of cellulose to release β-cellobiose, and β-glucosidases hydrolyze the terminal non-reducing β-D-glucosyl residues into glucose. These enzymes can be produced by many fungi (like *Aspergillus* and *Trichoderma*) and bacteria (like *Cellulomonas* and *Clostridium*) [2,3]. In any case, the rapid and complete conversion of cellulose is the limiting step during the process of bio-refinery [4].

Currently, *Trichoderma reesei* has been commonly used as a cellulase producer. Although two exoglucanases (CBH I and CBH II), eight endoglucanases (EGI-EGVIII), and seven β-glucosidase (BG I-BG VII) are produced by this fungus, the fairly low protein abundance (<1%) of β-glucosidase in the extracellular proteome limited the effective conversion of cellubiose to glucose [5]. The total cellulase activities of commercial preparations Viscozyme L and Celluclast 1.5L were as high as 33 and 95.2 FPU/mL, but the β-glucosidase activity was only 0.2 and 0.3 U/mL, respectively [6]. When β-glucosidase preparation Novezyme 188 was supplied to Celluclast 1.5L, glucose yield from sugarcane bagasse increased 1.5-fold than that of Celluclast 1.5L alone [7]. In addition, higher content of cellobiohydrolase in *Trichoderma reesei* extracellular proteome was detected, which caused the accumulation of cellobiose on account of the deficient amount of β-glucosidase, and then gave rise to severe product feedback inhibition of endoglucanase and cellobiohydrolase [8]. Some researchers found that hemicellulases, oxidoreductases, non-hydrolytic proteins and other auxiliary enzymes synergize to achieve an efficient enzymatic hydrolysis of cellulose [2]. Hemicellulase and pectinase could degrade the non-cellulose polysaccharide that covered cellulose and in turn increase the cellulose hydrolysis efficiency, and this synergistic effect could give rise to the cost reduction in deconstruction of lignocellulose [9]. Goldbeck et al. [10] found that when xylanase was added to the commercial cellulase preparation Accellerase 1500, the glucose yield of dilute acid pretreated sugarcane bagasse presented a 1.4-fold increase than that of Accellerase 1500 alone. Furthermore, hemicellulose hydrolysis products (xylose, arabinose, mannose, and galactose etc.) also have great potential in food and feed industry applications. Therefore, in terms of the whole component utilization of plant cell wall polysaccharides, not only lack of β-glucosidase restricted the efficient conversion of cellulose, the lack of hemicellulase in *Trichoderma reesei* extracellular proteome also made it difficult to utilize hemicellulose [11]. Hemicellulase activity of commercial cellulase production model strains *T. reesei* QM6a and *T. reesei* QM9414 has been determined by Li et al. [12], in which the xylanase activity was 5.42 and 1.27 U/mL, while xylosidase activity was only 0.4 and 0.005 U/mL, respectively. Recent studies showed that extracellular enzymes from *Penicillium* species perform better than cellulases from *Trichoderma* sp. On account of the higher hemicellulases activities of enzyme mixture secreted by *Penicillium* sp., greater fermentable sugar yield was obtained [13,14]. Yang et al. [15] found that when 50% of the commercial cellulase was replaced by *Penicillium chrysogenum* enzyme cocktail, release of reducing sugars was 78.6% higher than that with the cellulase alone, while the glucan and xylan conversion was increased by 37% and 106%, respectively. Additionally, the hydrolysis efficiency of cellulase increased more with the addition of multi-component hemicellulases cocktail than with xylanase alone. The artificial enzyme cocktail of pectinase, xylanase, arabinofuranosidase, acetyl xylan esterase, and ferulic acid esterase presented higher efficiency of bagasse degradation than the single component [16]. The appropriate dosage of auxiliary enzymes for efficient conversion of different lignocellulosic biomass is still uncertain. Therefore, exploring novel lignocellulosic degrading strains with complete and balanced enzyme cocktail for efficiently saccharification of lignocellulose is necessary for industrial enzyme preparation production.

Production of extracellular enzymes with high cellulase activity has been proved in members of *Trichoderma* [17]. *T. reesei* Rut-C30 mainly expresses cellulases including cellobiohydrolase and endoglucanase with a total abundance of 90–95% of the extracellular proteins [18]. Compared with the industrial cellulase producer *T. reesei*, *Trichoderma harzianum* harbors more comprehensive lignocellulosic degrading enzyme encoding genes in its genome [19]. To be specific, a total of 42 cellulase genes and 24 hemicellulase genes were annotated in genome of *T. harzianum* T6776, which were 1.5 and 1.7-fold of those of *T. reesei* Rut-C30 [20]. Meanwhile, the expression levels of xylanase, mannase, and various glycosidases of *T. harzianum* was significantly higher than that in *T. reesei* [19,21]. Compared with *T. reesei*, some *T. harzianum* strains produced a cellulolytic complex with higher β-glucosidase and endoglucanases activities, and the xylanase activity of some *T. harzianum* strains was higher than that of *T. reesei* [22–24]. However, efficiency of simultaneous cellulose and hemicellulose hydrolysis by the secretome of *T. harzianum* was still low [23,25]. Lignocellulose degrading enzymes in microbes are induced by substrates, and the low-cost carbon source and culture conditions had a great influence on the enzyme secretion and composition [26]. Nevertheless, the mechanism of lignocellulolytic enzyme inducing was still unclear [27]. So, searching and exploring highly effective enzyme systems for efficiently converting the whole component of plant cell wall polysaccharides is expected, and the comprehensive understanding of secretomes in the related microbes could help the development of efficiently tailor-made lignocellulolytic enzyme cocktails in vitro [28]. Isolated in our laboratory, *T. harzianum* EM0925 could secret high levels of cellulase and hemicellulase simultaneously, and its extracellular enzyme cocktail showed strong ability of lignocellulose degradation. The enzyme cocktail of *T. harzianum* EM0925 contained a great amount of complete lignocellulolytic enzymes as revealed by proteome analysis under the optimal inducing condition. Aiming at analyzing the degradation mechanism of *T. harzianum* EM0925, we performed the present study. The results provide a basis for tailor-made preparation of low-cost enzymes to effectively degrade the whole component of plant cell wall polysaccharides.

2. Results

2.1. Substrate Selection for Lignocellulosic Enzyme Production by T. Harzianum EM0925

In order to prepare low-cost lignocellulolytic enzymes cocktails, different kinds of lignocellulosic substrates were used separately as the carbon source for *T. harzianum* EM0925, including wheat bran, sunchoke (*Jerusalem artichoke*) stalks, corncob, miscanthus, giant juncao grass, switchgrass, corn stover, sugarcane bagasse, and straw of *Triarrhena lutarioriparia*. Measurement of the extracellular cellulase and xylanase revealed that the highest filter paper activity (1.54 U/mL) was induced by corn stover after seven days of cultivation (Figure 1). Simultaneously, high levels of endoglucanase activity (8.61 U/mL) and xylanase activity (54 U/mL) were induced (Figure 1), in which the xylanase activity was 91% of the maximum value (59.9 U/mL) detected in the culture with corncob after seven days of cultivation. In addition, when corn stover was used, rapid enzyme production capacity was observed in *T. harzianum* EM0925. The filter paper activity reached more than 70% of the maximum value, while the endoglucanase and xylanase activities reached more than 80% of their maximum values after two days of fermentation. Therefore, corn stover was selected as the optimal substrate for *T. harzianum* EM0925 to prepare the lignocellulolytic enzyme cocktail (EM0925) in further study.

2.2. Enzyme Activities of Enzyme Cocktail EM0925

Enzyme activities of commercial preparations (C 9748 and Celluclast 1.5L) and EM0925 (Enzyme preparation of *T. harzianum* EM0925) were measured by using the model substrates. The three enzyme preparations showed significant differences in their specific activities of cellulases, including endoglucanase, cellobiohydrolase, and β-glucosidase. The endoglucanase activity of EM0925 was only 56.9% of C 9748, and the cellobiohydrolase activity showed no significant difference with that of Celluclast 1.5L. Most notably, EM0925 displayed the highest level of β-glucosidase activity, which was 311 and 2.9 folds of that in C 9748 and Celluclast 1.5L, respectively. In addition, EM0925 displayed significantly higher hemicellulases activities than the two commercial preparations. The xylanase specific activitiy of EM0925 was 310.70 U/mg, which was 29.1 folds higher than that of Celluclast 1.5L. Xylosidase and arabinofuranosidase activities were as high as 203.60 and 11.72 U/mg, which was 333.8 and 37.8 folds higher than that of commercial enzyme preparations C 9748 and Celluclast 1.5L, respectively. Furthermore, amylase activities of EM0925 and C 9748 were 3.92 and 0.3 U/mg, respectively, and it was not detected in Celluclast 1.5L. EM0925 contained a more complete lignocellulosic enzyme system than the commercial enzyme preparations. It presented high levels of glycosidases that were necessary for cellulose and hemicellulose degradation, and showed a broader prospect in fermentable monosaccharide production from complex polysaccharide compounds (Table 1).

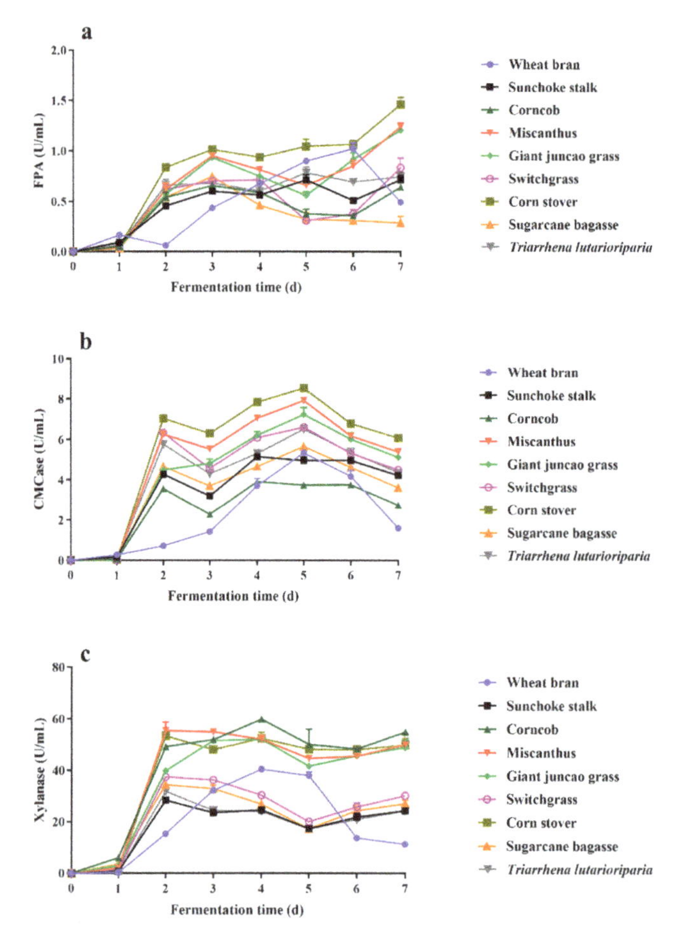

Figure 1. Activities of cellulase and hemicellulase secreted by *T. harzianum* EM0925 induced by different kinds of lignocellulosic biomass. (**a**) FPA in cultures; (**b**) CMCase activities in cultures; (**c**) hemicellulases activities in cultures.

Table 1. Cellulase and hemicellulase activities of EM0925 and commercial enzyme preparations (specific activities, U/mg protein).

Cocktail	FPA	EG	CBH	βG	Xyl	βX	Ara	Man	Amy
C 9748	1.60 ± 0.20 [a]	19.70 ± 2.90 [a]	52.20 ± 1.90 [a]	29.0 ± 2.70 [a]	26.60 ± 3.20 [a]	34.80 ± 1.40 [a]	0.03 ± 0.00 [a]	0.71 ± 0.00 [a]	0.30 ± 1.00 [a]
Celluclast 1.5 L	1.90 ± 0.02 [a]	8.33 ± 0.18 [b]	32.70 ± 1.80 [b]	0.27 ± 0.002 [b]	10.80 ± 0.14 [b]	0.61 ± 0.02 [b]	0.31 ± 0.00 [b]	3.90 ± 0.11 [b]	ND
EM0925	2.77 ± 0.08 [b]	11.20 ± 0.13 [c]	33.76 ± 2.40 [b]	84.97 ± 0.62 [c]	314.70 ± 3.23 [c]	203.60 ± 18.60 [c]	11.72 ± 0.33 [c]	6.70 ± 0.12 [c]	3.92 ± 0.02 [b]

FPA, filter paper activity; EG, CMCase activity; CBH, cellobiohydrolase activity; βG, β-glucosidase activity; Xyl, xylanase activity; βX, β-xylosidase activity; Ara, arabinofuranosidase activity; Man, mannase activity; Amy, amylase activity; C 9748, commercial cellulase preparation from Sigma; Celluclast 1.5 L, commercial cellulase preparation from Novozyme; EM0925, enzyme cocktail of *T. harzianum* EM0925; ND, not detected. [a, b, c] statistical significance is indicated by different letters in table as assessed by ANOVA ($p < 0.05$).

2.3. Extracellular Enzymes in the Proteome of *T. harzianum* EM0925

By using the label free quantitative proteomic approach, a total of 154 proteins were identified in the *T. harzianum* EM0925 secretome, which were common in all the three biological replicates. The predicted proteins in the secretome of *T. harzianum* EM0925 were grouped according to their biological function annotated in Pfam and dbCAN database. The 81 detected CAZymes contained 27 cellulases covering a complete cellulolytic enzyme system with 13 endoglucanases (GH5, GH7, GH16, GH17, GH30, GH55, GH64, GH81), 2 cellobiohydrolases (GH6, GH7), and 12 glucosidases (GH3, GH15, GH16, GH55) (Figure 2a, Figure S1). The detection of 39 hemicellulases revealed that this kind of enzymes was more abundant than the cellulases in *T. harzianum* EM0925 secretome, which included 8 xylanases (GH11, GH30), 1 xylosidases (GH3), 3 arabinofuranosidases (GH54, GH127), 7 galactosidases (GH2, GH22, GH27, GH30), 11 mannosidases (GH2, GH47, GH92, GH125), 1 mananase (GH26), and 8 carboxylesterases (CE1, CE5, CE10) (Figure 2b, Figure S1).

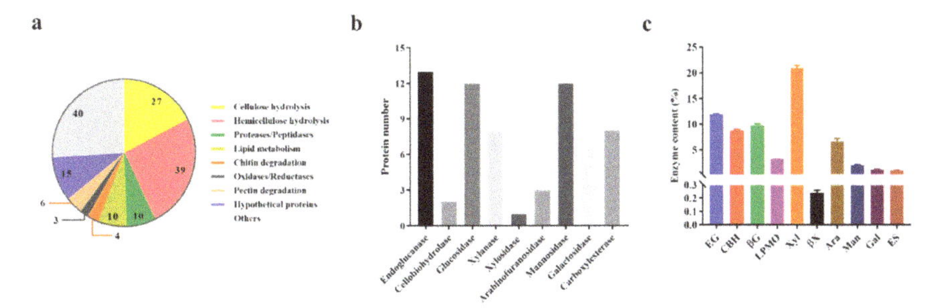

Figure 2. Functional classification of enzymes secreted by *T. harzianum* EM0925. (**a**) Functional classification of *T. harzianum* EM0925 enzyme cocktail; (**b**) numbers of different lignocellulosic enzyme in proteome of *T. harzianum* EM0925. (**c**) Abundance of lignocellulose degradading enzymes in the proteome of *T. harzianum* EM0925. EG, Endoglucanase; CBH, cellobiohydrolase; βG, β-glucosidase; LPMO, lytic polysaccharide monooxygenase; Xyl, xylanase; βX, β-xylosidase; Ara, arabinofuranosidase; Man, mannosidase; Gal, galactosidases; ES, carboxylesterase.

The quantitative proteomic data showed high ratios in the protein abundance of a core set of glycoside hydrolases secreted by *T. harzianum* EM0925, including total cellulase abundance of 31.5% and total hemicellulase abundance of 32.2% (Figure 2c). Cellulases, including endoglucanases, cellobiohydrolases and glucosidases accounted for 11.85%, 8.8%, and 9.76% of the total proteins, respectively (Figure 2c). Among the secreted cellulases and hemicellulases, some proteins in these sets took up a great proportion. For instance, the two GH5 endogluananses accounted for 4.54% and 5.30% of the total proteins, respectively, and both contained a CBM domain at each of the C and N ends of the protein. Cellobiohydrolases I (GH7) and cellobiohydrolases II (GH6) accounted for 7.30% and 1.50% of total protein abundance, respectively. Two GH3 β-glucosidases accounted for an outstanding proportion among the multiple glucosidases, with abundances of 6.00% and 1.76% (Figure 3a). Multiple abundant hemicellulases were also detected, including xylanase (21.01%), xylosidases (0.24%), arabinofuranosidases (6.66%), galactosidases (2.1%), mannosidases (1.18%), and carboxylesterase (1.01%). It was noteworthy that two highly expressed xylanases were relatively abundant in the secretome, with an abundance index of 14.69% and 4.51%, and accounted for 91.4% of the total xylanase abundance. A lytic polysaccharide monooxygenase (LPMO, AA9) secreted by *T. harzianum* EM0925 accounted for 3.23% of the total quantified CAZymes (Figure 2c). Moreover, the main glycosidases (glucosidases, xylosidases, arabinofuranosidases, and mannosidases) accounted for 17.2% of the total extracellular degradation enzymes secreted by *T. harzianum* EM0925 (Figure 3).

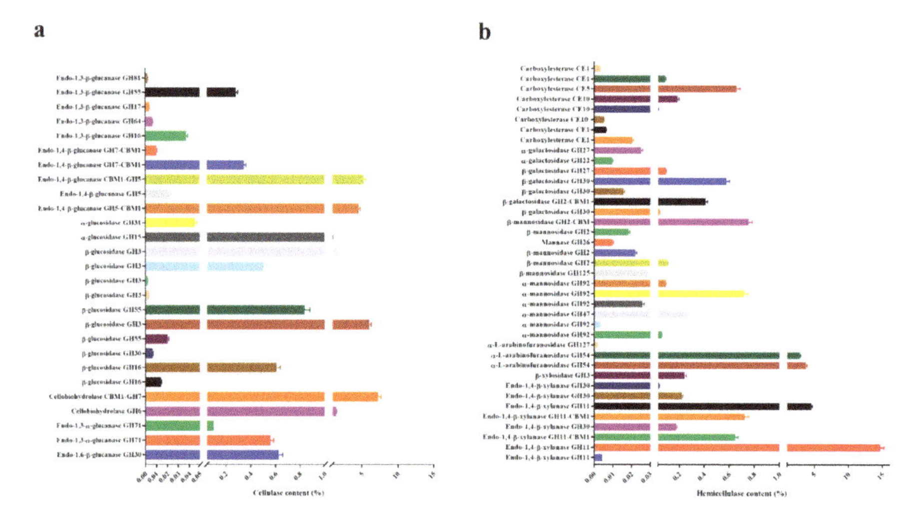

Figure 3. Cellulase and hemicellulase abundance in the proteome of *T. harzianum* EM0925. (**a**) Cellulase content in the proteome of *T. harzianum* EM0925; (**b**) hemicellulase content in the proteome of *T. harzianum* EM0925.

2.4. Properties of Enzyme Cocktail EM0925

The optimum temperatures of filter paper degrading enzymes and endoglucanases in the enzyme cocktail secreted by *T. harzianum* EM0925 were 60 °C, and that of xylanases were 50 °C. Filter paper activities and endoglucananse activities were 98.94% and 85.38% of their highest activities at 50 °C, respectively. When *T. harzianum* EM0925 enzymes were incubated at 50 °C for 1 h, the endoglucanases, total cellulases and xylanases still maintained 100%, 83.22%, and 85.72% relative activity, respectively.

The optimal pH of total cellulases, endoglucanases and xylanases was 4.5. The activity was above 75% of the highest activity for total cellulase at pH 3.0–6.0, above 70% for endoglucanase at pH 3.5–5.0, and above 90% for xylanase at pH 2.5–5.0. In addition, favorable pH stability was also observed for the enzymes: 75% total cellulase activity was maintained after 1 h of incubation at pH-values between 2.0 and 12.0, in which 90% activity was maintained at pH 3.0 to 9.0. In addition, about 98% endoglucanase activity and 80% xylanase activity were maintained at the pH range of 2.0–12.0 after 1 h incubation, which 95% xylanase activity was maintained at the pH range of 2.0–9.0 (Figure 4).

2.5. Hydrolysis of Lignocellulosic Biomass by Enzyme Cocktail EM0925

The structural carbohydrate and lignin contents of differently pre-treated fractions of corn stover were determined (Figure 5a). As expected, chlorite/acetic acid and NaOH pretreatment resulted a significant decrease in lignin content: only 2% of the initial lignin content remained in ALKCS, whereas the relative cellulose content was highly increased up to 60.7%, which was 1.4-fold of that in NTCS. Dilute acid and steam explosion treatment significantly decreased the content of hemicellulose, which decreased from 26.1% to 5% in DACS, and the highest cellulose content (62.5%) was observed at the same time (Figure 5a). At the enzyme dosage of 5 mg proteins/g substrate, hydrolysis with EM0925 generated 14.62, 19.00, 13.22, 12.43, and 8.83 mg reducing sugars per 20 mg of ultrafine grinding corn stover (UGCS), sodium hydroxide treated corn stover (ALKCS), sodium chlorite treated corn stover (DLCS), dilute acid treated corn stover (DACS), and stream explosion treated corn stover (SECS) after 72 h reaction, respectively. In contrast, only 4.31 mg reducing sugars were released from 20 mg of NTCS. Reducing sugar yield of UGCS and ALKCS was 96.6% and 100%, respectively. Furthermore, rapid saccharification was observed with these two substrates, which resulted in a sugar yield of 81.9% and 74.2% after 24 h of hydrolysis (Figure 5b).

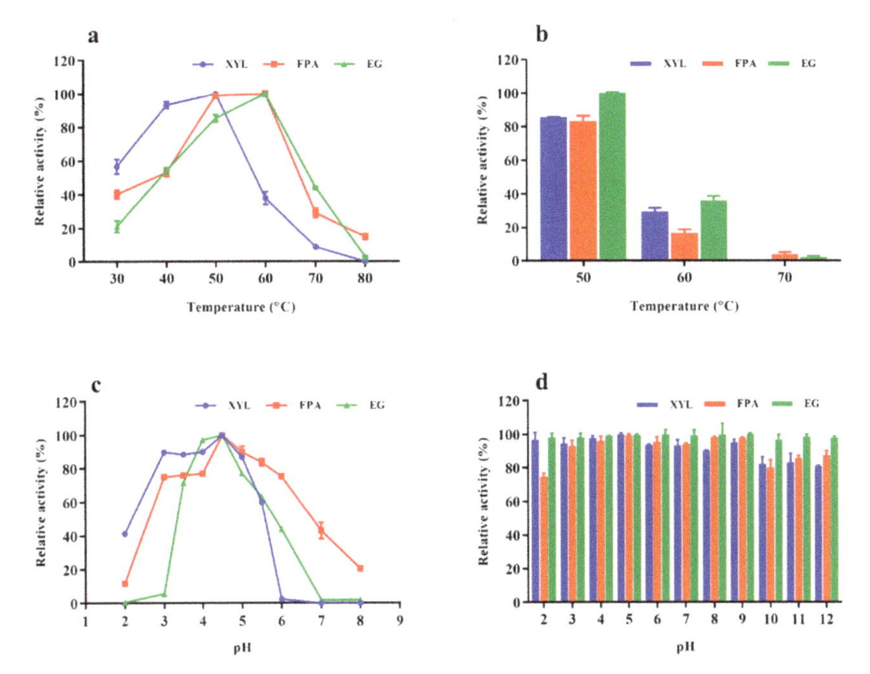

Figure 4. Characterization of FPA, CMCase, and xylanase of *T. harzianum* EM0925 enzyme cocktail. (**a**) The effect of tempertrure on FPA, CMCase, and xylanase of *T. harzianum* EM0925 enzyme cocktail; (**b**) thermostability of FPA, CMCase, and xylanase of *T. harzianum* EM0925 enzyme cocktail; (**c**) the effect of pH on FPA, CMCase, and xylanase of *T. harzianum* EM0925 enzyme cocktail; (**d**) pH stability of FPA, CMCase, and xylanase of *T. harzianum* EM0925 enzyme cocktail. pH stability was determined by measuring the residual activity after incubation of *T. harzianum* EM0925 enzyme cocktail in different buffers for 1 h. The relative activity was determined, and the maximum activity was defined as 100% (**a**,**c**). The initial activity of *T. harzianum* EM0925 enzyme cocktail not pre-incubated in different buffers was defined as 100% (**b**,**d**). Values are the mean of three replicates ± SD.

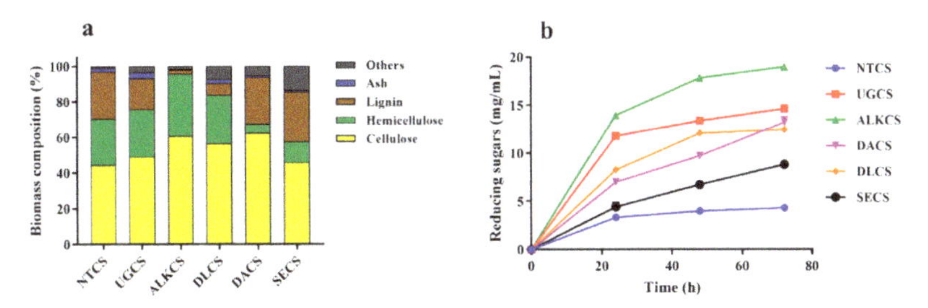

Figure 5. Reducing sugar yield from differently pre-treated fractions of corn stover by *T. harzianum* EM0925 enzyme cocktail. (**a**) Chemical compositon of pretreated corn stover. NTCS, matural corn stover without pretreatment; UGCS, ultrafine grinding corn stover; ALKCS, sodium hydroxide treated corn stover; DLCS, sodium chlorite treated corn stover; DACS, dilute acid treated corn stover; SECS, stream explosion treated corn stover. (**b**) Release of reducing sugars from differently pre-treated fractions of corn stover.

To compare the saccharification efficiency of EM0925 with those of two commercial enzyme preparations, an enzyme loading of 5, 10, and 30 FPU/g substrate was selected as the low, moderate and high enzyme dosage, respectively. Glucose and xylose yields from UGCS and ALKCS were measured after 72 h of hydrolysis reaction. Cellulose of UGCS was completely converted to glucose by EM0925 at low enzyme dosage, and a xylan conversion of 72.35% was obtained simultaneously. Meanwhile, only 45.68% of the cellulose and 57.15% of the xylan were converted by Celluclast 1.5L. The xylan conversion gradually increased with increasing enzyme dosages, xylose yield was 100% when 30 FPU EM0925 enzymes/g substrate was used. Meanwhile, only 85.59% of the cellulose and 66.18% of the xylan were converted by Celluclast 1.5L even at the highest enzyme dosage (Figure 6a,b). When ALKCS was used as substrate, 63.08% cellulose conversion and 70.47% xylan conversion were obtained at low dosage of EM0925, which were 1.55 and 1.40-fold of the same conversions by Celluclast 1.5L. The yields of glucose and xylose were all gradually improved with increasing enzyme dosages of the two preparations. 100% glucose and xylose were released when 30 FPU/g EM0925 was used, and only 73.78% of the cellulose and 81.28% of the xylan were converted by Celluclast 1.5L at the highest enzyme dosage (Figure 6c,d).

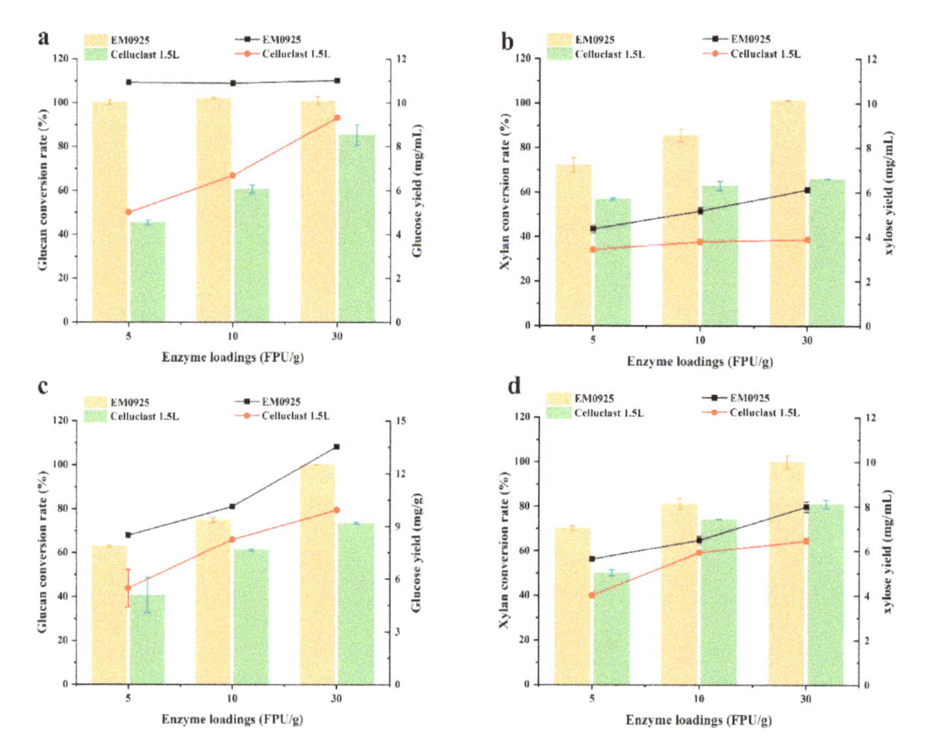

Figure 6. Conversion of glucan and xylan and sugar yield from UGCS and ALKCS by commercial enzyme preparation and *T. harzianum* EM0925 enzyme cocktails. (**a**) Glucan conversion and glucose yield from UGCS by commercial and *T. harzianum* EM0925 enzyme cocktails; (**b**) xylan conversion and xylose yield from UGCS by commercial and *T. harzianum* EM0925 enzyme cocktails; (**c**) glucan conversion and glucose yield from ALKCS by commercial and *T. harzianum* EM0925 enzyme cocktails; (**d**) xylan conversion and xylose yield from ALKCS by commercial and *T. harzianum* EM0925 enzyme cocktails. Celluclast 1.5 L, commercial cellulase preparation; UGCS, ultrafine grinding corn stover; ALKCS, sodium hydroxide treated corn stover.

When polysaccharides were completely converted to monosaccharides, 11.20 and 13.48 mg glucose were obtained from 20 mg of UGCS and ALKCS, respectively, while 6.11 and 7.98 mg pentose were released simultaneously. Monosaccharides yield was 865.5 mg/g of UGCS and 1073 mg/g of ALKCS, which was 1.29 and 1.31-fold of the highest monosaccharide yield obtained by Celluclast 1.5 L. In addition, the highest sugar yield from UGCS was obtained when EM0925 was applied at the dosage corresponding to 1/6 of the Celluclast 1.5 L dose (Figure 6a,b). These saccharification yields are among the highest yields reported in literature [29–34] (Table S1).

3. Discussion

Saprophytic fungi secrete a series of enzymes to degrade the plant cell walls in order to obtain nutrition for their growth. The main drawback of biofuels and other value-added bio-products generated from biomass are the high cost and the low saccharification efficiency of lignocellulolytic enzymes. In addition, the global market for industrial enzymes was expected to increase from nearly $5.0 billion in 2016 to $6.3 billion in 2021. Hence, lignocellulolytic enzymes which can be produced at low-cost and are capable of producing high yields of fermentable sugars require investigation. Using the agricultural residue corn stover as the sole carbon source, *T. harzianum* EM0925 secreted a quantitatively balanced enzyme system, which provides a low-cost enzyme preparation for biomass hydrolysis. High levels of endoglucanases and β-glucosidases could be secreted by *T. harzianum*, which has a high potential for application in industrial settings. According to Li et al. [12], the filter paper activity of the enzyme cocktail secreted by *T. harzianum* K223452 (CGMCC 17199) was 1.62 U/mL using sugarcane bagasse as the substrate supplied with 20 g/L glucose as carbon source. We observed a comparable activity for *T. harzianum* EM0925 when grown on corn stover without supplementation. These facts illustrate the possibility to develop low-cost degradation enzyme cocktails with high efficiency [2,26]. The high levels of lignocellulolytic enzymes and rapid enzyme production capacity of *T. harzianum* EM0925 made it a suitable candidate for enzyme cocktail production, since most of the previous studies on developing cost-effective cell wall degrading enzymes have focused on improving the activity of cellulase by adopting genetic operation and screening natural strains so as to obtain high yield of glucose [4,35]. Meng et al. [36] overexpressed the endoglucanase gene in *T. reesei* Rut-C30, which showed 90.0% and 132.7% increases in the activities of total cellulases and endoglucanases under flask culture conditions, respectively. Furthermore, hemicellulose is the second most abundant component of lignocellulose and is mainly composed of xylan, mannan, and xyloglucan. Hemicelluloses can be hydrolyzed into pentoses (xylose and arabinose) and hexoses (glucose, mannose and galactose) at high yield which is important for bio-refinery applications [37,38]. Besides the enzymes degrading the main chains of the structural polysaccharides in plant cell walls, various glycosidases have been found to be essential for obtaining fermentable sugars from lignocellulose [39]. Glucosidase, xylosidase, and arabinofuranosidase played key roles in the production of glucose, xylose, and arabinose, respectively. In addition, mannosidase, galactosidase, and esterase were also necessary to deconstruct the complex lignocellulose [40]. It was reported that *T. harzianum* LZ117 could secrete a more complete enzyme cocktail than *T. reesei*. This strain produced 29.24 U/mg xylanase after 144 h of fermentation, but its CMCase, *p*NPCase, and *p*NPGase were only 4.30, 0.19, and 1.81 U/mg, respectively, much lower than the other cellulase producers [12]. In our present study, CMCase, *p*NPCase, and *p*NPGase of EM0925 enzyme cocktail was 11.20, 33.76, and 84.97 U/mg, respectively. In addition, high levels of *p*NPXase and *p*NPAFase activities up to 203.60 and 11.72 U/mg were detected. Specific activities of glycosidases of EM0925 was the highest compared to the two extensively used inartificial enzyme cocktails under the conditions tested (Table 1). When using natural lignocellulose substrate as the sole carbon source, *T. harzianum* that secret cellulase and hemicellulase simultaneously, especially high levels of glycosidases has not been reported [27]. Therefore, the patterns of enzyme activities observed for *T. harzianum* EM0925 highlight its potential as efficient enzyme producer for industrial applications.

Different induction conditions have been shown to greatly influence the secretion of cellulolytic enzymes by fungi as revealed by Filho et al. [20], at the transcriptional level. In the transcriptome of *T. harzianum*, the GH7 endoglucanase gene was the highest expressed one when grown on cellulose, while expression level of GH10 xylanase was significantly increased when grown on the natural bagasse [20]. Secretome analysis of *T. harzianum* EM0925 demonstrated the multiple functions and the synergistic effects of the lignocellulolytic enzymes. The balanced abundance of three kinds of cellulases play a crucial role for their synergistic degradation effects [41]. The abundant endoglucanases increased the frequency for randomly degrading the main chain of cellulose, which provided more access points for cellobiohydrolases. The high abundance of glucosidases could quickly remove the cellobiose and release glucose efficiently. In addition, much higher abundance of glycosidases responsible for side chain degradation could also effectively release series of fermentable sugars from complex polysaccharides [42]. A total of 44 glycosidases among the 81 CAZymes were found in the secretome of EM0925, with GH3, GH5, GH11, GH30, and GH55 as the dominant components, which may play crucial role in lignocellulose degradation (Figure S1). Vanessa et al. [43] also reported that the glycoside hydrolases were the most abundant class of proteins secreted by *Trichoderma harzianum* IOC 3844, which represent 67% of the total proteins; however, pretreated (partially delignified) cellulolignin from sugarcane bagasse was used in their study as the inducer for lignocellulosic enzyme secretion. In addition, the most abundant groups in the secretome of *T. harzianum* IOC 3844 were GH3 (17%), GH5 (11%), GH 7 (10%), GH6 (6%), and GH55 (6%), which was slightly different from those verified for *T. harzianum* EM0925 in our present study and demonstrated the possible effects of the secretome.

For lignocellulose degradation, the amount of cellulases is not the most critical factor affecting sugar yield, since the synergistic effects between the cellulases and hemicellulases and between the enzymes for main- and side-chain degrading enzymes of the polysaccharides complex might be more important. Therefore, the high conversion efficiency of the secretome of *T. harzianum* EM0925 might be achieved from its numerously and quantitatively balanced cellulases and hemicellulases. Previously, partial replacement of cellulases with auxiliary enzymes has reduced the required amount of commercial cellulases to achieve high hydrolysis yields [11,44]. Yang et al. [45] reported a degree of synergy of 1.35 for the hydrolysis of delignified corn stover when studying the synergistic effect between a commercial enzyme preparation and a bifunctional enzyme consisting of an acetyl xylan esterase and a α-l-arabinofuranosidase. For glucan conversion of delignified corn stover, Zhu et al. [46] found a synergy of 1.96 between a hemicellulase preparation EMSD5 (including xylanases, β-xylosidases, α-l-arabinofuranosidases, α-glucuronidases, and acetyl xylan esterases) and the commercial cellulase from *T. reesei*. These findings revealed the significant role of glycosidases for efficient lignocellulose degradation. Moreover, 9 carbohydrate binding modules (CBMs) were also observed, which could help hydrolyze substrate more effectively [46]. In *T. harzianum* EM0925 enzyme enzyme cocktail, a LPMO with abundance of 3.23% was found. It was discovered that LPMOs promote degradation of the most recalcitrant crystalline cellulose by carrying out oxidative cleavage of polysaccharides [47]. Several studies showed synergetic effects between LPMOs and the lignocellulolytic enzymes during the saccharification process of cellulose and hemicellulose components [9,47]. All CAZymes and non-hydrolyzed protein detected in the secreted proteins of *T. harzianum* EM0925 are important for fermentable sugars production. In addition, some glycoside hydrolases such as GH30 xylanase and GH26 mannase secreted by *T. harzianum* EM0925 have been rarely studied in *Trichoderma*. Synergy between cellulolytic enzymes and these enzymes from *Acremonium alcalophilum* and *Aspergillus nidulans* has been described, which deserves more attention in the future [48,49]. The currently reported lignocellulosic degrading bacteria and fungi with industrial application prospects presented distinct optimal conditions for different enzymes even in the same induced enzyme system, and the stabilities against temperature and pH were also different, so it is necessary to determine the optimal conditions for enzymatic hydrolysis and saccharification with overall consideration [1,50,51]. In the present study, similar optimal temperature and pH for various functional enzymes in *T. harzianum* EM0925 enzyme cocktail were observed, which gave rise to better synergistic effects in efficient deconstruction of lignocellulose complex. Pretreatments could make the macroscopic and microscopic deconstruction and change chemical composition of lignocellulose, which improve the accessibility of enzymes and facilitate release of fermentable sugars. Among many pretreatment methods, dilute alkali treatment is widely used because of its remarkable effect, and physical crushing pretreatment without any pollution has been widely used as well [52].

When EM0925 was used, reducing sugar yields of 96.6% and 100% were obtained from UGCS and ALKCS, respectively, after 72 h of saccharification. The efficacy after 24 h of saccharification was 81.9% and 74.2%, respectively. (Figure 5b). So, the enzyme cocktail EM0925 was much more effective than the current commercial products under the conditions tested, since only 46.9% sugar yield was acquired from UGCS by Celluclast 1.5 L after 72 h hydrolysis [33] (Table S1). In general, saccharification efficiency of lignocellulose biomass by EM0925 was better than that in any other literature reported under the conditions tested [29–34]. To be specific, in our study, the maximum 85.59% glucose yield and 66.18% xylose yield were obtained from UGCS hydrolyzed with 30 FPU/g of Celluclast 1.5 L used; while 100% glucose yield and 100% xylose yield were obtained from UGCS and ALKCS with the same enzyme dosage of *T. harzianum* EM0925. To some extent, the glucose and xylose yields by EM0925 were lower than the total sugar yields when the same substrates were used, which indicated that not only monosaccharides but also oligosaccharides were obtained from UGCS and ALKCS. From these results, it could be estimated that various auxiliary enzyme including hemicellulase and amylase played important role to completely and effectively deconstruct the complex lignocellulose substrates [11].

Some accessory enzymes that assist biomass degradation could be used to improve the recovery of fermentable sugars for use in a biorefinery setting in order to improve total utilization of biomass. Supplementation of a commercial enzyme preparation with 30% crude enzyme complex from *Aspergillus oryzae* P21C3 increased the conversion of cellulose derived from pretreated sugarcane bagasse by 36%, which demonstrated the potential to use the supplementary enzymes in the total lignocellulose degradation, although 51.2% of cellulose conversion and 78.1% of xylan conversion were obtained [53]. With supplementary of β-glucosidase produced by *Aspergillus niger* and endoglucanase produced by *Talaromyces emersonii* to the currently used commercial enzyme cocktails, greater saccharification of alkaline-pretreated bagasse (87% glucose yield and 94% of xylose yield) was obtained, but its enzyme dosage was as high as 500 FPU/g substrate [54]. In the present study, only enzyme dosage of 5 mg EM0925/g substrate (13.85 FPU/g) was required for complete saccharification of lignocellulose, which provided sufficient sugars for the downstream fermentation of downstream products. Therefore, EM0925 presented great potential in industrial applications. To the best of our knowledge, it was the first example that enzyme cocktail from *T. harzianum* completely and simultaneously converted cellulose and xylan in natural biomaterials into fermentable sugars at the same time. To date, induction and saccharification mechanisms of lignocellulosic enzymes secreted by *T. harzianum* are still unknown. Differences in the composition of the *T. harzianum* secretome in response to different substrates was reported, which revealed the diversity of the fungus enzymatic toolbox [55]. Nevertheless, enzyme cocktails with excellent efficiency of complete saccharification of lignocellulose should be tailor-made, and degradation mechanisms remain to be elucidated in more detail. The expression of total extracellular proteins of *T. harzianum* EM0925 also could be optimized at the molecular level, in order to further expand the application scope of *T. harzianum* EM0925.

4. Materials and Methods

4.1. Strain and Culture Conditions

Thrichoderma harzianum EM0925 was isolated and screened for activities of lignocellulolytic enzyme in our laboratory previously. For enzyme production, the strain was incubated on potato agar for 3 days at 30 °C, then, 5 slices with diameter 0.5 cm of the agar plate were inoculated into each flask containing 200 mL Mandel's salt solution (4.0 KH_2PO_4, 2.8 $(NH_4)_2O_4$, 0.6 $MgSO_4 \cdot 7H_2O$, 0.01 $FeSO_4 \cdot 7H_2O$, 0.003 $MnSO_4 \cdot H_2O$, 0.003 $ZnSO_4 \cdot 7H_2O$, 0.004 $CoCl_2 \cdot 6H_2O$, and 2.0 peptone, in $g \cdot L^{-1}$) supplemented with 2% (*w/v*) lignocellulose substrate at 30 °C [15].

4.2. Preparation of Enzyme Cocktail EM0925

Different carbon sources were used to induce the enzyme production of *T. harzianum* EM0925. Lignocellulosic enzyme activities in the culture solution were measured each day, the optimal incubation time was estimated according to the filter paper activity (FPA), and the activities of endoglucanase (EG) and xylanase (XYL) [15]. Ammonium sulfate precipitation with a final concentration of 85% (w/v) was used to prepare the enzyme cocktails. After incubation at 4 °C for 4 h, the precipitate fraction containing the secreted enzymes was collected by centrifugation at 10,000× *g* for 20 min at 4 °C and then dissolved in sodium acetate buffer (50 mM, pH 5.0) and filtered through a 0.45 μm filter membrane (Millipore, Bedford, MA, USA). Proteins in the supe rnatant were dialyzed against sodium acetate buffer (50 mM, pH 5.0) and then freeze-dried. Commercial cellulase enzyme C 9748 derived from *Trichoderma longibrachiatum* (C 9748, Sigma Aldrich, St. Louis, MO, USA) and Celluclast 1.5 L from *Trichoderma reesei* ATCC 26921 (Sigma, USA, Novozyme products) were used as reference. The protein content of the enzyme cocktail was determined by a Bradford protein assay regent (Bio-rad, Hercules, CA, USA) [12].

4.3. Enzymatic Assays

FPA was measured according to the method reported by Urbánszki et al. [56]. EG and XYL were measured with carboxymethyl cellulose (CMC-Na, TCI, Tokyo, Japan) and Beechwood xylan (Sigma, USA) as substrates, respectively. Briefly, to measure the xylanase activity, an aliquot of 100 μL diluted crude enzyme was mixed with 100 μL 1% (*w*/*v*) of xylan. After 15 min of incubation at 40 °C, 150 μL of the 3,5-dinitrosalicylic acid (DNS) reagent was added. For CMCase activity measurement, an aliquot of 50 μL of diluted crude enzyme was mixed with 150 μL of 0.5% carboxymethyl cellulose. After incubation at 50 °C for 30 min, 50 μL of 1 M NaOH and the DNS reagent were sequentially added. After boiling the mixture at 100 °C for 5 min, it was cooled in cold water and the absorbance was read at 540 nm. Activities were calculated with glucose or xylose as the standard. One unit of enzyme activity was defined as the amount of enzyme catalyzing the release of 1 μmol of reducing sugars in 1 min from the substrate under the above conditions [57]. The cellobiohydrolase (CBH), β-d-glucosidase (βG), β-xylosidase (βX) and α-L-arabinofuranosidase (ABF) activities were determined according to the protocol of Zhu et al. [46] with minor modification: using *p*-nitrophenyl-β-D-cellobioside (*p*NPC), *p*-nitrophenyl-β-glucopyranoside (*p*NPG), *p*-nitrophenyl-β-d-xylopyranoside (*p*NPX), and *p*-nitrophenyl-α-L-arabinofuranoside (*p*NPAF) as substrate, respectively. The reaction mixture contained 50 μL of enzyme, 50 μL of 50 mM sodium acetate buffer (pH 5.0), and 50 μL of 10 mM substrate solution. After incubation at 50 °C for 10 min, the reaction was terminated by adding 50 μL of 1 M Na_2CO_3 and the absorbance of released *p*-nitrophenol was measured at 405 nm. Activities were calculated using p-nitrophenol as the standard. One unit of enzyme activity was defined as the amount of enzyme that produced 1 μmol of *p*NP in 1 min from the substrate under the above conditions. In addition, β-mannase activity (Man) was determined by the DNS method using mannose as the standard according to the methods of Li et al. [38]. The amylolytic activity (Amy) was assayed by the DNS method using maltose as the standard as reported by He et al. [58].

To determine the optimal temperature for FPA, EG, and XYL of the enzyme cocktail from *T. harzianum* EM0925, assays were performed in the temperature range from 30–70 °C at an interval of 10 °C. The optimal pH values for cellulases and hemicellulases were investigated at the optimal temperature in buffers ranging from pH 2–8 at intervals of 0.5 pH units, using 0.05 M different buffers: glycine-HCl (pH 2.0–3.0), sodium acetate (pH 4.0–5.0), and citric-Na_2HPO_4 (pH 6.0–8.0). Thermal stability of EM0925 was assessed by measuring the residual activity in the optimal conditions after incubation of the cocktail at different temperature (50, 60, and 70 °C) for 1 h. A dose of 1.0 mg/mL of enzyme cocktail was pre-incubated in 0.05 M sodium acetate, enzyme cocktail without pre-incubated was used as the control. The stabilities of various enzymes at different pH values were determined by measuring the residual activities in standard conditions after the enzyme incubation in 0.05 M buffers of different pH (pH 2.0, 3.0, 4.0, 5.0, 6.0, 7.0, 8.0, 9.0, 10.0, 11.0, and 12.0) at room temperature for 1 h. The buffers used were glycine-HCl (pH 2.0, 3.0), sodium acetate (pH 4.0, 5.0), citric-Na_2HPO_4 (pH 6.0–8.0) and glycine-NaOH (pH 9.0–12.0). One mg/mL of enzyme cocktail was pre-incubated in 0.05 M of the corresponding buffers, buffers, and enzyme cocktails without pre-incubated was used as the control [45].

4.4. Biomass Pretreatment and Component Analysis

The mature corn stover was collected from Shandong Province of China, and the other biomasses were available in our laboratory. The dried biomasses were ground and passed through a 50 mesh sieve before drying at 50 °C to a constant weight. Un-treated corn stover (NTCS) was subjected to different pretreatments as follows. Ultrafine grinding pretreatment corn stover (UGCS) was prepared by mixing NTCS and ZrO_2 balls (6–10 nm diameter) with a ratio of 1:2 and grinding for 0.5 h in a CJM-SY-B ultrafine vibration ball mill (Taiji Ring Nano Products Co., Hebei, China) [33]. Alkali-treated corn stover (ALKCS) was prepared by mixing the biomass and 1% (w/v) of NaOH at a solid-liquid ratio of 1: 10 and autoclaving the mixture at 121 °C for 1 h [59]. Sodium chlorite/acetic acid delignificated corn stover (DLCS) was obtained according to the protocol of Hu et al. [44]. Dilute acid pretreated corn stover (DACS) was treated with dilute acid according to the procedure of Ji et al. [59]. The steam explosion pretreated corn stover (SECS) was obtained by steam exploding the substrate according to the procedure of Liu et al. [57]. All slurries were filtered and washed until the pH reached 5.0 prior to use.

The structural carbohydrate and lignin contents of different pretreated biomasses were determined according to the laboratory analytical procedure of the National Renewable Energy Laboratory [60]. In brief, a sample of 0.5 g dry biomass was hydrolyzed at 30 °C with 3.0 mL H_2SO_4 (72%, w/v) for 1 h. Then 84 mL deionized water was added and a second hydrolysis was carried out in the autoclave at 121 °C for 1 h. All slurries were filtered and washed until reaching neutral pH. The glucose and xylose concentrations in the filter liquor were determined by HPLC. Acid-insoluble lignin was determined by subtracting the ash content from the solid residue dried at 60 °C. The ash content was determined by heating the solid residue at 575 °C for 12 h. The weight percentages of cellulose, xylan, arabinan, and lignin were calculated according to the method described previously [46].

4.5. Enzymatic Hydrolysis

Hydrolysis of all biomasses was conducted in 50 mM sodium citrate buffer (pH 4.5) in a 1 mL volume system containing 2% (*w/v*) substrate. The reaction was conducted in a 37 °C air bath with shaking at 200 rpm. Varying amounts (5 mg protein/g substrate and 5, 10, 30 FPU/g substrate) of enzyme loading were used for hydrolysis reactions. Samples were taken after 8 h, 12 h, 24 h, 48 h and 72 h of hydrolysis, and then heated at 100 °C for 10 min to terminate the reaction. Supernatants were collected by centrifugation at 10,000× *g* for 10 min. The content of total reducing sugars was determined by the DNS method. Monosaccharides in samples were determined by HPLC. Treatments containing substrate or enzyme alone were used as blank controls [15].

4.6. Extracellular Quantitative Proteome Analysis

The freeze-dried enzyme preparation of *T. harzianum* EM0925 dissolved in deionized water was separated by 12% (*w/v*) sodium dodecyl sulphate-poly-acrylamide gel electrophoresis (SDS-PAGE) at 120 V. The gel bands were excised and subjected to tryptic digestion and separated. The resulting peptides were reconstituted in 0.1% formic acid and then label-free quantitative proteomics identification by LC-MS/MS was performed. MaxQuant (version 1.4.1.2, https://www.maxquant.org/maxquant/) was used as the search engine in order to identify and quantify the extracellular proteins of *T. harzianum* EM0925. Raw data were preprocessed with Mascot Distiller 2.5 (Matrix Science, Boston, MA, USA) for peak picking, and resulting peaks were searched against the UniProt database (https://www.uniprot.org/) for *T. harzianum* strain CBS 226.95. The remaining parameter "intensity-based absolute quantification (iBAQ)" option was enabled. Only proteins identified with at least two peptides were considered for label-free quantification. Calculation of the protein label free quantification (LFQ) intensity was based on unique peptides using the built-in label-free quantification algorithm. The identified proteins were annotated using the Pfarm (https://pfam.xfam.org/search#tabview=tab1) and dbCAN database (http://bcb.unl.edu/dbCAN2/). Relative quantifications of extracellular proteins of *T. harzianum* EM0925 were analyzed. Three technical replicates were included in all the analyses [42,61].

5. Conclusions

Under induction of lignocellulosic biomass, *T. harzianum* EM0925 could rapidly secrete glycoside hydrolases, including numerically and quantitatively balanced cellulases and hemicellulases and especially high levels of glycosidases. *T. harzianum* EM0925 enzyme cocktail presented significantly higher enzyme activities than the commercial preparations under the conditions tested, and released 100% of glucose and xylose from UGCS and ALKCS simultaneously. These findings clearly indicate that *T. harzianum* EM0925 enzyme cocktail has great potential for industrial applications.

Supplementary Materials: The following are available online at http://www.mdpi.com/1422-0067/21/2/371/s1, Table S1: A comparison of different previous reported enzyme cocktails for lignocellulose saccharification. Figure S1: Distribution of glycoside hydrolase family in proteome of *T. harzianum* EM0925.

Author Contributions: Conceptualization, Y.Z. and H.Y.; methodology, Y.Z. and L.L. (Lijin Luo); software, Y.Z. and L.L. (Liang Liu); validation, J.Y., E.W., and H.Y.; formal analysis, R.W.; investigation, J.L.; data curation, Y.Z., R.W., and H.Y.; writing—original draft preparation, Y.Z.; writing—review and editing, E.W. and H.Y.; visualization, J.Y.; supervision, H.Y.; project administration, J.Y. All authors have read and agreed to the published version of the manuscript.

Funding: This research was funded by the project for extramural scientists of state key laboratory of agrobiotechnology, grant number 2018SKLAB6-28.

Conflicts of Interest: The authors declare no conflict of interest.

Abbreviations

CAZy	Carbohydrate active enzymes database
CBH	Cellobiohydrolase
CBM	Carbohydrate-binding module
EG	Endoglucanase
GH	Glycoside hydrolase
LPMO	Lytic polysaccharide monooxygenase
HPLC	High-performance liquid chromatography
FPA	Filter paper activity
CMC	Carboxymethylcellulose
βG	β-glucosidase
Xyl	Xylanase
βX	β-xylosidase
Ara	Arabinofuranosidase
Man	Mannase
Amy	Amylase
LC-MS/MS	Liquid chromatography-tandem mass spectrometry
NTCS	Non-pretreated corn stover
UGCS	Ultrafine grinding pretreated corn stover
ALKCS	Alkali pretreated corn stover
DACS	Dilute acid pretreated corn stover
DLCS	Sodium chlorite/acetic acid delignificated corn stover
SECS	Steam explosion pretreated corn stover
*p*NPC	*p*-nitrophenyl-β-d-cellobioside
*p*NPG	*p*-nitrophenyl-β-d-glucoside
*p*NPX	*p*-nitrophenyl-β-d-xyloside
*p*NPAF	*p*-nitrophenyl-α-l-arabinofuranoside
iBAQ	Intensity-based absolute quantification
LFQ	Label free quantification

References

1. Wang, J.; Chen, X.; Chio, C.; Yang, C.; Su, E.; Jin, Y.; Cao, F.; Qin, W. Delignification overmatches hemicellulose removal for improving hydrolysis of wheat straw using the enzyme cocktail from *Aspergillus niger*. *Bioresour. Technol.* **2019**, *274*, 459–467. [CrossRef]
2. Paz, A.; Outeirino, D.; Perez Guerra, N.; Dominguez, J.M. Enzymatic hydrolysis of brewer's spent grain to obtain fermentable sugars. *Bioresour. Technol.* **2018**, *275*, 402–409. [CrossRef] [PubMed]
3. Xin, F.; Dong, W.; Zhang, W.; Ma, J.; Jiang, M. Biobutanol production from crystalline cellulose through consolidated bioprocessing. *Trends Biotechnol.* **2019**, *37*, 167–180. [CrossRef] [PubMed]
4. Kun, R.S.; Gomes, A.C.S.; Hilden, K.S.; Cerezo, S.S.; Makela, M.R.; de Vries, R.P. Developments and opportunities in fungal strain engineering for the production of novel enzymes and enzyme cocktails for plant biomass degradation. *Biotechnol. Adv.* **2019**, *37*, 1–16. [CrossRef] [PubMed]
5. De Castro, A.M.; Pedro, K.C.; da Cruz, J.C.; Ferreira, M.C.; Leite, S.G.; Pereira, N., Jr. *Trichoderma harzianum* IOC-4038: A promising strain for the production of a cellulolytic complex with significant beta-glucosidase activity from sugarcane bagasse cellulignin. *Appl. Biochem. Biotechnol.* **2010**, *162*, 2111–2122. [CrossRef]
6. Gama, R.; Van Dyk, J.S.; Pletschke, B.I. Optimisation of enzymatic hydrolysis of apple pomace for production of biofuel and biorefinery chemicals using commercial enzymes. *3 Biotech* **2015**, *5*, 1075–1087. [CrossRef]
7. De Paula, R.G.; Antonieto, A.C.C.; Nogueira, K.M.V.; Ribeiro, L.F.C.; Rocha, M.C.; Malavazi, I.; Almeida, F.; Silva, R.N. Extracellular vesicles carry cellulases in the industrial fungus *Trichoderma reesei*. *Biotechnol. Biofuels* **2019**, *12*, 146. [CrossRef]
8. Jiang, Y.; Wu, R.; Zhou, J.; He, A.; Xu, J.; Xin, F.; Zhang, W.; Ma, J.; Jiang, M.; Dong, W. Recent advances of biofuels and biochemicals production from sustainable resources using co-cultivation systems. *Biotechnol. Biofuels* **2019**, *12*, 155. [CrossRef]
9. Sanhueza, C.; Carvajal, G.; Soto-Aguilar, J.; Lienqueo, M.E.; Salazar, O. The effect of a lytic polysaccharide monooxygenase and a xylanase from *Gloeophyllum trabeum* on the enzymatic hydrolysis of lignocellulosic residues using a commercial cellulase. *Enzym. Microb. Technol.* **2018**, *113*, 75–82. [CrossRef] [PubMed]
10. Goldbeck, R.; Gonçalves, T.A.; Damásio, A.R.L.; Brenelli, L.B.; Wolf, L.D.; Paixão, D.A.A.; Rocha, G.J.M.; Squina, F.M. Effect of hemicellulolytic enzymes to improve sugarcane bagasse saccharification and xylooligosaccharides production. *J. Mol. Catal. B Enzym.* **2016**, *131*, 36–46. [CrossRef]
11. Sun, F.F.; Hong, J.; Hu, J.; Saddler, J.N.; Fang, X.; Zhang, Z.; Shen, S. Accessory enzymes influence cellulase hydrolysis of the model substrate and the realistic lignocellulosic biomass. *Enzym. Microb. Technol.* **2015**, *79–80*, 42–48. [CrossRef] [PubMed]
12. Li, J.; Zhang, F.; Li, J.; Zhang, Z.; Bai, F.; Chen, J.; Zhao, X. Rapid production of lignocellulolytic enzymes by *Trichoderma harzianum* LZ117 isolated from Tibet for biomass degradation. *Bioresour. Technol.* **2019**, *292*, 122063. [CrossRef] [PubMed]
13. Liu, G.; Qin, Y.; Li, Z.; Qu, Y. Improving lignocellulolytic enzyme production with *Penicillium*: From strain screening to systems biology. *Biofuels* **2014**, *4*, 523–534. [CrossRef]
14. Gusakov, A.V.; Sinitsyn, A.P. Cellulases from *Penicillium* species for producing fuels from biomass. *Biofuels* **2014**, *3*, 463–477. [CrossRef]
15. Yang, Y.; Yang, J.; Liu, J.; Wang, R.; Liu, L.; Wang, F.; Yuan, H. The composition of accessory enzymes of *Penicillium chrysogenum* P33 revealed by secretome and synergistic effects with commercial cellulase on lignocellulose hydrolysis. *Bioresour. Technol.* **2018**, *257*, 54–61. [CrossRef]
16. Berlin, A.; Maximenko, V.; Gilkes, N.; Saddler, J. Optimization of enzyme complexes for lignocellulose hydrolysis. *Biotechnol. Bioeng.* **2007**, *97*, 287–296. [CrossRef]
17. Berrin, J.G.; Herpoel-Gimbert, I.; Ferreira, N.L.; Margeot, A.; Heiss-Blanquet, S. Use of cellulases from *Trichoderma reesei* in the Twenty-First Century-Part II: Optimization of cellulolytic cocktails for saccharification of lignocellulosic feedstocks. In *Biotechnology and Biology of Trichoderma*; Eisevier: Amsterdam, The Netherlands, 2014; pp. 263–280.
18. Chundawat, S.P.; Lipton, M.S.; Purvine, S.O.; Uppugundla, N.; Gao, D.; Balan, V.; Dale, B.E. Proteomics-based compositional analysis of complex cellulase-hemicellulase mixtures. *J. Proteome Res.* **2011**, *10*, 4365–4372. [CrossRef]

19. De Castro, A.M.; Ferreira, M.C.; da Cruz, J.C.; Pedro, K.C.; Carvalho, D.F.; Leite, S.G.; Pereira, N. High-Yield Endoglucanase Production by *Trichoderma harzianum* IOC-3844 cultivated in pretreated sugarcane mill byproduct. *Enzym. Res.* **2010**, *2010*, 854526. [CrossRef]
20. Ferreira Filho, J.A.; Horta, M.A.C.; Beloti, L.L.; Dos Santos, C.A.; de Souza, A.P. Carbohydrate-active enzymes in *Trichoderma harzianum*: A bioinformatic analysis bioprospecting for key enzymes for the biofuels industry. *BMC Genom.* **2017**, *18*, 779. [CrossRef]
21. Benoliel, B.; Torres, F.A.G.; de Moraes, L.M.P. A novel promising *Trichoderma harzianum* strain for the production of a cellulolytic complex using sugarcane bagasse in natura. *Springer Plus* **2013**, *2*, 656. [CrossRef]
22. Delabona Pda, S.; Farinas, C.S.; da Silva, M.R.; Azzoni, S.F.; Pradella, J.G. Use of a new *Trichoderma harzianum* strain isolated from the Amazon rainforest with pretreated sugar cane bagasse for on-site cellulase production. *Bioresour. Technol.* **2012**, *107*, 517–521. [CrossRef]
23. Zhao, X.; Xiong, L.; Zhang, M.; Bai, F. Towards efficient bioethanol production from agricultural and forestry residues: Exploration of unique natural microorganisms in combination with advanced strain engineering. *Bioresour. Technol.* **2016**, *215*, 84–91. [CrossRef]
24. Li, X.; Xie, X.; Liu, J.; Wu, D.; Cai, G.; Lu, J. Characterization of a putative glycoside hydrolase family 43 arabinofuranosidase from *Aspergillus niger* and its potential use in beer production. *Food Chem.* **2020**, *305*, 125382. [CrossRef] [PubMed]
25. Hu, J.; Arantes, V.; Pribowo, A.; Gourlay, K.; Saddler, J.N. Substrate factors that influence the synergistic interaction of AA9 and cellulases during the enzymatic hydrolysis of biomass. *Energy Environ. Sci.* **2014**, *7*, 2308–2315. [CrossRef]
26. Gusakov, A.V. Alternatives to *Trichoderma reesei* in biofuel production. *Trends Biotechnol.* **2011**, *29*, 419–425. [CrossRef]
27. De Souza, M.F.; da Silva, A.S.A.; Bon, E.P.S. A novel *Trichoderma harzianum* strain from the Amazon Forest with high cellulolytic capacity. *Biocatal. Agric. Biotechnol.* **2018**, *14*, 183–188. [CrossRef]
28. Li, C.; Pang, A.P.; Yang, H.; Lv, R.; Zhou, Z.; Wu, F.G.; Lin, F. Tracking localization and secretion of cellulase spatiotemporally and directly in living *Trichoderma reesei*. *Biotechnol. Biofuels* **2019**, *12*, 200. [CrossRef] [PubMed]
29. Boussaid, A.; Robinson, J.; Cai, Y.; Gregg, D.J.; Saddler, J.N. Fermentability of the hemicellulose-derived sugars from steam-exploded softwood (douglas fir). *Biotechnol. Bioeng.* **1999**, *64*, 284–289. [CrossRef]
30. Li, C.; Knierim, B.; Manisseri, C.; Arora, R.; Scheller, H.V.; Auer, M.; Vogel, K.P.; Simmons, B.A.; Singh, S. Comparison of dilute acid and ionic liquid pretreatment of switchgrass: Biomass recalcitrance, delignification and enzymatic saccharification. *Bioresour. Technol.* **2010**, *101*, 4900–4906. [CrossRef]
31. Herbaut, M.; Zoghlami, A.; Habrant, A.; Falourd, X.; Foucat, L.; Chabbert, B.; Paes, G. Multimodal analysis of pretreated biomass species highlights generic markers of lignocellulose recalcitrance. *Biotechnol. Biofuels* **2018**, *11*, 52. [CrossRef]
32. Chylenski, P.; Forsberg, Z.; Stahlberg, J.; Varnai, A.; Lersch, M.; Bengtsson, O.; Saebo, S.; Horn, S.J.; Eijsink, V.G.H. Development of minimal enzyme cocktails for hydrolysis of sulfite-pulped lignocellulosic biomass. *J. Biotechnol.* **2017**, *246*, 16–23. [CrossRef] [PubMed]
33. Li, J.; Zhang, H.; Lu, M.; Han, L. Comparison and intrinsic correlation analysis based on composition, microstructure and enzymatic hydrolysis of corn stover after different types of pretreatments. *Bioresour. Technol.* **2019**, *293*, 122016. [CrossRef]
34. Sills, D.L.; Gossett, J.M. Assessment of commercial hemicellulases for saccharification of alkaline pretreated perennial biomass. *Bioresour. Technol.* **2011**, *102*, 1389–1398. [CrossRef]
35. Liu, G.; Qu, Y. Engineering of filamentous fungi for efficient conversion of lignocellulose: Tools, recent advances and prospects. *Biotechnol. Adv.* **2018**, *37*, 519–529. [CrossRef] [PubMed]
36. Meng, Q.; Liu, C.; Zhao, X.; Bai, F. Engineering *Trichoderma reesei* Rut-C30 with the overexpression of *egl1* at the ace1 locus to relieve repression on cellulase production and to adjust the ratio of cellulolytic enzymes for more efficient hydrolysis of lignocellulosic biomass. *J. Biotechnol.* **2018**, *285*, 56–63. [CrossRef] [PubMed]
37. Ajijolakewu, K.A.; Leh, C.P.; Lee, C.K.; Wan Nadiah, W.A. Characterization of novel *Trichoderma* hemicellulase and its use to enhance downstream processing of lignocellulosic biomass to simple fermentable sugars. *Biocatal. Biotransfor.* **2017**, *11*, 166–175. [CrossRef]

38. Li, Y.; Yi, P.; Liu, J.; Yan, Q.; Jiang, Z. High-level expression of an engineered beta-mannanase (mRmMan5A) in *Pichia pastoris* for manno-oligosaccharide production using steam explosion pretreated palm kernel cake. *Bioresour. Technol.* **2018**, *256*, 30–37. [CrossRef]
39. Lyu, Y.; Zeng, W.; Du, G.; Chen, J.; Zhou, J. Efficient bioconversion of epimedin C to icariin by a glycosidase from *Aspergillus nidulans*. *Bioresour. Technol.* **2019**, *289*, 121612. [CrossRef]
40. Oliveira, D.M.; Mota, T.R.; Oliva, B.; Segato, F.; Marchiosi, R.; Ferrarese-Filho, O.; Faulds, C.B.; Dos Santos, W.D. Feruloyl esterases: Biocatalysts to overcome biomass recalcitrance and for the production of bioactive compounds. *Bioresour. Technol.* **2019**, *278*, 408–423. [CrossRef]
41. Kim, I.J.; Lee, H.J.; Kim, K.H. Pure enzyme cocktails tailored for the saccharification of sugarcane bagasse pretreated by using different methods. *Process Biochem.* **2017**, *57*, 167–174. [CrossRef]
42. Hwangbo, M.; Tran, J.L.; Chu, K.H. Effective one-step saccharification of lignocellulosic biomass using magnetite-biocatalysts containing saccharifying enzymes. *Sci. Total Environ.* **2019**, *647*, 806–813. [CrossRef] [PubMed]
43. Vanessa, A.L.R.; Roberto, N.M.; Pereira, N.; Marcelo, F.K.; Elias, L.; Rachael, S.; Clare, S.K.; Leonardo, D.G.; Simon, J.M.M. Characterization of the cellulolytic secretome of *Trichoderma harzianum* during growth on sugarcane bagasse and analysis of the activity boosting effects of swollenin. *Biotechnol. Prog.* **2016**, *32*, 327–336.
44. Hu, J.; Chandra, R.; Arantes, V.; Gourlay, K.; Van Dyk, J.S.; Saddler, J.N. The addition of accessory enzymes enhances the hydrolytic performance of cellulase enzymes at high solid loadings. *Bioresour. Technol.* **2015**, *186*, 149–153. [CrossRef] [PubMed]
45. Yang, Y.; Zhu, N.; Yang, J.; Lin, Y.; Liu, J.; Wang, R.; Wang, F.; Yuan, H. A novel bifunctional acetyl xylan esterase/arabinofuranosidase from *Penicillium chrysogenum* P33 enhances enzymatic hydrolysis of lignocellulose. *Microb. Cell Factories* **2017**, *16*, 166. [CrossRef] [PubMed]
46. Zhu, N.; Yang, J.; Ji, L.; Liu, J.; Yang, Y.; Yuan, H. Metagenomic and metaproteomic analyses of a corn stover-adapted microbial consortium EMSD5 reveal its taxonomic and enzymatic basis for degrading lignocellulose. *Biotechnol. Biofuels* **2016**, *9*, 243. [CrossRef]
47. Agger, J.W.; Isaksen, T.; Varnai, A.; Vidal-Melgosa, S.; Willats, W.G.T.; Ludwig, R.; Horn, S.J.; Eijsink, V.G.H.; Westereng, B. Discovery of LPMO activity on hemicelluloses shows the importance of oxidative processes in plant cell wall degradation. *Proc. Natl. Acad. Sci. USA* **2014**, *111*, 6287–6292. [CrossRef]
48. Šuchová, K.; Puchart, V.; Spodsberg, N.; Mørkeberg Krogh, K.B.R.; Biely, P. A novel GH30 xylobiohydrolase from *Acremonium alcalophilum* releasing xylobiose from the non-reducing end. *Enzym. Microb. Technol.* **2020**, *134*, 109484. [CrossRef]
49. Von Freiesleben, P.; Spodsberg, N.; Blicher, T.H.; Anderson, L.; Jorgensen, H.; Stalbrand, H.; Meyer, A.S.; Krogh, K.B. An *Aspergillus nidulans* GH26 endo-beta-mannanase with a novel degradation pattern on highly substituted galactomannans. *Enzym. Microb. Technol.* **2016**, *83*, 68–77. [CrossRef]
50. Aulitto, M.; Fusco, F.A.; Fiorentino, G.; Bartolucci, S.; Contursi, P.; Limauro, D. A thermophilic enzymatic cocktail for galactomannans degradation. *Enzym. Microb. Technol.* **2018**, *111*, 7–11. [CrossRef]
51. Yang, Z.; Liao, Y.; Fu, X.; Zaporski, J.; Peters, S.; Jamison, M.; Liu, Y.; Wullschleger, S.D.; Graham, D.E.; Gu, B. Temperature sensitivity of mineral-enzyme interactions on the hydrolysis of cellobiose and indican by beta-glucosidase. *Sci. Total Environ.* **2019**, *686*, 1194–1201. [CrossRef]
52. Gatt, E.; Rigal, L.; Vandenbossche, V. Biomass pretreatment with reactive extrusion using enzymes: A review. *Ind. Crop. Prod.* **2018**, *122*, 329–339. [CrossRef]
53. Braga, C.M.P.; Delabona, P.D.S.; Lima, D.; Paixao, D.A.A.; Pradella, J.; Farinas, C.S. Addition of feruloyl esterase and xylanase produced on-site improves sugarcane bagasse hydrolysis. *Bioresour. Technol.* **2014**, *170*, 316–324. [CrossRef] [PubMed]
54. Valadares, F.; Goncalves, T.A.; Goncalves, D.S.; Segato, F.; Romanel, E.; Milagres, A.M.; Squina, F.M.; Ferraz, A. Exploring glycoside hydrolases and accessory proteins from wood decay fungi to enhance sugarcane bagasse saccharification. *Biotechnol. Biofuels* **2016**, *9*, 110. [CrossRef] [PubMed]
55. Gomez-Mendoza, D.P.; Junqueira, M.; do Vale, L.H.; Domont, G.B.; Ferreira Filho, E.X.; Sousa, M.V.; Ricart, C.A. Secretomic survey of *Trichoderma harzianum* grown on plant biomass substrates. *J. Proteome Res.* **2014**, *13*, 1810–1822. [CrossRef]
56. Urbánszki, K.; Szakács, G.; Tengerdy, R.P. Standardization of the filter paper activity assay for solid substrate fermentation. *Biotechnol. Lett.* **2000**, *22*, 65–69. [CrossRef]

57. Liu, L.; Yang, J.; Yang, Y.; Luo, L.; Wang, R.; Zhang, Y.; Yuan, H. Consolidated bioprocessing performance of bacterial consortium EMSD5 on hemicellulose for isopropanol production. *Bioresour. Technol.* **2019**, *292*, 121965. [CrossRef]
58. He, L.; Mao, Y.; Zhang, L.; Wang, H.; Alias, S.A.; Gao, B.; Wei, D. Functional expression of a novel alpha-amylase from Antarctic psychrotolerant fungus for baking industry and its magnetic immobilization. *BMC Biotechnol.* **2017**, *17*, 22. [CrossRef]
59. Ji, G.; Han, L.; Gao, C.; Xiao, W.; Zhang, Y.; Cao, Y. Quantitative approaches for illustrating correlations among the mechanical fragmentation scales, crystallinity and enzymatic hydrolysis glucose yield of rice straw. *Bioresour. Technol.* **2017**, *241*, 262–268. [CrossRef]
60. Hussin, M.H.; Rahim, A.A.; Mohamad Ibrahim, M.N.; Yemloul, M.; Perrin, D.; Brosse, N. Investigation on the structure and antioxidant properties of modified lignin obtained by different combinative processes of oil palm fronds (OPF) biomass. *Ind. Crop. Prod.* **2014**, *52*, 544–551. [CrossRef]
61. Zhu, N.; Liu, J.; Yang, J.; Lin, Y.; Yang, Y.; Ji, L.; Li, M.; Yuan, H. Comparative analysis of the secretomes of *Schizophyllum commune* and other wood-decay basidiomycetes during solid-state fermentation reveals its unique lignocellulose-degrading enzyme system. *Biotechnol. Biofuels* **2016**, *9*, 42. [CrossRef]

International Journal of
Molecular Sciences

Article

Characterization of a Maltase from an Early-Diverged Non-Conventional Yeast *Blastobotrys adeninivorans*

Triinu Visnapuu, Aivar Meldre, Kristina Põšnograjeva, Katrin Viigand, Karin Ernits and Tiina Alamäe *

Department of Genetics, Institute of Molecular and Cell Biology, University of Tartu, Riia 23, 51010 Tartu, Estonia; triinu.visnapuu@ut.ee (T.V.); aivarmeldre@gmail.com (A.M.); kristina.poshnograjeva@gmail.com (K.P.); katrin.viigand@gmail.com (K.V.); karin.ernits@gmail.com (K.E.)
* Correspondence: tiina@alamae.eu

Received: 28 November 2019; Accepted: 30 December 2019; Published: 31 December 2019

Abstract: Genome of an early-diverged yeast *Blastobotrys* (*Arxula*) *adeninivorans* (*Ba*) encodes 88 glycoside hydrolases (GHs) including two α-glucosidases of GH13 family. One of those, the *rna_ARAD1D20130g*-encoded protein (*Ba*AG2; 581 aa) was overexpressed in *Escherichia coli*, purified and characterized. We showed that maltose, other maltose-like substrates (maltulose, turanose, maltotriose, melezitose, malto-oligosaccharides of DP 4-7) and sucrose were hydrolyzed by *Ba*AG2, whereas isomaltose and isomaltose-like substrates (palatinose, α-methylglucoside) were not, confirming that *Ba*AG2 is a maltase. *Ba*AG2 was competitively inhibited by a diabetes drug acarbose (K_i = 0.8 μM) and Tris (K_i = 70.5 μM). *Ba*AG2 was competitively inhibited also by isomaltose-like sugars and a hydrolysis product—glucose. At high maltose concentrations, *Ba*AG2 exhibited transglycosylating ability producing potentially prebiotic di- and trisaccharides. Atypically for yeast maltases, a low but clearly recordable exo-hydrolytic activity on amylose, amylopectin and glycogen was detected. *Saccharomyces cerevisiae* maltase MAL62, studied for comparison, had only minimal ability to hydrolyze these polymers, and its transglycosylating activity was about three times lower compared to *Ba*AG2. Sequence identity of *Ba*AG2 with other maltases was only moderate being the highest (51%) with the maltase MalT of *Aspergillus oryzae*.

Keywords: *Arxula adeninivorans*; glycoside hydrolase; α-glucosidase; maltose; panose; amylopectin; glycogen; inhibition by Tris; transglycosylation

1. Introduction

A non-conventional yeast *Blastobotrys adeninivorans* (syn. *Arxula adeninivorans*) belongs to a basal clade of Saccharomycotina subphylum and diverged in the evolution of fungi long before *Saccharomyces* [1–5]. A recent study states that the divergence of basal Saccharomycotina from *Saccharomyces cerevisiae* took place between 200 and 400 million years ago [4].

B. adeninivorans has several biotechnologically relevant properties: accumulation of lipids [6], salt tolerance, temperature-induced filamentation that promotes protein secretion and the ability to use a wide range of carbon and nitrogen sources, including purines, tannin and butanol, that are unusual nutrients for yeasts [2,7]. *B. adeninivorans* has been engineered for butanol production, applied in kits for the detection of hormones and dioxine in water as well as for manufacturing of tannase and cutinases [7]. Some other enzymes of *B. adeninivorans* have also been investigated, including alcohol dehydrogenase [8], extracellular glucoamylase [9] and invertase [10]. A highly active endo-inulinase of *B. adeninivorans* was cloned and recently characterized [11]. The genome of *B. adeninivorans* was sequenced in 2014 [2].

The genes potentially encoding α-glucosidases in the genomes of non-conventional yeasts were analysed in Viigand et al. [12]. The genes encoding two putative α-glucosidases designated as AG1 and AG2 were revealed in genome of *B. adeninivorans*. In the genomes of most yeasts addressed in Viigand et al. [12], the α-glucosidase genes resided in maltose utilization (*MAL*) clusters, whereas no *MAL* clusters were detected in *B. adeninivorans*.

α-Glucosidases have been popular objects to study protein evolution and phylogenesis [13–16], but they also have a biotechnological value. Thus, some of them have a high transglycosylating activity thanks to which they produce prebiotic oligosaccharides and potential functional food ingredients such as panose, melezitose, isomelezitose and isomalto-oligosaccharides [17–21]. For example, the α-glucosidase of *S. cerevisiae* produced isomelezitose from sucrose when the substrate concentration was high [21]. Transglycosylating ability of maltose by the α-glucosidase of *Xanthophyllomyces dendrorhous* (syn. *Phaffia rhodozyma*) has been studied in detail, and synthesis of tri- and tetrasaccharides with α-1,6 linkages was detected. This enzyme certainly has a biotechnological potential—it produced 3.6 times more transglycosylation products than the *S. cerevisiae* α-glucosidase studied at the same conditions [17,20].

Considering α-glucosidases of yeasts, they have mostly been studied in *S. cerevisiae* as these enzymes are crucial in baking and brewing [22]. *S. cerevisiae* has two types of α-glucosidases—maltases (EC 3.2.1.20) and isomaltases (EC 3.2.1.10)—that differ for substrate specificity. Maltases degrade maltose and maltose-like sugars, i.e., maltotriose, turanose and maltulose, but cannot degrade isomaltose and isomaltose-like sugars (α-1,6 linkages) such as palatinose. Both types of enzymes hydrolyze sucrose and a synthetic substrate *p*-nitrophenyl-α-D-glucopyranoside (*p*NPG) [15,16,23,24]. At the same time, some yeasts such as *Ogataea polymorpha* and *Scheffersomyces stipitis*, have promiscuous maltase-isomaltases that hydrolyze maltose- and isomaltose-like substrates [12,16,25].

In the current work, we expressed heterologously in *Escherichia coli* and biochemically characterized the α-glucosidase *Ba*AG2 of *B. adeninivorans* encoded by *rna_ARAD1D20130g*. We confirmed that *Ba*AG2 is a maltase with a considerable transglycosylating activity. Not typical for yeast maltases, *Ba*AG2 had exo-hydrolytic activity on amylose, amylopectin and glycogen. *Ba*AG2 is the first maltase characterized from *B. adeninivorans*.

2. Results

2.1. In Silico Analysis of BaAG2

According to annotations provided at the MycoCosm website [26], the genome of *Blastobotrys* (*Arxula*) *adeninivorans* [2] encodes 185 carbohydrate-active enzymes, including 88 glycoside hydrolases (GHs) assigned to different families. When mining the genome of *B. adeninivorans* [2] for the genes related to maltose hydrolysis, we found two genes encoding intracellular GH13 family proteins. Respective proteins were designated as AG1 and AG2 [12]. In MycoCosm, the AG1 was annotated as a protein similar to maltase Mal1 of *Schizosaccharomyces pombe* and the AG2 as similar to maltases of filamentous fungi *Aspergillus* and *Penicillium*. Both of these proteins were predicted to lack a signal peptide and to locate intracellularly. We confirmed this by using the SignalP program (see Materials and Methods). Aside from these two GH13 proteins, three putative extracellular α-glucosidases of GH31 family were detected in *B. adeninivorans* genome (see Table S1 of Supplementary Materials). Table S1 also includes two *B. adeninivorans* enzymes that have been experimentally studied: a secreted invertase AINV belonging to GH31 family [10] and a secreted glucoamylase [9] of GH15 family.

Substrate specificity of α-glucosidases can be *in silico* predicted based on so-called amino acid signature—a set of amino acids that locate in the vicinity of the substrate-binding pocket [12,15,27,28]. The upper panel of Figure 1 shows the amino acid signature of *O. polymorpha* maltase-isomaltase MAL1, *S. cerevisiae* maltase MAL62, isomaltase IMA1, and *B. adeninivorans* AG2. The amino acids of these proteins corresponding to Val216 of *Sc*IMA1 are shown inside a red frame as this position is considered of key importance in selective substrate binding [28].

Enzyme	Amino acid signature (numbering according to *Sc*IMA1)									Function
	158	216	217	218	219	278	279	307	411	
*Op*MAL1	F	T	A	G	L	V	G	D	N	Maltase-isomaltase
*Sc*MAL62	F	T	A	G	L	V	A	E	D	Maltase
*Sc*IMA1	Y	V	G	S	L	M	Q	D	E	Isomaltase
*Ba*AG2	Y	T	V	Q	I	G	S	R	N	Maltase or Maltase-isomaltase

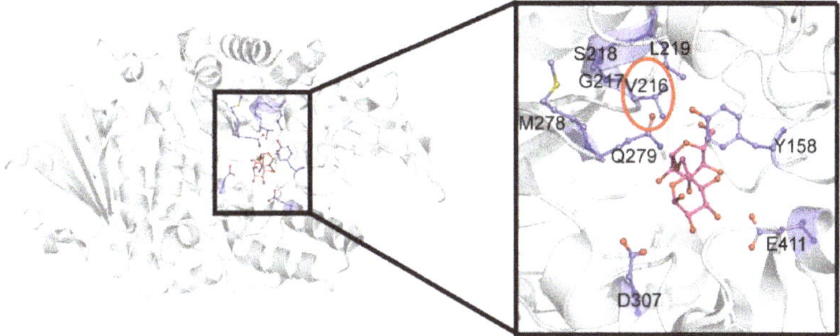

Figure 1. Amino acid signature of yeast α-glucosidases, including *B. adeninivorans* AG2 (upper panel) and their designation on the three-dimensional (3D) structure of *S. cerevisiae* isomaltase IMA1 in complex with isomaltose (RCSB Protein Data Bank, PDB: 3AXH [29]) (lower panel). Location of Val216 in the structure is marked with a red circle.

We then visualized the *S. cerevisiae* IMA1 structure in complex with isomaltose (PDB: 3AXH) [29] using PyMol [30] in order to display all these amino acids (Figure 1, lower panel). In *Ba*AG2, a Thr is present at position of Val216 and therefore we predicted that this enzyme is most probably a maltase. However, as maltase-isomaltases also have a Thr at that position (Figure 1, upper panel; [12,16]), based on the amino acid signature, *Ba*AG2 may also be a promiscuous enzyme with a wide substrate spectrum like *O. polymorpha* MAL1.

Figure 2 presents fragments of sequence comparison of *Ba*AG2 with those of five experimentally studied maltases from GH13 family: two from bacteria, two from yeasts and one from a filamentous fungus *Aspergillus*. The identity matrix of these proteins is presented in Supplementary Materials (Table S2). Though the proteins aligned sufficiently well over the entire sequence, *Ba*AG2 showed only moderate sequence identity with the other maltases ranging from 35% with *Halomonas* sp. H11 α-glucosidase to 51% with *Aspergillus oryzae* maltase MalT (Table S2). *In silico* assay of α-glucosidases of non-conventional yeasts [12] identified a putative α-glucosidase protein AG1 of *Lipomyces starkeyi* as the closest homologue (50% identity) of *Ba*AG2. The amino acid signature of the *Lipomyces* protein was YTVNKLSHE, and it was, therefore, predicted as a maltase [12].

The GH13 enzymes use an Asp (D) as a nucleophile and a Glu (E) as an acid-base catalyst [31]. Additionally, an Asp of the conserved 'NHD' motif serves as a transition state stabilizer [32]. In the *Ba*AG2 protein, Asp216 was predicted as a nucleophile, Glu274 as an acid-base catalyst and Asp348 as a stabilizer (Figure 2). Thr217 is located next to the catalytic nucleophile Asp216 in *Ba*AG2 (Figure 2). In maltases and maltase-isomaltases, either Thr or Ala is present at respective position, whereas in isomaltases, a Val is present [12,15,27,28], indicating that a Val residue at this position interferes with hydrolysis of maltose-like substrates. Indeed, if respective Thr was substituted with Val in *O. polymorpha* maltase-isomaltase, utilization of maltose-like sugars was severely hampered [16].

Furthermore, after substitution of Val216 in *S. cerevisiae* IMA1 with Thr, the isomaltase IMA1 gained the ability to hydrolyze maltose [28,29].

```
BaAG2     DAILFWLERG IDGFRIDTVQ 219
SpMal1    -ILRFWLDRG VDGFRLDAIN 215
ScMAL62   SAVGFWLDHG VDGFRIDTAG 217
GsAG      -MINWWLDKG IDGFRIDAIS 202
HaAG      -NMRFWLDLG VDGFRLDTVN 205
AoMalT    SAMEFWLQKG VDGFRVDTVN 213
Cc           : :**: *  :****:*:

BaAG2     SKYD-IMTVG EGS-PPSLEK 282
SpMal1    TEYD-AFSVG EMPYVLDTNE 278
ScMAL62   DGRE-IMRVG EVA--HGSDN 283
GsAG      ARYD-IMTVG EAN-GVTVDE 264
HaAG      DEYPGTTTVG EIGDDNPLER 280
AoMalT    AKYD-AMTVG ELPNTHTVDG 276
Cc                ** *        :

BaAG2     TTFFLENHDS GRSISRFASD 359
SpMal1    NASFIENHDQ TRTVSRYLSD 357
ScMAL62   ATTYIENHDQ ARSITRFADD 360
GsAG      NALFLENHDL PRSVSTWGND 337
HaAG      PCWATSNHDV VRSATRWGAD 344
AoMalT    STVFTENHDQ GRSVSRFGSE 356
Cc              .***   *: : :  :
```

Figure 2. Fragments of sequence alignment of six maltases. *Ba*AG2, *Blastobotrys adeninivorans* AG2 (580 aa); *Sp*Mal1, *Schizosaccharomyces pombe* Mal1 (579 aa, NP_595063.1) [33]; *Sc*MAL62, *Saccharomyces cerevisiae* MAL62 (584 aa, P07265) [23]; *Gs*AG, *Geobacillus stearothermophilus* exo-α-1,4-glucosidase (555 aa, BAA12704.1) [34]; *Ha*AG, *Halomonas* sp. H11 α-glucosidase (538 aa, BAL49684.1) [35]; *Ao*MalT, *Aspergillus oryzae* maltase MalT (574 aa, XP_001825184.1) [36]. Highlights: catalytic nucleophile (turquoise), acid-base catalyst (green), a transition state stabilizer (yellow) and a residue crucial for substrate specificity (red). Cc, Clustal consensus. Marking below the sequence alignment is according to Clustal consensus showing conservation: * positions with fully conserved residue; : positions with residues of strongly similar properties; . positions with residues of weakly similar properties.

2.2. Maltose-Like and Isomaltose-Like Sugars Are Growth Substrates for B. adeninivorans

According to the information present in the CBS-KNAW culture collection [37], *B. adeninivorans* CBS 8244 used in current work assimilates following α-glucosidic sugars: maltose, sucrose, melezitose, trehalose and α-MG. Of those, melezitose is a maltose-like sugar, and α-MG (a synthetic analogue of isomaltose [38]) is an isomaltose-like sugar. Glucose and many other monosaccharides are also assimilated. We asked if *B. adeninivorans* can also assimilate some other α-glucosidic sugars. We cultivated *B. adeninivorans* on Yeast Nitrogen Base (YNB) mineral medium containing 2 g/L of sugars indicated in Figure 3, and evaluated growth according to an optical density (OD) of 600 nm achieved by 24 h of growth (Figure 3). Our assay confirmed that five above-mentioned α-glucosidic sugars were indeed all assimilated. In addition, maltotriose, maltulose, turanose (maltose-like sugars) as well as isomaltose and palatinose (an isomaltose-like sugar) were identified as new growth substrates for this yeast. Thus, *B. adeninivorans* grows on both maltose-like and isomaltose-like sugars, meaning that it should possess enzymes for the hydrolysis of both types of sugars.

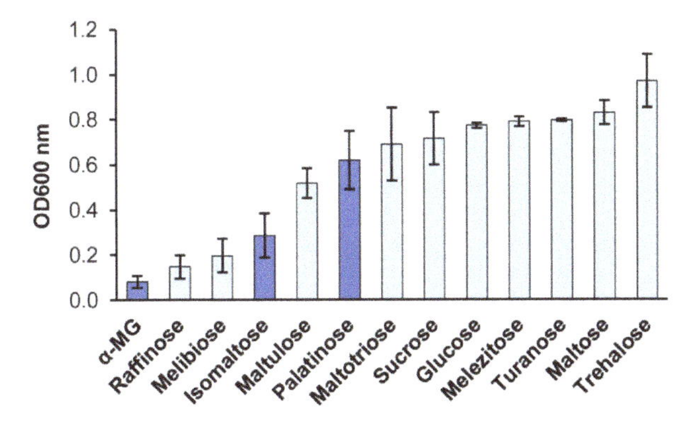

Figure 3. Growth of *B. adeninivorans* on sugars (supplemented at 2 g/L) evaluated by an optical density (OD) of the culture at 600 nm achieved by 24-h cultivation on a microplate at 37 °C. Isomaltose and isomaltose-like sugars are indicated by dark blue bars. Error bars are representing standard deviations (SD) of two independent experiments with two replicates.

2.3. *Cloning of the* Ba*AG2 Gene and Heterologous Expression of the* Ba*AG2 Enzyme*

The *Ba*AG2 protein deduced from the gene is 580 aa long. The protein was predicted as intracellular—no secretion signal was detected by the SignalP program v. 5.0 [39]. *Ba*AG2 was overexpressed in *E. coli* with the His_6-tag in its C-terminus that enabled purification of the protein using Ni^{2+}-affinity chromatography. The calculated molecular weight of the protein was 67.9 kDa and a prominent band of respective size was detected by electrophoresis of the lysate produced from induced *E. coli* cells overexpressing the *AG2* gene (Figure S1). The *E. coli* lysate exhibited catalytic activity of 1 mM *p*NPG hydrolysis at 30 °C (71 U/mg), which after purification of the protein increased 5.8 times, reaching 411.5 U/mg.

2.4. *Properties of* Ba*AG2*

2.4.1. Dependence of the *Ba*AG2 Activity on Temperature and pH. Thermal Stability of *Ba*AG2

The pH optimum of *Ba*AG2 was in moderately acidic region—from 5.5 to 6.5 (Figure S2). At pH 7.5, the activity was 81% from the maximum, and at pH 8.5, it was decreased to 52%. At pH 4.5 and 4.4, the respective values were 93 and 15%. Thus, the activity of *Ba*AG2 was significantly reduced at pH values below 4.5 and over 8.5 (Figure S2). The pH optimum of the *O. polymorpha* maltase is 6.0–6.5 [40], of *S. cerevisiae* maltase 6.5–6.8 [23,24] and of *Schizosaccharomyces pombe* maltase 6.0 [33]. In the current work, we routinely used the buffer with pH of 6.5 to characterize substrate specificity, kinetics and other properties of the enzyme.

As shown in Figure 4 (left panel), catalytic activity of *Ba*AG2 was the highest at 45 °C being 24% higher than activity measured at 30 °C—the temperature we routinely used for enzyme activity assay. Figure 4 shows that at temperature over 50 °C, the activity of the enzyme rapidly declined. Thermal stability assay of *Ba*AG2 indicated that the enzyme was rather thermolabile: if kept for 30 min at temperatures above 37 °C, its catalytic activity reduced significantly (Figure 4, right panel). Therefore, we routinely performed enzymatic assays at 30 °C since some reactions (e.g., transglycosylation and polysaccharide degradation assays) were conducted up to several days.

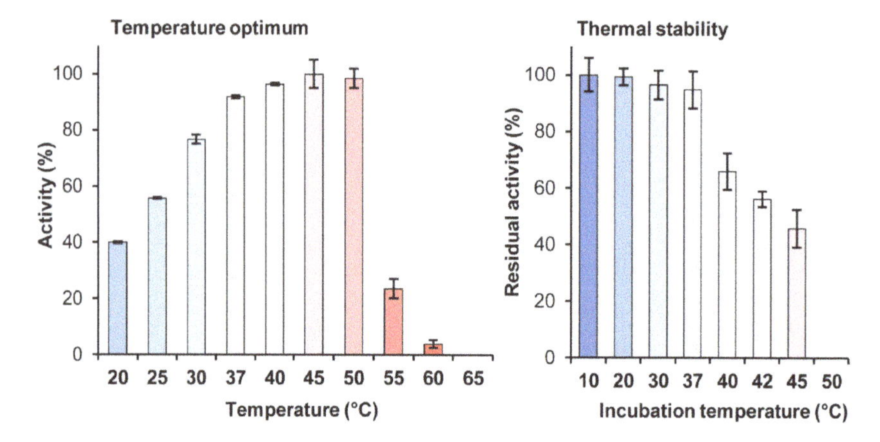

Figure 4. The effect of temperature on activity (left panel) and stability (right panel) of *Ba*AG2. For the thermostability assay, the enzyme was incubated for 30 min at the indicated temperature and residual activity was determined at 30 °C with *p*NPG as a substrate (see Materials and Methods, paragraph 4.4. for details). SDs of two to three replicates at each temperature are shown by error bars.

2.4.2. The Hydrolysis of Maltose and Maltose-Like Sugars

We predicted *in silico* that *Ba*AG2 is either maltase or maltase-isomaltase (see Figure 1). To find out which of the predictions was correct, the purified *Ba*AG2 protein was reacted with a selection of 100 mM sugars and 1 mM *p*NPG that serves as a substrate for maltases, isomaltases and maltase-isomaltases (Figure 5). According to our assay, *Ba*AG2 could hydrolyze universal substrates (*p*NPG and sucrose), maltose and maltose-like sugars such as turanose, maltotriose, melezitose and maltulose. Isomaltose and isomaltose-like substrates palatinose and α-methylglucoside were not hydrolyzed (Figure 5). Therefore, *Ba*AG2 should be considered as a maltase.

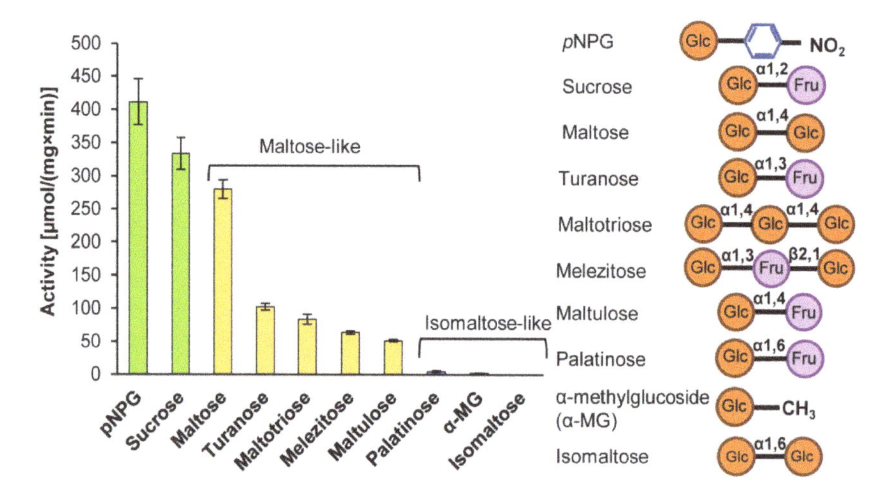

Figure 5. Activity of *Ba*AG2 on 100 mM sugars and 1 mM *p*NPG. Universal substrates are indicated by green, maltose and maltose-like sugars by yellow, and isomaltose and isomaltose-like sugars by blue bars. The composition and linkage types of the tested substrates are indicated. SD values of three to five replicates on each substrate are indicated.

2.4.3. The Kinetic Parameters of Hydrolysis of Maltose, Maltose-Like and Universal Substrates

We studied the kinetics of the hydrolysis of *p*NPG, maltose, sucrose, maltotriose, melezitose, maltulose and turanose to calculate K_m, V_{max}, k_{cat} and catalytic efficiency (k_{cat}/K_m) values for these substrates. Results are presented in Table 1.

Table 1. K_m, V_{max}, k_{cat} and k_{cat}/K_m values of hydrolysis of *p*NPG and sugars by *Ba*AG2.

Substrate *	K_m ± SD (mM)	V_{max} ± SD (μmol/(mg × min))	k_{cat} ± SD (1/s)	k_{cat}/K_m (1/(mM × s))
*p*NPG	0.76 ± 0.03	751.3 ± 14.5	850.2 ± 16.4	1106.1
Maltose	25.8 ± 1.6	336.4 ± 5.8	380.7 ± 6.5	14.8
Maltotriose	32.5 ± 3.3	117.9 ± 4.4	133.5 ± 5.0	4.1
Sucrose	35.9 ± 2.7	412.4 ± 8.9	466.8 ± 10.1	13.0
Turanose	45.2 ± 6.0	190.1 ± 10.4	215.1 ± 11.9	4.8
Maltulose	7.8 ± 1.0	52.2 ± 1.8	59.1 ± 2.1	7.6
Melezitose	238.3 ± 51.8	231.0 ± 31.8	261.4 ± 36.1	1.1

* For the structure and linkage type of the substrates, see Figure 5. SD, standard deviation.

Table 1 shows that natural sugars, maltose and sucrose (α-D-Glc-(1→2)-β-D-Fru) were hydrolyzed by *Ba*AG2 with the highest catalytic efficiency (k_{cat}/K_m). Additionally, *Ba*AG2 had a high affinity and activity towards a synthetic substrate *p*NPG—the K_m for *p*NPG was 0.76 mM and the V_{max} over 750 U/mg. Affinity of *Ba*AG2 for maltose was slightly higher than for sucrose. From this aspect, *Ba*AG2 differs from the maltases of *S. cerevisiae* [23,24] and *Candida albicans* [41], and also from the maltase-isomaltase of *O. polymorpha* [16,40,42], for which affinity for sucrose is about twice higher than for maltose. In contrast, the maltase of *Schizosaccharomyces pombe* prefers maltose to sucrose [33]. Among the substrates tested, the affinity and catalytic efficiency of *Ba*AG2 was the lowest for melezitose. Thin layer chromatography (TLC) analysis showed that the glycosidic bond of turanose moiety of melezitose was hydrolyzed first, yielding sucrose and glucose as products (Figure S3). Similar mode of melezitose hydrolysis was earlier shown for the maltase-isomaltase of *O. polymorpha* [16]. Interestingly, isomelezitose was hydrolyzed by *Ba*AG2 with release of palatinose (Figure S3).

We would like to emphasize that the activity of *Ba*AG2 (Table 1) was higher compared to some of other studied maltases of yeasts and filamentous fungi. For example, its V_{max} on maltose was 7.5 times higher compared to *S. cerevisiae* maltase MAL62 [23], and over two times higher compared to maltase-isomaltase of *O. polymorpha* [16]. On the other hand, α-glucosidases of *X. dendrorhous* and *A. niveus* preferred polysaccharides such as starch, amylopectin and glycogen, and their k_{cat} values on maltose were respectively 25 and 28 times lower compared to the value of *Ba*AG2 [43,44]. Catalytic constant of α-glucosidase of *A. niger* on maltose was 144 1/s (2.6 times lower than *Ba*AG2), but the affinity towards maltose was very high (0.75 mM) [45]. Based on the literature, only one α-glucosidase of yeast and filamentous fungi has much higher k_{cat} on maltose than that of *Ba*AG2—the extracellular α-glucosidase of *Schizosaccharomyces pombe* (k_{cat} = 709 1/s) [46].

2.4.4. The Inhibition Studies of Acarbose, Tris, Isomaltose-Like Sugars and Glucose

Having seen that isomaltose and isomaltose-like sugars are not hydrolyzed by *Ba*AG2 (Figure 5), we measured inhibition of *p*NPG hydrolysis reaction by these sugars as in [16]. Because maltases are usually strongly inhibited by glucose and acarbose, and much less by fructose [16], these substrates were also assayed as potential inhibitors of *Ba*AG2. When testing the effect of different buffers during the experiments, we noticed that the activity of *Ba*AG2 was lost in Tris-HCl buffer. More precise assay of the effect of tris(hydroxymethyl)aminomethane (Tris) on *Ba*AG2 revealed strong inhibition of the enzyme by this compound (Table 2). All tested compounds inhibited *p*NPG hydrolysis competitively, and the most powerful inhibitors of *Ba*AG2 were acarbose, Tris and glucose.

Table 2. K_i values and inhibition mode for *Ba*AG2 inhibitors of *p*NPG hydrolysis.

Inhibitor	$K_i \pm$ SD (mM)	Inhibition Mode
Palatinose	1.4 ± 0.1	Competitive
Isomaltose	22.7 ± 3.0	Competitive
α-MG	21.8 ± 1.4	Competitive
Acarbose	0.83 ± 0.01*	Competitive
Glucose	0.86 ± 0.05	Competitive
Fructose	36.9 ± 2.4	Competitive
Tris	70.5 ± 4.3*	Competitive

* These values are given in μM. SD, standard deviation.

We expected that binding of the substrates or competitive inhibitors of the enzyme should increase its thermostability. To confirm this, we conducted a differential scanning fluorimetry (DSF) assay of *Ba*AG2 in the presence and absence of competing inhibitors as in [16,47]. Acarbose, palatinose, Tris and glucose (strong inhibitors of *p*NPG hydrolysis by *Ba*AG2) and fructose that inhibited the reaction only weakly (see Table 2) were used as ligands. The melting temperature (T_m) values calculated from the DSF data are presented in Figure 6. The T_m of *Ba*AG2 without a ligand was similar to that of maltase-isomaltase MAL1 of *O. polymorpha*—51 °C [16]. Presence of acarbose (the strongest inhibitor of *Ba*AG2) increased the T_m of *Ba*AG2 by 11.4 °C. Presence of Tris raised the T_m value by 5.8 °C, and of glucose by 4.4 °C. Fructose and palatinose had only a minor effect on the T_m (Figure 6).

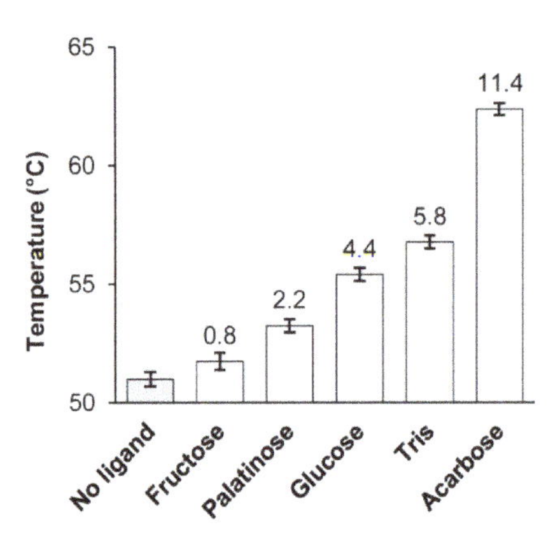

Figure 6. Thermostability of *Ba*AG2 in the presence and absence of indicated ligands. The T_m value of unliganded *Ba*AG2 (a grey bar) was 51 °C and the increase of T_m in the presence of a ligand is indicated above every bar. SDs of at least two independent experiments with two replicates at each condition are shown by error bars.

2.4.5. The Hydrolysis of Malto-Oligosaccharides of DP 3 to 7

We have earlier shown that *O. polymorpha* maltase-isomaltase MAL1 could use maltotriose and maltotetraose as a substrate, while malto-oligosaccharides (MOS) of higher degree of polymerization (DP) were not hydrolyzed [16]. We showed that MOS of DP3 (see Table 1), DP4, DP5, DP6 and DP7 served as substrates for *Ba*AG2 (Figure S4). Assay of reaction course indicated that exo-hydrolysis occurred—a glucose residue was stripped off from the oligomer. Activity on MOS was only moderate and a substantial proportion of it stayed unreacted even after extended (22 h) reaction time (see Figure S4). The MAL62 of *S. cerevisiae* that was assayed alongside could not hydrolyze MOS longer than maltotetraose (DP4) (Figure S4).

2.4.6. The Hydrolysis of Amylose, Amylopectin and Glycogen

Surprisingly, we detected the ability of *Ba*AG2 to hydrolyze polysaccharides that is a rather exceptional feature among maltases. After we noticed that *Ba*AG2 could hydrolyze soluble starch with the release of glucose, we performed a more detailed assay testing the hydrolysis of a set of polymeric α-glucans: amylose and amylopectin from potato, glycogen from oysters and dextrans of M_w 20 and 110 kDa. Commercial amylolytic enzymes amyloglycosidase (glucoamylase) from *Aspergillus niger*, and α-amylase from *Aspergillus oryzae* and *S. cerevisiae* MAL62 were used as reference enzymes.

Hydrolysis of the polymers was evaluated according to the release of glucose and by TLC analysis of reaction products. The commercial enzymes hydrolyzed amylose, amylopectin and glycogen rapidly and with the expected pattern of products (Figure 7). Dextrans were hydrolyzed only by the amyloglycosidase, and the release of glucose was minimal. *Ba*AG2 and *Sc*MAL62 had no activity on dextrans. However, *Ba*AG2 exhibited moderate, but clearly detectable and recordable exo-hydrolysis of amylose, amylopectin and glycogen. The activity was the highest with amylopectin, and the lowest with amylose (Figure 7). From amylopectin (5 g/L), 0.1 g/L of glucose was released by 24 h, and 0.3 g/L by 72 h of the reaction. In the case of *Sc*MAL62, no hydrolysis of amylose was detected, and hydrolysis of glycogen and amylopectin became detectable only by 72 h of the reaction (Figure 7).

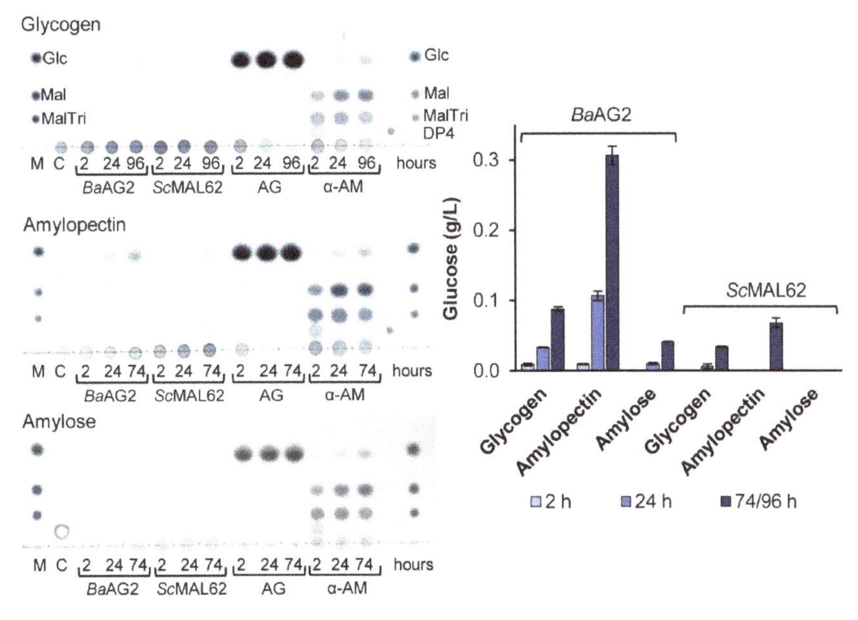

Figure 7. Hydrolysis of polysaccharides (5 g/L) by *Ba*AG2, *Sc*MAL62 and amyloglycosidase of *A. niger* (AG) and α-amylase of *A. oryzae* (α-AM). Samples withdrawn at indicated time points were analyzed using TLC. Reaction mixtures were spotted on TLC plates alongside with reference sugars (M): Glc (30 mM glucose); Mal (10 mM maltose); MalTri (10 mM maltotriose); DP4 (10 mM maltotetraose). The same marker sugars were used in all assays but are marked only on TLC plate of glycogen degradation. C—control sample without the enzyme but containing 5 g/L bovine serum albumin (BSA) incubated at the same conditions for 74/96 h. Glucose release was quantified enzymatically. See Materials and Methods, paragraphs 4.3 and 4.6 for details. SDs of two to three replicates are shown by error bars.

2.4.7. The Transglycosylation of Maltose

Many α-glucosidases can transglycosylate and produce short oligosaccharides, especially at high concentration of the substrate [17–21]. We assayed this possibility by incubating *Ba*AG2 with 250 and 500 mM maltose up to 72 h and analyzed the reaction products by HPLC. The maltase MAL62 of *S. cerevisiae* was used as a reference. Figure 8 and Table S3 show that already within 2 h at 250 mM (85.6 g/L) maltose concentration, *Ba*AG2 produced maltotriose (4.2 g/L) and panose, α-D-Glc-(1→6)-α-D-Glc-(1→4)-D-Glc (1.6 g/L), in addition to a maltose hydrolysis product—glucose. By 72 h of reaction, the maltotriose content was decreased and panose content increased to 2.6 g/L (Figure 8, Table S3). The *Sc*MAL62 produced only maltotriose under the same conditions, and its amount was considerably lower than in the case of *Ba*AG2—only 2.0 g/L produced by 2 h of reaction (Figure 8, Table S3). Transglycosylation of maltose was enhanced at 500 mM (171.2 g/L) concentration: up to 13.3 g/L of maltotriose was produced by 2 h and 10.4 g/L of panose by 72 h of reaction. Notably, a new transglycosylation product, isomaltose, emerged. It was produced by both enzymes, but its concentration was certainly higher in the case of *Ba*AG2—its concentration in the 72-h reaction sample was 5.2 g/L (Table S3). By 24 h of reaction with 500 mM maltose, the amount of transglycosylation products was 12.6% of total sugars in the reaction mixture, while the respective value for the MAL62 protein was about three times less—4.5% (see Table S3).

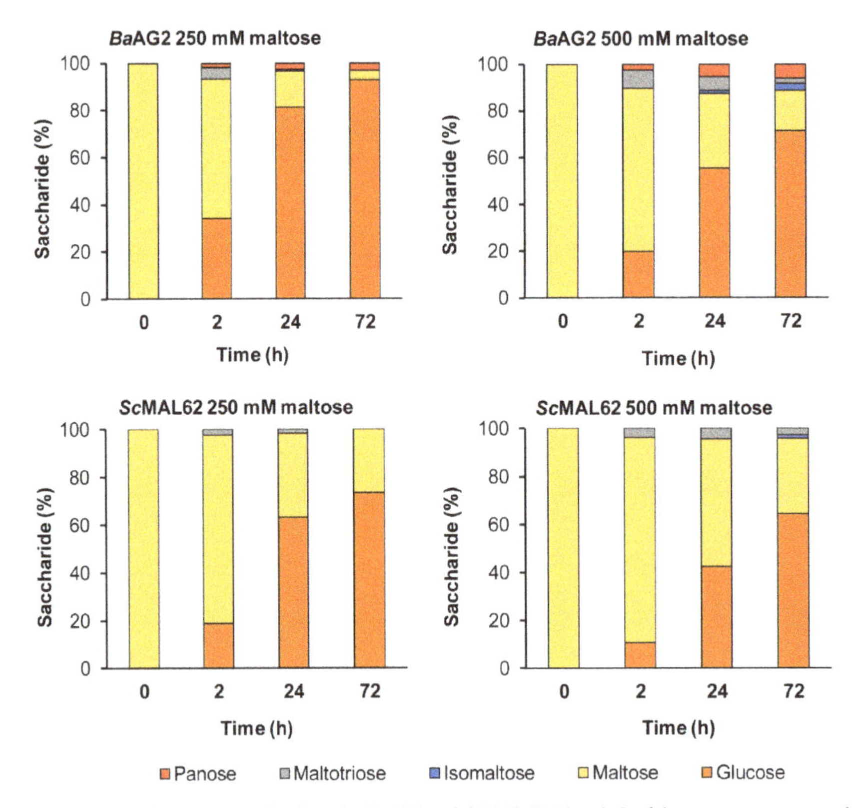

Figure 8. Transglycosylation of maltose by *Ba*AG2 and *Sc*MAL62. 20 μg/mL of the enzyme was reacted with 250 mM or 500 mM of maltose. Samples from the reaction mixtures were withdrawn at designated time points, heated to terminate the reaction and analyzed for sugar composition by HPLC as described in Materials and Methods, paragraphs 4.5. and 4.6. Total amount of detected saccharides at each time point was equalled to 100%. Products were identified using glucose, maltose, isomaltose, maltotriose and panose as references. SDs of two to three HPLC measurements at each time point were up to 20%.

We tested panose hydrolysis by *Ba*AG2 and conclude that it accumulated in transglycosylation reaction since it was not hydrolyzed by the enzyme even during extended (22 h) reaction time (Figure S3). In contrast, maltotriose was hydrolysed by *Ba*AG2 (Figure 5, Table 1), and due to that, its content decreased at prolonged transglycosylation reaction (Figure 8, Table S3).

3. Discussion

Utilization of α-glucosidic sugars by maltases and isomaltases has earlier been thoroughly studied in *S. cerevisiae* because metabolism of these sugars is crucial in brewing and baking [22,48]. However, transport and intracellular hydrolysis of α-glucosidic sugars have also been investigated in *Ogataea polymorpha* [25] and *Schizosaccharomyces pombe* [33,49]. A maltase has been characterized from *C. albicans* [41] and four maltase-isomaltases from *Scheffersomyces stipitis* [12]. Genes potentially encoding for either maltases, isomaltases or maltase-isomaltases were recently discovered in the genomes of many non-conventional yeasts [12]. Yeast species with deep phylogeny were expected to possess ancient-like α-glucosidases [12]. Figure 9 shows the phylogram of selected yeast species and *A. oryzae* based on sequence analysis of D1/D2 domains of large subunit ribosomal RNA to illustrate the evolutionary relationships between the species. *B. adeninivorans* and *Lipomyces starkeyi* belong to basal group of Saccharomycotina [50]. Based on phylogenetic analysis of orthologous proteins [4], this group diverged from *S. cerevisiae* lineage 200 to 400 million, and from the CTG clade 200 million years ago. The basal group is considered very heterogeneous, its most studied member is *Yarrowia lipolytica* (not presented in Figure 9) and the most basal lineage to this group and all Saccharomycotina is *Lipomyces starkeyi* [50].

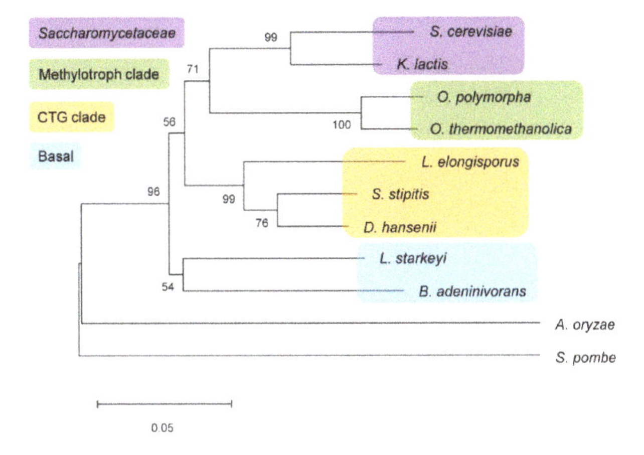

Figure 9. Phylogram based on analysis of D1/D2 domains of large subunit ribosomal RNA (rRNA) gene sequences of ten yeast species and of a filamentous fungus *Aspergillus oryzae*. The bootstrap values (1000 replicates) are shown at the nodes. The scale bar shows the number of base substitutions per site. The Saccharomycotina clades according to [50] are designated by background coloring.

On the phylogram (Figure 9), *B. adeninivorans* clusters with *Lipomyces starkeyi*. Genes for eight intracellular α-glucosidases (five maltases and three isomaltases) were predicted in the genome of *L. starkeyi*. In the phylogram of α-glucosidases from non-conventional yeasts, all eight *Lipomyces* enzymes clustered with those of *B. adeninivorans* [12]. However, these putative α-glucosidases of *L. starkeyi* have not been cloned for protein expression and characterization. In Kelly et al. an extracellular α-glucosidase from *L. starkeyi* was biochemically characterized [51]. Unfortunately, the sequence data of this protein is not available. The above-mentioned enzyme of *L. starkeyi* had equally high activity on maltose and isomaltose, but it also hydrolyzed maltotriose, isomaltotriose, panose, amylopectin and starch. Its activity with starch and amylopectin comprised 31 and 42% of that with maltose and isomaltose, and due to this feature, the authors considered it more similar to fungal rather than yeast enzymes [51]. Interestingly, this enzyme did not hydrolyze sucrose and had a quite low activity on *p*NPG. Typically, yeast α-glucosidases hydrolyze *p*NPG more rapidly than maltose. or sucrose [16,24] (Table 1) whereas the opposite is true for maltases of bacteria and archaea [52–54].

Our study showed that *Ba*AG2 is a maltase. *Ba*AG2 efficiently hydrolyzed maltose and maltose-like sugars, but could not hydrolyze isomaltose, palatinose nor α-MG that are specific substrates for yeast isomaltases (Figure 5, Table 1). Isomaltose, isomaltose-like sugars as well as acarbose and glucose competitively inhibited *p*NPG hydrolysis by *Ba*AG2 and increased thermostability of the enzyme (Table 2; Figure 6). We also discovered a very strong inhibition of *Ba*AG2 by Tris with the K_i of 70.5 μM (Table 2). Tris also considerably increased thermostability of *Ba*AG2 in a DSF assay (Figure 6). As thermostability of *S. cerevisiae* isomaltases was also elevated in the presence of Tris [55], it may bind to the active site of these enzymes. Literature mining revealed Tris as a competitive inhibitor of *Bacillus brevis* maltase with the K_i of 14.5 mM [56]. In a yeast *Brettanomyces bruxellensis* (former name *Br. lambicus*), both extra- and intracellular maltases were inhibited not only by acarbose (K_i values between 28.5 and 57 μM) but also by Tris (K_i values between 7.45 and 15.7 mM) [57]. Compared to *Br. bruxellensis* and *Bacillus brevis* maltases, *Ba*AG2 was much more sensitive to Tris. The pH optimum of *Ba*AG2 was in a moderate acidic region as in the case of other yeast maltases [23,24]. The temperature optimum for *Ba*AG2 was between 40–50 °C, with maximum activity (530 U/mg) achieved at 45 °C. Thermostability of the enzyme was not high—after keeping the enzyme at 45 °C for 30 min, the enzyme's activity dropped to 46% from the initial. Incubation for 30 min at 50 °C totally inactivated the enzyme (Figure 4). According to literature data, thermolability has been reported for some other yeast α-glucosidases. For example, the four isomaltases of *S. cerevisiae* had all low thermostability. Of those, the IMA1 was most stable, and IMA5 the least stable [55].

According to Hasegawa et al., the MalT protein of *A. oryzae* that has 51% of sequence identity to *Ba*AG2 (Table S2) exhibited *p*NPG-hydrolyzing activity, its expression was induced when grown on maltose, and thereby the MalT was defined by the authors as a maltase [36]. To date no additional data on MalT protein is available. The intracellular maltase (MAL1) protein of *Schizosaccharomyces pombe*, with 43.2% of sequence identity to *Ba*AG2 hydrolyzed *p*NPG and maltose, but had also activity on soluble starch, and some activity on sucrose [33].

Intriguingly, *Ba*AG2 had a detectable hydrolytic activity on MOS with DP up to 7, glycogen, amylose and amylopectin (Figure 7 and Table S4). We assume that the ability to hydrolyze MOS and to cleave polymeric α-glucans, at least to some extent, may be characteristic to maltases of early-diverged yeasts. As *Ba*AG2 is an intracellular enzyme, and this yeast possesses a secreted glucoamylase [9], the maltase *Ba*AG2 has most probably no role in starch degradation. However, yeasts synthesize glycogen as a reserve polysaccharide [58]. Considering that *Ba*AG2 had a remarkable activity on glycogen, the enzyme may contribute to glycogen catabolism in *B. adeninivorans*. We hypothesize that the ability to degrade malto-oligosaccharides and α-glycosidic polysaccharides may be characteristic for the maltases of yeasts with deep phylogeny. Isolation and study of α-glucosidases of the most basal lineages to Saccharomycotina, for example *Lipomyces starkeyi*, should verify the correctness of this hypothesis.

Several GHs of yeasts and filamentous fungi exhibit transglycosylating activity. For example, α-glucosidase of a yeast *X. dendrorhous* (syn. *Pfaffia rhodozyma*) produced from maltose a large proportion of transglycosylation products with α-1,6 linkage, among which panose was the most abundant [17]. It has been shown that the *S. cerevisiae* maltase also produced panose from maltose, yet transglycosylating activity of the *Saccharomyces* enzyme was more than three times smaller compared to the *X. dendrorhous* enzyme. Both enzymes also synthesized isomaltose and maltotriose, but the latter was rapidly used by the enzymes and was therefore not present among the final products [17]. We showed that maltotriose was produced from maltose also by *Ba*AG2 (Figure 8, Table S3), but as it serves as a substrate for the enzyme (Table 1), it was eventually hydrolyzed. α-Glucosidases of filamentous fungi have been suggested as feasible catalysts for transglycosylation. Thus, an *Aspergillus* enzyme with high transglycosylating activity was reported to produce panose and isomaltose from maltose [18,59,60]. When an *A. nidulans* α-glucosidase with strong transglycosylating activity was reacted with 5 g/L maltose during 6 h, approximately 50% of maltose was converted to transglycosylation products, 60% of which was isomaltose [60]. Notably, in addition to maltotriose and panose, isomaltose was also detected among the transglycosylation products produced from maltose by *Ba*AG2 (Figure 8, Table S3). Isomaltose synthesis from maltose was also confirmed for *Sc*MAL62, even though the content of it was only minimal (Figure 8, Table S3).

4. Materials and Methods

4.1. Yeast and Bacterial Strains, Cultivation of B. adeninivorans

Blastobotrys (*Arxula*) *adeninivorans* LS3 (CBS 8244) was kindly provided by Assoc. Prof. V. Passoth (SLU, Uppsala, Sweden). The yeast strain was grown on solid YPD medium (20 g/L peptone, 20 g/L glucose, 10 g/L yeast extract, 20 g/L agar) at 30 °C 24 h for harvesting the cells for genomic DNA extraction. *E. coli* DH5α (Thermo Fisher Scientific, Waltham, MA, USA) was used for DNA cloning and plasmid production. *E. coli* BL21 (DE3) [61] was used for heterologous expression of *Ba*AG2. The ability of *B. adeninivorans* to grow on sugars was assayed as in [12]. Yeast cells grown overnight on BD Difco YNB medium (Thermo Fisher Scientific, Waltham, MA, USA) without amino acids containing 2 g/L glucose were used as inoculum. The YNB medium was supplemented with 2 g/L of a filter-sterilized sugar (glucose, maltose, maltotriose, isomaltose, maltulose, α-methylglucoside, sucrose, raffinose, melibiose, turanose, palatinose, melezitose or trehalose). The cells were incubated on Greiner 96-well flat-bottom transparent polystyrene microplates (Greiner Bio-One, Frickenhausen, Germany) in 200 µl under agitation for 24 h at 37 °C. Optical density of the culture at 600 nm was measured against inoculated medium without sugar at the beginning and end of the experiment using an Infinite M200 PRO microplate reader (Tecan Group Ltd., Männedorf, Switzerland) equipped with Tecan i-control v. 1.7 software from the same provider. Two independent experiments with two parallel measurements were conducted.

4.2. Cloning, Heterologous Expression and Protein Purification

Genomic DNA of *B. adeninivorans* was extracted using PowerMax Soil DNA Isolation Kit (MoBio Laboratories, Carlsbad, CA, USA) and the standard protocol by manufacturer. The oligonucleotide primers Ba20130g_PURICter_Fw (5′ **TAACTTTAAGAAGGAGATATACAT** *ATGGTTCTAGGATTTTTCAAAAAG* 3′) and Ba20130g_PURICter_Rev (5′ **GCTATTAATGATGATG ATGATGAT***GGATTTCATAGATGACTGCCTCCA* 3′), designed according to the gene sequence of *rna_ARAD1D20130g*, were applied to amplify a 1789 bp fragment that represented the coding sequence of *Ba*AG2 [12]. In the primers, the nucleotides annealing with the pURI3Cter vector [62] are shown in bold and those annealing with the *BaAG2* gene sequence are shown in italics. The ATG start codon and the stop codon are underlined in the primer sequences. Recombinant *Pfu* polymerase (Thermo Fisher Scientific, Waltham, MA, USA) and 2.6 ng per µL of reaction mixture of genomic DNA were used in amplification. The PCR product was cloned into a pJET vector from CloneJET PCR Cloning Kit (Thermo Fisher Scientific, Waltham, MA, USA), yielding pJET-BaAG2. To produce the *Ba*AG2 protein with a C-terminal His_6-tag, an expression plasmid pURI3-AG2Cter was constructed by cloning the *BaAG2* gene into pURI3-Cter vector similarly as in [63]. Insertion of the *BaAG2* gene into the vector was confirmed by DNA sequencing. The cloning procedure was conducted with recombinant *Pfu* polymerase (Thermo Fisher Scientific, Waltham, MA, USA). DNA Clean & Concentrator-5 Kit (Zymo Research, Irvine, CA, USA) was used for purification and concentration of PCR products. Plasmid DNA was purified with FavorPrep Plasmid Extraction Mini Kit (Favorgen Biotech Corp., Ping-Tung, Taiwan). The *MAL62*-containing plasmid pURI-ScMAL62Cter [12] was used to produce the *S. cerevisiae* maltase protein that was analyzed as a reference.

The pURI3-BaAG2Cter and pURI-ScMAL62Cter containing the *BaAG2* gene or *MAL62* gene, respectively, were electroporated into *E. coli* BL21 (DE3) for heterologous expression. A simplified autoinduction medium as in [64] was used for protein overproduction: the LB-based medium (20 g/L tryptone, 5 g/L yeast extract, 5 g/L NaCl) was supplemented with 25 mM phosphate buffer (Na_2HPO_4/KH_2PO_4; pH 7.2) and 3 g/L glycerol to which filter-sterilized 0.25 g/L glucose and 1 g/L lactose were added. Medium for transformant selection contained ampicillin (Amp, 150 mg/L) for plasmid preservation. Bacterial cells were grown overnight in LB-Amp medium at 37 °C and diluted 100 times in autoinduction medium. At first, the cultures were incubated for 2 h at 37 °C followed by 20-h incubation at 22 °C. The cells were harvested by centrifugation (2400× *g*, 10 min) at 4 °C and stored at −20 °C until further use. Cells were disrupted by sonication with Ultrasonic Homogenizer (Cole-Parmer Instrument Company, Vernon Hills, IL, USA) in 100 mM K-phosphate buffer (pH 6.5) with the cOmplete, EDTA-free Protease Inhibitor Cocktail (Roche Diagnostics, Mannheim, Germany), and centrifuged 30 min at 2400 × *g* at 4 °C. The resulting supernatants were syringe-filtered (pore size 0.45 µm) and loaded onto an IMAC HisTrap FF column coupled with an ÄKTAprime plus chromatography system (GE Healthcare, Uppsala, Sweden). Further purification steps were conducted as described in [63]. Proteins were quantified by measuring the absorbance at 280 nm. The respective extinction coefficients of *Ba*AG2 [133,855 1/(M × cm)] and *Sc*MAL62 [148,990 1/(M × cm)] were computed at ExPASy Proteomics Server (http://expasy.org). Purified proteins were maintained in 100 mM K-phosphate buffer (pH 6.5) at 4 °C.

4.3. Determination of Substrate Specificity, Kinetic Parameters and Inhibition

Hydrolysis of *p*NPG was assayed as in [16,40] according to the release of *p*-nitrophenol. 100 mM K-phosphate buffer (pH 6.5) was used and reactions were conducted at 30 °C if not stated otherwise. The purified enzyme was unstable if diluted therefore 5 g/L bovine serum albumin (BSA) was added to the dilution buffer as a protein stabilizer to retain its full catalytic activity. For preliminary assay of substrate specificity of *Ba*AG2, the enzyme was reacted with 100 mM concentration of various potential substrates: maltose, sucrose, maltotriose, isomaltose, melezitose, maltulose, turanose, palatinose or α-MG. At fixed time points, aliquots were withdrawn, combined with three volumes of 200 mM Tris buffer (pH 8.3) and subsequently heated at 96°C for 5 min to stop the reaction. The content of the released glucose was determined spectrophotometrically as in [16,47]. The activities (µmoles of glucose released per minute per mg of protein; U/mg) were calculated from initial velocities of the reaction. For kinetic analysis, initial rates of *p*-nitrophenol or glucose release from substrates were measured at four to seven concentrations ranging from 0.1–3.0 mM for *p*NPG and 2.5–250 mM for di- and trisaccharides. At least three independent measurements for each substrate and concentration were made. *Ba*AG2 content in reaction mixtures ranged from 0.02 to 3.5 µg/mL. The initial velocity data analysis with enzyme kinetics module of the SigmaPlot (Systat Software, San Jose, CA, USA) yielded kinetic parameters (K_m, V_{max}) for the enzyme. k_{cat} and k_{cat}/K_m were calculated from these data. The theoretical M_w value of 67,901 Da for the k_{cat} calculation was computed in ExPASy Proteomics Server (http://expasy.org).

Inhibition of *Ba*AG2 was studied by incubating a suitable amount of enzyme (0.035–0.13 µg/mL) with 0.2–2.0 mM *p*NPG in the presence of following potential inhibitors: 5 µM acarbose, 0.1 mM Tris, 5 mM palatinose, 10 mM isomaltose, 10 mM glucose, 100 mM α-methylglucoside, 100 mM trehalose or 400 mM fructose. The K_i values were calculated using enzyme kinetics module of the SigmaPlot (Systat Software, San Jose, CA, USA).

Differential scanning fluorimetry (DSF) was used to evaluate the thermostability of *Ba*AG2 in the presence and absence of ligands. The concentration of *Ba*AG2 was used with 2 µM and for ligands: 100 mM fructose, 50 mM palatinose, 50 mM glucose, 5 mM Tris and 5 mM acarbose. The reaction was conducted in 50 mM 4-(2-hydroxyethyl)-1-piperazineethanesulfonic acid) (HEPES) buffer (pH 7.0) with 150 mM NaCl. The experiment was based on [16,47,65] with above-mentioned modifications. At least two independent experiments were performed with two technical replicates.

Degradation of glucose polymers, i.e., amylopectin-free amylose from potato, amylopectin from potato, glycogen from oysters and dextrans of M_W 20 kDa and 110 kDa, was assayed in K-phosphate buffer (100 mM, pH 6.5) containing 0.2 g/L Na-azide. Polysaccharide concentration in the reaction mixture was 5 g/L, and 20 µg/mL of the enzyme was used. At desired time points (2 h, 24 h, 74/96 h) aliquots were withdrawn and heated to stop the reaction. A negative control containing 20 µg/mL BSA instead of the enzyme was incubated alongside. Concentration of released glucose was measured as described above and 3 µl of the samples were analysed on TLC. Maltase from *S. cerevisiae* (*Sc*MAL62), amyloglycosidase from *Aspergillus niger* (Sigma-Aldrich, Merck, Darmstadt, Germany) and α-amylase from *Aspergillus oryzae* (Sigma-Aldrich, Merck, Darmstadt, Germany) were reacted with tested glucose polymers for comparison.

To assay the hydrolysis of malto-oligosaccharides (DP 4-7), panose and melezitose by *Ba*AG2 and *Sc*MAL62, 50 mM of the sugar was reacted with 2.6 µg/mL of the enzyme in K-phosphate buffer (100 mM, pH 6.5) containing 0.1 g/L of Na-azide and samples were withdrawn at fixed time points. 20 mM isomelezitose (transglycosylation product from sucrose of *Ogatae polymorpha* MAL1) was also tested as a potential substrate. The reaction samples were analyzed using TLC.

4.4. Determination of pH and Temperature Optima, Evaluation of Thermostability

Initial velocity of 1 mM *p*NPG hydrolysis by *Ba*AG2 was measured at 30 °C in BSA-supplemented buffers of varied pH (from 3.8 to 8.5) using McIlvaine's buffer (Na-phosphate/citrate buffer) [66] and 100 mM K-phosphate buffer to cover respective pH-interval, and data were plotted against the pH to determine the pH optimum. Hydrolysis of 1 mM *p*NPG was measured at varied temperatures from 20 to 65 °C. Initial velocity data were plotted against the temperature to reveal the temperature optimum. For thermostability determination, *Ba*AG2 was incubated in 100 mM K-phosphate buffer (pH 6.5) buffer containing 5 g/L BSA for 30 min at temperatures 10, 20, 30, 37, 45 and 50 °C. After cooling the samples on ice, residual activity of the enzyme was determined according to the hydrolysis of 1 mM *p*NPG at 30 °C. Every temperature point was assayed in triplicate. The activity measured after incubation at 10 °C was taken as 100%.

4.5. Study of Transglycosylation

20 µg/mL of the enzyme (*Ba*AG2 or *Sc*MAL62) was incubated in 100 mM K-phosphate buffer (pH 6.5) with 0.2 g/L Na-azide and 250 mM or 500 mM maltose at 30 °C up to 72 h. The samples with *Ba*AG2 also contained 5 g/L BSA. Samples were withdrawn at fixed intervals, heated at 95 °C to stop the reaction and analysed on TLC and by HPLC.

4.6. Chromatography of Substrates and Reaction Products

To visualize hydrolysis and polymerization products, the TLC analysis was conducted as in [16] on Silica Gel 60 F_{254} plates with concentrating zone (Merck, Darmstadt, Germany). 0.5 µl of the stopped reaction mixture were spotted onto the plate and sugars were separated with two runs in chloroform:acetic acid:water (6:7:1, v:v:v) [67]. For the analysis of products of polysaccharide degradation, 3 µl of the reaction mixtures were spotted. Sugars were visualized by immersion of the plates in aniline-diphenylamine reagent and subsequent heating of the dried plates at 100 °C [68].

HPLC analysis was performed similarly as in [65]. Glucose and fructose were used to calibrate the column. Fructose, maltose, sucrose, isomaltose, palatinose, turanose, maltotriose, panose and melezitose were used as reference sugars.

4.7. Alignment of RNA and Protein Sequences and Construction of the Phylogram

Gene sequences from domains 1 and 2 (D1/D2) of large subunit ribosomal RNA were aligned to build a neighbor-joining phylogenetic tree [69] of yeasts using MEGA v. 7.0 [70]. The maximum composite likelihood model [71] with 1000 bootstrap replicates was applied. Protein sequences were aligned using Clustal Omega [72] to calculate identity values of the proteins.

4.8. Extraction of Amino Acid Signature from the Alignments and Visualization of Respective Positions on the three-dimensional (3D) Model of S. cerevisiae Isomaltase IMA1

Protein sequences of *S. cerevisiae* maltase MAL62 (UniProtKB: P07265), *S. cerevisiae* isomaltase IMA1 (UniProtKB: YGR287C), *O. polymorpha* maltase-isomaltase MAL1 (UniProtKB: Q9P8G8) and *B. adeninivorans* AG2 were aligned using Clustal Omega [72] and amino acids corresponding to IMA1 signature positions determining the substrate specificity [12,15] were extracted from the alignment. The *S. cerevisiae* IMA1 structure in complex with isomaltose (PDB: 3AXH) [29] was visualized with PyMol [30] and amino acid signature was designated on the structure.

5. Conclusions

According to the literature data, a non-conventional yeast *Blastobotrys* (*Arxula*) *adeninivorans* belonging to the basal group of Saccharomycotina diverged in the evolution of yeasts hundreds of millions of years before *Saccharomyces* and can be considered as a yeast species with deep phylogeny. The genome of *B. adeninivorans* encodes two putative α-glucosidases. In current work, one of them, *Ba*AG2, was produced in *E. coli* and characterized in detail. *Ba*AG2 was proven to be a maltase—hydrolysing α-1,4 and α-1,3 but not α-1,6 linkages in glucose-containing substrates. Interestingly, *Ba*AG2 was strongly and competitively inhibited not only by acarbose, a diabetes drug and a well-known inhibitor of α-glucosidases, but also by Tris. Importantly, at high maltose concentrations, *Ba*AG2 exhibited transglycosylating ability producing potentially prebiotic di- and trisaccharides: isomaltose, panose and maltotriose. Thus, *Ba*AG2 may have a biotechnological value. In contrast to yeast maltases, *Ba*AG2 showed exo-hydrolytic activity on starch, amylose, amylopectin and glycogen. *S. cerevisiae* maltase MAL62 assayed for comparison had only minimal ability towards these polymers and its transglycosylating activity was much lower.

Supplementary Materials: Supplementary file can be found at http://www.mdpi.com/1422-0067/21/1/297/s1.

Author Contributions: Conceptualization, T.A. and T.V.; Methodology, T.V., K.E., A.M., K.P. and K.V.; Validation, T.V., K.E. and T.A.; Formal Analysis, T.V., A.M., K.P. and K.E.; Investigation, T.V. A.M., K.V., K.P., K.E. and T.A.; Resources, T.A. and T.V.; Writing—original draft preparation, T.A. and T.V.; Writing—review and editing, T.A., T.V., K.E. and K.P.; Visualization, T.V., K.E., A.M., K.P. and T.A.; supervision, T.A. and T.V.; funding acquisition, T.A and T.V. All authors have read and agreed to the published version of the manuscript.

Funding: This work was funded by the Estonian Research Council (grant number PUT1050) to T.A. The article processing charge was covered by University of Tartu Feasibility Fund grant PLTMRARENG13 to T.V.

Acknowledgments: We thank V. Passoth (SLU, Uppsala, Sweden) for providing the *B. adeninivorans* strain, B. Svensson (DTU, Kongens Lyngby, Denmark) for providing amylose and amylopectin and H. Vija (NICPB, Tallinn, Estonia) for the services in sugar quantification.

Conflicts of Interest: The authors declare no conflict of interest.

Abbreviations

*Ba*AG2	α-glucosidase 2 of *Blastobotrys adeninivorans*
BSA	bovine serum albumin
DP	degree of polymerization
DSF	differential scanning fluorimetry
GH	glycoside hydrolase
MOS	malto-oligosaccharides
α-MG	α-methylglucoside
PDB	RCSB Protein Data Bank
*p*NPG	*p*-nitrophenyl-α-D-glucopyranoside
OD	optical density
rRNA	ribosomal RNA
*Sc*MAL62	maltase MAL62 of *Saccharomyces cerevisiae*
SD	standard deviation
TLC	thin layer chromatography
T_m	melting temperature
Tris	tris(hydroxymethyl)aminomethane
YNB	Yeast Nitrogen Base

References

1. Kurtzman, C.P.; Robnett, C.J. Multigene phylogenetic analysis of the Trichomonascus, Wickerhamiella and Zygoascus yeast clades, and the proposal of *Sugiyamaella* gen. nov. and 14 new species combinations. *FEMS Yeast Res.* **2007**, *7*, 141–151. [CrossRef] [PubMed]
2. Kunze, G.; Gaillardin, C.; Czernicka, M.; Durrens, P.; Martin, T.; Böer, E.; Gabaldón, T.; Cruz, J.A.; Talla, E.; Marck, C.; et al. The complete genome of *Blastobotrys (Arxula) adeninivorans* LS3—A yeast of biotechnological interest. *Biotechnol. Biofuels* **2014**, *7*, 66. [CrossRef]
3. Shen, X.-X.; Zhou, X.; Kominek, J.; Kurtzman, C.P.; Hittinger, C.T.; Rokas, A. Reconstructing the backbone of the Saccharomycotina yeast phylogeny using genome-scale data. *G3 Genes Genomes Genet.* **2016**, *6*, 3927–3939. [CrossRef] [PubMed]
4. Correia, K.; Yu, S.M.; Mahadevan, R. AYbRAH: A curated ortholog database for yeasts and fungi spanning 600 million years of evolution. *Database* **2019**, *2019*, baz022. [CrossRef] [PubMed]
5. Hedges, S.B.; Marin, J.; Suleski, M.; Paymer, M.; Kumar, S. Tree of life reveals clock-like speciation and diversification. *Mol. Biol. Evol.* **2015**, *32*, 835–845. [CrossRef] [PubMed]
6. Thomas, S.; Sanya, D.R.A.; Fouchard, F.; Nguyen, H.V.; Kunze, G.; Neuvéglise, C.; Crutz-Le Coq, A.M. *Blastobotrys adeninivorans* and *B. raffinosifermentans*, two sibling yeast species which accumulate lipids at elevated temperatures and from diverse sugars. *Biotechnol. Biofuels* **2019**, *12*, 154. [CrossRef] [PubMed]
7. Malak, A.; Baronian, K.; Kunze, G. *Blastobotrys (Arxula) adeninivorans*: A promising alternative yeast for biotechnology and basic research. *Yeast* **2016**, *33*, 535–547. [CrossRef]
8. Kasprzak, J.; Rauter, M.; Riechen, J.; Worch, S.; Baronian, K.; Bode, R.; Schauer, F.; Kunze, G. Characterization of an *Arxula adeninivorans* alcohol dehydrogenase involved in the metabolism of ethanol and 1-butanol. *FEMS Yeast Res.* **2016**, *16*, fow018. [CrossRef]
9. Bui, D.M.; Kunze, I.; Förster, S.; Wartmann, T.; Horstmann, C.; Manteuffel, R.; Kunze, G. Cloning and expression of an *Arxula adeninivorans* glucoamylase gene in *Saccharomyces cerevisiae*. *Appl. Microbiol. Biotechnol.* **1996**, *44*, 610–619. [CrossRef]
10. Böer, E.; Wartmann, T.; Luther, B.; Manteuffel, R.; Bode, R.; Gellissen, G.; Kunze, G. Characterization of the *AINV* gene and the encoded invertase from the dimorphic yeast *Arxula adeninivorans*. *Antonie Van Leeuwenhoek* **2004**, *86*, 121–134. [CrossRef]
11. Bao, M.; Niu, C.; Xu, X.; Zheng, F.; Liu, C.; Wang, J.; Li, Q. Identification, soluble expression, and characterization of a novel endo-inulinase from *Lipomyces starkeyi* NRRL Y-11557. *Int. J. Biol. Macromol.* **2019**, *137*, 537–544. [CrossRef] [PubMed]
12. Viigand, K.; Põšnograjeva, K.; Visnapuu, T.; Alamäe, T. Genome mining of non-conventional yeasts: Search and analysis of MAL clusters and proteins. *Genes* **2018**, *9*, 354. [CrossRef] [PubMed]

13. Gabriško, M. Evolutionary history of eukaryotic α-glucosidases from the α-amylase family. *J. Mol. Evol.* **2013**, *76*, 129–145. [CrossRef] [PubMed]
14. Janeček, Š.; Gabriško, M. Remarkable evolutionary relatedness among the enzymes and proteins from the α-amylase family. *Cell. Mol. Life Sci.* **2016**, *73*, 2707–2725. [CrossRef]
15. Voordeckers, K.; Brown, C.A.; Vanneste, K.; van der Zande, E.; Voet, A.; Maere, S.; Verstrepen, K.J. Reconstruction of ancestral metabolic enzymes reveals molecular mechanisms underlying evolutionary innovation through gene duplication. *PLoS Biol.* **2012**, *10*, e1001446. [CrossRef]
16. Viigand, K.; Visnapuu, T.; Mardo, K.; Aasamets, A.; Alamäe, T. Maltase protein of *Ogataea (Hansenula) polymorpha* is a counterpart to the resurrected ancestor protein ancMALS of yeast maltases and isomaltases. *Yeast* **2016**, *33*, 415–432. [CrossRef]
17. Fernández-Arrojo, L.; Marín, D.; Gómez De Segura, A.; Linde, D.; Alcalde, M.; Gutiérrez-Alonso, P.; Ghazi, I.; Plou, F.J.J.; Fernández-Lobato, M.; Ballesteros, A. Transformation of maltose into prebiotic isomaltooligosaccharides by a novel α-glucosidase from *Xantophyllomyces dendrorhous*. *Process Biochem.* **2007**, *42*, 1530–1536. [CrossRef]
18. Mangas-Sánchez, J.; Adlercreutz, P. Enzymatic preparation of oligosaccharides by transglycosylation: A comparative study of glucosidases. *J. Mol. Catal. B Enzym.* **2015**, *122*, 51–55. [CrossRef]
19. Casa-Villegas, M.; Marín-Navarro, J.; Polaina, J. Amylases and related glycoside hydrolases with transglycosylation activity used for the production of isomaltooligosaccharides. *Amylase* **2018**, *2*, 17–29. [CrossRef]
20. Gutiérrez-Alonso, P.; Gimeno-Pérez, M.; Ramírez-Escudero, M.; Plou, F.J.; Sanz-Aparicio, J.; Fernández-Lobato, M. Molecular characterization and heterologous expression of a *Xanthophyllomyces dendrorhous* α-glucosidase with potential for prebiotics production. *Appl. Microbiol. Biotechnol.* **2016**, *100*, 3125–3135. [CrossRef]
21. Chiba, S.; Murata, M.; Matsusaka, K.; Shimomura, T. A new trisaccharide, 6F-α-D-glucosyl-sucrose, synthesized by transglucosylation reaction of brewer's yeast α-glucosidase. *Agric. Biol. Chem.* **1979**, *43*, 775–779. [CrossRef]
22. Stewart, G. *Saccharomyces* species in the production of beer. *Beverages* **2016**, *2*, 34. [CrossRef]
23. Needleman, R.B.; Marmur, J.; Federoff, H.J.; Eccleshall, T.R.; Buchferer, B. Purification and characterization of an α-glucosidase from *Saccharomyces carlsbergensis*. *Biochemistry* **1978**, *17*, 4657–4661. [CrossRef] [PubMed]
24. Krakenaĭte, R.P.; Glemzha, A.A. Some properties of two forms of alpha-glucosidase from *Saccharomyces cerevisiae*-II. *Biokhimiia* **1983**, *48*, 62–68.
25. Alamäe, T.; Viigand, K.; Põšnograjeva, K. Utilization of α-glucosidic disaccharides by *Ogataea (Hansenula) polymorpha*: Genes, proteins, and regulation. In *Non-Conventional Yeasts: From Basic Research to Application*; Sibirny, A., Ed.; Springer International Publishing: Cham, Switzerland, 2019; pp. 1–22.
26. Grigoriev, I.V.; Nikitin, R.; Haridas, S.; Kuo, A.; Ohm, R.; Otillar, R.; Riley, R.; Salamov, A.; Zhao, X.; Korzeniewski, F.; et al. MycoCosm portal: Gearing up for 1000 fungal genomes. *Nucleic Acids Res.* **2014**, *42*, D699–D704. [CrossRef]
27. Yamamoto, K.; Miyake, H.; Kusunoki, M.; Osaki, S. Crystal structures of isomaltase from *Saccharomyces cerevisiae* and in complex with its competitive inhibitor maltose. *FEBS J.* **2010**, *277*, 4205–4214. [CrossRef]
28. Yamamoto, K.; Nakayama, A.; Yamamoto, Y.; Tabata, S. Val216 decides the substrate specificity of α-glucosidase in *Saccharomyces cerevisiae*. *Eur. J. Biochem.* **2004**, *271*, 3414–3420. [CrossRef]
29. Yamamoto, K.; Miyake, H.; Kusunoki, M.; Osaki, S. Steric hindrance by 2 amino acid residues determines the substrate specificity of isomaltase from *Saccharomyces cerevisiae*. *J. Biosci. Bioeng.* **2011**, *112*, 545–550. [CrossRef]
30. Schrödinger, L.L.C. The {PyMOL} Molecular Graphics System, Version~1.8.6.0. 2015. Available online: https://pymol.org/2/ (accessed on 19 May 2017).
31. Janeček, Š.; Svensson, B.; MacGregor, E.A. A remote but significant sequence homology between glycoside hydrolase clan GH-H and family GH31. *FEBS Lett.* **2007**, *581*, 1261–1268. [CrossRef]
32. Uitdehaag, J.C.M.; Mosi, R.; Kalk, K.H.; Van der Veen, B.A.; Dijkhuizen, L.; Withers, S.G.; Dijkstra, B.W. X-ray structures along the reaction pathway of cyclodextrin glycosyltransferase elucidate catalysis in the α-amylase family. *Nat. Struct. Biol.* **1999**, *6*, 432–436. [CrossRef]
33. Chi, Z.; Ni, X.; Yao, S. Cloning and overexpression of a maltase gene from *Schizosaccharomyces pombe* in *Escherichia coli* and characterization of the recombinant maltase. *Mycol. Res.* **2008**, *112*, 983–989. [CrossRef] [PubMed]

34. Tsujimoto, Y.; Tanaka, H.; Takemura, R.; Yokogawa, T.; Shimonaka, A.; Matsui, H.; Kashiwabara, S.I.; Watanabe, K.; Suzuki, Y. Molecular determinants of substrate recognition in thermostable α-glucosidases belonging to glycoside hydrolase family 13. *J. Biochem.* **2007**, *142*, 87–93. [CrossRef] [PubMed]
35. Ojima, T.; Saburi, W.; Yamamoto, T.; Kudo, T. Characterization of *Halomonas* sp. strain H11 α-glucosidase activated by monovalent cations and its application for efficient synthesis of α-D-glucosylglycerol. *Appl. Environ. Microbiol.* **2012**, *78*, 1836–1845. [CrossRef] [PubMed]
36. Hasegawa, S.; Takizawa, M.; Suyama, H.; Shintani, T.; Gomi, K. Characterization and expression analysis of a maltose-utilizing (*MAL*) cluster in *Aspergillus oryzae*. *Fungal Genet. Biol.* **2010**, *47*, 1–9. [CrossRef] [PubMed]
37. CBS Database. Available online: http://www.wi.knaw.nl/Collections/DefaultInfo.aspx?Page=Home (accessed on 27 November 2019).
38. Teste, M.A.; Marie François, J.; Parrou, J.L. Characterization of a new multigene family encoding isomaltases in the yeast *Saccharomyces cerevisiae*, the IMA family. *J. Biol. Chem.* **2010**, *285*, 26815–26824. [CrossRef]
39. Almagro Armenteros, J.J.; Tsirigos, K.D.; Sønderby, C.K.; Petersen, T.N.; Winther, O.; Brunak, S.; von Heijne, G.; Nielsen, H. SignalP 5.0 improves signal peptide predictions using deep neural networks. *Nat. Biotechnol.* **2019**, *37*, 420–423. [CrossRef]
40. Liiv, L.; Pärn, P.; Alamäe, T. Cloning of maltase gene from a methylotrophic yeast, *Hansenula polymorpha*. *Gene* **2001**, *265*, 77–85. [CrossRef]
41. Geber, A.; Williamson, P.R.; Rex, J.H.; Sweeney, E.C.; Bennett, J.E. Cloning and characterization of a *Candida albicans* maltase gene involved in sucrose utilization. *J. Bacteriol.* **1992**, *174*, 6992–6996. [CrossRef]
42. Alamäe, T.; Liiv, L. Glucose repression of maltase and methanol-oxidizing enzymes in the methylotrophic yeast *Hansenula polymorpha*: Isolation and study of regulatory mutants. *Folia Microbiol.* **1998**, *43*, 443–452. [CrossRef]
43. Marín, D.; Linde, D.; Lobato, M.F. Purification and biochemical characterization of an α-glucosidase from *Xanthophyllomyces dendrorhous*. *Yeast* **2006**, *23*, 117–125. [CrossRef]
44. da Silva, T.M.; Michelin, M.; de Lima Damásio, A.R.; Maller, A.; Almeida, F.B.D.R.; Ruller, R.; Ward, R.J.; Rosa, J.C.; Jorge, J.A.; Terenzi, H.F.; et al. Purification and biochemical characterization of a novel α-glucosidase from *Aspergillus niveus*. *Antonie Van Leeuwenhoek* **2009**, *96*, 569–578. [CrossRef] [PubMed]
45. Kita, A.; Matsui, H.; Somoto, A.; Kimura, A.; Takata, M.; Chiba, S. Substrate specificity and subsite affinities of crystalline α-glucosidase from *Aspergillus niger*. *Agric. Biol. Chem.* **1991**, *55*, 2327–2335. [CrossRef]
46. Okuyama, M.; Tanimoto, Y.; Ito, T.; Anzai, A.; Mori, H.; Kimura, A.; Matsui, H.; Chiba, S. Purification and characterization of the hyper-glycosylated extracellular α-glucosidase from *Schizosaccharomyces pombe*. *Enzym. Microb. Technol.* **2005**, *37*, 472–480. [CrossRef]
47. Ernits, K.; Viigand, K.; Visnapuu, T.; Põšnograjeva, K.; Alamäe, T. Thermostability measurement of an α-glucosidase using a classical activity-based assay and a novel Thermofluor method. *Bio-Protocol* **2017**, *7*. [CrossRef]
48. Naumov, G.I.; Naumova, E.S.; Michels, C.A. Genetic variation of the repeated MAL loci in natural populations of *Saccharomyces cerevisiae* and *Saccharomyces paradoxus*. *Genetics* **1994**, *136*, 803–812.
49. Reinders, A.; Ward, J.M. Functional characterization of the α-glucoside transporter Sut1p from *Schizosaccharomyces pombe*, the first fungal homologue of plant sucrose transporters. *Mol. Microbiol.* **2001**, *39*, 445–454. [CrossRef]
50. Dujon, B.A.; Louis, E.J. Genome diversity and evolution in the budding yeasts (Saccharomycotina). *Genetics* **2017**, *206*, 717–750. [CrossRef]
51. Kelly, C.T.; Moriarty, M.E.; Fogarty, W.M. Thermostable extracellular α-amylase and α-glucosidase of *Lipomyces starkeyi*. *Appl. Microbiol. Biotechnol.* **1985**, *22*, 352–358. [CrossRef]
52. Egeter, O.; Bruckner, R. Characterization of a genetic locus essential for maltose-maltotriose utilization in *Staphylococcus xylosus*. *J. Bacteriol.* **1995**, *177*, 2408–2415. [CrossRef]
53. Schönert, S.; Buder, T.; Dahl, M.K. Identification and enzymatic characterization of the maltose-inducible α-glucosidase MalL (sucrase-isomaltase-maltase) of *Bacillus subtilis*. *J. Bacteriol.* **1998**, *180*, 2574–2578. [CrossRef]
54. Rolfsmeier, M.; Blum, P. Purification and characterization of a maltase from the extremely thermophilic crenarchaeote *Sulfolobus solfataricus*. *J. Bacteriol.* **1995**, *177*, 482–485. [CrossRef] [PubMed]

55. Deng, X.; Petitjean, M.; Teste, M.A.; Kooli, W.; Tranier, S.; François, J.M.; Parrou, J.L. Similarities and differences in the biochemical and enzymological properties of the four isomaltases from *Saccharomyces cerevisiae*. *FEBS Open Bio.* **2014**, *4*, 200–212. [CrossRef] [PubMed]
56. McWethy, S.J.; Hartman, P.A. Extracellular maltase of *Bacillus brevis*. *Appl. Environ. Microbiol.* **1979**, *37*, 1096–1102. [CrossRef] [PubMed]
57. Kumara, H.M.; De Cort, S.; Verachtert, H. Localization and characterization of alpha-glucosidase activity in *Brettanomyces lambicus*. *Appl. Environ. Microbiol.* **1993**, *59*, 2352–2358. [CrossRef]
58. Wilson, W.A.; Roach, P.J.; Montero, M.; Baroja-Fernández, E.; Muñoz, F.J.; Eydallin, G.; Viale, A.M.; Pozueta-Romero, J. Regulation of glycogen metabolism in yeast and bacteria. *FEMS Microbiol. Rev.* **2010**, *34*, 952–985. [CrossRef]
59. Casa-Villegas, M.; Marín-Navarro, J.; Polaina, J. Synthesis of isomaltooligosaccharides by *Saccharomyces cerevisiae* cells expressing *Aspergillus niger* α-glucosidase. *ACS Omega* **2017**, *2*, 8062–8068. [CrossRef]
60. Kato, N.; Suyama, S.; Shirokane, M.; Kato, M.; Kobayashi, T.; Tsukagoshi, N. Novel α-glucosidase from *Aspergillus nidulans* with strong transglycosylation activity. *Appl. Environ. Microbiol.* **2002**, *68*, 1250–1256. [CrossRef]
61. Studier, F.W.; Moffatt, B.A. Use of bacteriophage T7 RNA polymerase to direct selective high-level expression of cloned genes. *J. Mol. Biol.* **1986**, *189*, 113–130. [CrossRef]
62. Curiel, J.A.; de Las Rivas, B.; Mancheño, J.M.; Muñoz, R. The pURI family of expression vectors: A versatile set of ligation independent cloning plasmids for producing recombinant His-fusion proteins. *Protein Expr. Purif.* **2011**, *76*, 44–53. [CrossRef]
63. Visnapuu, T.; Mardo, K.; Mosoarca, C.; Zamfir, A.D.; Vigants, A.; Alamäe, T. Levansucrases from *Pseudomonas syringae* pv. tomato and *P. chlororaphis* subsp. aurantiaca: Substrate specificity, polymerizing properties and usage of different acceptors for fructosylation. *J. Biotechnol.* **2011**, *155*, 338–349. [CrossRef]
64. Ernits, K.; Eek, P.; Lukk, T.; Visnapuu, T.; Alamäe, T. First crystal structure of an endo-levanase—The BT1760 from a human gut commensal *Bacteroides thetaiotaomicron*. *Sci. Rep.* **2019**, *9*, 8443. [CrossRef] [PubMed]
65. Mardo, K.; Visnapuu, T.; Gromkova, M.; Aasamets, A.; Viigand, K.; Vija, H.; Alamäe, T. High-throughput assay of levansucrase variants in search of feasible catalysts for the synthesis of fructooligosaccharides and levan. *Molecules* **2014**, *19*, 8434–8455. [CrossRef] [PubMed]
66. Mcilvaine, T.C. A buffer solution for colorimetric comparison. *J. Biol. Chem.* **1921**, *49*, 183–186.
67. Stingele, F.; Newell, J.W.; Neeser, J.R. Unraveling the function of glycosyltransferases in *Streptococcus thermophilus* Sfi6. *J. Bacteriol.* **1999**, *181*, 6354–6360. [CrossRef]
68. Jork, H.; Funk, W.; Fischer, W.; Wimmer, H. *Thin-Layer Chromatography: Reagents and Detection Methods*; Ebel, H.F., Ed.; VCH Verlagsgesellschaft mbH: Weinheim, Germany, 1990; Volume 1a.
69. Saitou, N.; Nei, M. The neighbor-joining method: A new method for reconstructing phylogenetic trees. *Mol. Biol. Evol.* **1987**, *4*, 406–425.
70. Kumar, S.; Stecher, G.; Li, M.; Knyaz, C.; Tamura, K. MEGA X: Molecular evolutionary genetics analysis across computing platforms. *Mol. Biol. Evol.* **2018**, *35*, 1547–1549. [CrossRef]
71. Tamura, K.; Nei, M.; Kumar, S. Prospects for inferring very large phylogenies by using the neighbor-joining method. *Proc. Natl. Acad. Sci. USA* **2004**, *101*, 11030–11035. [CrossRef]
72. Sievers, F.; Higgins, D.G. Clustal Omega for making accurate alignments of many protein sequences. *Protein Sci.* **2018**, *27*, 135–145. [CrossRef]

International Journal of
Molecular Sciences

Article

The Structure of Sucrose-Soaked Levansucrase Crystals from *Erwinia tasmaniensis* reveals a Binding Pocket for Levanbiose

Ivan Polsinelli [1], Rosanna Caliandro [1], Nicola Demitri [2] and Stefano Benini [1,*]

[1] Bioorganic Chemistry and Bio-Crystallography laboratory (B2Cl), Faculty of Science and Technology, Free University of Bolzano, Piazza Università 5, 39100 Bolzano, Italy; ivan.polsinelli@unibz.it (I.P.); rosanna.caliandro@unibz.it (R.C.)

[2] Elettra-Sincrotrone Trieste, S.S. 14 Km 163.5 in Area Science Park, Basovizza, 34149 Trieste, Italy; nicola.demitri@elettra.eu

* Correspondence: stefano.benini@unibz.it; Tel.: +39-0471-017128

Received: 31 October 2019; Accepted: 17 December 2019; Published: 20 December 2019

Abstract: Given its potential role in the synthesis of novel prebiotics and applications in the pharmaceutical industry, a strong interest has developed in the enzyme levansucrase (LSC, EC 2.4.1.10). LSC catalyzes both the hydrolysis of sucrose (or sucroselike substrates) and the transfructosylation of a wide range of acceptors. LSC from the Gram-negative bacterium *Erwinia tasmaniensis* (EtLSC) is an interesting biocatalyst due to its high-yield production of fructooligosaccharides (FOSs). In order to learn more about the process of chain elongation, we obtained the crystal structure of EtLSC in complex with levanbiose (LBS). LBS is an FOS intermediate formed during the synthesis of longer-chain FOSs and levan. Analysis of the LBS binding pocket revealed that its structure was conserved in several related species. The binding pocket discovered in this crystal structure is an ideal target for future mutagenesis studies in order to understand its biological relevance and to engineer LSCs into tailored products.

Keywords: glycoside hydrolase; GH68; fructosyltransferase; fructooligosaccharides; FOS biosynthesis; prebiotic oligosaccharides

---

1. Introduction

In the last decade, interest has grown towards levan/inulin oligosaccharides. They have a wide range of applications, from personal care to packaging. These molecules are especially relevant for their medical applications and prebiotic activity [1–5].

Levansucrases (LSCs, EC: 2.4.1.10) and inulosucrases (INUs, EC: 2.4.1.9) are major fructosyltransferases employed as biocatalysts in the synthesis of fructans and fructooligosaccharides (FOSs). Both are members of glycosyl hydrolase family 68 (GH68) [6]. LSC catalyzes the transfructosylation of the fructose component of sucrose by using a variety of acceptor molecules, forming β-(2,6)-linked oligofructans. When a water molecule acts as an acceptor, the reaction results in the hydrolysis of sucrose into glucose and fructose [1].

LSCs are used in the fermentative production of microbial oligosaccharides and polysaccharides due to their ability to interact with low-cost substitutes of sucrose, e.g., syrups and molasses [7,8].

The existence of a wide spectrum of nonconventional fructosyl acceptors explains the biotechnological interest in LSCs. These enzymes can interact with nonconventional fructosyl acceptors and donors [9], such as monosaccharides, disaccharides, and sucrose homologs. For example, LSCs from *Pseudomonas syringae* pv. tomato DC3000 and *Pseudomonas aurantiaca* are able to transfructosylate deoxy sugars or alditols such as fucose, ribose, sorbitol, and xylitol [10]. Among nonconventional substrates, lactose has been one of the most extensively studied. This is due to its combined role with sucrose in a reaction catalyzed by LSCs from *Bacillus spp.* (*B. methylotrophicus* SK21.002 [11], *B. subtilis* NCIMB 11871 [12], and *B. licheniformis* [13]) to produce lactosucrose, which is a trisaccharide with prebiotic activity [14].

Aromatic alcohols such as phenol derivatives (e.g., hydroquinone) [15] and isoflavones (e.g., puerarin) [16] can also be transfructosylated by LSCs from *B. subtilis* (SacB, BsLSC) and *Gluconacetobacter diazotrophicus* (GdLSC), respectively. The improved physical, chemical, and bioactive properties (solubility, stability, availability, and activity) of these glycosides make them relevant to pharmaceutical applications.

In the last decade, several studies have been carried out to understand which residues are the most relevant in the reaction mechanism [2]. Thanks to these studies, engineered glycosyltransferases can be used to obtain specific compounds such as FOSs (e.g., 6-nystose) instead of high-molecular weight (HMW) levan [17].

To better describe the relevant residues, the active site of LSC is commonly divided into layers. Moving from the sucrose binding site outwards, there are three layers: the first, second, and third. Mutations S173A, S173G, and S422A (first layer) in the LSC of *Bacillus megaterium* (BmLSC, PDB ID: 3OM7) increase transfructosylate activity by 194%, 53%, and 42%, respectively [18]. In *P. syringae* pv. tomato, LSC3 with the E146Q mutation (second layer) exhibits increased production of FOSs compared to wildtype [19]. SacB of *B. subtilis* with a Y429N mutation (second layer) has mostly hydrolytic activity and can produce short-chain FOSs instead of HMW levan [20].

Due to the high-yield production of FOSs, the product spectrum and well-optimized production of recombinant enzyme in *Escherichia coli*, LSCs from *Erwinia tasmaniensis* (EtLSC) [21] and *Erwinia amylovora* (EaLSC) [22,23] are interesting candidates for engineered LSCs that produce tailor-made fructans.

The structure of LSC is known in *B. subtilis*, *B. megaterium*, *E. amylovora*, *E. tasmaniensis*, and *G. diazotrophicus*. LSCs have similar structures, and their active sites possess common structural features [24], such as the triad of amino acids involved in catalysis (Asp46, Asp203, and Glu287 in EaLSC) [25]. LSC has been successfully crystallized in complex with sucrose (*B. subtilis*, Protein Data Bank (PDB) ID: 1PT2), raffinose (*B. subtilis*, PDB ID: 3BYN), fructose, and glucose (*E. amylovora*, PDB ID: 4D47). While the sucrose binding site is conserved, superficial areas and volumes vary across species due to variability in the surrounding loops [2,25].

In this report, we present the first known crystal structure of an LSC, EtLSC, in complex with levanbiose (LBS). LBS is an intermediate in the synthesis of oligolevans in the LSC enzyme. The complex was obtained by soaking EtLSC apo crystals (PDB ID: 6FRW) in a concentrated solution of sucrose (0.5 M) in order to trap reaction intermediates/products in the crystals. We describe an unexplored plausible binding site for LBS. We analyzed conserved amino acids in the binding pocket of LBS and compared their structural arrangement to other LSCs from Gram-positive and -negative bacteria. The aims of these analyses were to understand the biological relevance of the binding pocket and explore possible implications for LSC engineering.

2. Results and Discussion

The structure of LSC from *E. tasmaniensis* in complex with LBS was determined with a maximum resolution of 1.58 Å (space group $P\ 4_1 2_1 2$). Data collection and structure-refinement statistics are summarized in Table 1. Atomic coordinates and experimental-structure factors were deposited in the PDB with PDB ID: 6RV5.

Overall, the protein structure showed great similarity with apo EtLSC (PDB ID: 6FRW) [21] and its closest homolog EaLSC in complex with glucose and fructose (PDB ID: 4D47) [25].

Table 1. Data collection and refinement statistics.

	6RV5 *Erwinia tasmaniensis* levansucrase
Wavelength (Å)	1.000
Temperature (K)	100
Resolution range (Å)	45.21–1.58 (1.64–1.58)
Space group	$P\,4_1 2_1 2$
a, b, c (Å) and α, β, γ (°)	127.886, 127.886, 58.268; 90, 90, 90
Total reflections	552,442 (86,048)
Unique reflections	162,091 (25,949)
Multiplicity	3.4 (3.3)
Completeness (%)	99.47 (99.95)
Mean I/sigma(I)	20.50 (2.91)
Wilson B-factor (Å^2)	18.23
R-merge	0.02668 (0.2867)
R-meas	0.03773 (0.4055)
R-pim	0.02668 (0.2867)
CC1/2	0.999 (0.888)
Reflections used in refinement	66,162 (6515)
Reflections used for R-free	3263 (343)
R-work	0.1456 (0.2640)
R-free	0.1929 (0.2902)
CC (work)	0.966 (0.827)
CC (free)	0.951 (0.716)
Ligands atoms	97
Solvent molecules	431
Protein residues	412
RMS (bonds) (Å)	0.017
RMS (angles) (°)	2.02
Ramachandran favored (%)	96.34
Ramachandran allowed (%)	3.41
Ramachandran outliers (%)	0.24
Rotamer outliers (%)	1.87
Clashscore	5.97
Average B-factor (Å^2)	26.33
Macromolecules	23.87
Ligands	42.23
Solvent	41.99

Statistics for highest-resolution shell shown in parentheses.

Both EtLSC and EaLSC act via a distributive (nonprocessive) mechanism. This mechanism is known to produce low-molecular weight (LMW) levan and a mixture of FOSs [2]. For example, EaLSC and EtLSC mainly produce short-chain FOSs with 3–6 degrees of polymerization (DP) [21,22], while the main product of GdLSC is 1-kestose [24,26]. Furthermore, BmLSC produces FOSs with a DP ranging from 2 to 20 [27]. Even SacB can catalyze the formation of LMW levan under conditions that favor a nonprocessive mechanism [28].

The complex process of polymerization is the main factor that determines the wide range of products synthesized by LSC. Specifically, the DP is increased via a cycle of fructosyl capture, transfer, and release of fructosylated intermediates. These intermediates belong to one of two main types of FOS, n fructose units with glucose moiety (GF_n) or exclusively n fructose units (F_n). LBS, a fructose dimer, is an intermediate that seems to be produced in the late phase of the reaction. It was described as a secondary intermediate in the LSC from *B. subtilis* (SacB) [28].

Both types of FOS (GF_n and F_n) have been found in the product mixture from EtLSC, and the following species were identified: Levanbiose (F_2), levantriose (F_3), 6-kestose (GF_2), 6-nystose (GF_3), and 6,6,6-kestopentaose (GF_4) [21].

A comparison of the structure of EtLSC (this report, PDB ID: 6RV5) and EaLSC (PDB ID: 4D47) was obtained by applying the same soaking procedure with sucrose but with different soaking times, revealing that different products bind to different locations on the enzyme. Although there was similarity between the products of hydrolysis (glucose and fructose) trapped in EaLSC [25] and LBS trapped in EtLSC (PDB ID: 6RV5), these molecules did not bind in similar pockets. In fact, LBS was bound to an exposed pocket on the surface of EtLSC (Figure 1A) while the products of hydrolysis were located inside the active site of EaLSC (Figure 1B). The active site was within the inner part of the β-propeller, which is also found in other LSCs.

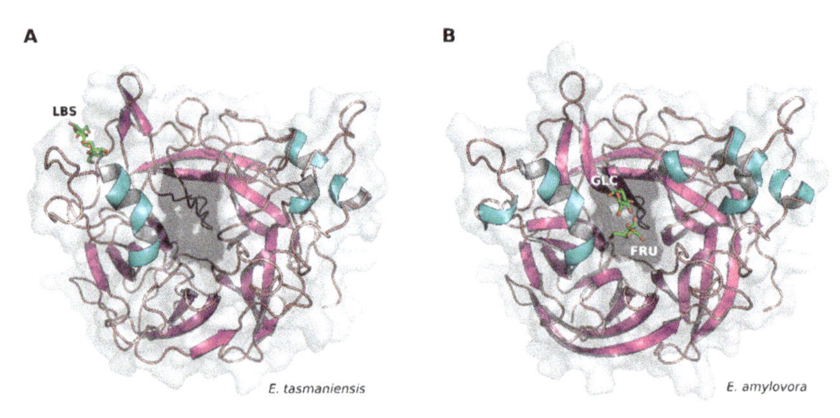

Figure 1. Comparison of ligand location in levansucrases (LSCs) from *Erwinia tasmaniensis* and *Erwinia amylovora*. (**A**) Cartoon representation of *E. tasmaniensis* LSC (EtLSC) structure with levanbiose (LBS) bound (PDB ID: 6RV5). Note: ligand molecule shown as green stick; active-site surface highlighted in black. (**B**) Cartoon representation of *E. amylovora* LSC (EaLSC) structure with fructose and glucose bound (PDB ID: 4D47). Note: ligands (hydrolysis products) shown as green sticks; active-site surface highlighted in black.

The complex of LBS with EtLSC was formed during the soaking of protein crystals with sucrose. It was located in a small pocket defined by residues Ala34, Phe35, Pro36, Val37, Arg73, Ile89, Trp371, Phe376, Arg377, and Ile378 (Figure 2A). LBS formed three hydrogen bonds with the Arg377 sidechain—LBS O1 interacted with Nη2, while LBS O2 and O3 interacted with Nε. LBS O4 formed another hydrogen bond with the N atom of the Ile378 backbone. Other residues in the pocket interacted with a nonpolar part of the molecule through hydrophobic bonds.

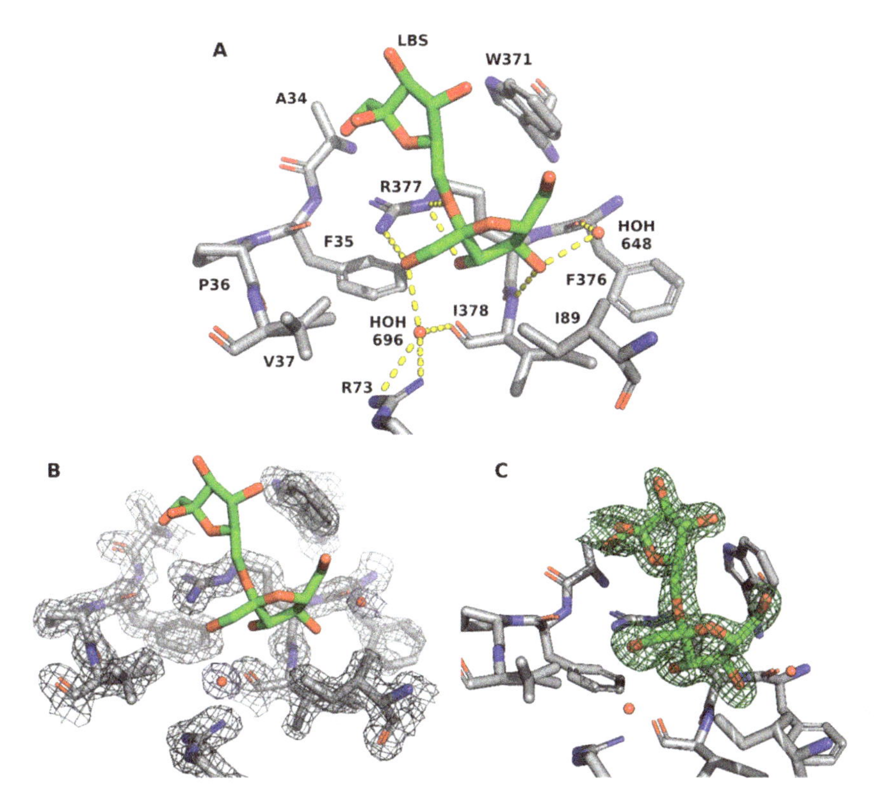

Figure 2. LBS binding pocket. (**A**) LBS interaction with EtLSC (PDB ID: 6RV5). Note: ligand molecule shown as green stick. (**B**) Representation of $2F_{obs}-F_{calc}$ electron density map of the LBS binding site. Note: electron-density map contoured at 1.5 σ. (**C**) Representation of polder map (omit map that excludes bulk solvent around omitted region) calculated with exclusion of LBS molecule. Note: polder map contoured at 4 σ.

Two water molecules were present in the pocket. The first water molecule (residue 648 in the PDB) formed hydrogen bonds with LBS O4 and the mainchain oxygen of Phe376. Backbone O of Ile378, Arg73 Nη1/Nη2, and LBS O1 formed bonds with a second water molecule (residue 696 in the PDB). After model building (see material and methods section) the residues and the waters of the binding pocket clearly fit the $2Fobs-Fcalc$ electron density map (Figure 2B) and the LBS perfectly fit the unbiased omit Polder map (Figure 2C)

LBS binding caused two noticeable movements in the sidechains of residues located in its binding pocket. This was remarkable when compared with the EtLSC apo structure (PDB ID: 6FRW) sharing the same crystallization conditions. The indole ring of the Trp371 sidechain tilted approximately 50°, and the benzyl moiety of Phe376 flipped 77.7°. Both residues moved toward the LBS moiety in the pocket (Figure 3). The movement of these two residues suggests that binding is mediated by one of the two fructose units in LBS.

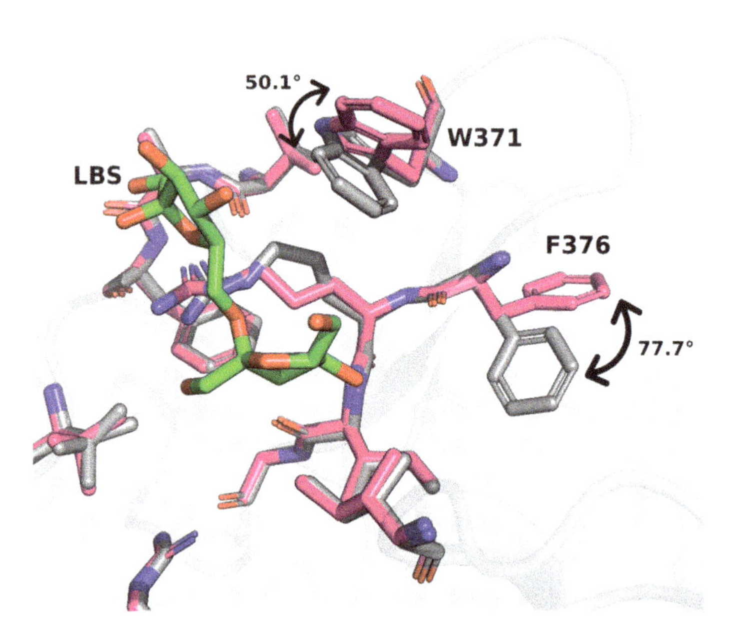

Figure 3. Residue movement upon LBS binding. Relevant conformational changes involved residues W371 and F376. Note: magenta, EtLSC structure with LBS; grey, apo enzyme (PDB ID: 6FRW).

Considering all the LSC studies to date, it can be concluded that LSCs from Gram-negative bacteria produce short-chain FOSs, while those from Gram-positive bacteria produce longer-chain oligosaccharides or levan (either LMW or HMW). Some structural features have been proven to be correlated with the length of the product, and are therefore different in Gram-negative and -positive bacteria [25]. For example, an arginine-to-histidine mutation at position 360 in SacB has been proposed as a switch in the production of either long- or short-chain FOSs [29]. Arginine is substituted by histidine in the LSCs of Gram-negative bacteria, including EaLSC and EtLSC, without losing the ability to perform the transfructosylation reaction.

It has previously been proposed that the loop formed by residues 366–380 contain residues that are able to shape the product spectrum of LSCs [25]. We compared the loop formed by residues 368–378 in EtLSC with other LSC structures available in the PDB (Figure 4A,B). From examining the available crystal structures, it is clear that the presence of LBS in the analyzed region could only be compatible with LSCs from Gram-negative bacteria. The loop conformation in structures of *E. amylovora* [25] and *G. diazotrophicus* [24] could allow for the presence of an LBS molecule (Figure 4A). In contrast, loop conformation is incompatible with LBS binding in the LSCs from Gram-positive bacteria *B. subtilis* [30] and *B. megaterium* [31] due to a ligand clash (Figure 4B). However, loop conformations could be influenced by crystal packing, and their variability could be limited. This may be explained by the involvement of loops in crystal contacts, as in the case of BmLSC. As a solution, loops could adopt slightly different conformations to allow for the required flexibility.

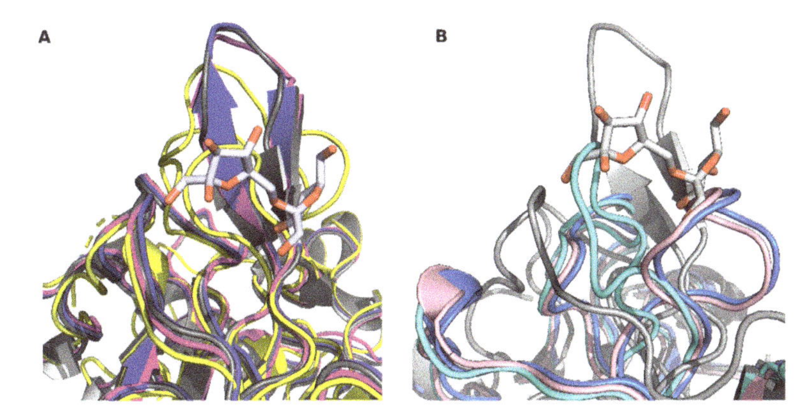

Figure 4. Comparison of LBS binding pocket in different LSC structures. (**A**) Proposed binding of LBS to structures with similar loop conformations. Note: grey and blue, *E. tasmaniensis* (PDB ID: 6RV5 and 6FRW respectively); magenta, *E. amylovora* (PDB ID: 4D47); and yellow, *Gluconacetobacter diazotrophicus* (PDB ID: 1W18). (**B**) Superimposition of LBS on structures with different loop conformations. Note: marine blue, *B. subtilis* (PDB ID: 1OYG); cyan and pink, *B. megaterium* (PDB ID: 3OM7 and 3OM2, respectively).

Nevertheless, conformations of loops surrounding the sucrose binding site [25], the presence of an LBS binding site, and other structural features may suggest a correlation with the synthesis of short-chain FOSs in Gram-negative bacteria. The chain length of fructans is determined by enzyme concentration and consequently by enzyme-product interactions [28]. Therefore, a stable LSC–LBS intermediate could favor the production of shorter-chain FOSs. The presence of LBS in the superficial pocket might also favor contacts between adjacent LSC molecules in the solution and increase enzyme density, thereby enhancing the probability of enzyme-product interactions in the active site. The ability to hold small oligofructans (e.g., LBS) in close vicinity to the active site of the enzyme may increase the likelihood that these molecules are used as fructose acceptors that therefore increase the production of small/medium DP oligosaccharides.

However, LSC structures currently available in the PDB belong to a restricted number of organisms, and further studies are required to gain a clear understanding of the differences between Gram-positive and -negative bacteria. Furthermore, the mechanism behind LBS migration towards or away from the active site, and its effect on the spectrum of generated products is still unclear.

3. Materials and Methods

The production of recombinant LSC from *E. tasmaniensis* (strain *Et1/99*) and its crystallization have been previously described [21]. In brief, the PCR-amplified gene was cloned into a pMCSG49 vector [32] and then expressed in *E. coli* BL21(DE3) star pLysS cells. Purified EtLSC was concentrated to 10 mg/mL (20 mM HEPES pH 7.5, 150 mM NaCl) and used for crystallization.

The crystals were grown by hanging drop vapor diffusion from drops consisting of 1 μL EtLSC solution and 1 μL precipitants (28% glycerol, 14% PEG4000, 2.5 mM manganese(II) chloride tetrahydrate, 2.5 mM cobalt(II) chloride hexahydrate, 2.5 mM nickel(II) chloride hexahydrate, and 2.5 mM zinc acetate dihydrate). The crystals were then soaked in a sucrose-containing solution corresponding to the crystallization drop to a final concentration of 0.5 M at different soaking times. Following this, crystals were flash-frozen in liquid N_2.

Diffraction data were collected on an XRD1 station of the synchrotron ELETTRA [33], Trieste, Italy (wavelength 1.000 Å, temperature 100 K, detector Pilatus 2 M), and processed with XDS [34]. Phase information was obtained by molecular replacement using the EtLSC apo structure (PDB ID: 6FRW) [21] as the input model for MOLREP [35]. The obtained starting model was iteratively refined with *Coot* [36], REFMAC5 [37], and PHENIX [38]. The quality of the model was assessed using MOLPROBITY [39]. The final refinement statistics for the structure are reported in Table 1.

The electron-density map and polder map [40] were calculated using REFMAC and PHENIX, respectively. Crystallographic figures were created using PyMOL (The PyMOL Molecular Graphics System, Version 2.20 Schroedinger, LLC).

4. Conclusions

In this study, we presented the crystal structure of EtLSC in complex with LBS. LBS, produced by EtLSC by soaking the crystals in a solution containing 0.5 M sucrose, binds to an unusual pocket that has not been previously reported. The pocket is a plausible site of interaction for LBS and fructose-like intermediates and could therefore be relevant to FOS production in EtLSC. This pocket contains residues conserved across *Erwinia spp.* and it may have a similar role in these bacteria.

Further studies (e.g., mutagenesis) are required to understand the possible relevance of the loop formed by residues 368–378, and the role of Arg377 in the determination of the product spectrum of EtLSC and EaLSC. This pocket could be a target for engineered LSCs with tuned specificity and/or increased yield of F_n and/or low-DP FOSs.

Author Contributions: Conceptualization, I.P., R.C., and S.B.; Methodology, I.P. and N.D.; Formal analysis, I.P., R.C., and S.B.; Funding acquisition, S.B.; Writing, all authors. All authors have read and agreed to the published version of the manuscript.

Funding: This research was funded by the Free University of Bolzano, project: MESCAL (grant number 1440).

Acknowledgments: We acknowledge Elettra Sincrotrone Trieste for providing access to the XRD1 beamline, to collect data under proposal 20180463.

Conflicts of Interest: The authors and Elettra Sincrotrone Trieste have no conflict of interest.

Abbreviations

LSC	Levansucrase
FOS	Fructooligosaccharide
LBS	Levanbiose
INU	Inulosucrase
GH68	Glycosyl hydrolase family 68
HMW	High-molecular weight
LMW	Low-molecular weight
DP	Degree of polymerization

References

1. Oner, E.T.; Hernandez, L.; Combie, J. Review of Levan Polysaccharide: From a Century of Past Experiences to Future Prospects. *Biotechnol. Adv.* **2016**, *34*, 827–844. [CrossRef] [PubMed]
2. Ortiz-Soto, M.E.; Porras-Domínguez, J.R.; Seibel, J.; López-Munguía, A. A Close Look at the Structural Features and Reaction Conditions that Modulate the Synthesis of Low and High Molecular Weight Fructans by Levansucrases. *Carbohydr. Polym.* **2019**, *219*, 130–142. [CrossRef]
3. Combie, J.; Öner, E.T. From Healing Wounds to Resorbable Electronics, Levan can Fill Bioadhesive Roles in Scores of Markets. *Bioinspir. Biomim.* **2018**, *14*. [CrossRef] [PubMed]
4. Xu, W.; Ni, D.; Zhang, W.; Guang, C.; Zhang, T.; Mu, W. Recent Advances in Levansucrase and Inulosucrase: Evolution, Characteristics, and Application. *Crit. Rev. Food Sci. Nutr.* **2018**, *59*, 3630–3647. [CrossRef] [PubMed]
5. González-Garcinuño, Á.; Tabernero, A.; Domínguez, Á.; Galán, M.A.; Martin del Valle, E.M. Levan and Levansucrases: Polymer, Enzyme, Micro-Organisms and Biomedical Applications. *Biocatal. Biotransfor.* **2018**, *36*, 233–244. [CrossRef]

6. Cantarel, B.L.; Coutinho, P.M.; Rancurel, C.; Bernard, T.; Lombard, V.; Henrissat, B. The Carbohydrate-Active EnZymes Database (CAZy): An Expert Resource for Glycogenomics. *Nucleic Acids Res.* **2009**, *37*, D233–D238. [CrossRef]
7. Özcan, E.; Öner, E.T. Microbial Production of Extracellular Polysaccharides from Biomass Sources. In *Polysaccharides*; Kishan, G.R., Jean-michel, M., Eds.; Springer International Publishing: Geneva, Switzerland, 2015; pp. 161–184.
8. Küçükaşik, F.; Kazak, H.; Güney, D.; Finore, I.; Poli, A.; Yenigün, O.; Nicolaus, B.; Öner, E.T. Molasses as Fermentation Substrate for Levan Production by Halomonas Sp. *Appl. Microbiol. Biotechnol.* **2011**, *89*, 1729–1740. [CrossRef]
9. Li, W.; Yu, S.; Zhang, T.; Jiang, B.; Mu, W. Recent Novel Applications of Levansucrases. *Appl. Microbiol. Biotechnol.* **2015**, *99*, 6959–6969. [CrossRef]
10. Visnapuu, T.; Mardo, K.; Mosoarca, C.; Zamfir, A.D.; Vigants, A.; Alamae, T. Levansucrases from Pseudomonas Syringae Pv. Tomato and P. Chlororaphis Subsp. Aurantiaca: Substrate Specificity, Polymerizing Properties and Usage of Different Acceptors for Fructosylation. *J. Biotechnol.* **2011**, *155*, 338–349. [CrossRef]
11. Wu, C.; Zhang, T.; Mu, W.; Miao, M.; Jiang, B. Biosynthesis of Lactosylfructoside by an Intracellular Levansucrase from Bacillus Methylotrophicus SK 21.002. *Carbohydr. Res.* **2015**, *401*, 122–126. [CrossRef]
12. Seibel, J.; Moraru, R.; Gotze, S.; Buchholz, K.; Na'amnieh, S.; Pawlowski, A.; Hecht, H.J. Synthesis of Sucrose Analogues and the Mechanism of Action of Bacillus Subtilis Fructosyltransferase (Levansucrase). *Carbohydr. Res.* **2006**, *341*, 2335–2349. [CrossRef] [PubMed]
13. Lu, L.; Fu, F.; Zhao, R.; Jin, L.; He, C.; Xu, L.; Xiao, M. A Recombinant Levansucrase from Bacillus Licheniformis 8-37-0-1 Catalyzes Versatile Transfructosylation Reactions. *Process Biochem.* **2014**, *49*, 1503–1510. [CrossRef]
14. Mu, W.; Chen, Q.; Wang, X.; Zhang, T.; Jiang, B. Current Studies on Physiological Functions and Biological Production of Lactosucrose. *Appl. Microbiol. Biotechnol.* **2013**, *97*, 7073–7080. [CrossRef] [PubMed]
15. Mena-Arizmendi, A.; Alderete, J.; Águila, S.; Marty, A.; Miranda-Molina, A.; López-Munguía, A.; Castillo, E. Enzymatic Fructosylation of Aromatic and Aliphatic Alcohols by Bacillus Subtilis Levansucrase: Reactivity of Acceptors. *J. Mol. Catal. B Enzym.* **2011**, *70*, 41–48. [CrossRef]
16. Núñez-López, G.; Herrera-González, A.; Hernández, L.; Amaya-Delgado, L.; Sandoval, G.; Gschaedler, A.; Arrizon, J.; Remaud-Simeon, M.; Morel, S. Fructosylation of Phenolic Compounds by Levansucrase from Gluconacetobacter Diazotrophicus. *Enzym. Microb. Technol.* **2019**, *122*, 19–25. [CrossRef] [PubMed]
17. Possiel, C.; Ortiz-Soto, M.; Ertl, J.; Münch, A.; Vogel, A.; Schmiedel, R.; Seibel, J. Exploring the Sequence Variability of Polymerization-Involved Residues in the Production of Levan- and Inulin-Type Fructooligosaccharides with a Levansucrase. *Sci. Rep.* **2019**, *9*, 7720. [CrossRef]
18. Ortiz-Soto, M.E.; Possiel, C.; Görl, J.; Vogel, A.; Schmiedel, R.; Seibel, J. Impaired Coordination of Nucleophile and Increased Hydrophobicity in the 1 Subsite Shift Levansucrase Activity Towards Transfructosylation. *Glycobiology* **2017**, *27*, 755–765. [CrossRef]
19. Mardo, K.; Visnapuu, T.; Gromkova, M.; Aasamets, A.; Viigand, K.; Vija, H.; Alamae, T. High-Throughput Assay of Levansucrase Variants in Search of Feasible Catalysts for the Synthesis of Fructooligosaccharides and Levan. *Molecules* **2014**, *19*, 8434–8455. [CrossRef]
20. Ortiz-Soto, M.E.; Rivera, M.; Rudiño-Piñera, E.; Olvera, C.; López-Munguía, A. Selected Mutations in Bacillus Subtilis Levansucrase Semi-Conserved Regions Affecting its Biochemical Properties. *Protein Eng. Des. Sel.* **2008**, *21*, 589–595. [CrossRef]
21. Polsinelli, I.; Caliandro, R.; Salomone-Stagni, M.; Demitri, N.; Rejzek, M.; Field, R.A.; Benini, S. Comparison of the Levansucrase from the Epiphyte Erwinia Tasmaniensis Vs its Homologue from the Phytopathogen Erwinia Amylovora. *Int. J. Biol. Macromol.* **2019**, *127*, 496–501. [CrossRef]
22. Caputi, L.; Nepogodiev, S.A.; Malnoy, M.; Rejzek, M.; Field, R.A.; Benini, S. Biomolecular Characterization of the Levansucrase of Erwinia Amylovora, a Promising Biocatalyst for the Synthesis of Fructooligosaccharides. *J. Agric. Food Chem.* **2013**, *61*, 12265–12273. [CrossRef] [PubMed]
23. Caputi, L.; Cianci, M.; Benini, S. Cloning, Expression, Purification, Crystallization and Preliminary X-Ray Analysis of EaLSC, a Levansucrase from Erwinia Amylovora. *Acta Crystallogr. Sect. F Struct. Biol. Cryst. Commun.* **2013**, *69*, 570–573. [CrossRef] [PubMed]
24. Martinez-Fleites, C.; Ortiz-Lombardia, M.; Pons, T.; Tarbouriech, N.; Taylor, E.J.; Arrieta, J.G.; Hernandez, L.; Davies, G.J. Crystal Structure of Levansucrase from the Gram-Negative Bacterium Gluconacetobacter Diazotrophicus. *Biochem. J.* **2005**, *390*, 19–27. [CrossRef] [PubMed]

25. Wuerges, J.; Caputi, L.; Cianci, M.; Boivin, S.; Meijers, R.; Benini, S. The Crystal Structure of Erwinia Amylovora Levansucrase Provides a Snapshot of the Products of Sucrose Hydrolysis Trapped into the Active Site. *J. Struct. Biol.* **2015**, *191*, 290–298. [CrossRef] [PubMed]
26. Hernandez, L.; Arrieta, J.; Menendez, C.; Vazquez, R.; Coego, A.; Suarez, V.; Selman, G.; Petit-Glatron, M.F.; Chambert, R. Isolation and Enzymic Properties of Levansucrase Secreted by Acetobacter Diazotrophicus SRT4, a Bacterium Associated with Sugar Cane. *Biochem. J.* **1995**, *309 (Pt 1)*, 113–118. [CrossRef]
27. Homann, A.; Biedendieck, R.; Goetze, S.; Jahn, D.; Seibel, J. Insights into Polymer Versus Oligosaccharide Synthesis: Mutagenesis and Mechanistic Studies of a Novel Levansucrase from Bacillus Megaterium. *Biochem. J.* **2007**, *407*, 189–198. [CrossRef]
28. Raga-Carbajal, E.; López-Munguía, A.; Alvarez, L.; Olvera, C. Understanding the Transfer Reaction Network Behind the Non-Processive Synthesis of Low Molecular Weight Levan Catalyzed by Bacillus Subtilis Levansucrase. *Sci. Rep.* **2018**, *8*, 15035. [CrossRef]
29. Chambert, R.; Petit-Glatron, M.F. Polymerase and Hydrolase Activities of Bacillus Subtilis Levansucrase can be Separately Modulated by Site-Directed Mutagenesis. *Biochem. J.* **1991**, *279*, 35–41. [CrossRef]
30. Meng, G.; Fütterer, K. Structural Framework of Fructosyl Transfer in Bacillus Subtilis Levansucrase. *Nat. Struct. Biol.* **2003**, *10*, 935–941. [CrossRef]
31. Strube, C.P.; Homann, A.; Gamer, M.; Jahn, D.; Seibel, J.; Heinz, D.W. Polysaccharide Synthesis of the Levansucrase SacB from Bacillus Megaterium is Controlled by Distinct Surface Motifs. *J. Biol. Chem.* **2011**, *286*, 17593–17600. [CrossRef]
32. Eschenfeldt, W.H.; Lucy, S.; Millard, C.S.; Joachimiak, A.; Mark, I.D. A Family of LIC Vectors for High-Throughput Cloning and Purification of Proteins. *Methods Mol. Biol.* **2009**, *498*, 105–115. [PubMed]
33. Lausi, A.; Polentarutti, M.; Onesti, S.; Plaisier, J.R.; Busetto, E.; Bais, G.; Barba, L.; Cassetta, A.; Campi, G.; Lamba, D.; et al. Status of the Crystallography Beamlines at Elettra. *Eur. Phys. J. Plus* **2015**, *130*, 43. [CrossRef]
34. Kabsch, W. XDS Acta Crystallogr. *D Biol. Crystallogr.* **2010**, *66*, 125–132. [CrossRef] [PubMed]
35. Vagin, A.; Teplyakov, A. Molecular Replacement with MOLREP. *Acta Crystallogr. D Biol. Crystallogr.* **2010**, *66*, 22–25. [CrossRef] [PubMed]
36. Emsley, P.; Lohkamp, B.; Scott, W.G.; Cowtan, K. Features and Development of Coot. *Acta Crystallogr. D Biol. Crystallogr.* **2010**, *66*, 486–501. [CrossRef]
37. Murshudov, G.N.; Skubak, P.; Lebedev, A.A.; Pannu, N.S.; Steiner, R.A.; Nicholls, R.A.; Winn, M.D.; Long, F.; Vagin, A.A. REFMAC5 for the Refinement of Macromolecular Crystal Structures. *Acta Crystallogr. D Biol. Crystallogr.* **2011**, *67*, 355–367. [CrossRef]
38. Adams, P.D.; Afonine, P.V.; Bunkoczi, G.; Chen, V.B.; Davis, I.W.; Echols, N.; Headd, J.J.; Hung, L.W.; Kapral, G.J.; Grosse-Kunstleve, R.W.; et al. PHENIX: A Comprehensive Python-Based System for Macromolecular Structure Solution. *Acta Crystallogr. D Biol. Crystallogr.* **2010**, *66*, 213–221. [CrossRef]
39. Williams, C.J.; Headd, J.J.; Moriarty, N.W.; Prisant, M.G.; Videau, L.L.; Deis, L.N.; Verma, V.; Keedy, D.A.; Hintze, B.J.; Chen, V.B. MolProbity: More and Better Reference Data for Improved All-atom Structure Validation. *Protein Sci.* **2018**, *27*, 293–315. [CrossRef]
40. Liebschner, D.; Afonine, P.V.; Moriarty, N.W.; Poon, B.K.; Sobolev, O.V.; Terwilliger, T.C.; Adams, P.D. Polder maps: Improving OMIT maps by excluding bulk solvent. *Acta Crystallogr. D Struct. Biol.* **2017**, *73*, 148–157. [CrossRef]

International Journal of
Molecular Sciences

Article

Cloning and Partial Characterization of an Endo-α-(1→6)-D-Mannanase Gene from *Bacillus circulans*

Shiva kumar Angala [1,*,†], Wei Li [1,†], Zuzana Palčeková [1], Lu Zou [2], Todd L. Lowary [2], Michael R. McNeil [1] and Mary Jackson [1,*]

[1] Mycobacteria Research Laboratories, Department of Microbiology, Immunology and Pathology, Colorado State University, Fort Collins, CO 80523-1682, USA; weili@colostate.edu (W.L.); Zuzana.Svetlikova@colostate.edu (Z.P.); M.Mcneil@ColoState.EDU (M.R.M.)

[2] Department of Chemistry, The University of Alberta, Edmonton, AB T6G 2G2, Canada; Lu.Zou@gilead.com (L.Z.); tlowary@ualberta.ca (T.L.L.)

* Correspondence: shivaka@colostate.edu (S.K.A.); Mary.Jackson@colostate.edu (M.J.); Tel.: +1-970-491-4067 (S.K.A.); +1-970-491-3582 (M.J.); Fax: +1-970-491-1815 (S.K.A.); +1-970-491-1815 (M.J.)

† These authors contributed equally to this work.

Received: 22 November 2019; Accepted: 6 December 2019; Published: 11 December 2019

Abstract: Mycobacteria produce two major lipoglycans, lipomannan (LM) and lipoarabinomannan (LAM), whose broad array of biological activities are tightly related to the fine details of their structure. However, the heterogeneity of these molecules in terms of internal and terminal covalent modifications and complex internal branching patterns represent significant obstacles to their structural characterization. Previously, an endo-α-(1→6)-D-mannanase from *Bacillus circulans* proved useful in cleaving the mannan backbone of LM and LAM, allowing the reducing end of these molecules to be identified as Man*p*-(1→6) [Man*p*-(1→2)]-Ino. Although first reported 45 years ago, no easily accessible form of this enzyme was available to the research community, a fact that may in part be explained by a lack of knowledge of its complete gene sequence. Here, we report on the successful cloning of the complete endo-α-(1→6)-D-mannanase gene from *Bacillus circulans* TN-31, herein referred to as *emn*. We further report on the successful production and purification of the glycosyl hydrolase domain of this enzyme and its use to gain further insight into its substrate specificity using synthetic mannoside acceptors as well as LM and phosphatidyl-*myo*-inositol mannoside precursors purified from mycobacteria.

Keywords: endo-α-(1→6)-D-mannase; mannoside; *Mycobacterium*; lipomannan; lipoarabinomannan; phosphatidylinositol mannosides

---

1. Introduction

Two complex α-(1→6)-linked D-mannose-containing lipoglycans, lipomannan (LM) and lipoarabinomannan (LAM) populate the cell envelope of mycobacteria [1]. They are essential to the integrity of the cell envelope [2] and play important roles in the immunopathogenesis of mycobacterial infections [1,3,4]. LM and LAM are believed to share a mannosylated phosphatidyl-*myo*-inositol lipid anchor and a mannan backbone composed of α-(1→6)-linked-D-mannopyranosyl residues occasionally branched with α-(1→2)-D-mannopyranosyl residues [1]. Despite recent advances in the biosynthetic pathway of these lipoglycans, the fine details of the structure of their mannan backbone, specifically, the precise distribution of the α-(1→2)-D-mannopyranosyl residues along the mannan backbone, is not known with certainty. To help in these analyses, two types of glycosyl hydrolases capable of hydrolyzing the mannan backbone of LM and LAM have traditionally been used: An exo-type α-(1→2,3,6)-mannosidase (EC # 3.2.1.24), which cleaves the terminal mannopyranosyl residues from the non-reducing end [5] and an endo-α-(1→6)-D-mannanase (EC # 3.2.1.101), which cleaves the glycosidic bond between two internal mannopyranosyl residues [6]. While the first enzyme can be purified from Jack bean and is commercially available, to the best of our knowledge, no commercial source of the second enzyme is available. The endomannanase hydrolyzing α-(1→6)-D-mannopyranosyl residues was discovered in 1974 by Ballou and co-workers [7] who then went on to purify this enzyme from the soil bacterium, *Bacillus circulans* [8], and subsequently shared this enzyme with other laboratories [6,9]. Some 24 years later, the gene encoding this enzyme, *aman6*, was identified by reverse genetics [10] but found to encode a protein only about half the size of that purified by Ballou et al., even though the recombinant product of the gene apparently displayed the expected catalytic activity on an (undefined) α-(1→6)-D-mannan substrate releasing 6α-mannotriose and 6α-mannobiose [10] [Figure S1]. Efforts by our laboratory to re-clone this gene by PCR based on its published sequence consistently failed, suggestive of potential rearrangements in the *Sau*3AI-digested chromosomal DNA libraries used in the isolation of *aman6* [10]. As a part of our continued effort to decipher the biosynthetic pathway of LM and LAM and establish the fine details of their structure, here, we report on the cloning of the full-size endo-α-(1→6)-D-mannanase gene (which we named *emn*) and its use to recombinantly express and purify the glycosyl hydrolase domain of this enzyme. The cleavage properties of the glycosyl hydrolase domain of Emn on synthetic mannoside acceptors, purified LM from *Mycobacterium smegmatis* and biosynthetic precursors, phosphatidyl-*myo*-inositol di- and hexa-mannosides, were investigated using mass spectrometry.

2. Results and Discussion

2.1. Cloning of the Full-Size Endo-α-(1→6)-D-Mannanase Gene from Bacillus circulans TN-31

Using reverse genetics, Maruyama et al. identified *aman6* as the gene encoding the endo-α-(1→6)-D-mannanase from *B. circulans* TN-31 by screening a *Sau*3A1-digested genomic DNA library of this bacterium. The *aman6* gene was reported to encode a 589-amino acid-long protein with a 36 amino-acid signal peptide (GenBank accession number AB024331) [10]. Attempts to PCR-amplify *aman6* in our laboratory using primers to the reported nucleotide sequence were unsuccessful. The nucleotide BLAST analysis of *aman6* yielded a gene, *gymC10_1685* from *Paenibacillus sp.* Y412MC10, with 88% DNA sequence identity to *aman6* on the first 1722-bp but a substantially larger coding sequence totaling 3249-bp instead of 1767 bp for *aman6*. Using primers specific to the 5′-end of *aman6* and to the downstream region of *gymC10_1685*, a 3522 bp-long PCR product was amplified from *B. circulans TN-31 genomic DNA*. The fragment encompasses a 3255 bp-long open reading frame encoding a 1084 amino acid protein with a 33 amino acid signal peptide (Figure S1). The molecular weight of the mature protein (131 kDa) matches that of the endo-α-(1→6)-D-mannanase purified by Ballou and collaborators from the same bacterium [8]. We named the full-size endomannanase gene *emn*. The mature Emn protein consists of a glycosyl hydrolase family 76 (GH-76) catalytic domain from amino acids 36 to 360 followed by three consecutive family six carbohydrate-binding modules from residues 392 to 522, 534 to 657 and 670 to 793, and a domain of unknown function (DUF4959) from residues 811 to 902. The GH-76 CaZy family of glycosyl hydrolases consists of α-(1→6)-mannanases (EC # 3.2.1.101) and α-glucosidases (EC # 3.2.1.20), whereas family six carbohydrate-binding modules are non-catalytic domains thought to facilitate the binding of the catalytic GH domain to its substrate(s).

2.2. Expression and Purification of the Glycosyl Hydrolase Domain of Emn

The production of the mature Emn protein from a pET14b expression plasmid in *E. coli* BL21 pLysS proved to be toxic to the cells and failed to yield any detectable protein product. We were, however, able to produce, using the same expression system, a 43 kDa N-terminally His_6-tagged recombinant protein encompassing the glycosyl hydrolase domain of Emn (hereafter referred to as GH-emn) corresponding to residues 36 to 400 of the full-size enzyme (Figure S1) and to purify it to near-homogeneity using a combination of nickel affinity chromatography and gel filtration (Figure S2). The purified GH-emn protein was stored at –80 °C in 20 mM Tris pH 7.5 buffer containing 150 mM NaCl and 20% glycerol for over 24 months without any appreciable loss of activity.

2.3. Digestion of Synthetic Mannoside Substrates

The purified GH domain of Emn was then tested for glycosyl hydrolase activity using two synthetic unbranched linear substrates consisting of octyl trimannoside (α-D-Man*p*-(1→6)-α-D-Man-(1→6)-α-D-Man-(1→octyl)) and octyl pentamannoside (α-D-Man*p*-(1→6)-α-D-Man-(1→6)-α-D-Man-(1→6)-α-D-Man-(1→6)-α-D-Man-(1→octyl)). These substrates are structurally similar to those originally used by Ballou and co-workers [8] to study the substrate specificity of the purified, native, full-size enzyme. Those used by Ballou were unsubstituted at the reducing end, whereas those we evaluated harbor an octyl aglycon at their reducing end to facilitate the identification of the hydrolytic products of the reaction and thus, exactly where the enzyme cleaves. Enzyme assays were carried out as described under Materials and Methods in liquid chromatography—mass spectrometry (LC/MS) sample vials at 50 °C for 16 h [8]—after which the reaction mixture was directly injected into the LC/MS instrument without further purification. Figure 1A–B shows the base peak chromatograms of octyl trimannoside and its GH-emn digestion products. The mass spectrum of the only major peak found for the substrate (peak (i)) showed a major [M–H] ion at *m/z* 615.29 as well as an ion at *m/z* 675.30 [M+HAc–H] and an ion at *m/z* 661.29 [M+HCOOH–H], all corresponding to the parent molecule, Manp_3-(1→octyl) (Figure 1C). In contrast, the base peak chromatogram of the enzymatic products revealed the loss of peak (i) with the concomitant appearance of two new peaks, (ii) and (iii). The mass spectrum of peak (ii) showed a strong ion at *m/z* 377.08 [M+Cl^-] corresponding to Manp_2 (Figure 1D), whereas peak (iii) yielded a signal at *m/z* 351.20 [M+HAc–H] corresponding to Man*p*-(1→octyl) (Figure 1E). These data show that the enzyme cleaves the trisaccharide at glycosidic bond of the middle Man*p* residue (labeled as M2 in Figure 1F) unit rather than the glycosidic bond of the Man*p* residue at either end of the trimannoside unit (labeled as M1 and M3 in Figure 1F).

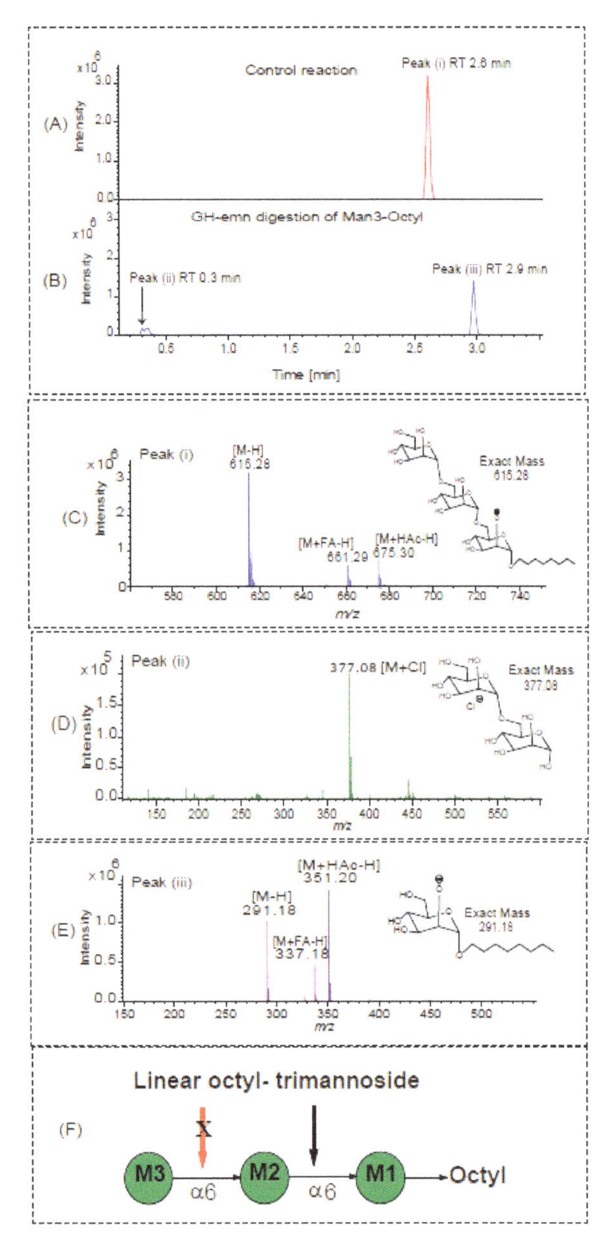

Figure 1. LC/MS analysis of the GH-emn-digested synthetic octyl trimannoside. (**A**,**B**) Base peak chromatograms showing the elution profiles of the substrate (**A**) and products of the enzymatic reaction (**B**). (**C**) Mass spectrum of the undigested synthetic octyl trimannoside at *m/z* 615.28 [M-H] as well as formate and acetate adducts. (**D**,**E**) Mass spectra of the products of the reaction: mannobiose (peak (ii)) (**D**) with a free reducing end at *m/z* 377.08 [M+Cl$^-$] and Manp_1-octyl (peak (iii)) (**E**) at *m/z* 291.18 [M-H], 337.18 (M+formate) and 351.2 (M+acetate). (**F**) Cartoon showing the location of the mannosyl cleavage (thicker arrow) of the octyl trimannoside at residue M2. The red arrow indicates the absence of cleavage by the enzyme.

A similar hydrolytic pattern was obtained upon digestion of the octyl pentamannoside by GH-emn (Figure 2). In general, Man$p_{1\text{-}4}$ as free reducing components were found as [M+Cl$^-$] ions and Man$p_{1,\,2,\,3,}$ and $_5$-(1→octyl) as [M+HAc–H] ions. The extracted ion chromatograms for these ions were integrated to obtain areas used to calculate the data presented in Figure 2A,B. The amounts of these ions varied with the enzyme concentration used (Figure 2A,B). Importantly, no Manp_4-(1→octyl) was detected with any amount of enzyme, as predicted from the data of Manp_3-(1→octyl) (Figure 1). Complete digestion occurred at the two highest enzyme concentrations (50 and 100 µg/mL) yielding only Manp_2-(1→octyl) and Manp_1-(1→octyl) in a ratio of approximately 2:1 and the free mannosides Manp_2 and Manp_1 in a 0.9:1 ratio. Traces of Manp_3-(1→octyl) (Figure 2A) along with oligomannosides Manp_4 and Manp_3 (Figure 2B) were only detected when less enzyme was used in the assay, suggesting that these products are further cleaved to form Manp_1-(1→octyl), and free Manp_2 and Manp_1. Collectively, these results indicate that GH-emn cleaves the octyl pentamannoside substrate at three different positions, as summarized in Figure 2C. At all three cleavage positions, the enzyme binds to three mannosyl residues and cuts between the middle Man and the reducing end Man. Consistently with that binding, the enzyme cannot further process Manp_2-(1→octyl) or Man$_2$, in agreement with previous studies from Dr. Ballou's laboratory [8].

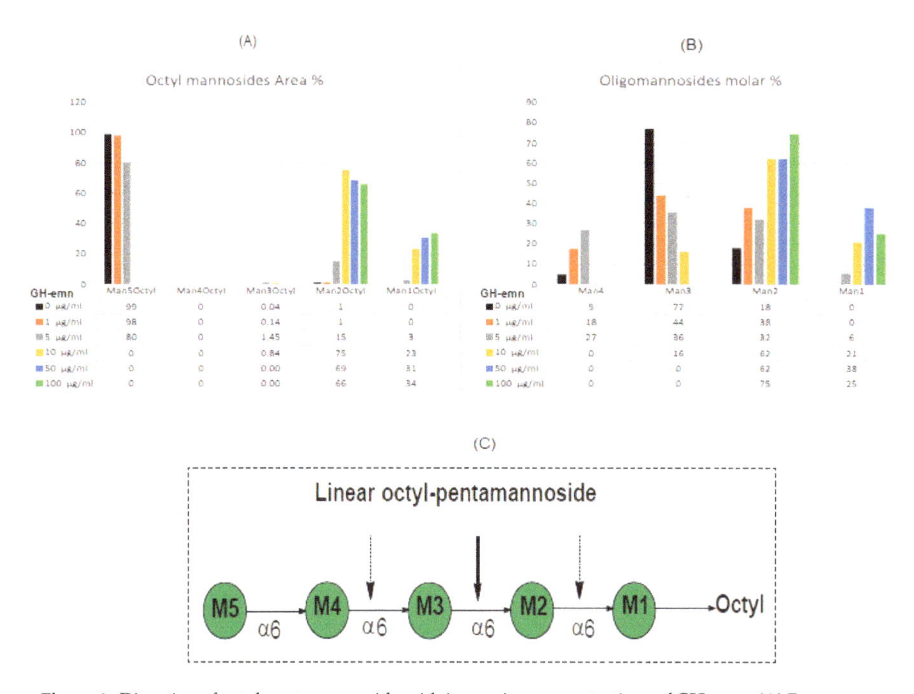

Figure 2. Digestion of octyl pentamannoside with increasing concentrations of GH-emn. (**A**) Percentage area of products containing an octyl chain. (**B**) Molar percentage of oligomannosides with a free reducing end. (**C**) Cartoon showing the locations of the three possible mannosyl cleavage sites. The thicker arrow denotes the dominant cleavage site having three mannosyl residues at the non-reducing end and the dotted arrows indicate other, less favored cleavage sites. The further cleavage of the released Manp_4 and Manp_3 is not shown.

2.4. Digestion of Phosphatidyl-Myo-Inositol-Mannosides

The reducing end of LM and LAM consists of phosphatidyl-*myo*-inositol where the *myo*-inositol residue is mannosylated at positions C-2 and C-6. This lipid anchor may be esterified with up to four acyl chains [1]. Phosphatidyl-*myo*-inositol mannosides, also known as PIMs, may also exist as free glycolipids populating the inner and outer membranes of mycobacteria [11]. Tri- and tetra-acylated phosphatidyl-*myo*-inositol dimannosides (Ac_1PIM_2 and Ac_2PIM_2) and tri- and tetra-acylated phosphatidyl-*myo*-inositol hexamannosides (Ac_1PIM_6 and Ac_2PIM_6) are the most abundant form of PIMs found in the mycobacterial cell envelope [11]. As expected from the results above, GH-emn showed no activity on deacylated PIM_2 (d-PIM_2) which lacks α-(1→6)-linked mannosyl residues (Figure S3).

Since earlier work has established that purified, native, Emn lacks activity on yeast mannan in which the backbone is substituted with Man*p*-α-(1→3)- and α-(1→2)-linked residues [7,8], we sought to determine whether GH-emn could cleave the pentamannoside extending from the six-position of *myo*-inositol in deacylated Ac_2PIM_6 (d-PIM_6). In this molecule, the first three mannosyl residues attached to *O*-6 of the inositol are linked via α-(1→6) glycosidic linkages and the last two mannosyl residues are present as the dimannoside α-Man*p*-(1→2)-α-Man*p* attached to position 2 of the third Man*p* from the inositol residue (Figure 3). PIM_6, as received from BEI, is a mixture containing both PIM_2 (shown above to not be a substrate for GH-emn) and PIM_6 in equal amounts. Hence, we needed to focus our analysis on the released free mannosides rather than the inositol-containing components. GH-emn activity on intact d-PIM_6 was analyzed by LC/MS in the negative ion mode. Upon enzymatic treatment, the amount of intact d-PIM_6 (MS and structure shown in Figure 3D) decreased dramatically (Figure 3A,B). Concomitantly, the chromatogram and mass spectrum of the enzyme-treated sample were dominated by a peak at *m/z* 701.19 [$M+Cl^-$] corresponding to the tetramannoside, D-Man*p*-(1→2)-α-D-Man*p*-(1→2)-α-D-Man*p*-(1→6)-α-D-Man*p*-(1→6) (Figure 3C,E). No ions corresponding to the mono-, di-, tri- or penta- mannosides were observed in the enzyme-treated sample. These results indicate that the GH-emn can tolerate an α-(1→2)-linked dimannoside substitution at the non-reducing terminus (M4 in Figure 3F) of the α-(1→6 trimannoside component (M4, M3, and M2) of d-PIM_6.

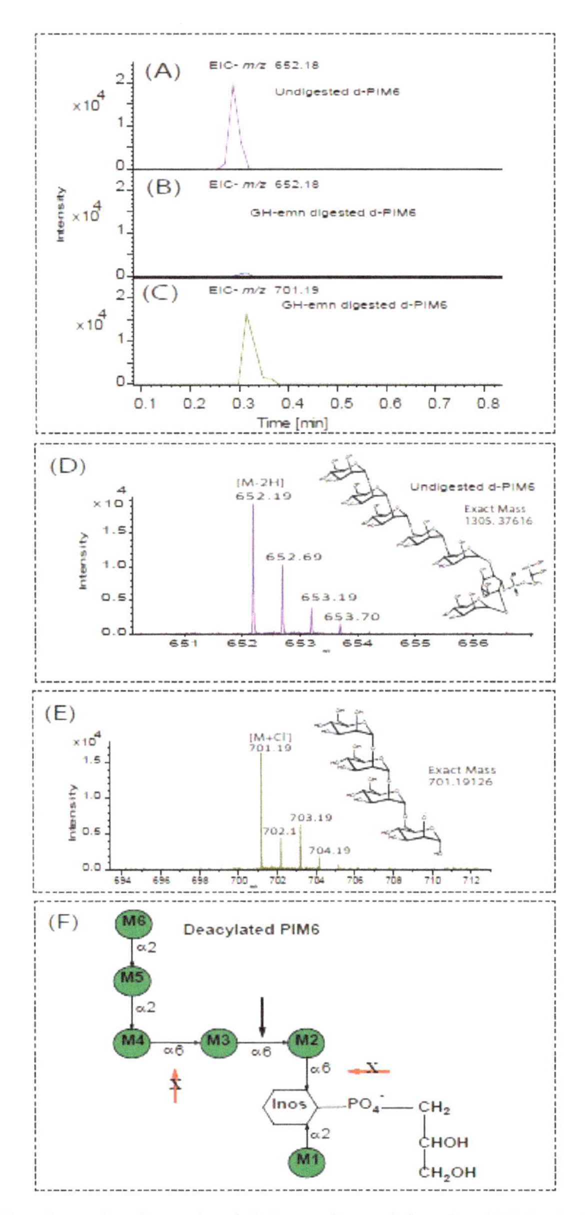

Figure 3. LC/MS analysis of undigested and GM-emn-digested deacylated PIM_6. (**A**,**D**) Extracted Ion Chromatogram (**A**), structure (**D**), and mass spectrum (**D**) of d-PIM_6 (*m/z* 652.18 [M–2H]) before enzyme treatment. (B) Extracted ion chromatograms (EIC) of d-PIM_6 (*m/z* 652.18 [M–2H]) after digestion with GH-emn. The EIC of *m/z* 701.19 [M+Cl^-] shown in panel (**C**) corresponds to the dominant tetramannoside product of d-PIM_6 after digestion with GH-emn. Its mass spectrum is shown in panel (**E**). (**F**) Cartoon showing the location of the mannosyl cleavage site of GH-emn to form Man_4. The red arrows indicate the absence of cleavage by the enzyme.

2.5. Analysis of Endomannanase-Digested LAM and LM

Next, we tested the endomannanase activity of GM-emn on the acylated forms of LM and LAM from *M. smegmatis*. Complete digestion of these substrates after an overnight incubation with GH-emn was evident from the analysis of the products of the reaction by SDS-PAGE followed by silver staining (Figure 4). Enzymatic digestion of LM and LAM further resulted in a low molecular weight product migrating slightly faster than acylated PIM_6 (Figure 4, lanes 3, 5 and 6), which we tentatively attribute to di- or tri-mannosylated forms of phosphatidyl-*myo*-inositol released from the reducing end of both lipoglycans.

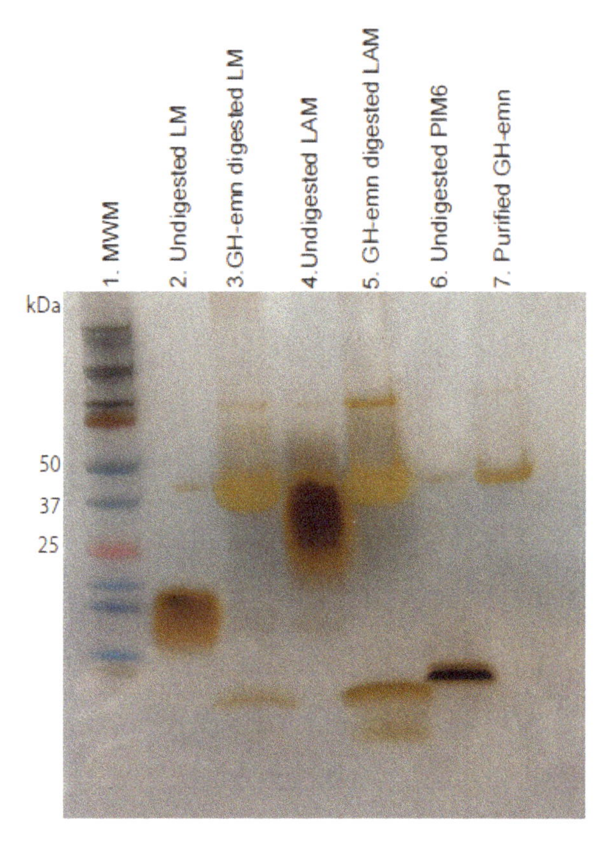

Figure 4. Silver-stained SDS PAGE showing the GH-emn digestion products of native *M. smegmatis* LM and LAM. MWM: Molecular weight marker.

To gain further insight into the nature of the fragments released by GH-emn from deacylated *M. smegmatis* LM (d-LM), the products of the reaction were also analyzed by LC/MS. The LC/MS profile of undigested d-LM was dominated by doubly, triply and quadruply charged, long-chain, high-molecular weight (range of 2,700 to 6,800 Da) mannans containing 15 to 40 mannosyl residues (Figure 5A). The most abundant components appeared as triply charged series [M–3H] of ions at *m/z* 1406.77, 1460.79, 1514.81, 1568.82, 1622.84, 1676.86 and 1730.88 corresponding to LM molecules with 24 to 34 mannosyl residues. The less abundant quadruply-charged series [M–4H] corresponded to mannans with 27 to 40 mannosyl residues, while the least abundant doubly-charged series contained 24 to 33 mannosyl residues. Together, these results confirm that the mannan backbone of *M. smegmatis* LM is composed of up to 40 mannosyl residues. GH-emn digestion of this heterogeneous population of d-LM resulted in two major ions at *m/z* 657.16 and 819.21, corresponding to the mass of d-PIM_2 and d-PIM_3 released from the reducing end (Figure 5B). Additionally, the presence of oligomannans (with free reducing ends) was deduced from the presence of singly, doubly and triply charged series of ions matching the mass of Manp_2 and higher oligomers with odd numbers of mannosyl residues from Manp_3 to Manp_{27} (Figure 5B). These oligomannans result from the presence of *O*-2 mannosyl substitutions at various positions of the mannan backbone of LM, preventing GH-emn from cutting. Their detailed structure awaits further characterization.

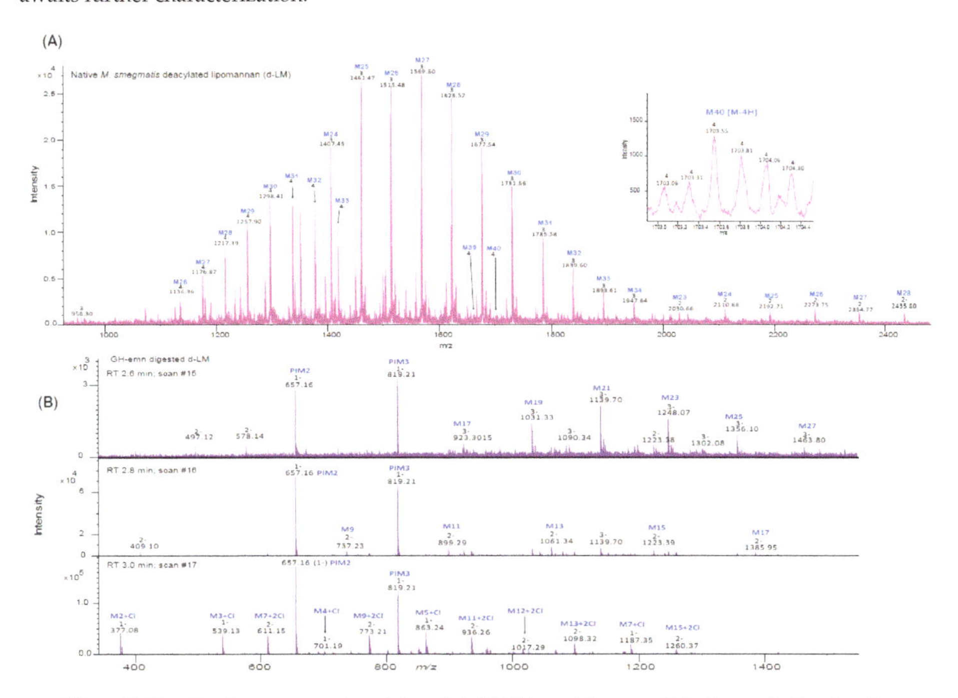

Figure 5. Negative ion mass spectra of deacylated LM from *M. smegmatis* before and after digestion with GH-emn. (**A**) Mass spectrum of undigested d-LM. The spectrum shows the mass distribution of the mannan backbone of d-LM which contains up to 40 mannosyl residues (differing by the mass of one hexose). Inset: The isotopic pattern for quadruply-charged d-LM with 40 mannosyl residues indicates the sensitivity of the mass spectrometer for high-molecular weight compounds. (**B**) Mass spectrum of d-LM after digestion by GH-emn. The mass spectra for three different retention times (RT) are shown. At all retention times, the spectra are dominated by singly-charged PIM_2 and PIM_3. RT 2.6 and 2.8 min show the presence of ions for Man17 to Man27 [M-3H] and Man9 to Man17 [M-2H]. RT 3.0 min shows low-molecular weight oligomannosides (Man_2 to Man_7) as singly charged and doubly charged chlorine adducts.

3. Materials and Methods

3.1. Cloning, Expression and Purification the Glycosyl Hydrolase Domain of Emn

B. circulans TN-31 (ATCC®290101™) was obtained from the American Type Culture Collection and grown on agar medium at 30 °C as recommended by the ATCC. Genomic DNA was isolated using a standard extraction protocol. The coding sequence of the full-length mature protein (3150 bp) was amplified using primers Emn1 (5′-CTCGAGTATACCGCATCAGATGGGG-3′) and Emn2 (5′-CTCGAGTTACTCCAAGCCATCCTGCC-3′). The glycosyl hydrolase domain sequence of *emn* (GH-emn; 1092 bp) was PCR-amplified using primers GH-emn1 (5′-CATATGTATACCGCATCAGAT GGGG-3′) and GH-emn2 (5′-CTCGAGTTAATTGTAGCGTTCCGCTTCGA-3′). Both PCR fragments were cloned into the *Nde*/ and *Xho*I sites of the pET14b expression plasmid and transformed into *E. coli* BL21 DE3 pLysS. For GH-emn production, freshly prepared LB broth containing 100 μg/mL ampicillin was inoculated with a preculture of pET14b-GH-emn transformant and incubated at 30 °C until optical density measured at 600 nm (OD_{600}) reached 0.2. Gene expression was induced by adding isopropyl-β-thio-galactopyranoside (1 mM final concentration) to the culture medium. After 4 h of induction at 30 °C, cells were collected by centrifugation and resuspended in lysis buffer consisting of Tris-HCl pH 7.5 containing 150 mM NaCl, 1 mM $MgCl_2$, 1 μg/mL DNase and protease inhibitor cocktail (Sigma, St. Louis, MO, USA). Cells were broken using a French press (1500 psi; 3 passages) and cell debris were removed by ultracentrifugation at 4 °C for 30 min at 110,000 × g. The supernatant was then applied to a His-trap column (GE Healthcare, Pittsburgh, PA, USA) and the protein eluted in a gradient of 10–500 mM imidazole, prior to gel filtration using a Superdex 200 column (GE Healthcare). Aliquots of the purified GH-emn protein were stored at –80 °C in 20 mM Tris pH 7.5 buffer containing 150 mM NaCl and 20% glycerol until further use. Protein concentrations were determined using the Pierce BCA protein assay kit as recommended by the manufacturer.

3.2. Assays Using Synthetic Mannosides Substrates

α-D-Man*p*-(1→6)-α-D-Man-(1→6)-α-D-Man-(1→octyl] (octyl trimannoside-) and α-D-Man*p*-(1→6) -α-D-Man-(1→6)-α-D-Man-(1→6)-α-D-Man-(1→6)-α-D-Man-(1→octyl (octyl pentamannoside) were chemically synthesized as described [12,13]. Endomannanase reaction mixtures contained the synthetic mannosides (100 μM) and purified GH-emn (0 to 500 μg/mL) in a total volume of 100 μL in 10 mM ammonium acetate buffer (pH 6). Control reactions contained buffer in place of the purified enzyme. Enzymatic reactions were carried out in LC/MS vials at 50 °C for 16 h at which point the enzyme was inactivated by heating at 65 °C for 30 min [8]. In the experiment where increasing concentrations of GH-emn were used, the reactions mixture were incubated at 50 °C for 2 h instead of 16 h.

3.3. Assays Using Purified Mycobacterial PIMs and Lipoglycans

Purified *M. tuberculosis* H37Rv phosphatidyl-*myo*-inositol hexamannosides (PIM_6) and *M. smegmatis* LAM were received from BEI resources. LM was extracted and purified from *M. smegmatis* whole cells as described previously [14]. Purified PIM_6 (20 μg) was deacylated using monomethylamine following an established procedure [15] and further digested with GH-emn (0.25 mg/mL) in 10 mM ammonium acetate buffer (pH 6) for 24 h at 50 °C. Native LM from *M. smegmatis* (50 μg) was deacylated with 0.2 N NaOH (200 μL) at 37 °C and then neutralized with 10% aqueous glacial acetic acid. Deacylated LM (d-LM) was separated from sodium salts using an Amicon ultra-0.5 mL centrifugal filter (3 kDa MWCO) and resuspended in 10 mM ammonium acetate buffer (pH 6) prior to digestion with GH-emn as described for d-PIM_6. D-LM and d-PIM_6 before and after enzymatic digestion were analyzed by LC/MS (see next section).

Native LAM (40 μg) and LM (40 μg) from *M. smegmatis* digested with GH-emn (50 μg) in 100 mM citrate phosphate buffer pH 6.6 for 16 h at 50 °C were also was analyzed by SDS-PAGE followed by silver staining alongside 10 μg of undigested starting material.

3.4. Analysis of Substrates and Reaction Products by Liquid Chromatography—Mass Spectrometry

Separation of the GH-emn-digested products was performed using a reverse-phase X-bridge C18 column (50 mm × 2.1 mm; 1.7 μM) on a Waters ACQUITY UPLC system. The mobile phases used were water (solvent A), acetonitrile (solvent B) and 0.5 M ammonium acetate (solvent C) under the following gradient conditions: 0–0.3 min (10% B), 0.3–3 min (70% B), 3–4.8 min (98% B), and 4.8–6.8 min (10% B) at a flow rate of 400 μL/min. A constant 2% solvent C was maintained throughout the LC run.

Mass spectrometry (MS) was performed using a Bruker maXis plus II high-resolution quadrupole time-of-flight (Q-TOF). The electrospray ionization (ESI) source settings were as follows: end plate offset voltage 500 V, capillary voltage 3500 V, nebulizer gas pressure 3.0 bar, dry gas flow rate 10 L/min. Two different tune parameters were used for analyzing native and GH-emn-digested products. For detecting low molecular compounds, the tune parameters were: funnel RF 300 Vpp, multipole RF 300 Vpp, ion energy 3.0 eV, low mass range for ion transmission *m/z* 100, collision energy 8.0 eV, collision RF 450 Vpp, pre-pulse storage 5.0 μs, ion cooler RF 800 Vpp and transfer time 80.0 μs. For detecting high molecular weight, d-LM tune parameters were: funnel RF 400 Vpp, multipole RF 400 Vpp, ion energy 3.0 eV, low mass range for ion transmission *m/z* 600, collision energy 8.0 eV, collision RF 2500 Vpp, pre-pulse storage 5.0 μs, ion cooler RF 800 Vpp and transfer time 140 μs. A data analysis was performed using Bruker compass data analysis 4.4 SR1.

4. Conclusions

The full-length endo-α-(1→6)-D-mannanase gene from *B. circulans* TN-31 was successfully cloned for the first time and its glycosyl hydrolase domain expressed and purified in active and stable form from *E. coli*. The analysis of the digestion pattern of synthetic mannosides, PIM and LM provided further insights into the substrate specificity of the endomannanase domain of the enzyme. The availability of this enzyme, together with a better understanding of its catalytic activity, should greatly facilitate the structural analysis of α-(1→6)-D-mannan-containing polysaccharides, including mycobacterial LM and LAM.

Supplementary Materials: Supplementary materials can be found at http://www.mdpi.com/1422-0067/20/24/6244/s1.

Author Contributions: Funding acquisition, M.J.; Conceptualization, M.J., M.R.M.; Investigation, S.K.A., W.L., Z.P., L.Z.; Data analysis, S.K.A., W.L., and M.R.M.; Writing—S.K.A., W.L., M.J., M.R.M.; Writing—reviewing & editing, M.J., M.R.M., and T.L.L.

Acknowledgments: This work was supported by the National Institute of Allergy and Infectious Diseases (NIAID)/National Institutes of Health (NIH) grant AI064798 (to MJ), and the Alberta Glycomics Centre. The content is solely the responsibility of the authors and does not necessarily represent the official views of the NIH. We thank Claudia M. Boot and Zachery C. Moen from CSU's Central Instrumentation Facility for their assistance with LC/MS analyses. The following reagents were obtained through BEI Resources, NIAID, NIH: *M. tuberculosis*, Strain H37Rv, Purified Phophatidylinositol Mannoside 6 (PIM_6) NR-14847 (lot 70009918), and *M. smegmatis*, lipoarabinomannan (LAM), NR-14849 (lot 61699475).

Conflicts of Interest: The authors declare no conflict of interest.

References

1. Angala, S.K.; Belardinelli, J.M.; Huc-Claustre, E.; Wheat, W.H.; Jackson, M. The Cell Envelope Glycoconjugates of Mycobacterium Tuberculosis. *Crit. Rev. Biochem. Mol. Biol.* **2014**, *49*, 361–399. [CrossRef] [PubMed]
2. Fukuda, T.; Matsumura, T.; Ato, M.; Hamasaki, M.; Nishiuchi, Y.; Murakami, Y.; Maeda, Y.; Yoshimori, T.; Matsumoto, S.; Kobayashi, K.; et al. Critical Roles for Lipomannan and Lipoarabinomannan in Cell Wall Integrity of Mycobacteria and Pathogenesis of Tuberculosis. *MBio* **2013**, *4*, e00472-12. [CrossRef] [PubMed]
3. Vergne, I.; Gilleron, M.; Nigou, J. Manipulation of the Endocytic Pathway and Phagocyte Functions by Mycobacterium Tuberculosis Lipoarabinomannan. *Front. Cell. Infect. Microbiol.* **2014**, *4*, 187. [CrossRef] [PubMed]
4. Ishikawa, E.; Mori, D.; Yamasaki, S. Recognition of Mycobacterial Lipids by Immune Receptors. *Trends Immunol.* **2017**, *38*, 66–76. [CrossRef] [PubMed]
5. Li, Y.T. Presence of Alpha-D-Mannosidic Linkage in Glycoproteins. Liberation of D-Mannose from Various Glycoproteins by Alpha-Mannosidase Isolated from Jack Bean Meal. *J. Biol. Chem.* **1966**, *241*, 1010–1012. [PubMed]
6. Chatterjee, D.; Hunter, S.W.; McNeil, M.; Brennan, P.J. Lipoarabinomannan. Multiglycosylated Form of the Mycobacterial Mannosylphosphatidylinositols. *J. Biol. Chem.* **1992**, *267*, 6228–6233. [PubMed]
7. Nakajima, T.; Ballou, C.E. Structure of the Linkage Region between the Polysaccharide and Protein Parts of Saccharomyces Cerevisiae Mannan. *J. Biol. Chem.* **1974**, *249*, 7685–7694. [PubMed]
8. Nakajima, T.; Maitra, S.K.; Ballou, C.E. An Endo-alpha1 leads to 6-D-Mannanase from a Soil Bacterium. Purification, Properties, and Mode of Action. *J. Biol. Chem.* **1976**, *251*, 174–181. [PubMed]

9. Hunter, S.W.; Brennan, P.J. Evidence for the Presence of a Phosphatidylinositol Anchor on the Lipoarabinomannan and Lipomannan of Mycobacterium Tuberculosis. *J. Biol. Chem.* **1990**, *265*, 9272–9279. [PubMed]
10. Maruyama, Y.; Nakajima, T. The Aman6 Gene Encoding a Yeast Mannan Backbone Degrading 1,6-Alpha-D-Mannanase in Bacillus Circulans: Cloning, Sequence Analysis, and Expression. *Biosci. Biotechnol. Biochem.* **2000**, *64*, 2018–2020. [CrossRef] [PubMed]
11. Guerin, M.E.; Kordulakova, J.; Alzari, P.M.; Brennan, P.J.; Jackson, M. Molecular Basis of Phosphatidyl-Myo-Inositol Mannoside Biosynthesis and Regulation in Mycobacteria. *J. Biol. Chem.* **2010**, *285*, 33577–33583. [CrossRef] [PubMed]
12. Tam, P.H.; Besra, G.S.; Lowary, T.L. Exploring the Substrate Specificity of a Mycobacterial Polyprenol Monophosphomannose-Dependent Alpha-(1–>6)-Mannosyltransferase. *ChemBioChem* **2008**, *9*, 267–278. [CrossRef] [PubMed]
13. Angala, S.K.; McNeil, M.R.; Zou, L.; Liav, A.; Zhang, J.; Lowary, T.L.; Jackson, M. Identification of a Novel Mycobacterial Arabinosyltransferase Activity Which Adds an Arabinosyl Residue to Alpha-D-Mannosyl Residues. *ACS Chem. Biol.* **2016**, *11*, 1518–1524. [CrossRef] [PubMed]
14. Kaur, D.; Obregon-Henao, A.; Pham, H.; Chatterjee, D.; Brennan, P.J.; Jackson, M. Lipoarabinomannan of Mycobacterium: Mannose Capping by a Multifunctional Terminal Mannosyltransferase. *Proc. Natl. Acad. Sci. USA* **2008**, *105*, 17973–17977. [CrossRef] [PubMed]
15. Brearley, C.A.; Hanke, D.E. Assaying Inositol Phospholipid Turnover in Plant Cells. *Signal. Inosit. A Pract. Approach* **1997**, *176*, 1–32.

International Journal of
Molecular Sciences

Article

Crystal Structure of a GH3 β-Glucosidase from the Thermophilic Fungus *Chaetomium thermophilum*

Imran Mohsin [1], Nirmal Poudel [1], Duo-Chuan Li [2,*] and Anastassios C. Papageorgiou [1,*]

1 Turku Bioscience, University of Turku and Åbo Akademi University, 20520 Turku, Finland; dr.mohsinimran@hotmail.com (I.M.); nirpou@utu.fi (N.P.)
2 Department of Mycology, Shandong Agricultural University, Taian 271018, China
* Correspondence: lidc20@sdau.edu.cn (D.-C.L.); anapap@utu.fi (A.C.P.); Tel.: +358-29-4503752 (A.C.P.)

Received: 29 October 2019; Accepted: 25 November 2019; Published: 27 November 2019

Abstract: Beta-glucosidases (β-glucosidases) have attracted considerable attention in recent years for use in various biotechnological applications. They are also essential enzymes for lignocellulose degradation in biofuel production. However, cost-effective biomass conversion requires the use of highly efficient enzymes. Thus, the search for new enzymes as better alternatives of the currently available enzyme preparations is highly important. Thermophilic fungi are nowadays considered as a promising source of enzymes with improved stability. Here, the crystal structure of a family GH3 β-glucosidase from the thermophilic fungus *Chaetomium thermophilum* (*Ct*BGL) was determined at a resolution of 2.99 Å. The structure showed the three-domain architecture found in other β-glucosidases with variations in loops and linker regions. The active site catalytic residues in *Ct*BGL were identified as Asp287 (nucleophile) and Glu517 (acid/base). Structural comparison of *Ct*BGL with Protein Data Bank (PDB)-deposited structures revealed variations among glycosylated Asn residues. The enzyme displayed moderate glycosylation compared to other GH3 family β-glucosidases with similar structure. A new glycosylation site at position Asn504 was identified in *Ct*BGL. Moreover, comparison with respect to several thermostability parameters suggested that glycosylation and charged residues involved in electrostatic interactions may contribute to the stability of the enzyme at elevated temperatures. The reported *Ct*BGL structure provides additional insights into the family GH3 enzymes and could offer new ideas for further improvements in β-glucosidases for more efficient use in biotechnological applications regarding cellulose degradation.

Keywords: glycoside hydrolase; cellulose degradation; thermophilic fungus; β-glucosidases; *Chaetomium thermophilum*; protein structure; fungal enzymes

1. Introduction

Beta-glucosidases (β-glucosidases) are key cellulolytic enzymes that catalyze the hydrolysis of cellobiose to glucose. They are terminal enzymes of the cellulase system as they act during the final step of the cellulose degradation. Although beta-glucosidases do not act directly on cellulose, their activity is important to circumvent the inhibitory effect of cellobiose on endoglucanases and cellobiohydrolases. Consequently, inclusion of β-glucosidases in cellulase preparations can synergistically increase the hydrolytic efficiency of other cellulolytic enzymes [1].

Beta-glucosidases (β-glucoside glucohydrolases, EC 3.2.1.21; BGL) have been found in the glycoside hydrolase (GH) families GH1, GH3, GH5, GH9, GH30, and GH116 of the CAZy database [2]. They are anomeric configuration-retaining enzymes that operate through the canonical double-displacement glycosidase mechanism, except those of the GH9 family.

Despite their extensive use in biotechnological application, structural data on fungal β-glucosidases are still scarce. Many fungal BGLs are classified as part of the GH3 family, one of the largest CAZy families with over 13,600 annotated protein sequences. Two fungal β-glucosidase structures currently available are produced by the mesophilic fungus *Aspergilus aculeatus* [3] and the moderate thermophilic *Rasamsonia emersonii* [4], respectively. A third structure produced by the mesophilic fungus *Trichoderma reesei* of BGL1 has been deposited in the Protein Data Bank (PDB), but no paper has been published yet. Recently, the structure of the filamentous fungus *Neurospora crassa* Cel3A (*Nc*Cel3A) was reported at 2.25 Å resolution [5], as well as structures of GH3 β-glucosidases from *Hypocrea jecorina* (a teleomorph of *Trichoderma reesei*) [6], *Aspergilus niger* [7], *Aspergillus fumigatus*, and *Aspergillus oryzae* [8].

The first structure of a β-glucosidase, that of barley (*Hordeum vulgare*) exo-1,3-1,4-glucanase ExoI (*Hv*ExoI) [9] showed a two-domain architecture: An N-terminal (α/β)8 TIM-barrel that hosts the catalytic nucleophile aspartate and an $(\alpha/\beta)_6$ sandwich domain where the acid/base glutamate resides. Subsequent studies, however, revealed variations in the modular architecture of GH3 β-glucosidases with the first domain to adopt a collapsed/incomplete triosephosphate isomerase (TIM)-barrel domain and the presence of a third domain with a fibronectin FnIII-like fold. The function of the third domain is still unknown, although it has been suggested to stabilize the incomplete TIM-barrel domain [4]. Notably, a fourth domain has also been identified in *Kluyveromyces marxianus* and *Pseudoalteromonas* sp. BB1 β-glucosidases [10,11].

Thermophilic fungi have attracted considerable attention in recent years as an alternative reservoir of thermostable cellulases for cellulose degradation [12,13]. Importantly, thermophilic fungi can produce thermostable enzymes that can be used at temperatures up to 70 °C, whereas enzymes from mesophilic organisms are typically active up to 50 °C. The importance of using thermophilic cellulases in cellulose degradation stems from the fact that, at higher temperatures, cellulose swells and becomes more susceptible to breaking. Various thermophilic fungi have been studied in recent years and their β-glucosidases have been characterized [12]. Understanding the structure–function–stability relationships in fungal β-glucosidases is therefore important for finding new and better alternatives for industrial biocatalysts. Hydrolysis rate, inhibitors, and stability are considered critical factors for the efficient use of β-glucosidases in complex biomass hydrolysis [14]. Moreover, β-glucosidases have been suggested for use in the synthesis of various glycoconjugates, and a GH3 β-glucosidase from the thermophilic fungus *Myceliophthora thermophila* has been found to act as an efficient biocatalyst in alkyl glycoside synthesis [15].

Here, we report the crystal structure of a β-glucosidase from the thermophilic filamentous fungus *Chaetomium thermophilum* (*Ct*BGL) at a resolution of 2.99 Å, and compare it with other β-glucosidases of the GH3 family. This is the second structure of a β-glucosidase produced by a thermophilic fungus, and therefore, it is expected to provide further insights into the structure–function relationship of this family of enzymes in high-temperature settings.

2. Results

2.1. Quality of the Structure

The crystal structure of *Ct*BGL was refined using data up to 2.99 Å resolution. The final R_{cryst} and R_{free} (5% of the reflections excluded from refinement) were 0.201 and 0.252, respectively (Table 1). The structure exhibits good stereochemistry despite the limited resolution with root mean square deviation (rmsd) in bond lengths and bond angles of 0.007 Å and 0.94°, respectively. The Ramachandran plot shows 92.8% of the residues in the most favorable regions and 1.0% in disallowed regions. Residues that fall into the disallowed regions belong to flexible parts of the structure with weak electron density. Structure quality statistics for *Ct*BGL fall within the distribution found in other crystal structures of similar resolution as analyzed and displayed by POLYGON [15]. The refined structure contains 836 residues of the mature protein, starting at residue Trp32. The full-length cDNA encodes for a 867-residue enzyme and the first 16 residues have been suggested to act as a secretion signal peptide [16]. Residues 17–31 at the N-terminal were not modeled, owing to lack of adequate electron density, possibly due to high flexibility. The enzyme crystallized with one molecule in the crystallographic asymmetric unit in contrast to other β-glucosidases that crystallize either with two molecules such as the β-glucosidases from *Aspergillus oryzae* (*Ao*βG) and *A. fumigatus* (*Af*βG) [8], *A. aculeatus* (*Aa*βG) [8] and *Nc*Cel3A [5], or with four molecules in the asymmetric unit such as *Re*Cel3A [4]. The solvent content is unusually high (~77%) for a protein crystal, which could explain the low resolution of the data and the disintegration of the crystals during manipulation prior to mounting. The NCS dimer found in other β-glucosidases is seen in *Ct*BGL as a crystallographic dimer that involves the −Y, −X, −Z+ $\frac{1}{2}$ symmetry-related molecule.

Table 1. Data collection and refinement statistics.

Data Collection	
Beamline	EMBL-DESY X13
Wavelength (Å)	0.8123
Data Processing	
Space group	$P4_12_12$
Unit cell dimensions a, b, c (Å)	121.9, 121.9, 264.9
No. of molecules/asymmetric units	1
Mosaicity (°)	0.070
Resolution range (Å)	20.0–2.99 (3.11–2.99) *
Total measurements/unique reflections	222508 (22974)/41051 (4513)
Completeness (%)	99.3 (98.6)
$\langle I/\sigma(I)\rangle$	7.8 (1.4)
R_{meas}	0.287 (1.452)
R_{pim}	0.122 (0.846)
$CC_{1/2}$	0.978 (0.492)
Wilson B-factor (Å^2)	46.2
Refinement	
Reflections (working/test)	38861/2047
R_{cryst}/R_{free}	0.201/0.249
No. of protein atoms	6457
Ligand atoms	12
No. of Sugar Atoms	355
No. of Water Molecules	6
Geometry	
rmsd in bond lengths (Å)	0.007
rmsd in bond angles (°)	0.94
Most favorable regions (%)	92.8
Additional allowed regions (%)	6.2
Outliers (%)	1.0
Clashscore	10.4
Average B-Factors (Å^2)	
Protein	50.4
Ligand	63.8
Sugars	72.4
Water molecules	32.8
Pyranose Conformations (Total/Percentage)	
Lowest energy conformations	23/85.2
Highest energy conformations	4/14.8

* Values for the outermost resolution shell are shown in parentheses.

2.2. Overall Structure Comparisons

Structure-based sequence alignment (Figure 1) showed that *Ct*BGL has the highest structural similarity with *Nc*Cel3A (PDB id 5nbs), as indicated by the root mean square deviation (rmsd) of 0.60 Å for 831 aligned residues (sequence identity 72%), followed by *Aspergilus fumigatus Af*βG (PDB id 5fji; rmsd 0.81 Å; seq. identity 62%), *Re*Cel3A (PDB id 5ju6; rmsd 0.84 Å; 62.0%) and *Aspergillus aculeatus Aa*Bgl1 (4iig; 0.83 Å; 62.0%). Lower similarities were found with *Hypocrea jecorina Hj*Cel3A (3zyz; 1.21 Å; 47%).

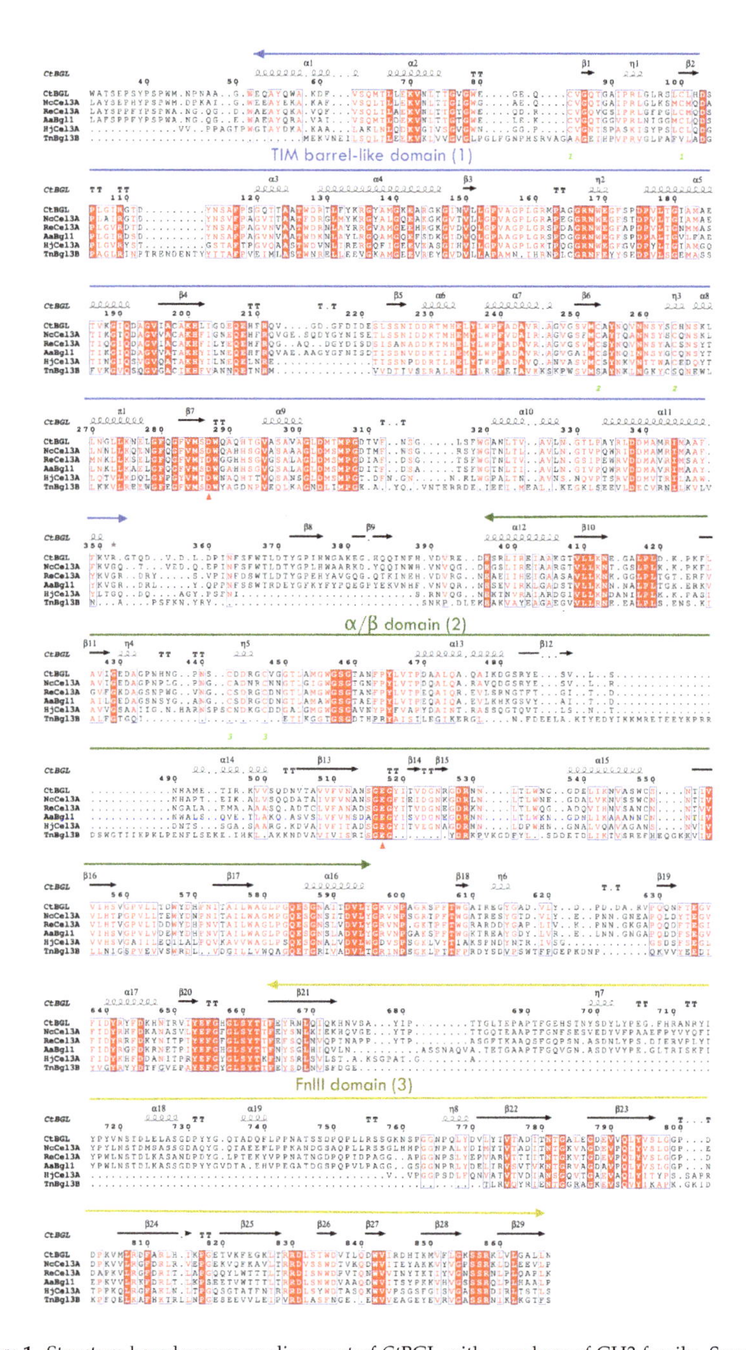

Figure 1. Structure-based sequence alignment of *Ct*BGL with members of GH3 family. Secondary structure elements are shown on the top of the alignment, while the red triangles below the alignment indicate the catalytic conserved residues (Asp287 and Glu517 in *Ct*BGL). Disulfide bonds are depicted with green numbers. The three domains are indicated. A column is framed if more than 70% of its residues are similar according to physicochemical properties. Frames in red background with white letters depict strict identity. The figure was constructed with ESPript [16].

2.3. Description of the Structure

The structure of *Ct*BGL (Figure 2) consists of three distinct domains, similar to other GH3 family members: A catalytic triose phosphate isomerase (TIM) barrel-like domain (Leu51–Ser354), an α/β sandwich domain (His396–Gly596) and a FnIII (fibronectin type III) domain (Thr663–Gln867) with a prominent insertion region (677–766). A linker region (residues 355–394) connects domain 1 to domain 2 and a second linker region (600–662) connects domain 2 to domain 3. The 39-residue linker between domain 1 and domain 2 is comparable to that in *Nc*Cel3A and significantly longer than the linker in other GH3 β-glucosidases such as *Hj*Cel3A and *Hv*ExoI with only 18 and 16 residues in length, respectively. In *Nc*Cel3A, the linker comprises 42 residues (residues 341–383) with a 25-residue (residues 351–376) insertion previously described as a hydrophobic linker responsible for the activation of *Re*Cel3A and *Aa*Bgl in organic solvents [5]. This insertion has been coined as loop II and contains several aromatic residues, of which *Ct*GBL Phe364 (Phe352 in *Nc*Cel3A) and *Ct*GBL Trp367 (Trp355 in *Nc*Cel3A) are the most conserved. Moreover, loop II contains residues Phe366, Trp367, and Trp377 that line up one side of the substrate-binding site, and they are also found in *Nc*Cel3A (Phe354, Trp355, and Trp365, respectively) with variations in other β-glucosidases, such as *Re*Cel3A and *Aa*Bgl1. It has been suggested that this loop is stabilized by interactions with the N-glycans from a neighboring glycosylated Asn residue (Asn57 in *Nc*Cel3A) [5]. In *Ct*BGL, the equivalent residue is Asn72, which is also glycosylated (see below).

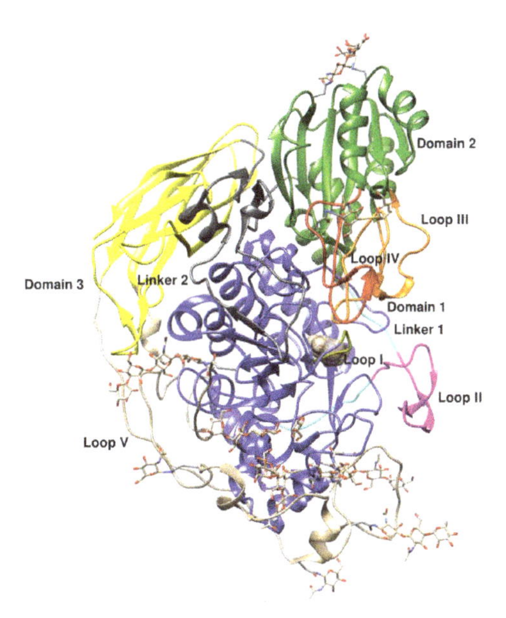

Figure 2. Overall crystal structure of *Ct*BGL in ribbon representation. The domains, linkers, and loops are shown in different colors and labeled: Catalytic TIM barrel-like domain (blue), α/β sandwich domain (green), FnIII domain (yellow), loop I (gold), linker 1 (cyan), loop II or insertion region (magenta), loop III (orange), loop IV (orange–red), linker 2 (gray), loop V (brown). The N-glycans and glucose in the active site are shown as gray sticks. The figure was created with UCSF Chimera [17].

*Ct*BGL second domain, an $(\alpha/\beta)_6$ sandwich fold, is structurally well conserved among all the GH3 enzymes. It consists of residues His396 to Gly596 and includes loops III and IV that encompass residues Gly433–Val467 and Ala513–Asn536, respectively (Figure 3). Loop III houses the conserved cysteine residues Cys442 and Cys447, which form a disulfide bridge that stabilizes the folded loop III architecture. Ser458 and Asp444 are also highly conserved in all compared structures. This domain hosts the catalytic acid Glu517 in loop IV, which is found to be conserved in all of the compared β-glucosidases, *Ct*BGL, *Nc*Cel3A, *Re*Cel3A, *Hj*Cel3A, *Aa*Bgl1, and *Thermotoga neapolitana* β-glucosidase 3B (*Tn*Bgl3B). Moreover, the architecture and location of loops III and IV, which constitute one site of the active site cavity in between loops I and II, are also found conserved. Ser458, Asp444, and Glu517 are found pointing downwards from the loop regions and towards the active site.

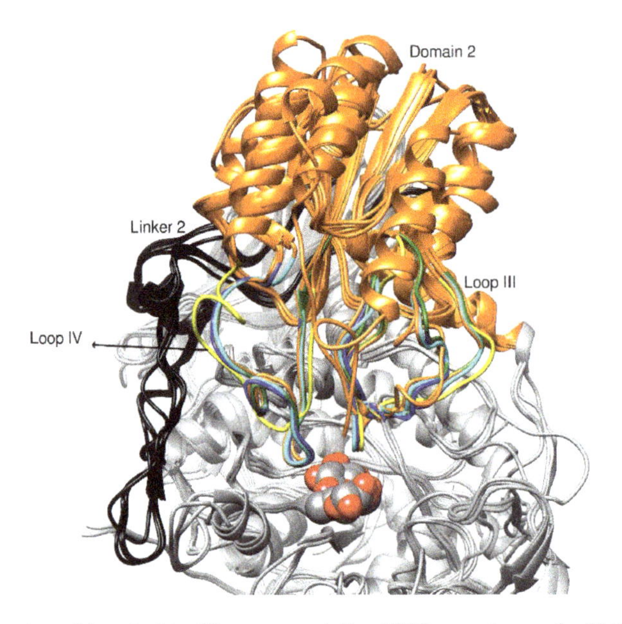

Figure 3. Overview of domain 2 in ribbon representation *Ct*BGL superimposed with the compared structures. Domain 2 is colored orange; loops III and IV in *Ct*BGL, *Nc*Cel3A, *Re*Cel3A, and *Tn*Bgl3B are colored in green, blue, cyan, and yellow, respectively, for structural comparisons. Linker 2 loop is colored in black, and the active site is depicted by the bound BGC in sphere representation.

Domain 2 of the *Ct*BGL structure is followed by a second linker region (Linker 2; residues Lys597–Ser660). This linker region exhibits an extended structure, which almost resembles a boundary that separates domain I and domain II from domain III. The extended linker region probably plays a role in stabilizing loops I and IV.

The FnIII-like domain 3 was first observed in the GH3 family in the *Tn*Bgl3B structure [18]. It consists of residues Tyr661–Asn867 that form a beta sandwich composed of a total of nine β-strands arranged in two layers of β-sheets with three and four β-strands, respectively. Loop V takes an extended structure and encompasses domain I. Importantly, loop V is present in *Nc*Cel3A, *Re*Cel3A, and *Aa*Bgl1, but it is absent in *Tn*Bgl3B and *Hj*Cel3A (Figure 4). Several conserved aromatic residues, namely, Tyr716, Tyr718, Tyr733, and Phe740, are found in loop V. Tyr716 and Tyr733 are found to be conserved in all structures except *Hj*Cel3A and *Tn*Bgl3B, where loop V is absent, while Tyr718 is replaced by Trp in *Re*Cel3A and *Aa*Bgl1. Moreover, Phe740 is replaced by a Tyr residue in *Re*Cel3A and a His residue in *Aa*Bgl1. These conserved aromatic residues form π-stack interactions with *N*-acetyl-*β*-D-glucosamine (GlcNAc) residues. Also, conserved in loop V is Asn720, which was found to be N-glycosylated.

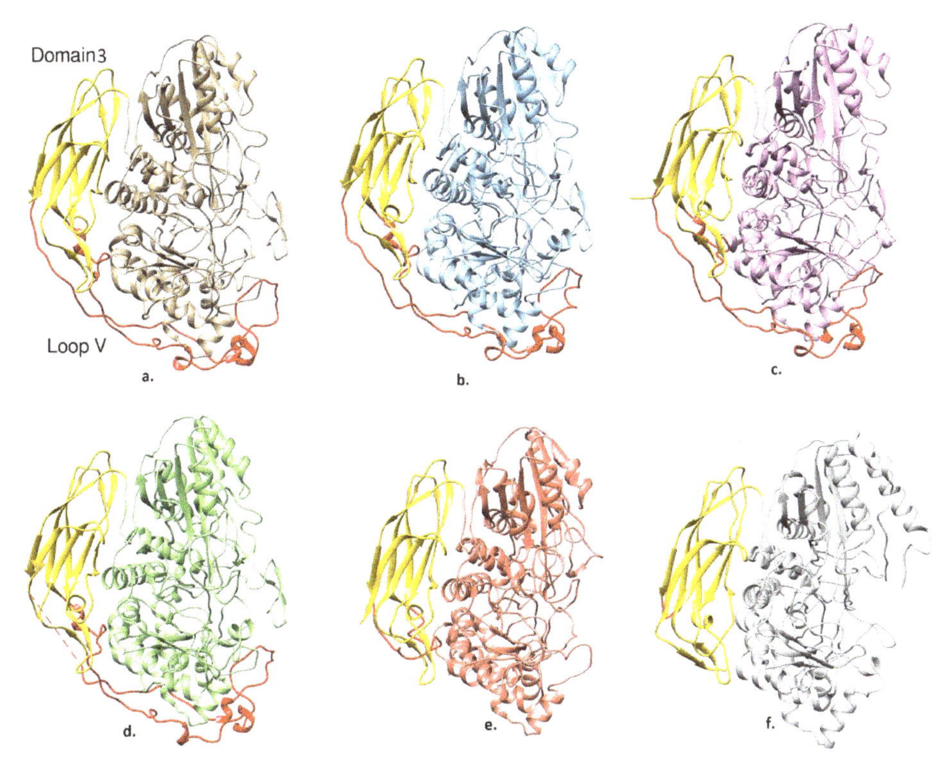

Figure 4. Loop V comparisons in (**a**) *Ct*BGL, (**b**) *Nc*Cel3A, (**c**) *Re*Cel3A, (**d**) AaBgl1, (**e**) *Hj*Cel3A, and (**f**) *Tn*Bgl3B. Domain 3 is colored in yellow and the loop V is represented by red coloration. Domains 1 and 2 are colored in the same color in each enzyme and differently amongst the enzymes.

2.4. Glycosylation Sites

Potential glycosylation sites were observed in the *Ct*BGL structure in 10 positions, all in Asn residues (Table 2). The glycosylation sites found in *Ct*BGL when compared with other structures showed some variations. A unique glycosylation site was observed in *Ct*BGL at Asn504 on the surface of the molecule, where the corresponding aligned structures had either a different residue or a gap in the structure-based alignment (Table 3). Asn259 was found highly conserved and glycosylated in all structures. Moreover, *Re*Cel3A possess 16 glycosylation sites [4], similarly to *Aspergillus* β-glucosidases [8]. On the other hand, *Hj*Cel3A showed the lowest number of glycosylation sites, with only two glycosylated sites (Asn208 and Asn310 in chains A and B, respectively). As observed before in this class of β-glucosidases, all glycosylation sites in *Ct*BGL were also located on one face of the molecule (Figure 5).

A total of 27 glycan moieties were found in *Ct*BGL. The glycans ranged in length from single GlcNAc to longer chain. The longest glycosylation chain was composed of eight residues. The overall degree of glycosylation in *Ct*BGL can be considered as moderate when compared against other glycosylation chains, such as those found in *Re*Cel3A and *Aa*BGL1, with the longest chain composed of 10 residues and 45–50 glycosylation residues in total per chain in the crystallographic asymmetric unit. Interestingly, the glycosylation pattern in *Ct*BGL showed high GlcNAc-type N-glycans, while *Re*Cel3A and *Aa*BGL1 displayed high mannose-type N-glycans. A total of 15 GlcNAc residues were identified in *Ct*BGL. Out of the total 10 glycosylation sites in *Ct*BGL, three of them were found to consist of a single GlcNAc monosaccharide. Single GlcNAc monosaccharides in other enzymes, such as *Aa*BGL1, were obtained possibly as a result of a treatment with endoglycosidase H prior to crystallization. In the case of *Ct*BGL, no endoglycosidase treatment was employed. Similarly, single GlcNAc molecules were found in *Af*βG (in Asn543 and Asn715) without any attempt of enzymatic cleavage. It is possible that, in these cases, the enzymes were subjected to endoglycosidase activity during the expression stage. The other seven N-glycosylation sites in *Ct*BGL ranged from two to eight monosaccharides. The N-glycan at *Ct*BGL Asn329 was located in domain I of the enzyme and consisted of eight monosaccharides (two GlcNAc, five α-D-Man, and one β-D-Man). Extra densities that could accommodate another two monosaccharide molecules were present in the electron-density map, but they were not modeled owing to lack of clarity. It was, nevertheless, the largest N-glycan structure in *Ct*BGL and was involved in forming multiple H-bonds with the enzyme residues at Thr293, Asp730, Val295, Gly734, and Tyr733 and a pi-sigma interaction with Tyr733 residue. The structural equivalent Asn residue in *Aa*Bgl1 (Asn322) had a long-length glycan with eight Man and two GlcNAc that were also involved in extensive stabilizing interactions with domain I. N-glycans at positions Asn72 and Asn720 were located in domains I and III, respectively, with loop V residues in between them. They both participated in interactions with protein residues. The N-glycan at Asn72 was found to interact in conventional H-bonding interactions with Ala90, Tyr716, Tyr718, and in pi-sigma interaction with Tyr715 from domain III. The N-glycan at Asn720 interacted with Arg710 via H-bonding and Tyr704 via weak van der Waals interactions. A long chain glycan moiety at Asn259 at the outer surface of the enzyme was found to interact via H-bonding with Ser35, Glu36, and Asp229.

Notably, the protein glycans were also shown to exhibit a potential binding affinity for polysaccharides such as cellulose [19] and aromatic compounds [20,21]. This suggests a possible role of N-glycans in promoting cellulose binding and also in protein–glycan interactions. Glycans in β-glucosidases can stabilize the crystal packing contacts and also participate in hydrogen-bonding interactions at the dimer interface [8]. There are two glycosylated Asn residues at the dimer interface of β-glucosidases that provide protein–glycan interactions between the two chains. One of those Asn residues corresponds to Asn531 in *Ct*BGL. This Asn bears only two monosaccharide molecules, whereas, for comparison, the equivalent Asn residue in *Aa*GL1 has seven monosaccharides with the terminal ones able to reach the adjacent subunit. The second Asn residue that participates in dimer interface contacts is not present in *Ct*BGL. The lack of these contacts could therefore contribute to reduced strength of intermolecular interactions in the crystal lattice, resulting in further instability of the *Ct*BGL crystals.

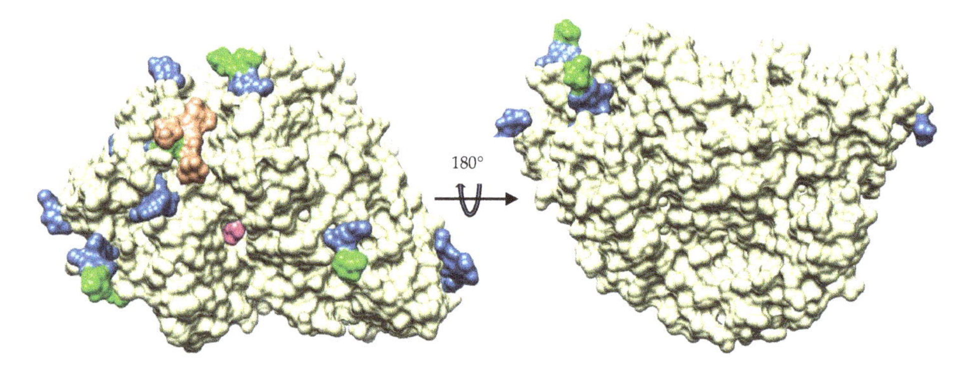

Figure 5. Distribution of glycosylation sites in the *Ct*BGL structure. GlcNAc is shown in blue, β-D-mannose in green, α-D-mannose in orange, and BGC in magenta.

Table 2. Glycosylation sites and glycan descriptions in *Ct*BGL.

Residue	Corresponding Glycan Structure
Asn72	Man-β1,4–GlcNAc-β1,4–GlcNAc
Asn259	Man-β1,4–Man-β1,4–GlcNAc-β1,4–GlcNAc
Asn322	GlcNAc-β1,4–GlcNAc
Asn329	Man-α1,2–Man-α1,6–(α1,3-Man)-α1,6–Man-(α1,3-Man)Man-β1,4–GlcNAc-β1,4–GlcNAc
Asn504	GlcNAc
Asn531	Man-β1,3–GlcNAc
Asn572	GlcNAc-β1,4–GlcNAc
Asn698	GlcNAc
Asn720	Man-β1,4–GlcNAc-β1,4–GlcNAc
Asn744	GlcNAc

Table 3. Comparison of glycosylation sites [%].

*Ct*BGL	*Nc*Cel3A(5nbs)	*Aa*Bgl1 (4iif)	*Re*Cel3A (5ju6)	*Hj*Cel3A (3zyz) [#]
Asn72 (+)	Asn57 (+)	Asn61 (+)	Asn61 (+)	Gly30 (−)
Asn259 (+)	Asn248 (+)	Asn252 (+)	Asn249 (+)	Asn208 (+)
Asn322 (+)	Asn311 (+)	Asn315 (+)	Asn312 (+)	Ala270 (−)
Asn329 (+)	Asn318 (+)	Asn322 (+)	Asn319 (+)	Ser277 (−)
Asn504 (+)	Asp492 (−)	Ala496 (−)	Ala492(−)	Lys428 (−)
Asn531 (+)	Asn519 (+)	Asn523 (+)	Asn519 (+)	Asn455 (−)
Asn572 (+)	Asn560 (+)	Asn564 (+)	Asn560 (+)	Gln496 (−)
Asn698 (+)	Ser687 (−)	Asn690 (+)	Asn685 (+)	
Asn720 (+)	Asn710 (+)	Asn712 (+)	Asn707 (+)	
Asn744(+)	Lys734 (−)	Gly736 (−)	Asn731 (+)	

[%] (+) depicts glycosylation, (−) absence of glycosylation [#] In subunit A.

2.5. Active Site

The active site is located in a shallow pocket near the interface of the first and second domain. A molecule of β-D-glucose (BGC) was fitted at the active site based on residual electron density observed in electron density F_o–F_c difference maps. The source of the β-D-glucose is most likely the growth medium used as no β-D-glucose was used during crystallizations or soaking. Two catalytic residues were identified, Asp287 (nucleophile) at the N-terminal TIM-barrel domain and Glu517 (acid/base) at the sandwich α/β domain II. Both catalytic residues were found conserved in the GH3 family members (Figure 6). The corresponding catalytic residues were Asp276/Glu505 in *Nc*Cel3A, Asp277/Glu505 in *Re*Cel3A, Asp280/Glu509 in *Aa*Bgl1, Asp236/Glu441 in *Hj*Cel3A, and Asp242/Glu458 in *Tn*Bgl3B. The collapsed TIM-barrel model of domain I is vital for the proper accession of the active site. It was found that near the active site, the second barrel β strand (Gly87 to Thr89) was much shorter and antiparallel, which creates an active site much wider and accessible when compared with GH3 enzyme structures with complete TIM-barrel fold [6]. An additional electron density present at the active site next to BGC was not interpretable and may suggest a bound buffer molecule, MPD from the crystallization mother liquor, or a partially bound glucose molecule.

2.6. CtBGL Thermostability

*Ct*BGL has been found to be thermostable at 50 °C and to retain half of its activity after incubation at 65 °C for 55 min [22]. The enzyme also retains 29.7% of its activity after incubation at 70 °C for 10 min. In regard to most other β-glucosidases from thermophilic fungi, *Ct*BGL exhibits comparable thermostability, as previously reported [22]. In contrast, *A. fumigatus* β-glucosidase has been found highly thermostable and able to retain most of its activity for at least 19 h at 65 °C [23].

Protein thermostability is usually hard to predict and there is no a common mechanism yet available [24,25]. Several factors of protein thermostability have been proposed that could provide some clues (Table 4). Solvent-accessible surface (SAS), charged residues, and glycosylation patterns are some of the key indicators. The structure of *Hj*Cel3A from the mesophilic fungus *Hypocrea jecorina* had the lowest SAS (22812 Å^2), owing to the smaller number of residues and the lack of loop V. Similarly, loop V was absent in *Tn*Bgl3A and the SAS value was reduced. The SAS values for the other β-glucosidases, including *Ct*BGL, were quite similar, i.e., around 28100 Å^2.

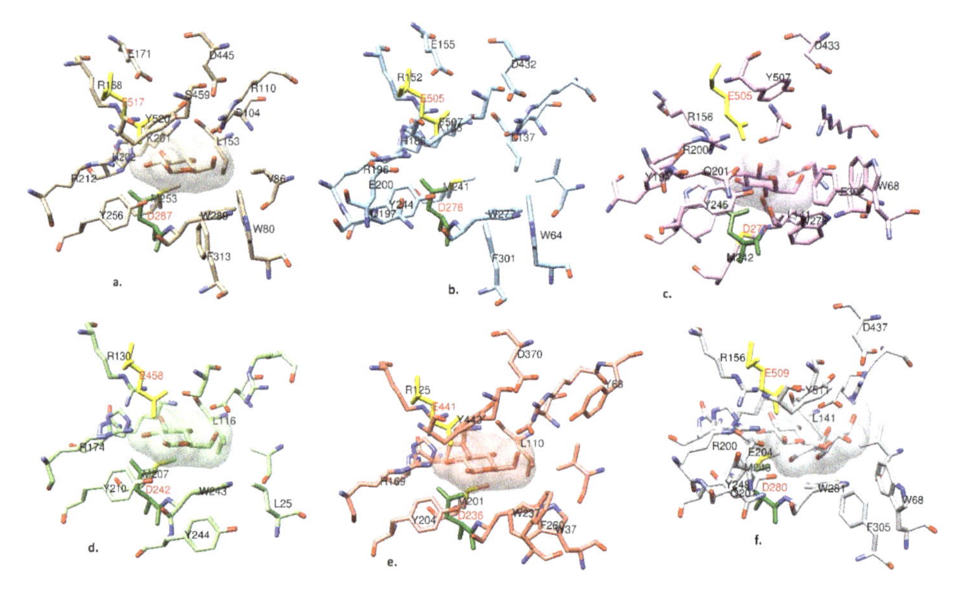

Figure 6. Active site structure comparisons of (**a**) *Ct*BGL, (**b**) *Nc*Cel3A, (**c**) *Re*Cel3A, (**d**) *Tn*Bgl3B, (**e**) *Hj*Cel3A, and (**f**) *Aa*BGL1. The corresponding conserved nucleophile Asp is colored in dark green and the acid/base Glu is colored in yellow. Both residues are labeled in red. Carbon atoms are colored differently in each enzyme. The active site BGC is displayed with its solvent-excluded surface for visual clarity.

Charged residues in the structures can contribute to structural integrity and, in turn, to thermostability. The negatively (Asp and Glu) and positively (Arg and Lys) charged residues may provide a stability profile of the structure [26]. The numbers showed significant variations when the proteins were compared. In particular, *Hj*CelA showed reduced numbers of positively and negatively charged residues, despite its high thermostability with an optimum temperature at 90 °C and an unfolding temperature of ~88 °C depending on the enzyme concentration [27]. The number was increased for the β-glucosidases from other thermophilic and mesophilic β-glucosidases. In addition, *Afβ*G, despite its high thermostability, was also characterized by a similar content of charged residues, suggesting that the strength of individual ion–pair interactions may play a key role.

Finally, the glycosylation pattern found may also contribute to the thermostability of the enzyme by promoting interactions with amino acid residues. It has been shown that glycosylation enhances the solubility, reduces the aggregation, and increases the thermal stability of proteins [28,29]. The exact mechanism by which the enzyme glycosylation pattern affects the overall function and structure of proteins is not yet well understood. Analysis of protein structures deposited in the Protein Data Bank has suggested that N-glycosylation causes no significant local or global structural changes; however, it decreases the protein dynamics, thus leading to increased stability [26]. Aglycosylated proteins are, in general, found to be less stable, and therefore aggregate more easily than its glycosylated counterparts at certain temperatures. Thus, glycosylation has been suggested as a potential factor to enhance enzyme solubility, stability, and function [30]. Thermostability measurements in *Hj*Cel3A samples with different degree of N-glycosylation revealed the same melting temperature (74.0 ± 0.2 °C), thus suggesting that the effect of glycosylation may be case-specific and other factors could play a role. Interestingly, *Tn*Bgl3B that lacks glycosylation exhibited thermostability, most likely as a result of the high numbers of charged residues. Nevertheless, the moderate glycosylation in *Ct*BGL could not be ruled out as a contributing factor to the limited stability of the enzyme compared to other more thermostable β-glucosidases characterized by extensive glycosylation. Further studies are, however, required to better understand the role of glycosylation in β-glucosidases and in protein stability.

Table 4. Comparison of thermostability parameters.

	*Ct*BGL	*Af*βG (5FJI)	*Aa*Bgl1 (4IIF)	*Re*Cel3A (5JU6)	*Tn*Bgl3B (2X40)	*Hj*Cel3A (3ZYZ)
Thermal stability characteristics	$T_{1/2}$ 65 °C after 55 min	19 h at 65 °C [23]	$T_{1/2}$ 62 °C after 30 min [31]	T_m 87.3 °C [4]	Unfolding temperature 86–89 °C [27]	T_m 77.6 °C [4]
Asp + Glu (−)	93	95	92	87	113	55
Arg + Lys (+)	66	67	65	59	93	51
Pro: Gly	0.51	0.55	0.51	0.61	0.58	0.52
Val (%)	7.0	7.9	8.0	7.5	9.2	8.8
Amino acid residues	867	844	841	857	721	714
SAS (Å^2)	28,502.9	28,092.1	28,074.9	28,481.7	26,430.0	22,551.0
Intra-chain salt Bridges	46	49	42	44	39	34
H-bonding interactions	12 ∞	22	25	22	23	8
Interface area, Å^2	1544.2 ∞	1324.3	1445.3	1322.5	1253.1	699.3
Helical content (%)	21.2	21.8	21.8	20.5	26.5	23.7
Sugars	27	45 (chain A)	44 %	50 (chain C)	NA $	2

∞ In the crystallographic dimer. % Prior to crystallization, the enzyme was subjected to endoglycosidase H treatment with ~50% of the N-glycans removed [3]. $ The enzyme was produced in *Escherichia coli*.

3. Materials and Methods

3.1. Protein Expression, Purification, and Crystallization

Protein expression and purification of the β-glucosidase from *Chaetomium thermophilum* CT2 was carried out as previously described [22]. Briefly, the enzyme (Uniprot id A6YRT4) was produced in *Pichia pastoris* GS115 cells and purified by ion-exchange chromatography on a DEAE-Sepharose (Pharmacia, Uppsala, Sweden) to homogeneity as judged by SDS-PAGE. β-Glucosidase activity was assayed with salicin using the Miller's method and also detected in native polyacrylamide gel using 4-methylumbeliferyl-β-D-glucopyranoside [22]. The activity of the enzyme was 1.62 ± 0.20 U/mg (1 U corresponds to the release of 1 μM of glucose per min) from three independent measurements at optimum conditions of pH and temperature.

3.2. Protein Crystallization

Prior to crystallization, the enzyme solution was concentrated to ~10 mg/mL with Amicon® Ultra Centrifugal Filters (10,000 MW cut-off) (Millipore, MA, USA) in 10 mM HEPES–NaOH, pH 7.0 buffer. Crystals were obtained by the hanging-drop vapor diffusion method at 16 °C using a well solution of 35–45% *v*/*v* MPD (Sigma-Aldrich, St. Louis, MO, USA). The drops were prepared by mixing 2 μL of protein solution with an equal volume of well solution. The crystals grew as octahedra to a maximum size of approximately 0.06 × 0.06 × 0.08 mm^3 within a period of 1 month.

3.3. Data Collection and Processing

Data were collected on the X13 beamline at EMBL-Hamburg (c/o DESY) from a single crystal under cryogenic (100 K) temperature using a MARCCD detector. The presence of MPD in the crystallization was sufficient for cryoprotection, thus no additional cryoprotectant was needed. One hundred and fifty diffraction images were collected from a single crystal with a rotation range of 0.45° per image. Data processing was carried out with XDS [32]. The crystal was found to belong to the tetragonal space group $P4_12_12/P4_32_12$. Assuming one molecule in the asymmetric unit, the Matthews coefficient V_M [33] is 5.4 Å^3/Da corresponding to a solvent content of ~77%.

3.4. Structure Determination and Refinement

Initial phases were obtained with molecular replacement using Phaser [34], as implemented in Phenix 1.15.2_3472 [35]. The crystal structure of *Aspergillus aculeatus* β-glucosidase in complex with castanospermine (PDB id 4iif; sequence identity 61.5%) was used as a search model after pruning side chains with Sculptor [36] based on sequence alignment considerations. Initially, the search was carried out assuming two molecules in the asymmetric unit, but no solution was produced. Based, however, on the statistics for one molecule (TFZ = 30.9), the search was limited to one molecule and a single solution was obtained in space group $P4_12_12$. Refinement was carried out using simulated annealing (1000 K) in Phenix with maximum likelihood as target function. The refinement was alternated with model visualization and rebuilding using Coot 0.8.9 [37]. Tight restraints were used to avoid overfitting of the structure to the data, owing to the low resolution. The progress of the refinement was monitored by the R_{free} with 5% of the reflections used for the calculations [38]. High-resolution structures were used to assist the rebuilding in places with poor electron density and ambiguities in atom positions. Structure validation was performed with tools implemented in Phenix and Coot. The stereochemistry and conformation of the sugars were tested with Privateer [39]. Data collection and final refinement statistics are shown in Table 1. The atomic coordinates and the structure factors have been deposited to the Protein Data Bank under the accession code 6SZ6.

3.5. Structure Analysis

PDBeFold [40] was employed to identify structural similarities between *Ct*BGL and deposited PDB structures. Structure-based sequence alignment was performed with UCSF Chimera 1.13.1 [41]. ESPript 3 [16] was used in rendering sequence similarities and secondary structure information from aligned sequences. Salt bridges were calculated using ESBRI (http://bioinformatica.isa.cnr.it/ESBRI/). SAS area was calculated using PDBePISA [42]. The values for charged residues were calculated using the ExPASy server (http://web.expasy.org/protparam/) ProtParam tool [43].

4. Conclusions

The *Ct*BGL structure was determined at 2.99 Å resolution and refined to good stereochemical and refinement numbers, despite the limited resolution. The structure exhibited a three-domain architecture with two linker regions as other β-glucosidase structures. Variations in the length of the first linker regions and the third domain were identified. *Ct*BGL showed the highest structural similarity with *Nc*Cel3A. The catalytic residues at the active site were identified as Asp287 (nucleophile) and Glu517 (acid/base). *Ct*BGL showed a low number of glycans at glycosylation sites compared to other heavily glycosylated GH3 β-glucosidases. Charged residues and the glycosylation pattern are suggested as potential contributing factors for the thermostability properties of *Ct*BGL. The analysis presented in this study could offer new ideas towards further improvements in β-glucosidases for better use in biotechnological applications.

Author Contributions: Conceptualization, D.-C.L. and A.C.P.; data analysis, I.M., N.P., and A.C.P.; writing—original draft preparation, I.M. and A.C.P.; writing—review and editing, I.M., N.P., D.-C.L., and A.C.P.

Funding: The work was supported by the Chinese National Nature Science Foundation (31571949).

Acknowledgments: The authors thank Teemu Haikarainen for his initial contribution to the project. The authors are also grateful to Biocenter Finland for infrastructure support. Access to EMBL Hamburg (c/o DESY) was provided by the European Community's Seventh Framework Programme (FP7/2007-2012) under grant agreement no. 226716 and Biostruct-X (283570).

Conflicts of Interest: The authors declare no conflicts of interest.

Abbreviations

BGC	β-D-glucose
GH3	Glycoside hydrolase family 3
MPD	2-methyl-2,4-pentanediol
PDB	Protein Data Bank
NCS	Non-crystallographic symmetry
rmsd	Root mean square deviation
SAS	Solvent-accessible surface
$T_{1/2}$	Half-life inactivation
T_m	Melting temperature

References

1. Davies, G.J.; Gloster, T.M.; Henrissat, B. Recent structural insights into the expanding world of carbohydrate-active enzymes. *Curr. Opin. Struc. Biol.* **2005**, *15*, 637–645. [CrossRef]
2. Levasseur, A.; Drula, E.; Lombard, V.; Coutinho, P.M.; Henrissat, B. Expansion of the enzymatic repertoire of the CAZy database to integrate auxiliary redox enzymes. *Biotechnol. Biofuels* **2013**, *6*, 41. [CrossRef] [PubMed]
3. Suzuki, K.; Sumitani, J.-I.; Nam, Y.-W.; Nishimaki, T.; Tani, S.; Wakagi, T.; Kawaguchi, T.; Fushinobu, S. Crystal structures of glycoside hydrolase family 3 β-glucosidase 1 from *Aspergillus aculeatus*. *Biochem. J.* **2013**, *452*, 211–221. [CrossRef]
4. Gudmundsson, M.; Hansson, H.; Karkehabadi, S.; Larsson, A.; Stals, I.; Kim, S.; Sunux, S.; Fujdala, M.; Larenas, E.; Kaper, T.; et al. Structural and functional studies of the glycoside hydrolase family 3 β-glucosidase Cel3A from the moderately thermophilic fungus *Rasamsonia emersonii*. *Acta Crystallogr. Sect. D Struct. Biol.* **2016**, *72*, 860–870. [CrossRef]
5. Karkehabadi, S.; Hansson, H.; Mikkelsen, N.E.; Kim, S.; Kaper, T.; Sandgren, M.; Gudmundsson, M. Structural studies of a glycoside hydrolase family 3 β-glucosidase from the model fungus *Neurospora crassa*. *Acta Crystallogr. Sect. F Struct. Biol. Commun.* **2018**, *74*, 787–796. [CrossRef]
6. Karkehabadi, S.; Helmich, K.E.; Kaper, T.; Hansson, H.; Mikkelsen, N.-E.; Gudmundsson, M.; Piens, K.; Fujdala, M.; Banerjee, G.; Scott-Craig, J.S.; et al. Biochemical Characterization and Crystal Structures of a Fungal Family 3 β-Glucosidase, Cel3A from *Hypocrea jecorina*. *J. Biol. Chem.* **2014**, *289*, 31624–31637. [CrossRef]
7. Lima, M.A.; Oliveira-Neto, M.; Kadowaki, M.S.; Rosseto, F.R.; Prates, E.T.; uina, F.; Leme, A.F.; Skaf, M.S.; Polikarpov, I. *Aspergillus niger* β-glucosidase has a cellulase-like tadpole molecular shape: Insights into glycoside hydrolase family 3 (GH3) β-glucosidase structure and function. *J. Biol. Chem.* **2013**, *288*, 32991–33005. [CrossRef]
8. Agirre, J.; Ariza, A.; Offen, W.A.; Turkenburg, J.P.; Roberts, S.M.; McNicholas, S.; Harris, P.V.; McBrayer, B.; Dohnalek, J.; Cowtan, K.D.; et al. Three-dimensional structures of two heavily N-glycosylated *Aspergillus* sp. family GH3 *β*-D-glucosidases. *Acta Crystallogr. D Struct. Biol.* **2016**, *72*, 254–265. [CrossRef]
9. Varghese, J.; Hrmova, M.; Fincher, G. Three-dimensional structure of a barley beta-D-glucan exohydrolase, a family 3 glycosyl hydrolase. *Structure* **1999**, *7*, 179–190. [CrossRef]
10. Nakatani, Y.; Cutfield, S.M.; Cowieson, N.P.; Cutfield, J.F. Structure and activity of exo-1, 3/1, 4-β-glucanase from marine bacterium *Pseudoalteromonas* sp. BB1 showing a novel C-terminal domain. *FEBS J.* **2012**, *279*, 464–478. [CrossRef]
11. Yoshida, E.; Hidaka, M.; Fushinobu, S.; Koyanagi, T.; Minami, H.; Tamaki, H.; Kitaoka, M.; Katayama, T.; Kumagai, H. Role of a PA14 domain in determining substrate specificity of a glycoside hydrolase family 3 β-glucosidase from *Kluyveromyces marxianus*. *Biochem. J.* **2010**, *431*, 39–49. [CrossRef] [PubMed]
12. Li, D.-C.; Papageorgiou, A.C. Cellulases from Thermophilic Fungi: Recent Insights and Biotechnological Potential. In *Fungi in Extreme Environments: Ecological Role and Biotechnological Significance*; Tiquia-Arashiro, S., Grube, M., Eds.; Springer: Cham, Switzerland, 2019.
13. de Pereira, C.J.; Marques, P.N.; Rodrigues, A.; de Oliveira, B.T.; Boscolo, M.; da Silva, R.; Gomes, E.; Martins, B.D. Thermophilic fungi as new sources for production of cellulases and xylanases with potential use in sugarcane bagasse saccharification. *J. Appl. Microbiol.* **2015**, *118*, 928–939. [CrossRef] [PubMed]

14. Sørensen, A.; Lübeck, M.; Lübeck, P.; Ahring, B. Fungal Beta-Glucosidases: A Bottleneck in Industrial Use of Lignocellulosic Materials. *Biomolecules* **2013**, *3*, 612–631. [CrossRef] [PubMed]
15. Karnaouri, A.; Topakas, E.; Paschos, T.; Taouki, I.; Christakopoulos, P. Cloning, expression and characterization of an ethanol tolerant GH3 β-glucosidase from *Myceliophthora thermophila*. *PeerJ* **2013**, *1*, e46. [CrossRef]
16. Robert, X.; Gouet, P. Deciphering key features in protein structures with the new ENDscript server. *Nucleic Acids Res.* **2014**, *42*, W320–W324. [CrossRef]
17. Pettersen, E.F.; Goddard, T.D.; Huang, C.C.; Couch, G.S.; Greenblatt, D.M.; Meng, E.C.; Ferrin, T.E. UCSF Chimera—A visualization system for exploratory research and analysis. *J. Comput. Chem.* **2004**, *25*, 1605–1612. [CrossRef]
18. Pozzo, T.; Pasten, J.; Karlsson, E.; Logan, D.T. Structural and functional analyses of beta-glucosidase 3B from *Thermotoga neapolitana*: A thermostable three-domain representative of glycoside hydrolase. *J. Mol. Biol.* **2010**, *397*, 724–739. [CrossRef]
19. Payne, C.M.; Jiang, W.; Shirts, M.R.; Himmel, M.E.; Crowley, M.F.; Beckham, G.T. Glycoside hydrolase processivity is directly related to oligosaccharide binding free energy. *J. Am. Chem. Soc.* **2013**, *135*, 18831–18839. [CrossRef]
20. Yamaguchi, H.; Nishiyama, T.; Uchida, M. Binding Affinity of N-Glycans for Aromatic Amino Acid Residues: Implications for Novel Interactions between N-Glycans and Proteins. *J. Biochem.* **1999**, *126*, 261–265. [CrossRef]
21. Hudson, K.L.; Bartlett, G.J.; Diehl, R.C.; Agirre, J.; Gallagher, T.; Kiessling, L.L.; Woolfson, D.N. Carbohydrate–Aromatic Interactions in Proteins. *J. Am. Chem. Soc.* **2015**, *137*, 15152–15160. [CrossRef]
22. Xu, R.; Teng, F.; Zhang, C.; Li, D. Cloning of a Gene Encoding β-Glucosidase from *Chaetomium thermophilum* CT2 and Its Expression in Pichia pastoris. *J. Mol. Microb. Biotechnol.* **2011**, *20*, 16–23. [CrossRef] [PubMed]
23. Kim, K.-H.; Brown, K.M.; Harris, P.V.; Langston, J.A.; Cherry, J.R. A Proteomics Strategy To Discover β-Glucosidases from *Aspergillus fumigatus* with Two-Dimensional Page In-Gel Activity Assay and Tandem Mass Spectrometry. *J. Proteome Res.* **2007**, *6*, 4749–4757. [CrossRef] [PubMed]
24. Sadeghi, M.; Naderi-Manesh, H.; Zarrabi, M.; Ranjbar, B. Effective factors in thermostability of thermophilic proteins. *Biophys. Chem.* **2006**, *119*, 256–270. [CrossRef] [PubMed]
25. Unsworth, L.D.; van der Oost, J.; Koutsopoulos, S. Hyperthermophilic enzymes-stability, activity and implementation strategies for high temperature applications. *FEBS J.* **2007**, *274*, 4044–4056. [CrossRef] [PubMed]
26. Lee, H.; Qi, Y.; Im, W. Effects of N-glycosylation on protein conformation and dynamics: Protein Data Bank analysis and molecular dynamics simulation study. *Sci. Rep.* **2015**, *5*, 8926. [CrossRef] [PubMed]
27. Turner, P.; Svensson, D.; Adlercreutz, P.; Karlsson, E. A novel variant of *Thermotoga neapolitana* β-glucosidase B is an efficient catalyst for the synthesis of alkyl glucosides by transglycosylation. *J. Biotechnol.* **2007**, *130*, 67–74. [CrossRef] [PubMed]
28. Ioannou, Y.; Zeidner, K.; Grace, M.; Desnick, R. Human alpha-galactosidase A: Glycosylation site 3 is essential for enzyme solubility. *Biochem. J.* **1998**, *332*, 789–797. [CrossRef]
29. Kayser, V.; Chennamsetty, N.; Voynov, V.; Forrer, K.; Helk, B.; Trout, B.L. Glycosylation influences on the aggregation propensity of therapeutic monoclonal antibodies. *Biotechnol. J.* **2011**, *6*, 38–44. [CrossRef]
30. Meldgaard, M.; Svendsen, I. Different effects of N-glycosylation on the thermostability of highly homologous bacterial (1, 3-1, 4)-beta-glucanases secreted from yeast. *Microbiology* **1994**, *140*, 159–166. [CrossRef]
31. Baba, Y.; Sumitani, J.-I.; Tani, S.; Kawaguchi, T. Characterization of *Aspergillus aculeatus* β-glucosidase 1 accelerating cellulose hydrolysis with *Trichoderma cellulase* system. *AMB Express* **2015**, *5*, 3. [CrossRef]
32. Kabsch, W. XDS. *Acta Crystallogr. Sect. D Biol. Crystallogr.* **2010**, *66*, 125–132. [CrossRef] [PubMed]
33. Matthews, B. Solvent content of protein crystals. *J. Mol. Biol.* **1968**, *33*, 491–497. [CrossRef]
34. McCoy, A.J.; Grosse-Kunstleve, R.W.; Adams, P.D.; Winn, M.D.; Storoni, L.C.; Read, R.J. Phaser crystallographic software. *J. Appl. Crystallogr.* **2007**, *40*, 658–674. [CrossRef] [PubMed]
35. Adams, P.D.; Afonine, P.V.; Bunkóczi, G.; Chen, V.B.; Davis, I.W.; Echols, N.; Headd, J.J.; Hung, L.-W.; Kapral, G.J.; Grosse-Kunstleve, R.W.; et al. PHENIX: A comprehensive Python-based system for macromolecular structure solution. *Acta Crystallogr. Sect. D Biol. Crystallogr.* **2010**, *66*, 213–221. [CrossRef]
36. Bunkoczi, G.; Read, R. Improvement of molecular-replacement models with Sculptor. *Acta Crystallogr. Sect. D Biol. Crystallogr.* **2011**, *67*, 303–312. [CrossRef]

37. Emsley, P.; Cowtan, K. Coot: Model-building tools for molecular graphics. *Acta Crystallogr. Sect. D Biol. Crystallogr.* **2004**, *60*, 2126–2132. [CrossRef]
38. Brünger, A. Free R value: A novel statistical quantity for assessing the accuracy of crystal structures. *Nature* **1992**, *355*, 472–475. [CrossRef]
39. Agirre, J. Strategies for carbohydrate model building, refinement and validation. *Acta Crystallogr. Sect. D Struct. Biol.* **2017**, *73*, 171–186. [CrossRef]
40. Krissinel, E.; Henrick, K. Secondary-structure matching (SSM), a new tool for fast protein structure alignment in three dimensions. *Acta Crystallogr. Sect. D Biol. Crystallogr.* **2004**, *60*, 2256–2268. [CrossRef]
41. Meng, E.C.; Pettersen, E.F.; Couch, G.S.; Huang, C.C.; Ferrin, T.E. Tools for integrated sequence-structure analysis with UCSF Chimera. *BMC Bioinform.* **2006**, *7*, 339.
42. Krissinel, E. Macromolecular complexes in crystals and solutions. *Acta Crystallogr. Sect. D Biol. Crystallogr.* **2011**, *67*, 376–385. [CrossRef] [PubMed]
43. Gasteiger, E.; Hoogland, C.; Gattiker, A.; Duvaud, S.; Wilkins, M.R.; Appel, R.D.; Bairoch, A. Protein Identification and Analysis Tools on the ExPASy Server. In *The Proteomics Protocols Handbook*; Walker, J.M., Ed.; Humana Press: Totowa, NJ, USA, 2005; pp. 571–607.

International Journal of
Molecular Sciences

Article

Two Homologous Enzymes of the GalU Family in *Rhodococcus opacus* 1CP—*Ro*GalU1 and *Ro*GalU2

Antje Kumpf [1,2,3,*], Anett Partzsch [1], André Pollender [1], Isabel Bento [2] and Dirk Tischler [3,*]

1. Environmental Microbiology, Institute of Biosciences, TU Bergakademie Freiberg, Leipziger Str. 29, 09599 Freiberg, Germany; anett.partzsch@web.de (A.P.); andre.pollender@ioez.tu-freiberg.de (A.P.)
2. EMBL Hamburg, Notkestr. 85, 22607 Hamburg, Germany; ibento@embl-hamburg.de
3. Microbial Biotechnology, Faculty of Biology & Biotechnology, Ruhr University Bochum, Universitätsstr. 150, 44780 Bochum, Germany

* Correspondence: antje.kumpf@rub.de (A.K.); dirk.tischler@rub.de (D.T.); Tel.: +49-234-32-22082 (A.K.); +49-234-32-22656 (D.T.)

Received: 28 October 2019; Accepted: 16 November 2019; Published: 19 November 2019

Abstract: Uridine-5′-diphosphate (UDP)-glucose is reported as one of the most versatile building blocks within the metabolism of pro- and eukaryotes. The activated sugar moiety is formed by the enzyme UDP-glucose pyrophosphorylase (GalU). Two homologous enzymes (designated as *Ro*GalU1 and *Ro*GalU2) are encoded by most *Rhodococcus* strains, known for their capability to degrade numerous compounds, but also to synthesize natural products such as trehalose comprising biosurfactants. To evaluate their functionality respective genes of a trehalose biosurfactant producing model organism—*Rhodococcus opacus* 1CP—were cloned and expressed, proteins produced (yield up to 47 mg per L broth) and initially biochemically characterized. In the case of *Ro*GalU2, the V_{max} was determined to be 177 U mg^{-1} (uridine-5′-triphosphate (UTP)) and K_m to be 0.51 mM (UTP), respectively. Like other GalUs this enzyme seems to be rather specific for the substrates UTP and glucose 1-phosphate, as it accepts only dTTP and galactose 1-phoshate in addition, but both with solely 2% residual activity. In comparison to other bacterial GalU enzymes the *Ro*GalU2 was found to be somewhat higher in activity (factor 1.8) even at elevated temperatures. However, *Ro*GalU1 was not obtained in an active form thus it remains enigmatic if this enzyme participates in metabolism.

Keywords: glycosylation; UDP-glucose pyrophosphorylase; UDP-glucose; nucleotide donors; *Rhodococcus*, Actinobacteria, gene redundancy; Leloir glycosyltransferases; activated sugar; UTP

---

1. Introduction

Uridine-5′-diphosphate (UDP)-glucose is a key metabolite in most organisms and thus used in a variety of reactions of the sugar and starch metabolism, sugar interconversions, amino and nucleotide sugar metabolism, biosynthesis of antibiotics and cell envelope components, and as precursor for different primary and secondary metabolites [1–3]. Being an interesting and valuable compound, UDP-glucose attracted more and more attention for biotechnological applications. As an example, in glycosylation reactions, their production and application in fine chemical scale could be shown [1–3]. It was presented that whole cell catalysis by sucrose synthase (SuSy) with UDP as precursor led to large-scale production of 100 $g_{UDP\text{-}glucose}$/L with a yield of 86% [3]. Another report with 0.1 g/L of free SuSy could achieve 144 $g_{UDP\text{-}glucose}$/L with a comparable conversion rate of 85% based on the precursor UDP [1]. The enzymatic production of various sugar-nucleotides can be realized via several UDP-glucose synthesizing enzymes (Scheme 1). Glycosyltransferases, like sucrose synthase, are the most common used representatives [1,4–7]. Nucleotidyl transferases, like UDP-glucose pyrophosphorylase have been known for a long time but were seldom employed as biocatalysts to produce UDP-glucose, maybe due to low specific activities. Those enzymes are more often studied for their metabolic role.

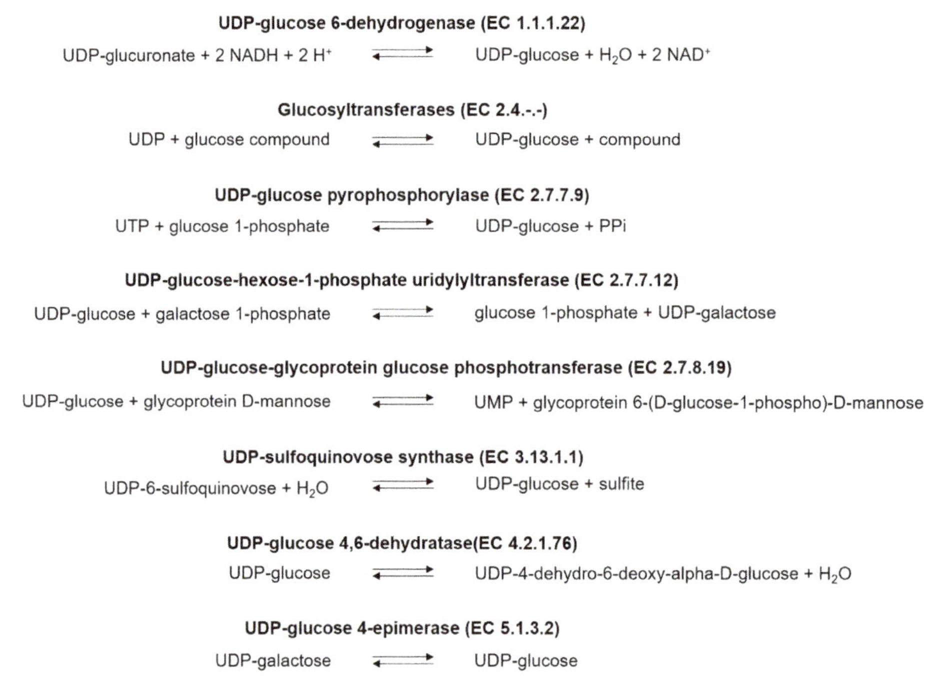

Scheme 1. Reactions in which UDP-glucose is consumed or formed. UDP: Uridine-5′-diphosphate, UMP: Uridine-5′-monophosphate, UTP: Uridine-5′-triphosphate, NAD^+/NADH: Nicotinamide adenine dinucleotide in the oxidized and reduced form.

UDP-glucose pyrophosphorylases (GalU or UGPase; EC 2.7.7.9) catalyze the reversible reaction of glucose 1-phosphate and UTP into UDP-glucose and inorganic pyrophosphate (PP_i). Enzymes of the GalU family are ubiquitous and can be found among the tree of life [8]. With respect to bacteria mostly proteobacterial GalUs have been studied. Like many other nucleotidyl transferases, also GalU requires divalent cations to promote the reaction (Scheme 2). In most cases magnesium ions are employed, and so far only magnesium chloride has been investigated [9–16]. The reaction mechanism follows a sequential bi-bi-mechanism starting with the binding of UTP to the active site, in presence of a magnesium ion [17], followed by the binding of glucose 1-phosphate. The octahedral coordination sphere of the magnesium positions the substrates in the right way and enables the nucleophilic attack of glucose 1-phosphate on UTP [18]. A lysine, an aspartate and several water molecules within the active site help to stabilize the position of the substrates and cofactor for the proper nucleophilic attack of the phosphoryl oxygen of glucose 1-phosphate towards the α-phosphor atom of UTP [17]. Finally, PP_i is released from the GalU/Mg^{2+}/UDP-glucose complex [17].

GalU enzymes have been shown to play crucial roles in galactose fermentation [19]. Other metabolic pathways in which this enzyme family is active are similar like the ones UDP-glucose itself plays an important role. It produces UDP-glucose as precursor for sucrose, glucan and amylose in the sugar and starch metabolism. Furthermore, it plays a role in galactose metabolism and glucuronate interconversion, where galactose 1-phosphate is over all converted into UDP-glucuronate with the help of a hexose 1-phosphate uridylyl transferase and a UDP-glucose 6-dehydrogenase. In amino sugar and nucleotide sugar metabolism, GalU is active in interconverting different UDP-sugars via the intermediate UDP-glucose. Additionally, it can be found in glycolipid metabolism and the synthesis of antibiotics. Respectively, many GalU enzymes from several organisms have been investigated to date, e.g., GalUs of plants [20–24], of mammals [25–27], including human [28], parasites [29], and many different bacteria [9,10,12,13,30–33], only to name a few groups. Fields of study included the determination of structures [17,30,34], biochemical characterization [14,31,35,36] and biological function [15,33,37].

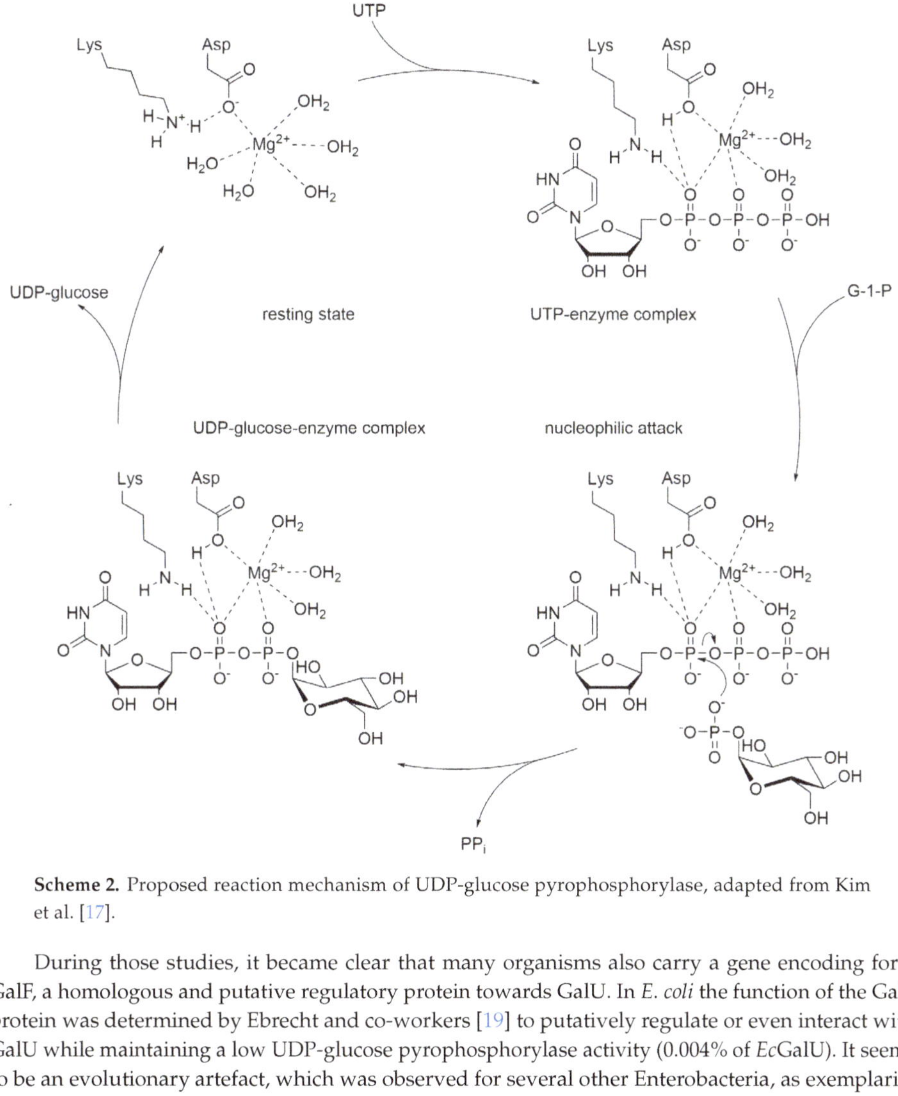

Scheme 2. Proposed reaction mechanism of UDP-glucose pyrophosphorylase, adapted from Kim et al. [17].

During those studies, it became clear that many organisms also carry a gene encoding for a GalF, a homologous and putative regulatory protein towards GalU. In *E. coli* the function of the GalF protein was determined by Ebrecht and co-workers [19] to putatively regulate or even interact with GalU while maintaining a low UDP-glucose pyrophosphorylase activity (0.004% of *Ec*GalU). It seems to be an evolutionary artefact, which was observed for several other Enterobacteria, as exemplarily different species and strains of *Klebsiella, Yersinia, Pectobacterium, Shigella, Escherichia, Pantoea, Salmonella, Serratia,* and some others [19]. Both proteins, GalU and GalF, seem to have evolved independently from a common ancestor, but only GalU kept the initial enzymatic role. Apparently, GalF underwent several mutations, which led to a much lower activity compared to GalU. By means of mutagenesis of identified key residues, the activity of GalF was restored and comparable to GalU and even a physiological complementation of a GalU-knockout strain was possible [19]. GalUs are known to have two important conserved motifs: The *N*-terminal one, which is involved in the binding of the uracil ring with the sequence G-X-G-T-R-X-L-P-X-T-K (X stands for any amino acid) [8,38] and a V-E-K-P-motif with an important lysine for glucose 1-phosphate binding [38–40]. The GalF of *E. coli* was shown to have the same V-E-K-P-motif, but instead of G-L-G-T-R, GalF has G-L-G-M-H residues in the *N*-terminal motif [19]. By means of mutagenesis it could be shown that the alteration of T-R into M-H was possibly the reason for the reduced UDP-glucose pyrophosphorylase activity of *E. coli* GalF. But, the actual function of GalF among Enterobacteria still remains unclear. If gene redundancy and physiological flexibility here between *E. coli* GalU and GalF is of relevance is not yet clarified. Furthermore, it remains enigmatic why this presence of homologous GalU/GalF-proteins has so far only been reported for Enterobacteria among the bacteria. However, in the genomes of *Rhodococcus* species, gene redundancy plays a crucial role [20,22,27,41–43]. Interestingly, most *Rhodococcus* species also carry two *galU* genes which we will highlight herein for the first time. Thus, *R. opacus* 1CP encodes for two isoforms of the GalU enzyme—*Ro*GalU1 and *Ro*GalU2, which have not been characterized yet. In some cases, it could be shown that enzymatic isoforms derived from rhodococci have the same or only a slightly different function, so that it was assumed that the coding isogenes have evolved independently from a common ancestor [44]. However, the functionality of respective proteins has not been investigated so far. Furthermore, to the best of our knowledge studies on UDP-glucose pyrophosphorylases in *Rhodocccus* species have not been carried out. Herein we describe the redundant GalU enzymes of *R. opacus* 1CP whereas one was found to be active as recombinant enzyme and might participate in several metabolic pathways, which is discussed in detail. Phylogenetic analysis of the amino acid sequences of GalU proteins of several rhodococci, Enterobacteria, and Actinobacteria helps to further understand the role of *Ro*GalU1 and *Ro*GalU2. This will also give more information on gene redundancy in Actino- and Enterobacteria in general.

2. Results

2.1. Genome Mining and Phylogenetic Analyses

Via a genome mining approach two genes encoding for two UDP-glucose pyrophosphorylases (GalUs) were identified in *Rhodococcus opacus* 1CP. The respective genes and amino acid sequences used for codon usage optimization were derived from the NCBI protein accession numbers ANS26426 (*Ro*GalU1, theoretical size: 33.2 kDa, GenBank accession number of the codon usage optimized nucleotide sequence: MN617759) and ANS26629 (*Ro*GalU2, theoretical size: 33.9 kDa, GenBank accession number of the codon usage optimized nucleotide sequence: MN617760). They have a sequence identity of more than 70% on amino acid level to each other when compared over full length.

Having a look at the genomic environment of both genes, various genes were found flanking *RogalU1*, which were involved in the sugar- and nucleotide metabolism, e.g., a mannose 6-phosphate isomerase, a D-glycero-D-manno-heptose 1-phosphate guanosyltransferase or a undecaprenyl-phosphate galactose phosphotransferase, whereas *RogalU2* is flanked by cation channels, heat shock proteins or putative regulatory proteins.

Analyzing the amino acid sequences of the GalUs originating from *R. opacus* 1CP, both contain the same motifs for uracil binding, G-X-G-T-R-F-L-P (start: amino acid 17 in *Ec*GalU sequence; 18 and 21 in *Ro*GalU1 and *Ro*GalU2, respectively). Equally, both display the V-E-K-P motif (start: amino acid 200 in *Ec*GalU sequence; 199 and 202 in *Ro*GalU1 and *Ro*GalU2, respectively) with the important lysine (K202 in *Ec*GalU) for binding of glucose 1-phosphate (see Figure 1; the subsequent numbering of amino acids is according to *Ec*GalU; [19]). Interestingly, only the GalF of *E. coli* shows an alteration from T20-R21 to M-H in the first motif, which was shown before to be the reason for the reduced UDP-glucose pyrophosphorylase activity of *E. coli* GalF. Large parts of *Ro*GalU1 and *Ro*GalU2 are identical and seem to be conserved. Several of those conserved amino acids can be found in *Ec*GalU and/or *Ec*GalF, respectively. Furthermore, all four sequences share the same residues for the hydrogen bonding of the uracil ring (A16-G17 and Q109), and at least the *Rhodococcus* GalUs and *Ec*GalU have the same amino acids for the phosphoryl binding of UDP-glucose (G179, K202 and G-A-G-D, start: amino acid: 234). The overall alignment indicates that the structure of all four proteins should be highly similar, as the secondary elements presented by α-helical and β-sheet areas of *Ec*GalU are present in *Ro*GalU1 and *Ro*GalU2, as well.

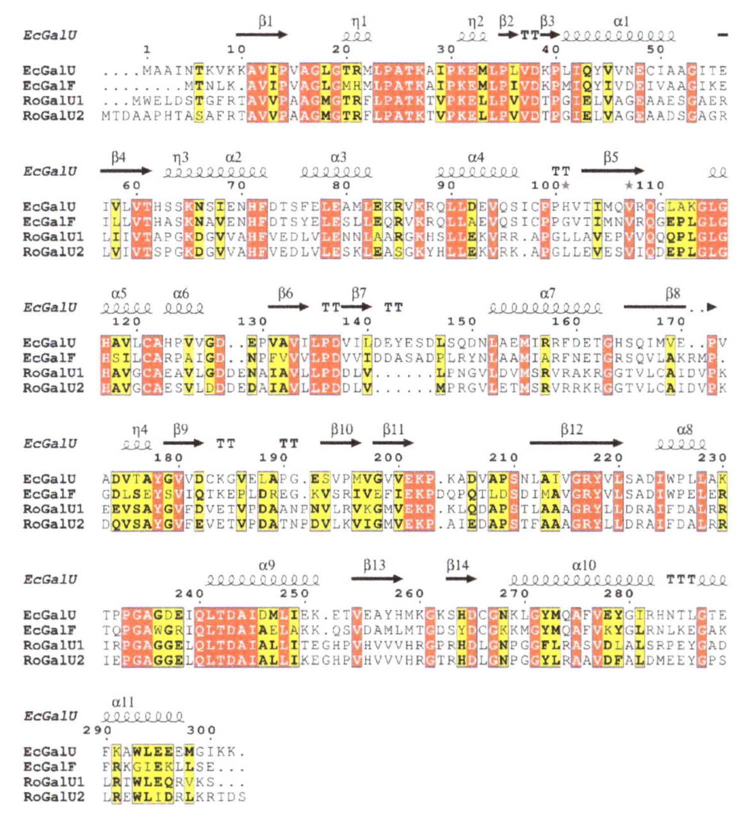

Figure 1. Alignment of *Ro*GalU1 and *Ro*GalU2 with the GalU and GalF amino acid sequences of *E. coli* K-12 [19], executed with the MUSCLE algorithm [45,46] in the program MEGA X [47] for multiple sequence alignment with default settings and imaged with ESPript 3.0 [48]. Conserved amino acids are color-coded. Red box with white letters: strictly identical amino acids, yellow box with black bold letters: similar amino acids, which are conserved among at least two sequences, black squiggles with α or η signs: α-helical structures, black arrows with β sign: β-sheet structures, **TT**: strict β-turns, **TTT**: strict α-turns, grey stars: residues with alternate conformations.

The alignment (Figure 1; part of the complete alignment with all 48 amino acid sequences) that underlies the phylogenetic tree (Figure 2) shows remarkable sequence similarities and again huge conserved areas between the GalU sequences of different actinobacterial species and especially strains of *Rhodococcus*, *Mycobacterium*, *Corynebacterium,* and *Gordonia*. As they are close relatives, it is not surprising that those sequences cluster together. It is striking that there are two sequence sections around the G-X-G-T-R-F-L-P (start: amino acid 17) motif and the V-E-K-P (start: amino acid 200) motif, which are highly conserved among all chosen sequences. Besides those and the highly conserved motifs for binding of the substrates and products, there are also three more conserved motifs: One G-L-G-H sequence (start: amino acid 114) before and two sequences between the V-E-K-P motif and the *C*-terminus, namely G-R-Y-L-L (start: amino acid 216) and Q-L-T-D-A-I (start: amino acid 240) within all analyzed sequences. In addition, those seem to be highly conserved, not only among Actinobacteria, but also Enterobacteria. Furthermore, the sequences contain single conserved amino acids, like G53, F72, E79, D137, Y178, S210, G261, D265, and G267 (Figure 1).

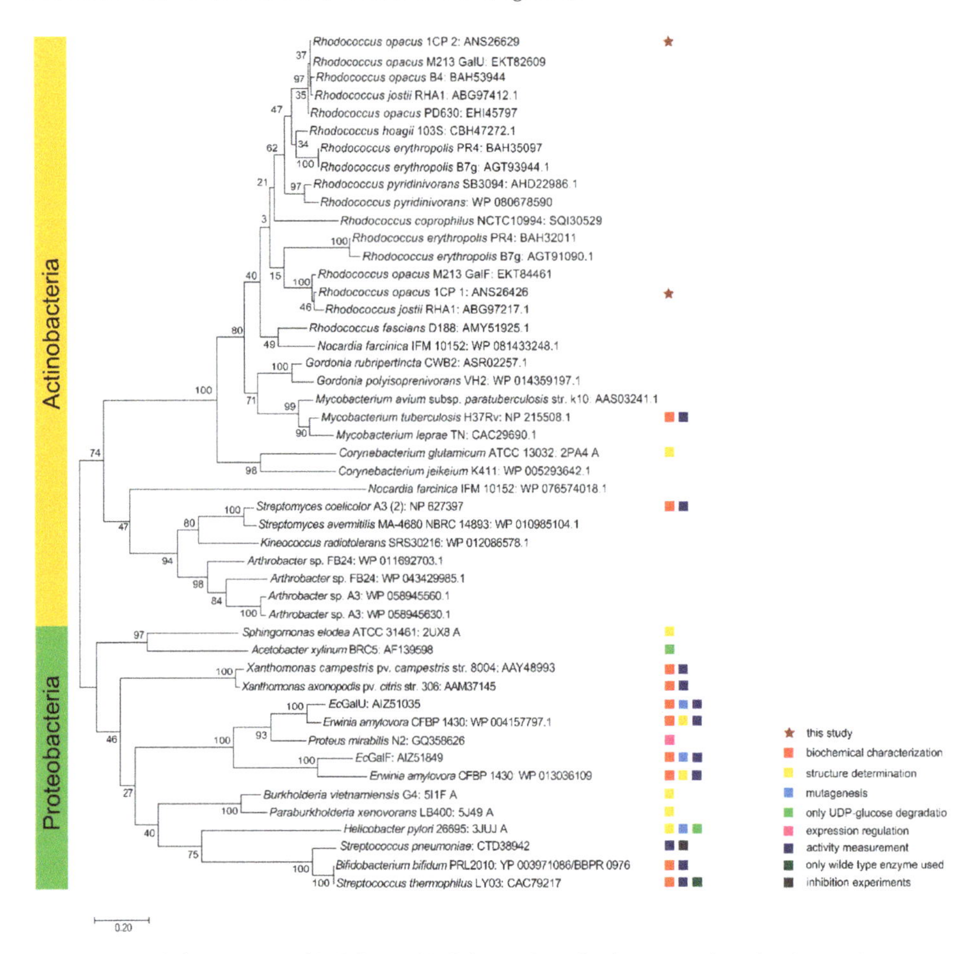

Figure 2. Phylogenetic tree of *Ro*GalU1 and *Ro*GalU2 with similar characterized or related UDP-glucose pyrophosphorylase sequences. The evolutionary history was inferred by using the Maximum Likelihood method and JTT matrix-based model [49]. The tree with the highest log likelihood (−13775.30) is shown. The percentage of trees in which the associated taxa clustered together is shown next to the branches. Initial tree(s) for the heuristic search were obtained automatically by applying Neighbor-Join and BioNJ algorithms to a matrix of pairwise distances estimated using a JTT model, and then selecting the topology with superior log likelihood value. The tree is drawn to scale, with branch lengths measuring the number of substitutions per site. This analysis involved 48 amino acid sequences. There was a total of 375 positions in the final dataset. Evolutionary analyses were conducted in MEGA X [47].

The above described alignment was used and completed by numerous GalU protein sequences in order to generate a distance tree for a phylogenetic analysis (Figure 2). The first big branch of the tree contains all actinobacterial proteins, which are very close relatives. Within those, also *Rhodococcus* enzymes form one big cluster. Interestingly, the enzymes from the same strains do not cluster together, but form two different smaller clusters. The *Rhodococcus* strains containing only one GalU protein are mapped within the branch of our tested and active *Ro*GalU2. The second big branch of the tree contains Actinobacteria in the upper part, Firmicutes, and Proteobacteria in the lower part.

2.2. Recombinant Expression of *RogalU1* in *E. coli*, Purification, and Renaturation

For characterization purposes in more detail *RogalU1* was cloned into the expression vector pET16bP. After transformation of chemically competent *E. coli* BL21(DE3) pLysS with pET16bP-*RogalU1* overexpression and respective protein production and purification were carried out as described in the material and methods section. The SDS-PAGE in Figure 3a shows the expected molecular weight of *Ro*GalU1 of 33 kDa, but most of the produced enzyme is present as insoluble protein in the cells. The formation of those inclusion bodies was determined from the greyish to white color of the pellet after crude cell extract preparation. Hence, it was tried to renature the misfolded protein with the Thermo Scientific Pierce Protein Refolding Kit according to the manufacturer's instruction. Unfortunately, it was not possible to obtain *Ro*GalU1 in an active and soluble form. Also, protein within the inclusion bodies only had a marginal residual enzyme activity. This was determined to be <0.001 U mg^{-1} by means of the standard test which is described in the materials and methods section (4.6). In this work, we determined throughout enzyme activity with respect to product formation (1 U corresponds to the formation of 1 μmol UDP-glucose per minute). In order to turn inclusion bodies into native protein, the expression was altered with respect to various media and cultivation conditions. All efforts failed and thus this enzyme was not characterized further.

Figure 3. SDS-PAGE gels of *Ro*GalU1 (**a**) and *Ro*GalU2 (**b**). Arrows indicate the expected molecular weight of 33 kDa for *Ro*GalU1 and 34 kDa for *Ro*GalU2. SP: soluble proteins, IB: inclusion bodies, PP: purified protein, M: Marker.

2.3. Recombinant Expression in *E. coli*, Purification and Production of Active *Ro*GalU2

The gene *RogalU2* was treated as described in the materials and methods and similarly as the *RogalU1* described above. In this case, it was possible to produce about 47 mg of pure, soluble protein with an expected molecular weight of 34 kDa (see Figure 3b) from 25 g wet biomass. The maximum enzyme activity observed was 3.6 U mg^{-1} in the direction of UDP-glucose formation, using the standard reaction conditions. In addition, different media for gene expression were tested (LB, TB, or NB), but no significant difference in terms of protein yield or activity was found. TB medium was the fastest medium for cell growth and led to most biomass, therefore using TB auto induction medium for later large-scale production was an obvious choice.

In order to validate the oligomeric state of the enzyme, a size exclusion chromatography under non-denaturizing conditions was performed in duplicates with two different concentrations of the protein (0.2 and 2.0 mg mL^{-1}), each. The elution profiles were similar in those cases (not shown). The respective calibration was also measured twice and revealed the occurrence of this protein mostly in a hexameric state (188 kDa elution volume vs. subunit size of 34 kDa; Figures 3 and 4). Thus, we can state that under these conditions the protein was present mostly in a single hydrodynamic state and therefore protein characterization can be performed.

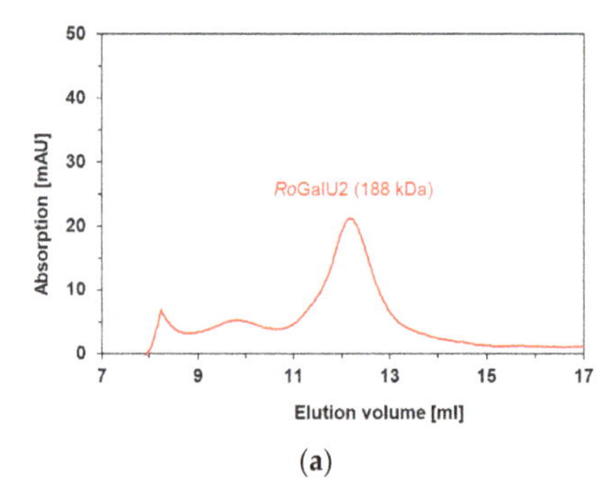

(a)

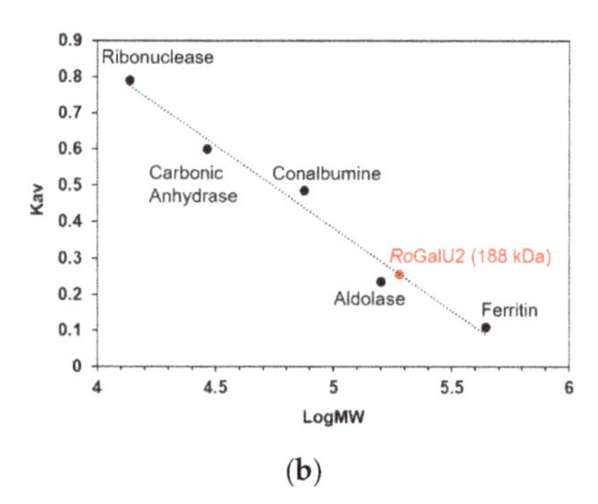

(b)

Figure 4. Size exclusion chromatography of *Ro*GalU2 (**a**) and the corresponding calibration measurements (**b**). The calibration curve was prepared with the following proteins of known molecular weight (given in brackets): Ribonuclease (13.7 kDa), Carbonic Anhydrase (29 kDa), Conalbumin (75 kDa), Aldolase (158 kDa) and Ferritin (440 kDa). All elutions were performed in a buffer containing 50 mM Hepes, pH 7.0 and 1 mM magnesium chloride. See methods Section 4.5. The protein concentration was either 0.2 or 2.0 mg mL^{-1} whereas the latter is presented herein.

2.4. Determining Optimum Reaction Conditions for RoGalU2

In order to collect proper kinetic data for the enzyme activity of *Ro*GalU2 in the direction of UDP-glucose formation, it was necessary to find optimum reaction conditions. In addition, the product detection method and fitting procedure had to be established. UDP-glucose formation by *Ro*GalU2 was determined by HPLC as described in materials and methods section. A typical result and a respective rate determination is presented in Figure 5.

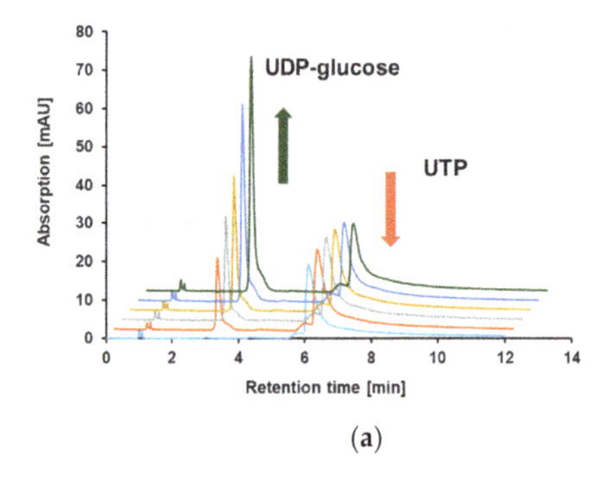

(a)

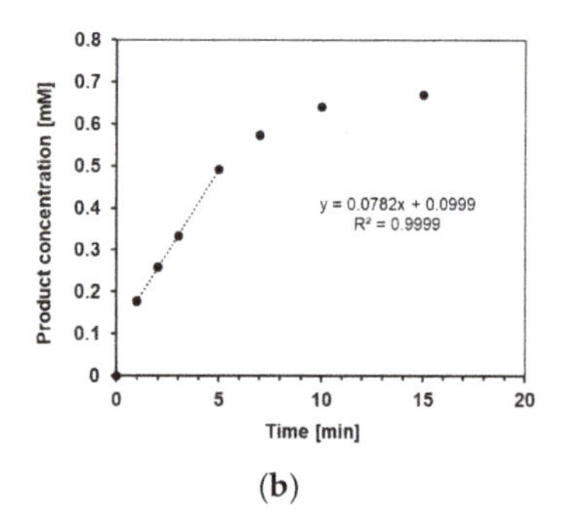

(b)

Figure 5. Enzymatic conversion of UTP and glucose 1-phosphate into UDP-glucose and PP_i. In (**a**) the HPLC chromatograms of a time course from 0 to 7 min are presented and the determined concentration of UDP-glucose was used to calculate a product formation rate as shown in (**b**). For this analysis the standard enzyme assay and determination methods as described in materials and methods section have been used.

Firstly, the reaction temperature was varied from 4 °C to 87 °C. There was an activity plateau determined between 40 °C and 45 °C, with a maximum observed enzyme activity of 6.7 U mg^{-1} (Figure 6a). To monitor the stability of *Ro*GalU2 at different temperatures, the enzyme was incubated for 30 min at temperatures ranging from 0 °C to 70 °C. The activity assay was then carried out at 45 °C. *Ro*GalU2 shows at least 80% of activity when incubated for 30 min between 19 °C and 45 °C (see Figure 6b). Because of the higher activity that could be achieved at 42 °C *versus* 45 °C, it was decided to perform all subsequent experiments at 42 °C.

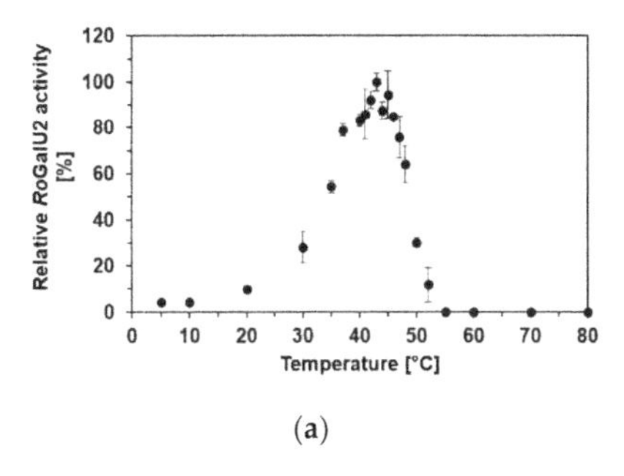

(a)

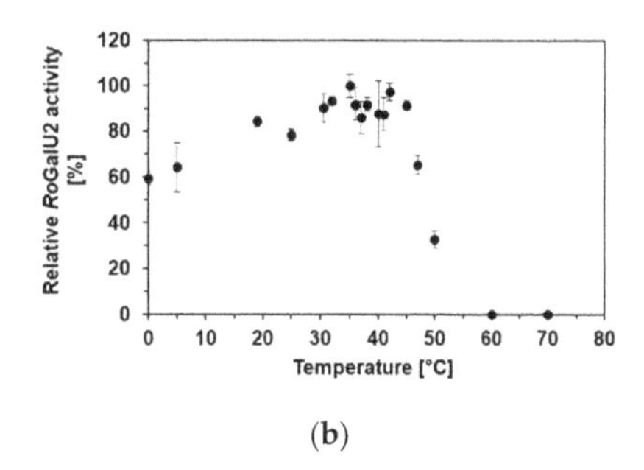

(b)

Figure 6. Temperature-dependent activity (**a**) and stability (**b**) of purified *Ro*GalU2 enzyme was measured in triplicates. Activity was measured with pre-incubation of the assay mixture (2 mM UTP, 2 mM glucose 1-phosphate, 4 mM $MgCl_2$, 50 mM Hepes buffer, pH 7.0, 30 µg *Ro*GalU2, 1 mL reaction volume) at indicated temperatures. The enzyme was stored on ice and directly used for measurements. For stability measurements the enzyme was incubated shaking for 30 min at indicated temperatures and directly used for activity assay at 45 °C. Means and standard deviations are shown. 100% relative *Ro*GalU2 activity correspond to 6.7 U mg^{-1} ((**a**), activity) and 5.0 U mg^{-1} ((**b**), stability), respectively.

To verify the dependence of *Ro*GalU2 on magnesium the concentration of magnesium chloride and magnesium sulfate present in the standard reaction setup at 42 °C was varied between 0–10 mM. For magnesium salt concentrations below 1 mM, the standard buffer of the enzyme had to be exchanged by a buffer without magnesium chloride freshly prior to respective experiments since *Ro*GalU2 is not stable in a buffer without magnesium. The activity was also measured after every buffer exchange. Figure 7 shows the absolute dependence of the activity on the concentration of magnesium ions. There is no activity and thus UDP-glucose formation measurable without magnesium salt. The highest activity was determined to be 5.2 U mg^{-1} with magnesium chloride and 9.5 U mg^{-1} with magnesium sulfate, in the presence of only 1 mM of the respective salt. When increasing the magnesium concentration, the *Ro*GalU2 activity decreases significantly, falling below 60% with 10 mM magnesium salt. Thus, it can be concluded that magnesium is a crucial cofactor for the enzyme as presented in Scheme 2, respectively.

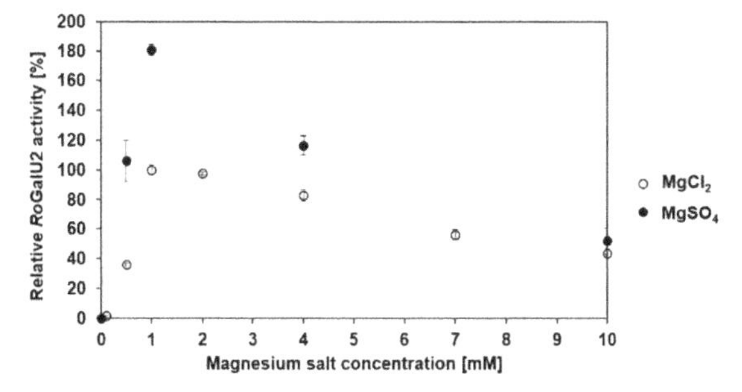

Figure 7. Relative *Ro*GalU2 enzyme activity depending on the $MgCl_2$ and $MgSO_4$ concentration between 0 mM and 10 mM. Assay mixture contained 2 mM UTP, 2 mM glucose 1-phosphate, 0–10 mM $MgCl_2$ or $MgSO_4$, 50 mM Hepes buffer, pH 7.0, 30 µg *Ro*GalU2. Reaction was carried out at 42 °C in 1 mL reaction volume. Means with standard deviations of triplicate measurements are shown. 100% relative *Ro*GalU2 activity correspond to 5.2 U mg^{-1}.

The activity of *Ro*GalU2 with respect to magnesium salts was most similar at a concentration of 4 mM. Thus, further divalent cations were tested as metal chloride salts at 4 mM each like manganese, nickel, cobalt, calcium, and zinc. In addition, trivalent metal cations like iron and aluminum were tested. No activity was observed with the trivalent metal chlorides. For all other divalent chlorides, the relative activity was below 20%. Thus, a magnesium dependency was clearly determined.

To test whether masking of the magnesium by means of EDTA results in a reduced enzyme activity, we applied 1 mM EDTA to the reaction solution and could not observe any activity. Therefore, the standard assay was used but at a lower magnesium chloride concentration, which was set to 1 mM as well. We also tested the reaction in presence of acetonitrile as solvent and received the same result. This was of importance since acetonitrile was used to stop the reactions prior to HPLC analysis as well as to adjust samples towards elution conditions. Thus, it was proven that acetonitrile was a proper reagent to stop the reaction.

Different buffers like Bis-Tris, MOPS, sodium phosphate and imidazole were also tested to see whether the enzymatic activity can be further increased. The pH of the reaction solution was also varied ranging from 6.6–9.3 in total and split into the following ranges of different buffers: 6.6–7.4 (Imidazole), 7.4–8.4 (Hepes), and 7.6–9.3 (Tris-HCl). Figure 8 shows a graph with the comparison of *Ro*GalU2 activities measured in those reaction solutions.

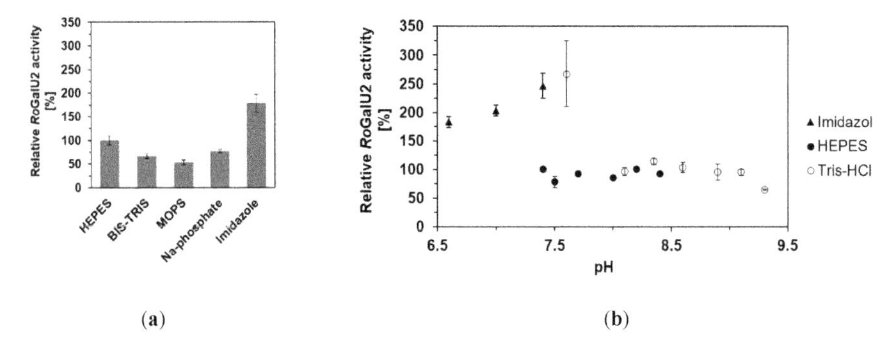

(a) (b)

Figure 8. *Ro*GalU2 enzyme activity depending on the buffer (**a**) and pH (**b**) of the reaction solution. (**a**) Assay mixtures contained 1.5 mM UTP, 250 mM glucose 1-phosphate, 1 mM $MgCl_2$, 50 mM buffer, pH 7.4, 1.7 µg *Ro*GalU2. Reaction was carried out at 42 °C in 1 mL volume. (**b**) Assay mixtures contained 1.5 mM UTP, 250 mM glucose 1-phosphate, 1 mM $MgCl_2$, 50 mM Imidazole, pH 6.6–7.4 (black triangle) or 50 mM Hepes, pH 7.4–8.4 (black circle) or 50 mM Tris-HCl, pH 7.6–9.3 (white circle), 1.7 µg *Ro*GalU2. Reaction was carried out at 42 °C in 1 mL volume. Means with standard deviations are shown out of a minimum set of three independent measurements. 100% relative activities correspond to 126 U mg^{-1} (**a**) and 109 U mg^{-1} (**b**).

In none of the used buffers the relative *Ro*GalU2 activity dropped below 60%. But it was possible to increase the relative activity up to 180% with Imidazole at pH 7.4. The pH profile even shows an increase up to 250% in activity when Imidazole or Tris-HCl are used at pH values of around 7.5. Interestingly, there is no significant difference when Hepes buffer is used at pH values between 7.4 and 8.3 or when Tris-HCl is used between pH 8.1 and 9.1. Here we only observed relative *Ro*GalU2 activities around 80%.

The above described results allowed to formulate an improved enzyme assay with the following conditions: 1.5 mM UTP, 250 mM glucose 1-phosphate, 1 mM magnesium chloride, 50 mM Hepes, pH 7.4, 42 °C in 1 mL reaction volume. Even though the activity of *Ro*GalU2 is higher with Imidazole pH 6.6–7.4 or Tris-HCl pH 7.6, we decided to continue using Hepes pH 7.4, because the standard deviation with Hepes buffer was found to be much smaller than with other buffers. This might be due to the weaker buffering properties of Tris-HCl or Imidazole in the applied range.

2.5. Enzyme Kinetics with RoGalU2

In order to collect proper kinetic data, the above described improved enzyme assay was employed. Thus, we could now determine the common kinetic constants K_m, V_{max}, k_{cat}, and k_{cat}/K_m by means of varying the substrate concentrations. Already here we like to state the K_m and thus k_{cat} values have to be considered as apparent values as later indicated in Table 1. This is necessary as a saturation of the enzyme by both substrates cannot be secured as the later on presented data show. Pre-experiments defined 1.5 mM UTP and 250 mM of glucose 1-phosphate as most suitable fixed concentrations for the variation of the opposite substrate, respectively. As shown in Figure 9 UTP was varied between 10 µM and 5 mM with a fixed concentration of 250 mM glucose 1-phosphate. The data were fitted to the model of Yano and Koga [50] (Equation (2), see Methods section) for substrate inhibition, because above 1.5 mM UTP a strong substrate inhibition became obvious. Below this concentration, the *Ro*GalU2-activity reaches a maximum observed activity of about 122 U mg^{-1}. The calculated V_{max} was determined to be 177 U mg^{-1} according to the inhibition fit of Yano and Koga [50].

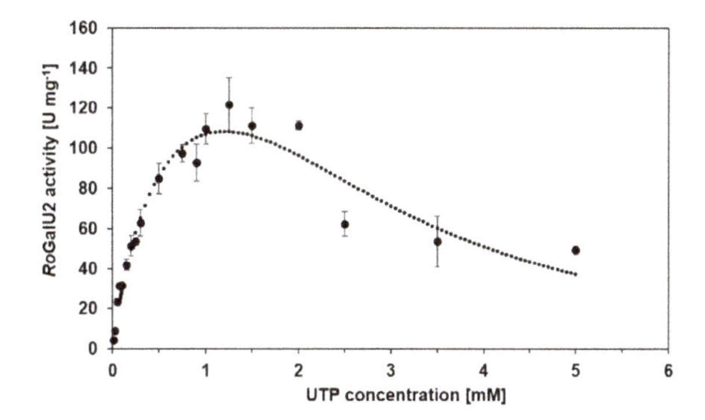

Figure 9. *Ro*GalU2 enzyme activity depending on the UTP concentration between 10 µM and 5 mM at a fixed glucose 1-phosphate concentration of 250 mM. Fit is calculated by the Model of Yano and Koga [50], for enzyme kinetics with substrate inhibition. The assay mixture contained 0.01–5 mM UTP, 250 mM glucose 1-phosphate, 1 mM $MgCl_2$, 50 mM Hepes buffer, pH 7.4, 1.7 µg *Ro*GalU2. Reaction was carried out at 42 °C in 1 mL volume. Mean values with standard deviations from triplicates are shown.

Figure 10 shows the graphs for substrate variation of glucose 1-phosphate between 50 µM and 500 mM with 1.5 mM UTP in each reaction. Figure 10a shows a Michaelis–Menten-like kinetic at concentrations between 50 µM and 10 mM. This is common for many GalU enzymes as discussed later. But with a closer look at the data, it became clear that the model does not describe the values in a proper way. In contrast to common Michaelis–Menten kinetics, the *Ro*GalU2 activity still increased, beyond the concentration of 10 mM glucose 1-phosphate applied. Therefore, the collection of data for higher glucose 1-phosphate concentrations was necessary and showed a kinetic behavior that was different from general models that describe enzyme kinetics (Figure 10b), with a maximum observed activity of about 119 U mg^{-1}. No model allowed to fit the obtained data properly. Further, we repeated the experiments with other batches of protein and the results were similar. To achieve a maximum formation rate of UDP-glucose, a high concentration of glucose 1-phosphate was mandatory, resulting in those data.

The kinetic properties of *Ro*GalU2 are summarized in Table 1. Further, to describe the substrate scope of *Ro*GalU2, we used UTP, ATP, GTP, CTP, and dTTP as nucleotides and glucose 1-phosphate, galactose 1-phosphate, ribose 5-phosphate and glucose as sugars in different combinations to each other. Only with sugar 1-phosphates and UTP or dTTP, product formation was detectable. Nevertheless, the relative activity of the reaction of dTTP and glucose 1-phosphate could not be evaluated due to unavailability of a product standard. A low relative *Ro*GalU2 activity of 2% was observable when galactose 1-phosphate and UTP were tested (100% relative *Ro*GalU2 activity corresponded to 8.1 U mg^{-1}). With other nucleotides or sugars no activity or product formation was observable.

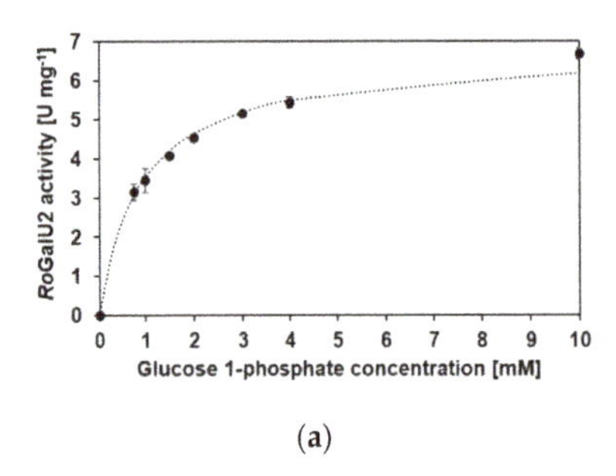

(a)

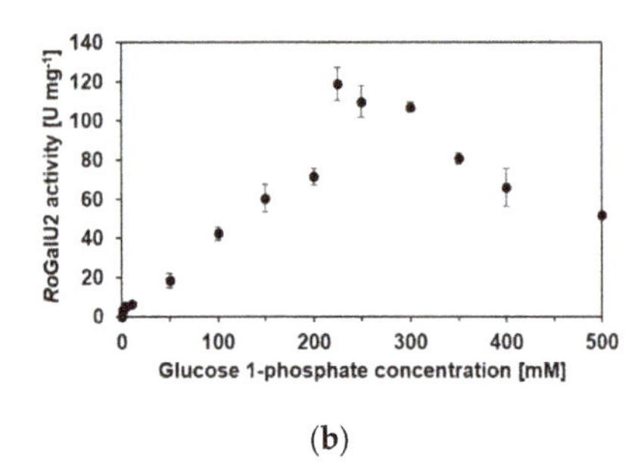

(b)

Figure 10. *Ro*GalU2 enzyme activity depending on the glucose 1-phosphate concentration between 50 µM and 500 mM at a fixed UTP concentration of 1.5 mM. Data for glucose 1-phosphate concentrations between 0 and 10 mM were fit to the kinetic model of Michaelis–Menten (**a**). The data obtained over the total glucose 1-phosphate variation are presented in (**b**), whereas no fit was able to describe the data. Assay mixture contained 0.05–500 mM glucose 1-phosphate, 1.5 mM UTP, 1 mM $MgCl_2$, 50 mM Hepes buffer, pH 7.4, 1.7 µg *Ro*GalU2. Reaction was carried out at 42 °C in 1 mL volume. Means with standard deviations are shown from a minimum of triplicates.

Table 1. Kinetic properties of *Ro*GalU2.

Varied Substrate	UTP [1]	Glucose 1-Phosphate [1]	
conc. [mM]	0.01–5	0.05–10	0.05–500
apparent K_m [mM]	0.51	0.9	150
V_{max} [U mg^{-1}] [2]	177	6.75	-
$V_{max,ob}$ [U mg^{-1}] [3]	122	6.68	119
apparent k_{cat} [s^{-1}]	96.2	3.7	65
K_I [mM]	2.61	-	225
k_{cat}/K_m [µM^{-1} s^{-1}]	0.19	0.004	0.0004

[1] Assay mixture contained substrates at varied concentrations given whereas the other substrate was set constant at 250 mM glucose 1-phosphate or 1.5 mM UTP, respectively, 1 mM $MgCl_2$, 50 mM Hepes buffer, pH 7.4, 1.7 µg *Ro*GalU2. Reaction was carried out at 42 °C in 1 mL volume. All experiments were performed at least in triplicates. Fits were according Michaelis–Menten or Yano and Koga [50]. [2] The specific activity was obtained from fitting the data. [3] The maximal observed activity was taken as a read out from the plot.

3. Discussion

UDP-glucose pyrophosphorylases are important enzymes for the general metabolism of many organisms as highlighted in the introduction. However, among the bacterial representatives mostly GalU enzymes from Proteobacteria have been studied. To a lesser extent, GalUs of Actinobacteria were described, as for example some information are available for the ones originating of *Mycobacterium* or *Streptomyces* species [10,13,51]. Here we wanted to add knowledge and studied the enzyme family from rhodococci, which is interesting from a phylogenetic as well as mechanistic point of view. In a model organism, *Rhodococcus opacus* 1CP, relevant for trehalose biosurfactant production or aromatic compound degradation [52,53], two isogenes encoding for UDP-glucose pyrophosphorylases were identified and described for the first time herein.

A phylogenetic tree based on an alignment of 48 protein sequences of partially characterized UDP-glucose pyrophosphorylases was generated (Figure 2). The tree is separated into two groups; proteobacterial and actinobacterial representatives. It should be stressed that some protein sequences have been annotated as GalU and others as GalF enzymes or regulatory proteins, respectively, but in most cases this has not been experimentally verified.

The first group contains GalU- or GalF-designated protein sequences of partially characterized UDP-glucose pyrophosphorylases or related sequences from Actinobacteria, the other contains GalUs and GalFs from Proteobacteria. The branch of actinobacterial enzymes is separated again into two subgroups; Corynebacteriales and other orders of Actinobacteria. Within the branch of corynebacterial enzymes different *Rhodococcus* species cluster together. Gene products derived from soil or wastewater populating rhodococci (which encode for two GalU enzymes) cluster into two subbranches, respectively. Within those, GalUs of the zoonotic pathogen or animal infecting *Rhodococcus* species, such as *R. coprophilus* and *R. hoagie* (*equii*), as well as those *Rhodococcus* strains that only comprise a single GalU gene form one cluster. Herein, the protein *Ro*GalU2 from our model strain is localized. An exception is the plant pathogenic *R. fascians*, which also encodes for only one GalU protein. It seems more related to a pathogenic *Nocardia* then to other rhodococci enzymes. From those *Rhodococcus* species carrying two GalUs, a second branch is formed in which the protein *Ro*GalU1 clusters. Another branch within corynebacterial representatives is formed by GalUs of the genera *Gordonia*, *Mycobacterium* and *Corynebacterium*. And another more distant group of the actinobacterial GalUs contains enzymes of the genera *Streptomyces*, *Kineococcus* and *Arthrobacter*. Interestingly, *Arthrobacter* also has strains carrying two isoforms of GalU proteins, but they form a group far from the *Rhodococcus* proteins. A similar behavior is present in the group of proteobacterial GalUs including some annotated but also experimentally verified GalFs. Here the enzymes are arranged according to their phylogenetic relation as expected. However, *E. coli* and *Erwinia amylovora* each encode two related proteins which form distinct branches.

As the branch of proteins named GalU2 also contains other actinobacterial GalUs only having one GalU isoform, it is possible that GalU2 is the older protein with the initial function in the genus *Rhodococcus*. Thus, GalU1 might have evolved from GalU2 and gained the function of a UDP-glucose pyrophosphorylase. But the genetic environment of the respective coding genes does not support this hypothesis. The genetic environment of *RogalU1* is more related to the sugar metabolism with sugar-phospho transferases and sugar-phospho isomerases, whereas the genetic environment of *RogalU2* comprises many putative and for the sugar or amino metabolism unspecific genes. It consists of different genes coding for cation and phosphate channels, transmembrane proteins, conductance and mechanosensitive channels, acting as an osmotic release valve in response to osmotic stress, as well as ATP converting enzymes and heat shock proteins. Interestingly, the gene cluster of *RogalU1* is only available in *R. jostii* RHA1, whereas the gene cluster surrounding *RogalU2* is present in many other Actinobacteria, like *N. farcinica*, *M. smegmatis*, *M. tuberculosis*, *C. glutamicum*, among others. Therefore, the *RogalU2* cluster seems to be the more common one for Actinobacteria.

Interestingly, *Ro*GalU2 has 99% amino acid sequence identity to the GalU regulator GalF of *R. opacus* PD630. However, this GalF designation for strain PD630 is based only on a bioinformatic annotation and no other *galU* gene is present in the genome of *R. opacus* PD630. Thus, this protein must also be an UDP-glucose pyrophosphorylase and not a regulatory element. *Ro*GalU1 only has 83% sequence identity to this mentioned protein of *R. opacus* PD630. Thus, it seems likely that the protein more related to our *Ro*GalU2 is of importance for rhodococci, which will need to be verified by further studies and representatives. GalF of *E. coli* only has a slight residual UDP-glucose pyrophosphorylase activity. *E. coli* GalF showed a drastically increased activity and substrate specificity for glucose 1-phosphate when exchanging the methionine in position 15 into a threonine and the histidine in position 16 into an arginine (both within the G-X-G-T-R-F-P-L motif of GalUs), which are the corresponding residues in *E. coli* GalU [19]. In the sequences of *Ro*GalU1 and *Ro*GalU2, there is no alteration in the G-X-G-T-R-F-L-P motif. Thus, a regulatory role among the *Rhodococcus* proteins is excluded at this stage.

The underlying alignment of the distance tree shows that GalU enzymes in general are very conserved among bacteria, and that the degree of conservation within Actinobacteria is even higher. The first 100 amino acids are almost fully conserved with only some exceptions. As the two largest conserved areas of all shown sequences are around the motifs for substrate and product binding, it is obvious that the reaction mechanism is very likely the same in all bacterial GalU enzymes. It is striking that there are also four more motifs and several single amino acids which are highly conserved among bacterial GalU sequences. The tyrosine of the G-R-Y-L-L motif and the leucine in the Q-L-T-D-A motif are also mentioned by Aragão et al. to form a hydrophobic cap to the base of the sugar ring in the active site [9]. It could be possible that the other motifs and amino acids are not only important for the structure formation, like establishing the subunit interaction, but also for the construction of the active site, the binding of substrate and product or even the binding of magnesium ions as cofactors. They could help to stabilize the nucleophilic attack of glucose 1-phosphate to UTP or to bring the active site in the right conformation when the substrates are bound.

Hence, both genes *RogalU1* and *RogalU2* were successfully cloned into expression systems and protein production for a subsequent biochemical characterization was studied. It was not possible to produce active and soluble *Ro*GalU1. In addition, the inclusion bodies obtained had only a low residual enzyme activity. Neither the optimization of medium or expression conditions nor the renaturation of wrongly folded protein lead to soluble and active *Ro*GalU1. Inclusion bodies can be avoided or reduced by different methods. Reduction of the expression temperature to 14–16 °C or even lower, reduction of the expression time or ITPG concentration could possibly increase the amount of soluble active protein [54]. Also, co-expression of the gene of interest with chaperones, are reported to help folding the protein in the right way [55]. The production of a fusion protein with mCherry or other suitable proteins can increase the protein solubility as well [2]. The sole production of inclusion bodies of a GalU from *M. tuberculosis* could be overcome by using an expression system which was more related to the donor organism, so gene expression was achieved in competent *M. smegmatis* cells and the mycobacterial specific expression vector pMIP12 for improved gene expression [51].

*Ro*GalU2 was successfully overproduced and purified in reasonable amounts compared to literature, but with a higher yield of insoluble recombinant *Ro*GalU2, as well [13]. SDS-PAGE revealed the expected molecular weight of about 34 kDa. Within bacteria this molecular weight corresponds to that of other GalUs [12,13,56]. The oligomeric state of the protein was determined to be hexameric by size-exclusion chromatography. But the other known bacterial GalUs were shown to be dimers [10,31] or tetramers [8,9,12,17,30]. Only Lai et al. reported the GalU of *M. tuberculosis* strain H37Rv to be dimeric and hexameric in solution [13]. As the latter one is closely related to our target *Ro*GalU2 this fits well to our findings (Figure 2). But we did observe a hexameric state under applied conditions, which can be studied in more detail by altering the buffer or presence of divalent ions through the size exclusion chromatography. Respectively, the presence of a protein in a single hydrodynamic state allowed us to study the biochemical properties in more detail.

*Ro*GalU2 with a V_{max} of 177 U mg^{-1} (UTP) and 119 U mg^{-1} (G1P), respectively, has a high activity compared to other bacterial GalUs. Many other bacterial GalUs show activities between 0.1 U mg^{-1} for *M. tuberculosis* GalU [13] and 90 U mg^{-1} for *Xantomonas campestris* [31]. Similar or slightly higher activities for bacterial GalUs are known as well, like 270 U mg^{-1} for *Streptomyces coelicolor* GalU [10] and 340 U mg^{-1} for *E. coli* GalU [19]. Higher activities are only known from eukaryotes [24,57,58]. *Ro*GalU2 also showed a high apparent turnover frequency (app. k_{cat}) with 96 s^{-1}. But this value is very heterogeneous among bacteria, showing lower, similar and higher turnover frequencies [13,19,30]. An overview on bacterial GalU enzymes is provided in Table 2.

Table 2. Kinetic properties of UDP-glucose pyrophosphorylases used for the phylogenetic analysis. Units are given in brackets in the table head, unless otherwise indicated.

Organism and Protein	V_{max} [U mg^{-1}]		K_m [mM]		k_{cat} [s^{-1}]		c [mM]		Ref.
	UTP	G1P	UTP	G1P	UTP	G1P	UTP	G1P	
R. opacus 1CP GalU2	177	6.75[1] 119[2]	0.51	0.9[1] 150[2]	96.2	3.7[1] 65[2]	0.01–5	0.05–500	This study
E. coli K-12 GalU	340		0.17	0.035	187		2	2	[19]
E. coli K-12 GalF	0.015		0.36	0.52	0.008		2	2	[19]
M. tuberculosis H37Rv GalU	2.5–2.7[3] 9.8 U[4]	5.8 U[4]	0.1[3] 0.012[4]	0.13[3] 0.045[4]	93.44[4]	55.03[4]	0.001–0.01[4]	0.008–0.05[4]	[13,51]
E. amylovora CFBP 1430 GalU	14.4	32.5	0.027	0.007	7.9	17.2	0.006–0.1	0.003–0.1	[30]
S. coelicolor A3 (2) GalU	270		>10	0.06	149		≤20	n.s.	[10]
B. bifidum PRL2010 GalU	13		0.042	0.098	12		0–2	0–2	[59]
X. capestris pv. *campestris* str. 8004 GalU	60–90		0.21	0.06	40		n.s.	0-8	[31]
X. axonopodis pv. *citris* str. 306 GalU	60–90		0.11	n.s.	29		n.s.	n.s.	[31]
S.thermophilus LY03 GalU	0.2		n.s.	n.s.	0.11		1.25	1	[32]

[1] for glucose-1-phosphate concentrations up to 10 mM; [2] for glucose-1-phosphate concentrations between 10–500 mM; [3] [51]; [4] [13]; n.s.: not specified; The K_m and k_{cat} values in case of *Ro*GalU2 are apparent values, see Table 1.

The maximum activity was observed between 40 °C and 45 °C and the protein was stable up to 45 °C. The enzymatic activity drastically decreased after an incubation for 30 min at higher temperatures. Other GalUs have temperature optima around only 37 °C [60,61]. Only nucleotidyltransfrases of thermophilic bacteria and Archaea have a higher temperature optimum [62,63] and a half-life of 30 min at 95 °C [36]. However, *Ro*GalU2 originates from a mesophilic Actinobacterium, *R. opacus* 1CP, which grows best at 30 °C. Thus, this temperature optimum was not expected for a metabolically relevant enzyme. Only two other enzymes of this strain showed a higher temperature activity or stability so far; a flavin-dependent monooxygenase [64] and membrane-linked isomerase [65].

The use of magnesium sulfate with respect to UDP-glucose pyrophosphorylases was not described in literature before. Here we show the definite dependency on magnesium ions and an activity increase to 180% when using sulfate as anion instead of chloride. The total electronegativity of sulfate is higher than of chloride, which could be advantageous for charge neutralization and coordination of the phosphoryl oxygen. This could then help binding UTP and implementing the nucleophilic attack, which is the base for the proposed reaction mechanism (Scheme 2) [17,18]. Another possibility is that the sulfate variant of magnesium salt forms a more stable complex with UTP than the chloride. As Kleczkowski already reported, those complexes could be the actual substrates for GalUs and that free UTP inhibits the reaction when present in too high concentrations [66]. This is in agreement to our kinetic study in which substrate inhibition was determined at higher UTP concentrations (Figure 9). Furthermore, this is supported by the results obtained by the variation of magnesium salt concentration. There the activity of *Ro*GalU2 decreased rapidly when the concentration of the magnesium salt was higher than 1–2 mM. It is likely that the activity increases when both, the cofactor and UTP, are fed in equimolar concentrations. But an experimental set up will need to verify this hypothesis. Other groups did not report the occurrence of substrate inhibition. Here, we describe a substrate inhibition for both, glucose 1-phosphate and UTP. The only report showing a similar behavior of a GalU enzyme at increasing glucose 1-phosphate concentrations is for the enzyme from potato by Gupta and colleagues [57]. But it has to be noted that the tested glucose 1-phosphate concentration range was quite small (0.05–1 mM) and the activity still increased at 1 mM [57].

GalUs show activities with the same divalent metal ions that were tested with *Ro*GalU2 and also the same degree of inhibition by EDTA [12,36,63,67]. Thus, a clear preference for Mg^{2+} is demonstrated for all GalU enzymes.

The literature reported pH range tolerated by GalUs is huge, between pH 5.5 and pH 10.0, but the optimum is often found around pH 7.5 [13,36,63,67]. *Ro*GalU2 behaves expectedly with a pH optimum of around 7.5, depending on the buffer. However, it maintains a comparable high activity up to a pH of about 9 and thus behaves as many GalU enzymes.

*Ro*GalU2 accepts UTP and dTTP as nucleotides, as well as glucose 1-phosphate and galactose 1-phosphate as sugar phosphates. Substrate promiscuity is known for some GalUs, whereas some are very specific [39]. Indeed, it has been reported that GalUs from *S. coelicolor*, *Salmonella enterica*, *Sphingomonas elodea*, and *E. coli* are able to accept both UTP and dTTP as substrates [10,40,68,69], whereas for example the GalU from *Helicobacter pylori* is specific only for UDP [17]. It has been shown for *H. pylori* that configuration of the active site prevents a thymine from binding due to steric clashes with a methionine residue (M105) [17]. Superposition of a *Ro*GalU2 homology model, produced using SWISS-MODEL [70,71], with the crystal structure from *H. pylori* in complex with UDP-glucose (pdbID 3JUK, [17]) shows that *Ro*GalU2 does not have a methionine but a proline residue in the designated position. This observation is similar to what has been reported for *S. elodea* and *E. coli* GalUs having a proline and an alanine, respectively, at that position and also being active towards dTTP [39]. Furthermore, *Ro*GalU2 is also able to take galactose 1-phosphate as a substrate, but the enzyme is less active (only 2% residual activity), as was observed for *S. elodea* and *E. coli* GalUs [30]. It has been suggested that this less favorable binding of galactose 1-phosphate is due to the loss of the H-bond formed between the glucose 1-phosphate and the main-chain nitrogen atom of a glycine residue (Gly179 *E. coli* and Gly180 in *Ro*GalU2 homology model) [39].

Having all this in mind and the optimal conditions for *Ro*GalU2 experimentally verified, we could use this knowledge to obtain a maximum observed activity value for this UDP-glucose pyrophosphorylases in the direction of UDP-glucose formation (Figure 11). Under optimal conditions the activity of *Ro*GalU2 was 270 U mg^{-1}, which is an increase in activity of about 37%. Thus this enzyme is among the most active GalUs of bacteria and might be interesting to be studied for various biotechnological applications described recently [72].

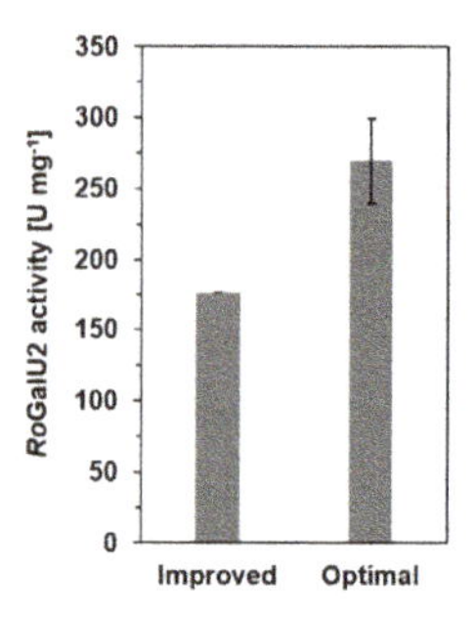

Figure 11. *Ro*GalU2 activity of the improved activity assay compared to an assay with optimal reaction conditions. Reaction solution for improved test contained 1.5 mM UTP, 250 mM glucose 1-phosphate, 1 mM $MgCl_2$, 50 mM Hepes, pH 7.4, 1.7 µg *Ro*GalU2. Reaction was carried out at 42 °C in 1 mL scale. Reaction solution for optimal test contained 1.5 mM UTP, 250 mM glucose 1-P, 1 mM $MgSO_4$, 50 mM Imidazol, pH 7.4, 1.7 µg *Ro*GalU2. Reaction was carried out 43 °C in 1 mL scale. Means with standard deviations are shown (triplicates).

4. Materials and Methods

4.1. Bacterial Strains, Plasmids, and Gene Synthesis

Protein sequences of the UDP-glucose pyrophosphorylases *Ro*GalU1 and *Ro*GalU2 of *Rhodococcus opacus* 1CP were taken from the NCBI accessions ANS26426 (*Ro*GalU1) and ANS26629 (*Ro*GalU2), respectively. The corresponding genes *RogalU1* (914 bp) and *RogalU2* (932 bp) used in this study were codon usage optimized to increase the expression level in *E. coli* and synthesized by Eurofins Genomics (Ebersberg, Germany) with flanking restriction sites of *Nde*I and *Not*I (GenBank accession numbers of the codon usage optimized nucleotide sequences of *RogalU1*: MN617759 and *RogalU2*: MN617760). Both genes were delivered in separate vectors (pEX-A2). They were cloned into the expression vector pET16bP (5740 bp) carrying a resistance against ampicillin, a DNA sequence that allowed the production of the GalU proteins with an *N*-terminal $Histidine_{10}$-tag and an additional

DNA sequence for the gene expression induction with isopropyl β-D-1-thiogalactopyranoside (IPTG) (see Table 3).

Table 3. Strains, plasmids, and primers used in this study.

Sample	Relevant Characteristics	Source, Reference
	Strains	
R. opacus 1 CP	Benzoate$^+$, 4-hydroxybenzoate$^+$, 3-chlorobenzoate$^+$, phenol$^+$, 4-chlorophenol$^+$, 2,4-dichlorophenol$^+$, 2-chlorophenol$^+$, 3-methylphenol$^+$, 4-methylphenol$^+$, phthalate$^+$, isophthalate$^+$, *n*-alkanes$^+$ (C10-C16), styrene$^+$	[64,73]
E. coli DH5α	fhuA2 Δ(argF-lacZ) U169 phoA glnV44 Φ80 Δ(lacZ)M15 gyrA96 recA1 relA1 endA1 thi-1 hsdR17	NEB [1]
E. coli BL21(DE3) pLysS	fhuA2 [lon] ompT gal (λ DE3) [dcm] ΔhsdS λ DE3 = λ sBamHIo ΔEcoRI-B int::(lacI::PlacUV5::T7 gene1) i21 Δnin5	NEB [1]
	Plasmids	
pEX-A2	multiple cloning site, Lac-Promoter, pUC origin, Amp^r	Eurofins [2]
pET16bP	pET16b with additional multiple cloning site; allows production of recombinant proteins with N-terminal Histidine$_{10}$-Tag and gene expression induction with IPTG	Wehmeyer [3]
pEX-A2-*RogalU1*	pEX-A2 vector with recombinant UDP-glucose-pyrophosphorylase gene 1 of *R. opacus* 1 CP (*RogalU1*) 914 bp, NCBI protein accession: ANS26426	This study
pEX-A2-*RogalU2*	pEX-A2 vector with recombinant UDP-glucose-pyrophosphorylase gene 2 of *R. opacus* 1 CP (*RogalU2*) 932 bp, NCBI protein accession: ANS26629	This study
pET16bP-*RogalU1*	pET16bP vector with recombinant UDP-glucose-pyrophosphorylase gene 1 of *R. opacus* 1 CP (*RogalU1*)	This study
pET16bP-*RogalU2*	pET16bP vector with recombinant UDP-glucose-pyrophosphorylase gene 2 of *R. opacus* 1 CP (*RogalU2*)	This study
	Primer	
pET16bP-fw	5′ - CATCACAGCAGCGGCCATATCGAAG - 3′	This study
pET16bP-rev	5′ - CAGCTTCTTTTCGGGCTTTGTTAG - 3′	This study

[1] New England Biolabs Inc. or [2] Eurofins Genomics as a commercial source. [3] personal communication by U. Wehmeyer.

4.2. Protein Production

Transformation of *E. coli* BL21(DE3) pLysS was carried out as recommended by New England Biolabs Inc. (Ipswich, Massachusetts, USA).

Protein production was realized in a 1 L scale in a Fernbach flask in standard LB medium (lysogenic broth: 10 g L^{-1} trypton, 5 g L^{-1} yeast extract, 10 g L^{-1} sodium chloride) with 100 mg L^{-1} ampicillin as well as 50 mg L^{-1} chloramphenicol. Expression cultures were inoculated 1:50 with an overnight pre-culture at 37 °C of *E. coli* BL21(DE3) pLysS-pET16bP-*RogalU1* or *E. coli* BL21(DE3) pLysS-pET16bP-*RogalU2* in the same medium used for the expression culture. The main culture was incubated for about 2 h at 37 °C until an OD_{600} of 0.2–0.3 could be observed. After cooling down to 20 °C, the gene expression and thus protein production started after induction with 0.5 mM isopropyl β-D-1-thiogalactopyranoside (IPTG) at an OD_{600} of 0.3–0.4.

The large-scale protein production was performed in an analog manner in a 10 L Eppendorf bioreactor in TB autoinduction medium [74] with the same concentrations of antibiotics and inoculation culture as described above.

After 22 h the protein production was stopped in both cases by harvesting the cells at 4 °C and 5000× *g* for 30 min. After washing with 25 mM sodium phosphate buffer, pH 7.1, pelleted cells were frozen and stored at −80 °C in portions of about 25 g of wet biomass.

4.3. Purification of Recombinant Proteins

For purification 25 g of pelleted cells were thawed as fast as possible in a warm water bath and mixed gently with 25 mL of a buffer containing the following compounds: 25 mM sodium phosphate buffer, pH 7.1, 100 mM sodium chloride, 1 mM magnesium chloride, 240 U of DNase I and 20 mg of lysozyme. After incubation at 30 °C for 45 to 60 min the cells were sonicated 10 times for 30 s with an intensity of 70% (Bandelin Sonoplus HD 2070, MS 72) and centrifuged at 12,000× *g* at 4 °C for 20 min. Another two centrifugation steps with the supernatant followed at 4 °C, 50,000× *g* for

30 min, each. The clear supernatant was filtered through 0.45 μm and 0.2 μm filter, respectively, before purification with an Äkta Prime Plus FPLC system with 5 mL HisTrap HP nickel column (GE Healthcare) with a flow of 5 mL min^{-1}. For equilibration of the affinity chromatography column a buffer containing 25 mM sodium phosphate buffer, pH 7.1, 300 mM sodium chloride and 25 mM imidazole was used. Protein loading was performed with equilibration buffer containing 25 mM imidazole. Unspecific proteins were washed off the column with equilibration buffer containing 40 mM imidazole. Protein purification was then performed by applying a linear gradient from 40–500 mM imidazole in this buffer in a course of total 10–20 mL elution, depending on the injection volume. The protein of interest was eluted at 500 mM imidazole.

Fractions that showed UDP-glucose pyrophosphorylase activity were pooled and precipitated with 80% saturated ammonium sulfate solution. The precipitated protein was dissolved and stored in a buffer of 50 mM Hepes, pH 7.0, 100 mM sodium chloride and 1 mM magnesium chloride. Protein aliquots were stored at –80 °C for long term storage or at 4 °C for short term storage and had concentrations of about 2–8 mg mL^{-1}.

Here, it needs to be mentioned that *Ro*GalU2 has a very low extinction coefficient due to the low amount of aromatic amino acids in the polypeptide chain and thus purifying by following the UV/VIS trace was somewhat difficult.

4.4. Renaturation of RoGalU1

Isolation of inclusion bodies and renaturation of the wrongly folded *Ro*GalU1 was realized with the Thermo Scientific Pierce Protein Refolding Kit according to the manufacturer's instruction, but without any EDTA in the buffers or solutions. The *Ro*GalU1 concentrations for the renaturation were 1 mg mL^{-1} and 10 mg mL^{-1}, respectively.

4.5. Protein Determination

Determination of the protein concentration was done as described before by Bradford [75] with a bovine serum albumin (BSA) standard. Identification of the proteins was performed by determination of the molecular weight with sodium dodecyl sulfate polyacrylamide gel electrophoresis (SDS-PAGE) and coomassie staining [76]. Furthermore, the oligomeric state of the proteins was determined by size exclusion chromatography using a 24 mL Superdex 200 10/300 GL column (GE Healthcare), the manufacturers protocol and a calibration standard mix containing Ferritin, Conalbumin, Carbonic Anhydrase, Ribonuclease, and Aprotinin. Dextran Blue was used to determine the void volume of the size exclusion column. The buffer that was used for this method contained 50 mM Hepes, pH 7.0 and 1 mM magnesium chloride and was the same as used for the most characterization experiments.

4.6. Enzyme Activity Assay

The specific enzyme activity of *Ro*GalU1 and *Ro*GalU2 was measured only in the direction of UDP-glucose formation in a reaction volume of 1 mL. The following assay composition was used in this work mostly and thus designated as standard test. The reaction solution for initial activity measurements contained 2 mM UTP, 2 mM glucose 1-phosphate, 4 mM magnesium chloride, and 50 mM Hepes, pH 7.0. Pre-incubation of the reaction samples was done for 15 min at 30 °C and the reaction was started by addition of 30 μg of enzyme, unless otherwise indicated. Samples of 100 μL were taken after defined time points to determine the initial reaction rates. Therefore, the reaction was stopped by adding 100 μL acetonitrile and vortexing in order to denature proteins. After centrifugation for 2 min at 20,000× *g* 100 μL of the clear supernatant were used for HPLC analysis.

In order to characterize GalU enzymes the above described standard assay was altered with respect to buffer, pH, temperature and various other additives during experimentation and to find optimal reaction conditions.

The enzyme activity is expressed as U mg^{-1}. Respectively, 1 U corresponds to the formation of 1 µmol UDP-glucose per minute. This value is then referred to the amount of enzyme in the assay.

4.7. Product Determination by HPLC for Specific Enzyme Activity Evaluation

HPLC measurement was performed with a Thermo Scientific Dionex Ultimate 3000 with UV/VIS detector and Macherey-Nagel EC 150/4.6 Nucleoshell HILIC column with a particle size of 2.7 µm. For determination of UDP-glucose formation an isocratic chromatography program was used with 70% acetonitrile and 30% 134 mM ammonium acetate, pH 5.35 at a flow of 1.3 mL min^{-1} and an oven temperature of 30 °C. Detection of UDP-glucose was done at a wave length of 260 nm with an injection volume of 5 µL. Calibration was carried out with appropriate concentrations of a UDP-glucose standard under the same conditions.

For data evaluation and calculation of product formation the area of the UDP-glucose peak was used and referred to the time point of the enzyme assay to obtain an initial rate. Those rates were plotted according to the enzyme kinetic models for analysis.

The following two equations according to Michaelis–Menten (1) and Yano und Koga [50] (2);

$$\mathrm{V} = \frac{V_{\mathrm{max}} \cdot c_{\mathrm{S}}}{K_{\mathrm{m}} + c_{\mathrm{S}}} \tag{1}$$

$$\mathrm{V} = \frac{V_{\mathrm{max}} \cdot c_{\mathrm{S}}}{K_{\mathrm{m}} + c_{\mathrm{S}} + \frac{c_{\mathrm{S}}^{3}}{K_{\mathrm{i}}^{2}}} \tag{2}$$

have been used. Equation (1) was used for non-limiting conditions and 2 for the cases of substrate inhibition.

5. Conclusions

The biochemical characterization of UDP-glucose pyrophosphorylases has been reported for several organisms. Here, we described the first characterization of a UDP-glucose pyrophosphorylase from a *Rhodococcus* strain that contained two isogenes coding for GalUs. Based on phylogenetic analyses and activity data obtained it seems obvious that *Ro*GalU2 represents the metabolically active enzyme, whereas the role of *Ro*GalU1 remains enigmatic and needs to be investigated. Furthermore, the activity of *Ro*GalU2 is higher than that of some other reported GalUs, but the use of conventional kinetic models is limited. Therefore, further biochemical investigation will be necessary. In addition, this is the first report of the use of magnesium sulfate as metal cofactor. The sulfate salt of magnesium was able to double the activity of *Ro*GalU2. A maximum activity of *Ro*GalU2 of about 270 U mg^{-1} was determined which renders it a candidate for further biocatalytic investigations.

Author Contributions: Conceptualization, A.K., D.T. and I.B.; methodology, A.K., A.P. (Anett Partzsch) and A.P. (André Pollender); formal analysis, A.K. and A.P. (André Pollender); investigation, A.K. and A.P. (Anett Partzsch); data curation, A.K.; writing—original draft preparation, A.K. and D.T.; writing—review and editing, all authors.; visualization, A.K.; funding acquisition, D.T.

Funding: This research was funded by European Union and Saxonian Government, grant number 100263899 related to the ERA-IB project No-P, grant ERA-IB-15-110. The APC was partially funded by Ruhr-Universität Bochum.

Acknowledgments: We thank the collaborator of the ERA-IB project No-P for fruitful discussion and among those especially Katarzyna Szymańska, Andrzej Jarzębski, Daria Kowalczykiewicz and Marta Przypis from the Silesian University of Technology (Poland), Ulf Hanefeld, Peter-Leon Hagedoorn and Luuk Mestrom from the TU Delft (The Netherlands), as well as Arkadiusz Chruściel from MEXEO (Poland) and Rob Schoevaart from Chiral Vision (The Netherlands). In addition, we thank Daniel Eggerichs and Carolin Mügge for graphical support and constructively reviewing.

Conflicts of Interest: The authors declare no conflict of interest. The funders had no role in the design of the study; in the collection, analyses, or interpretation of data; in the writing of the manuscript, or in the decision to publish the results.

References

1. Gutmann, A.; Nidetzky, B. Unlocking the potential of Leloir glycosyltransferases for applied biocatalysis: Efficient synthesis of Uridine 5′-diphosphate-glucose by sucrose synthase. *Adv. Synth. Catal.* **2016**, *358*, 3600–3609. [CrossRef]
2. Mestrom, L.; Marsden, S.R.; Dieters, M.; Achterberg, P.; Stolk, L.; Bento, I.; Hanefeld, U.; Hagedoorn, P.-L. Artificial fusion of mCherry enhances trehalose transferase solubility and stability. *Appl. Environ. Microbiol.* **2019**, 85. [CrossRef] [PubMed]
3. Schmölzer, K.; Lemmerer, M.; Gutmann, A.; Nidetzky, B. Integrated process design for biocatalytic synthesis by a Leloir glycosyltransferase: UDP-glucose production with sucrose synthase. *Biotechnol. Bioeng.* **2017**, *114*, 924–928. [CrossRef] [PubMed]
4. Bungaruang, L.; Gutmann, A.; Nidetzky, B. Leloir glycosyltransferases and natural product glycosylation: Biocatalytic synthesis of the C-glucoside Nothofagin, a major antioxidant of redbush herbal tea. *Adv. Synth. Catal.* **2013**, *355*, 2757–2763. [CrossRef] [PubMed]
5. Dai, L.; Li, J.; Yao, P.; Zhu, Y.; Men, Y.; Zeng, Y.; Yang, J.; Sun, Y. Exploiting the aglycon promiscuity of glycosyltransferase Bs-YjiC from *Bacillus subtilis* and its application in synthesis of glycosides. *J. Biotechnol.* **2017**, *248*, 69–76. [CrossRef]
6. Diricks, M.; Gutmann, A.; Debacker, S.; Dewitte, G.; Nidetzky, B.; Desmet, T. Sequence determinants of nucleotide binding in sucrose synthase: Improving the affinity of a bacterial sucrose synthase for UDP by introducing plant residues. *Protein Eng. Des. Sel.* **2017**, *30*, 141–148. [CrossRef]
7. Schmölzer, K.; Gutmann, A.; Diricks, M.; Desmet, T.; Nidetzky, B. Sucrose synthase: A unique glycosyltransferase for biocatalytic glycosylation process development. *Biotechnol. Adv.* **2016**, *34*, 88–111. [CrossRef]
8. Thoden, J.B.; Holden, H.M. The molecular architecture of glucose-1-phosphate uridylyltransferase. *Protein Sci.* **2007**, *16*, 432–440. [CrossRef]
9. Aragão, D.; Fialho, A.M.; Marques, A.R.; Mitchell, E.P.; Sa-Correia, I.; Frazao, C. The complex of *Sphingomonas elodea* ATCC 31461 glucose-1-phosphate uridylyltransferase with glucose-1-phosphate reveals a novel quaternary structure, unique among nucleoside diphosphate-sugar pyrophosphorylase members. *J. Bacteriol.* **2007**, *189*, 4520–4528. [CrossRef]
10. Asención Diez, M.D.; Peirú, S.; Demonte, A.M.; Gramajo, H.; Iglesias, A.A. Characterization of recombinant UDP- and ADP-glucose pyrophosphorylases and glycogen synthase to elucidate glucose-1-phosphate partitioning into oligo- and polysaccharides in *Streptomyces coelicolor*. *J. Bacteriol.* **2012**, *194*, 1485–1493. [CrossRef]
11. Ebrecht, A.C.; Asencion Diez, M.D.; Piattoni, C.V.; Guerrero, S.A.; Iglesias, A.A. The UDP-glucose pyrophosphorylase from *Giardia lamblia* is redox regulated and exhibits promiscuity to use galactose-1-phosphate. *Biochim. Biophys. Acta* **2015**, *1850*, 88–96. [CrossRef] [PubMed]
12. Koo, H.M.; Yim, S.-W.; Lee, C.-S.; Pyun, Y.R.; Kim, Y.S. Cloning, sequencing, and expression of UDP-glucose pyrophosphorylase gene from *Acetobacter xylinum* BRC5. *Biosci. Biotechnol. Biochem.* **2014**, *64*, 523–529. [CrossRef] [PubMed]
13. Lai, X.; Wu, J.; Chen, S.; Zhang, X.; Wang, H. Expression, purification, and characterization of a functionally active *Mycobacterium tuberculosis* UDP-glucose pyrophosphorylase. *Protein Expr. Purif.* **2008**, *61*, 50–56. [CrossRef] [PubMed]
14. Meng, M.; Wilczynska, M.; Kleczkowski, L.A. Molecular and kinetic characterization of two UDP-glucose pyrophosphorylases, products of distinct genes, from *Arabidopsis*. *Biochim. Biophys. Acta* **2008**, *1784*, 967–972. [CrossRef]
15. Okazaki, Y.; Shimojima, M.; Sawada, Y.; Toyooka, K.; Narisawa, T.; Mochida, K.; Tanaka, H.; Matsuda, F.; Hirai, A.; Hirai, M.Y.; et al. A chloroplastic UDP-glucose pyrophosphorylase from *Arabidopsis* is the committed enzyme for the first step of sulfolipid biosynthesis. *Plant Cell* **2009**, *21*, 892–909. [CrossRef]
16. Roeben, A.; Plitzko, J.M.; Korner, R.; Bottcher, U.M.; Siegers, K.; Hayer-Hartl, M.; Bracher, A. Structural basis for subunit assembly in UDP-glucose pyrophosphorylase from *Saccharomyces cerevisiae*. *J. Mol. Biol.* **2006**, *364*, 551–560. [CrossRef]
17. Kim, H.; Choi, J.; Kim, T.; Lokanath, N.K.; Ha, S.C.; Suh, S.W.; Hwang, H.-Y.; Kim, K.K. Structural basis for the reaction mechanism of UDP-glucose pyrophosphorylase. *Mol. Cells* **2010**, *29*, 397–405. [CrossRef]

18. Thoden, J.B.; Holden, H.M. Active site geometry of glucose-1-phosphate uridylyltransferase. *Protein Sci.* **2007**, *16*, 1379–1388. [CrossRef]
19. Ebrecht, A.C.; Orlof, A.M.; Sasoni, N.; Figueroa, C.M.; Iglesias, A.A.; Ballicora, M.A. On the Ancestral UDP-Glucose Pyrophosphorylase Activity of GalF from *Escherichia coli*. *Front. Microbiol.* **2015**, *6*, 1253. [CrossRef]
20. Balan, D.; Tokas, J.; Singal, H.R. UDP-glucose pyrophosphorylase: Isolation, purification and characterization from developing thermotolerant wheat (*Triticum aestivum*) grains. *Protein Expr. Purif.* **2018**, *148*, 68–77. [CrossRef]
21. Chen, R.; Zhao, X.; Shao, Z.; Zhu, L.; He, G. Multiple isoforms of UDP-glucose pyrophosphorylase in rice. *Physiol Plant* **2007**, *129*, 725–736. [CrossRef]
22. Cotrim, C.A.; Soares, J.S.M.; Kobe, B.; Menossi, M. Crystal structure and insights into the oligomeric state of UDP-glucose pyrophosphorylase from sugarcane. *PLoS ONE* **2018**, *13*, e0193667. [CrossRef] [PubMed]
23. McCoy, J.G.; Bitto, E.; Bingman, C.A.; Wesenberg, G.E.; Bannen, R.M.; Kondrashov, D.A.; Phillips, G.N., Jr. Structure and dynamics of UDP-glucose pyrophosphorylase from *Arabidopsis thaliana* with bound UDP-glucose and UTP. *J. Mol. Biol.* **2007**, *366*, 830–841. [CrossRef] [PubMed]
24. Meng, M.; Fitzek, E.; Gajowniczek, A.; Wilczynska, M.; Kleczkowski, L.A. Domain-specific determinants of catalysis/substrate binding and the oligomerization status of barley UDP-glucose pyrophosphorylase. *Biochim. Biophys. Acta* **2009**, *1794*, 1734–1742. [CrossRef] [PubMed]
25. Dutta, S.K. UDPglucose pyrophosphorylase from Ehrlich ascites carcinoma cell-purification and characterization. *Indian J. Biochem. Biophys.* **1985**, *22*, 203–207. [PubMed]
26. Granzow, C.; Kopun, M.; Zimmermann, H.P. Role of nuclear glycogen synthase and cytoplasmic UDP glucose pyrophosphorylase in the biosynthesis of nuclear glycogen in HD33 Ehrlich-Lettré ascites tumor cells. *J. Cell Biol.* **1981**, *89*, 475–484. [CrossRef]
27. Reynolds, T.H.; Pak, Y.; Harris, T.E.; Manchester, J.; Barrett, E.J.; Lawrence, J.C. Effects of insulin and transgenic overexpression of UDP-glucose pyrophosphorylase on UDP-glucose and glycogen accumulation in skeletal muscle fibers. *J. Biol. Chem.* **2005**, *280*, 5510–5515. [CrossRef]
28. Führing, J.I.; Cramer, J.T.; Schneider, J.; Baruch, P.; Gerardy-Schahn, R.; Fedorov, R. A quaternary mechanism enables the complex biological functions of octameric human UDP-glucose pyrophosphorylase, a key enzyme in cell metabolism. *Sci. Rep.* **2015**, *5*, 9618. [CrossRef]
29. Dickmanns, A.; Damerow, S.; Neumann, P.; Schulz, E.-C.; Lamerz, A.-C.; Routier, F.H.; Ficner, R. Structural basis for the broad substrate range of the UDP-sugar pyrophosphorylase from *Leishmania major*. *J. Mol. Biol.* **2011**, *405*, 461–478. [CrossRef]
30. Benini, S.; Toccafondi, M.; Rejzek, M.; Musiani, F.; Wagstaff, B.A.; Wuerges, J.; Cianci, M.; Field, R.A. Glucose-1-phosphate uridylyltransferase from *Erwinia amylovora*: Activity, structure and substrate specificity. *Biochim. Biophys. Acta Proteins Proteom.* **2017**, *1865*, 1348–1357. [CrossRef]
31. Bosco, M.B.; Machtey, M.; Iglesias, A.A.; Aleanzi, M. UDPglucose pyrophosphorylase from *Xanthomonas* spp. Characterization of the enzyme kinetics, structure and inactivation related to oligomeric dissociation. *Biochimie* **2009**, *91*, 204–213. [CrossRef] [PubMed]
32. Degeest, B.; de Vuyst, L. Correlation of activities of the enzymes alpha-phosphoglucomutase, UDP-galactose 4-epimerase, and UDP-glucose pyrophosphorylase with exopolysaccharide biosynthesis by *Streptococcus thermophilus* LY03. *Appl. Environ. Microbiol.* **2000**, *66*, 3519–3527. [CrossRef] [PubMed]
33. Padilla, L.; Morbach, S.; Krämer, R.; Agosin, E. Impact of heterologous expression of *Escherichia coli* UDP-glucose pyrophosphorylase on trehalose and glycogen synthesis in *Corynebacterium glutamicum*. *Appl. Environ. Microbiol.* **2004**, *70*, 3845–3854. [CrossRef] [PubMed]
34. Toccafondi, M.; Cianci, M.; Benini, S. Expression, purification, crystallization and preliminary X-ray analysis of glucose-1-phosphate uridylyltransferase (GalU) from *Erwinia amylovora*. *Acta Crystallogr. F Struct. Biol. Commun.* **2014**, *70*, 1249–1251. [CrossRef] [PubMed]
35. Kawano, Y.; Sekine, M.; Ihara, M. Identification and characterization of UDP-glucose pyrophosphorylase in cyanobacteria *Anabaena* sp. PCC 7120. *J. Biosci. Bioeng.* **2014**, *117*, 531–538. [CrossRef]
36. Kim, J.S.; Koh, S.; Shin, H.J.; Lee, D.S.; Lee, S.Y. Biochemical characterization of a UDP-sugar pyrophosphorylase from *Thermus caldophilus* GK24. *Biotechnol. Appl. Biochem.* **1999**, *29 (Pt 1)*, 11–17.
37. Holden, H.M.; Rayment, I.; Thoden, J.B. Structure and function of enzymes of the Leloir pathway for galactose metabolism. *J. Biol. Chem.* **2003**, *278*, 43885–43888. [CrossRef]

38. Marques, A.R.; Ferreira, P.B.; Sá-Correia, I.; Fialho, A.M. Characterization of the *ugpG* gene encoding a UDP-glucose pyrophosphorylase from the gellan gum producer *Sphingomonas paucimobilis* ATCC 31461. *Mol. Genet. Genomics* **2003**, *268*, 816–824. [CrossRef]
39. Berbis, M.; Sanchez-Puelles, J.; Canada, F.; Jimenez-Barbero, J. Structure and function of prokaryotic UDP-glucose pyrophosphorylase, a drug target candidate. *CMC* **2015**, *22*, 1687–1697. [CrossRef]
40. Silva, E.; Marques, A.R.; Fialho, A.M.; Granja, A.T.; Sá-Correia, I. Proteins encoded by *Sphingomonas elodea* ATCC 31461 *rmlA* and *ugpG* genes, involved in gellan gum biosynthesis, exhibit both dTDP- and UDP-glucose pyrophosphorylase activities. *Appl. Environ. Microbiol.* **2005**, *71*, 4703–4712. [CrossRef]
41. Gröning, J.A.D.; Eulberg, D.; Tischler, D.; Kaschabek, S.R.; Schlömann, M. Gene redundancy of two-component (chloro)phenol hydroxylases in *Rhodococcus opacus* 1CP. *FEMS Microbiol. Lett.* **2014**, *361*, 68–75. [CrossRef] [PubMed]
42. Letek, M.; González, P.; Macarthur, I.; Rodríguez, H.; Freeman, T.C.; Valero-Rello, A.; Blanco, M.; Buckley, T.; Cherevach, I.; Fahey, R.; et al. The genome of a pathogenic *Rhodococcus*: Cooptive virulence underpinned by key gene acquisitions. *PLoS Genet.* **2010**, *6*, e1001145. [CrossRef] [PubMed]
43. Mizanur, R.M.; Pohl, N.L. A thermostable promiscuous glucose-1-phosphate uridyltransferase from *Helicobacter pylori* for the synthesis of nucleotide sugars. *J. Mol. Catal. B Enzymatic* **2008**, *50*, 13–19. [CrossRef]
44. Tischler, D.; Gröning, J.A.D.; Kaschabek, S.R.; Schlömann, M. One-component styrene monooxygenases: An evolutionary view on a rare class of flavoproteins. *Appl. Biochem. Biotechnol.* **2012**, *167*, 931–944. [CrossRef]
45. Edgar, R.C. MUSCLE: A multiple sequence alignment method with reduced time and space complexity. *BMC Bioinform.* **2004**, *5*, 113. [CrossRef]
46. Edgar, R.C. MUSCLE: Multiple sequence alignment with high accuracy and high throughput. *Nucleic Acids Res.* **2004**, *32*, 1792–1797. [CrossRef]
47. Kumar, S.; Stecher, G.; Li, M.; Knyaz, C.; Tamura, K. MEGA X: Molecular evolutionary genetics analysis across computing platforms. *Mol. Biol. Evol.* **2018**, *35*, 1547–1549. [CrossRef]
48. Robert, X.; Gouet, P. Deciphering key features in protein structures with the new ENDscript server. *Nucleic Acids Res.* **2014**, *42*, W320-4. [CrossRef]
49. Jones, D.T.; Taylor, W.R.; Thornton, J.M. The rapid generation of mutation data matrices from protein sequences. *Comput. Appl. Biosci.* **1992**, *8*, 275–282. [CrossRef]
50. Yano, T.; Koga, S. Dynamic behavior of the chemostat subject to substrate inhibition. *Biotechnol. Bioeng.* **1969**, *11*, 139–153. [CrossRef]
51. Asención Diez, M.D.; Demonte, A.M.; Syson, K.; Arias, D.G.; Gorelik, A.; Guerrero, S.A.; Bornemann, S.; Iglesias, A.A. Allosteric regulation of the partitioning of glucose-1-phosphate between glycogen and trehalose biosynthesis in *Mycobacterium tuberculosis*. *Biochim. Biophys. Acta* **2015**, *1850*, 13–21. [CrossRef] [PubMed]
52. Patil, H.I.; Pratap, A.P. Production and quantitative analysis of trehalose lipid biosurfactants using high-performance liquid chromatography. *J. Surfactants Deterg.* **2018**, *21*, 553–564. [CrossRef]
53. Tischler, D.; Niescher, S.; Kaschabek, S.R.; Schlömann, M. Trehalose phosphate synthases OtsA1 and OtsA2 of *Rhodococcus opacus* 1CP. *FEMS Microbiol. Lett.* **2013**, *342*, 113–122. [CrossRef] [PubMed]
54. Pacheco, B.; Crombet, L.; Loppnau, P.; Cossar, D. A screening strategy for heterologous protein expression in *Escherichia coli* with the highest return of investment. *Protein Expr. Purif.* **2012**, *81*, 33–41. [CrossRef] [PubMed]
55. Strandberg, L.; Enfors, S.O. Factors influencing inclusion body formation in the production of a fused protein in *Escherichia coli*. *Appl. Environ. Microbiol.* **1991**, *57*, 1669–1674. [PubMed]
56. Aragão, D.; Marques, A.R.; Frazão, C.; Enguita, F.J.; Carrondo, M.A.; Fialho, A.M.; Sá-Correia, I.; Mitchell, E.P. Cloning, expression, purification, crystallization and preliminary structure determination of glucose-1-phosphate uridylyltransferase (UgpG) from *Sphingomonas elodea* ATCC 31461 bound to glucose-1-phosphate. *Acta Crystallogr. Sect. F Struct. Biol. Cryst. Commun.* **2006**, *62*, 930–934. [CrossRef] [PubMed]
57. Gupta, S.K.; Sowokinos, J.R.; Hahn, I.-S. Regulation of UDP-glucose pyrophosphorylase isozyme UGP5 associated with cold-sweetening resistance in potatoes. *J. Plant Physiol.* **2008**, *165*, 679–690. [CrossRef]
58. Steiner, T.; Lamerz, A.C.; Hess, P.; Breithaupt, C.; Krapp, S.; Bourenkov, G.; Huber, R.; Gerardy-Schahn, R.; Jacob, U. Open and closed structures of the UDP-glucose pyrophosphorylase from *Leishmania major*. *J. Biol. Chem.* **2007**, *282*, 13003–13010. [CrossRef]
59. De Bruyn, F.; Beauprez, J.; Maertens, J.; Soetaert, W.; de Mey, M. Unraveling the Leloir pathway of *Bifidobacterium bifidum*: Significance of the uridylyltransferases. *Appl. Environ. Microbiol.* **2013**, *79*, 7028–7035. [CrossRef]

60. Ma, Z.; Fan, H.-j.; Lu, C.-p. Molecular cloning and analysis of the UDP-glucose pyrophosphorylase in *Streptococcus equi* subsp. *zooepidemicus*. *Mol. Biol. Rep.* **2011**, *38*, 2751–2760. [CrossRef]
61. Weissborn, A.C.; Liu, Q.; Rumley, M.K.; Kennedy, E.P. UTP: Alpha-D-glucose-1-phosphate uridylyltransferase of *Escherichia coli*: isolation and DNA sequence of the *galU* gene and purification of the enzyme. *J. Bacteriol.* **1994**, *176*, 2611–2618. [CrossRef] [PubMed]
62. Honda, Y.; Zang, Q.; Shimizu, Y.; Dadashipour, M.; Zhang, Z.; Kawarabayasi, Y. Increasing the thermostable sugar-1-phosphate nucleotidylyltransferase activities of the archaeal ST0452 protein through site saturation mutagenesis of the 97th amino acid position. *Appl. Environ. Microbiol.* **2017**, 83. [CrossRef] [PubMed]
63. Zhang, Z.; Tsujimura, M.; Akutsu, J.-i.; Sasaki, M.; Tajima, H.; Kawarabayasi, Y. Identification of an extremely thermostable enzyme with dual sugar-1-phosphate nucleotidylyltransferase activities from an acidothermophilic archaeon, *Sulfolobus tokodaii* strain 7. *J. Biol. Chem.* **2005**, *280*, 9698–9705. [CrossRef] [PubMed]
64. Tischler, D.; Eulberg, D.; Lakner, S.; Kaschabek, S.R.; van Berkel, W.J.H.; Schlömann, M. Identification of a novel self-sufficient styrene monooxygenase from *Rhodococcus opacus* 1CP. *J. Bacteriol.* **2009**, *191*, 4996–5009. [CrossRef] [PubMed]
65. Oelschlägel, M.; Gröning, J.A.D.; Tischler, D.; Kaschabek, S.R.; Schlömann, M. Styrene oxide isomerase of *Rhodococcus opacus* 1CP, a highly stable and considerably active enzyme. *Appl. Environ. Microbiol.* **2012**, *78*, 4330–4337. [CrossRef] [PubMed]
66. Kleczkowski, L.A. Glucose activation and metabolism through UDP-glucose pyrophosphorylase in plants. *Phytochemistry* **1994**, *37*, 1507–1515. [CrossRef]
67. Li, Q.; Huang, Y.-Y.; Conway, L.P.; He, M.; Wei, S.; Huang, K.; Duan, X.-C.; Flitsch, S.L.; Voglmeir, J. Discovery and biochemical characterization of a thermostable glucose-1-phosphate nucleotidyltransferase from *Thermodesulfatator indicus*. *Protein Pept. Lett.* **2017**, *24*, 729–734. [CrossRef]
68. Bernstein, R.L.; Robbind, P.W. Control aspects of Uridine 5′-diphosphate glucose and Thymidine 5′-diphosphate glucose synthesis by microbial enzymes. *J. Biol. Chem.* **1965**, *240*, 391–397.
69. Lindquist, L.; Kaiser, R.; Reeves, P.R.; Lindberg, A.A. Purification, characterization and HPLC assay of *Salmonella* glucose-1-phosphate thymidylyltransferase from the cloned *rfbA* gene. *Eur. J. Biochem.* **1993**, *211*, 763–770. [CrossRef]
70. Waterhouse, A.; Bertoni, M.; Bienert, S.; Studer, G.; Tauriello, G.; Gumienny, R.; Heer, F.T.; de Beer, T.A.P.; Rempfer, C.; Bordoli, L.; et al. SWISS-MODEL: Homology modelling of protein structures and complexes. *Nucleic Acids Res.* **2018**, *46*, W293–W303. [CrossRef]
71. Kumpf, A.; Tischler, D. Structural investigations of GalU enzymes from Actinobacteria, Crystals. Unpublished work.
72. Mestrom, L.; Przypis, M.; Kowalczykiewicz, D.; Pollender, A.; Kumpf, A.; Marsden, S.R.; Bento, I.; Jarzębski, A.B.; Szymańska, K.; Chruściel, A.; et al. Leloir glycosyltransferases in applied biocatalysis: A multidisciplinary approach. *Int. J. Mol. Sci.* **2019**, *20*, 5263. [CrossRef] [PubMed]
73. Moiseeva, O.V.; Solyanikova, I.P.; Kaschabek, S.R.; Gröning, J.; Thiel, M.; Golovleva, L.A.; Schlömann, M. A new modified ortho cleavage pathway of 3-chlorocatechol degradation by *Rhodococcus opacus* 1CP: Genetic and biochemical evidence. *J. Bacteriol.* **2002**, *184*, 5282–5292. [CrossRef]
74. Studier, F.W. Protein production by auto-induction in high-density shaking cultures. *Protein Expr. Purif.* **2005**, *41*, 207–234. [CrossRef] [PubMed]
75. Bradford, M.M. A rapid and sensitive method for the quantitation of microgram quantities of protein utilizing the principle of protein-dye binding. *Anal. Biochem.* **1976**, *72*, 248–254. [CrossRef]
76. Laemmli, U.K. Cleavage of structural proteins during the assembly of the head of bacteriophage T4. *Nature* **1970**, *227*, 680–685. [CrossRef]

International Journal of
Molecular Sciences

Article

The C-Type Lysozyme from the upper Gastrointestinal Tract of *Opisthocomus hoatzin*, the Stinkbird

Edward J. Taylor [1,†,‡], Michael Skjøt [2,†,§], Lars K. Skov [2], Mikkel Klausen [2], Leonardo De Maria [2,||], Garry P. Gippert [2], Johan P. Turkenburg [1], Gideon J. Davies [1] and Keith S. Wilson [1,*]

1 Structural Biology Laboratory, Department of Chemistry, The University of York, York YO10 5DD, UK; etaylor@lincoln.ac.uk (E.J.T.); Johan.Turkenburg@york.ac.uk (J.P.T.); Gideon.Davies@york.ac.uk (G.J.D.)
2 Novozymes A/S, Biologiens Vej 2, 2800 Kongens Lyngby, Denmark; mskj@novonordisk.com (M.S.); LaKS@novozymes.com (L.K.S.); MLKL@novozymes.com (M.K.); leonardo.demaria@astrazeneca.com (L.D.M.); GPGI@novozymes.com (G.P.G.)
* Correspondence: keith.wilson@york.ac.uk
† These authors contributed equally to this work.
‡ Current address: School of Life Sciences, Joseph Banks Laboratories, University of Lincoln, Green Lane, Lincoln, LN6 7DL, UK.
§ Current address: Novo Nordisk A/S, 2760 Måløv, Denmark.
|| Current address: AstraZeneca AB R & D, Pepparedsleden 1, 431 50 Mölndal, Sweden.

Received: 21 October 2019; Accepted: 1 November 2019; Published: 6 November 2019

Abstract: Muramidases/lysozymes are important bio-molecules, which cleave the glycan backbone in the peptidoglycan polymer found in bacterial cell walls. The glycoside hydrolase (GH) family 22 C-type lysozyme, from the folivorous bird *Opisthocomus hoazin* (stinkbird), was expressed in *Aspergillus oryzae*, and a set of variants was produced. All variants were enzymatically active, including those designed to probe key differences between the Hoatzin enzyme and Hen Egg White lysozyme. Four variants showed improved thermostability at pH 4.7, compared to the wild type. The X-ray structure of the enzyme was determined in the apo form and in complex with chitin oligomers. Bioinformatic analysis of avian GH22 amino acid sequences showed that they separate out into three distinct subgroups (chicken-like birds, sea birds and other birds). The Hoatzin is found in the "other birds" group and we propose that this represents a new cluster of avian upper-gut enzymes.

Keywords: lysozyme; peptidoglycan cleavage; avian gut GH22; crystal structure

---

1. Introduction

Peptidoglycans are unique to prokaryotic organisms and consist of a glycan backbone of muramic acid and glucosamine (both N-acetylated), cross-linked with peptide chains. In Gram-positive bacteria (e.g., *Staphylococcus aureus*) the glycan backbone is highly cross-linked, while it is only partially cross-linked in Gram-negative bacteria, such as *Escherichia coli*. The cross-linking amino acid chain contains L-alanine, D-glutamic acid, meso-diaminopimelic acid, and D-alanine in *E. coli*, or L-alanine, D-glutamine, L-lysine, and D-alanine, with a five-glycine interbridge between tetrapeptides, in the case of *S. aureus* [1]. The unique composition of both the carbohydrate polymer and the peptide cross-linker means that only specialised enzymes can hydrolyse peptidoglycans. Lysins (amidases, lysozymes/muramidases and peptidases) are such specialised bio-molecules. The term lysozyme, or muramidase, is broadly used to describe the enzymes that cleave the β-1,4-glycosidic bond between *N*-acetylglucosamine (NAG) and *N*-acetylmuramic acid (NAM) (or vice versa) in the carbohydrate backbone of peptidoglycan, Figure 1.

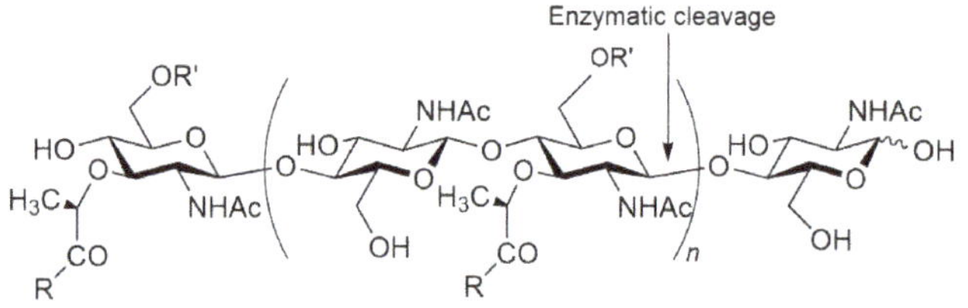

Figure 1. The cleavage site of the cell wall glycan by lysozyme/muramidase.

In nature, the β-1,4 bonds of peptidoglycan are cleaved by a structurally diverse set of enzymes. Lysozymes/muramidases (EC 3.2.1.17) are found in several glycoside hydrolase families in the Carbohydrate-Active enZYmes Database (CAZy, www.cazy.org [2]), including GH18, 19, 22, 23, 24, 25, 73 and 108, some of which, such as GH25, remain largely uncharacterized biochemically. While the first three families have very low sequence identities, they do have some common structural features, consisting of a constant core of two helices and a three-stranded β-sheet that accommodates the substrates in the inter-domain cleft [3]. Higher organisms typically have enzymes from several GH families, e.g., *Gallus gallus* has at least three from GH18, GH22 and GH23. An excellent review of "Lysozymes in the animal kingdom" [4] summarises a wealth of information on the enzymes, and their classification into subfamilies—lysozymes C (chicken-type, the archetypal lysozymes), G (goose-type) and I (invertebrate). Subfamilies C and I are both grouped in CAZy family GH22, while G is in GH23.

CAZy has over 700 entries for GH22, almost all from Eukaryota, but the 3D structure has only been determined for about 25 species. The most well-known is the C-type lysozyme from *G. gallus* (chicken), commonly called Hen Egg White Lysozyme (HEWL), which is almost synonymous with lysozyme. The deposited structures are dominated by the enzymes from chickens and *Homo sapiens*. While the overall amino acid sequences identity is quite low, the structural similarity of the C-type lysozymes is very high. A partial explanation for this is assumed to be due to the four conserved disulphide bridges, that ensure a compact and rather rigid 3D arrangement.

The structure of HEWL revealed the GH22 fold [5] to be a $\alpha + \beta$ motif, made up of five α-helical regions and five containing β-strands, with two catalytic groups, Glu35 and Asp52. The active site consists of six subsites (originally termed A, B, C, D, E and F, but now more generally named −4, −3, −2, −1, +1 and +2) [6], which bind up to six consecutive sugar residues. The glycosidic bond between the N-acetyl muramic acid (NAM) at subsite −1, and the N-acetyl -glucosamine (NAG) at subsite +1, is weakened by steric distortion of the sugar ring in subsite −1, and is the target of the hydrolytic cleavage. In 2001, experimental evidence for the correct working mechanism of HEWL was finally established [7], with the hydrolysis of the β-(1,4)-glycosidic bond occurring through a double displacement reaction. This mechanism is believed to apply to all members of this very broad class of enzymes.

HEWL has over 25 years of recorded use in wine and cheese making [8], and, classically C-type GH22 lysozymes have been known to act as antimicrobials, or at least as microbial growth inhibitors [9,10]. In addition, some mammalian (typically ruminant) GH22s have been proposed to have a digestive role in the stomach, where they could degrade bacteria after the front-gut fermentation process [11]. More recently, lysozymes have been proposed to modulate the bacterial flora and to digest bacterial cell wall debris, thereby affecting the immune system [12].

If selected lysozymes could be expressed in a heterologous host, suitable for industrial production, these could be used in applications where peptidoglycans are present and their elimination would be useful (such as biofilms, washing and nutritional supplements). For a long time, the literature indicated that lysozymes were difficult to express in such hosts. In particular, the group of David Archer at the University of Nottingham published more than 10 papers on the heterologous expression of HEWL and human lysozyme in *Aspergilli* [13] and, indeed, in *Pichia,* albeit with limited yields being obtained. Furthermore, the lack of gastric stability has been a hindrance for commercializing HEWL as an animal feed additive.

With the aim of expressing a digestive lysozyme in a suitable fungal host, the literature was scanned for a GH22 C-type lysozyme with high stability at low pH (gastric conditions). Earlier reports had indicated the extraordinary stability of the digestive lysozyme from *Opisthocomus hoazin* at low pH [14], and the corresponding cDNA sequence (Genbank entry AAA73935.1) had been published [15]. *O. hoazin*, known as the Hoatzin, or stinkbird, lives in parts of the rainforest in South America. The Hoatzin is unique in being the only known bird with crop fermentation in the foregut [15,16]. A recent review of avian crop function covers the importance of digestion in bird species [17]. In some respects, digestion in the Hoatzin gut seems more like that of ruminants than other folivorous birds, and is derived from the morphological and microbiological environment in the digestion tract [14]. Both ruminants and Hoatzins express high levels of gastric lysozyme and use fermentation in their foregut. This enables the Hoatzin to take advantage of energy from both the cellular content and the cell wall polysaccharides of hydrolysed bacteria [18]. Lysozyme is an important element of the Hoatzin's digestion system, wherein the digestive tract of the Hoatzin can also be found. Based on the predicted amino acid sequence, HEWL and *O. hoazin* lysozyme (henceforth *Oh*Lys) have very different isoelectric points. As a result, quite different properties for their selectivity might be expected.

We here report the successful cloning and expression in *Aspergillus oryzae* (NCBI:txid90341) of a synthetic gene, corresponding to *Oh*Lys. Mutational studies were applied to modify the activity and/or stability of *Oh*Lys. In addition, we have determined the crystal structure of the apo enzyme and of complexes with reaction products (chito-oligosaccharides). This is the first structure of an avian GH22 lysozyme/muramidase outside the chicken-like sub-group.

2. Results

2.1. Expression of OhLys and Variants in Aspergillus

HEWL, the best characterized GH22, shows a surprising level of promiscuous chitinolytic activity [19]. As the *Aspergillus* fungal cell wall consists mainly of chitin, expression of an enzyme with chitinase activity may be expected to counter-select high-level expressing transformants. Indeed, Archer et al. previously concluded that proteolysis occurs in HEWL between Gly49 and Ser50 when it is expressed in *A. niger* [20]. In contrast, our results show that *Oh*Lys can be expressed at a reasonable level (about 1 g/L) in *A. oryzae*, with about 30 variants being produced and shown to be active in the turbidity assay.

2.2. Crystal Structure of the Apo Enzyme and Its Chitotriose Complex

Crystals of the apo enzyme belong to the orthorhombic space group $P2_12_12_1$ with one molecule in the asymmetric unit, Table 1. As expected, *Oh*Lys has a typical GH22 lysozyme fold (Figure 2). The chain was traced from Glu1 through to Cys126, with a glycerol from the cryoprotectant, four Cl^- ions and 200 water molecules. There are four disulphide bridges (Cys6–Cys126, Cys30–Cys114, Cys63–Cys79 and Cys75–Cys93) in the structure.

Table 1. Crystallographic statistics.

Data Collection [(a)]		
Data set	Apo-Native	Mutant Ligand complex
Beamline	I02	I04
Wavelength (Å)	0.9795	0.9795
PDB code	6T5S	6T6C
Space group	$P2_12_12_1$	$P2_12_12_1$
Cell parameters (Å)	a = 36.97 b = 54.90 c = 65.32	a = 37.05 b = 55.01 c = 65.83

Table 1. *Cont.*

Data Collection [(a)]		
Resolution range (Å)	42.02-1.5 (1.53–1.50)	42.21-1.25 (1.27–1.25)
Number of reflections	109904	304278
Unique reflections	21950	38043
Monomers in asymmetric unit	1	1
Completeness (%)	99.8 (97.1)	99.9 (99.1)
<I/σ(I)>	12.3 (1.9)	13.4 (2.8)
$CC_{1/2}$ [(b)]	0.998 (0.646)	0.998 (0.535)
Multiplicity	5.0 (3.2)	8.0 (4.5)
R_{meas} [(c)]	0.082 (0.637)	0.111 (1.013)
Refinement statistics		
Fraction of free reflections	0.050	0.050
Final R_{cryst}	0.116	0.108
Final R_{free}	0.166	0.134
R.m.s. deviations from ideal geometry (target values are given in parentheses)		
Bond distances (Å)	0.015(0.013)	0.018(0.013)
Bond angles (°)	1.74 (1.64)	2.19 (1.71)
Average main chain B values ($Å^2$)	12.5	9.1
Average side chain B values ($Å^2$)	16.1	13.3
Average B values for Ligand ($Å^2$)	-	25.5
Molprobity score	1.85	1.94
Ramachandran favoured (%) [(d)]	95.16	96.03
Ramachandran outliers (%) [(d)]	0.0	0.0
Clash score	6.34	7.77

[(a)] values in parentheses correspond to the highest resolution shell. [(b)] $CC_{1/2}$ values for Imean are calculated by splitting the data randomly in half. [(c)] R_{meas} is defined as $\Sigma \sqrt{(N/N-1)}|I - <I>|/\Sigma I$, where I is the intensity of the reflection. [(d)] Ramachandran plot analysis was carried out by MOLPROBITY [21].

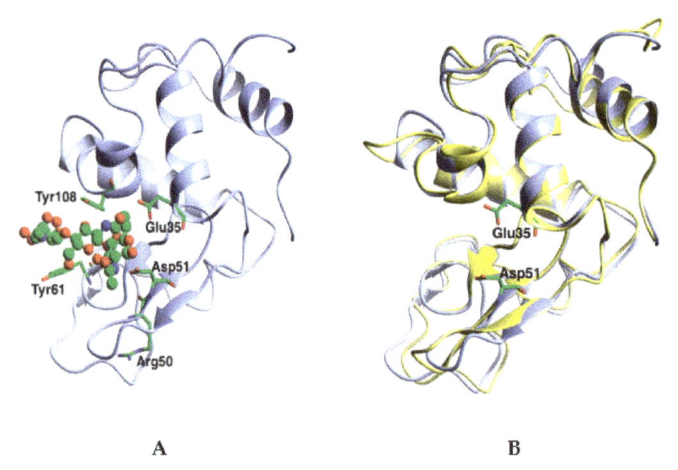

Figure 2. The 3D structure of *O. hoazin* lysozyme (*Oh*Lys). (**A**) *Oh*Lys shown as ribbons with the chitotriose from the complex as ball and stick binding in sites -1 to -3. The position of the catalytic Glu35 and Asp51 (cylinders) are taken from the apo structure. Three key residues which were subsequently mutated to resemble those in Hen Egg White Lysozyme (HEWL) are also shown as cylinders: Arg50, Tyr61 and Tyr108. (**B**) Superposition of the *Oh*Lys (blue) on HEWL (yellow).

The GH22 fold is highly conserved over a range of organisms, as can be seen in Table 2, where the r.m.s. difference in Cα position over between 112 and 125 residues is between 0.79 and 1.4Å, from rainbow trout to mouse. The 3D structure is highly conserved; the chains are of very similar lengths in all species, with almost no deletions or insertions. The structures show a remarkably high level of similarity, greater than might be expected for the sequence identity, with the r.m.s. difference in Cα positions showing little correlation with the evolutionary tree for the GH22 lysozymes discussed below. In part, this likely reflects the conserved set of disulphide bridges in these enzymes. The only variation is in the so-called calcium loop, at the bottom of the structure in Figure 1, with residues in the range 45–51, which is displaced in a couple of the structures. This loop is occupied by a sodium ion in several deposited PDB files. The last two GH22 structures in the Table, from bivalves, show somewhat more extensive differences in several loops, hence the reduced number of equivalent Cα atoms.

Table 2. The structures of GH22 lysozymes in the PDB and their similarity to *Oh*Lys.

Species	Common Name	PDB	Resn. Å	RMS Å	No. Cα
Oncorhynchus mykiss	Rainbow trout	1lmq	1.6	0.78	125
Bos taurus	Cow	2z2f	1.5	0.86	125
Homo sapiens	Human	133l	1.77	0.87	125
Pelodiscus sinensis	Chinese soft-shelled turtle	2gv0	1.9	0.92	125
Tachyglossus aculeatus	Australian echidna: Spiny anteater	1jug	1.9	0.93	124
Phasianus colchicus	Ring necked pheasant	1ghl	2.1	0.94	125
Coturnix japonica	Japanese quail	2ihl	1.4	0.95	124
Mus musculus	Mouse	4yf2	2.1	0.97	123
Canis lupus familiaris	Dog	1qqy	1.85	1.02	122
Gallus gallus	Chicken	3lzt	0.94	1.05	125
Anas platyrhynchos	Duck/Mallard	5v8g	1.2	0.98	125
Colinus virginianus	Virginia quail: Northern bobwhite	1dkj	2	1.09	125
Equus caballus	Horse	2eql	2.5	1.14	123
Bombyx mori	Domestic silk worm	1GD6	2.5	1.17	112
Meleagris gallopavo	Turkey	135l	1.3	1.18	125
Musca domestica	House Fly Lys1	2fbd	1.9	1.19	108
Antheraea mylitta	Silkworm	1IIZ	2.4	1.27	112
Numida meleagris	Helmeted guidea fowl	1hhl	1.9	1.31	125
Musca domestica	House Fly Lys2	3cb7	1.9	1.4	111
Meretrix lusoria	Bivalve	3ab6	1.78	2.33	80
Tapes japonica	Bivalve	2dqa	1.6	2.36	81

The overall charge of *Oh*Lys differs somewhat from that of other C-type lysozymes (Table 3). This increase in negative charge of *Oh*Lys is evident in the surface electrostatics of the enzyme (Figure 3). While the significance of this is not clear, it should be noted that the peptidoglycan substrate is also a charged molecule, and that peptidoglycan from different bacterial sources have different pI. So, charge interactions between the enzyme and the substrate at working pH are probably functionally important.

Table 3. The number of charged residues and pI for four representative lysozymes. Histidines have been excluded from the positive set.

	*Oh*Lys	Bovine	House Fly	HEWL
No. Residues	126	129	122	129
Negative Charge: Asp and Glu	20	15	9	9
Positive Charge: Arg and Lys	13	14	10	17
Ratio (negative/positive)	1.5	≈1	≈1	0.5
Total Charged	33	29	19	26
Theoretical pI	5.1	6.5	7.7	9.3

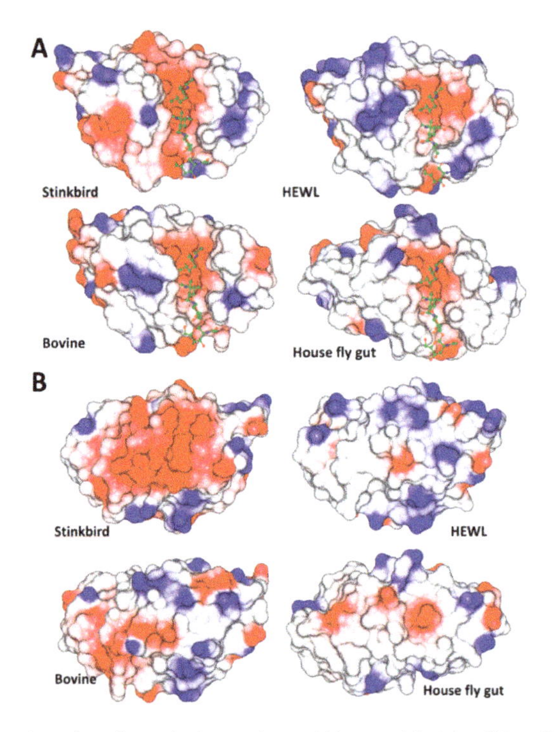

Figure 3. Electrostatic surface charge (red: negative and blue: positive) for *Oh*Lys, HEWL, Bovine gut lysozyme and house fly gut lysozyme, calculated at pH 7.0, within CCP4mg. (**A**) viewed from the active site side of the enzymes, (**B**) from the opposite side. The single clear observation is that the rear (**B**) of *Oh*Lys carries a substantially higher negative charge. The structures were superposed using the Gesamt option in CCP4mg. The ligand is taken from the *Oh*Lys complex.

Co-crystallisation of the inactive variants of the enzyme with chito-oligosaccharides was partially successful. Screening for ligand complexes was carried out with the supposed inactive mutants, E35A and D51A. E35A itself crystallised readily in INDEX screen no. 7 (Hampton Research) in Falcon 24 well plates, producing large, well-diffracted crystals, while it was not possible to obtain crystals from the D51A mutant. Co-crystallisations of E35A were set up with chitobiose, chitotriose, chitotetraose, chitopentaose and chitohexaose, and were successful with chitobiose, chitotetraose, and chitohexaose. The resulting electron density clearly confirmed the Glu to Ala mutation.

The electron density for the crystal soaked in chitohexaose is shown in Figure 4, which shows excellent density for the −2 and −3 subsite sugars, and good density for −1, with poorly ordered density for −4, with a significant difference in density around this subsite. Unfortunately, there is no density in the +1 subsite—a key aim of the experiment had been to observe sugar bound across the point of catalysis between −1 and +1. It is evident that the mutant retains a sufficient level of residual activity to hydrolyse; at the high protein concentrations of a prolonged crystallization, the chitohexaose substrate to a mixture of the two, three and four membered sugars. The density for the chitotetraose ligand is essentially identical to this, while the chitotriose only shows binding in sites −1 to −3. Sites +1 and +2 do appear to be accessible in the structure, and binding of non-hydolysable chitohexaose analogues, for example, with sulphur replacing the glycosidic oxygens, might prove successful if such ligands should become available.

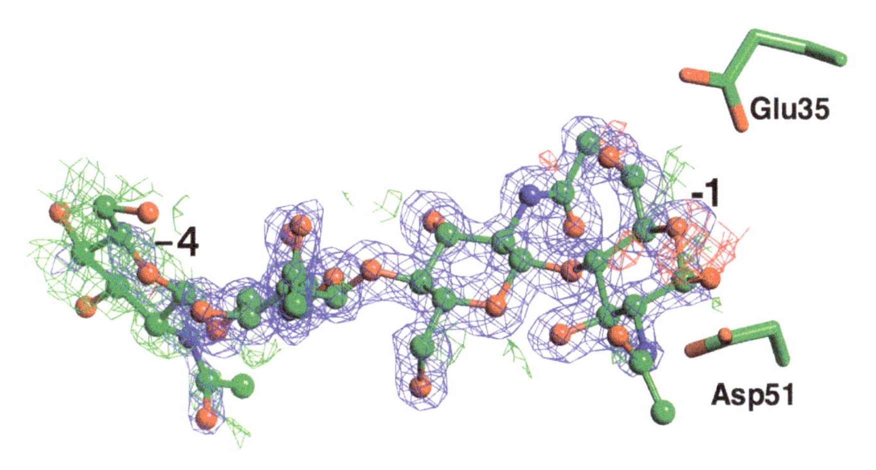

Figure 4. Electron density maps for the ligand from the crystal co-crystallised with chitohexaose. The REFMAC maximum likelihood weighted map, contoured at 1 σ, is shown in blue, the difference map, contoured at 2.5 σ in green (positive), and red (negative), with phases calculated prior to the incorporation of any ligand atoms in refinement. The catalytic residues are superposed from the apo structure.

2.3. Position of OhLys in the GH22 Family Tree: the Avian Gut Enzymes

A bioinformatic analysis of GH22 amino acid sequences shows that vertebrates and invertebrates separate into different subgroups, and in the vertebrates, avian species separate out into three distinct subgroups (chicken-like birds, sea birds and other birds), represented in a tree (Figure 5). *Oh*Lys is found in the "other birds" group. We propose that this cluster represents avian upper-gut enzymes and that there is a physiological reason for this differentiation; one observation is that seabirds and "other birds" feed their chicks, while the chicken-like group does not. In the chicken-like group, high amounts of lysozyme are found in the egg. There is, at present, no conclusion as to which avian lysozymes are digestive gut/crop enzymes. While, for a few (Hoatzin, for example) it has it been firmly established that the lysozyme functions in the crop, it is of note that the avian "upper gut" group (including the Hoatzin, Zebra finch, dove and others, Figure 5) lies on a separate branch of the tree from the avian-egg enzymes, and has conserved changes relative to HEWL. A number of the produced variants addresses the conserved differences between HEWL and the "upper gut" group (for example, Arg50, discussed below). We propose that the "upper gut" lysozymes are utilized differently in these species and that this division will allow for further work regarding the digestive system in birds (and other organisms) by comparing their GH22s. The sequences of the proposed avian gut lysozymes are aligned in Figure 6A and the Hoazin sequence compared to HEWL in Figure 6B.

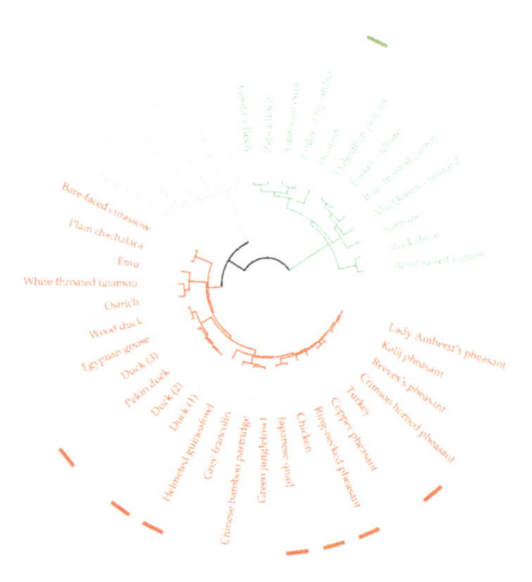

Figure 5. Phylogenetic tree of avian GH22 lysozymes, marked with different colours for different types of birds (red: chicken-like, violet: seabirds, and green: "other birds"/"upper gut"). Truncated sequences were omitted from the analysis and sequences that are more than 98% identical are shown as one entry. Known 3D structures in the PDB are indicated by coloured squares (green: *Oh*Lys and red: pdb). A list of sequence identifiers (common name, scientific name, database ID and cluster sizes can be found in Supplementary Table S1.

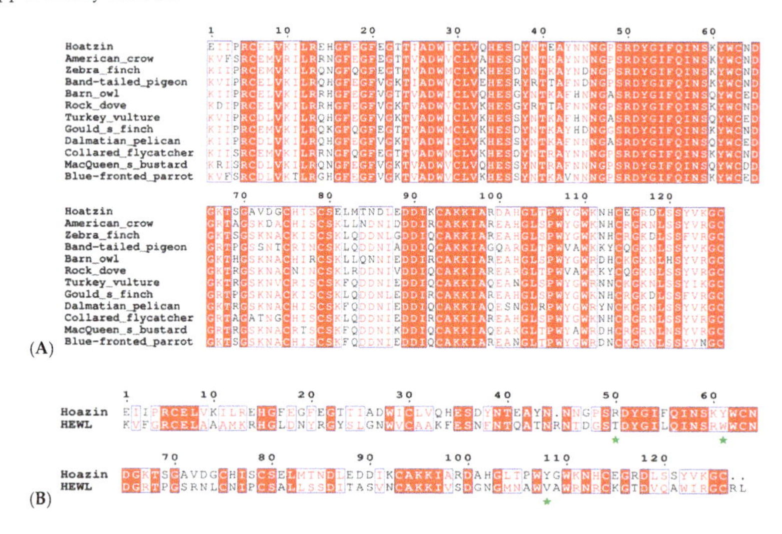

Figure 6. (**A**) Sequences of the proposed avian gut set of GH22 lysozymes: the SWISSPROT/PDB codes and full species names are listed in Supplementary Table S1. (**B**) For comparison, the alignment of *Oh*Lys with HEWL. The sequence numbers correspond to *Oh*Lys. Conserved residues are highlighted in red; conservative changes are shown in red text. There is a single deletion at residue 45 in the *Oh*Lys sequence, and a loss of two amino acids at the C-terminus. The Hoatzin sequence is that reported from the gene sequencing [15], and does not include the two amino acid insertion after residue 1 in the *Oh*Lys variant. Figure produced using ESPrpt [22]. Three key differences are indicated by * (R50/T51, Y61/W62 and Y108/V109): each of these residues was mutated from its *Oh*Lys amino acid to that in HEWL, as part of the mutational studies. The first two of these are totally conserved in the avian gut family in (**A**), the latter is either Y or V.

2.4. Properties of OhLys and Variants

2.4.1. The Extended Set of OhLys Variants

In these variants, a kexin-like protease B (kexB) site was introduced between the signal and the mature region, inserting a lysine and arginine (*1aK *1bR) between residues Glu1 and Ile2 of the wild type sequence. The variants listed in Table S2 were originally designed to probe a number of properties, as shown, and, while these were not all characterised in detail, the results of some mutations are briefly described below.

A selected set of variants was produced in larger amounts, purified and tested for pH optimum, and thermostability, T_m, was measured using Nano differential scanning fluorimetry (Table 4) (all variants derived from *Oh*Lys-KexB). The data show that R50T is destabilized at both pH 4.7 and 8.0, while D90A and Y108V are stabilized, especially at pH 8.0. In summary, Arg at position 50 makes *Oh*Lys more stable, while the D90A and Y108V changes to HEWL residues make the enzyme more alkaline-stable. The latter pair would make these two constructs good candidates for applications requiring high pH stability.

Table 4. Thermostability, as examined by nanoDSF.

	T_m (°C) at pH 4.7	T_m (°C) at pH 8.0
*Oh*Lys-KexB wt	65.2	55.4
R50T	59.8	53.3
K60R Y61W	64.8	54.2
S78P	66.9	57.4
D90A	69.0	64.0
I97V	62.7	54.3
Y108V	69.7	63.8
K124R	67.4	54.1

2.4.2. Retention of Activity of OhLys at Low pH

The pH activity profile on peptidoglycan from *Micrococcus lysodeikticus* was measured for the wild-type and a set of mutants. It cannot be guaranteed that this corresponds directly to the activity on the peptidoglycan found in the real in vivo situation. The pH optimum of the WT *Oh*Lys is 4.4, while the KexB insertion increases this to ~5.0. The optimum remains at 5.0 for the other mutants, only reverting to ~4.4 for the R50T and Y61W variants. These values should be compared with those reported earlier for a set of digestive and more typical lysozymes [11], which showed that the optima for three ruminants were in the range 4.5–5.0, while those for the typical chicken and pig lysozymes had a broad peak between 6 and 8, with the leaf-eating monkey lysozyme around 5.2. This supports the acquisition of a low pH optimum for *Oh*Lys, as a crop digestive enzyme, which has evolved independently of the ruminant set.

From the pH curves of the variants (Figure 7) it was clear that almost all constructs with the kexB insertion had a shift in the pH optimum, from pH 4.4 to about pH 5. It seems likely that this is caused by the additional positive residues, which increase the pI. Five of the seven variants had higher activity than the WT, while two had lower. The two variants with the Y61W mutation had the highest activity compared to HEWL. Variant R50T had significantly lower activity and stability than the WT, and it seems that activity at pH 5 has almost been eliminated. Arg50 is adjacent to the catalytic Asp51 in the chain (Figure 8) and its sidechain is hydrogen-bonded to the sidechains of Asp65 and Thr68, and to the backbone O of Gly47. Both Asp65 and Thr68 are in the loop region from Cys63 to Cys79. Removal of the Arg50 sidechain will remove these hydrogen bond possibilities and could, therefore, lead to a slight rearrangement of this part of the structure, leading to a new position of the catalytic Asp51 and decreased stability of this part of the enzyme.

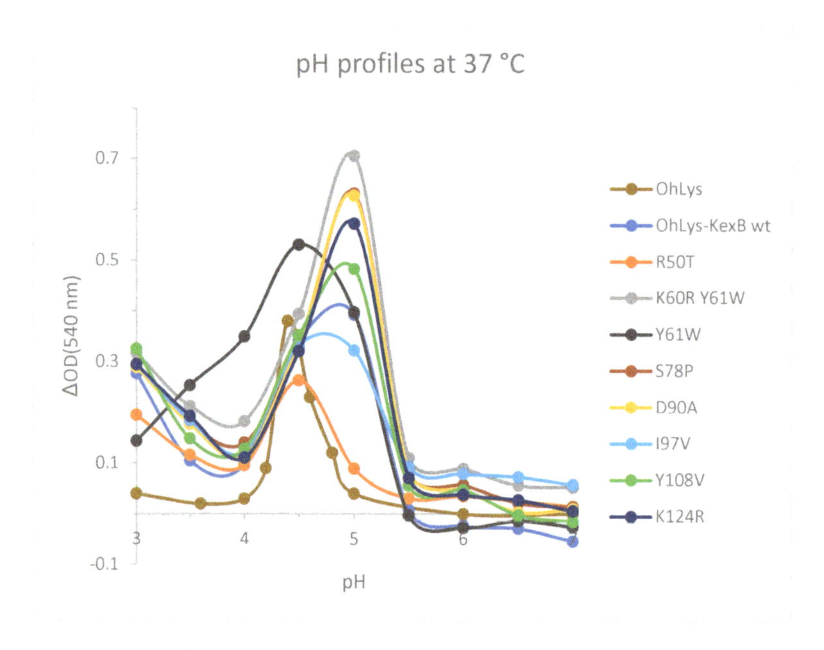

Figure 7. The pH profiles of wild-type *Oh*Lys, the *Oh*Lys-KexB wt, and a selected set of variants, all derived from *Oh*Lys–KexB.

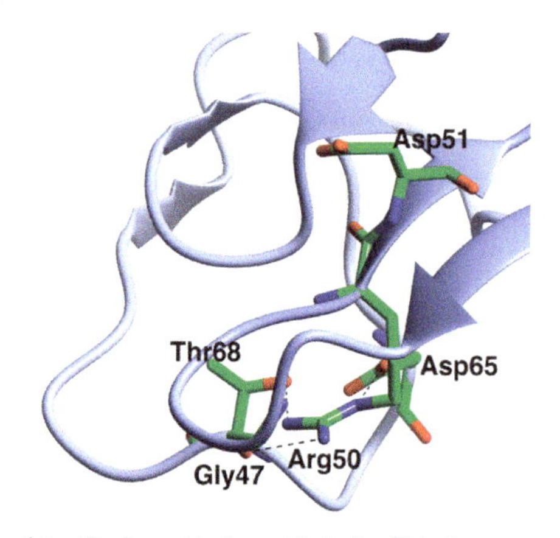

Figure 8. The position of Arg50 adjacent to the catalytic Asp51 in the sequence, showing the residues with which its side chain forms H-bonds, shown as dashed lines.

The two variants with the Y61W mutation had the highest activity. Kumagai and co-workers mutated Trp62 to His in HEWL, and observed an altered substrate-binding mode [23]. In *Oh*Lys, the corresponding residue is Tyr61, which is found frequently among GH22s, with the Trp almost exclusively occurring in HEWL. Variant R50T had significantly lower activity and stability.

The low pH optimum and pepsin resistance reported earlier for *Oh*Lys (Kornegay et al., 1994 and Ruiz et al., 1994) were confirmed for the enzyme expressed in *A. oryzae*. A comparison of activity after incubation under gastric conditions, between HEWL and *Oh*Lys, is shown in Figure 9.

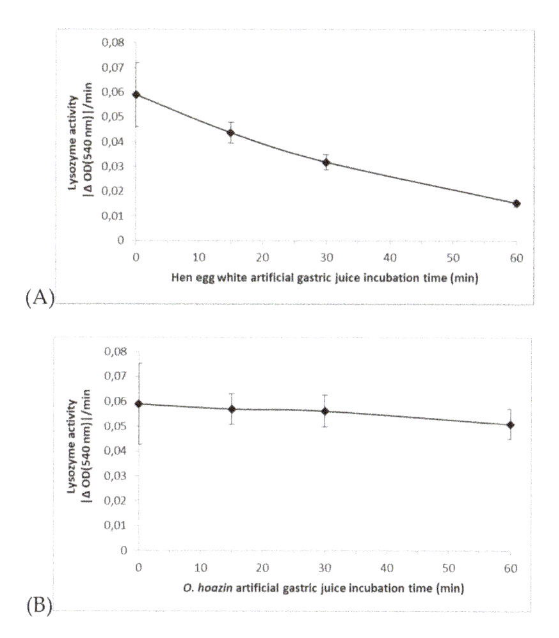

Figure 9. Activity of (**A**) HEWL and (**B**) *Oh*Lys measured by the turbidity assay after incubation, as described above under "In vitro stability of OhLys vs. HEWL under gastric conditions". *Oh*Lys retains a high level of activity after one hour (close to 100%), while that of HEWL has dropped substantially. These results refer to WT *Oh*Lys.

3. Discussion

The major aim of this study, which was successfully achieved, was to establish a possible alternative lysozyme to HEWL for application in animal feed, where the low gastric stability of HEWL posed a major problem for survival in the stomach, so the enzyme could act in the intestine (where most peptidoglycan is present, either dead or alive). The Hoatzin had been previously described as having a gastric stable GH22 lysozyme, and, in addition, it is a creature in which the genome had been sequenced (although more common now, this information is still only available for a limited number of birds). Expression in a micro-organism is essential for an economically viable commercial product, so successful overexpression in an *Aspergillus oryzae* host presents an important advance in this, and enables protein engineering for modifying properties, such as pH-optimum or T-stability. The results confirmed that the enzyme showed considerably better gastric stability, key for its potential application, compared to that of HEWL, reported previously in [11]. The pH optimum of wild type *Oh*Lys was 4.5. While the chicken feed application has been superseded by the development of a GH25 muramidase into a successful commercial product (Balancius^{TM} from DSM nutritional products, Kaiseraugst, Switzerland) [24,25], the observations made for *Oh*Lys could be useful in other applications.

With regard to the variants, possibly the most important observation is that a range of mutations could be made and expressed quite easily, while retaining stability and catalytic activity. The quality of the experimental analysis of the properties of the mutants was limited by the OD assay, which, while being a well-established lysozyme analytical tool, is not highly reproducible. The introduction of the KexB mutation, involving the insertion of two residues *1aK *1bR, increased the pH optimum by one pH unit, and this remained essentially unchanged in the set of mutants based on this variant. The kexB mutation was not successful in allowing removal of the N-terminal Glu. Finally, some mutations appeared to have higher specific activity, and four variants had higher thermal stability.

Bioinformatic analysis of avian GH22 amino acid sequences shows that they separate into three distinct subgroups (chicken-like birds, sea birds and other birds). *Oh*Lys is found in the "other birds" group, and we propose that this cluster represents avian upper-gut enzymes and that there is a physiological reason for this differentiation.

The structure presented here is the first to be determined for the avian upper-gut subset of GH22 lysozymes, and is, as expected, very similar overall to other C-type lysozymes. The close similarity in the 3D structure for this family—more than might be expected for the sequence similarity—is assumed to be at least in part due to the disulfide bridges. Indeed, it is notable that the enzymes can be different, with, for example, low sequence ID and different pH-optima, despite having such similar structural frameworks. The binding of the ligand in subsites −4, −3, −2 and −1 is typical for this family. Unfortunately, the "inactive" mutant retained sufficient activity to degrade longer oligosaccharide substrates, and so the complexes did not span the −1 and +1 sites as planned. However, the long active site cleft is accessible to ligand in the crystals, which may prove beneficial in future.

In summary, we present the first structure and a ligand complex of a digestive lysozyme from a new sub-group of GH22s from the avian upper gut. *Oh*Lys can be expressed in a production-relevant host and protein engineering can be performed to optimize the performance.

4. Materials and Methods

4.1. Cloning, Expression and Purification of OhLys

4.1.1. Heterologous Expression of OhLys in *Aspergillus oryzae*

According to the published Genbank entry (AAA73935.1, [15]) genomic DNA was extracted from a frozen hoatzin tissue sample provided by the Louisiana State University Museum of Vertebrate Zoology collection (*Opisthocomus hoazin* (GenBank® accession number L36032) and the sequence determined. This was used as a starting point to design an artificial coding region (CDR), which was adapted to *Aspergillus oryzae* codon usage. The CDR was amplified by PCR, and cloned into an expression vector using InFusion cloning. Resulting clones were sequenced on both strands and transformed into an *A. oryzae* (NCBI:txi 90341) expression host. Transformants were grown for four days in yeast extract/Peptone/Mannitol 5 g/3 g/25 g per liter supplemented media in a microtiter plate, and the expression in the resulting supernatants was evaluated by SDS-PAGE. The best transformant was spore purified twice.

4.1.2. Fermentation and Purification

The transformant was fermented in shaker flasks at 30 °C for 72 h. Culture broths were filtered through filtration cloth, and, subsequently, through a 0.2 μm filtration unit (Nalgene, Thermo Fisher Scientific, Waltham, MA, USA) to remove the *Aspergillus* host. Solid NaCl was added to a final concentration of 200 mM and the pH was adjusted to pH 5.5 with 20% CH_3COOH. The adjusted enzyme solution was applied to a SP-sepharose FF column (GE Healthcare (Brondby, Denmark)), which was equilibrated in 50 mM CH_3COOH/NaOH, 200 mM NaCl, pH 5.5. The column was thoroughly washed with an equilibration buffer to remove loosely bound protein. The enzyme was eluted using a linear NaCl gradient (200 mM to 1000 mM) in 50 mM CH_3COOH/NaOH, pH 5.5 over five column volumes. *Oh*Lys eluted as a single peak, and the purity was analysed by SDS-PAGE. The resultant protein solution was concentrated to 26 mg mL^{-1} and buffer exchanged into 25 mM HEPES pH 7.5, using a Vivaspin ((Sartorius, Goettingen, Germany) 10-kDa cut-off concentrator. Edman degradation on the purified protein indicated an N-terminus, corresponding to the expected (EIIPRCELVK-), and intact molecule mass spectrometry gave a molecular weight of 14260.4 g/mol (theoretical value 14,260.9 g/mol).

4.2. Expression of *Oh*Lys Variants

In the first variant, a kexin-like protease B (kexB) site was introduced between the signal and the mature region, inserting a lysine and arginine (*1aK *1bR) between residues Glu1 and Ile2 of the wild-type sequence. The aim had been to remove the N-terminal glutamate, to avoid pyro-glutamate formation, but unfortunately no cleavage was observed with kexB, and the result was a construct with an N-terminal sequence of E-1 K0 R1 I2 I3 P4 R5 C6 E7 L8 V9- (the residues are numbered from -1, so as to retain the normal mature sequence for the bulk of the chain), instead of the hoped for IIPRCELV-. This is henceforth referred to as *Oh*Lys-kexB.

Subsequent variants were made by established methods (see WO 2012/035103 [26]) starting from the *Oh*Lys-kexB variant, rather than the true wild-type. Variants were transformed into an *A. oryzae* host and the expression estimated by SDS-page. Where appropriate, the activity was evaluated using a turbidity assay, with *Micrococcus luteus* as substrate (see below). The variants fell into two groups, described in the following sections. The first aimed to prepare enzymatically inactive constructs for co-crystallisation with ligands, and the second group targeted changing a number of properties, as listed in Table S1, including three aimed at key differences between *Oh*Lys and HEWL: R50T, Y61W and Y108V.

4.2.1. Inactive Catalytic Site Variants

Based on the *Oh*Lys-kexB variant, four variants (E35A, E35Q, D51A and D51N) of the catalytic residues E35 and D51 were produced and purified with the methods described above. The aim was to express an inactive enzyme for co-crystallisation with ligand.

4.2.2. The Extended Set of Variants

An extensive set of variants was created, all starting from the *Oh*Lys-kexB variant, and are briefly summarised in Supplementary Table S1. A key observation was that that all these variants were shown to retain enzyme activity using the turbidity assay below, supporting the robustness of the *Oh*Lys fold.

4.2.3. Preparation of *Micrococcus Lysodeikticus* Substrate

Before use, *M. lysodeikticus* cells were resuspended in citric acid/phosphate buffer pH 4.4 in a concentration of 0.5 mg cells/mL and the optical density (OD) was measured at 540 nm. The cell suspension was adjusted so the cell concentration equaled an OD_{540} of 1.0, and the adjusted cell suspension was stored cold before use. Resuspended cells were used within 4 h.

4.2.4. Turbidity Assay of Activity

The OD-drop assay measures lysozyme activity through a reduction in OD (light scattering) caused by cell lysis (cell wall hydrolysis), as described in many papers on HEWL [11,27]. Here, the activities of *Oh*lLys and variants were determined by measuring the decrease (drop) in absorbance/optical density of a solution of resuspended *Micrococcus lysodeikticus* ATTC No. 4698 (Sigma-Aldrich M3770), measured in a spectrophotometer at 540 nm (https://www.sigmaaldrich.com/technical-documents/protocols/biology/enzymatic-assay-of-lysozyme.html and [28]).

4.2.5. Measurement of Lysozyme Activity in the Turbidity Assay

The lysozyme sample was diluted to a concentration in the range 100–200 mg enzyme protein/L in citric acid/phosphate buffer pH 4.4 and kept on ice until use. In a 96 well microtiterplate (Nunc), 200 µL of the substrate was added to each well, and the plate incubated at 37 °C for 5 min in a VERSAmax microplate reader (Molecular Devices, Wokingham, United Kingdom). Following incubation, the absorbance of each well was measured at 540 nm (start value). To start, the activity measurement 20 µL of the diluted lysozyme samples was added to the 200 µL substrate in each well and kinetic measurement of absorbance at 540 nm was initiated for a minimum of 30 min, up to 24 h, at 37 °C.

The measured absorbance at 540 nm was monitored for each well, and over time, and a drop-in absorbance was taken as the drop-in lysozyme activity. To compare results from the turbidity assay, the samples to be compared were tested in the same experimental run, using the same buffer and substrate batch.

4.3. Biophysical Properties of OhLys and Variants

4.3.1. pH Optimum of *Oh*Lys

pH activity curves between pH 3 and 7 were recorded using the turbidity assay described above, but with the pH of the citrate/phosphate buffer adjusted to the relevant value. All samples were diluted to an enzyme concentration of 50 µg/mL. The wild type enzyme, without the KR insertion, had a pH optimum of pH 4.4.

4.3.2. In Vitro Stability of *Oh*Lys vs. HEWL under Gastric Conditions

*Oh*Lys and HEWL were incubated for 0, 15, 30 and 60 min in artificial gastric juice (HCl pH 2, 1 mg/mL pepsin, 0.1 M NaCl) at 37 °C, and their activity determined using the turbidity assay described above.

4.3.3. Thermostability of Variants Using Nano Differential Scanning Fluorescence

Nano Differential Scanning Fluorescence (NanoDSF) was performed with a Prometheus NT.48 instrument (NanoTemper Technologies GmbH, München, Germany). Purified *Oh*Lys variants (in either 250 mM Na-acetate, pH 4.7, or 20 mM tris(hydroxymethyl)aminomethane, pH 8.0) were loaded into nanoDSF standard grade capillaries (NanoTemper Technologies GmbH; catalogue number PR-C002) through capillary action. Three capillaries were filled for each sample. The capillaries were then placed into the instrument (up to 48 single capillaries can be loaded in a single run) and the laser intensity required for optimum signal generation was determined. The samples were run with the following experimental setting: temperature slope 2 °C/minute, start temperature 20 °C and end temperature 95 °C. The data were analysed using the software supplied with the instrument (PR.ThermControl v2.0.4, NanoTemper Technologies GmbH) and the Tm (for the ratio 350 nm/330 nm).

4.4. Apo-Enzyme Crystallisation and Structure Solution

Crystals of apo-*Oh*Lys, the wild-type, not the kexB variant, were grown in 96-well MRC crystallisation Plates™ (Molecular Dimensions Ltd.), set up by a Mosquito Nanodrop crystallisation robot (Molecular Dimensions Ltd.). A total of 150 nl of protein was mixed with 150 nl of mother liquor solution. Crystals grew in 0.1 M MIB system (malonic acid, imidazole, boric acid), with pH 4.0 and 25% Peg 3350, corresponding to conditions B1–B4 of the PACT premier™ screen (Molecular Dimensions Ltd.). Crystals were cryoprotected in a mother liquor solution, incorporating 25% glycerol prior to flash freezing in liquid nitrogen. Diffraction data were collected at the European Synchrotron Radiation Facility beamline ID23-1, at 100 K to 1.5 Å resolution.

All computations were carried out using programs from the *CCP4* suite [29], unless otherwise stated. Data were processed with MOSFLM [30,31], and scaled and merged with AIMLESS. The apo structure was solved by molecular replacement, using PHASER [32], with the coordinates of a human mutant lysozyme (PDB code: 1gft) as a search model. The starting model was improved manually using COOT [33], alternating with cycles of REFMAC [34]. The structure was validated using MOLPROBITY [35] prior to deposition.

4.5. Ligand Complexes Crystallization and Structure Solution

Screening for ligand complexes was carried out with the supposedly inactive mutants, E35A and D51A, both of which contained the kexB dipeptide insertion. E35A itself crystallised readily in INDEX screen no. 7 (Hampton Research) in Falcon 24 well plates, producing large, well-diffracted crystals, while it proved impossible to obtain crystals of the D51A mutant. Co-crystallisations of E35A were set up with chitobiose, chitotriose, chitotetraose, chitopentaose and chitohexaose, and were successful with chitobiose, chitotetraose, and chitohexaose. Crystals were harvested and frozen, and data were collected at the Diamond Light Source beam line IO4 for all three co-crystals. They were all isomorphous with those of the apo-enzyme, which was therefore used as the starting model. There was electron density in the active site of all three structures, but the density did not span the active site, and was restricted to the −1, −2 and −3 subsites, for a maximum of three monosaccharides. As the three structures were so similar, only that resulting from the chitohexaose sample was fully refined. It is assumed that the "inactive" E35A mutant had retained a low level of activity, sufficient to hydrolyse the longer oligosaccharides during complex formation and/or crystal growth.

For both the apo enzyme and the ligand complex, data collection and refinement statistics are given in Table 1. Structural figures were drawn with CCP4mg [36].

4.6. Patents

The patent application "Lysozymes" (WO 2012/035103) [26] was based on the protein engineering work described in this paper.

Supplementary Materials: Supplementary Materials can be found at http://www.mdpi.com/1422-0067/20/22/5531/s1.

Author Contributions: Conceptualization, M.S., L.K.S. and L.D.M.; Data curation, J.P.T.; Formal analysis, K.S.W., M.S., L.K.S. and G.P.G.; Investigation, E.J.T., M.S., L.K.S., M.K. and J.P.T.; Methodology, G.P.G.; Project administration, K.S.W. and G.J.D.; Software, L.D.M., G.P.G. and J.P.T.; Validation, J.P.T.; Writing—Original draft, E.J.T.; Writing—Review and editing, K.S.W., L.K.S., M.K. and G.J.D.

Funding: This research received no external funding. We note that L.K.S., M.S., and G.P.G. are employees of Novozymes (Denmark).

Acknowledgments: Edward J. Taylor gratefully acknowledges financial support from the Royal Society and the Biotechnology and Biological Sciences Research Council (BBSRC). Gideon J. Davies is the Royal Society "Ken Murray" Research Professor. We thank Diamond Light Source for access to beamlines I02 (proposal number mx-306) and I04 (proposal number mx-7864) that contributed to the results presented here. The authors thank Sam Hart for assistance during data collection.

Conflicts of Interest: The authors declare no conflict of interest, but we note that the Novozymes authors declare the following competing financial interest(s): Novozymes is a commercial enzyme supplier.

Abbreviations

CAZy	Carbohydrate-Active enZYmes Database
GH	glycoside hydrolase
HEWL	HEWL: hen egg white lysozyme
NAG	*N*-acetylglucosamine
NAM	*N*-acetylmuramic acid
*Oh*Lys	*Opisthocomus hoazin* lysozyme

References

1. Malanovic, N.; Lohner, K. Antimicrobial Peptides Targeting Gram-Positive Bacteria. *Pharmaceuticals (Basel)* **2016**, *9*, 59. [CrossRef]
2. Cantarel, B.L.; Coutinho, P.M.; Rancurel, C.; Bernard, T.; Lombard, V.; Henrissat, B. The Carbohydrate-Active EnZymes database (CAZy): An expert resource for Glycogenomics. *Nucleic Acids Res.* **2009**, *37*, D233–D238. [CrossRef]
3. Monzingo, A.F.; Marcotte, E.M.; Hart, P.J.; Robertus, J.D. Chitinases, chitosanases, and lysozymes can be divided into procaryotic and eucaryotic families sharing a conserved core. *Nat. Struct. Biol.* **1996**, *3*, 133–140. [CrossRef] [PubMed]
4. Callewaert, L.; Michiels, C.W. Lysozymes in the animal kingdom. *J. Biosci.* **2010**, *35*, 127–160. [CrossRef]

5. Blake, C.C.; Koenig, D.F.; Mair, G.A.; North, A.C.; Phillips, D.C.; Sarma, V.R. Structure of hen egg-white lysozyme. A three-dimensional Fourier synthesis at 2 Angstrom resolution. *Nature* **1965**, *206*, 757–761. [CrossRef] [PubMed]
6. Davies, G.J.; Wilson, K.S.; Henrissat, B. Nomenclature for sugar-binding subsites in glycosyl hydrolases. *Biochem. J.* **1997**, *321*, 557–559. [CrossRef] [PubMed]
7. Vocadlo, D.J.; Davies, G.J.; Laine, R.; Withers, S.G. Catalysis by hen egg-white lysozyme proceeds via a covalent intermediate. *Nature* **2001**, *412*, 835–838. [CrossRef]
8. Gao, Y.C.; Zhang, G.P.; Krentz, S.; Darius, S.; Power, J.; Lagarde, G. Inhibition of spoilage lactic acid bacteria by lysozyme during wine alcoholic fermentation. *Aust. J. Grape Wine R* **2002**, *8*, 76–83. [CrossRef]
9. Proctor, V.A.; Cunningham, F.E. The chemistry of lysozyme and its use as a food preservative and a pharmaceutical. *Crit. Rev. Food Sci. Nutr.* **1988**, *26*, 359–395. [CrossRef]
10. Hughey, V.L.; Johnson, E.A. Antimicrobial activity of lysozyme against bacteria involved in food spoilage and food-borne disease. *Appl. Environ. Microbiol.* **1987**, *53*, 2165–2170. [PubMed]
11. Dobson, D.E.; Prager, E.M.; Wilson, A.C. Stomach lysozymes of ruminants. I. Distribution and catalytic properties. *J. Biol. Chem.* **1984**, *259*, 11607–11616. [PubMed]
12. Ragland, S.A.; Criss, A.K. From bacterial killing to immune modulation: Recent insights into the functions of lysozyme. *PLoS Pathog.* **2017**, *13*, e1006512. [CrossRef] [PubMed]
13. Jeenes, D.J.; Mackenzie, D.A.; Archer, D.B. Transcriptional and post-transcriptional events affect the production of secreted hen egg white lysozyme by Aspergillus niger. *Transgenic Res.* **1994**, *3*, 297–303. [CrossRef] [PubMed]
14. Ruiz, M.C.; Dominguezbello, M.G.; Michelangeli, F. Gastric Lysozyme as a Digestive Enzyme in the Hoatzin (Opisthocomus Hoazin), a Ruminant-Like Folivorous Bird. *Experientia* **1994**, *50*, 499–501. [CrossRef]
15. Kornegay, J.R.; Schilling, J.W.; Wilson, A.C. Molecular adaptation of a leaf-eating bird: Stomach lysozyme of the hoatzin. *Mol. Biol. Evol.* **1994**, *11*, 921–928.
16. Grajal, A.; Strahl, S.D.; Parra, R.; Gloria Dominguez, M.; Neher, A. Foregut fermentation in the hoatzin, a neotropical leaf-eating bird. *Science* **1989**, *245*, 1236–1238. [CrossRef]
17. Kierończyk, B.; Rawski, M.; Długosz, J.; Świątkiewicz, S.; Józefiak, D. Avian crop function–A review. *Ann. Anim. Sci.* **2016**, *16*, 653–678. [CrossRef]
18. Godoy-Vitorino, F.; Ley, R.E.; Gao, Z.; Pei, Z.; Ortiz-Zuazaga, H.; Pericchi, L.R.; Garcia-Amado, M.A.; Michelangeli, F.; Blaser, M.J.; Gordon, J.I.; et al. Bacterial community in the crop of the hoatzin, a neotropical folivorous flying bird. *Appl. Environ. Microbiol.* **2008**, *74*, 5905–5912. [CrossRef]
19. Flach, J.; Pilet, P.E.; Jolles, P. Whats New in Chitinase Research. *Experientia* **1992**, *48*, 701–716. [CrossRef]
20. Archer, D.B.; Mackenzie, D.A.; Jeenes, D.J.; Roberts, I.N. Proteolytic Degradation of Heterologous Proteins Expressed in Aspergillus-Niger. *Biotechnol. Lett.* **1992**, *14*, 357–362. [CrossRef]
21. Chen, V.B.; Arendall, W.B., 3rd; Headd, J.J.; Keedy, D.A.; Immormino, R.M.; Kapral, G.J.; Murray, L.W.; Richardson, J.S.; Richardson, D.C. MolProbity: All-atom structure validation for macromolecular crystallography. *Acta Crystallogr. D Biol. Crystallogr.* **2010**, *66*, 12–21. [CrossRef] [PubMed]
22. Robert, X.; Gouet, P. Deciphering key features in protein structures with the new ENDscript server. *Nucleic Acids Res.* **2014**, *42*, W320–W324. [CrossRef] [PubMed]
23. Kumagai, I.; Maenaka, K.; Sunada, F.; Takeda, S.; Miura, K. Effects of subsite alterations on substrate-binding mode in the active site of hen egg-white lysozyme. *Eur. J. Biochem.* **1993**, *212*, 151–156. [CrossRef] [PubMed]
24. Goodarzi Boroojeni, F.; Manner, K.; Rieger, J.; Perez Calvo, E.; Zentek, J. Evaluation of a microbial muramidase supplementation on growth performance, apparent ileal digestibility, and intestinal histology of broiler chickens. *Poult. Sci.* **2019**, *98*, 2080–2086. [CrossRef] [PubMed]
25. Sais, M.; Barroeta, A.C.; Lopez-Colom, P.; Nofrarias, M.; Majo, N.; Lopez-Ulibarri, R.; Perez Calvo, E.; Martin-Orue, S.M. Evaluation of dietary supplementation of a novel microbial muramidase on gastrointestinal functionality and growth performance in broiler chickens. *Poult. Sci.* **2019**, pez466. [CrossRef] [PubMed]
26. Klausen, M.; Allesen-Holm, M.; de Maria, L.; Skjoet, M. Lysozymes. Patent WO/2012/035103, 22 March 2012.
27. Parry, R.M., Jr.; Chandan, R.C.; Shahani, K.M. A Rapid and Sensitive Assay of Muramidase. *Proc. Soc. Exp. Biol. Med.* **1965**, *119*, 384–386. [CrossRef] [PubMed]
28. Shugar, D. The measurement of lysozyme activity and the ultra-violet inactivation of lysozyme. *Biochim. Biophys. Acta* **1952**, *8*, 302–309. [CrossRef]

29. Winn, M.D.; Ballard, C.C.; Cowtan, K.D.; Dodson, E.J.; Emsley, P.; Evans, P.R.; Keegan, R.M.; Krissinel, E.B.; Leslie, A.G.; McCoy, A.; et al. Overview of the CCP4 suite and current developments. *Acta Crystallogr D Biol. Crystallogr.* **2011**, *67*, 235–242. [CrossRef]
30. Leslie, A.G.W. Recent changes to the MOSFLM package for processing film and image plate data. *Jt. CCP4 ESF-EAMCB Newsl. Protein Cryst.* **1992**, *26*.
31. Leslie, A.G. The integration of macromolecular diffraction data. *Acta Crystallogr D Biol. Crystallogr* **2006**, *62*, 48–57. [CrossRef]
32. McCoy, A.J.; Grosse-Kunstleve, R.W.; Adams, P.D.; Winn, M.D.; Storoni, L.C.; Read, R.J. Phaser crystallographic software. *J. Appl. Crystallogr.* **2007**, *40*, 658–674. [CrossRef] [PubMed]
33. Emsley, P.; Lohkamp, B.; Scott, W.G.; Cowtan, K. Features and development of Coot. *Acta Crystallogr D Biol. Crystallogr.* **2010**, *66*, 486–501. [CrossRef] [PubMed]
34. Murshudov, G.N.; Vagin, A.A.; Dodson, E.J. Refinement of macromolecular structures by the maximum-likelihood method. *Acta Crystallogr D Biol. Crystallogr.* **1997**, *53*, 240–255. [CrossRef] [PubMed]
35. Davis, I.W.; Leaver-Fay, A.; Chen, V.B.; Block, J.N.; Kapral, G.J.; Wang, X.; Murray, L.W.; Arendall, W.B., 3rd; Snoeyink, J.; Richardson, J.S.; et al. MolProbity: All-atom contacts and structure validation for proteins and nucleic acids. *Nucleic Acids Res.* **2007**, *35*, W375–W383. [CrossRef] [PubMed]
36. McNicholas, S.; Potterton, E.; Wilson, K.S.; Noble, M.E. Presenting your structures: The CCP4mg molecular-graphics software. *Acta Crystallogr D Biol. Crystallogr.* **2011**, *67*, 386–394. [CrossRef] [PubMed]

International Journal of
Molecular Sciences

Article

Deep Eutectic Solvents as New Reaction Media to Produce Alkyl-Glycosides Using Alpha-Amylase from *Thermotoga maritima*

Alfonso Miranda-Molina †,‡, Wendy Xolalpa †, Simon Strompen §, Rodrigo Arreola-Barroso, Leticia Olvera, Agustín López-Munguía, Edmundo Castillo and Gloria Saab-Rincon *

Departamento Ingeniería Celular y Biocatálisis, Instituto de Biotecnología, Universidad Nacional Autónoma de México, Apartado Postal 510-3, Cuernavaca, Morelos 62250, Mexico; alfonso_itz@yahoo.com.mx (A.M.-M.); wxolalpa@ibt.unam.mx (W.X.); simon.strompen@basf.com (S.S.); rarreolb@ibt.unam.mx (R.A.-B.); lolvera@ibt.unam.mx (L.O.); agustin@ibt.unam.mx (A.L.-M.); edmundo@ibt.unam.mx (E.C.)

* Correspondence: gsaab@ibt.unam.mx

† These authors contributed equally to this work.

‡ Present address: Centro de Investigaciones Químicas, Universidad Autónoma del Estado de Morelos. Av. Universidad 1001, Col. Chamilpa, C.P. 62209, Cuernavaca, Morelos 62210, Mexico

§ Present address: BASF Personal Care and Nutrition GmbH, 40589 Duesseldorf-Holthausen, Germany

Received: 17 August 2019; Accepted: 25 September 2019; Published: 31 October 2019

Abstract: Deep Eutectic Solvents (DES) were investigated as new reaction media for the synthesis of alkyl glycosides catalyzed by the thermostable α-amylase from *Thermotoga maritima* Amy A. The enzyme was almost completely deactivated when assayed in a series of pure DES, but as cosolvents, DES containing alcohols, sugars, and amides as hydrogen-bond donors (HBD) performed best. A choline chloride:urea based DES was further characterized for the alcoholysis reaction using methanol as a nucleophile. As a cosolvent, this DES increased the hydrolytic and alcoholytic activity of the enzyme at low methanol concentrations, even when both activities drastically dropped when methanol concentration was increased. To explain this phenomenon, variable-temperature, circular dichroism characterization of the protein was conducted, finding that above 60 °C, Amy A underwent large conformational changes not observed in aqueous medium. Thus, 60 °C was set as the temperature limit to carry out alcoholysis reactions. Higher DES contents at this temperature had a detrimental but differential effect on hydrolysis and alcoholysis reactions, thus increasing the alcoholyisis/hydrolysis ratio. To the best of our knowledge, this is the first report on the effect of DES and temperature on an enzyme in which structural studies made it possible to establish the temperature limit for a thermostable enzyme in DES.

Keywords: Enzymatic glycosylation; alkyl glycosides (AG)s; Deep eutectic solvents (DES); Amy A; alcoholysis; hydrolysis; methanol; circular dichroism; protein stability; alpha-amylase

---

1. Introduction

Alkyl glycosides (AGs) are a class of non-ionic surfactants prepared from renewable agricultural resources, namely starch and fats or their derivatives [1]. AGs contain a carbohydrate head group such as glucose, galactose, maltose, xylose, α-cyclodextrin, and a hydrocarbon tail, usually a primary alcohol of different chain length of saturated or unsaturated nature [2,3]. Their high biodegradability and a low toxicity make them attractive for many applications such as cosmetics, foods, the extraction of organic dyes, membrane protein research, and pharmaceuticals [4]. The industrial production of alkyl glycosides has been typically achieved by traditional organic chemical procedures. However, this strategy requires multiple protection, deprotection and activation steps, the preparation of an anomerically-pure AG requiring high temperature and pressure, as well as the use of toxic catalysts, harmful chemicals, dangerous solvents, and other harsh conditions that adversely impact the environment and human health [4,5].

Considering that one of the most interesting properties of enzymes is their regio- and stereo-selectivity, a biocatalytic approach represents an interesting alternative to cope with this limitation. Moreover, the reactions are conducted under mild temperature and pH conditions which minimize side reactions [6–13]. In this context, AGs have been successfully prepared enzymatically, by applying glycosidases, such as β-galactosidases, β-glucosidases [14], or β-xylosidases [15], among others. Nevertheless, reports dealing with α-amylases are scarce [1,16–18].

The natural reaction of α-amylases (E.C. 3.2.1.1) is the hydrolysis of internal α-1,4-glycosidic bonds in starch through a double-displacement mechanism in which a covalent intermediate glycosyl enzyme is deglycosylated by water [19]. Like all retaining glycosidases, α-amylases can also catalyze transfer reactions which result from the affinity of alternative molecules other than water as glucosyl acceptors. Sugars are the most efficient group of acceptors studied (transglycosylation reactions), but glucosyl residue may also be transferred to alcohols (alcoholysis reactions) [20]. As water is the natural medium in these reactions, the transfer yield is always limited by the competition with hydrolysis. Deep eutectic solvents (DES) usually consist of a salt (most frequently quaternary ammonium salts such as choline chloride) and a complexing agent acting as a hydrogen bond donor (e.g. urea). Replacing the conventional aqueous solvent by alternative solvents such as DES offers the possibility to favor transglycosylation activity. Therefore, together with enzyme engineering, a strategy based on solvent engineering is a promising alternative to improve the synthesis yield of industrially-relevant alkyl-glycosides.

The aim of this work was to study the effect of DES as alternative reaction media for the α-amylase catalyzed synthesis of alkyl-glycosides. DES share many characteristics of ionic liquids, including their capacity to dissolve compounds of different polarities, their availability, and their environmental friendliness. Recent studies indicate that pure DESs or cosolvents represent "green" alternatives for reaction media in lipase-catalyzed synthesis [21–23]. Few reports exist, however, on enzymes reactions other than lipases in DES, as they generally have high viscosities and require temperatures higher than 50 °C, either for melting, or simply to reduce the medium viscosity. Amy A, an α-amylase from *T. maritima*, is a stable enzyme, active at temperatures above 80 °C, and thus considered particularly suited for application in DES.

2. Results

In the present work, a series of 12 deep eutectic solvents (DES) were prepared. The selection of DES was based on viscosity and literature data reporting biocatalytic reactions in the corresponding solvents. Notably, natural deep eutectic solvents (NADES) without water or choline chloride as a salt are highly viscous, and therefore, are unsuitable for enzymatic reactions. On the other hand, Cholinium (Ch+), a natural Generally Recognized as Safe (GRAS) compound, commonly used as a food additive, is the preferred hydrogen bond acceptor (HBA) [24]. Thus, choline chloride was the most common HBA selected in combination with hydrogen bond donors (HBD) of different structure: amides, carboxylic acids, alcohols, and polyols (sugars). The composition of all the assayed DES and their preparation method are reported in Table S1 in Supplementary Information.

2.1. Screening of DES for Application in α-amylase Catalyzed Reactions and Effect of Water Content on Enzyme Activity and Stability

While lipases can be used in essentially water-free media or systems with low water activity [25,26], most enzymes require water for activity [27–31]. In this context, a screening in 2 stages was designed to evaluate the viability of DESs in Amy A catalyzed reactions. In a first stage, DESs were assayed as pure solvent and cosolvent in order to define the minimal amount of water required for the reaction.

In the second stage, the DES or water: DES mixtures were assayed as reaction media for transfer reactions with methanol as the acceptor substrate, using maltotriose as a glucosyl donor.

To identify the required amount of water for enzyme activity, the 12 selected DES were used as solvents in reactions containing increasing water concentrations of 0, 5, 10, 25, 50, 75, and 100% (*v/v*), with maltotriose as a substrate. Substrate conversion was measured as the amount of maltotriose hydrolyzed to maltose and glucose. Further hydrolysis of maltose to glucose was not considered. In the absence of water (0%), methanol (5% *v/v*) was included in the reaction to provide a reactive nucleophile. In Figure 1, substrate conversion after 4 h of reaction is shown as a function of the water content. DES-containing acidic groups were not suitable solvents at any water concentration assayed, and were therefore not included in the figure.

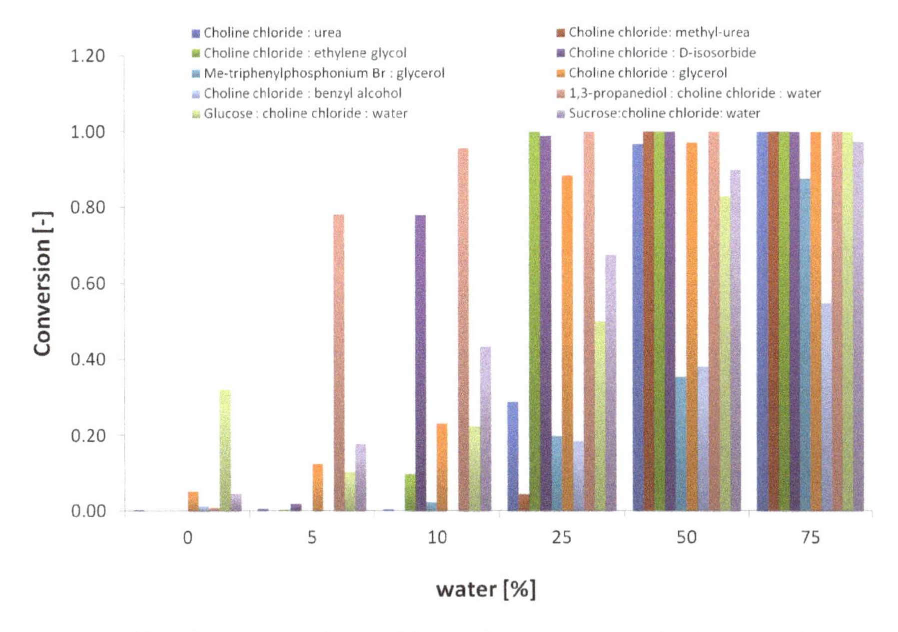

Figure 1. Effect of water content in water:DES cosolvent systems on maltotriose conversion in α-amylase-catalyzed hydrolysis reactions. No reaction was observed on choline chloride:malonic acid (1:1) and choline chloride:levulinic acid (1:2).

As shown in Figure 1, significant α- amylase hydrolytic activity takes place in solvents containing only 10% water or less for several DES, except for choline chloride:malonic acid (1:1) and choline chloride:levulinic acid (1:2), for which no activity was observed. In order to include enzyme stability in the assay, samples were taken from the same reactions after 4 h and checked for residual amylase activity on starch. Figure 2 shows the residual amylase activity after 4 h reaction in DES, based on the I^-/I_2 staining assay. Residual amylose produces a blue stain, indicating that the amylase was deactivated by the corresponding DES at some time during the reaction. Clear wells indicate the presence of active amylase. The yellow color (column 7) was formed, probably from a side reaction of the DES with iodine.

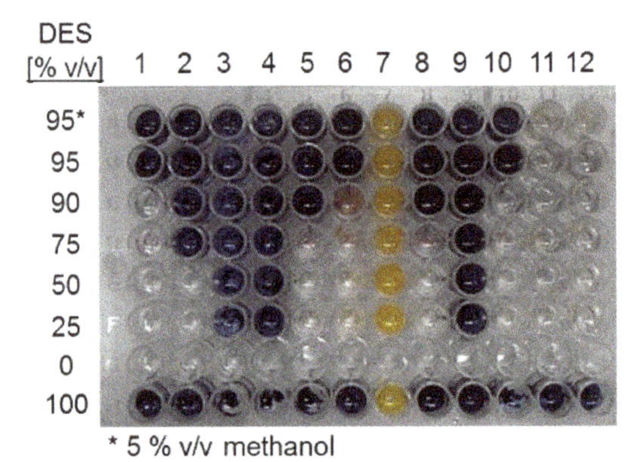

Figure 2. Screening for α-amylase stability in DES:water cosolvent mixtures as detected by the I^-/I_2 staining assay with amylose, after 4h reactions in the corresponding DES. **1**. choline chloride:urea (1:2); **2**. choline chloride:methyl urea (1:2), **3**. choline chloride:malonic acid (1:1), **4**. choline chloride:levulinic acid (1:2), **5**. choline chloride:ethylene glycol (1:2), **6**. choline chloride:D-isosorbide (1:2); **7**. MPh_3PBr:glycerol (1:2), **8**. choline chloride:glycerol (1:3); **9**. choline chloride:benzyl alcohol (1:2); **10**. choline chloride:1,3-propanediol:water (1:1:1); **11**. choline chloride:glucose:water (4:1:4); **12**. choline chloride:sucrose:water (5:2:5).

In the last row of Figure 2, it may be observed that the enzyme was fully inactive after 4 h of reaction when 100% DESs were used as solvent. Also, the enzyme inactivated in those reactions containing acidic DESs (columns 3–4), probably as a result of a severe pH shift, as reported for acidic DES [32,33]. A similar inactivation profile was observed with DES 9 (choline chloride:benzyl alcohol (1:2)). In the case of choline chloride:urea (1:2) and choline chloride:1,3-propanediol:water (1:1:1), the activity was retained with 90% of DES. No information could be obtained on amylase stability in methyl (triphenyl) phosphoniumbromid (MPh_3PBr):glycerol (column 7), probably due to a side reaction of the iodine-staining solution with the DES (yellow color). Amylase activity was retained in all reactions containing choline chloride:glucose:water (column 11) and choline chloride:sucrose:water (column 12). From this analysis, choline chloride:urea (1:2), choline chloride:1,3-propanediol:water (1:1:1), choline chloride:glucose:water (4:1:4); and choline chloride:sucrose:water (5:2:5) were defined as the most adequate DESs for amylase reactions. However, this is a preliminary qualitative analysis based on a staining procedure so further exploration of the enzyme stability is required.

2.2. Screening for Transfer to Alcohols (Alcoholysis) in Selected DES

The initial screening evaluating the hydrolytic activity of the α-amylase in DES/water systems allowed us to identify DESs in which the α-amylase activity could be retained after several hours. The six most promising DESs were selected for a second screening with methanol as the acceptor to screen for alcoholytic activity (see materials and methods). As shown in Figure 1, water was required for Amy A stability, but as water also competes as a nucleophile, both water and methanol concentrations were varied in order to identify the conditions for an optimal alcoholysis:hydrolysis ratio.

Methanol was assayed in a 5–30% *v/v* of total volume concentration range, with the remaining volume corresponding to the DES:water solvent. In the latter, water concentration ranged between 5 to 40% (*v/v*), while DES completed the remaining 60–95% (*v/v*). The same stability assay where blue wells indicate enzyme deactivation was performed. A general trend may be observed in Figure 3 for all the assayed DESs, as low water and high methanol concentrations tend to deactivate the enzyme. In effect, when the assay was carried out in a DES:buffer solvent containing 90% DES (10% buffer), no activity was observed in all cases except for choline chloride:sucrose:water with low methanol concentrations of 5 or 10% (*v/v*). All other solvents required a minimum of 20% (*v/v*) water as a cosolvent. Although one would expect Amy A to tolerate low concentrations of water, the presence of water in some DES, like choline chloride:1, 3-propanediol:water (1:1:1) required an additional 10% more buffer in the DES:water solvent, as well as the choline chloride:glucose:water (4:1:4) DES that required 60%.

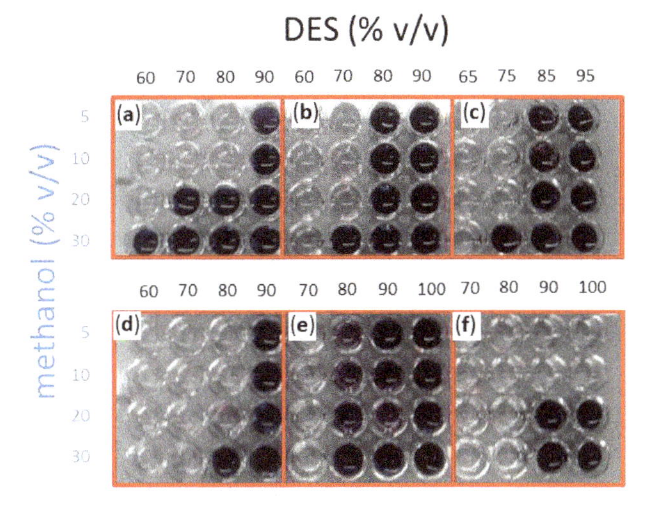

Figure 3. Retained Amy A activity as evaluated after 12 h of reaction in DES:water:MeOH systems at 75 °C. A reaction aliquot was taken to see retaining activity. The assay shown here was carried out with 1% starch (*w/v*) as a substrate at 50 °C. Six different DESs were screened: (**a**) choline chloride:urea (1:2); (**b**) choline chloride:D-isosorbide (1:2); (**c**) choline chloride:glycerol (1:3); (**d**) choline chloride:1,3-propanediol:water (1:1:1); (**e**) choline chloride:glucose:water (4:1:4); (**f**) choline chloride:sucrose:water (5:2:5). Methanol was tested in a range between 5–30% (*v/v*) of the total volume as indicated. The remaining volume was completed by a DES:buffer solution in which DES concentration was varied as indicated in each panel.

The alcoholysis reactions were analyzed by HPLC using an ELSD detector. However, the high concentration of the DESs components in some cases interfered with certain product peaks. Additionally, some DESs contained alcohols such as 1, 3-propanediol, glycerol, sucrose, and D-isosorbide. These alcohols represent potential nucleophiles in addition to methanol or water that could compete as acceptors in the transfer reactions. Accordingly, the unidentified peaks probably correspond to products formed from such alcoholysis reactions. However, reference compounds for these substances are not available, making identification and quantification difficult. From these results, we decided to select chloride:urea (1:2) DES as cosolvent. In the first experiment, maltotriose was used as the glycosyl donor. The results are summarized in Figure 4a–c. Conversion was based on the remaining maltotriose concentration after reaction.

Figure 4a shows that higher DES concentrations in the reaction result in lower conversion levels, most likely due to both low activity and stability of the enzyme (for stability see Figures 1 and 3). Similarly, conversions levels also decrease in the presence of high methanol concentrations. It must be considered, however, that the conversion of maltotriose can be achieved by hydrolysis, alcoholysis, or transglycosylation. Therefore, in order to evaluate the specific production of the targeted methyl-glycosides, the concentrations of products obtained from alcoholysis were quantified by HPLC. Methyl-glucoside, methyl-maltoside, and interestingly, also methyl-maltotrioside were detected. The formation of the latter indicated a transfer reaction of oligosaccharides, as observed earlier by our group using Amy A [16]. In order to reduce the number of poly-glycosylated methanol products, a glucoamylase (GA) digestion treatment of the reaction products obtained with Amy A was carried out. It was found that this enzyme was capable not only of transforming the residual dextrins to glucose through the hydrolysis of glucose units linked to amylose non-reducing ends, but also of hydrolyzing the poly-glucosylated methanol forms to the simplest mono-glucosylated form [16,17]. The highest methyl-glycoside concentrations of approximately 39 mM (9.51 g/L) were obtained using methanol at 10% (*v/v*) in 60% choline chloride:urea (30% water). This concentration was higher than previously reported (7.5 g/L) [16]. Figure 4c shows the percentage of transfer to methanol as compared to hydrolysis. Higher methanol concentrations generally lead to a higher methanolysis vs. hydrolysis ratio, as expected. However, above 10% methanol (*v/v*), no further increase could be observed in methyl glycoside production in all reactions. The overall conversion was low with high DES concentrations (Figure 4a).

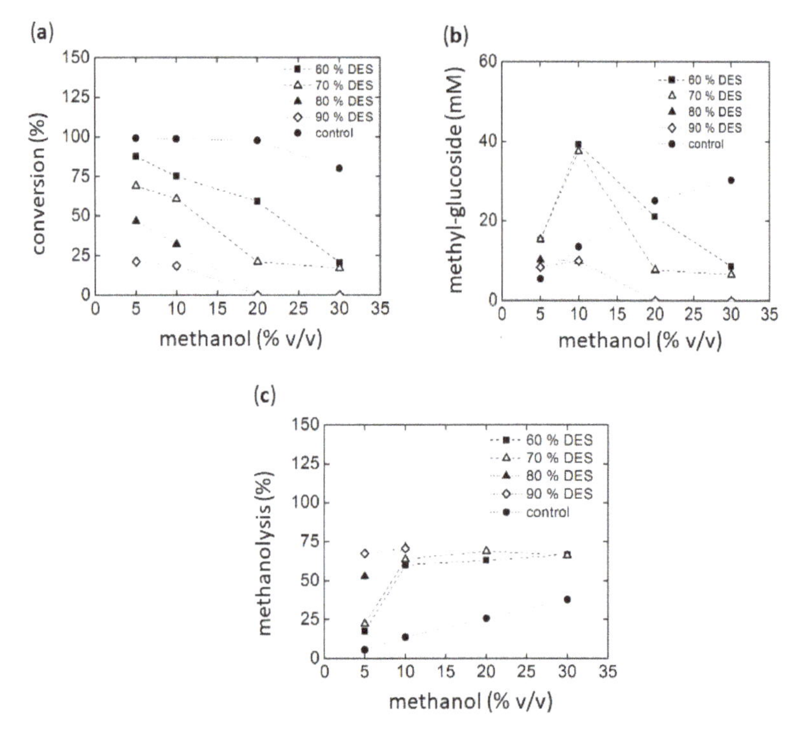

Figure 4. Methanolysis with maltotriose as substrate using Amy A as catalyst in choline chloride:urea (1:2):water as a solvent with increasing amounts of MeOH. (**a**) Maltotriose conversion, (**b**) Alcoholysis evaluated as methyl-glucoside synthesis after glucoamylase treatment of the glucosylated products (see text), (**c**) final yield of methanolysis reaction.

2.3. Exploration of Alcoholysis Activity in Choline Chloride:Urea (1:2):Water DES Using Starch as Glucosyl Donor

The use of maltotriose as a glycosyl donor showed promising results; however, this is an expensive substrate for industrial purposes. We therefore decided to use starch, as it is not only an economic alternative, but it also offers a stabilizing effect on amylases. In previous works, it was shown that with 15% (*w/v*) starch, conversions higher than 90% were obtained in methanol concentrations up to 20% (*v/v*) [16,18].

In a preliminary experiment, the alcoholysis reaction products using different concentrations of methanol (10, 20 and 30% *v/v*) and DES (0%–50% *v/v*) were quantified during the reaction in periods of 6, 12, and 24 h. Overall, the amount of methyl-glucoside (MG) obtained did not change much after 6 h of reaction (data not shown). DES produced an increment in alcoholysis yield when methanol was used at 10%; however, there was no further yield increase with higher concentrations of methanol in the DES system. In contrast, in a buffer system, higher alcoholysis yields were observed up to 30% methanol concentrations (data not shown). These and the results with maltotriose as a glycosyl donor prompted us to investigate whether DES could have a destabilizing effect on the enzyme, especially because the reaction was carried out at a high temperature (85 °C).

2.4. Stability of Amy A in Mixtures of DES-Methanol

The far-UV region of circular dichroism spectra is sensitive to changes in the secondary structure of proteins, and can be used to monitor the stability of the protein under different conditions. Figure 5 shows the CD spectra of the protein at DES concentrations from 60 to 80%. Light scattering due to the high DES concentration precluded the comparison of spectra below 210 nm. However, Figure 5 shows the negative signal around 220 nm, characteristic of proteins that present secondary structures. This negative signal looks similar for all the DES mixtures containing water at 25 °C, suggesting minimal changes in secondary structure. In contrast, Amy A in pure DES showed a near 50% decrease in the signal at this wavelength, suggesting the loss of the secondary structure, and in good agreement with the lack of activity observed under this condition.

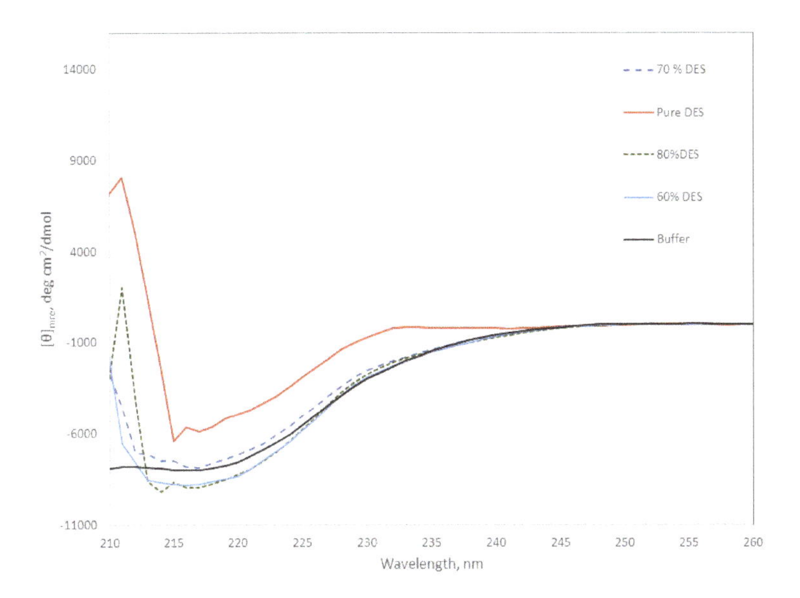

Figure 5. Far-UV region of the CD spectra of Amy A in pure phosphate buffer and in choline chloride:urea (2:1) DES/buffer mixtures (20 to 100% (*v/v*) of DES) at 25 °C. Pure DES solid red line; 80% DES dashed green line; 70% DES blue dashed line; 60% DES blue solid line; buffer, black solid line.

The activity assay, however, was carried out at 85 °C; therefore, we decided to carry out temperature scans of Amy A registering the signal at 220 nm at different concentrations of DES and MeOH. Figure 6 shows the change of signal by increasing and then decreasing the temperature from 25 °C to 90 °C; the insets show the CD spectra before and after heating to 90 °C and cooling back the protein solutions to 25 °C. Interestingly, the proteins in DES/water mixtures showed minimal differences from the proteins in the buffer, even though the concentrations of DES were as high as 80% (Figure 6a–c). The protein did not show significant changes in the mean residue ellipticity at 220 nm ([θ] mre) upon heating in any of the DES-buffer mixtures, showing a final spectrum that was practically unchanged. However, methanol turned out to be the major destabilizing factor. Even with only 10% methanol, a cooperative decrease in the signal was observed between 60 and 80 °C (Figure 6d–f). After reaching a minimum, the signal increased and the refolding process did not follow the same curve as the unfolding, showing considerable loss of the secondary structure. It is noteworthy that although the sample containing 80% DES lost about 35% of its [θ]mre at 220 nm, even at 25 °C (Figure 6f), after heating and cooling, it had a signal that was similar to those of the samples containing less DES, suggesting that the secondary structure was disturbed more by the increase of methanol concentration than by that of DES. When the methanol concentration increased to 20%, the sample containing 80% DES lost about 50% of the secondary structure, even at 25 °C (Figure 6i), but there was only a limited further loss of signal upon heating, and the refolding signal stayed steady. In contrast, the samples with 60% DES-20% MeOH and 70%DES-20% MeOH showed a high secondary structure content at 25 °C, and went through an irreversible unfolding when heated to 90 °C (Figure 6g,h). Again, the signal after cooling was similar for all samples containing 20% MeOH, around −1000 mdeg*cm^2/dmol, implying a major dependence on MeOH concentration.

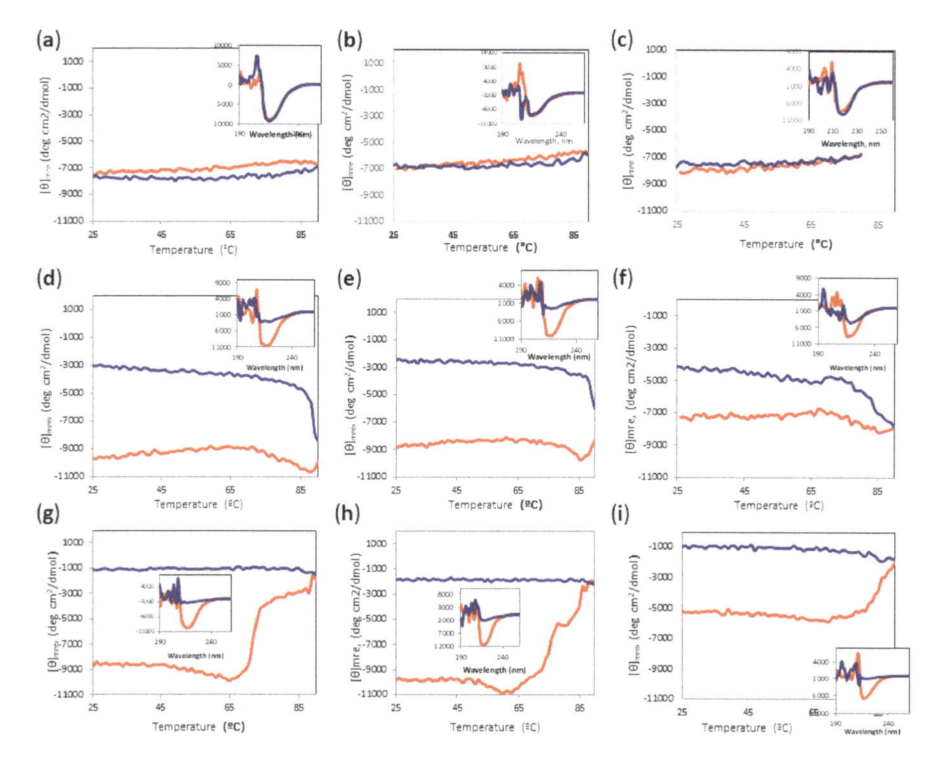

Figure 6. CD temperature scans of Amy A from 25 °C to 90 °C in different choline chloride:urea (1:2) DES and MeOH concentrations. Main chart: Amy A heating signal (red) and cooling signal (blue) at 220 nm. Inset: Far UV-CD spectra before heating (red) and after cooling (blue). (**a**) 60% DES; (**b**) 70% DES; (**c**) 80% DES; (**d**) 60% DES, 10% MeOH; (**e**) 70% DES, 10% MeOH; (**f**) 80% DES, 10% MeOH; (**g**) 60% DES, 20% MeOH; (**h**) 70% DES, 20% MeOH; (**i**) 80% DES, 20% MeOH.

Since major irreversible changes can occur in most of the conditions at temperatures above 60 °C, we decided to check the reversibility of heating the protein to only 60 °C for all conditions. As can be observed in Figure 7, even when there were subtle changes in [θ]mre at 220 nm upon heating, the final protein spectrum looked very similar to the initial one. Thus, we established 60 °C as the limit temperature to be able to compare the yield of alcoholysis reaction for all conditions.

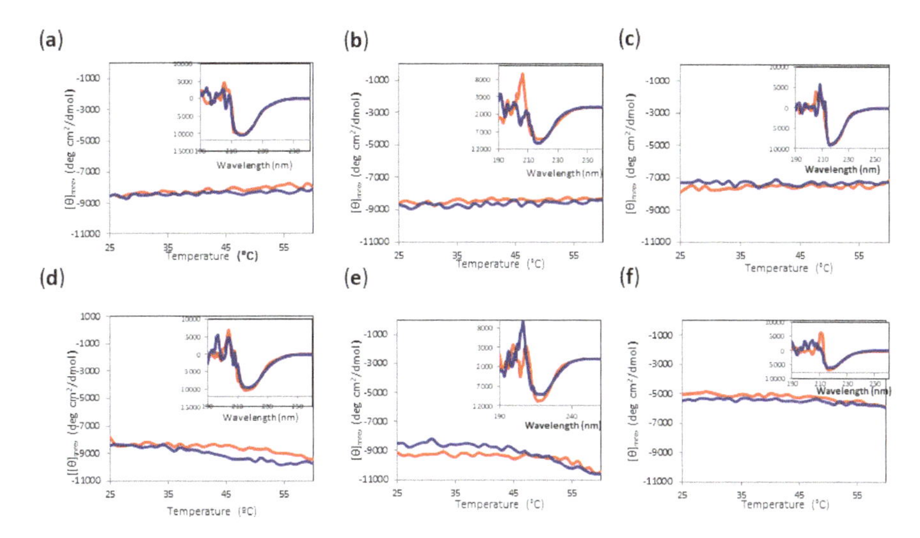

Figure 7. CD temperature scans of Amy A from 25 °C to 60 °C in different choline chloride:urea (1:2) DES and MeOH concentrations. Main chart: Amy A heating signal (red) and cooling signal (blue) at 220 nm. Inset: Far UV-CD spectra before heating (red) and after cooling (blue). (**a**) 60% DES, 10% MeOH; (**b**) 70% DES, 10% MeOH; (**c**) 80% DES, 10% MeOH; (**d**) 60% DES, 20% MeOH; (**e**) 70% DES, 20% MeOH; (**f**) 80% DES, 20% MeOH.

2.5. Activity of Amy A at 60 °C

Amy A is a thermostable enzyme with an optimal temperature at 85 °C, an advantage under the high temperature required for starch solubilization and for DES viscosity reduction. However, the structural information we obtained established a temperature limit for the alcoholysis reaction which was well below the optimal temperature. We measured activity at 60 °C under the different solvent conditions, including with an aqueous buffer, which is not reported at this temperature. Table 1 shows the V_{max} and K_M values at 60°C for the hydrolysis reaction.

Table 1. Kinetic constants of amylase Amy A.

Condition	K_M (mg/mL)	V_{max} (µmol/mg *min)
No DES	2.8 ± 0.5	196 ± 10
60% DES	5.7± 1.7	27 ± 4
60% DES 10% MeOH	1.7± 0.6	3.6 ± 0.4

Measurements at 60 °C, in buffer 50 mM Tris pH7 containing 150 mM NaCl and 2 mM $CaCl_2$. The results are the average of three replicates and the errors represent one standard deviation.

Hydrolytic activity was reduced almost three-fold by decreasing the temperature to 60 °C, relative to that at T_{opt} (85 °C) in aqueous buffer in good agreement with the observations by Liebl, et al. [32]. At 60 °C, in the absence of DES, Amy A follows a typical M-M model (Figure S1). However, in the presence of 60% DES with or without 10% methanol, an inhibitory effect was observed as the starch concentration increased (Figure S1). The apparent inhibition might be explained by the high viscosity in the resultant reaction medium as a result of the high concentration of both DES and starch, making diffusion a limiting factor for the catalytic efficiency as observed in other enzymes by the addition of trehalose [34–36]. For this reason, initial velocities up to 8 mg/mL were fit to the Michelis-Menten equation to estimate catalytic parameters (Figure S2). Table 1 shows that hydrolytic activity decreases by almost an order of magnitude in the presence of DES, and that the addition of 10% MeOH decreases this value by another order of magnitude, while K_M increased twice in DES, reflecting the lower availability of substrate in the active site. The presence of methanol in DES, on the other hand, decreased slightly K_M, probably as a result of reduced viscosity. Nevertheless, relevant for our purposes is the ratio of alcoholysis to hydrolysis reaction, which defines the final product yield and purity.

2.6. Alcoholysis Yield by Amy A at 60 °C

Alcoholysis initial rates are difficult to measure, since product quantification requires the hydrolysis of methyl-oligosaccharides and quantification of the resultant methyl-glucoside by HPLC (Figure S3). Instead, as a point of comparison, we decided to measure the final yield of both hydrolysis (measured by reducing sugars) and alcoholysis (measured by the amount of methyl-glucoside) after 18 h reaction, when presumably both reactions had reached equilibrium. We carried out alcoholysis with *T. maritima* α-amylase using 10% of methanol and varying the DES concentration from 10 to 80%. The product concentration after 18 h of incubation at 60 °C is compared to that obtained under the same conditions in the absence of DES in Figure 8. The systems with higher DES concentration (70 and 80%) have a very high viscosity when mixed with starch, making the medium and the measurement unmanageable. We excluded these experiments from the plots.

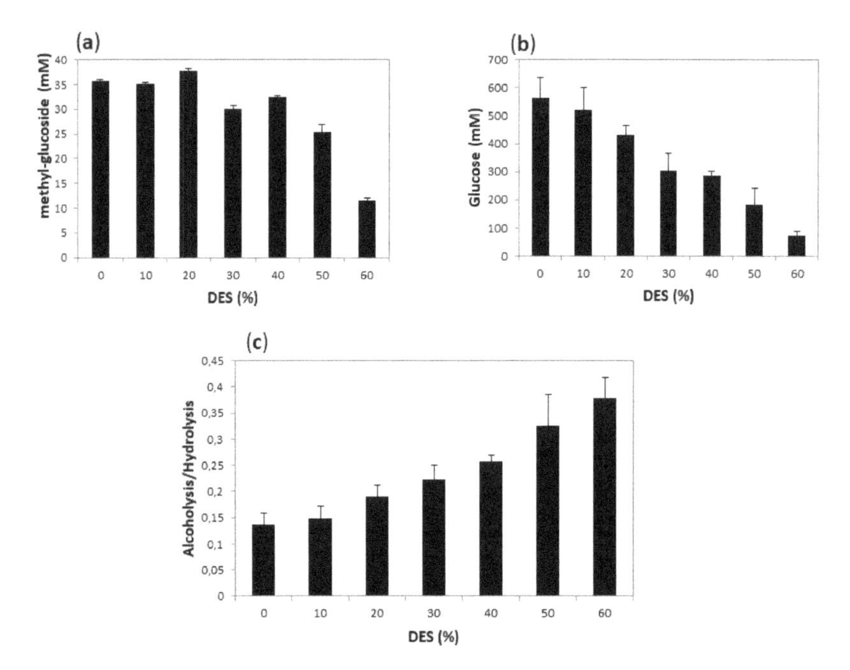

Figure 8. Methanolysis of starch after 18 h using Amy A as a catalyst, and 10% MeOH in different concentrations of choline chloride:urea (1:2):water DES as a solvent. (**a**) shows the yield of alcoholysis reaction (expressed as mM of methyl-glucoside); (**b**) the yield of hydrolysis (expressed as equivalent to dextrose in mM) and (**c**) the alcoholysis/hydrolysis ratio. The values were calculated as described in Materials and Methods.

As shown in Figure 8, alcoholysis yield remained practically unchanged up to 20% DES, and still reasonably high up to 40% DES, while hydrolysis yield showed a constant decrease as DES concentration increased, so that the alcoholysis:hydrolysis ratio increased with the increase of DES concentration.

3. Discussion

Water activity is a key parameter affecting glycosidase-catalyzed synthesis. On one hand, a sufficiently high-water activity (aw = 0.6) is required by glycosidases to retain activity, but on the other hand, a high-water content promotes hydrolysis, thus decreasing the overall transfer reaction to alcohols [4]. In each catalytic event an α-amylase-glycan intermediate is formed whose sugar moiety may be transferred to water or methanol. The selectivity of this transfer depends on the concentration of nucleophiles and their specificities defined by their corresponding rate constants [27,30]. For most α-amylases, transglycosylation reactions are unfavorable under normal conditions. To become significant, they require a high concentration of substrates and low water activities [3,4]. Amy A has an important transglycosidic activity; however, its α-1,4 hydrolytic activity seems to be also higher, so a rather limited transference of glycosyl residues to methanol occurs [16,20]. Although DES was envisioned as a way to reduce water content in the reaction mixture, water is always required in small amounts to ensure activity and stability. Thus, achieving an anhydrous medium that would eliminate the hydrolysis reaction that competes with the production of alkyl glycosides is not possible, indicating that some hydration is needed to form an aqueous shell around the protein that allows it to fold and the necessary lubrication in the active site to work, as has also been observed by others [33]; this contrasts with lipases for which DES seem sufficient to mimic the water layer covering active enzymes in alternative reaction media to water [21,22].

Despite the need for water in the reaction medium, further studies with the choline chloride:urea DES demonstrated that DES can improve the reaction selectivity to the desired production of alkyl glucosides. Increasing concentrations of DES reduced both hydrolysis and alcoholysis, but this reduction was higher for the undesired hydrolysis reaction, reducing its production more than that of the alcoholysis reaction, resulting in an increase in selectivity.

Circular dichroism (CD) structural studies revealed that the loss of activity might result from thermal denaturation in reactions media with high DES content. Surprisingly, MeOH disturbed the structure of Amy A more than DES. This loss of structure was prevented by performing the reaction at temperatures below the transition temperature observed at the thermal denaturation curves. This decrease in temperature reduced the enzyme activity; as a result, the reaction time should be elongated to increase the yield of methyl glycosides.

4. Materials and Methods

4.1. Materials

Glucoamylase from *A<.spergillus niger*, choline chloride, soluble starch, maltotriose, urea, methyl urea, malonic acid, levulinic acid, ethylene glycol, D-isosorbide, MPh_3PBr, glycerol, benzyl alcohol, 1,3-propanediol, glucose, sucrose, α-methyl-glucoside, and methanol were purchased from Sigma Chemical Co. (St. Louis, MO, USA).

4.2. Analytical Procedures

Amylase from *T. maritima*, Amy A, was obtained from the heterologous expression in *E. coli* as described previously [16,20]. Cells containing the overexpressed protein were lysed by sonication and partially purified via heat precipitation. The protein concentration of the applied α-amylase stock was 15 mg mL^{-1} for initial screening. To determine catalytic parameters, the protein extracts were further purified through Ni^{2+} affinity columns as previously described [16]

4.3. Synthesis of Deep Eutectic Solvents (DES)

The DES used in this work were synthesized in two different ways. In the first method, a thermal approach was used, which consisted in mixing and heating the components of eutectic solvent until a liquid was formed. A vacuum was applied in order to avoid water entering the system, particularly in cases where hygroscopic compounds such as choline chloride were used. The second method consisted of the dissolution of the components in water followed by freeze-drying [37]. Table S1 in supplementary information, summarizes the preparation of each DES used in this work.

4.4. Screening of Deep Eutectic Solvents for Application in α-amylase Catalyzed Reactions

Twelve DESs were used from the list presented in Table S1 (see supplementary data) for analysis in α-amylase-catalyzed reactions: 1. choline chloride:urea (1:2); 2. choline chloride:methyl urea (1:2), 3. choline chloride:malonic acid (1:1), 4. choline chloride:levulinic acid (1:2), 5. choline chloride:ethyleneglycol (1:2), 6. choline chloride:D-isosorbide (1:2); 7. MPh_3PBr:glycerol (1:2), 8. choline chloride:glycerol (1:3); 9. choline chloride:benzyl alcohol (1:2); 10. choline chloride:1,3-propanediol:water (1:1:1); 11. choline chloride:glucose:H_2O (4:1:4); 12. choline chloride:sucrose:H_2O (5:2:5). The 12 DES were tested in 96-well PCR plates each at different water concentrations of 0, 5, 10, 25, 50, 75, and 100% (*v/v*) to determine the minimum amount of water needed for enzyme activity, using 50 mM maltotriose as substrate. In order to provide a reactive nucleophile, methanol (5% *v/v*) was added to those reactions not containing water (0%). The reaction containing 0% water contained methanol as a nucleophile instead of water, and lyophilized Amy A was used. The PCR plate was sealed with a silicone microplate sealing mat to avoid water or methanol evaporation. After 4 h incubation in a PCR cycler at 75 °C, samples of each reaction were quenched (20 µl of sample + 180 µl of 0.225 M HCl) and analyzed via HPLC to determine remaining maltotriose.

4.5. Stability of Amy A in the Different Deep Eutectic Solvents

In order to determine the stability of Amy A during the reaction time that took place in the different DES systems, 20 µl samples from each reaction after 4 h incubation was diluted in 180 µl of a 50 mM Tris-HCl, 2 mM $CaCl_2$ buffer at pH 7.0, containing 1% (*w/v*) starch and incubated at 50 °C for 1 h. To visualize the remaining enzyme activity, 100 µl of this reaction were diluted 1:2 in water and stained with 20 µl of a 5 mM I^-/I_2 solution.

4.6. Screening for Alcoholytic Activity in Selected DES

Based on the first screening, only six different DES were further screened for alcoholytic activity: choline chloride:urea (1:2); choline chloride:D-isosorbide (1:2); choline chloride:glycerol (1:3); choline chloride:1,3-propanediol:water (1:1:1); choline chloride:glucose:water (4:1:4); choline chloride:sucrose:water (5:2:5). Methanol was tested in a range between 5–30% *v/v* of the total volume. The remaining volume was completed by a DES:buffer solution in which DES concentration ranged between 60–95% *v/v*. Thus, the six DES under the 16 different conditions could be evaluated in a 96-well plate. Then, 50 mM maltotriose was used as substrate, 5 µl of a 15 mg/mL Amy A solution obtained after purification by heat precipitation were added as described above. The total reaction volume was 150 µl in all reactions. Reactions were carried out at 75 °C in a PCR cycler. The same stability test as described above was carried out for each sample after 4 h of reaction. The reactions were additionally analyzed by HPLC. The percentage of conversion was estimated from the remaining maltotriose.

4.7. Product Quantification by HPLC

When maltotriose was used as substrate, the reaction products were analyzed by HPLC in an Ultimate 3000 equipped with a 300 RS Pump, a 3000 RS Autosampler and a RS Column Compartment, using an ELSD 2000ES Alltech detector. Both maltooligosaccharides and methyl-glucosides could be separated and detected using a Prevail Carbohydrate column (acetonitrile:water 70:30 as mobile phase, flow: 1 mL/min, column T: 30 °C, ELSD T: 80 °C, nitrogen Flow: 1.7 L/min). When starch was used as substrate, reaction products were analyzed in a Waters-Millipore 510 HPLC system equipped with an automatic sampler (model 717 Plus) (Waters Corp., MA, USA), a refractive-index detector (Waters 410) (Waters Corp., MA, USA) and a Hypersil GOLD™ Amino column (Thermo Scientific, Paisley, UK) using acetonitrile:water (80:20) as the mobile phase at a flow rate of 1.0 mL/min. The peak areas were measured and compared against those of a standard curve containing known amounts of methyl-glucoside.

To calculate the % conversion of the reaction, the following formula was used:

$$\%Conversion = \frac{[Maltotriose]_{inicial} - [Maltotriose]_{final}}{[Maltotrios]_{inicial}} \times 100$$

4.8. Hydrolysis Activity Assay

The enzyme activity of Amy A from *T. maritima* was determined by measuring the initial velocities of reducing sugars released at saturation starch concentration by the 3,5-dinitrosalicylic acid method, as reported previously [38]. The reaction was carried out in 1 mL of 10 mg/mL soluble starch dissolved in 50 mM Tris, 150 mM NaCl, 2 mM $CaCl_2$ buffer, pH 7, at 85 °C. A unit of enzyme activity was defined as the amount of enzyme required to release 1 µmol of glucose equivalents per minute. For the characterization of Amy A at 60 °C, the activity measurements were carried out varying the starch concentration from 1 mg/mL to 20 mg/mL and using as a solvent mixtures containing 50 mM Tris, 150 mM NaCl, 2 mM $CaCl_2$ buffer, pH 7, and variable *v/v* concentrations of DES and methanol. The initial velocities were plotted against starch concentration and the data fit to the Michaelis-Menten equation to obtain the catalytic parameters.

4.9. Alcoholysis Reactions with Methanol and Starch in Choline Chloride:Urea 1:2 DES as co Solvent

Alcoholysis reactions were performed in 2 mL Eppendorf tubes which were tightly sealed to avoid methanol and water evaporation. DES concentration was varied from 0 to 80%, using 10% methanol and 10% starch as a glycosyl donor, and 50 mM Tris, 150 mM NaCl, 2 mM $CaCl_2$ buffer at pH 7 to complete the 100% volume. The solvent mixture was homogenized and pre-equilibrated at the reaction assay temperature before adding 20 U of Amy A to start the reaction, which was incubated during 18 h at the defined temperature according to the experiment.

4.10. Hydrolysis Quantification at the End Point of Reaction

The determination of reducing sugars was evaluated as described above from a 1:50 dilution of the final reaction. The hydrolysis events were estimated by subtracting the equivalents of the alcoholysis reaction from the equivalent of the reducing sugars.

4.11. Alcoholysis Reaction Measurement

The measurement of alcoholysis is complicated by the diversity of products possible. To reduce the complexity, the final reaction products were hydrolyzed for 12 h at 50 °C with *A. niger* glucoamylase (25 U/mL), which is an exoglycosidase that hydrolyses α-1,4-glycosidic bonds from the non-reducing end, transforming any alcoholysis product in glucose plus methyl-glucoside for each alcoholysis event. At the end of incubation period, all the samples were centrifuged at 13,000 rpm for 15 min and filtered through a 0.22 µm nylon membrane before the analyses by HPLC as described above, using a standard curve of methyl-glucoside to quantify the amount in each sample.

4.12. Alcoholysis Yield and Alcoholysis/Hydrolysis Ratio

When a hydrolysis reaction takes place, for each molecule of a reducing substrate cleaved, one additional reducing end is formed, increasing in one its contribution to the solution reducing power; in contrast, when an alcoholysis reaction occurs, there is no real gain in reducing power. Therefore, a measure of the increasing equivalent glucose concentration measured by the DNS method, as described above, reflects the number of hydrolysis events that occurred during the reaction, while the number of alcoholysis events can be independently determined through the quantification of the increase in alkyl glucoside concentration in the same period of time. From these data it is possible to determine the number of hydrolysis and alcoholysis events in the reaction, as well as an alcoholysis/hydrolysis ratio and an alcoholysis yield:

$$\text{Alcoholysis/hydrolysis ratio} = \frac{AlcoholysisEvents}{HydrolysisEvents}$$

$$\text{Alcoholysis efficiency} = \frac{AlcoholysisEvents}{AlcoholysisEvents + HydrolysisEvents}$$

4.13. Structural Characterization of Amy A in Choline Chloride:Urea 1:2 DES as co Solvent

Circular dichroism spectra (CD) were recorded on a J-710 spectropolarimeterTM (Jasco, USA) equipped with a Peltier temperature control using a 1.00 mm path length cell. Three CD scans were recorded and averaged from 190 to 260 nm for each sample containing 0.3 mg/mL of Amy A in 10 mM Phosphate, 0.5 mM $CaCl_2$ buffer at pH 7 at 25 °C and varying the DES (60%–80% *v/v*) and methanol (0, 10 and 20 % (*v/v*)) concentrations.

4.14. Thermal Stability of Amy A in Choline Chloride:Urea 1:2 DES as co Solvent

Thermal stability of Amy A was determined by following the CD signal at 220 nm during a temperature scan from 25 °C to 90° at a rate of 1 °C/min of all the samples described above. The reversibility of any change observed was also evaluated by following the CD signal at 220 nm in the return to 25 °C and recording a final spectrum for comparison with the initial one.

5. Conclusions

We demonstrated that Amy A can perform both hydrolysis and alcoholysis in media containing deep eutectic solvents (DES) using starch or maltotriose as substrates. However, the enzyme activity and stability in each medium depends on the nature of the DES. The possibility of reducing water activity using alternative reaction media as the DES cosolvent was successful to a certain degree, as shown by the increase in the alcoholysis/hydrolysis ratio; however, it was found that using pure DES had a destabilizing effect. The alcholysis yield measured as methyl-glucoside with the optimal DES cosolvent was 20% greater than that previously reported in an aqueous buffer with the same methanol concentration.

Our results highlight the importance of performing variable temperature structural studies to determine the temperature limit at which an enzyme is operative.

Supplementary Materials: Supplementary materials can be found at http://www.mdpi.com/1422-0067/20/21/5439/s1.

Author Contributions: Conceptualization, G.S.-R., E.C. and A.L.-M.; methodology, A.M.-M. W.X. R.A.-B. S.S., L.O.; validation, W.X.; writing—original draft preparation, A.M.-M. W.X. R.A.-B. G.S.-R. funding acquisition, G.S.-R.

Funding: This research was funded by Dirección General de Asuntos del Personal Académico, Universidad Nacional Autónoma de México through PAPIIT Grants No. IT200617 to G.S.R. and IA202619 to W.X.

Acknowledgments: The authors thank Humberto Flores and Fernando González-Muñoz for technical assistance. Juan Manuel Hurtado and the following facilities of the Instituto de Biotecnología, Universidad Nacional Autónoma de México for their services at different stages of this work: Unidad de Cómputo, Unidad de Síntesis y Secuenciación and Unidad de Biblioteca.

Conflicts of Interest: The authors declare no conflict of interest. The funders had no role in the design of the study; in the collection, analyses, or interpretation of data; in the writing of the manuscript, or in the decision to publish the results.

References

1. Larsson, J.; Svensson, D.; Adlercreutz, P. α-Amylase-catalysed synthesis of alkyl glycosides. *J. Mol. Catal. B Enzym.* **2005**, *37*, 84–87. [CrossRef]

2. Matsumura, S.; Imai, K.; Yoshikawa, S.; Kawada, K.; Uchibori, T. Surface activities, biodegradability and antimicrobial properties of n-alkyl glucosides, mannosides and galactosides. *J. Am. Oil Chem. Soc.* **1990**, *67*, 996–1001. [CrossRef]
3. Rather, M.Y.; Mishra, S. β-glycosidases: An alternative enzyme based method for synthesis of alkyl-glycosides. *Sustain. Chem. Process.* **2013**, *1*, 7. [CrossRef]
4. Iglauer, S.; Wu, Y.; Shuler, P.; Tang, Y.; Goddard, W.A. Analysis of the influence of alkyl polyglycoside surfactant and cosolvent structure on interfacial tension in aqueous formulations versus n-octane. *Tenside Surfactants Deterg.* **2010**, *47*, 87–97. [CrossRef]
5. Ismail, A.; Soultani, S.; Ghoul, M. Enzymatic-catalyzed synthesis of alkylglycosides in monophasic and biphasic systems. I. The transglycosylation reaction. *J. Biotech.* **1999**, *69*, 135–143. [CrossRef]
6. Fink, M.J.; Syrén, P.-O. Redesign of water networks for efficient biocatalysis. *Curr. Opin. Chem. Biol.* **2017**, *37*, 107–114. [CrossRef] [PubMed]
7. Fuchs, M.; Farnberger, J.E.; Kroutil, W. The industrial age of biocatalytic transamination. *Eur. J. Org. Chem.* **2015**, *2015*, 6965–6982. [CrossRef]
8. Gotor, V. Biocatalysis applied to the preparation of pharmaceuticals. *Org. Proc. Res. Dev.* **2002**, *6*, 420–426. [CrossRef]
9. Kulishova, L.M.; Zharkov, D.O. Solid/gas biocatalysis. *Biochemistry* **2017**, *82*, 95–105. [CrossRef]
10. Polakovič, M.; Švitel, J.; Bučko, M.; Filip, J.; Neděla, V.; Ansorge-Schumacher, M.B.; Gemeiner, P. Progress in biocatalysis with immobilized viable whole cells: Systems development, reaction engineering and applications. *Biotechnol. Lett.* **2017**, *39*, 667–683. [CrossRef]
11. Sheldon, R.A.; Pereira, P.C. Biocatalysis engineering: The big picture. *Chem. Soc. Rev.* **2017**, *46*, 2678–2691. [CrossRef] [PubMed]
12. Sheldon, R.A.; Woodley, J.M. Role of biocatalysis in sustainable chemistry. *Chem. Rev.* **2018**, *118*, 801–838. [CrossRef] [PubMed]
13. Zaks, A. Industrial biocatalysis. *Curr. Opin. Chem. Biol.* **2001**, *5*, 130–136. [CrossRef]
14. Panintrarux, C.; Adachi, S.; Matsuno, R. β-glucosidase-catalyzed condensation of glucose with 2-alcohols in buffer-saturated alcohols. *Biotechnol. Lett.* **1997**, *19*, 899–902. [CrossRef]
15. Kamiyama, Y.; Yasui, T. Enzymatic synthesis of alkyl β-xylosides from xylobiose by application of the transxylosyl reaction of aspergillus niger β-xylosidase AU—Shinoyama, Hirofumi. *Agric. Biol. Chem.* **1988**, *52*, 2197–2202. [CrossRef]
16. Damian-Almazo, J.Y.; Moreno, A.; Lopez-Munguia, A.; Soberon, X.; Gonzalez-Munoz, F.; Saab-Rincon, G. Enhancement of the alcoholytic activity of α-amylase Amy A from Thermotoga maritima MSB8 (DSM 3109) by site directed mutagenesis. *Appl. Environ. Microbiol.* **2008**, *74*, 5168–5177. [CrossRef]
17. Saab-Rincon, G.; del-Rio, G.; Santamaria, R.I.; Lopez-Munguia, A.; Soberon, X. Introducing transglycosylation activity in a liquefying α-amylase. *FEBS Lett.* **1999**, *453*, 100–106. [CrossRef]
18. Santamaria, R.I.; Del, R.G.; Saab, G.; Rodriguez, M.E.; Soberon, X.; Lopez, M.A. Alcoholysis reactions from starch with alpha-amylases. *FEBS Lett.* **1999**, *452*, 346–350. [CrossRef]
19. Vihinen, M.; Ollikka, P.; Niskanen, J.; Meyer, P.; Suominen, I.; Karp, M.; Holm, L.; Knowles, J.; Mantsala, P. Site-directed mutagenesis of a thermostable alpha-amylase from *Bacillus stearothermophilus*: Putative role of three conserved residues. *J. Biochem.* **1990**, *107*, 267–272. [CrossRef]
20. Moreno, A.; Damian-Almazo, J.Y.; Miranda, A.; Saab-Rincon, G.; Gonzalez, F.; Lopez-Munguia, A. Transglycosylation reactions of *Thermotoga maritima* α-amylase. *Enzym. Microb. Tech.* **2010**, *46*, 331–337. [CrossRef]
21. Durand, E.; Lecomte, J.; Baréa, B.; Piombo, G.; Dubreucq, E.; Villeneuve, P. Evaluation of deep eutectic solvents as new media for *Candida antarctica* B lipase catalyzed reactions. *Process Biochem.* **2012**, *47*, 2081–2089. [CrossRef]
22. Juneidi, I.; Hayyan, M.; Hashim, M.A.; Hayyan, A. Pure and aqueous deep eutectic solvents for a lipase-catalysed hydrolysis reaction. *Biochem. Eng. J.* **2017**, *117*, 129–138. [CrossRef]
23. Khodaverdian, S.; Dabirmanesh, B.; Heydari, A.; Dashtban-moghadam, E.; Khajeh, K.; Ghazi, F. Activity, stability and structure of laccase in betaine based natural deep eutectic solvents. *Int. J. Biol. Macromol.* **2018**, *107*, 2574–2579. [CrossRef] [PubMed]

24. Mbous, Y.P.; Hayyan, M.; Hayyan, A.; Wong, W.F.; Hashim, M.A.; Looi, C.Y. Applications of deep eutectic solvents in biotechnology and bioengineering—Promises and challenges. *Biotechnol. Adv.* **2017**, *35*, 105–134. [CrossRef] [PubMed]
25. Anderson, E.M.; Larsson, K.M.; Kirk, O. One biocatalyst–many applications: The use of Candida Antarctica B-lipase in organic synthesis. *Biocatal. Biotransform.* **1998**, *16*, 183–204. [CrossRef]
26. Kumar, A.; Dhar, K.; Kanwar, S.S.; Arora, P.K. Lipase catalysis in organic solvents: Advantages and applications. *Biol. Proced. Online* **2016**, *18*, 2. [CrossRef]
27. Bell, G.; Halling, P.J.; Moore, B.D.; Partridge, J.; Rees, D.G. Biocatalyst behaviour in low-water systems. *Trends Biotechnol.* **1995**, *13*, 468–473. [CrossRef]
28. Chahid, Z.; Montet, D.; Pina, M.; Graille, J. Effect of water activity on enzymatic synthesis of alkylglycosides. *Biotechnol. Lett.* **1992**, *14*, 281–284. [CrossRef]
29. Klibanov, A.M. Enzymatic catalysis in anhydrous organic solvents. *Trends Biochem. Sci.* **1989**, *14*, 141–144. [CrossRef]
30. Montiel, C.; Bustos-Jaimes, I.; Bárzana, E. Enzyme-catalyzed synthesis of heptyl-β-glycosides: Effect of water coalescence at high temperature. *Bioresour. Technol.* **2013**, *144*, 135–140. [CrossRef]
31. Vulfson, E.N.; Patel, R.; Law, B.A. Alkyl-β-glucoside synthesis in a water-organic two-phase system. *Biotechnol. Lett.* **1990**, *12*, 397–402. [CrossRef]
32. Qin, H.; Hu, X.; Wang, J.; Cheng, H.; Chen, L.; Qi, Z. Overview of acidic deep eutectic solvents on synthesis, properties and applications. *Green Energy Environ.* **2019**. [CrossRef]
33. Skulcova, A.; Russ, A.; Jablonsky, M.; Sima, J. The pH behavior of seventeen deep eutectic solvents. *Bioresources* **2018**, *13*, 5042–5051. [CrossRef]
34. Hernandez-Meza, J.M.; Sampedro, J.G. Trehalose mediated inhibition of lactate dehydrogenase from rabbit muscle. The application of Kramers' theory in enzyme catalysis. *J. Phys. Chem. B* **2018**, *122*, 4309–4317. [CrossRef] [PubMed]
35. Sampedro, J.G.; Uribe, S. Trehalose-enzyme interactions result in structure stabilization and activity inhibition. The role of viscosity. *Mol. Cell Biochem.* **2004**, *256–257*, 319–327. [CrossRef] [PubMed]
36. Uribe, S.; Sampedro, J.G. Measuring solution viscosity and its effect on enzyme activity. *Biol. Proced. Online* **2003**, *5*, 108–115. [CrossRef]
37. Gutierrez, M.C.; Ferrer, M.L.; Yuste, L.; Rojo, F.; del Monte, F. Bacteria incorporation in deep-eutectic solvents through freeze-drying. *Angew. Chem. Int. Ed. Engl.* **2010**, *49*, 2158–2162. [CrossRef]
38. Summer, J.B.; Howell, S.F. A method for determination of saccharase activity. *J. Biol. Chem.* **1935**, *108*, 51–54.

International Journal of
Molecular Sciences

Article

A Genome-Centric Approach Reveals a Novel Glycosyltransferase from the GA A07 Strain of *Bacillus thuringiensis* Responsible for Catalyzing 15-*O*-Glycosylation of Ganoderic Acid A

Te-Sheng Chang [1,†], Tzi-Yuan Wang [2,†], Tzu-Yu Hsueh [1], Yu-Wen Lee [1], Hsin-Mei Chuang [1], Wen-Xuan Cai [1], Jiumn-Yih Wu [3], Chien-Min Chiang [4] and Yu-Wei Wu [5,6,*]

1. Department of Biological Sciences and Technology, National University of Tainan, Tainan 70005, Taiwan; mozyme2001@gmail.com (T.-S.C.); vuvu99983@gmail.com (T.-Y.H.); s10458017@gm2.nutn.edu.tw (Y.-W.L.); tiffany170420@gmail.com (H.-M.C.); amy19990630@yahoo.com.tw (W.-X.C.)
2. Biodiversity Research Center, Academia Sinica, Taipei 11529, Taiwan; tziyuan@gmail.com
3. Department of Food Science, National Quemoy University, Kinmen County 892, Taiwan; wujy@nqu.edu.tw
4. Department of Biotechnology, Chia Nan University of Pharmacy and Science, No. 60, Erh-Jen Rd., Sec. 1, Jen-Te District, Tainan 71710, Taiwan; cmchiang@mail.cnu.edu.tw
5. Graduate Institute of Biomedical Informatics, College of Medical Science and Technology, Taipei Medical University, Taipei 11031, Taiwan
6. Clinical Big Data Research Center, Taipei Medical University Hospital, Taipei 11031, Taiwan

* Correspondence: yuwei.wu@tmu.edu.tw; Tel.: +886-2-66382736 (ext. 1505)
† These authors contributed equally to the work.

Received: 9 September 2019; Accepted: 18 October 2019; Published: 20 October 2019

Abstract: Strain GA A07 was identified as an intestinal *Bacillus* bacterium of zebrafish, which has high efficiency to biotransform the triterpenoid, ganoderic acid A (GAA), into GAA-15-*O*-β-glucoside. To date, only two known enzymes (BsUGT398 and BsUGT489) of *Bacillus subtilis* ATCC 6633 strain can biotransform GAA. It is thus worthwhile to identify the responsible genes of strain GA A07 by whole genome sequencing. A complete genome of strain GA A07 was successfully assembled. A phylogenomic analysis revealed the species of the GA A07 strain to be *Bacillus thuringiensis*. Forty glycosyltransferase (GT) family genes were identified from the complete genome, among which three genes (*FQZ25_16345*, *FQZ25_19840*, and *FQZ25_19010*) were closely related to BsUGT398 and BsUGT489. Two of the three candidate genes, *FQZ25_16345* and *FQZ25_19010*, were successfully cloned and expressed in a soluble form in *Escherichia coli*, and the corresponding proteins, BtGT_16345 and BtGT_19010, were purified for a biotransformation activity assay. An ultra-performance liquid chromatographic analysis further confirmed that only the purified BtGT_16345 had the key biotransformation activity of catalyzing GAA into GAA-15-*O*-β-glucoside. The suitable conditions for this enzyme activity were pH 7.5, 10 mM of magnesium ions, and 30 °C. In addition, BtGT_16345 showed glycosylation activity toward seven flavonoids (apigenein, quercetein, naringenein, resveratrol, genistein, daidzein, and 8-hydroxydaidzein) and two triterpenoids (GAA and antcin K). A kinetic study showed that the catalytic efficiency (k_{cat}/K_M) of BtGT_16345 was not significantly different compared with either BsUGT398 or BsUGT489. In short, this study identified BtGT_16345 from *B. thuringiensis* GA A07 is the catalytic enzyme responsible for the 15-*O*-glycosylation of GAA and it was also regioselective toward triterpenoid substrates.

Keywords: Nanopore sequencing; ganoderic acid; *Bacillus thuringiensis*; biotransformation; glycosyltransferase; whole genome sequencing

1. Introduction

Glycosyltransferase (GT, EC 2.4.x.y) exists in all living beings and is able to catalyze the glycosylation of molecules such as proteins, nucleic acids, polysaccharides, and lipids. Most GTs use a nucleotide-activated sugar donor, such as uridine diphosphate (UDP)-glucose, in the catalytic reaction. According to a carbohydrate-activating enzyme (CAZy) database, GTs are classified into 107 families [1]. Among them, GTs that use small molecules (such as flavonoids or triterpenoids) as sugar acceptors are classified into the GT1 family. Many members of the GT1 family with activities toward flavonoids have been identified [2,3], however, very few GT1 family members with glycosylation activities toward triterpenoids were reported until recently [4].

Biotransformation of xenobiotics by either a microorganism's whole cells or purified enzymes may form more-bioactive metabolites than the precursor molecules [5–9]. Among different biotransformations, glycosylation was shown to improve water solubility, stability, and bioactivities of flavonoids, such as anti-oxidant and anti-allergic activities [10–12]. Glycosylation of triterpenoids to form saponins can also improve some bioactivities of the triterpenoid precursors. For examples, dozens of reports showed that triterpenoid glycosides, ginseng saponins, from the medicinal plant ginseng, possess more bioactivities involved in the central nervous system, cardiovascular system, immune system, anticarcinogenic activities, and diabetes mellitus, than do ginseng triterpenoid aglycones [13]. Therefore, using GT to biotransform xenobiotics to new glycoside compounds is a worthy field of study.

Ganoderic acid A (GAA) is a triterpenoid isolated from the medicinal fungus, *Ganoderma lucidum* [14]. In addition to GAA, more than 300 different kinds of triterpenoids have been isolated from *Ganoderma* spp. [15], and studies suggested that these triterpenoids may possess many bioactivities [15–17]. Despite numerous kinds of triterpenoids having been identified from *G. lucidum*, very few natural *Ganoderma* triterpenoids exist in the form of glycosides (saponins) [15]. In addition, only five microbial GTs were found to biotransform triterpenoids into new bioactive derivatives [1–4]. Taken together, finding GTs that target *Ganoderma* triterpenoids could potentially expand the diversities of both GT enzymes and *Ganoderma* triterpenoids.

Our previous study identified an intestinal bacterium of zebrafish, *Bacillus* sp. GA A07 strain, which could biotransform GAA into GAA-15-*O*-β-glucoside [18]. In order to identify the GTs of the GA A07 strain responsible for this triterpenoid biotransformation, the complete genome of the strain was resolved using both Nanopore long-read and BGI short-read sequencing technologies. Candidate GT genes were then discovered by comparing them to the CAZy database [1] and also subsequently searching potential sequences against five triterpenoid-glycosylation genes [4,19–24]. These candidate genes were then subcloned and overexpressed in *Escherichia coli*; the biotransformation activities of the purified recombinant GTs were also determined.

2. Results

2.1. Comparison of GAA-15-O-β-Glucoside Production between B. subtilis ATCC 6633 and Bacillus sp. GA A07

Our previous study identified two *Bacillus* strains with the ability to biotransform GAA to GAA-15-*O*-β-glucoside, *B. subtilis* ATCC 6633 [4] and *Bacillus* sp. GA A07 [18]. To compare the biotransformation activity between the two strains, fermentation broths of the two strains fed with GAA were analyzed by ultra-performance liquid chromatography (UPLC) during cultivation. Results showed that *Bacillus* sp. GA A07 possessed 12.5-fold higher GAA biotransformation activity than *B. subtilis* ATCC 6633 after a 24 h incubation of GAA (Figure 1). Based on these results, the complete genome of the GA A07 strain was resolved to identify the GTs responsible for the triterpenoid biotransformation.

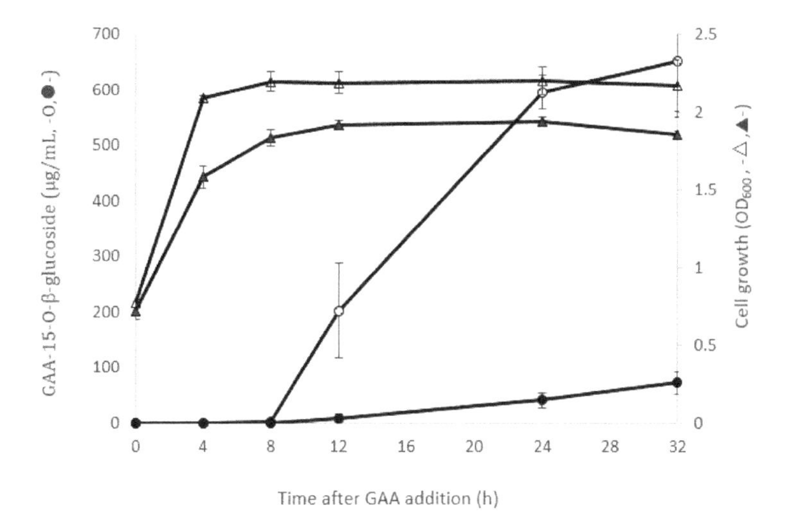

Figure 1. Time course of ganoderic acid A (GAA)-15-*O*-β-glucoside production (circle symbols) and cell growth (triangle symbols) by either *Bacillus subtilis* ATCC 6633 (solid) or *Bacillus sp.* GA A07 (open). The two strains were cultivated in Luria-Bertani (LB) media with shaking at 180 rpm and 30 °C. GAA at 1 mg/mL was added to the fermentation broth as the optical density (OD) at 600 nm of the culture reached 0.6, and cultivation continued for another 32 h. During cultivation, the fermentation broth was analyzed by ultra-performance liquid chromatography (UPLC). The UPLC operating conditions are described in the Materials and Methods section.

2.2. Genome Sequencing, Assembly, Annotation, and Reclassification of the GA A07 Strain

Genome sequencing of *Bacillus* sp. GA A07 was performed in order to determine the enzymes that contribute to GAA glycosylation. Totally 1,217,502,092 base pairs (bps) were sequenced from 142,886 Nanopore reads. The average read length was 8521 bps. The assembly process (outlined in the Materials and Methods section) yielded a complete circular genome along with four circular plasmids (GenBank BioProject accession no. PRJNA557365; Genome accession no. CP042270). The genome size was 5,272,357 bps with a G+C percentage of 35.33%. Totally, 5094 putative protein-coding genes, 106 transfer RNA (tRNA) genes, and 42 ribosomal RNA (rRNA) genes were annotated for the GA A07 strain, as shown in the circular genome map (Figure 2).

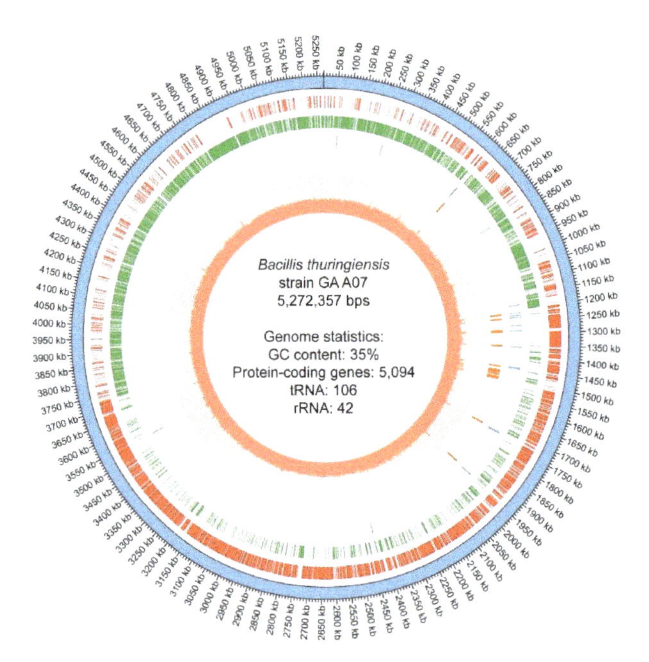

Figure 2. Circular genome map of the *Bacillus thuringiensis* GA A07 strain. The six rings from the outer to inner side represent (1) assembled scaffolds, (2) genes in the forward strand, (3) genes in the reverse-complement strand, (4) transfer RNA (tRNA) genes, (5) ribosomal RNA (rRNA) genes, and (6) the GC content distribution per 1000 bps.

The 16S gene tree was only able to group the GA A07 strain with other *Bacillus* species (including *B. subtilis*, *B. thuringiensis*, *B. cereus*, and *B. anthracis*) [18]; however, the actual species to which it belonged could not be determined using only the 16S rRNA gene. Thus, we employed a two-step method to identify which species the GA A07 strain belongs to. First, we used three different approaches [25–27], namely average nucleotide identity (ANI), average amino acid identity (AAI), and a tetra correlation search (TCS), to find the most closely related species to the GA A07 strain. All methods identified *B. thuringiensis* as the closest species (*B. thuringiensis* serovar *canadensis* identified by both the ANI and AAI approaches and *B. thuringiensis* BMB171 identified by the TCS approach). We then built a phylogenetic tree (Figure 3) from 250 single-copy marker genes (the full list of marker genes can be found in Table S1). Figure 3 revealed the GA A07 strain indeed belongs to the group of *B. thuringiensis*.

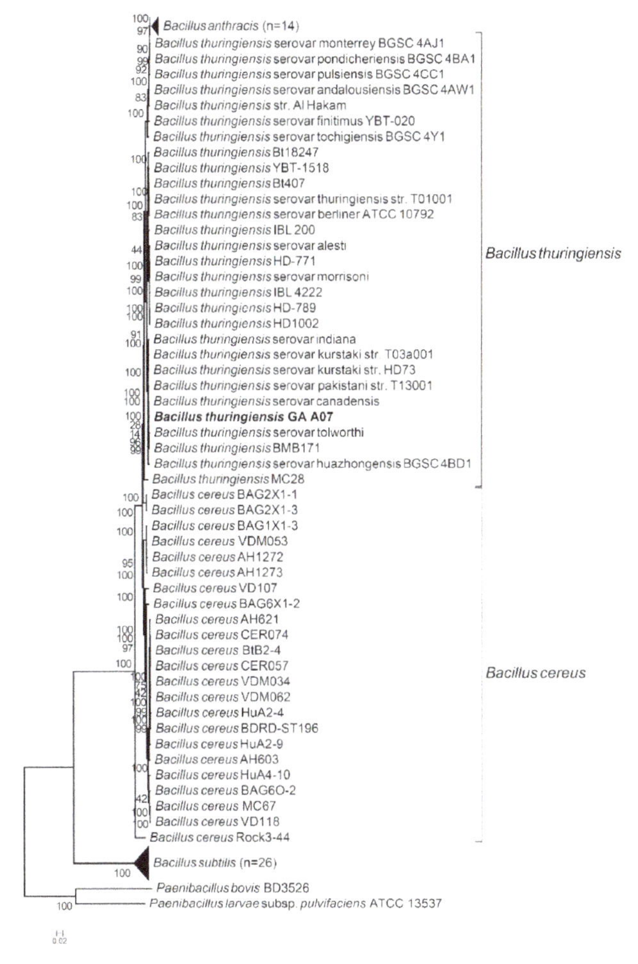

Figure 3. Phylogenetic tree for the *Bacillus thuringiensis* GA A07 strain. The tree was built from 250 single-copy marker genes. Two *Paenibacillus* genomes were included as outgroups. The GA A07 strain is marked in bold font. See the Materials and Methods section for details.

2.3. Phylogenetic Analysis of GTs from the GA A07 Strain

Previous studies showed that five microbial GTs were validated to have triterpenoid glycosylation activity, including BsYjiC (GenBank Protein accession no. NP_389104) from *B. subtilis* 168 [19–23], UGT109A1 (GenBank Protein accession no. ASY97769) from *B. subtilis* CTCG 63501 [24,28], BsGT1 (GenBank Protein accession no. ANP92054) from *B. subtilis* KCTC 1022 [29], and two GTs, BsUGT398 and BsUGT489, from *B. subtilis* ATCC 6633 (GenBank Protein accession nos. WP_003225398 and WP_003220489, respectively) [4]. To classify which genes were responsible for the biotransformation of GAA, GT genes were first annotated from the GA A07 genome. The 40 identified GT genes were then used to build a phylogenetic tree using the five validated genes with triterpenoid glycosylation activities (Figure 4). Among the 40 GTs, one GT1 (*FQZ25_19010*) and two GT28 (*FQZ25_16345*, *FQZ25_19840*) family genes were most closely related to the five validated genes (marked by stars in Figure 4), and were considered putative gene candidates.

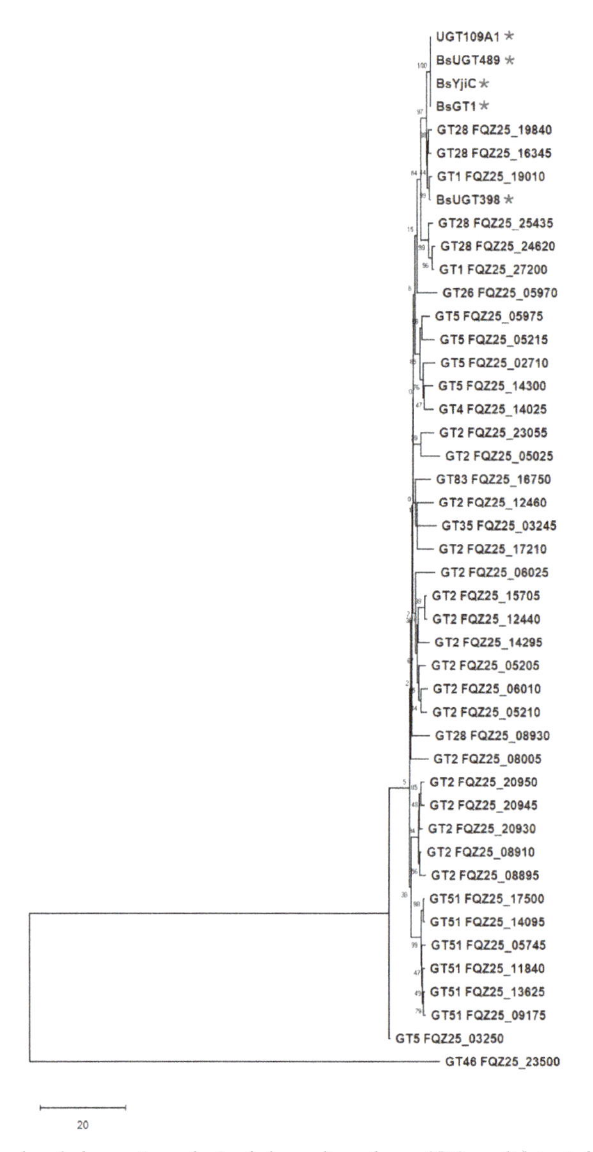

Figure 4. Molecular phylogenetic analysis of glycosyltransferase (GT) candidates inferred from the maximum likelihood (ML) method. The best-fit ML model selection was mtREV24+I [30], and the tree with the highest log likelihood (−21419.90) is shown. The percentage of trees in which the associated taxa clustered together is shown next to the branches. Initial trees for the heuristic search were automatically obtained by applying the Neighbor-joining and BioNJ algorithms to a matrix of pairwise distances estimated using a Jones–Taylor–Thornton (JTT) model and then selecting the topology with the superior log likelihood value. The rate variation model allowed for some sites to be evolutionarily invariable ([+I], 0.00% sites). The tree is drawn to scale, with branch lengths measured in the number of substitutions per site. This analysis included 45 amino acid sequences. All positions with less than 95% site coverage were eliminated, i.e., fewer than 5% alignment gaps, missing data, and ambiguous bases were allowed at any position (partial deletion option). There were 184 positions in the final dataset. Evolutionary analyses were conducted using MEGA X [31].

2.4. Cloning, Overexpression, and Purification of GT from the GA A07 Strain in E. coli

To obtain the pure GT for the assay of the GAA biotransformation, the three candidate genes (*FQZ25_16345*, *FQZ25_19010*, and *FQZ25_19840*) were subcloned into the pETDuet-1™ expression vector (Figure S1a) and overexpressed with a fusion of His-tag in the amino-terminal in *E. coli* BL21 (DE3), and the produced GT proteins, respectively designated BtGT_16345, BtGT_19010, and BtGT_19840, were purified with Ni^{2+} chelate affinity chromatography. Among them, BtGT_16345 (Figure S1b) and BtGT_19010 (Figure S1d) were successfully purified (shown as a single band in the sodium dodecylsulfate polyacrylamide gel electrophoresis (SDS-PAGE) analysis). In contrast, BtGT_19840 could not be purified due to the insoluble form of the expressed proteins (Figure S1c).

2.5. Activity Assays of Recombinant GT Proteins toward GAA

The purified enzymes were incubated with 0.02 mg/mL of GAA, 10 mM of Mg^{2+}, and 1 mM of UDP-glucose at pH 8 and 40 °C for 30 min. After incubation, the reaction mixtures were assayed by UPLC. Results showed that BtGT_16345 catalyzed GAA to GAA-15-*O*-β-glucoside (Figure 5a), while BtGT_19010 did not catalyze GAA (Figure 5b).

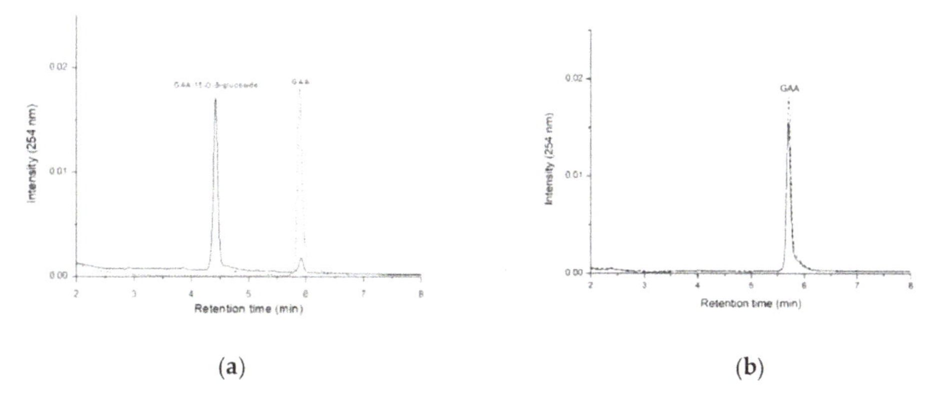

Figure 5. Biotransformation of ganoderic acid A (GAA) by purified BtGT_16345 (**a**) and BtGT_19010 (**b**). Purified enzymes (10 μg/mL) were incubated with 1 mM UDP-glucose and 0.02 mg/mL of GAA in the presence of 50 mM Tris at pH 8.0 and 10 mM $MgCl_2$. Before (the 0-min curve) or after (the 30-min curve) 30 min of incubation at 40 °C, the mixtures, together with the standard GAA or GAA-15-*O*-β-glucoside, were analyzed by UPLC. UPLC conditions are described in the Materials and Methods section.

2.6. Catalytic Conditions for BtGT_16345

The activity of purified BtGT_16345 was determined at different pH values and temperatures, and with different metal ions. Many GTs utilize divalent metal ion cofactors, and Mg^{2+} was found to be present in native crystals of some GTs [32]. Results showed that the suitable catalytic conditions for BtGT_16345 protein were at pH 7.5 and 30 °C, with 10 mM of Mg^{2+} (Figure 6).

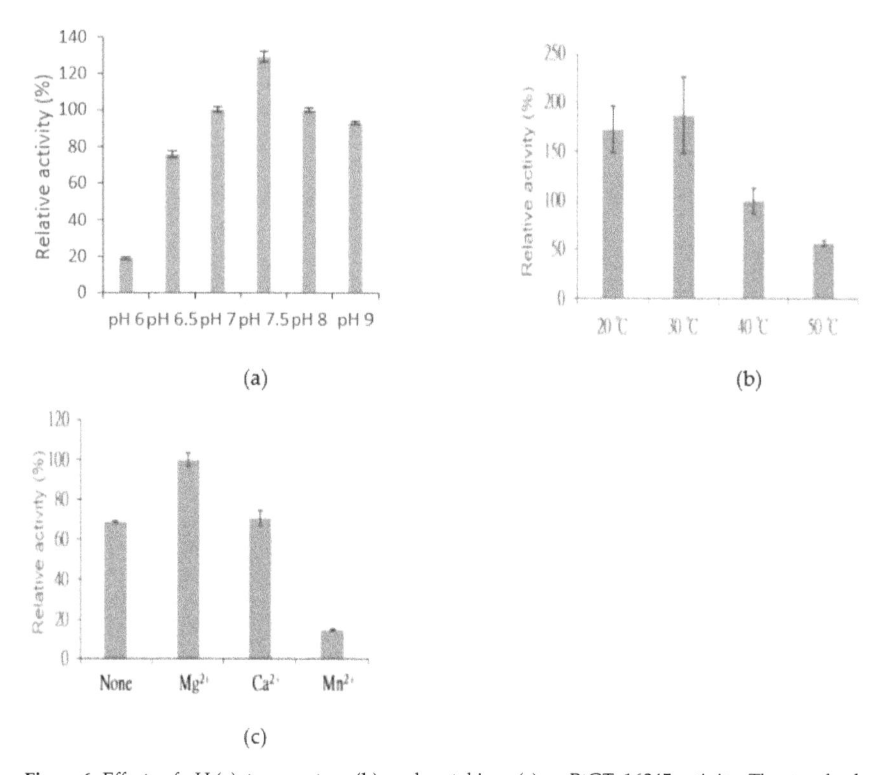

Figure 6. Effects of pH (**a**), temperature (**b**), and metal ions (**c**) on BtGT_16345 activity. The standard condition was set to 10 μg/mL of the purified enzyme, 1 mg/mL of ganoderic acid A (GAA), 10 mM of $MgCl_2$, and 10 mM of UDP-glucose in 50 mM of Tris at pH 8.0 and 40 °C. To determine suitable reaction conditions, different pH values, temperatures, and metal ions in the standard condition were replaced with the tested condition. Relative activities were obtained by dividing the area of the product peak of the reaction in the UPLC profile by that of the reaction at the standard condition and are presented as mean values (n = 3) along with error bars representing standard deviations.

2.7. Substrate Specificity of BtGT_16345

To determine the substrate specificity of BtGT_16345, the GAA triterpenoid, two additional non-GAA triterpenoids, and seven flavonoids were used as substrates for biotransformation assays (Figure 7a). Conversion of each product was calculated by dividing the peak area of each product by that of the initial input substrate in the UPLC chromatogram. Thus, the calculation of the conversion was only based on the UPLC area due to the different extinctions coefficients of the various products. Our results showed that BtGT_16345 exhibited glycosylation activity toward all tested flavonoids as well as GAA and antcin K, however no activity was detected toward another triterpenoid, celastrol (Figure 7b). In contrast, the purified recombinant BtGT_19010 was not functional on any tested compounds, including GAA.

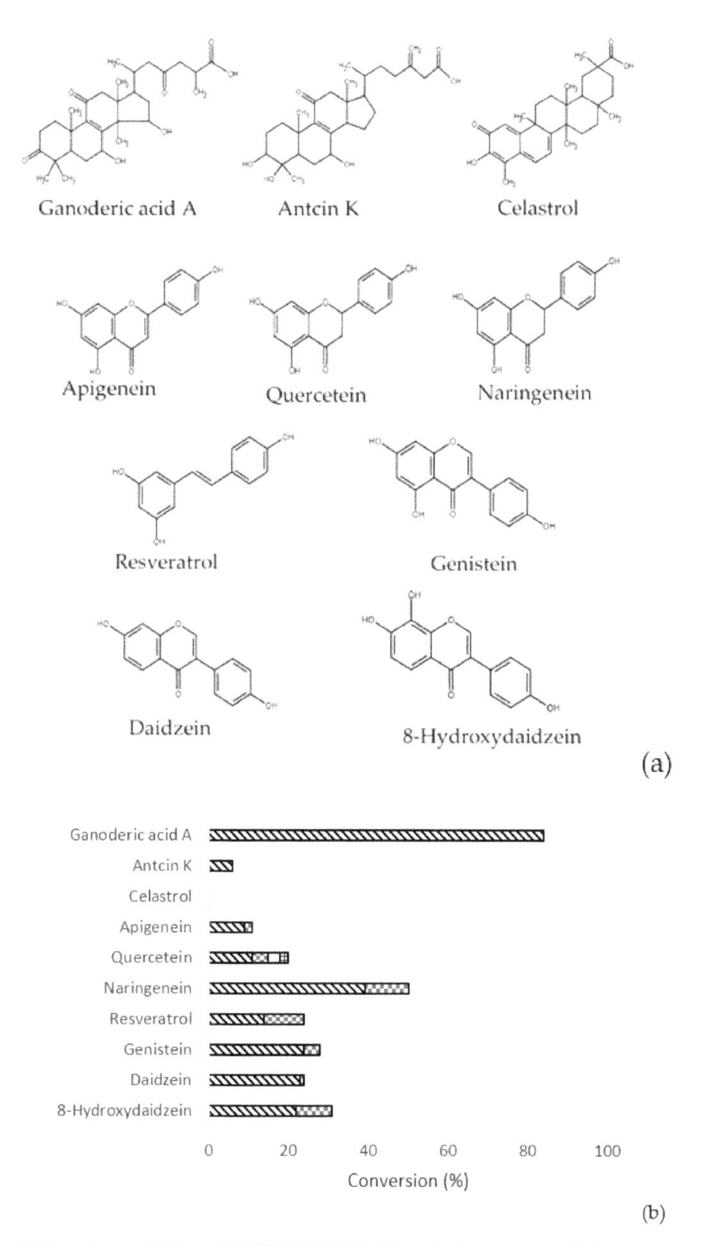

Figure 7. Substrate specificity of BtGT_16345. (**a**) Chemical structures of the test compounds. (**b**) Conversion (%) of the test compounds by BtGT_16345. For the substrate specificity assay, 1 mg/mL of different compounds was mixed with 10 μg/mL of BtGT_16345, 10 mM of UDP-glucose, 10 mM of $MgCl_2$, and 50 mM of PB at pH 7.0 and incubated at 30 °C for 30 min. After incubation, the reaction mixture was analyzed by UPLC. UPLC conditions are described in the Materials and Methods section. Conversion of each product was calculated by dividing the peak area of each product by that of the input substrate before biotransformation in the UPLC chromatogram. For biotransformation with multiple products, each product is presented by a bar with a different background pattern.

2.8. Kinetic Study of BtGT_16345 toward GAA

Figure 1 shows that *B. thuringiensis* GA A07 possessed higher GAA biotransformation activity than *B. subtilis* ATCC 6633. Moreover, the GTs of *B. subtilis* ATCC 6633 responsible for catalyzing the biotransformation were identified as BsUGT398 and BsUGT489 in a previous study [4]. To compare the catalytic efficiency of GTs toward GAA from the two strains, a kinetic study was performed on the three GTs: BsUGT398, BsUGT489, and BtGT_16345. Both recombinant purified BsUGT398 and BsUGT489 were obtained from a previous study [4]. The kinetic study was performed by using different concentrations of GAA as the substrate for the individual testing GT enzyme and the reaction velocity at each concentration of GAA was obtained from the slope of the plot of time versus the amount of the product (Figure 8). The kinetic parameters were calculated by nonlinear regression analysis applied to Michaelis–Menten equation (Table 1). Results showed that BsUGT398 exhibited the significant highest GAA-binding affinity with a K_M value of 90.71 ± 14.86 μM, while BsUGT489 exhibited the highest turnover number with a k_{cat} value of 0.9336 ± 0.0626 s^{-1}. However, the catalytic efficiency (k_{cat}/K_M) of BtGT_16345 did not show significantly different compared with either BsUGT398 or BsUGT489 (Table 1).

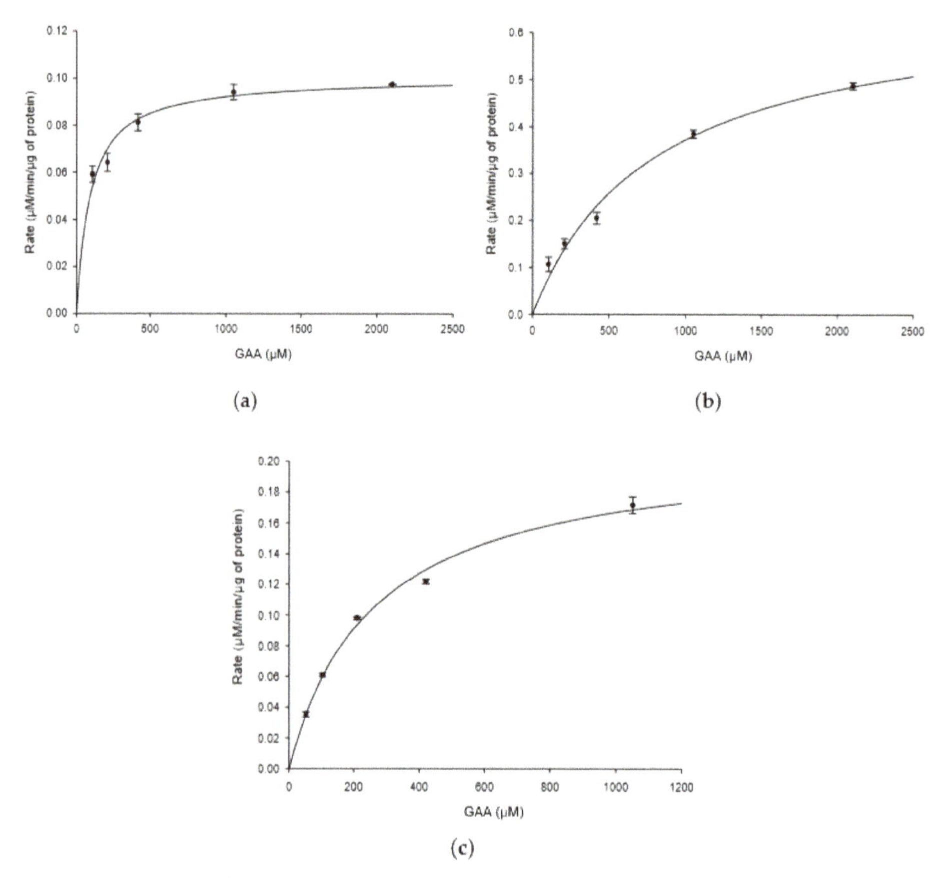

Figure 8. Kinetic study of BsUGT398 (**a**); BsUGT489 (**b**); and BtGT_16345 (**c**). Different concentrations of GAA were mixed with 10 μg purified GT protein, 10 mM UDP-glucose, 10 mM $MgCl_2$, and 50 mM PB (pH 7.0) for BtGT_16345, or 50 mM of Tris (pH 8.0) for BsUGT398 and BsUGT489 in 1 mL reaction mixture and incubated at 30 °C for BtGT_16345, or 40 °C for BsUGT398 and BsUGT489 for 20 min. During the incubation, samples from each reaction were removed and analyzed by UPLC every 2 min.

The reaction rate for each concentration of GAA was obtained from the slope of the plot of the amount of product over time. The amount of GAA-15-*O*-β-glucoside produced from the reaction was calculated from the peak area of the UPLC analysis normalized to a standard curve. The reaction velocity at each concentration of GAA was obtained from the slope of the plot of time versus the amount of the product and presented as mean values ($n = 2$) along with error bars representing standard deviations. The UPLC operation procedure was described in the Materials and Methods section. The kinetic parameters were calculated by nonlinear regression analysis applied to Michaelis-Menten equation as the description in the Materials and Methods section.

Table 1. Kinetic parameters of GT toward GAA.

GT	K_M (μM)	k_{cat} (s^{-1})	k_{cat}/K_M (s^{-1} mM^{-1})
BsUGT398	90.71 ± 14.86	0.1401 ± 0.0051	1.5445 ± 0.2592
BsUGT489	793.96 ± 124.09	0.9336 ± 0.0626	1.1759 ± 0.2000
BtGT_16345	263.82 ± 24.78	0.2944 ± 0.0109	1.1159 ± 0.1127

3. Discussion

This study sequenced and assembled the complete genome of the GA A07 strain for strain classification and GT identification. The circular map in Figure 2 not only contained critical information about the genome (including the numbers of protein-coding genes, ribosomal RNA genes, and tRNA genes, and the GC content distribution on the genome) but also revealed several interesting characteristics. For example, the GC proportions were higher on the rRNA regions, and that the rRNA genes were grouped in a portion of the genome (especially between positions 1.2 M to 1.5 M) instead of distributed evenly on the genome. The distribution of the protein-coding genes were also uneven for both forward and reverse-complement strands, in which genes in forward strand were more abundant in half of the genome (from positions 1.2 M to 3.75 M) while genes in reverse-complement strand were in greater number in another half of the genome. We consider these characteristics outside the scope of this manuscript but may warrant future genome analysis to find their underlying meanings.

The phylogenetic tree built from 16S genes could only identify that the phylogenetic placement of the GA A07 strain was very closely related to a group of *Bacilli*, including *B. thuringiensis*, *B. anthracis*, *B. cereus*, and *B. subtilis* [18]. Herein, a three-step approach was applied to identify strain GA A07: (1) downloading all available *Bacillus* genomes; (2) applying ANIs, AAIs, and a TCS to pinpoint the most closely related species/strain among *Bacillus*; and (3) building a phylogenetic tree using 250 single-copy marker genes. This approach helped us reclassify strain GA A07 as a part of *B. thuringiensis* (Figure 3). Figure 4 further shows putative GTs identified from the novel genome sequence and three GT candidates, one GT1 (BtGT_19010) and two GT28s (BtGT_16345, BtGT_19840), were grouped with the five triterpenoid-glycosylation GTs (BsYjiC, UGT109A1, BsGT1, BsUGT398, and BsUGT489).

Previous studies identified *B. subtilis* ATCC 6633 and *B. thuringiensis* GA A07 were able to biotransform GAA into GAA-15-*O*-β-glucoside [4,18]. Herein *B. thuringiensis* GA A07 exhibited over 10-fold higher biotransformation activity than *B. subtilis* ATCC 6633 (Figure 1). Through a genome-centric analysis, we further identified that BtGT_16345, which belongs to the GT28 family, exhibited good glycosylation activity toward GAA (Figure 5). This is the first report that a GT28 enzyme, not only GT1, can catalyze GAA triterpenoids. From the results of the kinetics study, the catalytic efficiency (k_{cat}/K_M) of BtGT_16345 for *B. thuringiensis* GA A07 did not show significantly different compared with either BsUGT398 or BsUGT489 for *B. subtilis* ATCC 6633 (Table 1). Therefore, BtGT_16345 identified in *B. thuringiensis* GA A07 might not be the major contributor to 10-fold higher biotransformation activity. BtGT_19840, the activity of which could not be evaluated in this study owing to an inability to obtain its soluble expression, may contribute to the higher GAA biotransformation of *B. thuringiensis*. Other possible reasons include: (1) higher expression of BtGT with GAA biotransformation activity, (2) a higher uptake rate of GAA into cells; (3) a higher UDP-glucose concentration accumulating in cells, and (4) other enzymes and/or coenzymes cooperating in cells, may account for the contribution of the higher catalytic efficiency of *B. thuringiensis* GA A07.

Although over 500,000 GTs were identified in the CAZy database, only six microbial GTs were found to exhibit glycosylation activities toward triterpenoids, including BtGT_16345 (this study), BsYjiC [19–23], UGT109A1 [24,28], BsGT1 [29], BsUGT398, and BsUGT489 [4]. BtGT_16345, which was classified into the GT28 family, is the only exception out of the other five GTs that belong to the GT1 family [1]. Previous studies showed that these GT1 enzymes catalyzed the glycosylation of several flavonoids at multiple positions to form mixture of flavonoid glycosides [19–24,28,29]. In the present study, BtGT_16345 exhibited similar catalytic activities toward seven flavonoids to form multiple products (Figure 7). These results revealed that microbial GTs potentially accept a broader range of different flavonoids as catalytic substrates and exhibit less regioselectivity toward flavonoid substrates than do plant GTs [2,3].

As to triterpenoid substrates, both BsYjiC [19–23] and UGT109A1 [24,28] catalyzed *O*-glycosylation toward triterpenoids at multiple positions (C-3, C-6, C-12, and C-20). As a result, one triterpenoid substrate can potentially be biotransformed into many types of triterpenoid glycosides. The above reports revealed that the two GT1s were less regioselective toward triterpenoid substrates. In contrast, BtGT_16345 specifically catalyzed glycosylation at the C-15 position out of the three hydroxyl groups capable of the *O*-glycosylation (C-7, C-15, and C-26) of GAA and no other glycosylated products were detected during the biotransformation using the analytical techniques herein (Figure 5). Moreover, only one product was produced by BtGT_16345 from the biotransformation toward antcin K, another triterpenoid containing four sites capable of *O*-glycosylation (C-3, C-4, C-7, and C-26) (Figure 7). BtGT_16345 was regioselective toward triterpenoid substrates, and could be further used for industrial applications or stepwise biosynthesis for structure-activity studies.

4. Materials and Methods

4.1. Microorganism and Chemicals

The *Bacillus* sp. GA A07 strain was isolated from intestinal bacteria of zebrafish in our previous study [18]. Both purified recombinant BsUGT398 and BsUGT489 were obtained in our previous study [4]. Antcin K was obtained by a procedure in our previous study [33]. GAA and celastrol were bought from Baoji Herbest Bio-Tech (Xi-An, Shaanxi, China). 8-OHDe was prepared according to Wu et al.'s [34] method. Other flavonoids were purchased from Sigma (St. Louis, MO, USA) or Tokyo Chemical Industry (Tokyo, Japan). UDP-glucose was purchased from Cayman Chemical (Ann Arbor, MI, USA). All materials needed for the polymerase chain reaction (PCR), including primers, deoxyribonucleotide triphosphate, and Taq DNA polymerase, were purchased from MDBio (Taipei, Taiwan). The pETDuet-1 plasmid was purchased from Novagen (Madison, WI, USA). Restriction enzymes and DNA ligase were obtained from New England Biolabs (Ipswich, MA, USA). Other reagents and solvents used were of high quality, and were purchased from commercially available sources.

4.2. Whole-Genome Sequencing

Bacillus sp. GA A07 cells were harvested from Luria-Bertani (LB) plates pre-cultured in a 25 °C incubator for 3 days. Genomic DNA of cells was extracted with a ZR Soil Microbe DNA Kit™ (D6001, Zymo Research, Irvine, CA, USA). Genomic DNA (1~3 μg) was end-repaired and ligated with NB12 barcode sequencing adaptors (EXP-NBD104, Native Barcoding Expansion 1-12) via a KAPA Hyper Prep Kit (Cat#KR0961, Kapa Biosystems, Wilmington, MA, USA), following the manufacturer's instructions. The barcoded genomic DNA library was premixed with LB and SQB buffer of the Ligation Sequencing Kit (SQK-LSK109, Oxford Nanopore Technologies, UK) loaded in a Flow Cell (R9.4.1; FLO-MIN106), and sequenced by MinION devices for 24 h. For short-read sequencing, the genomic DNA library was fragmented by Covaris S220 with a 350-bp size peak, a peak incident power of 140 W, a duty factor of 10, 200 cycles per burst, and a treatment time of 100 s. Sheared DNA fragments were then used for library construction with the MGIEasy DNA Library Prep Kit v1.1. The library was sequenced with the BGISEQ-500RS sequencer by Tri-I Biotech (New Taipei, Taiwan).

4.3. Genome Assembly and Annotation

Nanopore reads were first error-corrected using canu v1.8 [35] (parameters: -correct -nanopore-raw); the error-corrected reads were then assembled using wtdbg2 v2.4 [36]. The yielded assembly was mapped to BGISEQ-500RS paired-end sequences (trimmed using SOAPnuke v2.0.7 [37]; parameters: filter -l 10 -q 0.5 -n 0.1 -M 2 –adaMR 0.5 -f AAGTCGGAGGCCAAGCGGTCTTAGGAAGACAA -r AAGTCGGATCGTAGCCATGTCGTTCTGTGAGCCAAGGAGTTG) using bowtie2v 2.2.3 [38]. The mapping SAM file was first converted to a sorted indexed BAM file using SAMtools v1.9-45 [39] (samtools sort; samtools index) and then used as input for the assembly error correction tool, Pilon v1.23 [40].

4.4. Reclassification of GA A07 Strain

In total, 882 *Bacillus* genomes were downloaded to identify the most closely related species. FastANI v1.2 [41] was leveraged to identify the most closely related species (i.e., the ones with the highest average nucleotide identity compared to the GA A07 genome). Average amino acid identity was also checked by the following steps: (1) Prodigal v2.6.3 [25] was used to in batch-predict protein-coding genes from the genomes, and genes were converted into amino acid sequences; (2) amino acid identities between the GA A07 genome and all downloaded *Bacillus* genomes were compared using BLASTP [26] (parameters: -max_target_seqs 1 -evalue 1e-10); and (3) the mean value of the best-hit identities was calculated and was regarded as the average amino acid identity between the two genomes. The Tetra Correlation Search (TCS) function implemented in the JSpeciesWS webserver [27] was also pursued to find the most closely related bacterial genome based on tetra-nucleotide composition evidence.

Only genomes with better assembly quality (defined as genomes with at most ten scaffolds) were used to build the phylogenetic tree. The tree was generated using ezTree [42], which is capable of identifying single-copy marker genes among input genomes, thereby creating a concatenated alignment of all marker genes, and using FastTree 2 [43] with the Jones–Taylor–Thornton (JTT) evolutionary model and 1000 resampling tests to construct a reliable tree. The nwk file of the phylogenetic tree was then visualized using Molecular Evolutionary Genetics Analysis (MEGA) X software [31].

4.5. Identification and Analysis of GT Genes

The dbCAN2 webserver [44] was employed to identify potential GTs from the *B. thuringiensis* GA A07 genome. An unrooted phylogenetic tree of all extracted GT protein sequences was constructed using MEGA X software [31] with the maximum-likelihood method, 500 bootstrap replications, the general reversible mitochondrial model [30], and partial deletion.

4.6. Fermentation and Biotransformation of GAA

Bacillus subtilis ATCC 6633 or *B. thuringiensis* GA A07 was cultivated in a 250-mL baffled Erlenmeyer flask containing 20 mL of LB medium with 5% of glucose at 180 rpm and 28 °C. When the OD_{600} of the cell culture reached 0.6, 1 mg/mL of GAA was added to the broth. Cultivation was carried out for another 32 h, and fermentation broth (0.5 mL) of the culture was taken at predicted time intervals and used for the UPLC analysis to measure the biotransformation activity.

4.7. UPLC Analysis

The UPLC system (Acquity UPLC H-Class, Waters, Milford, MA, USA) was equipped with an analytic C18 reversed-phase column (Kinetex® C18, 1.7 μm, 2.1 i.d. × 100 mm, Phenomenex, Torrance, CA, USA). The operating conditions of UPLC for analysis of GAA, antcin K, and celastrol were consistent with those of our previous study [4] except for the 430-nm absorbance detection of celastrol. Operating conditions for flavonoids were from our previous study [11].

4.8. Expression and Purification of GT from GA A07 Strain

Genomic DNA of the GA A07 strain was isolated using the commercial kit Geno *Plus*™ (Viogene, Taipei, Taiwan). Candidate GT genes were amplified from genomic DNA using a PCR with specific primer sets (Table S2). The amplified GT genes were subcloned into the pETDuet-1™ vector through suitable restriction enzyme sites (Table S2) to obtain the expression vector, pETDuet-BtGT (Figure S1a). Expression vectors were transformed into *E. coli* BL21 (DE3) via electroporation to obtain recombinant *E. coli*.

Recombinant BtGT_16345, BtGT_19840, and BtGT_19010 were produced and purified from the recombinant *E. coli*, and analyzed by SDS-PAGE (Figure S1b–d). The protein concentration was determined by a Bradford assay using bovine serum albumin as the standard. The experimental procedures were the same as those in our previous study [4].

4.9. In Vitro Biotransformation Assay

In vitro biotransformation was performed using purified GT proteins. In a 0.1mL standard reaction mixture, 1 μg of purified GT protein, 0.02 mg/mL of GAA, 1 mM of UDP-glucose, 10 mM of $MgCl_2$, and 50 mM of Tris at pH 8.0 were added. The reaction was carried out at 40 °C for 30 min, stopped by adding 0.9 mL of methanol, and analyzed by UPLC.

For optimization experiments, different pH values, temperatures, and metal ions were replaced in the standard reaction, where 1 mg/mL of GAA was used. For pH testing, PB at pH 6.0 to 7.5, and Tris buffer at pH 8.0 and pH 9.0 were used. For metal ion testing, 10 mM of $MgCl_2$, $CaCl_2$, or $MnCl_2$ was used. The relative activity was obtained by dividing the area of the product peak of the reaction in the UPLC profile by that of the reaction with Tris pH 8.0, at 40 °C, and with 10 mM of $MgCl_2$.

For the substrate specificity assay, 25 mg/mL of the substrate soluble in dimethyl sulfoxide conditions, all tested substances were soluble in the reaction buffer. 1 mg/mL of different test compounds was mixed with 1 μg of purified GT protein, 10 mM of UDP-glucose, 10 mM of $MgCl_2$, and 50 mM of PB pH 7.0 in a 0.1mL reaction mixture and incubated at 30 °C for 30 min. After incubation, the reaction mixture was analyzed by UPLC.

For the kinetic experiments, different concentrations of GAA were mixed with 10 μg of purified GT protein, 10 mM of UDP-glucose, 10 mM of $MgCl_2$, and 50 mM of PB at pH 7.0 for BtGT_16345, or 50 mM of Tris at pH 8.0 for BsUGT398 and BsUGT489 in a 1-mL reaction mixture and incubated at 30 °C for BtGT_16345, or 40 °C for BsUGT398 and BsUGT489 for 20 min. During incubation, samples from each reaction were taken out and analyzed by UPLC every 2 min. The amount of GAA-15-*O*-β-glucoside produced from the reaction was calculated from the peak area of the UPLC analysis normalized to a standard curve. The reaction velocity at each concentration of GAA was obtained from the slope of the plot of time versus the amount of the product. The kinetic parameters were calculated by nonlinear regression analysis applied to Michaelis-Menten equation using SigmaPlot 14.0 software (Systat Software, San Jose, CA, USA). The k_{cat} values were calculated using the predicted molecular mass for each recombinant enzyme.

5. Conclusions

A novel GT28 family enzyme, BtGT_16345, from a new genome assembly of the *B. thuringiensis* GA A07 strain, was identified that can biotransform GAA into GAA-15-*O*-β-glucoside. To our knowledge, BtGT_16345 is the first GT28 family enzyme with triterpenoid glycosylation activity.

Supplementary Materials: Supplementary materials can be found at http://www.mdpi.com/1422-0067/20/20/5192/s1. The following materials are available online. Table S1: Single copy marker genes used for building the phylogenetic tree; Table S2: Nucleotide sequences of the primers used for amplification of glycosyltransferase (GT) genes in the present study; Figure S1: Expression and purification of glycosyltransferases (GTs) from *Bacillus thuringiensis* GA A07 in *E. coli*.

Author Contributions: Conceptualization, T.-S.C., T.-Y.W., and Y.-W.W.; data curation, Y.-W.W., T.-Y.H., Y.-W.L., H.-M.C., W.-X.C., and T.-S.C.; methodology, Y.-W.W., C.-M.C., T.-Y.W., J.-Y.W., and T.-S.C.; project administration, T.-S.C.; Writing—original draft, Y.-W.W., C.-M.C., T.-Y.W., J.-Y.W., and T.-S.C.; writing—review and editing, Y.-W.W., C.-M.C., T.-Y.W., J.-Y.W., and T.-S.C.

Funding: This research was financially supported by grants mainly from the Ministry of Science and Technology of Taiwan (Project No. MOST108-2221-E-024-008-MY2 to T.-S.C.) and partially from MOST-108-2628-E-038-002-MY3 to Y.-W.W. and Academia Sinica to T.-Y.W.

Conflicts of Interest: The authors declare no conflicts of interest.

Abbreviations

GAA	Ganoderic acid A
GT	Glycosyltransferase
UDP	Uridine diphosphate

References

1. Cantarel, B.; Coutinho, P.M.; Rancurel, C.; Bernard, T.; Lombard, V.; Henrissat, B. The Carbohydrate-Active EnZymes database (CAZy): An expert resource for Glycogenomics. *Nucleic Acids Res.* **2009**, *37* (Suppl. 1), D233–D238. [CrossRef]
2. Hofer, B. Recent developments in the enzymatic *O*-glycosylation of flavonoids. *Appl. Microbiol. Biotechnol.* **2016**, *100*, 4269–4281. [CrossRef] [PubMed]
3. Tiwari, P.; Sangwan, R.S.; Sangwan, N.S. Plant secondary metabolism linked glycosyltransferases: An update on expanding knowledge and scopes. *Biotechnol. Adv.* **2016**, *34*, 716–739. [CrossRef] [PubMed]
4. Chang, T.S.; Wu, J.J.; Wang, T.Y.; Wu, K.Y.; Chiang, C.M. Uridine diphosphate-dependent glycosyltransferases from *Bacillus subtilis* ATCC 6633 catalyze the 15-*O*-glycosylation of ganoderic acid A. *Int. J. Mol. Sci.* **2018**, *19*, 3469. [CrossRef] [PubMed]
5. Cao, H.; Chen, X.; Jassbi, A.R.; Xiao, J. Microbial biotransformation of bioactive flavonoids. *Biotechnol. Adv.* **2015**, *33*, 214–223. [CrossRef] [PubMed]
6. Shah, S.A.A.; Tan, H.L.; Sultan, S.; Faridz, M.A.B.M.; Shah, M.A.B.M.; Nurfazilah, S.; Hussain, M. Microbial-catalyzed biotransformation of multifunctional triterpenoids derived from phytonutrients. *Int. J. Mol. Sci.* **2014**, *15*, 12027–12060. [CrossRef] [PubMed]
7. Sultana, N.; Saify, Z.S. Enzymatic biotransformation of terpenes as bioactive agents. *J. Enzym. Inhib. Med. Chem.* **2013**, *28*, 1113–1128. [CrossRef] [PubMed]
8. Muffler, K.; Leipold, D.; Scheller, M.C.; Haas, C.; Steingroewer, J.; Bley, T.; Neuhaus, H.E.; Mirata, M.A.; Schrader, J.; Ulber, R. Biotransformation of triterpenes. *Process Biochem.* **2011**, *46*, 1–15. [CrossRef]
9. Parra, A.; Rivas, F.; Garcia-Granados, A.; Martinez, A. Microbial transformation of triterpenoids. *Mini-Rev. Org. Chem.* **2009**, *6*, 307–320. [CrossRef]
10. Shimoda, K.; Hamada, H.; Hamada, H. Synthesis of xylooligosaccharides of daidzein and their anti-oxidant and anti-allergic activities. *Int. J. Mol. Sci.* **2011**, *12*, 5616–5625. [CrossRef]
11. Chiang, C.M.; Wang, T.Y.; Yang, S.Y.; Wu, J.Y.; Chang, T.S. Production of new isoflavone glucosides from glycosylation of 8-hydroxydaidzein by glycosyltransferase from *Bacillus subtilis* ATCC 6633. *Catalysts* **2018**, *8*, 349. [CrossRef]
12. Chang, T.S.; Wang, T.Y.; Yang, S.Y.; Kao, Y.H.; Wu, J.J.; Chiang, C.M. Potential industrial production of a well-soluble, alkaline-stable, and anti-inflammatory isoflavone glucoside from 8-hydroxydaidzein glucosylated by recombinant amylosucrase of *Deinococcus geothermalis*. *Molecules* **2019**, *24*, 2236. [CrossRef] [PubMed]
13. Shi, Z.Y.; Zeng, J.Z.; Wong, A.S.T. Chemical structures and pharmacological profiles of ginseng saponins. *Molecules* **2019**, *24*, 2443. [CrossRef] [PubMed]

14. Kubota, T.; Asaka, Y. Structures of ganoderic acid A and B, two new lanostane type bitter triterpenes from *Ganoderma lucidum* (FR.) Karst. *Helv. Chim. Acta* **1982**, *65*, 611–619. [CrossRef]
15. Xia, Q.; Zhang, H.; Sun, X.; Zhao, H.; Wu, L.; Zhu, D.; Yang, G.; Shao, Y.; Zhang, X.; Mao, X.; et al. A comprehensive review of the structure elucidation and biological activity of triterpenoids from *Ganoderma* spp. *Molecules* **2014**, *19*, 17478–17535. [CrossRef] [PubMed]
16. Wu, J.W.; Zhao, W.; Zhong, J.J. Biotechnological production and application of ganoderic acids. *Appl. Microbiol. Biotechnol.* **2010**, *87*, 457–466.
17. Liang, C.; Tian, D.; Liu, Y.; Li, H.; Zhu, J.; Li, M.; Xin, M.; Xia, J. Review of the molecular mechanisms of *Ganoderma lucidum* triterpenoids: Ganoderic acids A, C2, D, F, DM, X and Y. Eur. *J. Med. Chem.* **2019**, *174*, 130–141. [CrossRef]
18. Chang, T.S.; Chiang, C.M.; Wang, T.Y.; Lee, C.H.; Lee, Y.W.; Wu, J.Y. New triterpenoid from novel triterpenoid 15-*O*-glycosylation on ganoderic acid A by intestinal bacteria of zebrafish. *Molecules* **2018**, *23*, 2345. [CrossRef]
19. Dai, L.; Li, J.; Yao, P.; Zhu, Y.; Men, Y.; Zeng, Y.; Yang, J.; Sun, Y. Exploiting the aglycon promiscuity of glycosyltransferase Bs-YjiC from *Bacillus subtilis* and its application in synthesis of glycosides. *J. Biotechnol.* **2017**, *248*, 69–76. [CrossRef]
20. Dai, L.; Li, J.; Yang, J.; Zhu, Y.; Men, Y.; Zeng, Y.; Cai, Y.; Dong, C.; Dai, Z.; Zhang, X.; et al. Use of a promiscuous glycosyltransferase from *Bacillus subtilis* 168 for the enzymatic synthesis of novel protopanaxtriol-type ginsenosides. *J. Agric. Food Chem.* **2017**, *66*, 943–949. [CrossRef]
21. Dai, L.; Li, J.; Yang, J.; Men, Y.; Zeng, Y.; Cai, Y.; Sun, Y. Enzymatic synthesis of novel glycyrrhizic acid glucosides using a promiscuous *Bacillus glycosyltransferase*. *Catalysts* **2018**, *8*, 615. [CrossRef]
22. Li, K.; Feng, J.; Kuang, Y.; Song, W.; Zhang, M.; Ji, S.; Qiao, X.; Ye, M. Enzymatic synthesis of bufadienolide O-glycosides as potent antitumor agents using a microbial glycosyltransferase. *Adv. Syn. Cat.* **2017**, *359*, 3765–3772. [CrossRef]
23. Chen, K.; He, J.; Hu, Z.; Song, W.; Yu, L.; Li, K.; Qiao, X.; Ye, M. Enzymatic glycosylation of oleanane-type triterpenoids. *J. Asia. Nat. Prod. Res.* **2018**, *20*, 615–623. [CrossRef] [PubMed]
24. Liang, H.; Hu, Z.; Zhang, T.; Gong, T.; Chen, J.; Zhu, P.; Li, Y.; Yang, J. Production of a bioactive unnatural ginsenoside by metabolically engineered yeasts based on a new UDP-glycosyltransferase from *Bacillus subtilis*. *Metab. Eng.* **2017**, *44*, 60–69. [CrossRef]
25. Hyatt, D.; Chen, G.L.; Locascio, P.F.; Land, M.L.; Larimer, F.W.; Hauser, L.J. Prodigal: Prokaryotic gene recognition and translation initiation site identification. *BMC Bioinformatics* **2010**, *11*, 119. [CrossRef]
26. Altschul, S.F.; Madden, T.L.; Schaffer, A.A.; Zhang, J.; Zhang, Z.; Miller, W.; Lipman, D.J. Gapped BLAST and PSI-BLAST: A new generation of protein database search programs. *Nucleic Acids Res.* **1997**, *25*, 3389–3402. [CrossRef]
27. Richter, M.; Rossello-Mora, R.; Oliver Glockner, F.; Peplies, J. JSpeciesWS: A web server for prokaryotic species circumscription based on pairwise genome comparison. *Bioinformatics* **2016**, *32*, 929–931. [CrossRef]
28. Zhang, T.T.; Gong, T.; Hu, Z.F.; Gu, A.D.; Yang, J.L.; Zhu, P. Enzymatic synthesis of unnatural ginsenosides using a promiscuous UDP-glucosyltransferase from *Bacillus subtilis*. *Molecules* **2018**, *23*, 2797. [CrossRef]
29. Wang, D.D.; Jin, Y.; Wang, C.; Kim, Y.J.; Perez, J.E.J.; Baek, N.I.; Mathiyalagan, R.; Markus, J.; Yang, D.C. Rare ginsenoside Ia synthesized from F1 by cloning and overexpression of the UDP-glycosyltransferase gene from Bacillus subtilis: Synthesis, characterization, and in vitro melanogenesis inhibition activity in BL6B16 cells. *J. Gingeng Res.* **2018**, *42*, 42–49. [CrossRef]
30. Adachi, J.; Hasegawa, M. Model of amino acid substitution in proteins encoded by mitochondrial DNA. *J. Mol. Evol.* **1996**, *42*, 459–468. [CrossRef]
31. Kumar, S.; Stecher, G.; Li, M.; Knyaz, C.; Tamura, K. MEGA X: Molecular Evolutionary Genetics Analysis across computing platforms. *Mol. Biol. Evol.* **2018**, *35*, 1547–1549. [CrossRef] [PubMed]
32. Lairson, L.L.; Henrissat, B.; Davies, G.J.; Withers, S.G. Glycosyltransferases: Structures, functions, and mechanisms. *Annu. Rev. Biochem.* **2008**, *77*, 25.1–25.35. [CrossRef] [PubMed]
33. Chang, T.S.; Chiang, C.M.; Siao, Y.Y.; Wu, J.Y. Sequential biotransformation of antcin K by *Bacillus subtilis* ATCC 6633. *Catalysts* **2018**, *8*, 349. [CrossRef]
34. Wu, S.C.; Chang, C.W.; Lin, C.W.; Hsu, Y.C. Production of 8-hydroxydaidzein polyphenol using biotransformation by *Aspergillus oryzae*. *Food Sci. Technol. Res.* **2015**, *21*, 557–562. [CrossRef]

35. Koren, S.; Walenz, B.P.; Berlin, K.; Miller, J.R.; Bergman, N.H.; Phillippy, A.M. Canu: Scalable and accurate long-read assembly via adaptive k-mer weighting and repeat separation. *Genome Res.* **2017**, *27*, 722–736. [CrossRef]
36. Ruan, J.; Li, H. Fast and accurate long-read assembly with wtdbg2. *bioRxiv* **2019**. [CrossRef]
37. Chen, Y.; Chen, Y.; Shi, C.; Huang, Z.; Zhang, Y.; Li, S.; Li, Y.; Ye, J.; Yu, C.; Li, Z.; et al. SOAPnuke: A MapReduce acceleration-supported software for integrated quality control and preprocessing of high-throughput sequencing data. *Gigascience* **2018**, *7*, 1–6. [CrossRef]
38. Langmead, B.; Salzberg, S.L. Fast gapped-read alignment with Bowtie 2. *Nat. Methods* **2012**, *9*, 357–359. [CrossRef]
39. Li, H.; Handsaker, B.; Wysoker, A.; Fennell, T.; Ruan, J.; Homer, N.; Marth, G.; Abecasis, G.; Durbin, R.; Genome Project Data Processing, S. The Sequence Alignment/Map format and SAMtools. *Bioinformatics* **2009**, *25*, 2078–2079. [CrossRef]
40. Walker, B.J.; Abeel, T.; Shea, T.; Priest, M.; Abouelliel, A.; Sakthikumar, S.; Cuomo, C.A.; Zeng, Q.; Wortman, J.; Young, S.K.; et al. Pilon: An integrated tool for comprehensive microbial variant detection and genome assembly improvement. *PLoS ONE* **2014**, *9*, e112963. [CrossRef]
41. Jain, C.; Rodriguez, R.L.; Phillippy, A.M.; Konstantinidis, K.T.; Aluru, S. High throughput ANI analysis of 90K prokaryotic genomes reveals clear species boundaries. *Nat. Commun.* **2018**, *9*, 5114. [CrossRef]
42. Wu, Y.W. ezTree: An automated pipeline for identifying phylogenetic marker genes and inferring evolutionary relationships among uncultivated prokaryotic draft genomes. *BMC Genom.* **2018**, *19* (Suppl. 1), 921. [CrossRef] [PubMed]
43. Price, M.N.; Dehal, P.S.; Arkin, A.P. FastTree 2–approximately maximum-likelihood trees for large alignments. *PLoS ONE* **2010**, *5*, e9490. [CrossRef] [PubMed]
44. Zhang, H.; Yohe, T.; Huang, L.; Entwistle, S.; Wu, P.; Yang, Z.; Busk, P.K.; Xu, Y.; Yin, Y. dbCAN2: A meta server for automated carbohydrate-active enzyme annotation. *Nucleic Acids Res.* **2018**, *46*, W95–W101. [CrossRef] [PubMed]

International Journal of
Molecular Sciences

Article

Structural and Functional Characterization of Three Novel Fungal Amylases with Enhanced Stability and pH Tolerance

Christian Roth [1,2,†], Olga V. Moroz [1,†], Johan P. Turkenburg [1], Elena Blagova [1], Jitka Waterman [1,3], Antonio Ariza [1,4], Li Ming [5], Sun Tianqi [5], Carsten Andersen [6], Gideon J. Davies [1] and Keith S. Wilson [1,*]

1. York Structural Biology Laboratory, Department of Chemistry, University of York, Heslington, York YO10 5DD, UK; Christian.Roth@mpikg.mpg.de (C.R.); olga.moroz@york.ac.uk (O.V.M.); Johan.turkenburg@york.ac.uk (J.P.T.); lena.blagova@york.ac.uk (E.B.); jitka.waterman@diamond.ac.uk (J.W.); antonio.ariza@path.ox.ac.uk (A.A.); gideon.davies@york.ac.uk (G.J.D.)
2. Carbohydrates: Structure and Function, Biomolecular Systems, Max Planck Institute of Colloids and Interfaces, 14195 Berlin, Germany
3. Diamond Light Source, Diamond House, Harwell Science and Innovation Campus, Fermi Ave, Didcot OX11 0DE, UK
4. Sir William Dunn School of Pathology, University of Oxford, Oxford OX1 3RE, UK
5. Novozymes (China) Investment Co. Ltd., 14 Xinli Road, Haidian District, Beijing 100085, China; MLIX@novozymes.com (L.M.); TQSU@novozymes.com (S.T.)
6. Novozymes (Denmark), Krogshojvej 36, DK-2880 Bagsvaerd, Denmark; CarA@novozymes.com

* Correspondence: keith.wilson@york.ac.uk; Tel.: +44-1904-328262

† These authors contributed equally to this work.

Received: 16 September 2019; Accepted: 24 September 2019; Published: 3 October 2019

Abstract: Amylases are probably the best studied glycoside hydrolases and have a huge biotechnological value for industrial processes on starch. Multiple amylases from fungi and microbes are currently in use. Whereas bacterial amylases are well suited for many industrial processes due to their high stability, fungal amylases are recognized as safe and are preferred in the food industry, although they lack the pH tolerance and stability of their bacterial counterparts. Here, we describe three amylases, two of which have a broad pH spectrum extending to pH 8 and higher stability well suited for a broad set of industrial applications. These enzymes have the characteristic GH13 α-amylase fold with a central $(\beta/\alpha)_8$-domain, an insertion domain with the canonical calcium binding site and a C-terminal β-sandwich domain. The active site was identified based on the binding of the inhibitor acarbose in form of a transglycosylation product, in the amylases from *Thamnidium elegans* and *Cordyceps farinosa*. The three amylases have shortened loops flanking the nonreducing end of the substrate binding cleft, creating a more open crevice. Moreover, a potential novel binding site in the C-terminal domain of the *Cordyceps* enzyme was identified, which might be part of a starch interaction site. In addition, *Cordyceps farinosa* amylase presented a successful example of using the microseed matrix screening technique to significantly speed-up crystallization.

Keywords: α-amylase; starch degradation; biotechnology; structure

---

1. Introduction

The use of enzymes in industrial processes is a multi-billion-dollar market. One of the first enzymes discovered in 1833 was diastase, an enzyme able to hydrolyze starch [1]. Nowadays, amylases, also able to hydrolyze starch, constitute up to 25% of the market for enzymes and have virtually replaced chemical methods for degrading starch in the industrial sector (reviewed in [2]). Amylases are the most important class of enzymes for degrading starch and can be subdivided into three subclasses: α-, β-, and gluco-amylases based on their reaction specificity and product profiles. α-amylases degrade the α- 1,4 linkage between adjacent glucose units and are extensively used for example in bioethanol production or in washing powder and detergents [3] (and reviewed in [4]). One of the most widely used α-amylases is that from *Bacillus licheniformis*, known under the tradename "Termamyl". Microbial amylases are generally used in detergent applications and other industrial processes, including bioethanol production, with new amylases, in particular those from hyperthermophilic organisms, offering further improvement in the production process (reviewed in [5]).

α-amylases belong to glycoside hydrolase family 13 (GH13) in the CAZy database classification [6]. They have a $(\beta/\alpha)_8$ barrel domain harboring the active site, a subdomain which includes the canonical calcium binding site inserted between the third β-strand and the third α-helix and a C-terminal β-sandwich domain, thought to be important for the interaction with raw starch (reviewed in [7]) [8,9]. Amylases follow a retaining mechanism with an aspartate as nucleophile and one glutamate as general acid/base [10,11]. Up to ten consecutive sugar subsites forming the active site cleft have been identified in bacterial amylases [12].

To date, recombinant fungal amylases have been isolated from mesophilic hosts such as *Aspergillus oryzae* and are of particular interest to the food industry as they match the temperature and pH range used in typical applications in the baking process, where they are active in the dough but inactivated during baking. Due to the widespread use of fungal enzymes for the production of food and food ingredients (such as citric acid), they are classified as GRAS (generally recognized as safe) organisms by organizations including the FDA (US Food and Drug Administration) [13].

Up till now, fungal enzymes with a higher pH-tolerance and thermostability have not been reported. Here, we describe the structure and function of three novel α-amylases from *Cordyceps farinosa* (CfAM), *Rhizomucor pusillus* (RpAM) and *Thamnidium elegans* (TeAM) with a higher stability and pH-tolerance with the potential to act as novel biocatalysts for various industrial processes. The sequence of all three enzymes groups them in the GH13 sub-family 1 along with, for example, the amylase from *Aspergillus oryzae* (also known as TAKA amylase). However, unlike other fungal amylases, the enzymes in this study have been shown to have a broad pH profile with an optimum around pH 5 while retaining activity at pH 8. Furthermore, their more open crevice leads to the production of longer oligomers compared to TAKA amylase.

The native RpAM and TeAM have a four-domain fold with a carbohydrate binding domain (CBM20) at the C-terminus and a short serine-rich linker in between, while native CfAM lacks this CBM20 domain. In this study, only the core of the amylases including the A, B and C domains was cloned and expressed. In addition, crystallization of *Cordyceps farinosa* amylase again demonstrates the power of the microseed matrix screening technique [14].

2. Results

2.1. Biochemical Characterization

The pH, temperature and product profiles were characterized for all three amylases. Of great desire are amylases with a broader pH-tolerance compared to TAKA amylase. Our analysis showed that all three amylases have a pH optimum around 5. Whereas TeAM has no significant activity above pH 7, RpAM and CfAM retain significant activity at pH 7 extending up to a pH of 9 (Figure 1a). In particular, CfAM shows the highest pH tolerance, retaining 70% of its activity at pH 8. RpAM and TeAM both show a pronounced shoulder, suggesting the involvement of more titratable residues in the substrate recognition and catalysis process. The temperature profiles reveal that RpAM and CfAM also have a considerably higher thermotolerance compared to TAKA and TeAM (Figure 1b). In particular, RpAM retains full activity even at 80 °C, making it an attractive enzyme for industrial high temperature starch saccharification processes. Compared to TAKA amylase, all three amylases show a tendency to produce higher amounts of oligomers with a degree of polymerization (dp) of three, with trace amounts of oligomers with a dp of up to seven for TeAM (Figure 1c).

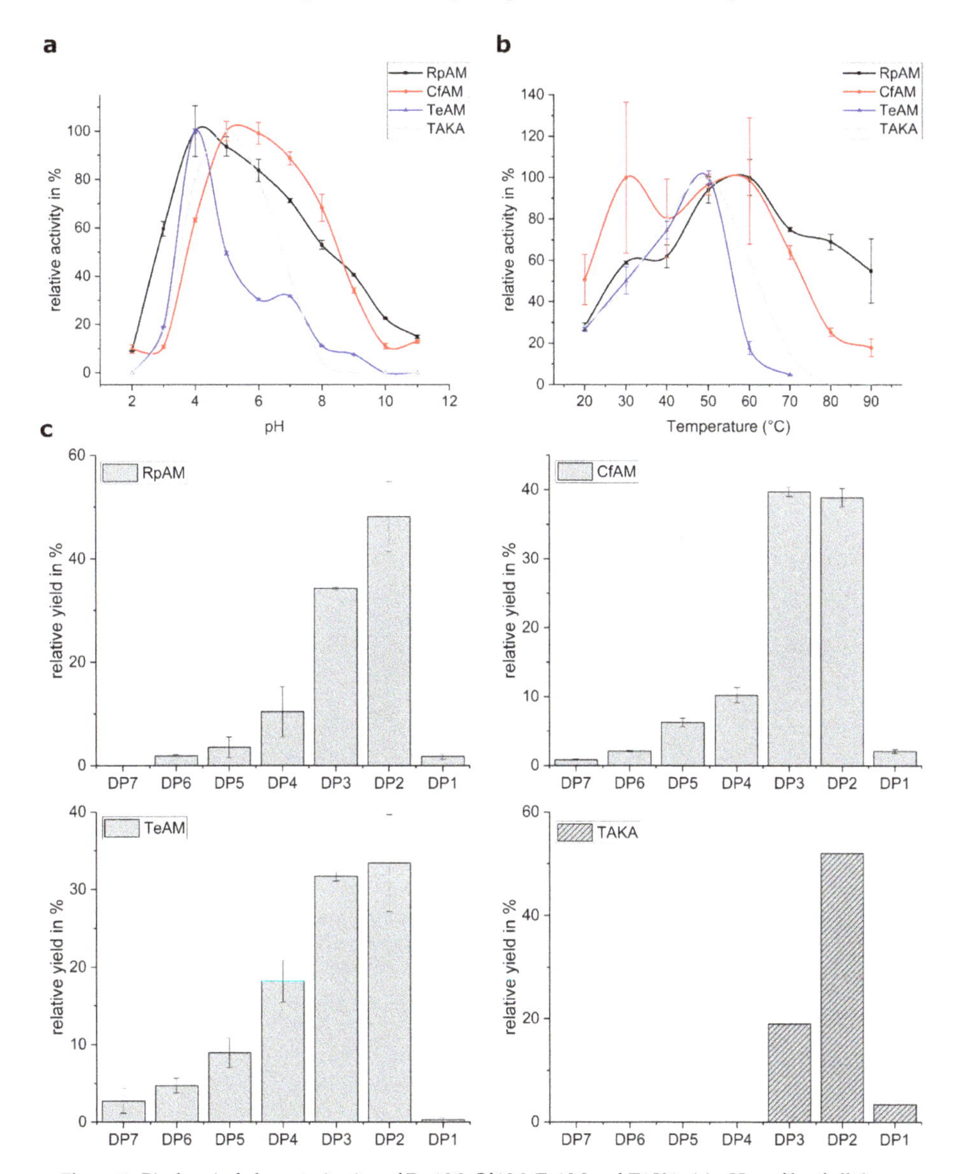

Figure 1. Biochemical characterization of RpAM, CfAM. TeAM and TAKA. (**a**) pH-profile of all three amylases in comparison with TAKA amylase; (**b**) temperature profile of all three amylases in comparison with TAKA amylase; (**c**) product profile of all three amylases and the abundance of oligomers with a degree of polymerization (dp) of 1 to 7 after hydrolysis of starch.

2.2. Overall Fold

The structures were solved using molecular replacement starting from the *A. oryzae* amylase as template (pdb-ID: 7taa and 3vx0) to a resolution of 1.4 Å for RpAM, 1.2 Å for TeAM and 1.35 Å for CfAM, respectively. The final model of RpAM includes two monomers in the asymmetric unit comprising residues 1 to 438 in both chains, which superpose on each other with an r.m.s.d. of 0.54 Å. The model of TeAM contains one monomer in the asymmetric unit including residues 1 to 438. For CfAM, there are two monomers in the asymmetric unit comprising residues 19 to 459 for chain A and 19 to 460 for chain B, which superpose with an r.m.s.d. of 0.3 Å. All three amylases have the classical domain structure with a central $(\beta/\alpha)_8$-barrel with the active site located on its C-terminal face, together with a small subdomain, inserted between the third strand and helix and a C-terminal β-sandwich (Figure 2a). All three superpose with each other (Figure 2b) and with TAKA-amylase with an r.m.s.d. between 0.6 to 0.9 Å for up to 423 residues. Two conserved disulphide bridges stabilize flexible loops in subdomains A and B. There is an additional disulphide bridge in CfAM, located in the C-terminal domain. All three α-amylases have the conserved canonical calcium binding site located between the $(\beta/\alpha)_8$-barrel and the insertion domain B.

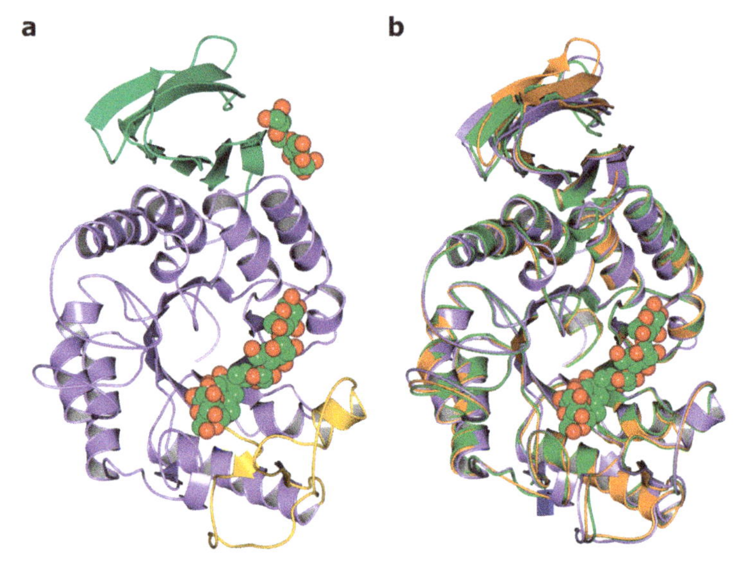

Figure 2. Structural overviews. (**a**) ribbon representation of the structure of CfAM amylase in ribbon representation. The domains are colored separately with the central barrel in purple. subdomain B in yellow and the C-terminal β-sandwich in green. The bound ligands acarbose transglycosylation product (ATgp) and maltose are shown as spheres; (**b**) structural superposition of CfAM (purple) TeAM (orange) and RpAM (green).

2.3. Ligand Binding Site

Although all three amylases were co-crystallized with acarbose, a well-known inhibitor for amylases, a complex with acarbose bound was only obtained for TeAM and CfAM. The reason why acarbose was not bound to RpAM is not clear. As expected, the acarbose was found in the substrate binding cleft in each monomer of TeAM and CfAM, with the acarviosine unit sitting in subsites -1 and +1, (Figure 3a–d). In both enzymes, the binding mode is conserved, and the ligands superpose with each other (Figure 3e), except for the monomer in subsite -4. The distorted pseudosugar valieneamine in subsite 1 with its $^2H_3$ half chair conformation mimics the conformation of the putative transition state along the catalytic itinerary of α-amylases. Additional density in subsites -2 and -3 and -4 was modelled as a second acarbose unit, covalently attached to the first acarbose. The catalytic nucleophile D190/D192(CfAM/TeAM) is in a near attack conformation poised to react with the anomeric carbon, whilst the catalytic acid/base E214/E216(CfAM/TeAM) forms a hydrogen bond with the bridging nitrogen of the glycosidic bond with the 4-deoxyglucose in subsite +1. In addition, a hydrogen bond with H194/H196 stabilizes the 4-deoxyglucose in that subsite. The +3 subsite is formed by the sugar tong, composed of Y142/144 of subdomain B and F216/218 of the central domain, sandwiching the glucose between them. The reducing end of acarbose is stabilized by a hydrophobic platform interaction with Y240/F242 and a hydrogen bond with the main chain nitrogen of G218/G220. Interestingly, additional density at the non-reducing end was observed and was modelled as an additional acarbose unit in subsites −2 and −3 and −4. The glucose in subsite −2 is stabilized by multiple hydrogen bonds with D323/325, R327/329 and W375/377. The glucose in subsite −3 is held in place by only one hydrogen bond with D323/325. The last visible part of the acarbose molecule is the acarviosine unit in subsite −4, which is not stabilized by direct interactions with the protein. Furthermore, the acarviosine unit is in two different positions in the two structures, reflecting the lack of strong stabilizing interactions between the ligand and the protein beyond subsite −3 (Figure 3e).

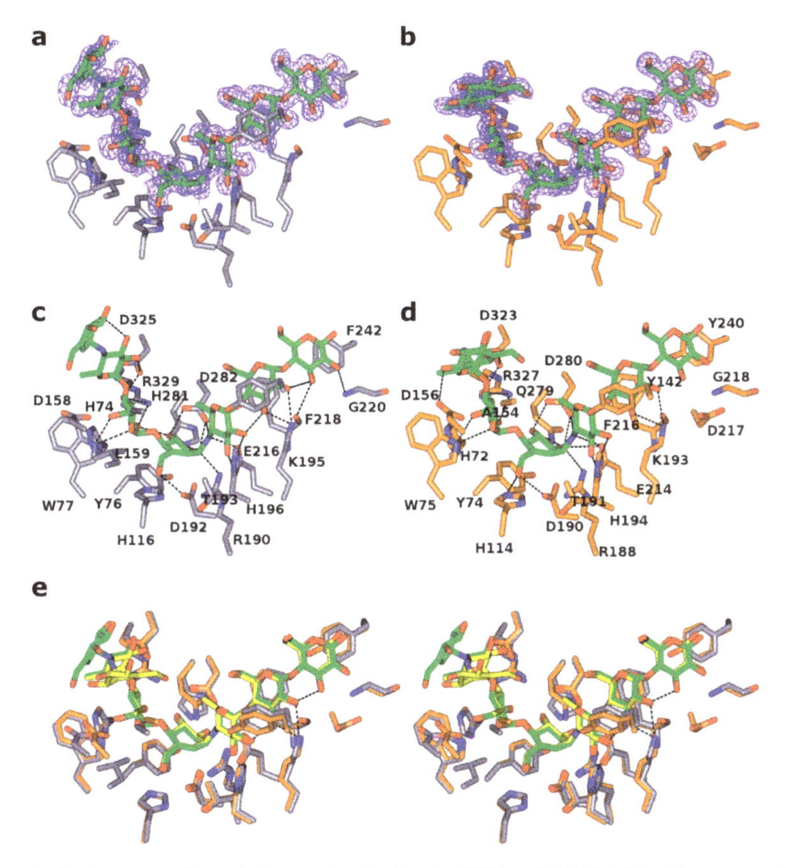

Figure 3. Acarbose transglycosylation product binding in CfAM and TeAM. (**a**,**b**) stick representation of the acarbose derived transglycosylation product in the substrate binding crevice of CfAM and TeAM, respectively. The 2Fo-Fc electron density around the ligands is contoured at 0.3 e/Å^3. The interacting residues are shown as cylinders. (**c**,**d**) hydrogen bonding pattern between ATgp and CfAM and TeAM in the active site. (**e**) stereo view of the overlay of the binding crevice of CfAM (purple) and TeAM (orange). The residues and the ligands overlap very closely with the only major difference being the orientation of the acarviosine subunit in subsite -4.

2.4. Secondary Glucose Binding Site

In CfAM, a secondary binding site in domain C was identified and modelled as maltose located at the edge of the β-sandwich (Figure 4). The glucose units are held in place mainly via hydrogen bonds without the usual stacking interactions with aromatic side chains.

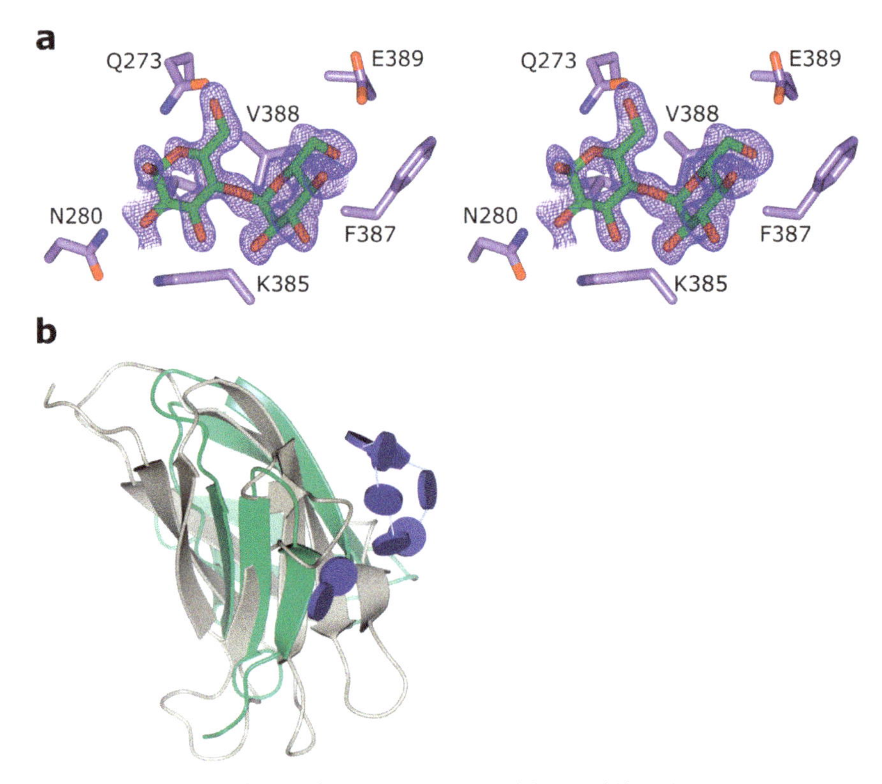

Figure 4. The secondary maltose binding site in the C-terminal domain of CfAM. (**a**) stereo view showing the maltose in cylinder representation with the corresponding 2Fo-Fc electron density contoured at 0.4 e/Å^3. The interacting residues are shown as blue cylinders; (**b**) superposition of the C-terminal domain (green) with the CBM20 domain from *A. niger* glucoamylase (pdb-ID: 1ac0) in beige. The bound β-cyclodextrin of CBM20 and the maltose unit are shown as glycoblocks [15].

2.5. N-Glycosylation

There are three N-glycosylation sites, one at N144 in RpAM and two at N180 and 412 in TeAM. We observed only the core GlcNAc residue in all three enzymes. In the case of TeAM, this is due to the deglycosylation procedure with EndoH.

2.6. Isoasparate Formation

We observed the formation of an isoaspartate by succinimide formation and deamidation of N120 in chain B of RpAM. The same asparagine in chain A shows high flexibility and the resulting density suggest partial isoaspartate formation, but a model could not be built with confidence.

3. Discussion

We have analyzed structurally and functionally three novel fungal α-amylases with potential to be used in the food industry and other industrial processes. All three structures determined show the canonical amylase fold and overlap with each other with an r.m.s.d. of 0.54 Å (Figure 1b). Further analysis of the sequence showed that both RpAM and CfAM have a slightly lower number of charged residues and a higher number of hydrophobic residues compared to TeAm and TAKA amylase, which might contribute to the higher thermostability of these two variants. Increased internal hydrophobicity while keeping external hydrophilicity was found to correlate well with the thermostability of *Bacillus* α-amylases [16]. Furthermore, the shortened loops in these enzymes may also contribute to the overall rigidity of the enzymes and therefore the thermostability as observed for other enzymes as well [17,18].

The substrate crevice in all three amylases, if defined on the basis of protein carbohydrate interactions, spans from subsite -3 to +3. Having only three defined subsites for the non-reducing end is common for amylases and is in line with the number of donor subsites described for the TAKA-amylase. Potentially, there could be more subsites for additional carbohydrate units at the reducing end, which might connect the active site crevice with the observed second binding site (see below).

The observed complexes are most likely the result of limited transglycosylation, an unusual side reaction previously reported *in crystallo* for several amylases—for example, TAKA-amylase [19]. Though this reaction is common in the closely related CGTases (GH13_2) and amylomaltases (GH77), it was not observed in solution for α-amylases. However, in crystals, transglycosylation products with 10 or more units have been reported as a result of multiple transglycosylation events. Interestingly, the final complex always has the pseudosaccharide unit, thought to mimic the transition state, in the -1 subsite, rendering the enzyme inactive. Other binding modes are clearly possible as evidenced by the final product and a pre-Michaelis complex observed for GH77 *Thermus aquaticus* amylomaltase with acarbose [20].

All three amylases have as their hallmark a shortened loop between β2/α3 and two shorter loops in subdomain B located between β3 and α4 of the central $(\beta/\alpha)_8$-barrel, compared to structures of other fungal amylases, e.g., TAKA-amylase (Figure 5a). The importance of subdomain B for the physicochemical properties—for example, pH-stability, as well as substrate and product specificity, is well known [21–24]. Indeed, the shorter loops open up the substrate crevice on the non-reducing end (Figure 5b), which might explain the shift in the product profile for all three amylases towards oligomers with a higher dp compared to TAKA amylase (Figures 1c and 5c).

The C-terminal domain in α-amylases is implicated in starch binding and shows structural similarity to classic CBM domains, based on an analysis using PDBeFOLD [25]. The additional binding site in this domain in CfAM strengthen the role of this domain in substrate binding. Additional carbohydrate binding sites have been observed as well for example in barley α-amylase 1 [26]. While none of these sites overlap with the binding site seen in CfAM, a structure of a CBM20 in complex with β-cyclodextrin revealed two binding sites, with the site termed SB1 in close proximity to the binding site in CfAM (Figure 4b) [27]. This was confirmed to be the primary binding site for the interaction with raw starch, and it is likely that the observed binding site in CfAM is a genuine carbohydrate binding site. Furthermore, it is intriguing to speculate about a potential path from the primary substrate crevice to the secondary glucose binding site, which could be rather easily thought as a simple extension of the acarbose from the reducing end.

Only limited information about the influence of glycosylation on amylase activity is available. It was shown that, for α-amylase, Amy1 from the yeast *Cryptococcus flavus* N-glycosylation enhances thermostability and resistance to proteolytic degradation [28]. The same effect is observed for *Trichoderma reesei* Cel7a [29]. Indeed, N144 is located in an extended loop and N-glycosylation might help to shield the loop against proteolytic attack. The other two glycosylation sites are located in or at the beginning of secondary structure elements, with N412 being located in the C- domain.

The observed isoaspartate formation is thought usually to be an age-related side effect of protein decomposition, but a functional role cannot be ruled out [30]. Indeed, it was shown in GH77 enzymes that such unusual posttranslational rearrangement might play a functional role in glycoside hydrolases [31,32]. The observed isoaspartate is located in one of the shortened loops in subdomain B close to the substrate binding cleft, suggesting a functional role in CfAM as well.

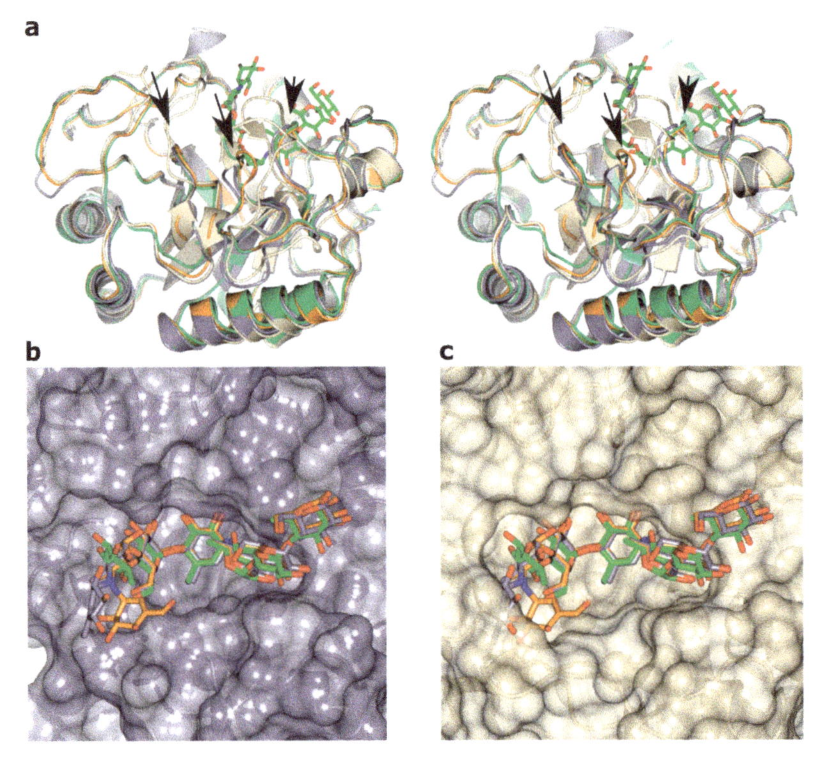

Figure 5. (**a**) Stereo view of all three amylases compared to TAKA-amylase with the three shortened loops in the front marked with arrows. The ligand in CfAM is shown as sticks to identify the active site; (**b**) surface representation of CfAM with the bound ligand. The substrate is more open on the donor subsite; (**c**) surface representation of TAKA-amylase. The elongated loops create a more restricted active site crevice precluding the binding mode observed in CfAM and TeAM due to steric clashes.

4. Materials and Methods

4.1. Macromolecule Production

The coding sequence of CfAM for the A, B and C domains was amplified from *Cordyceps farinosa* gDNA by the polymerase chain reaction (PCR). The PCR fragment was obtained using primer pairs: 5′-ACACAACTGGGGATCCACCATGAAGCTTACTGCGTCCCTC-3′ and 5′-GATGGTGATGGGATCCTTACTGCGCAACAAAAACAATGGG-3′. The fragment was then ligated in the expression vector pSUN515 using *Bam*HI and *Xho*I restriction sites. The ligation protocol was performed according to the IN-FUSION™ Cloning Kit instructions. A transformation of TOP10 competent *E. coli* cells (Tiangen, Beijing China) with the plasmid, containing the CfAM gene, was performed and positive clones confirmed by sequencing. The transformation of *Aspergillus oryzae* (strain *MT3568*) with the expression vector comprising CfAM gene was performed according to patent application WO95/002043 [33]. After incubation for 4–7 days at 37 °C, spores of four transformants were inoculated into 3 mL of YPM medium. After 3-day cultivation at 30 °C, the culture broths were analyzed by SDS-PAGE to identify the transformant producing the largest amount of recombinant mature amylase with an estimated size of 48 kDa. Spores from the best expressing transformant were cultivated in YPM medium in shake flasks for 4 days at a temperature of 30 °C. The culture broth was harvested by filtration using a 0.2 μm filter device, and the filtered fermentation broth was used for purification and further assays.

RpAM was cloned and expressed in a similar manner as CfAM while TeAM was expressed in *Pichia pastoris* with a similar protocol to that described for the lipase from *Gibberella zeae* [34]. The entire coding sequence of TeAM was amplified from cDNA by the polymerase chain reaction and transformation into ElectroMax DH10B competent cells (Invitrogen, Waltham, MA, USA) by electroporation. Transformed cells were plated on LB plates containing 100 mM ampicillin. After overnight incubation at 27 °C, a positive clone was selected by colony PCR and confirmed by sequencing. The plasmid DNA of the positive clone was linearized with PmeI (NEB, Ipswich, MA, USA) and transformed into *Pichia pastoris* KM71 (Invitrogen, Waltham, MA, USA) following the manufacturer's instructions. An amylase positive clone was inoculated into 3 mL buffered minimal sorbitol complex medium and incubated at 28 °C for 3 days until the OD600 reached 20. Methanol was added to the culture daily to a final concentration of 0.5% for the following 4 days. On day 4 of induction, the culture supernatant was separated from the cells by centrifugation and the pH of the supernatant was adjusted to 7.0.

The CfAM culture broth was precipitated with ammonium sulphate (80% saturation), then dialyzed with 20 mM Na-acetate at pH 5.0. The solution was loaded on to a Q Sepharose Fast Flow column (GE Healthcare, Brondby, Denmark) equilibrated with 20 mM Na Acetate at pH 5.0. Protein was eluted with a salt gradient from zero to 1 M NaCl Fractions were analyzed for amylase activity and pooled accordingly. The flow-through fraction, containing the bulk of amylase activity was supplemented with ammonium sulphate to a final concentration of 1.2 M and then loaded on to Phenyl Sepharose 6 Fast Flow column (GE Healthcare, Brondby, Denmark). The activity was eluted by a linear gradient of decreasing salt concentration. The fractions with activity were analyzed by SDS-PAGE and then concentrated for further use.

Amylase activity was detected by Azo dyed and azurine cross-linked hydroxyethyl-amylose (AZCL-HE-amylose) (Megazyme International Ireland Ltd., Bray, Ireland) as substrate. In addition, 10µL enzyme sample and 120 µL 0.1% substrate at pH 7 were mixed in a microtiter plate and incubated at 50 °C for 30 min. Then, 70 µL supernatant was transferred to a new microtiter plate and the absorption at 595 nm determined. All reactions were done as duplicates.

4.2. Biochemical Characterisation

4.2.1. pH Optimum

To determine the pH Optimum, each enzyme (3 µL of a 0.5 mg/mL solution) was incubated with 40 µL 1% substrate (AZCL-HE-amylose) (Megazyme International Ireland Ltd., Bray, Ireland). The pH between 2 and 11 was adjusted using 100 µL of B&R buffer (Britton–Robinson buffer: 0.1 M boric acid, 0.1 M acetic acid, and 0.1 M phosphoric acid, adjusted to pH-values 3.0, 4.0, 5.0, 6.0, 7.0, 8.0, 9.0, 10.0 and 11.0 with HCl or NaOH) [35]. The reactions were incubated at 30 °C for 30 min and afterwards 60 µL were transferred in a new microtiter plate and the absorption was measured at 595 nm.

4.2.2. Temperature Optimum

To determine the Temperature Optimum, each enzyme was incubated with 100 µL 0.1% substrate (AZCL-HE-amylose) (Megazyme International Ireland Ltd., Bray, Ireland) in 50 mM Na Acetate pH 4.3. The substrate solution was preincubated at 20–90 °C for 5 min and the reaction was started by addition of 3 µL of enzyme solution (0.5 mg/mL). The reaction mixture was further incubated at the respective temperature for 30 min at 950 rpm. The reaction was stopped by rapid cooling on ice. Afterwards, 60 µL of each reaction was transferred in a microtiterplate and the absorption was measured at 595 nm. Each reaction was performed in triplicate.

4.2.3. Product Profile

For product profile determination, each enzyme (15 µL) was incubated with 120 µL 0.1% substrate (AZCL-HE-amylose) (Megazyme International Ireland Ltd., Bray, Ireland) at pH 5 and 62 °C for 14 h.

70 µL of each reaction was mixed with equal amounts of Acetonitril. The mixture was centrifuged for 30 min at 16.000× *g* and the supernatant was analyzed using HPAEC with pulsed amperometric detection.

4.3. Crystallisation

4.3.1. RpAM

The concentrated protein was mixed with acarbose in a molar ration of 4:1 before the initial screening in 96 well format using commercially available screens. An initial hit (0.2 M NaCl, 0.1 M Na-acetate pH 4.6, 30% MPD) was further refined in 24-well format using the initial crystals as seeds. Crystals suitable for data collection were cryoprotected using 25% glycerol and flash frozen in liquid nitrogen prior data collection.

4.3.2. TeAM

Prior to providing the sample to York, the protein was deglycosylated using Endo-H treatment. The protein was concentrated using Amicon (Merck, Germany) filter units and stored at −80 °C for later use. For the crystallization, the protein was mixed with 5 mM acarbose prior to setting up the screen. Initial screens were set up in a 96-well sitting drop format using commercially available screens. Initial hits were further refined in a 24-well hanging drop format. The best crystals grew in 0.1 M di-hydrogen phosphate, 1.8 M ammonium sulphate. Crystals were cryoprotected by addition of ethylene glycol to a final concentration of 15%. The crystals were flash frozen in liquid nitrogen prior to data collection.

4.3.3. CfAM

Prior to crystallization, the protein was concentrated to 22.5 mg/mL by ultrafiltration in an Amicon centrifugation filter unit (Millipore), aliquoted to 50 µL; aliquots that were not immediately set up for crystallization were flash frozen in liquid nitrogen and stored at −80°C to use later in optimizations. Initial crystallization experiments were carried out in the presence or absence of 4 mM $CaCl_2$ and 40 mM acarbose. An initial hit was obtained for an acarbose complex, in just one condition (H3, Bis-tris 5.5, 25% w/v PEG3350) of JCSG screen (Figure 6a), out of total 192 conditions in two initial screens set up – JCSG and PACT premier™ HT-96 (Molecular Dimensions (Suffolk, UK)). The crystals were imperfect and were used to make the seeding stock. The seeding stock was prepared and microseed matrix screening (MMS, recent review in [14]) carried out using an Oryx robot (Douglas Instruments (Hungerford, UK)) according to the published protocols [36,37]. Briefly, crystals were crushed, and diluted with ~50 µL of mother liquor. The solution was transferred into a seed bead containing reaction tube and vortexed for three minutes. The seeding stock was used straightaway, and the remaining seeds were frozen and kept at ™20 °C. MMS was carried out in the PACT screen, giving a significant number of hits (Figure 6b). Crystals from condition A11 were used to make a seeding stock for the next seeding round. This time it was not a "classical" MMS-seeding into a random screen, but rather seeding into an optimization screen based on the initial conditions, but with different pH, salts and PEGs/PEG concentrations. The crystallization drops contained 150 nl protein + 50 nl seeding stock + 100 nl mother liquor from a new random screen. The final, good quality crystal was obtained in 12% PEG 3350 0.2 M $NaNO_3$, CAPS pH 11.0 (Figure 6c).

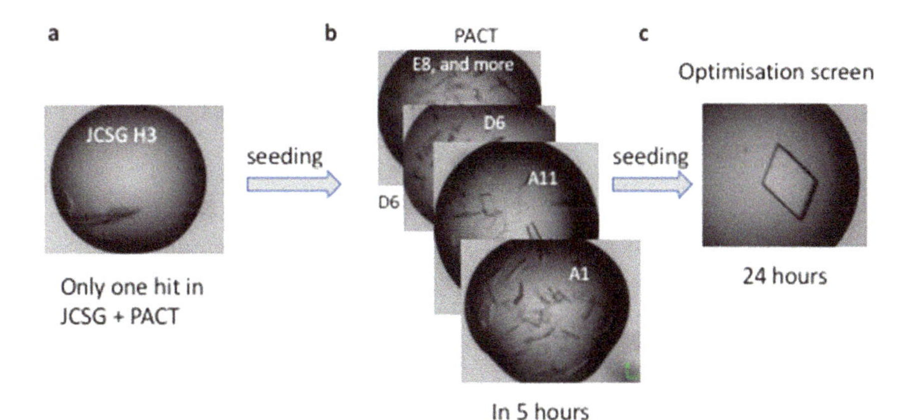

Figure 6. Crystal optimization using microseed matrix screening.

4.4. Data Collection and Processing

The data were collected at Diamond on beam line I02, processed by XDS [38], and scaled with Aimless [39]. The statistics are shown in Table 1.

Table 1. Data collection and processing statistics.

	CfAM	TeAM	RpAM
Diffraction source	Diamond I02	Diamond I02	ESRF ID29
Wavelength (Å)	0.9795	0.9795	1.0004
Temperature (K)	100	100	100
Space group	P1	$P2_12_12_1$	P1
a, *b*, *c* (Å)	56.88, 61.97, 70.40	51.02, 56.63, 166.01	51.22, 62.60, 66.81
α, β, γ (°)	79.33, 82.88, 67.99	90, 90, 90	77.03, 81.04, 89.62
Resolution range (Å)	33.1–1.35 (1.37–1.35)	48.76–1.20 (1.22-1.20)	43.21–1.4 (1.42–1.40)
Total No. of reflections	342,708	1,149,540	315,876
No. of unique reflections	163,777	150,529	146,177
Completeness (%)	85.3 (38.2)	99.8 (96.7)	92.9 (61.9)
Redundancy	2.1 (2.1)	7.6 (4.6)	2.2 (2.1)
$\langle I/\sigma(I)\rangle$	13.7 (10.3)	14.1(1.7)	9.6 (2.3)
$R_{r.i.m.}$	0.076 (0.129)	0.021 (0.446)	0.030 (0.225)
CC1/2	0.983 (0.970	0.999 (0.615)	0.998 (0.892)
Overall *B* factor from Wilson plot (Å^2)	6.8	8.7	8.1

Values for the outer shell are given in parentheses.

4.5. Structure Solution and Refinement

The structure of RpAM was solved by molecular replacement with Molrep [40], using TAKA amylase as template (PDB-ID:7taa). The structure of TeAM was solved with Molrep using the model of RpAM. The CfAM structure was solved using Molrep [40] with 3vx0 α-amylase from *Aspergillus oryzae* as a model. The final models were built using automated chain tracing with Buccaneer [41], followed by manual building in Coot [42], iterated with reciprocal space refinement using Refmac [43]. The statistics are summarized in Table 2.

Table 2. Structure solution and refinement.

	CfAM	TeAM	RpAM
PDB-ID	6SAV	6SAO	6SAU
Resolution range (Å)	33.1–1.35 (1.385–1.35)	48.76–1.20 (1.22–1.20)	39.99–1.4 (1.42–1.40)
Completeness (%)	85.3 (39.7)	99.8 (96.7)	92.8 (89.2)
No. of reflections, working set	155,488	143,033	138,848
No. of reflections, test set	8289	7574	7328
Final R_{cryst}	0.113 (0.09)	0.110 (0.27)	0.136 (0.22)
Final R_{free}	0.150 (0.17)	0.134 (0.29)	0.164 (0.26)
Cruickshank DPI	0.051	0.027	0.056
No. of subunits in the asymmetric unit	2	1	2
No. of non-H atoms	Chain A/B	Chain A	Chain A/B
Protein	3557/3609	3570	3662/3592
Ion	1/2	1	1/1
Ligand	99/120	133	14/36
Water	875	568	943
Total	8263	4272	8306
R.m.s. deviations			
Bonds (Å)	0.0191	0.0163	0.0147
Angles (°)	2.06	1.937	1.875
Average B factors (Å^2)	Chain A/B	Chain A	Chain A/B
Protein	10/8.7	12.9	12.9/12.3
Ions			
Ca^{2+}	6.7/5.8	9.5	7.39/7.5
Na^{2+}	N/A/10.9		
Ligand	19.6/18.0	22.2	19.2/21.4
Water	19.0	28.8	24.21
Ramachandran plot			
Most favoured (%)	98.6	97.7	97.2
Allowed (%)	1.4	2.3	2.7

Values for the outer shell are given in parentheses.

5. Conclusions

Taken together, we describe the structural and functional characterization of three novel fungal α- amylases with enhanced stability, of which two, CfAM and RpAM, have a higher pH optimum and greater temperature tolerance, well suited for usage in the detergent or saccharification industry. The structures reveal that these amylases follow the canonical domain structure of α-amylases, and that three shortened loops between β_2/α_3 and in subdomain B are likely to be responsible for the altered enzymatic properties of the amylases compared to TAKA-amylase. For the first time, we have unambiguously identified up to three different N-glycosylation sites in α-amylases in the structures. Furthermore, the observed formation of an isoaspartate from an asparagine in one of the shortened loops might play a functional role. The complexes with acarbose derived transglycosylation products define seven subsites of the substrate binding crevice and helped to identify the catalytic residues unambiguously. In addition, a new previously unobserved carbohydrate binding site was revealed in the C-terminal β-sandwich domain of CfAM, which might be important for the initial interaction with its polymeric substrate.

6. Patents

The *Rhizomucor pusillus* amylase and the use of this amylase in various industrial applications have been claimed in patent application WO2006065579. A close homologue of the *Thamnidium elegans* amylase was claimed in patent application WO2006069290 including the use in industrial applications.

Author Contributions: C.R. and O.V.M. analyzed the data, built and refined the structure and prepared the original draft. J.P.T. collected and analyzed the X-ray data. O.V.M., E.B., A.A. and J.W. crystallized the amylases and solved the initial structures. L.M. and S.T. cloned, produced, purified and characterized the amylases biochemically, C.R., C.A., G.J.D. and K.S.W. wrote, analyzed, and reviewed all stages of the manuscript. G.J.D. C.A. and K.S.W. planned and supervised the work.

Funding: This research received no external funding.

Acknowledgments: The authors are grateful for financial support by Novozymes. We thank ESRF for the access to beamline ID29 and Diamond Light Source for access to beamline I02 (proposal numbers mx-1221 and mx-9948) that contributed to the results presented here. The authors also thank Sam Hart for assistance during data collection.

Conflicts of Interest: The authors declare no conflict of interest, but we note that the Novozymes authors declare the following competing financial interest(s): Novozymes are a commercial enzyme supplier. Novozymes, provided the enzyme samples used for crystallization, did the functional characterization and provided the financial support for the project.

Abbreviations

CfAM.	*Cordyceps farinosa* amylase
RpAM	*Rhizomucor pusillus* amylase
TeAM	*Thamnidium elegans* amylase
TAKA	*Aspergillus oryzae* amylase
dp	Degree of polymerization

References

1. Payen, A.P.J.F. Memoire sur la diastase, les principaux produits de ses réactions et leurs applications aux arts industriels" (Memoir on diastase, the principal products of its reactions, and their applications to the industrial arts). *Annal. Chim. Phys.* **1833**, *2*, 73–92.
2. Gurung, N.; Ray, S.; Bose, S.; Rai, V. A broader view: Microbial enzymes and their relevance in industries, medicine, and beyond. *Biomed Res. Int.* **2013**, *2013*, 329121. [CrossRef] [PubMed]
3. Roy, J.K.; Manhar, A.K.; Nath, D.; Mandal, M.; Mukherjee, A.K. Cloning and extracellular expression of a raw starch digesting alpha-amylase (Blamy-I) and its application in bioethanol production from a non-conventional source of starch. *J. Basic Microbiol.* **2015**, *55*, 1287–1298. [CrossRef] [PubMed]
4. Gupta, R.; Gigras, P.; Mohapatra, H.; Goswami, V.K.; Chauhan, B. Microbial α-amylases: A biotechnological perspective. *Process Biochem.* **2003**, *38*, 1599–1616. [CrossRef]
5. Niehaus, F.; Bertoldo, C.; Kahler, M.; Antranikian, G. Extremophiles as a source of novel enzymes for industrial application. *Appl. Microbiol. Biotechnol.* **1999**, *51*, 711–729. [CrossRef]
6. Lombard, V.; Golaconda Ramulu, H.; Drula, E.; Coutinho, P.M.; Henrissat, B. The carbohydrate-active enzymes database (CAZy) in 2013. *Nucleic Acids Res.* **2014**, *42*, D490–D495. [CrossRef]
7. Janecek, S.; Svensson, B.; MacGregor, E.A. Structural and evolutionary aspects of two families of non-catalytic domains present in starch and glycogen binding proteins from microbes, plants and animals. *Enzyme Microb. Technol.* **2011**, *49*, 429–440. [CrossRef] [PubMed]
8. Liu, Y.; Yu, J.; Li, F.; Peng, H.; Zhang, X.; Xiao, Y.; He, C. Crystal structure of a raw-starch-degrading bacterial alpha-amylase belonging to subfamily 37 of the glycoside hydrolase family GH13. *Sci. Rep.* **2017**, *7*, 44067. [CrossRef]
9. Mehta, D.; Satyanarayana, T. Domain C of thermostable alpha-amylase of Geobacillus thermoleovorans mediates raw starch adsorption. *Appl. Microbiol. Biotechnol.* **2014**, *98*, 4503–4519. [CrossRef]
10. Sogaard, M.; Kadziola, A.; Haser, R.; Svensson, B. Site-directed mutagenesis of histidine 93, aspartic acid 180, glutamic acid 205, histidine 290, and aspartic acid 291 at the active site and tryptophan 279 at the raw starch binding site in barley alpha-amylase 1. *J. Biol. Chem.* **1993**, *268*, 22480–22484.
11. Kadziola, A.; Sogaard, M.; Svensson, B.; Haser, R. Molecular structure of a barley alpha-amylase-inhibitor complex: Implications for starch binding and catalysis. *J. Mol. Biol.* **1998**, *278*, 205–217. [CrossRef] [PubMed]
12. Brzozowski, A.M.; Lawson, D.M.; Turkenburg, J.P.; Bisgaard-Frantzen, H.; Svendsen, A.; Borchert, T.V.; Dauter, Z.; Wilson, K.S.; Davies, G.J. Structural analysis of a chimeric bacterial alpha-amylase. High-resolution analysis of native and ligand complexes. *Biochemistry* **2000**, *39*, 9099–9107. [CrossRef] [PubMed]

13. Pritchard, P.E. Studies on the bread-improving mechanism of fungal alpha-amylase. *J. Biol. Educ.* **1992**, *26*, 12–18. [CrossRef]
14. D'Arcy, A.; Bergfors, T.; Cowan-Jacob, S.W.; Marsh, M. Microseed matrix screening for optimization in protein crystallization: What have we learned? *Acta Crystallogr F Struct Biol Commun* **2014**, *70*, 1117–1126. [CrossRef]
15. McNicholas, S.; Agirre, J. Glycoblocks: A schematic three-dimensional representation for glycans and their interactions. *Acta Crystallogr. D Struct. Biol.* **2017**, *73*, 187–194. [CrossRef] [PubMed]
16. Janeček, Š. Does the increased hydrophobicity of the interior and hydrophilicity of the exterior of an enzyme structure reflect its increased thermostability? *Int. J. Biol. Macromol.* **1993**, *15*, 317–318. [CrossRef]
17. Mok, S.C.; Teh, A.H.; Saito, J.A.; Najimudin, N.; Alam, M. Crystal structure of a compact alpha-amylase from Geobacillus thermoleovorans. *Enzyme Microb. Technol.* **2013**, *53*, 46–54. [CrossRef]
18. Mazola, Y.; Guirola, O.; Palomares, S.; Chinea, G.; Menendez, C.; Hernandez, L.; Musacchio, A. A comparative molecular dynamics study of thermophilic and mesophilic beta-fructosidase enzymes. *J. Mol. Model.* **2015**, *21*, 228. [CrossRef]
19. Brzozowski, A.M.; Davies, G.J. Structure of the Aspergillus oryzae alpha-amylase complexed with the inhibitor acarbose at 2.0 A resolution. *Biochemistry* **1997**, *36*, 10837–10845. [CrossRef]
20. Przylas, I.; Terada, Y.; Fujii, K.; Takaha, T.; Saenger, W.; Strater, N. X-ray structure of acarbose bound to amylomaltase from Thermus aquaticus. Implications for the synthesis of large cyclic glucans. *Eur. J. Biochem.* **2000**, *267*, 6903–6913. [CrossRef]
21. Juge, N.; Rodenburg, K.W.; Guo, X.J.; Chaix, J.C.; Svensson, B. Isozyme hybrids within the protruding third loop domain of the barley alpha-amylase (beta/alpha)8-barrel. Implication for BASI sensitivity and substrate affinity. *FEBS Lett.* **1995**, *363*, 299–303. [CrossRef]
22. Rodenburg, K.W.; Juge, N.; Guo, X.J.; Sogaard, M.; Chaix, J.C.; Svensson, B. Domain B protruding at the third beta strand of the alpha/beta barrel in barley alpha-amylase confers distinct isozyme-specific properties. *Eur. J. Biochem.* **1994**, *221*, 277–284. [CrossRef] [PubMed]
23. Penninga, D.; Strokopytov, B.; Rozeboom, H.J.; Lawson, C.L.; Dijkstra, B.W.; Bergsma, J.; Dijkhuizen, L. Site-directed mutations in tyrosine 195 of cyclodextrin glycosyltransferase from Bacillus circulans strain 251 affect activity and product specificity. *Biochemistry* **1995**, *34*, 3368–3376. [CrossRef] [PubMed]
24. Nakamura, A.; Haga, K.; Yamane, K. Four aromatic residues in the active center of cyclodextrin glucanotransferase from alkalophilic Bacillus sp. 1011: Effects of replacements on substrate binding and cyclization characteristics. *Biochemistry* **1994**, *33*, 9929–9936. [CrossRef]
25. Krissinel, E. On the relationship between sequence and structure similarities in proteomics. *Bioinformatics* **2007**, *23*, 717–723. [CrossRef] [PubMed]
26. Robert, X.; Haser, R.; Gottschalk, T.E.; Ratajczak, F.; Driguez, H.; Svensson, B.; Aghajari, N. The structure of barley alpha-amylase isozyme 1 reveals a novel role of domain C in substrate recognition and binding: A pair of sugar tongs. *Structure* **2003**, *11*, 973–984. [CrossRef]
27. Sorimachi, K.; Le Gal-Coeffet, M.F.; Williamson, G.; Archer, D.B.; Williamson, M.P. Solution structure of the granular starch binding domain of Aspergillus niger glucoamylase bound to beta-cyclodextrin. *Structure* **1997**, *5*, 647–661. [CrossRef]
28. de Barros, M.C.; do Nascimento Silva, R.; Ramada, M.H.; Galdino, A.S.; de Moraes, L.M.; Torres, F.A.; Ulhoa, C.J. The influence of N-glycosylation on biochemical properties of Amy1, an alpha-amylase from the yeast Cryptococcus flavus. *Carbohydr. Res.* **2009**, *344*, 1682–1686. [CrossRef]
29. Amore, A.; Knott, B.C.; Supekar, N.T.; Shajahan, A.; Azadi, P.; Zhao, P.; Wells, L.; Linger, J.G.; Hobdey, S.E.; Vander Wall, T.A.; et al. Distinct roles of N- and O-glycans in cellulase activity and stability. *Proc. Natl. Acad. Sci. USA* **2017**, *114*, 13667–13672. [CrossRef]
30. Reissner, K.J.; Aswad, D.W. Deamidation and isoaspartate formation in proteins: Unwanted alterations or surreptitious signals? *Cell. Mol. Life Sci.* **2003**, *60*, 1281–1295. [CrossRef]
31. Barends, T.R.; Bultema, J.B.; Kaper, T.; van der Maarel, M.J.; Dijkhuizen, L.; Dijkstra, B.W. Three-way stabilization of the covalent intermediate in amylomaltase, an alpha-amylase-like transglycosylase. *J. Biol. Chem.* **2007**, *282*, 17242–17249. [CrossRef] [PubMed]
32. Roth, C.; Weizenmann, N.; Bexten, N.; Saenger, W.; Zimmermann, W.; Maier, T.; Strater, N. Amylose recognition and ring-size determination of amylomaltase. *Sci. Adv.* **2017**, *3*, e1601386. [CrossRef] [PubMed]

33. Dalboege, H.; Christgau, S.; Andersen, L.N.; Kofod, L.V.; Kauppinen, M.S. DNA encoding an enxyme with endoglucanase activity from Trichoderma harzianum. World Patent WO/1995/002043, 1995.
34. Sun, Y.; Li, M.; Zhang, Y.; Liu, L.; Liu, Y.; Liu, Z.; Li, X.; Lou, Z. Crystallization and preliminary crystallographic analysis of Gibberella zeae extracellular lipase. *Acta Crystallogr. Sect. F Struct. Biol. Cryst Commun.* **2008**, *64*, 813–815. [CrossRef] [PubMed]
35. Britton, H.T.S.; Robinson, R.A. CXCVIII.—Universal buffer solutions and the dissociation constant of veronal. *J. Chem. Soc.* **1931**, 1456–1462. [CrossRef]
36. Shaw Stewart, P.D.; Kolek, S.A.; Briggs, R.A.; Chayen, N.E.; Baldock, P.F.M. Random Microseeding: A Theoretical and Practical Exploration of Seed Stability and Seeding Techniques for Successful Protein Crystallization. *Cryst. Growth Des.* **2011**, *11*, 3432–3441. [CrossRef]
37. Shah, A.K.; Liu, Z.-J.; Stewart, P.D.; Schubot, F.D.; Rose, J.P.; Newton, M.G.; Wang, B.-C. On increasing protein-crystallization throughput for X-ray diffraction studies. *Acta Crystallogr. Sect. D* **2005**, *61*, 123–129. [CrossRef] [PubMed]
38. Kabsch, W. Xds. *Acta Crystallogr. D Biol. Crystallogr.* **2010**, *66*, 125–132. [CrossRef] [PubMed]
39. Evans, P.R.; Murshudov, G.N. How good are my data and what is the resolution? *Acta Crystallogr. D Biol. Crystallogr.* **2013**, *69*, 1204–1214. [CrossRef]
40. Vagin, A.; Teplyakov, A. Molecular replacement with MOLREP. *Acta Crystallogr. D Biol. Crystallogr.* **2010**, *66*, 22–25. [CrossRef]
41. Cowtan, K. The Buccaneer software for automated model building. 1. Tracing protein chains. *Acta Crystallogr. D Biol. Crystallogr.* **2006**, *62*, 1002–1011. [CrossRef]
42. Emsley, P.; Lohkamp, B.; Scott, W.G.; Cowtan, K. Features and development of Coot. *Acta Crystallogr. D Biol. Crystallogr.* **2010**, *66*, 486–501. [CrossRef] [PubMed]
43. Murshudov, G.N.; Vagin, A.A.; Dodson, E.J. Refinement of macromolecular structures by the maximum-likelihood method. *Acta Crystallogr. D Biol. Crystallogr.* **1997**, *53*, 240–255. [CrossRef] [PubMed]

International Journal of
Molecular Sciences

Article

Directed Evolution of *Clostridium thermocellum* β-Glucosidase A Towards Enhanced Thermostability

Shahar Yoav [1,2], Johanna Stern [2], Orly Salama-Alber [2], Felix Frolow [3,†], Michael Anbar [2], Alon Karpol [4], Yitzhak Hadar [1], Ely Morag [2] and Edward A. Bayer [2,*]

1. Department of Plant Pathology and Microbiology, Robert H. Smith Faculty of Agriculture, Food and Environment, the Advanced School for Environmental Studies, The Hebrew University of Jerusalem, Rehovot 76100, Israel; shaharyoav@gmail.com (S.Y.); hadar@agri.huji.ac.il (Y.H.)
2. Department of Biomolecular Sciences, The Weizmann Institute of Science, Rehovot 7610001, Israel; jostern@rcip.co.il (J.S.); orlysala@gmail.com (O.S.-A.); michaelanbar@gmail.com (M.A.); elymorag11@gmail.com (E.M.)
3. Department of Molecular Microbiology and Biotechnology, Tel Aviv University, Tel Aviv 6997801, Israel; hanbayer@gmail.com
4. CelDezyner, 2 Bergman St, Tamar Science Park, Rehovot 7670504, Israel; alonkarpol@gmail.com

* Correspondence: ed.bayer@weizmann.ac.il; Tel.: +972-8-934-2373

† Deceased, 29 August 2014.

Received: 4 September 2019; Accepted: 20 September 2019; Published: 23 September 2019

Abstract: β-Glucosidases are key enzymes in the process of cellulose utilization. It is the last enzyme in the cellulose hydrolysis chain, which converts cellobiose to glucose. Since cellobiose is known to have a feedback inhibitory effect on a variety of cellulases, β-glucosidase can prevent this inhibition by hydrolyzing cellobiose to non-inhibitory glucose. While the optimal temperature of the *Clostridium thermocellum* cellulosome is 70 °C, *C. thermocellum* β-glucosidase A is almost inactive at such high temperatures. Thus, in the current study, a random mutagenesis directed evolutionary approach was conducted to produce a thermostable mutant with K_{cat} and K_m, similar to those of the wild-type enzyme. The resultant mutant contained two mutations, A17S and K268N, but only the former was found to affect thermostability, whereby the inflection temperature (T_i) was increased by 6.4 °C. A17 is located near the central cavity of the native enzyme. Interestingly, multiple alignments revealed that position 17 is relatively conserved, whereby alanine is replaced only by serine. Upon the addition of the thermostable mutant to the *C. thermocellum* secretome for subsequent hydrolysis of microcrystalline cellulose at 70 °C, a higher soluble glucose yield (243%) was obtained compared to the activity of the secretome supplemented with the wild-type enzyme.

Keywords: Cellulase; random mutagenesis; cellulose degradation; structural analysis

---

1. Introduction

Cellulose, the major polymer in the plant cell wall, is the most abundant organic resource on Earth, and cellulosic is a primary feedstock for the production of ethanol-based biofuels. Cellulose is a highly crystalline homopolymer composed of individual chains of glucose, which form a planar structure, reinforced by inter and intra-chain hydrogen bond interactions [1]. Depending on the source, each chain contains from 100 to more than 10,000 glucose units, with the disaccharide cellobiose (composed of two glucose units oriented at 180° along the chain axis) being its smallest repetitive unit [2]. In order to utilize cellulose as a resource for biofuel production, the chain must first be enzymatically hydrolyzed into its primary monomeric glucose units. The glucose is then used as a carbon source for alcoholic fermentation to produce bioethanol. Alternative fermentation processes can also be used for the production of various other biochemicals such as butanol, acetone, lactic acid, succinic acid and more. Thus, efficient enzymatic hydrolysis of the cellulose is crucial for increasing the cost-effectiveness of the bioethanol and biochemical production process [3,4].

In the plant cell wall, cellulose is encompassed by the hemicelluloses and lignin, which together create a chemically complex and recalcitrant structure [5,6]. The plant cell wall structure and the rigid nature of cellulose itself render the cellulose degradation process extremely difficult. For efficient degradation, a diverse set of plant cell wall-degrading enzymes is therefore required [7,8]. Cellulose hydrolysis is mediated by three major types of enzymes: endoglucanases, exoglucanases and β-glucosidases that work in synergy [9]. Endoglucanases can cleave the cellulosic chains in the middle, creating reducing and non-reducing ends. Exoglucanases hydrolyze the cellulosic chains from the newly formed chain ends in a "processive" (sequential) manner, leading to the formation of the soluble disaccharide cellobiose. Finally, β-glucosidases are capable of cleaving cellobiose into soluble glucose units. Cellobiose is known to serve as a strong feedback inhibitor (mainly for exoglucanases) [10], thus highlighting the significance of the β-glucosidases not only in providing the end product (glucose) but also in preventing feedback inhibition. For example, near-complete inhibition of the cellulosome of *Clostridium thermocellum* was observed at a concentration of only 2% cellobiose [11,12]. The addition of its native β-glucosidase A (BglA) was able to relieve inhibition, and thus to enhance the rate and degree of solubilization of crystalline cellulose [13,14]. The important role of BglA was also demonstrated by manipulating *C. thermocellum* 27405 to overexpress the BglA gene in vivo. The resultant strain demonstrated increased total cellulolytic activity during growth [15].

Extensive efforts have been made during the last decades for the development and assembly of efficient cellulolytic enzymatic cocktails. However, cellulose degradation is still not efficient enough to be cost effective [3,16]. One of the key bottlenecks for achieving cost-effective degradation of plant cell wall biomass is the requirement for large amounts of cellulases (about 100–200 g of cellulase per gallon of cellulosic ethanol) [17]. In this context, thermostable enzymes are gaining wide interest in the industry, since they are better suited for harsh process conditions, such as those used for the bioethanol production [18]. Thermostability of enzymes can be increased by genetic modification. Increasing the thermostability of an enzyme, while retaining its activity, is thought to enhance its overall performance, especially for the extended time periods necessary for degradation of cellulosic substrates [19,20]. Moreover, thermostable enzymes can be recycled more efficiently, thereby lowering overall production costs [21,22]. Finally, thermostable processes can reduce contamination [23]. Taken together, engineering thermostable enzymes is important to achieve the relatively low-cost biodegradation of biomass for the production of cellulosic ethanol [24]. Indeed, a wide range of bacterial or fungal cellulases was subjected to genetic modification to increase their thermostability [24–28].

C. thermocellum is a thermophilic bacterium, and its cellulosome is considered to be one of the most efficient natural systems for cellulose conversion [29]. Our group and others have previously reported the design of highly thermostable mutants derived from *C. thermocellum* cellulolytic enzymes, such as the endoglucanase Cel8A [30–32] and the exoglucanase Cel48S [33], which have proved to be stable at very high temperature ranges (around 80 °C). In nature, *C. thermocellum* utilizes cell-surface-bound cellulosomes to hydrolyze the cellulose into soluble cellobiose and other cellodextrins, which are then actively transported into the cells and hydrolyzed to glucose by periplasmic β-glucosidase [10,34]. The maximal cellulose degradation capacity mediated by the *C. thermocellum* cellulosome could be achieved at 70 °C [35]. However, the results reported here demonstrate inactivation of the recombinant *C. thermocellum* BglA (Clo1313_2020) at such high temperatures. Thus, and due to the important role of β-glucosidases, enhancing the thermostability of *C. thermocellum* BglA is of great significance and was the goal of the current study.

Directed evolution, which consists of random mutagenesis and high-throughput screening approaches, is a powerful technique which does not require prior functional, structural, or mechanistic knowledge. Only suitable and effective screening strategies for the desired activity are required [36,37]. The chromogenic product released from *p*-nitrophenyl-β-D-1,4-glucopyranoside (*p*NPG, an analogue of the natural substrate of β-glucosidase: cellobiose), enables efficient and rapid screening. Indeed, directed evolution methodologies have already been used in the past for enhanced thermostability of β-glucosidases from other (mostly mesophilic) organisms [36,38,39], which resulted in the creation of mutants stable at temperature ranges of 50–60 °C (the natural wild-type range of *C. thermocellum* recombinant enzymes). In the current study, we used the directed evolution strategy based on the substrate analogue *p*NPG, in order to create and reveal thermostable mutants of *C. thermocellum* BglA. The functionality of this mutant was also validated under near-natural conditions, namely by examining the contribution of the thermostable BglA to the hydrolyses of microcrystalline cellulose by *C. thermocellum* cellulase mixtures.

2. Results

2.1. Construction and Screening of C. Thermocellum BglA Clones Library

In order to generate thermostable mutants of *C. thermocellum* BglA, in vitro directed evolution was applied on the full-length open reading frame (ORF) of the Clo1313_2020 gene. High numbers of mutation events per clone enable broad screening, but, in contrast, too many mutation events per clone might mask the desired mutation events. Here, a frequency leading to ~20% active clones was chosen. Mutation frequency is determined by two parameters: the template amount and the number of polymerase reaction cycles. Hence, by using different template amounts and different thermal cycle numbers, appropriate mini-libraries were created. The amount of 100 ng DNA template and 23 PCR thermal cycles led to 23% active colonies, which were used to create a library of ~40,000 clones. Sequencing eight active and non-active clones revealed an average of three and seven mutation events per active and non-active clones, respectively. In the next step, the library was screened for thermostable clones. About 8000 clones were screened, revealing 40 thermostable (red) clones.

2.2. Characterization of the Thermostable Mutants

In order to verify and quantify the thermostability of the detected clones, their residual activity after heat shock was calculated (residual activity was defined as the activity of the heat-shocked lysate × 100/activity of non-heated lysate). The residual activities of the two most thermostable clones were 149% and 140% higher than the wild type. The most thermostable clone was sequenced, revealing one silent mutation and two mutation events: A17S and K268N (herein referred to as Mut 1). The second thermostable clone revealed a single mutation event: S39T (referred here as Mut 2).

For further characterization, the two thermostable mutants and the wild-type BglA enzyme were recombinantly expressed in *E. coli* and purified. Purified enzymes were subjected to heat-shock treatment (66–72 °C, 1 h) and residual activities were calculated (Figure 1). Indeed, both mutants were more thermostable than the wild-type enzyme, with Mut 1 being more thermostable than Mut 2. Mut 1 lost only ~40% of its activity (residual activity ~60%) after heat-shock at 68.4 °C. Under the same conditions, Mut 2 lost ~90% of its activity, and the wild-type BglA totally lost its activity. After heat-shock at 70 °C, both the wild-type and Mut 2 enzymes totally lost their activity, while Mut 1 still exhibited residual activity of ~9.5%.

The kinetic parameters of the wild type and Mut 1 were measured (Table 1). Mut 1 revealed relatively similar and only slightly higher catalytic efficiency (K_{cat}/K_m), with no statistical differences between the K_{cat} and K_m values compared to the wild type enzyme. The effect of each mutation event (namely A17S and K268N) on the thermostability of Mut 1 was examined. For this purpose, two recombinant enzymes were constructed and purified, one containing only the A17S mutation, and the other only K268N. The enzymes were subjected to heat-shock (66–72 °C, 1 h), and the residual activities were calculated (Figure 1). A17S demonstrated similar thermostability to Mut 1, indicating the major contribution of this substitution to the thermostability. In contrast, K268S did not demonstrate improvement in its thermostability (compared to that of the wild type), indicating that the improved thermostability derived only from the A17S mutation event. Using site-direct mutagenesis, A17 was substituted with other polar amino acids, namely, threonine, glutamine and asparagine. The resultant enzymes (A17T, A17Q and A17N) demonstrated much lower activity and thermostability, compared to the wild-type enzyme (Figure S1).

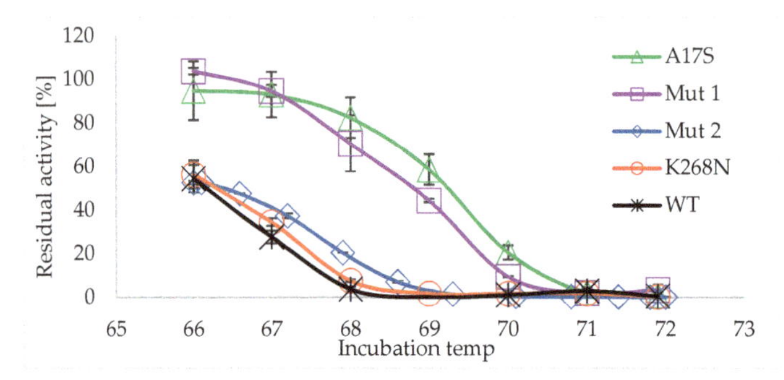

Figure 1. Thermostability of the various *C. thermocellum* BglA mutants. Wild-type (WT) *C. thermocellum* BglA and the different mutants were incubated at 66–72 °C for 1 h, followed by activity assay (using *p*-nitrophenyl-β-D-1,4-glucopyranoside [*p*NPG], an analogue of the natural β-glucosidase substrate). Residual activity of the different mutants (defined as the activity of the heat-shocked enzyme × 100/activity of the non-heated enzyme) was calculated. Mut 1 and A17S demonstrated similar thermostability.

Table 1. Kinetic parameters of *p*NPG hydrolysis by wild-type *C. thermocellum* BglA and the thermostable mutant (Mut 1).

Title	Wild Type	Mut 1
V_{max} [$M{\cdot}s^{-1}$]	$9.92 \times 10^{-7} \pm 6.57 \times 10^{-8}$	$9.1 \times 10^{-7} \pm 3.97 \times 10^{-8}$
K_{cat} [s^{-1}]	76 ± 5.036	70 ± 3.05
K_m [mM]	6.7 ± 1.111	5 ± 0.59
K_{cat}/K_m [$s^{-1}{\cdot}M^{-1}$]	11,282 ± 900	14,018 ± 867

Kinetic parameters of *C. thermocellum* BglA and Mut 1 were measured by *p*NPG assay, and calculated by nonlinear fit by GraphPadPrism software.

The inflection temperatures (T_i) of the wild-type, Mut 1 and A17S enzymes were measured using a NanoTemper Tycho NT.6 instrument. This assay is based on the changes in the intrinsic fluorescence from the aromatic amino acid residues tryptophan and tyrosine (measured at 350 nm and 330 nm). During the assay, the temperature of the protein solution is ramped from 35 °C to 95 °C for a 3 min period, accompanied by continuous measurement of the fluorescence. Changes in the fluorescence signal indicate transitions in the folding state of a protein. The midpoint temperature at which a transition occurs is called the inflection temperature (T_i) [40–42]. The T_i of wild-type BglA, Mut 1 and A17S were 79.3 ± 0.08, 85.7 ± 0.15 and 85.7 ± 0.16 °C, respectively, demonstrating an increase of ~6.4 °C in the T_i (Figure 2). The effect of cellobiose (the natural substrate of β-glucosidases) added to the reaction mixture (at 1 mM) on the T_i was also tested. The results were similar (80 ± 0.48 °C, 85.9 ± 0.53 °C and 85.8 ± 0.1 °C for the wild-type, Mut 1 and A17S, respectively), indicating no stability effect in the presence of cellobiose.

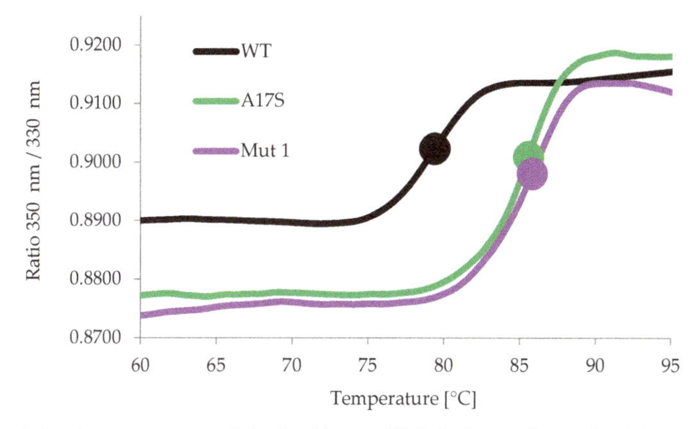

Figure 2. Inflection temperatures (T_i) of wild-type (WT) *C. thermocellum* BglA, Mut 1 and A17S. The inflection temperatures were measured using a NanoTemper Tycho NT.6 instrument as described in the Methods section.

The amino acid sequence of *C. thermocellum* BglA was BLASTed against the NCBI nonredundant protein database. BLAST results for the residues adjacent to the A17S mutation event are represented schematically by the diagrams of amino acids frequencies in Figure 3. The residues near position 17 were found to be relatively conserved. Interestingly, the only amino acid replacing the alanine in position 17 in the different homologues was serine.

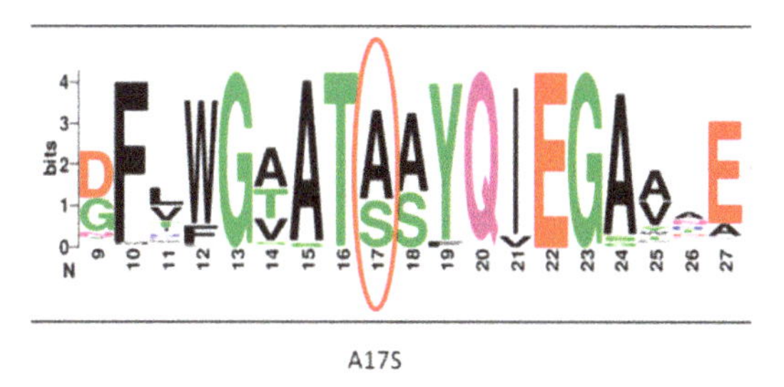

Figure 3. Amino acid frequencies in the residues surrounding the A17S mutation event. The relevant amino acid sequence of *C. thermocellum* BglA was BLASTed against the NCBI nonredundant protein database. The top 1000 hits were used to create the distribution scheme using WEBLOGO. Position 17 is marked by the red circle.

2.3. Structural Aspects of BglA

The crystal structure of wild-type *C. thermocellum* BglA was recently determined in our lab (PDB code 5OGZ, Table S1). BglA adopts the expected $(\beta/\alpha)_8$ TIM barrel fold, typically observed for clan-A β-glucosidases, with two active-site glutamates, Glu166 on strand β4 and Glu355 on strand β7, presumed to act as catalytic acid/base and nucleophile, respectively. Structure alignment of the Bgl1A structure with those of four other family-1 β-glucosidases revealed that the glutamate residues are positioned very similarly with respect to the aligned structures. In addition, the distance between their Cδ atoms is 5.33 Å, consistent with the properties of a retaining β-glycosidase. These highly conserved motifs are responsible for substrate binding and enzymatic hydrolysis of the glycosidic bond within the active site.

Figure 4A shows the position of A17, which is located in a loop near the central cavity of the enzyme. The nearby coding region of K268, however, which is located in an outer α-helix, was relatively non-conserved (data not shown).

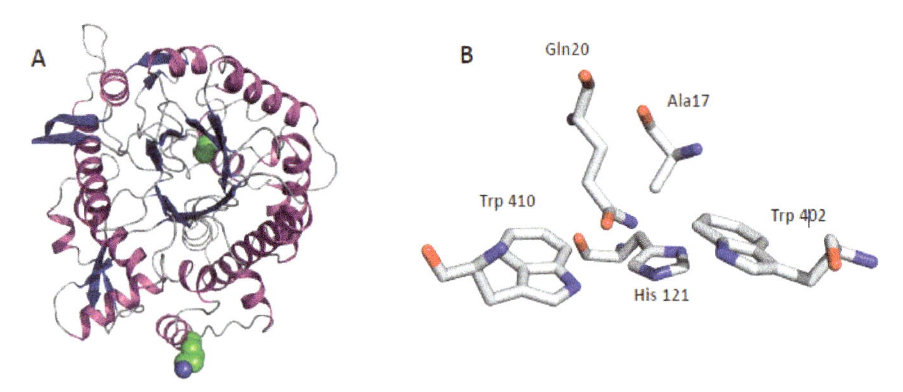

Figure 4. Structural analysis of *C. thermocellum* BglA. (**A**) Crystal structure of *C. thermocellum* BglA. α-Helices are colored purple. β-sheets are colored blue. Ala17 and Lys268 (the two mutated positions in Mut 1) are displayed by spheres. (**B**) Ala17 and its adjacent residues. Carbon atoms are shown in white, oxygen atoms in red and nitrogen atoms in blue. Analysis was performed using PyMol software.

2.4. *Advantage of Thermostable BglA in the Cellulose Hydrolysis Process*

The contribution of thermostable BglA (Mut 1) to the cellulose hydrolysis process was examined. For this purpose, microcrystalline cellulose (Avicel) was hydrolyzed by the *C. thermocellum* secretome, supplemented with either Mut 1 or wild-type BglA at 60 and 70 °C, and the concentration of the released soluble glucose was measured (Figure 5). Final glucose concentrations in the Mut 1 samples were higher than those of the wild-type samples at both temperatures. However, this advantage was much higher while working at 70 °C (57.4 mM in Mut 1 versus 21.7 mM in the wild-type samples), and lower while working at 60 °C (47.1 mM in Mut 1 versus 40.1 mM in the wild-type).

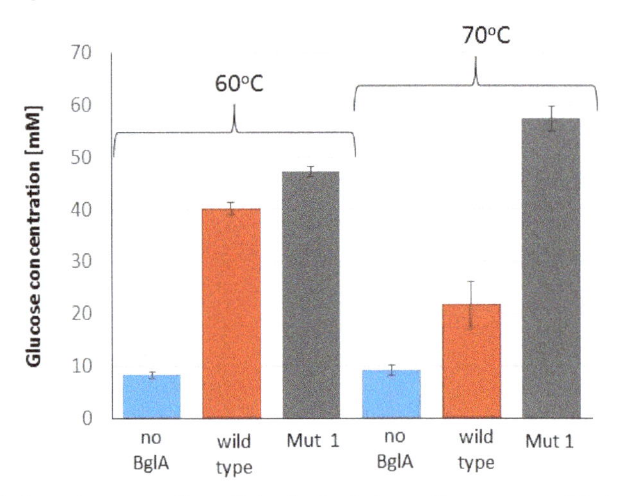

Figure 5. Hydrolysis of microcrystalline cellulose by the *C. thermocellum* secretome. Solutions containing the *C. thermocellum* secretome were applied on Avicel with or without the addition of BglA (wild-type or mutant), followed by incubation at either 60 or 70 °C for 48 h. The concentration of the released soluble glucose was measured by HPLC. Mut 1 showed higher glucose yields at both temperatures, with significant advantage at 70 °C. The experiment was conducted in triplicate. Bars indicate standard deviation.

In addition, Mut 1 was integrated in our lab into a thermostable designer cellulosome [43]. To do so, a plasmid was created containing clone 1 attached to the gene segment coding for the dockerin module from *Clostridium clariflavum*. The expressed recombinant enzyme was successfully integrated into an artificial thermostable designer scaffoldin, which also contained appropriate dockerin-bearing thermostable *Clostridium thermocellum* mutants of both exoglucanase Cel48S and endoglucanase Cel8A. The resulting thermostable designer cellulosome exhibited a 1.7-fold enhancement in cellulose degradation (compared to the action of conventional designer cellulosomes that contain the respective wild-type enzymes). The results were published by Moraïs et al. in 2016 [43].

3. Discussion

Developing thermostable and highly active cellulase preparations is critical for achieving cost-effective enzymatic deconstruction of cellulosic biomass [24]. In this report, directed evolution was conducted on the β-glucosidase A gene of the thermophilic bacterium, *C. thermocellum*, to produce a potent thermostable mutant (Mut 1), which contained two mutations: Alanine at position 17 was substituted with serine, and lysine at position 258 was substituted with asparagine. However, only the A17S mutation was found to be responsible for the observed thermostability. The residues near position 17 were found to be conserved among the top 1000 homologous sequences of *C. thermocellum* β-glucosidase A. The observed conservation can indicate the important role of this position and on the importance of alanine 17 to the functionality of the enzyme [44,45]. The fact that serine was the only amino acid replacing alanine in position 17 in the different homologues is fully consistent with our results, in which the alanine-to-serine mutation produced a functional and thermostable mutant enzyme. Indeed, substitution of A17 with threonine, glutamine and asparagine resulted in almost inactive enzymes. According to Daniel et al. (1996), the enhanced stability of proteins can be achieved by an additional stabilizing force which is equivalent to only a few weak interactions [31,46]. Indeed, single mutation events were found to increase the thermostability of various cellulases [31,36,47]. In the case of *C. thermocellum* BglA, substitution of serine for alanine would likely result in hydrogen bonding with His 121 (Figure 4B), both of which are located near the active site. The newly created hydrogen bond would presumably help stabilize the enzyme. Our results also demonstrate that a single point mutation can increase the thermostability of the already naturally thermostable *C. thermocellum* BglA, thereby increasing its T_i by 6.4 °C.

Mut 1 showed increased activity over the wild-type enzyme, with a much higher effect at 70 °C versus 60 °C (the optimum growth temperature for *C. thermocellum*). These results, together with the fact that Mut 1 has similar K_{cat} and K_m, indicate that the mutation affects mainly the thermostability, rather than the activity of the enzyme. Directed evolution conducted on the mesophilic bacterium *Paenibacillus polymyxa* BglA revealed a more thermostable mutant containing the same A17S mutation event [36]. This mutation event increased the half-life of thermoinactivation by 11-fold, when applied at 50 °C. In addition, the authors obtained a lower K_m and a higher K_{cat}, resulting in doubling the catalytic efficiency. Considering the crystal structure of *P. polymyxa* BglA, a possible explanation for those effects was suggested [36]. The alanine at position 17 occurred in the internal cavity, buried among Gln20, His121, Trp398 and Trp406, near the active site. It was suggested that the alanine-to-serine substitution increased the residue volume, which in turn was assumed to be important for enzyme thermostability [48]. In addition, the proximity of the newly substituted polar residue (serine) to the above-mentioned amino acids, all of which participate in the ligand binding, was suggested to increase the binding affinity [49]. The crystal structure of *C. thermocellum* BglA revealed a similar structure (Figure 4B) to that of *P. polymyxa* BglA, with the Ala17 buried in the internal cavity among Gln20, His121, Trp402 and Trp410. Thus, a very similar mechanism may well be valid for *C. thermocellum* BglA although no significant effect on K_m was measured.

Chromogenic substrates are often used for the screening of enzyme activity, since they enable rapid visual detection of the desired phenotype. However, improved hydrolysis of the synthetic chromogenic substrate does not necessarily correlate with that of the natural substrate of the enzyme [38,50,51]. Thus, a verification step of the enzymatic activity on the natural substrate is required following screening on substrate analogues. Thus, the contribution of Mut 1 to cellulose hydrolysis was tested in this study, thereby revealing its advantage at higher temperatures.

In nature, the assembly of the catalytic units of *C. thermocellum* into cellulosomes containing cellulose-binding modules (CBMs) resulted in the formation of higher local cellobiose concentrations at particular sites. However, *C. thermocellum* BglA does not possess a CBM module, and is not targeted towards the increased local cellobiose concentration. Instead, cellulosome-generated cellobiose is transported directly into the cell [34], and hydrolyzed to glucose by periplasmic β-glucosidases. In cell-free enzymatic systems, such as that reported here, removal of inhibitory cellobiose can be performed by adding BglA to the assay. Thus, targeting the recombinant BglA to the increased local cellobiose concentration might improve hydrolysis efficiency. In a former study conducted by our lab, we reported the design of a recombinant form of the wild-type *C. thermocellum* BglA, which possessed the ability to directly bind to the cellulosome via cohesin-dockerin interaction [52]. Integration of BglA into the *C. thermocellum* cellulosome led to higher degradation levels of microcrystalline cellulose and pretreated switchgrass, compared to cellulosomes supplemented with the soluble wild-type form of the enzyme. By using the same technique, the thermostable Mut 1 was incorporated into thermostable designer cellulosomes, which now demonstrated a 1.7-fold enhancement in cellulose degradation, compared to a non-thermostabilized designer cellulosome preparation [43]. These results further emphasize the advantage of thermostable mutants for improving lignocellulosic biomass conversion.

Several approaches can be applied to design thermostable cellulases. In the current study, the directed evolution approach was used for the improvement of *C. thermocellum* BglA. This strategy is based on random mutagenesis and sequential screening rather than a rational hypothesis-based approach. This powerful method does not require preliminary knowledge about the structure of the enzyme, and can reveal mutations that would not be revealed by knowledge-dependent approaches. However, in the future, the recently solved crystal structure of *C. thermocellum* BglA (Table S1) can be used for rational design, in order to further improve the thermostability and activity of Mut 1. [53].

Overall, the current study demonstrates that natural thermostable cellulases can be further improved. Exposing the "hidden" potential of plant cell wall-degrading enzymes is thus an important step towards cost-effective conversion of plant biomass into bioethanol or other biochemicals.

4. Materials and Methods

4.1. Random Mutagenesis and Library Construction

A library of *C. thermocellum* BglA mutant clones was created as previously described [30,31] with minor modifications. Different quantities (20, 100 or 400 ng) of the *C. thermocellum* BglA open reading frame Clo1313_2020, cloned in a pET28a plasmid, were used as a template for error prone PCR according to the manufacturer's instructions, using Gene-Morph II Random Mutagenesis Kit (Stratagene, La Jolla, CA, USA). Thermal cycling parameters were 95 °C for 2 min followed by 18, 23 or 27 cycles of 95 °C for 1 min, 55 °C for 45 s and 72 °C for 2 min, followed by a final step of 72 °C for 10 min. T7 promoter primer and T7 terminator primer were used for amplification. The resulting PCR products were applied on 0.75% agarose gel. Extracted bands were treated with DpnI and diluted 100 times. The solution was then used as a template for sequential PCR reaction using ReadyMix™ Taq PCR Reaction Mix (Sigma-Aldrich, Rehovot, Israel) with primers: CAGTC**CATGGC**AAAGATAAC (NcoI restriction site) and CACG**CTCGAG**GAAACCGTTGTTTTGATTAC (XhoI restriction site). The thermal cycling parameters followed the manufacturer's instruction, with an annealing temperature of 55 °C and elongation time of 2 min. The amplified products were purified, restricted with NcoI and XhoI according to the manufacturer's instructions and ligated to the pET28a-based plasmid treated with NcoI/XhoI and shrimp alkaline phosphatase (SAP). The ligated plasmids were treated with SacI in order to remove unrestricted vector. (New England Biolabs, UK enzymes were used in the restriction and ligation process). Plasmids were then electrotransformed into *E. coli* XL1 electro-competent cells and purified using a miniprep kit (QIAprep Spin Miniprep Kit, Qiagen, Redwood City, CA, USA), creating minilibraries. The minilibraries were transformed into *E. coli* BL21 competent cells and plated on LB plates, containing 1.5% agar, 4 μM isopropyl β-D-1-thiogalactopyranoside (IPTG) and 50 μg/mL kanamycin. The plates were incubated overnight at 37 °C. A solution of 25 mM citrate buffer, pH = 6.1, containing 0.75% agar, was then boiled, cooled to 45 °C and supplemented with Magenta GlcA (5-Bromo-6-chloro-3-indolyl β-D-glucuronide cyclohexylammonium salt, Sigma Aldrich) to a final concentration of 0.02%, and applied onto the plates, creating an additional layer. The plates were dried for 1 h at room temperature and then incubated at 60 °C for 1.5 h (until red colonies appeared). The percentage of the red colonies was calculated and the parameters leading to 20–30% active clones were further used to increase library.

4.2. Screening for Thermostable Clones

Library screening was performed as detailed in the previous section with two additional steps: Plasmids were transformed into *E. coli* BL21 competent cells and plated on LB plates containing 1.5% agar, 4 μM IPTG and 50 μg/mL kanamycin. The plates were incubated overnight at 37 °C. They were then replicated on fresh LB agar plates using silk snippets, heat-shocked at 70 °C for 50 min and cooled at 4 °C. A layer of 25 mM citrate buffer, pH 6.1, 0.75% agar and 0.02% Magenta GlcA, was added to the plates (as detailed above), which were dried for 1 h at room temperature and then incubated at 60 °C for 1.5 h. The replicates were used to purify the plasmids of selected red colonies, and selected clones were sequenced. A plate containing wild-type *C. thermocellum* BglA was used as a control.

4.3. Residual Activity of Overexpressing BglA Colonies

Colonies overexpressing thermostable mutants (as indicated by the appearance of red color) were grown on liquid LB medium (0.5 mL), containing 0.1 mM IPTG and 50 μg/mL kanamycin, overnight at 37 °C in 96-deep-well plates. In order to extract the proteins, each well was supplemented with 20 μL of Popculture (Novagene, Darmstadt, Germany), DNaseI and lysozyme, and the plate was incubated at 37 °C for 20 min. The lysate was diluted 30 times in 50 mM citrate buffer, pH 6.1. Diluted lysate (100 μL) was incubated at 66 °C for 75 min and cooled on ice. Then, 15 μL of heated and non-heated samples were added to 1 mM *p*-nitrophenyl-β-D-glucopyranoside (*p*NPG, Sigma Aldrich, St. Louis, MO, USA) solution and incubated for 45 min at 60 °C. The reaction was terminated by adding 85 μl of 1 M carbonate buffer, pH 9.5, and optical densities of the samples were measured at a wavelength of 405 nm. Residual activity was calculated by comparing the activity of the heated versus non-heated samples.

4.4. Protein Expression and Purification

The resulting *C. thermocellum* BglA mutants and wild-type enzymes were produced by expression of relevant plasmids into *E. coli* BL21 (lDE3) pLysS cells. The proteins were extracted and purified on an Ni-nitrilotriacetic acid (NTA) column (Qiagen, Hilden, Germany), as reported earlier [30]. Purity of the recombinant proteins was assessed by SDS-PAGE on 12% acrylamide gels, and fractions, containing the pure recombinant protein, were pooled and concentrated using AmiconUltra 15 mL 50,000 MWCO concentrators (Millipore, Bedford, MA, USA). Protein concentration was estimated from the absorbance at 280 nm, based on the known amino acid composition of the protein, using the Protparam tool (http://www.expasy.org/tools/protparam.html). Proteins were stored in 50% (*v*/*v*) glycerol at −20 °C.

4.5. Stability Assay

Solutions of 50 mM citrate buffer, pH 6, containing 7 μg/mL of the recombinant enzymes, were incubated at 66–72 °C for 1 h and then cooled on ice. Heated and non-heated samples were diluted to a final enzymatic concentration of 1.05 μg/mL in a solution of 50 mM citrate buffer, pH 6.1, containing 1 mM *p*NPG, incubated for 10 min at 60 °C, cooled on ice, and supplemented with equal amounts of 1 M carbonate buffer, pH 9.5. Optical densities of the samples were measured at a wavelength of 405 nm. Residual activity was calculated by comparing the activity of the heated versus non-heated enzymes.

4.6. Kinetic Parameters Measurements

A solution of 50 mM citrate buffer, pH 6.1, containing 0–25 mM *p*NPG, was supplemented with a concentration of 13 nM enzyme (wild-type or mutant) and incubated at 60 °C for 8 min in a preheated 96-well plate, accompanied by continued measurements at OD_{405}. The concentrations of end product (*p*-nitrophenol) were calculated using known concentrations of *p*-nitrophenol. Kinetic parameters were calculated by nonlinear fit using the GraphPadPrism software (GraphPad Software, Inc., San Diego, CA, USA)

4.7. Sequence Analysis

The protein sequence of *C. thermocellum* BglA (ADU75064.1) was BLASTed against the NCBI non-redundant protein database. The top 1000 hits with *E*-value < 0.001 were further aligned. Frequency of amino acids was visualized using WEBLOGO version 2.8.2.

4.8. Purification of the C. Thermocellum Secretome

C. thermocellum DSM1313 was grown on GS-2 medium (0.5 g/L K_2HPO_4, 0.5 g/L $MgCl_2{\cdot}6H_2O$, 0.5 g/L KH_2PO_4, 1.3 g/L $(NH_4)_2SO_4$, 0.002 g/L resazurin, 10.5 g/L MOPS buffer, 5 g/L yeast extract, 1.25 mg/L iron(II) sulfate and 0.5 mM $CaCl_2$, adjusted with 10 M NaOH to a final pH of 7.2) with 0.5% microcrystalline cellulose (Avicel, Sigma Aldrich, St. Louis, MO, USA) in batch culture. Nitrogen flushing was used to achieve anaerobic conditions. After 48 h of incubation at 60 °C, growth medium was centrifuged (10,808× *g*, 10 min). Soluble proteins were precipitate by 80% ammonium sulfate and re-suspended in Tris-buffered saline (TBS) buffer, pH 7.4. Protein concentration was measured by Bradford assay, using Bio-Rad protein assay solution (Bio-Rad) [54].

4.9. Cellulose Hydrolysis Assay

A quantity of 0.6 mg/mL of *C. thermocellum* secretome solution was applied to a suspension of 250 mg/mL of microcrystalline cellulose in 20 mM citrate buffer, pH 6.1, with or without the addition of 2 µg/mL BglA (wild-type or mutant) in a reaction volume of 2 mL. Samples were incubated at 60 and 70 °C for 48 h, and centrifuged (16,100× *g*, 5 min). Released soluble sugars were analyzed by high-pressure liquid chromatography (HPLC, Agilent Infinity 1260 system, Agilent Technologies, Santa Clara, CA, USA) using an Aminex®HPX-87H Ion Exclusion column (Bio-Rad, Hercules, CA, USA) with a guard column, mobile phase of 5 mM H_2SO_4 (flow-through of 0.6 mL/min at 45 °C) in an Agilent 1260 Infinity LC system with RID detector (G1362A). Experiments were performed in triplicate.

4.10. T_i Measurements

Recombinant BglA enzymes in TBSx1 buffer (with and without the addition of 1 mM cellobiose) were used for T_i measurements, using a NanoTemper Tycho NT.6 instrument Agentek (1987) Ltd., Tel Aviv, Israel), according to the manufacturer's instructions [42].

Supplementary Materials: Supplementary material can be found at http://www.mdpi.com/1422-0067/20/19/4701/s1.

Author Contributions: S.Y., Y.H., E.M. and E.A.B. conceived and designed the experiments; S.Y. performed the experiments and analyzed the data, M.A. and A.K. supported and advised the kinetic-parameter measurements, directed evolution and the screening method; O.S.-A. and F.F. resolved the 3D structure, S.Y., Y.H., J.S. and E.A.B. wrote the paper.

Funding: This research was supported by the United States–Israel Binational Science Foundation (BSF grant No. 2013284), Jerusalem, Israel; the Israel Science Foundation (ISF grant no. 1349/13); the European Union NMP.2013.1.1-2: CellulosomePlus Project number 604530, and by a research grant from the Yotam Project via the Sustainability and Energy Research Initiative (SAERI) at the Weizmann Institute of Science.

Acknowledgments: The authors are pleased to acknowledge the advice and assistance of Yoav Barak, Department of Chemical Research Support, The Weizmann Institute of Science. We also thank Itamar Kass (Ben-Gurion

University of the Negev, Beer Sheva, Israel and Amai Proteins, Ltd., Rehovot, Israel) and Oren Yaniv (Beckman Coulter, Rehovot, Israel) for their insight into the structural aspects of BglA.

Conflicts of Interest: The authors declare no conflict of interest.

Abbreviations

BglA	β-glucosidase A
*p*NPG	*p*-nitrophenyl-β-D-glucopyranoside
CBM	cellulose binding module

References

1. Bayer, E.A.; Shoham, Y.; Lamed, R. Lignocellulose-decomposing bacteria and their enzyme system. In *The Prokaryotes*, 4th ed.; Rosenberg, E., Ed.; Springer: Berlin/Heidelberg, Germany, 2013; pp. 216–266.
2. Varrot, A.; Frandsen, T.P.; Von Ossowski, I.; Boyer, V.; Cottaz, S.; Driguez, H.; Schülein, M.; Davies, G.J. Structural Basis for Ligand Binding and Processivity in Cellobiohydrolase Cel6A from Humicola insolens. *Structure* **2003**, *11*, 855–864. [CrossRef]
3. Viikari, L.; Vehmaanperä, J.; Koivula, A. Lignocellulosic ethanol: from science to industry. *Biomass Bioenergy* **2012**, *46*, 13–24. [CrossRef]
4. Klein-Marcuschamer, D.; Oleskowicz-Popiel, P.; Simmons, B.A.; Blanch, H.W. The challenge of enzyme cost in the production of lignocellulosic biofuels. *Biotechnol. Bioeng.* **2012**, *109*, 1083–1087. [CrossRef] [PubMed]
5. Pettolino, F.A.; Walsh, C.; Fincher, G.B.; Bacic, A. Determining the polysaccharide composition of plant cell walls. *Nat. Protoc.* **2012**, *7*, 1590–1607. [CrossRef] [PubMed]
6. Pauly, M.; Keegstra, K. Cell-wall carbohydrates and their modification as a resource for biofuels. *Plant J.* **2008**, *54*, 559–568. [CrossRef] [PubMed]
7. Himmel, M.E.; Xu, Q.; Luo, Y.; Ding, S.-Y.; Lamed, R.; Bayer, E.A. Microbial enzyme systems for biomass conversion: emerging paradigms. *Biofuels* **2010**, *1*, 323–341. [CrossRef]
8. Lombard, V.; Ramulu, H.G.; Drula, E.; Coutinho, P.M.; Henrissat, B. The carbohydrate-active enzymes database (CAZy) in 2013. *Nucleic Acids Res.* **2014**, *42*, 490–495. [CrossRef] [PubMed]
9. Horn, S.J.; Vaaje-Kolstad, G.; Westereng, B.; Eijsink, V.G. Novel enzymes for the degradation of cellulose. Biotechnol. *Biofuels* **2012**, *5*, 45. [CrossRef]
10. Strobel, H.J. Growth of the thermophilic bacterium Clostridium thermocellum in continuous culture. *Curr. Microbiol.* **1995**, *31*, 210–214. [CrossRef]
11. Lamed, R.; Kenig, R.; Setter, E.; Bayer, E.A. Major characteristics of the cellulolytic system of Clostridium thermocellum coincide with those of the purified cellulosome. *Enzyme Microb. Technol.* **1985**, *7*, 37–41. [CrossRef]
12. Andrić, P.; Meyer, A.S.; Jensen, P.A.; Dam-Johansen, K. Effect and modeling of glucose inhibition and in situ glucose removal during enzymatic hydrolysis of pretreated wheat straw. *Appl. Biochem. Biotechnol.* **2010**, *160*, 280–297. [CrossRef] [PubMed]
13. Kadam, S.K.; Demain, A.L. Addition of cloned β-glucosidase enhances the degradation of crystalline cellulose by the Clostridium thermocellum cellulase complex. *Biochem. Biophys. Res. Commun.* **1989**, *161*, 706–711. [CrossRef]
14. Lamed, R.; Kenig, R.; Morgenstern, E.; Calzada, J.F.; De Micheo, F.; Bayer, E.A. Efficient cellulose solubilization by a combined cellulosome-β-glucosidase system. *Appl. Biochem. Biotechnol.* **1991**, *27*, 173–183. [CrossRef]
15. Maki, M.L.; Armstrong, L.; Leung, K.T.; Qin, W. Increased expression of β-glucosidase A in Clostridium thermocellum 27405 significantly increases cellulase activity. *Bioengineered* **2013**, *4*, 15–20. [CrossRef] [PubMed]
16. Mittal, A.; Decker, S.R. Special issue: Application of biotechnology for biofuels: transforming biomass to biofuels. *3 Biotech* **2013**, *3*, 341–343. [CrossRef] [PubMed]
17. Zhang, Y.H.P.; Lynd, L.R. A functionally based model for hydrolysis of cellulose by fungal cellulase. *Biotechnol. Bioeng.* **2006**, *94*, 888–898. [CrossRef] [PubMed]
18. Heinzelman, P.; Snow, C.D.; Wu, I.; Nguyen, C.; Villalobos, A.; Govindarajan, S.; Minshull, J.; Arnold, F.H. A family of thermostable fungal cellulases created by structure-guided recombination. *Proc. Natl. Acad. Sci. USA* **2009**, *106*, 5610–5615. [CrossRef] [PubMed]

19. Doi, R.H.; Kosugi, A. Cellulosomes: plant-cell-wall-degrading enzyme complexes. *Nat. Rev. Microbiol.* **2004**, *2*, 541–551. [CrossRef]
20. Blumer-Schuette, S.E.; Brown, S.D.; Sander, K.B.; Bayer, E.A.; Kataeva, I.; Zurawski, J.V.; Conway, J.M.; Adams, M.W.W.; Kelly, R.M. Thermophilic lignocellulose deconstruction. *FEMS Microbiol. Rev.* **2014**, *38*, 393–448. [CrossRef]
21. Cristina, A.; Felby, C.; Gama, M. Cellulase stability, adsorption/desorption profiles and recycling during successive cycles of hydrolysis and fermentation of wheat straw. *Bioresour. Technol.* **2014**, *156*, 163–169.
22. Skovgaard, P.A.; Jørgensen, H. Influence of high temperature and ethanol on thermostable lignocellulolytic enzymes. *J. Ind. Microbiol. Biotechnol.* **2013**, *40*, 447–456. [CrossRef] [PubMed]
23. Abdel-banat, B.M.A.; Hoshida, H.; Ano, A.; Nonklang, S.; Akada, R. High-temperature fermentation: How can processes for ethanol production at high temperatures become superior to the traditional process using mesophilic yeast? *Appl. Microbiol. Biotechnol.* **2010**, *85*, 861–867. [CrossRef] [PubMed]
24. Wu, I.; Arnold, F.H. Engineered thermostable fungal Cel6A and Cel7A cellobiohydrolases hydrolyze cellulose efficiently at elevated temperatures. *Biotechnol. Bioeng.* **2013**, *110*, 1874–1883. [CrossRef] [PubMed]
25. Nakazawa, H.; Okada, K.; Onodera, T.; Ogasawara, W.; Okada, H.; Morikawa, Y. Directed evolution of endoglucanase III (Cel12A) from Trichoderma reesei. *Appl. Microbiol. Biotechnol.* **2009**, *83*, 649–657. [CrossRef] [PubMed]
26. Liu, W.; Zhang, X.; Zhang, Z.; Zhang, Y.P. Engineering of Clostridium phytofermentans endoglucanase Cel5A for improved thermostability. *Appl. Environ. Microbiol.* **2010**, *76*, 4914–4917. [CrossRef] [PubMed]
27. Voutilainen, S.P.; Boer, H.; Alapuranen, M.; Jänis, J.; Vehmaanperä, J.; Koivula, A. Improving the thermostability and activity of Melanocarpus albomyces cellobiohydrolase Cel7B. *Appl. Microbiol. Biotechnol.* **2009**, *83*, 261–272. [CrossRef]
28. Trudeau, D.L.; Lee, T.M.; Arnold, F.H. Engineered thermostable fungal cellulases exhibit efficient synergistic cellulose hydrolysis at elevated temperatures. *Biotechnol. Bioeng.* **2014**, *111*, 2390–2397. [CrossRef]
29. Bayer, E.A.; Belaich, J.P.; Shoham, Y.; Lamed, R. The cellulosomes: multienzyme machines for degradation of plant cell wall polysaccharides. *Annu. Rev. Microbiol.* **2004**, *58*, 521–554. [CrossRef] [PubMed]
30. Anbar, M.; Gul, O.; Lamed, R.; Sezerman, U.O.; Bayer, E.A. Improved thermostability of Clostridium thermocellum endoglucanase Cel8A by using consensus-guided mutagenesis. *Appl. Environ. Microbiol.* **2012**, *78*, 3458–3464. [CrossRef]
31. Anbar, M.; Lamed, R.; Bayer, E.A. Thermostability enhancement of Clostridium thermocellum cellulosomal endoglucanase Cel8A by a single glycine substitution. *ChemCatChem* **2010**, *2*, 997–1003. [CrossRef]
32. Yi, Z.; Pei, X.; Wu, Z. Introduction of glycine and proline residues onto protein surface increases the thermostability of endoglucanase CelA from Clostridium thermocellum. *Bioresour. Technol.* **2011**, *102*, 3636–3638. [CrossRef] [PubMed]
33. Smith, M.A.; Rentmeister, A.; Snow, C.D.; Wu, T.; Farrow, M.F.; Mingardon, F.; Arnold, F.H. A diverse set of family 48 bacterial glycoside hydrolase cellulases created by structure-guided recombination. *FEBS J.* **2012**, *279*, 4453–4465. [CrossRef] [PubMed]
34. Nataf, Y.; Yaron, S.; Stahl, F.; Lamed, R.; Bayer, E.A.; Scheper, T.; Sonenshein, A.; Shoham, Y. Cellodextrin and laminaribiose ABC transporters in Clostridium thermocellum. *J. Bacteriol.* **2009**, *191*, 203–209. [CrossRef] [PubMed]
35. Xu, C.; Qin, Y.; Li, Y.; Ji, Y.; Huang, J.; Song, H.; Xu, J. Factors influencing cellulosome activity in Consolidated Bioprocessing of cellulosic ethanol. *Bioresour. Technol.* **2010**, *101*, 9560–9569. [CrossRef] [PubMed]
36. Liu, W.; Hong, J.; Bevan, D.R.; Zhang, Y.-H.P. Fast identification of thermostable beta-glucosidase mutants on cellobiose by a novel combinatorial selection/screening approach. *Biotechnol. Bioeng.* **2009**, *103*, 1087–1094. [CrossRef] [PubMed]
37. Cobb, R.E.; Sun, N.; Zhao, H. Directed evolution as a powerful synthetic biology tool. *Methods* **2013**, *60*, 81–90. [CrossRef] [PubMed]
38. McCarthy, J.K.; Uzelac, A.; Davis, D.F.; Eveleigh, D.E. Improved catalytic efficiency and active site modification of 1,4-beta-D-glucan glucohydrolase A from Thermotoga neapolitana by directed evolution. *J. Biol. Chem.* **2004**, *279*, 11495–11502. [CrossRef] [PubMed]
39. Arrizubieta, M.J. Increased thermal resistance and modification of the catalytic properties of a beta -glucosidase by random mutagenesis and in vitro recombination. *J. Biol. Chem.* **2000**, *275*, 28843–28848. [CrossRef]

40. Sierla, M.; Hõrak, H.; Overmyer, K.; Waszczak, C.; Yarmolinsky, D.; Maierhofer, T.; Vainonen, J.P.; Salojärvi, J.; Denessiouk, K.; Laanemets, K.; et al. The Receptor-like Pseudokinase GHR1 Is Required for Stomatal Closure[OPEN]. *Plant Cell* **2018**, *30*, 2813–2837.
41. Nilsen, J.; Bern, M.; Sand, K.M.K.; Grevys, A.; Dalhus, B.; Sandlie, I.; Andersen, J.T. Human and mouse albumin bind their respective neonatal Fc receptors differently. *Sci. Rep.* **2018**, *8*, 14648. [CrossRef]
42. Mohamadi, M.; Tschammer, N.; Breitsprecher, D. Quick protein binding analysis by label-free thermal shift analysis on the Tycho NT. 6. Available online: https://www.accela.eu/files/produc (accessed on 20 August 2019).
43. Moraïs, S.; Stern, J.; Kahn, A.; Galanopoulou, A.P.; Yoav, S.; Shamshoum, M.; Smith, M.A.; Hatzinikolaou, D.G.; Arnold, F.H.; Bayer, E.A. Enhancement of cellulosome-mediated deconstruction of cellulose by improving enzyme thermostability. *Biotechnol. Biofuels* **2016**, *9*, 323. [CrossRef] [PubMed]
44. Rodina, A.; Godson, G.N. Role of conserved amino acids in the catalytic activity of Escherichia coli primase. *J. Bacteriol.* **2006**, *188*, 3614–3621. [CrossRef] [PubMed]
45. Dokholyan, N.V.; Mirny, L.A.; Shakhnovich, E.I. Understanding conserved amino acids in proteins. *Physica* **2002**, *314*, 600–606. [CrossRef]
46. Daniel, R.M.; Dines, M.; Petach, H.H. The denaturation and degradation of stable enzymes at high temperatures. *Biochem. J.* **1996**, *317*, 1–11. [CrossRef] [PubMed]
47. Heinzelman, P.; Snow, C.D.; Smith, M.A.; Yu, X.; Kannan, A.; Boulware, K.; Villalobos, A.; Govindarajan, S.; Minshull, J.; Arnold, F.H. SCHEMA Recombination of a Fungal Cellulase Uncovers a Single Mutation That Contributes Markedly to Stability*. *J. Boil. Chem.* **2009**, *284*, 26229–26233. [CrossRef]
48. Haney, P.J.; Badger, J.H.; Buldak, G.L.; Reich, C.I.; Woese, C.R.; Olsen, G.J. Thermal adaptation analyzed by comparison of protein sequences from mesophilic and extremely thermophilic Methanococcus species. *Proc. Natl. Acad. Sci. USA* **1999**, *96*, 3578–3583. [CrossRef]
49. Sanz-Aparicio, J.; Hermoso, J.A.; Martínez-Ripoll, M.; Lequerica, J.L.; Polaina, J. Crystal structure of beta-glucosidase A from Bacillus polymyxa: insights into the catalytic activity in family 1 glycosyl hydrolases. *J. Mol. Biol.* **1998**, *275*, 491–502. [CrossRef] [PubMed]
50. González-Candelas, L.; Aristoy, M.C.; Polaina, J.; Flors, A. Cloning and characterization of two genes from Bacillus polymyxa expressing beta-glucosidase activity in Escherichia coli. *Appl. Environ. Microbiol.* **1989**, *55*, 3173–3177. [PubMed]
51. Kaur, J.; Chadha, B.S.; Kumar, B.A.; Ghatora, S.K.; Saini, H.S. Purification and characterization of ß-glucosidase from Melanocarpus sp. MTCC 3922. *Electron. J. Biotechnol.* **2007**, *10*, 260–270. [CrossRef]
52. Gefen, G.; Anbar, M.; Morag, E.; Lamed, R.; Bayer, E.A. Enhanced cellulose degradation by targeted integration of a cohesin-fused β-glucosidase into the Clostridium thermocellum cellulosome. *Proc. Natl. Acad. Sci. USA* **2012**, *109*, 10298–10303. [CrossRef]
53. Anbar, M.; Bayer, E.A. Approaches for improving thermostability characteristics in cellulases. *Methods Enzymol.* **2012**, *510*, 261–271. [PubMed]
54. Bradford, M.M. A rapid and sensitive method for the quantitation of microgram quantities of protein utilizing the principle of protein-dye binding. *Anal. Biochem.* **1976**, *72*, 48–254. [CrossRef]

International Journal of
Molecular Sciences

Article

Active Site Architecture and Reaction Mechanism Determination of Cold Adapted β-D-galactosidase from *Arthrobacter* sp. 32cB

Maria Rutkiewicz [1], Anna Bujacz [1,*], Marta Wanarska [2], Anna Wierzbicka-Wos [3] and Hubert Cieslinski [2]

1 Institute of Technical Biochemistry, Faculty of Biotechnology and Food Sciences, Lodz University of Technology, Stefanowskiego 4/10, 90-924 Lodz, Poland
2 Department of Molecular Biotechnology and Microbiology, Faculty of Chemistry, Gdansk University of Technology, Narutowicza 11/12, 80-233 Gdansk, Poland
3 Department of Microbiology, Faculty of Biology, University of Szczecin, Felczaka 3c, 71-412 Szczecin, Poland
* Correspondence: anna.bujacz@p.lodz.pl; Tel.: +48-42-631-34-94

Received: 31 July 2019; Accepted: 30 August 2019; Published: 3 September 2019

Abstract: *Arth*βDG is a dimeric, cold-adapted β-D-galactosidase that exhibits high hydrolytic and transglycosylation activity. A series of crystal structures of its wild form, as well as its *Arth*βDG_E441Q mutein complexes with ligands were obtained in order to describe the mode of its action. The *Arth*βDG_E441Q mutein is an inactive form of the enzyme designed to enable observation of enzyme interaction with its substrate. The resulting three-dimensional structures of complexes: *Arth*βDG_E441Q/LACs and *Arth*βDG/IPTG (ligand bound in shallow mode) and structures of complexes *Arth*βDG_E441Q/LACd, *Arth*βDG/ONPG (ligands bound in deep mode), and galactose *Arth*βDG/GAL and their analysis enabled structural characterization of the hydrolysis reaction mechanism. Furthermore, comparative analysis with mesophilic analogs revealed the most striking differences in catalysis mechanisms. The key role in substrate transfer from shallow to deep binding mode involves rotation of the F581 side chain. It is worth noting that the 10-aa loop restricting access to the active site in mesophilic GH2 βDGs, in *Arth*βDG is moved outward. This facilitates access of substrate to active site. Such a permanent exposure of the entrance to the active site may be a key factor for improved turnover rate of the cold adapted enzyme and thus a structural feature related to its cold adaptation.

Keywords: galactosidase; hydrolysis; reaction mechanism; complex structures; cold-adapted; GH2

1. Introduction

Glycosyl hydrolases (GHs) are sugar processing enzymes, divided into families based on their structures. This is the reason why the lactose processing β-D-galactosidases (βDGs) belong to different GH families: GH1, GH2, GH35, GH42, and GH59. Their common structural feature is presence of a TIM-barrel type catalytic domain followed by a variety of β-architecture domains, which nature and occurrence differ among GH subfamilies [1].

The most studied β-D-galactosidase in the GH2 family is bacterial *lacZ* βDG from *Escherichia coli* (*Ecol*βDG) [2–4]. It is a large homotetramer, where each monomer consists of 1023 amino acids. Its primary mode of action is to catalyze the hydrolysis of lactose to D-galactose and D-glucose. To achieve its full catalytic efficiency *Ecol*βDG requires the presence of divalent ions such as Mg^{2+} or Mn^{2+}, which can result in 5–100-fold increase of activation depending on the substrate. Within the *Ecol*βDG catalytic site, two subsites can be distinguished: the first exhibits high specificity for binding the galactose moiety, whereas the second provides a platform for weak binding of different moieties [5]. If an excess of galactose occurs, *Ecol*βDG exhibits transglycosylation activity that results in formation of allolactose, a disaccharide composed of D-galactose and D-glucose moieties linked through a β-(1,6)-glycosidic bond [6,7].

Lactose processing enzymes are commonly used in the dairy industry for production of lactose-free products. Keeping dairy products in refrigerated conditions results in crystallization of lactose that leads to an undesirable grainy texture. That is why lactose removal is used for improving the quality of final products, such as ice-creams and some types of cheeses [8,9].

Another example of an enzyme from GH2 family, commonly used in dairy industry, is yeast βDG from *Klyvuromyces lactis* (*Klyv*βDG) [10]. Similar to *Ecol*βDG, it consists of 1032 aa and its functional form is a homotetramer. Not only it catalyzes lactose hydrolysis, but also transglycosylation reaction that results in formation of galactose derivatives, among other alkyl-galactosides [11], gal-mannitol, [12], and bionic acids [13].

However, the usage of a cold-adapted enzyme, exhibiting similar catalytic efficiency to ones already implemented but at lower temperatures (4–18 °C), is highly sought after. Especially for the food industry, removing the need for a heating step not only brings the costs of production down, but it also prevents potential mesophilic contamination and loss of nutritional value of food products due to heating, and production of unwanted products by thermal conversion [14,15].

*Arth*βDG is an interesting candidate for industrial use. Not only can it hydrolyze lactose at a rate comparable to βDG from *Klyvuromyces lactis* but it exhibits additional transglycosylation activity [16]. Galactooligosaccharides (GOS) and heterooligosaccharides (HOS) are prebiotics, which are important for human health. That is why, with constantly increasing evidence of their consumption benefits, they found their way into infant nutrition and special nutrition, and more recently have become increasingly present in everyday food products [17–26].

The modification of transglycosylase activity specificity and efficiency may be achieved by controlling reaction equilibrium or by enzyme engineering. Studies concentrated on introducing mutations into subsites of GHs showed that the modulation of hydrolysis and transglycosylation activities can be achieved by means of knowledge-based enzyme engineering. However, the role of individual amino acids in the active site must be discovered as a basis for successful design of an enzyme with specific, desired activities [27].

*Arth*βDG is a five-domain protein of molecular weight 110 kDa. The catalytic domain, in form of TIM barrel, is surrounded by three IG-like domains and, as typical for the GH2 family, an N-terminal super β-sandwich domain. Regardless of low sequence identity, this monomer's architecture is strikingly similar to *Klyv*βDG and *Ecol*βDG, which enabled determination of catalytic residues E441 and E517. Cold-adapted *Arth*βDG possesses a functional dimer (not typical for the GH2 family). The dimer is stabilized by head-to-tail interactions between Domains 1 and Domains 5 from neighboring molecules [28]. The same oligomerization state, though shaped differently, was previously described by us for another cold-adapted β-D-galactosidase from *Paracoccus* sp. 32d for which we had determined crystal structure [29]. Thanks to comparative analysis, the structural features that may play a key part in its cold-adaptation were described. Most interesting was maximization of energy gain from the surface residue–solvent interactions, that was obtained by reduction of oligomerization state and formation of hydrophobic patches on the protein's surface [30].

The comprehensive structural study of cold-adapted *Arth*βDG reaction mechanism is a first necessary stage for the knowledge-based enzyme engineering that could lead to creation of an enzyme that would not only hydrolyze lactose, but also effectively convert it to the beneficial for human health GOS and HOS at cold conditions. Usage of native protein would limit us to analysis of substrate binding using substrate analogs which could not by hydrolyzed by the enzyme, such as isopropyl β-D-1-thiogalactopyranoside (IPTG), or less preferable substrates such as ortho-nitrophenyl-β-galactoside (ONPG). However, thanks to designing an inactive mutant in which catalytic E441, acting as acid catalyst, was substituted with the structurally isomorphous glutamate residue, we were able to obtain complexes with the natural substrate lactose bound in both deep and shallow binding modes.

2. Results

2.1. Crystal Structures of ArthβDG and ArthβDG_E441Q Complexes

During soaking of *Arth*βDG crystals with ONPG and X-gal (5-bromo-4-chloro-3-indolyl-β-D-galactopyranoside) a coloration of crystals was observed. It was rapid for crystals soaked with X-gal, as they turned intensely blue within 3 min (Figure 1B).

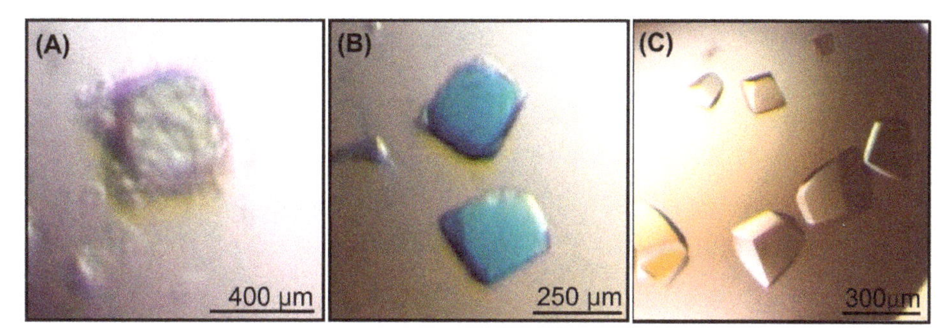

Figure 1. The crystals of *Arth*βDG: after addition of ONPG (**A**); after 2 h of soaking in X-gal (**B**). The crystals of *Arth*βDG_E441Q mutein soaked 24 h with mixture of lactose and galactose (**C**).

Different short soaking times (in a range of 10 s to 10 min, 22 crystals tested) and excess of X-gal (from 3 to 20 molar access in respect to protein concentration) gave intensely blue coloration of soaked crystals. This indicates that the enzyme in the crystal was in its active form and performed hydrolysis of X-gal. The structure solved using diffraction data collected from blue crystals was identical with the *Arth*βDG/GAL obtained by soaking with lactose. It means that X-gal, similarly to lactose, was hydrolyzed by enzyme during soaking of native crystals in ligand solution. Blue color came from 5,5′-dibromo-4,4′-dichloro-indigo—a dimer of the second product (5-bromo-4-chloro-3-hydroxyindol) of X-gal hydrolysis. The blue dye is either deposited in solvent channels of the crystal or randomly bound to the surface of protein. If the dimerized product of X-gal hydrolysis would be bound specifically, it should be detectible in the resulting crystal structure, because it contains two Br atoms, which give strong picks on the electron density maps.

Soaking of *Arth*βDG crystals with ONPG resulted in a yellow halo around the soaked crystals (Figure 1A); however, the rate of ONPG hydrolysis was probably lower than X-gal as the coloration was visually observed only after more than an hour soaking. For the crystal soaked in ONPG for 15 min, the color was not observed and the complex of *Arth*βDG with ONPG bound was formed - an intact ONPG was visible in the active site after determining the crystal structure.

Crystals of *Arth*βDG soaked with lactose underwent fast deterioration, and a very short soaking time was required to obtain crystals still suitable for diffraction experiments. The diffraction data collected after only 1 min of soak with lactose resulted in the crystal structure with galactose. This is evidence that the active enzyme performed hydrolysis of lactose in crystal, in such fast rate, and product of reaction–galactose, bound in active site was visible in electron density maps.

The soaking of *Arth*βDG_E441Q mutein crystals with lactose yielded no complex in solved crystal structures, as for soaking times up to 6 h no ligand was found in the structure and for longer soaking times crystals were destroyed. We have succeeded in determining the crystal structure of *Arth*βDG_E441Q complex with lactose after soaking crystals of *Arth*βDG_E441Q with mixtures of lactose-galactose (Figure 1C) and lactose-fructose for 24 h. After this time, crystals of the mutein complexes were still without visible signs of deterioration and resulted in very good resolution data.

The complex structures of lactose bound in shallow and deep mode were obtained while subjecting those crystals to diffraction experiments.

The crystal structures of *Arth*βDG complexes with galactose (2.1 Å), IPTG (2.2 Å), ONPG (2.8 Å), and its mutein *Arth*βDG_E441Q (1.8 Å) in complexes with lactose (LAC) bound in shallow mode *Arth*βDG_E441Q/LACs (1.9 Å) and in deep mode *Arth*βDG_E441Q/LACd (1.8 Å) were processed in trigonal space group $P3_121$, the same as *Arth*βDG native structure. Matthew's volume calculation [31] revealed that no changes in crystal packing were detected, and protein monomer was present in each asymmetric unit. Crystal structures of these complexes were solved in PHENIX by isomorphous replacement using rigid body procedure and the native structure of *Arth*βDG (PDB ID: 6ETZ) as a model. This allowed us to obtain the molecule in the same position and orientation in all the analyzed structures, which facilitated comparison of not only the structures, but also electron densities. The details for the diffraction data collection and processing are presented in Table 1. Each structure was further refined in PHENIX.REFINE, including TLS parameters [32] defined for each domain. The resulting refinement statistics are given in Table 1.

Table 1. Diffraction data collection, processing, and refinement statistics for crystal structures of investigated *Arth*βDG complexes.

	***Arth*βDG_E441Q** **PDB ID: 6SE8**	***Arth*βDG_LACs** **PDB ID: 6SE9**	***Arth*βDG_LACd** **PDB ID: 6SEA**	***Arth*βDG_IPTG** **PDB ID: 6SEB**	***Arth*βDG_ONPG** **PDB ID: 6SEC**	***Arth*βDG_GAL** **PDB ID: 6SED**
Diffraction source	P13 PETRA, Hamburg, Germany	BL 14.1 BESSY, Berlin, Germany	BL 14.1 BESSY, Berlin, Germany	BL 14.2 BESSY, Berlin, Germany	BL 14.2 BESSY, Berlin, Germany	BL 14.2 BESSY, Berlin, Germany
Wavelength (Å)	0.976250	0.918400	0.918400	0.918400	0.918400	0.918400
Temperature (K)	100 K	100 K	100 K	100 K	100 K	100 K
Detector	PILATUS 6M	PILATUS 3S 2M	PILATUS 3S 2M	PILATUS 3S 2M	PILATUS 3S 2M	PILATUS 3S 2M
Rotation range per image (°)	0.05	0.1	0.1	0.2	0.2	0.2
Total rotation range (°)	160	180	180	180	180	180
Exposure time per image (s)	0.1	0.2	0.2	0.3	0.3	0.3
Space group	P3121	P3121	P3121	P3121	P3121	P3121
a, b, c (Å)	136.8, 136.8, 127.0	138.9, 138.9, 127.9	138.6, 138.6, 127.4	137.1, 137.1, 126.9	137.4, 137.4, 126.8	136.8, 136.8, 126.8
α, β, γ (°)	90, 90, 120	90, 90, 120	90, 90, 120	90, 90, 120	90, 90, 120	90, 90, 120
Mosaicity (°)	0.133	0.084	0.72	0.115	0.247	0.130
Resolution range (Å)	50.0–1.8 (1.9–1.8)	47.0–2.0 (2.1–2.0)	46.9–1.9 (2.0–1.9)	46.6–2.2 (2.3–2.2)	46.6–2.6 (2.7–2.6)	50.0–2.1 (2.2–2.1)
No. of unique reflections	118,383	95,079	109,583	75,085	43,328	80,260
Completeness (%)	98.8 (92.5)	99.4 (97.0)	99.9 (99.5)	99.9 (99.7)	99.4 (98.1)	99.8 (99.3)
Redundancy	7.75 (6.33)	10.09 (10.23)	9.68 (9.85)	5.60 (5.58)	6.63 (6.32)	10.08 (9.66)
I/σ (I)	13.94 (2.12)	15.23 (1.17)	13.86 (1.12)	11.86 (1.01)	11.75 (1.03)	16.07 (1.72)
R_{meas} (%)	8.6 (64.5)	10.8 (193.7)	9.8 (177.8)	14.2 (170.6)	16.8 (165.0)	13.6 (140.6)
Overall B factor: Wilson plot/refinement (Å2)	37.8/30.3	47.3/38.8	43.7/36.1	47.2/41.7	60.2/58.3	43.1/37.7
No. of reflections: working/test set	118,348/2101	100,463/2091	116,171/2088	63,550/2101	67,872/2112	66,731/2101
R/R_{free}	0.135/0.165	0.203/0.238	0.182/0.205	0.159/0.205	0.174/0.240	0.166/0.204
No. of non-H atoms: Protein/Ligand/Water	7794/140/826	7619/133/400	7652/97/671	7649/64/617	7624/30/135	7672/65/605
R.m.s. deviations: Bonds (Å)/Angles (°)	0.008/0.964	0.003/0.602	0.010/1.004	0.007/0.873	0.008/1.008	0.002/0.563
Ramachandran plot: Most favored/allowed (%)	97.4/2.6	96.8/3.2	97.6/2.4	97.1/2.9	94.9/5.1	97.2/2.8

2.2. Active Center of *Arth*βDG and *Arth*βDG_E441Q Mutein

Catalytic site of *Arth*βDG, located at the bottom of a relatively wide funnel on the top of catalytic Domain 3, is complemented with parts of the chain from Domain 1 and Domain 5 at the entrance of the active cavity (Figure 2). The funnel leading to the active site has a strongly acidic character, which is beneficial for binding of carbohydrate substrates (Figure 3).

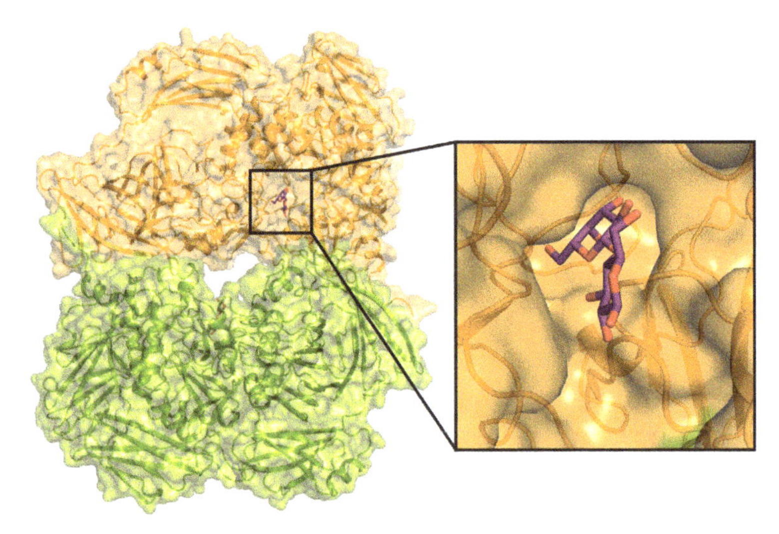

Figure 2. The dimer of *Arth*βDG_E441Q and the zoom of one of the active site cavities with lactose.

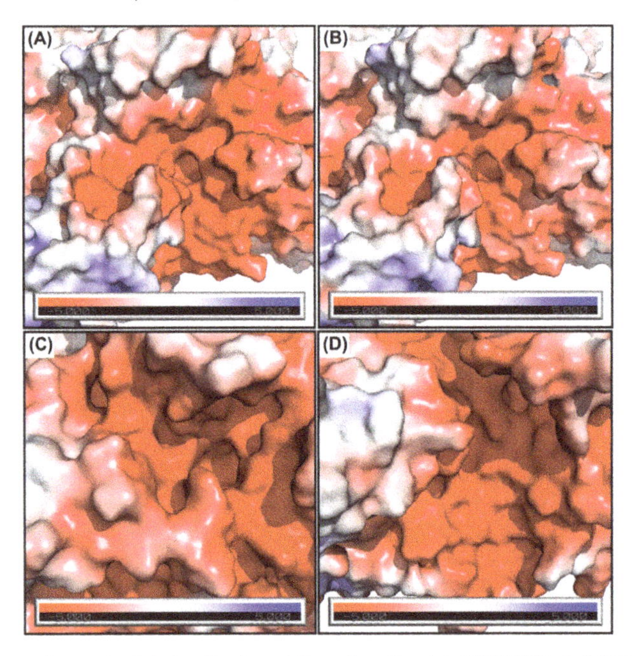

Figure 3. The surface potential visualization at the active site of *Arth*βDG (**A**), *Arth*βDG_E441Q (**B**), *Ecol*βDG (**C**), and *Klyv*βDG (**D**).

The active site cavity has an acidic character throughout, which facilitates the binding of the saccharide substrate, which is typically lactose. Such a shape of the active site cavity is observed for other βDGs with transglycosylation activities: *Ecol*βDG and *Klyv*βDG [4,7,10] as it facilitates the binding of a larger acceptor of galactosyl group, such as galactose, fructose, or salicin. *Arth*βDG forms a widely open entrance to its catalytic site, which makes it more accessible for the substrate but also promotes product dissociation. Both, easier product dissociation and active sites not being shielded or restricted from solvent may be considered as structural cold-adaptation as it influences the enzymes' turnover rate.

2.3. Structural Analysis of Reactions' Mechanism Catalyzed by *Arth*βDG

All determined crystal structures showed precisely the changes in the active site of the enzyme in different stages of catalyzed reaction. Visualizing these structural changes helps to understand and explain a classical Koshland double-displacement mechanism occurring during hydrolysis of (1,4)-β-O-glycosidic bond catalyzed by *Arth*βDG (Figure 4).

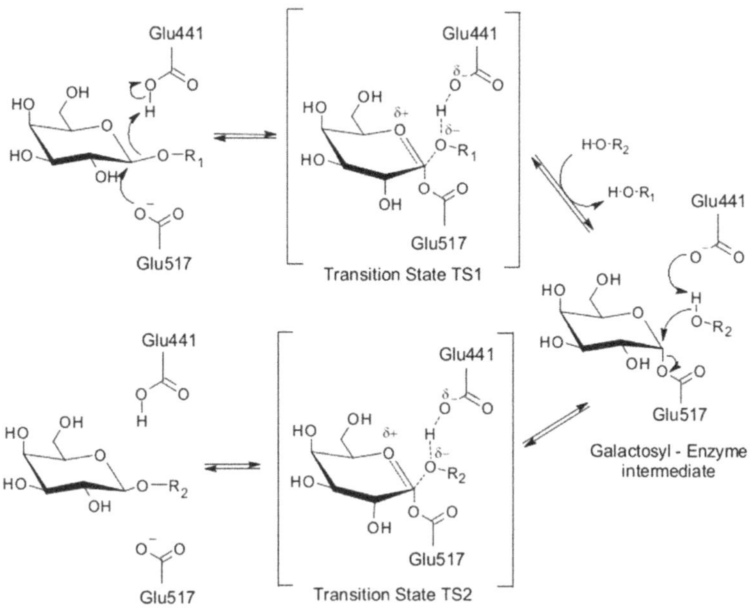

Figure 4. The reaction mechanism of Koshland double displacement with the catalytic residues numbered as for *Arth*βDG [33].

Determined crystal structures of *Arth*βDG complexes with specific ligands may be divided into three groups: complexes with substrates and their analogues (early complexes) *Arth*βDG_E441Q/LACs, *Arth*βDG/IPTG; the second group are intermediates complexes *Arth*βDG_E441Q/LACd *Arth*βDG/ONPG; and the third one is the complex with product *Arth*βDG/GAL.

The early complexes with substrate show LAC and IPTG binding in shallow mode, intermediate complexes depicts deep binding of substrate that directly precedes formation of galactosyl-enzyme covalent bond, and the product complex allows description of the product release process.

At the early stage of the reaction, the substrate is bound in the shallow binding site where the galactosyl group is stabilized by a number of H-bonds between its hydroxyl groups and residues N110, E441, E517, M481, H520, and H368 via water molecules and by an interaction with a sodium ion. Additionally, the glucosyl moiety of lactose is stabilized by H395 and E398 via a water molecule, even though there is already an interaction between substrate and catalytic residues (E441 and E517), and the position of the substrate does not allow access to O-glycosidic bond (Figure 5).

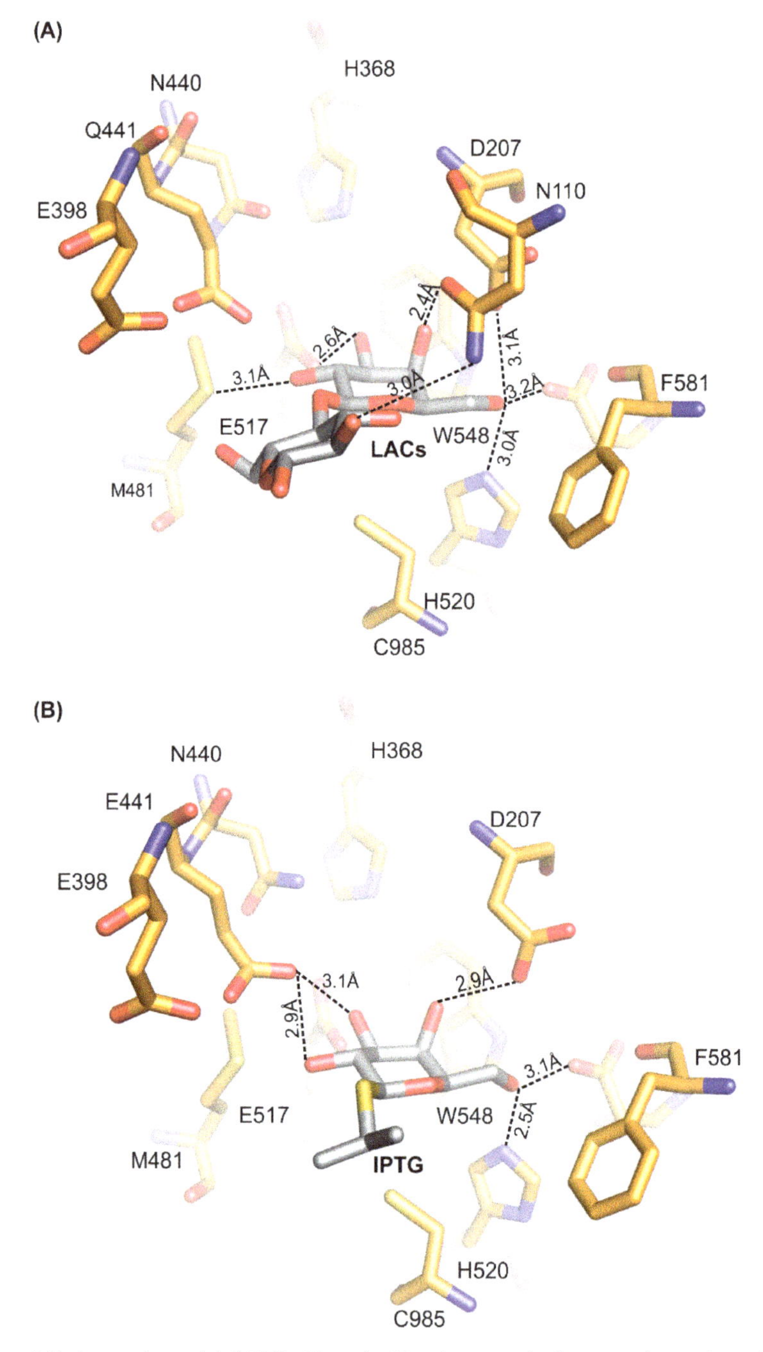

Figure 5. Early complexes of *Arth*βDG with saccharide substrate and substrate analogue: the molecule of lactose (**A**) and IPTG (**B**) bound at shallow binding site.

The insertion of substrate into deep binding is associated with movement of F581 phenyl ring, which rotates around Cα-Cβ bond causing shift of aromatic ring by 2.9 Å in the direction of the active center, reducing the volume of the shallow binding site (Figure 6). Surprisingly, no movement of backbone is observed during the transfer of ligand from shallow to deep binding site. During substrate transfer into the deep binding site, the galactosyl moiety, properly positioned in shallow binding stage, is moved deeper into the active site by approximately 2.4 Å being at the same time rotated around an axis perpendicular to the sugar ring by approximately 60°.

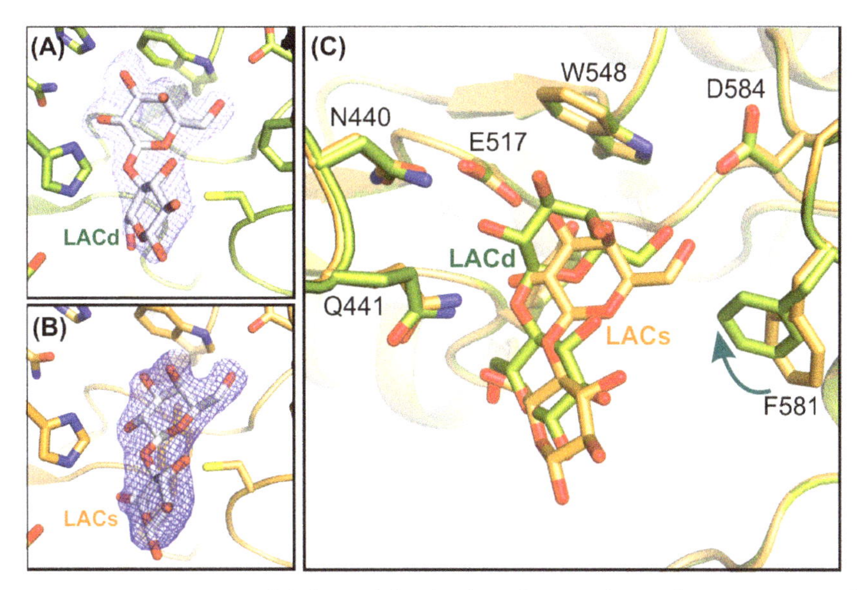

Figure 6. Enzyme active site of shallow and deep binding of lactose. Electron density $2F_o$-F_c map of lactose in deep (**A**) and shallow (**B**) binding mode (contoured at 1σ). Superposition of enzyme active site in both structures (**C**).

In the deep binding site of *Arth*βDG, the galactosyl ring is stabilized by direct interactions of its hydroxyl groups with E441 (Q441 in mutein), E517, H368, D207, sodium ion and N440, D584, H520 (latter observed by deeper bound lactose in complex of *Arth*βDG_E441Q_LACd). Now in *Arth*βDG_E441Q_LACd, the NH_2 from the amide group of Q441 interacts directly with oxygen from the glycosidic bond of lactose, which is 2.6Å away (Figure 7A). In *Arth*βDG/ONPG, the carboxyl group of E441 interacts directly with oxygen from the glycosidic bond of ONPG, which is at a distance of 2.8 Å (Figure 7B).

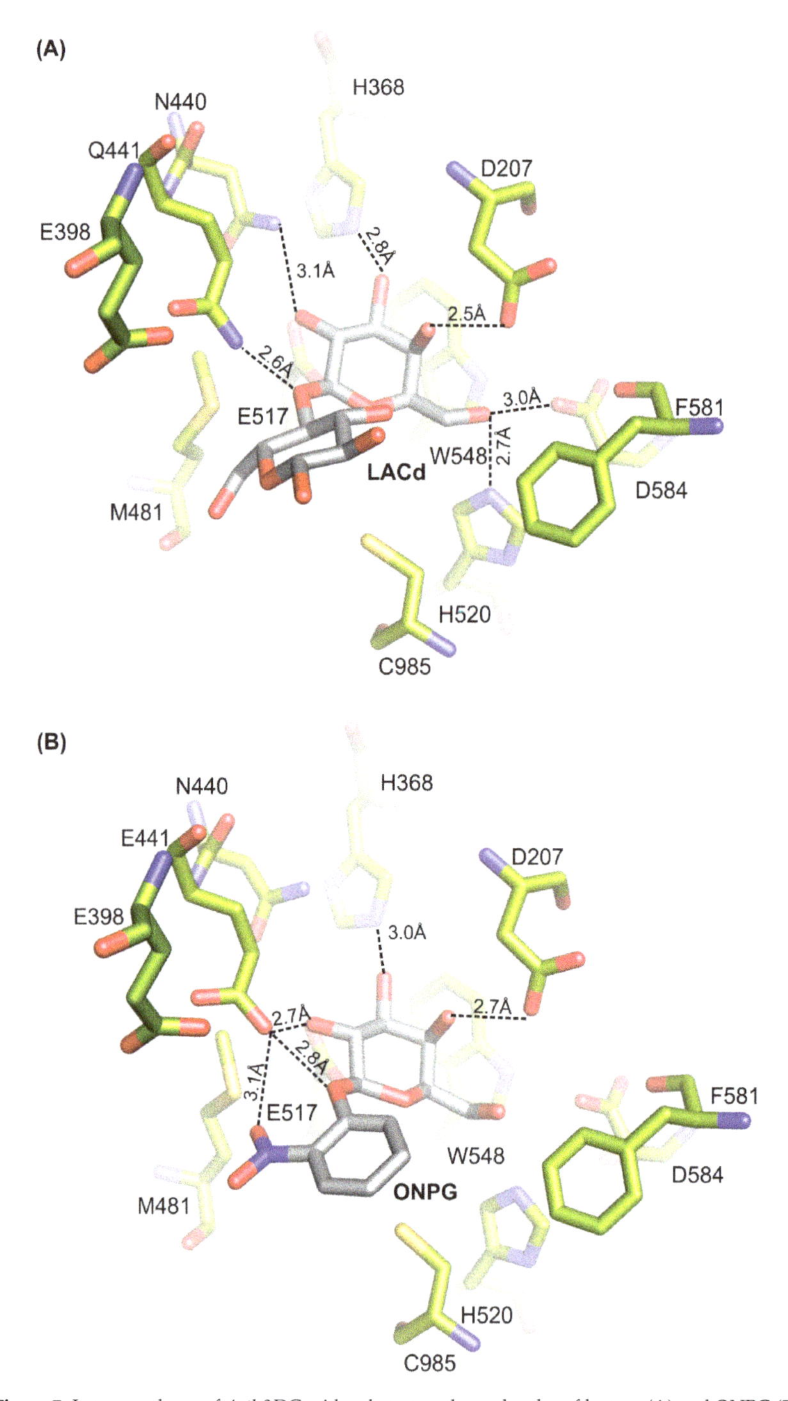

Figure 7. Late complexes of *Arth*βDG with substrates: the molecules of lactose (**A**) and ONPG (**B**).

The superposition of active sites of *Arth*βDG with *Ecol*βDG shows the conservation of amino acids involved in stabilization of the galactosyl moiety. However, these two enzymes differ in the stabilization of a second moiety of β-D-galactoside (Figure 8). If we consider binding of the natural substrate, lactose, the second moiety is glucopyranose. In the active site of *Ecol*βDG, the glucopyranose ring is stabilized by π-stacking interaction with W999. However, in the *Arth*βDG active site, W999 is substituted by C985. This cysteine residue does not influence stabilization of substrate in shallow binding mode. However, when lactose is bound in deep mode, the center of the glucopyranose ring is at a distance of 4.4 Å from C985 making π-sulphur interaction possible. Thus, substitution of W999 with C985 reduces stabilization of the second moiety of β-D-galactoside during shallow binding mode, but is still creating stabilizing interactions when the substrate is bound in deep mode. Furthermore, this results in creating more space in the close vicinity of the active site by substituting the bulky indol group with a smaller cysteine residue side chain.

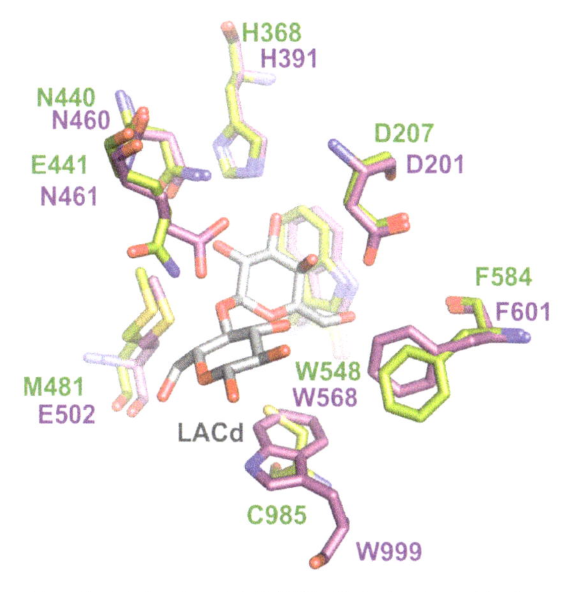

Figure 8. Superposition of catalytic sites of *Arth*βDG with lactose bound in deep mode (green) and *Ecol*βDG (purple).

After the hydrolysis reaction is completed, the F581 side chain moves back into its previous position, opening the way for galactose molecule evacuation from the active site (Figure 9).

The product, now in half-chair conformation, is still stabilized by a number of interactions: D207, H368, N440, E441, Y482, E517, H520, and C985 (Figure 7). The 'open' position of F581 is also observed for unliganded structures of *Arth*βDG (PDB IDs: 6ETZ) and its mutant *Arth*βDG_E441Q, suggesting that its movement is dependent upon substrate presence at the shallow binding site.

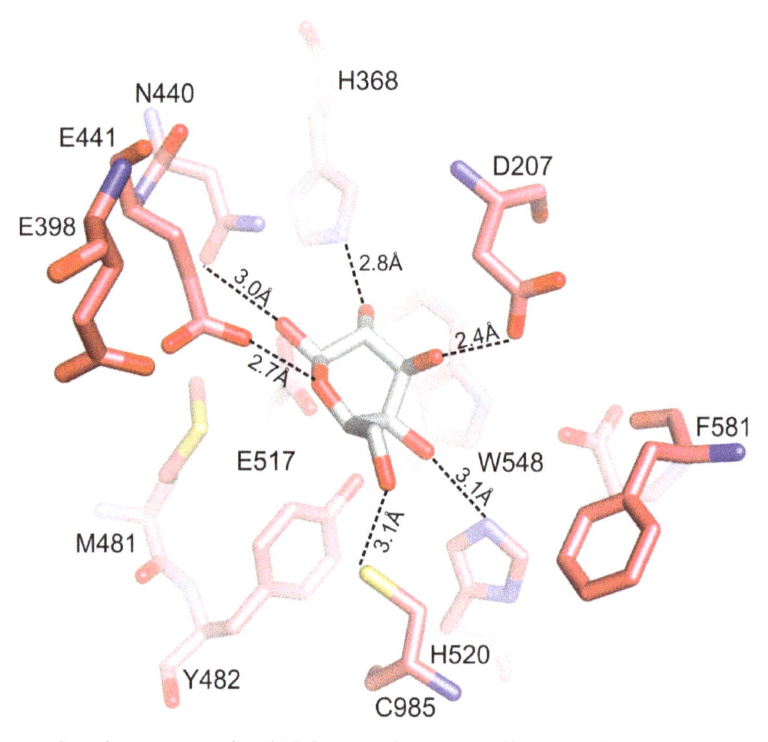

Figure 9. Complex structure of *Arth*βDG with galactose in half-chair conformation bound in the active center.

3. Discussion

GH2 family β-D-galactosidases are sugar configuration-retaining enzymes that follow a classical Koshland double-displacement mechanism. These crystal structures of *Arth*βDG complexes with ligands enabled characterization of the active site and determined which residues take part in two modes of substrate binding: deep and shallow.

The large rotation of the galactosyl residue during deep binding would most probably result in forming π-stacking interaction with W548. Such a form of intermediate stabilization was described for *Ecolβ*DG [7]. It should be noted that a tryptophan residue is the preferred aromatic amino acid for binding of carbohydrates [34,35] and is frequently present in carbohydrate binding domains of proteins. In the case of *Arth*βDG, only one tryptophan, W548, is located in the bottom of the active site. Additionally, three tryptophan residues are present at the entrance to the catalytic pocket (W402, W470, and W773), where they may form platforms for initial sugar binding. Another amino acid which is considered to play an important role in carbohydrate binding is histidine. The active site of *Arth*βDG contains several histidine residues: H334, H368, H395, H520, and H553. Among them, H368 and H520 are directly involved in stabilizing the galactosyl moiety. H520 is primarily involved in stabilization of hydroxyl group O6 of the galactosyl moiety during shallow binding of substrate. When the substrate is moved deeper into catalytic site, H368 stabilizes the position of hydroxyl group O3. It must be noted that the catalytic site architecture of *Arth*βDG is composed such a way that only a sugar moiety with a proper conformation of hydroxyls O2, O3, and O4 can be effectively bound in the active site. Hence, residues forming H-bonds with hydroxyl groups in these positions H368, N440, and D207, play a crucial role in enzyme's specificity.

It is worth noting that a typical chair conformation of the galactosyl ring in substrate ($^1C_4$) is changing to a half-chair ($^3H_4$)^{hkkkkHhhh} conformation in the still bound product of the half-reaction.

There are many conformations of pyranose ring possible in solution; however, some of them are more stable than others. In the case of lactose, it usually has a relaxed chair conformation in solution. The double displacement mechanism, in which lactose is hydrolyzed by retaining galactosidases, such as *Arth*βDG, undergoes formation of two oxocarbenium ion-like transition states (Figure 4). Such transition states must be formed with sp^2 hybridization and formation of a positive charge on anomeric carbon atom of the substrate. Only a few conformations of galactosyl moiety allow sp^2 hybridization on anomeric carbon, one of which is half-chair conformation $^3H_4$, observed for the galactose bound in active site of *Arth*βDG [36].

The rotation of F581 (F601 in *Ecol*βDG) was described as one of the factors associated with the deep binding mode, together with 10 Å movement of a 10-aa loop from Domain 5. However, in the case of *Arth*βDG, the movement of F581 and D207 are the only conformational changes accompanying the reaction mechanism. In fact, the 10-aa loop in *Arth*βDG is stabilized by a number of strong interactions with other parts of Domain 5, in a position allowing better access of the substrate to the active site. It should be noted that it is one of the regions of *Arth*βDG in which the backbone differed significantly from homologous structures. These facts lead us to consider this permanent exposure of the entrance to the active site as a structural adaptation towards activity in cold conditions. Fewer structural hindrances for substrate entering and product leaving the active site can result in a higher turnover rate. Analysis of these obtained crystal structures shows that *Arth*βGD forms a widely open entrance to its catalytic site, which makes it more accessible for the saccharide substrate and promotes product dissociation.

Both galactosyl binding sites, shallow and deep ones, form a net of H-bonds that stabilize this part of the substrate. On the other hand, the glucosyl moiety of lactose, or the non-galactose moieties of IPTG and ONPG, are hardly stabilized by any interactions during shallow binding. The *Ecol*βDG W999 is substituted at *Arth*βDG with a cysteine residue which may stabilize the second sugar ring of the substrate by π-sulphur interactions during deep binding, however such a substitution would render the enzyme less specific toward binding a sugar moiety at this position. Thus, not only disaccharides, but also other galactosides are processed by *Arth*βDG. The enzyme's lack of preference for the second moiety in galactoside may be the main reason for its ability to hydrolyze a wide variety of substrates, as well as for its ability to transfer galactosyl group to a variety of acceptors [16] resulting in an interesting range of potentially useful heterooligosaccharides.

4. Materials and Method

4.1. Site-Directed Mutagenesis of Gene Encoding ArthβDG

The gene encoding the *Arth*βDG enzyme, which was previously cloned into the pBAD/Myc-His A expression vector [16], has been mutated in a site-specific manner using the Q5 Site-Directed Mutagenesis Kit (NEB, Ipswich, MA, USA) following the manufacturer's protocol. For this purpose, a pair of mutagenic primers was designed and synthesized (Genomed, Warszawa, Poland). Primer ForBglAr32cBm441: 5′GTCCCTGGGCAACCAGGCGGCACCGG3′ and primer RevBglAr32m441: 5′CACATGACCACCGAGGCGTGGTTCTTGTCGCGC3′ allowed us to introduce a point mutation at 1321 nucleotide position in the gene substituting G with C resulting in the substitution of glutamic acid (E) residue with glutamate (Q) residue in the 441 position of the amino acid chain of *Arth*βDG. Hence, the product of mutated gene expression has been called *Arth*βDG_E441Q. In theory, this amino acid change should abolish β-D-galactosidase activity of mutein *Arth*βDG_E441Q.

PCR cycling conditions were as follows: (1) Initial DNA denaturation at 98 °C for 30 s; then (2) 25 cycles of PCR product amplification consisting of 10 s of DNA denaturation at 98 °C, 20 s of mutagenic primers annealing at 70 °C, and 3 min 20 s of PCR product extension at 72 °C; and (3) the final PCR product extension at 72 °C for 7 min. After PCR, the amplified DNA product was directly added to unique Kinase-Ligase-DpnI (KLD) enzymes mix. Then the product of KLD reaction (5 min at room temperature) was directly used to transform NEB 5-alpha chemically competent

E.coli cells (the *lacZ* deletion mutant, Δ (lacZ) M15). After that, transformants were spread on Luria–Bertani agar plates (10 g L^{-1} of peptone K, 5 g L^{-1} of yeast extract, 10 g L^{-1} of NaCl, and 15 g L^{-1} of agar) supplemented with ampicillin (100 μg mL^{-1}), X-Gal (40 μg mL^{-1}) and L-arabinose (200 μg mL^{-1}). After plate incubation—firstly at 37 °C for 12 h, and then at 22 °C for next 12 h—a few recombinant colonies without β-D-galactosidase activity were chosen for further studies. Plasmids isolated using the ExtractMe Plasmid DNA Kit (Blirt, Gdansk, Poland) from selected recombinants were sequenced (Genomed, Warszawa, Poland) and analyzed (blast2go on-line tool). Recombinant plasmid pBAD-Bgal32cB_E441Q(A) harboring the properly mutated *Arthrobacter* sp. 32cB β-D-galactosidase gene under the control of the P_{BAD} promoter was used for effective production of *Arth*βDG_E441Q mutein in *E. coli* host [16].

4.2. Expression and Purification of ArthβDG and ArthβDG_E441Q

Heterologous expressions of recombinant *Arth*βDG and *Arth*βDG_E441Q proteins were performed in the *E. coli* LMG 194 cells transformed with pBAD-Bgal32cB and pBAD-Bgal32cB_E441Q plasmids, respectively, as previously described. [25] Both proteins were purified by two ion-exchange chromatography steps (weak anion exchanger and strong anion exchanger), followed by a size-exclusion chromatography step.

The fractions containing *Arth*βDG were identified by SDS-PAGE electrophoresis run on 10% SDS-polyacrylamide gel and by enzymatic activity assay with ONPG as a substrate [25], whereas the fractions containing *Arth*βDG_E441Q were identified by SDS-PAGE only, due to lack of enzymatic activity. The sample buffer was changed into 0.05 M HEPES pH 7.0 and the samples were concentrated using 50 kDa cut-off membrane Vivaspin filters (Sartorius, Goettingen, Germany) up to the protein concentration of 15 mg/mL.

4.3. ArthβDG Crystallization and Diffraction Data Collection

Crystals of *Arth*βDG and *Arth*βDG_E441Q mutein were grown using the same optimization matrix of 25–45% Tacsimate^{TM} and pH ranges between 6.0–8.0. All the drops were set up using a seed stock prepared from crystals of *Arth*βDG grown at 35% Tacsimate^{TM} pH 7.0 and diluted 10,000 times. Numerous attempts of co-crystallization with ligands were undertaken but no crystal structures of desired complexes were obtained. Furthermore, addition of natural substrate, lactose, prevented formation of *Arth*βDG_E441Q crystals even at very low concentration of added ligand. Crystal structures of investigated *Arth*βDG and *Arth*βDG_E441Q complexes were obtained by soaking of native and mutant crystals with desired ligand or ligands mixture. Soaking was performed by adding powder of ligand directly to the crystallization drop. The soaking experiments were performed for 15 min, 30 min, 1 h, 2 h, 6 h, 14 h, and 24 h prior to flash-freezing. The crystals, prior to mounting and flash-freezing, were protected with 60% Tacsimate^{TM} of pH corresponding to crystallization conditions [37].

High-resolution diffraction data were collected using synchrotron sources on beamlines 14.1 and 14.2 at BESSY, Berlin, Germany and P13 beamline at PETRA, DESY Hamburg, Germany. The diffraction images were collected with fine slicing 0.1° and diffraction data were processed using XDSapp [38]. Crystal structures were solved and refined using the PHENIX program suite [39]. As a model, the structure of *Arth*βDG (PDB ID: 6ETZ) was used.

Author Contributions: M.R. performed crystallization; M.R. and A.B. performed synchrotron diffraction data collection, processing, structure solving, and carried out structural analysis; M.R. purified enzyme, refined the structures; M.R. and A.B. prepared the manuscript; M.W. performed native *Arth*βDG enzyme expression in *E. coli* and determined the purification protocol; H.C. designed and performed site-direct mutagenesis experiment resulted in a gene encoding *Arth*βDG_E441Q mutein; A.W.-W. performed *Arth*βDG-E441Q mutein expression in *E.coli*; A.B. coordinated the project.

Funding: This research was funded by National Science Centre of Poland grant number 2016/21/B/ST5/00555 (A.B.) and 2018/28/T/ST5/00233 scholarship (M.R.).

Acknowledgments: We thank HZB for the allocation of synchrotron radiation beamtime at BL 14.1, BL 14.2, and PETRA synchrotron at P13.

Conflicts of Interest: The authors declare that they have no conflict of interest.

Abbreviations

*Arth*βDG	β-D-galactosidase from *Arthrobacter* sp. 32cB
*Ecol*βDG	β-D-galactosidase from *Escherichia coli*
*Klyv*βDG	β-D-galactosidase from *Klyvuromyces lactis*
GAL	galactose
GOS	galactooligosaccharides
HOS	heterooligosaccharides
IPTG	isopropyl β-D-1-thiogalactopyranoside
Lac	Lactose
ONPG	ortho-nitrophenyl-β-galactoside
X-gal	5-bromo-4-chloro-3-indolyl-β-D-galactopyranoside

References

1. Talens-Perales, D.; Górska, A.; Huson, D.H.; Polaina, J.; Marín-Navarro, J. Analysis of Domain Architecture and Phylogenetics of Family 2 Glycoside Hydrolases (GH2). *PLoS ONE* **2016**, *11*, e0168035. [CrossRef] [PubMed]
2. Cohn, M.; Monod, J. Purification and properties of the beta-galactosidase (lactase) of *Escherichia coli*. *Biochim. Biophys. Acta* **1951**, *7*, 153–174. [CrossRef]
3. Jacobson, R.H.; Zhang, X.J.; DuBose, R.F.; Matthews, B.W. Three-dimensional structure of beta-galactosidase from *E. coli*. *Nature* **1994**, *369*, 761–766. [CrossRef] [PubMed]
4. Juers, D.H.; Matthews, B.W.; Huber, R.E. LacZ beta-galactosidase: Structure and function of an enzyme of historical and molecular biological importance. *Protein Sci.* **2012**, *21*, 1792–1807. [CrossRef] [PubMed]
5. Brás, N.F.; Fernandes, P.A.; Ramos, M.J. QM/MM Studies on the β-Galactosidase Catalytic Mechanism: Hydrolysis and Transglycosylation Reactions. *J. Chem. Theory Comput.* **2010**, *6*, 421–433. [CrossRef] [PubMed]
6. Juers, D.H.; Rob, B.; Dugdale, M.L.; Rahimzadeh, N.; Giang, C.; Lee, M.; Matthews, B.W.; Huber, R.E. Direct and indirect roles of His-418 in metal binding and in the activity of beta-galactosidase (*E. coli*). *Protein Sci.* **2009**, *18*, 1281–1292. [CrossRef] [PubMed]
7. Juers, D.H.; Heightman, T.D.; Vasella, A.; McCarter, J.D.; Mackenzie, L.; Withers, S.G.; Matthews, B.W. A structural view of the action of *Escherichia coli* (lacZ) beta-galactosidase. *Biochemistry* **2001**, *40*, 14781–14794. [CrossRef] [PubMed]
8. Harju, M. Milk sugars and minerals as ingredients. *Int. J. Dairy Technol.* **2001**, *54*, 61–63. [CrossRef]
9. Harju, M.; Kallioinen, H.; Tossavainen, O. Lactose hydrolysis and other conversions in dairy products: Technological aspects. *Int. Dairy J.* **2012**, *22*, 104–109. [CrossRef]
10. Pereira-Rodriguez, A.; Fernandez-Leiro, R.; Gonzalez-Siso, M.I.; Cerdan, M.E.; Becerra, M.; Sanz-Aparicio, J. Structural basis of specificity in tetrameric *Kluyveromyces lactis* beta-galactosidase. *J. Struct. Biol.* **2012**, *177*, 392–401. [CrossRef]
11. Stevenson, D.E.; Stanley, R.A.; Furneaux, R.H. Optimization of alkyl beta-D-galactopyranoside synthesis from lactose using commercially available beta-galactosidases. *Biotechnol. Bioeng.* **1993**, *42*, 657–666. [CrossRef]
12. Klewicki, R.; Belina, I.; Wojciechowska, A.; Klewicka, E.; Sójka, M. Synthesis of Galactosyl Mannitol Derivative Using β-Galactosidase from *Kluyveromyces lactis*. *Pol. J. Food Nutr. Sci.* **2017**, *67*, 33–39. [CrossRef]
13. Wojciechowska, A.; Klewicki, R.; Sójka, M.; Klewicka, E. Synthesis of the Galactosyl Derivative of Gluconic Acid with the Transglycosylation Activity of β-Galactosidase. *Food Technol. Biotechnol.* **2017**, *55*, 258–265. [CrossRef]
14. Cavicchioli, R.; Charlton, T.; Ertan, H.; Omar, S.M.; Siddiqui, K.S.; Williams, T.J. Biotechnological uses of enzymes from psychrophiles. *Microb. Biotechnol.* **2011**, *4*, 449–460. [CrossRef]
15. Cavicchioli, R.; Siddiqui, K.S.; Andrews, D.; Sowers, K.R. Low-temperature extremophiles and their applications. *Curr. Opin. Biotechnol.* **2002**, *13*, 253–261. [CrossRef]

16. Pawlak-Szukalska, A.; Wanarska, M.; Popinigis, A.T.; Kur, J. A novel cold-active β-d-galactosidase with transglycosylation activity from the Antarctic *Arthrobacter* sp. 32cB–Gene cloning, purification and characterization. *Process Biochem.* **2014**, *49*, 2122–2133. [CrossRef]
17. Boehm, G.; Fanaro, S.; Jelinek, J.; Stahl, B.; Marini, A. Prebiotic concept for infant nutrition. *Acta Paediatr.* **2003**, *92*, 64–67. [CrossRef]
18. Lee, L.Y.; Bharani, R.; Biswas, A.; Lee, J.; Tran, L.A.; Pecquet, S.; Steenhout, P. Normal growth of infants receiving an infant formula containing *Lactobacillus reuteri*, galacto-oligosaccharides, and fructo-oligosaccharide: A randomized controlled trial. *Matern. Health Neonatol. Perinatol.* **2015**, *1*, 9. [CrossRef]
19. Li, M.; Monaco, M.H.; Wang, M.; Comstock, S.S.; Kuhlenschmidt, T.B.; Fahey, G.C., Jr.; Miller, M.J.; Kuhlenschmidt, M.S.; Donovan, S.M. Human milk oligosaccharides shorten rotavirus-induced diarrhea and modulate piglet mucosal immunity and colonic microbiota. *ISME J.* **2014**, *8*, 1609–1620. [CrossRef]
20. Hughes, C.; Davoodi-Semiromi, Y.; Colee, J.C.; Culpepper, T.; Dahl, W.J.; Mai, V.; Christman, M.C.; Langkamp-Henken, B. Galactooligosaccharide supplementation reduces stress-induced gastrointestinal dysfunction and days of cold or flu: A randomized, double-blind, controlled trial in healthy university students. *Am. J. Clin. Nutr.* **2011**, *93*, 1305–1311. [CrossRef]
21. Kunz, C.; Rudloff, S. Biological functions of oligosaccharides in human milk. *Acta Paediatr.* **1993**, *82*, 903–912. [CrossRef]
22. Torres, D.P.; Gonçalves, M.; Teixeira, J.A.; Rodrigues, L.R. Galacto-Oligosaccharides: Production, Properties, Applications, and Significance as Prebiotics. *Compr. Rev. Food Sci. Food Saf.* **2010**, *9*, 438–454. [CrossRef]
23. McVeagh, P.; Miller, J.B. Human milk oligosaccharides: Only the breast. *Acta Paediatr.* **1997**, *33*, 281–286. [CrossRef]
24. Musilova, S.; Rada, V.; Vlkova, E.; Bunesova, V. Beneficial effects of human milk oligosaccharides on gut microbiota. *Benef. Microbes.* **2014**, *5*, 273–283. [CrossRef]
25. Mussatto, S.I.; Mancilha, I.M. Non-digestible oligosaccharides: A review. *Carbohydr. Polym.* **2007**, *68*, 587–597. [CrossRef]
26. Oliveira, D.L.; Wilbey, R.A.; Grandison, A.S.; Roseiro, L.B. Milk oligosaccharides: A review. *Diary Technol.* **2015**, *68*, 305–321. [CrossRef]
27. Manas, N.H.A.; Illias, R.M.; Mahadi, N.M. Strategy in manipulating transglycosylation activity of glycosyl hydrolase for oligosaccharide production. *Crit. Rev. Biotechnol.* **2018**, *38*, 272–293. [CrossRef]
28. Rutkiewicz-Krotewicz, M.; Pietrzyk-Brzezinska, A.J.; Wanarska, M.; Cieslinski, H.; Bujacz, A. In Situ Random Microseeding and Streak Seeding Used for Growth of Crystals of Cold-Adapted beta-D-Galactosidases: Crystal Structure of beta DG from *Arthrobacter* sp. 32cB. *Crystals* **2018**, *8*, 13. [CrossRef]
29. Rutkiewicz-Krotewicz, M.; Pietrzyk-Brzezinska, A.J.; Sekula, B.; Cieśliński, H.; Wierzbicka-Woś, A.; Kur, J.; Bujacz, A. Structural studies of a cold-adapted dimeric β-D-galactosidase from *Paracoccus* sp. 32d. *Acta Cryst. D Struct. Biol.* **2016**, *72*, 1049–1061. [CrossRef]
30. Rutkiewicz, M.; Bujacz, A.; Bujacz, G. Structural features of cold-adapted dimeric GH2 β-D-galactosidase from *Arthrobacter* sp. 32cB. *Biochim. Biophys. Acta (BBA)-Proteins Proteom.* **2019**, *1867*, 776–786. [CrossRef]
31. Matthews, B.W. Solvent content of protein crystals. *J. Mol. Biol.* **1968**, *33*, 491–497. [CrossRef]
32. Winn, M.D.; Isupov, M.N.; Murshudov, G.N. Use of TLS parameters to model anisotropic displacements in macromolecular refinement. *Acta Cryst. D* **2001**, *57*, 122–133. [CrossRef]
33. Sinnott, M.L.; Souchard, I.J.L. The mechanism of action of β-galactosidase. Effect of aglycone nature and α-deuterium substitution on the hydrolysis of aryl galactosides. *Biochem. J.* **1973**, *133*, 89–98. [CrossRef]
34. Hudson, K.L.; Bartlett, G.J.; Diehl, R.C.; Agirre, J.; Gallagher, T.; Kiessling, L.L.; Woolfson, D.N. Carbohydrate-Aromatic Interactions in Proteins. *J. Am. Chem. Soc.* **2015**, *137*, 15152–15160. [CrossRef]
35. Malecki, P.H.; Vorgias, C.E.; Petoukhov, M.V.; Svergun, D.I.; Rypniewski, W. Crystal structures of substrate-bound chitinase from the psychrophilic bacterium *Moritella marina* and its structure in solution. *Acta Cryst. D Biol. Cryst.* **2014**, *70*, 676–684. [CrossRef]
36. Ardevol, A.; Rovira, C. Reaction Mechanisms in Carbohydrate-Active Enzymes: Glycoside Hydrolases and Glycosyltransferases. Insights from ab Initio Quantum Mechanics/Molecular Mechanics Dynamic Simulations. *J. Am. Chem. Soc.* **2015**, *137*, 7528–7547. [CrossRef]
37. Bujacz, G.; Wrzesniewska, B.; Bujacz, A. Cryoprotection properties of salts of organic acids: A case study for a tetragonal crystal of HEW lysozyme. *Acta Cryst. D* **2010**, *66*, 789–796. [CrossRef]

38. Sparta, K.M.; Krug, M.; Heinemann, U.; Mueller, U.; Weiss, M.S. XDSAPP2.0. *J. Appl. Crystallogr.* **2016**, *49*, 1085–1092. [CrossRef]
39. Adams, P.D.; Afonine, P.V.; Bunkoczi, G.; Chen, V.B.; Davis, I.W.; Echols, N.; Headd, J.J.; Hung, L.-W.; Kapral, G.J.; Grosse-Kunstleve, R.W.; et al. PHENIX: A comprehensive Python-based system for macromolecular structure solution. *Acta Crystallogr. Sect. D* **2010**, *66*, 213–221. [CrossRef]

International Journal of
Molecular Sciences

Article

Structural Comparison of a Promiscuous and a Highly Specific Sucrose 6^F^-Phosphate Phosphorylase

Jorick Franceus [1,*], Nikolas Capra [2], Tom Desmet [1] and Andy-Mark W.H. Thunnissen [2]

[1] Centre for Synthetic Biology (CSB), Department of Biotechnology, Ghent University, Coupure Links 653, 9000 Ghent, Belgium

[2] Groningen Biomolecular Sciences and Biotechnology Institute, University of Groningen, Nijenborgh 4, 9747 AG Groningen, The Netherlands

* Correspondence: jorick.franceus@ugent.be; Tel.: +32-9264-9920

Received: 17 July 2019; Accepted: 8 August 2019; Published: 11 August 2019

Abstract: In family GH13 of the carbohydrate-active enzyme database, subfamily 18 contains glycoside phosphorylases that act on α-sugars and glucosides. Because their phosphorolysis reactions are effectively reversible, these enzymes are of interest for the biocatalytic synthesis of various glycosidic compounds. Sucrose 6^F-phosphate phosphorylases (SPPs) constitute one of the known substrate specificities. Here, we report the characterization of an SPP from *Ilumatobacter coccineus* with a far stricter specificity than the previously described promiscuous SPP from *Thermoanaerobacterium thermosaccharolyticum*. Crystal structures of both SPPs were determined to provide insight into their similarities and differences. The residues responsible for binding the fructose 6-phosphate group in subsite +1 were found to differ considerably between the two enzymes. Furthermore, several variants that introduce a higher degree of substrate promiscuity in the strict SPP from *I. coccineus* were designed. These results contribute to an expanded structural knowledge of enzymes in subfamily GH13_18 and facilitate their rational engineering.

Keywords: GH13_18; sucrose phosphorylase; glycoside phosphorylase; *Ilumatobacter coccineus*; *Thermoanaerobacterium thermosaccharolyticum*; crystallography

---

1. Introduction

Glycoside phosphorylases (GPs) are carbohydrate-active enzymes that catalyze the reversible cleavage of glycosidic bonds using inorganic phosphate [1–3]. In their physiological context, they provide a degradative route for carbohydrates and glycosides that is more energy-efficient when compared to hydrolysis. Their glycosyl phosphate reaction products can enter various pathways, such as glycolysis, without prior activation by a kinase, saving one molecule of ATP [4]. Because of the reversibility of the phosphorolytic reaction, GPs are also attractive enzymes for the synthesis of glycosidic bonds in vitro. One of the most well-studied GPs in that regard is sucrose phosphorylase (SP; EC 2.4.1.7), which catalyzes the phosphorolysis of sucrose into α-D-glucose 1-phosphate (Glc1P) and D-fructose. Thanks to the high energy content of sucrose and the enzyme's remarkable substrate promiscuity, SP is often applied as a catalyst in cost-effective transglucosylation reactions using the cheap and renewable substrate sucrose as a glucosyl donor [5–10]. Moreover, SP has been a target in various engineering studies to further alter its specificity, selectivity, and thermostability [11–15].

Sucrose phosphorylase is found in subfamily 18 of glycoside hydrolase family GH13 (GH13_18) according to the Carbohydrate-Active Enzyme database (http://www.cazy.org) [16,17]. While searching this subfamily for more thermostable SPs, we previously came across a peculiar putative SP from the thermophile *Thermoanaerobacterium thermosaccharolyticum*. Although the enzyme did show significant activity on sucrose, as was expected, its kinetic parameters revealed that sucrose is in fact not its native substrate. Instead, a clear preference for sucrose 6^F-phosphate was observed, making it the first sucrose

6^F-phosphate phosphorylase (SPP; EC 2.4.1.329) ever reported [18]. A few key residues responsible for the difference in specificity were identified through homology modeling and mutagenesis. Furthermore, the *T. thermosaccharolyticum* SPP (TtSPP) was shown to accept an even wider range of substrates than the known SPs, offering opportunities for biotechnological applications and further engineering. For instance, a mutant of TtSPP was rationally designed to achieve high activity towards bulky phenolics such as resveratrol, enabling the quantitative production of a resveratrol glucoside [19]. The same mutant could also be applied for the synthesis of glucosylated 3-hydroxy-β-lactams [20].

The discovery of TtSPP hinted at a broader functional diversity in the family left to uncover and triggered the search for other novel GH13_18 enzymes in recent years. By characterizing proteins from unexplored branches of the phylogenetic tree, phosphorylases from *Meiothermus silvanus*, *Spirochaeta thermophila*, and *Escherichia coli* were found to be dedicated to the substrate 2-*O*-α-D-glucosylglycerate (EC 2.4.1.352) [21]. Further, a comparison of distinctive active site sequence motifs in different branches of the tree led to the discovery of a *Marinobacter adhaerens* enzyme that acts as a configuration-retaining 2-*O*-α-D-glucosylglycerol phosphorylase (EC 2.4.1.359) [22]. In this work, we report the expression and characterization of an SPP from *Ilumatobacter coccineus* YM16–304 (IcSPP) with active site sequence motifs that are different from the motifs found in TtSPP or in the phosphorylases with other specificities in the GH13_18 subfamily. The enzyme shows high activity on sucrose 6^F-phosphate, but unlike TtSPP, it does not catalyze reactions with alternative substrates. Crystal structures of the two SPPs were determined to allow for a comparison of their active site. Automated docking and mutational analyses were carried out to obtain variants of IcSPP with a less stringent specificity.

2. Results and Discussion

2.1. Choice of the Target Sequence

As knowledge on the diversity of substrate specificities in subfamily GH13_18 expanded, a few sequence motifs that are highly conserved among enzymes sharing the same specificity could be identified. Such signature motifs were first detected in sucrose 6^F-phosphate phosphorylases and glucosylglycerate phosphorylases and were then utilized to predict that the putative sucrose phosphorylase from *M. adhaerens* is actually a glucosylglycerol phosphorylase [18,21,22]. Perhaps the most distinct specificity-determining region is the so-called loop A, found in domain B′ between strand β7 and helix α7 of the catalytic domain (Figure 1). For example, one position in that loop carries a glutamine residue in all enzymes except glucosylglycerate phosphorylases, where glutamate is present instead. Likewise, glucosylglycerol phosphorylases possess a conserved VGA motif that is absent in all other enzymes.

Another clade of interesting enzymes was encountered by thoroughly searching the subfamily's phylogenetic tree for more sequences with an aberrant motif in loop A. These atypical sequences feature several notable conservations that are not found in any other clades, such as a YYQ motif and a lysine at a position where others hold glycine, valine, or asparagine. The enzymes in the target clade primarily originate from the microbial classes *Clostridia*, *Flavobacteriia*, and *Acidimicrobiia*. We set out to elucidate their properties and selected a candidate from *Ilumatobacter coccineus* YM16–304 for expression and characterization. Very recently, Tauzin et al. already reported the characterization of an enzyme from the gut bacterium *Ruminococcus gnavus* that shows the same characteristic residues in loop A [23]. That enzyme, which has 52% sequence identity to the homologue from *Ilumatobacter coccineus* presented here, was found to be an SPP with a very strict specificity.

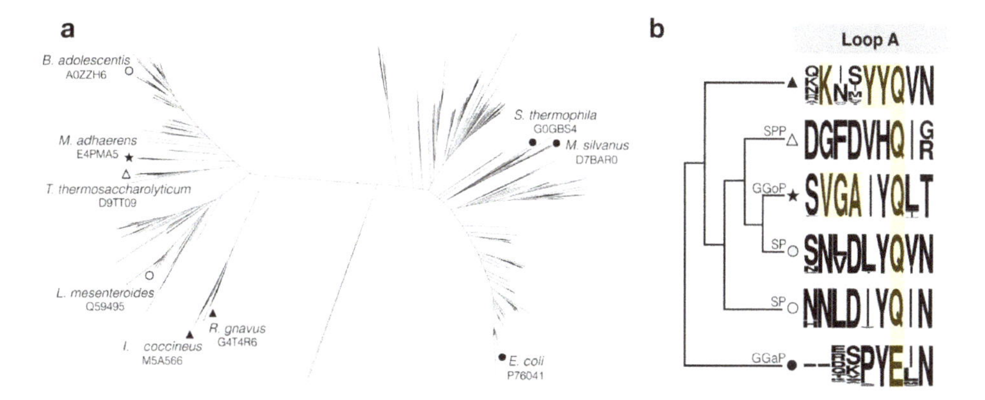

Figure 1. (**a**) Phylogenetic tree of protein sequences classified in subfamily GH13_18. A selection of representatives that have been characterized and reported in the literature are indicated with their UniProt ID. (**b**) Simplified phylogenetic tree showing the sequence logo of an acceptor site loop for the proteins in the branch. Motifs mentioned in the main text are highlighted. Black triangle, IcSPP-like sucrose 6^F-phosphate phosphorylases (target clade in this study); white triangle, TtSPP-like sucrose 6^F-phosphate phosphorylases; star, glucosylglycerol phosphorylases (GGoP); grey circle, *Bifidobacterium*-like sucrose phosphorylases; white circle, lactic acid bacteria-like sucrose phosphorylases; black circle, glucosylglycerate phosphorylases (GGaP).

2.2. Expression and Characterization of IcSPP

IcSPP, provided with a C-terminal His_6-tag, was recombinantly expressed in *E. coli* and purified to apparent homogeneity by affinity chromatography. Approximately 12 mg of purified protein was obtained from a 250 mL culture. The protein migrated in sodium dodecyl sulfate-polyacrylamide gel electrophoresis as a single band with an apparent molecular mass that is in accordance with the theoretical mass deduced from the amino acid sequence (59.9 kDa; Figure S1).

The activity in synthetic direction was determined for more than 30 different putative acceptor substrates, using Glc1P as glucosyl donor (Table S1). When fructose 6-phosphate was added, the measured specific activity (45 ± 3 U/mg) was in the same range as reported wild-type activities of other GH13_18 phosphorylases. However, no activity was observed on any of the other compounds. These results indicate that the enzyme is a sucrose 6^F-phosphate phosphorylase with strict specificity (Figure 2), consistent with the earlier characterization of the homologous *R. gnavus* SPP by Tauzin et al. [23]. Reactions in the physiologically relevant degradative direction confirmed the high activity on sucrose 6^F-phosphate (53 ± 4 U/mg) and the lack of activity on other compounds. IcSPP and the *R. gnavus* SPP described previously thus behave very differently from the promiscuous *T. thermosaccharolyticum* SPP (TtSPP) that resides in a different clade of the subfamily's phylogenetic tree.

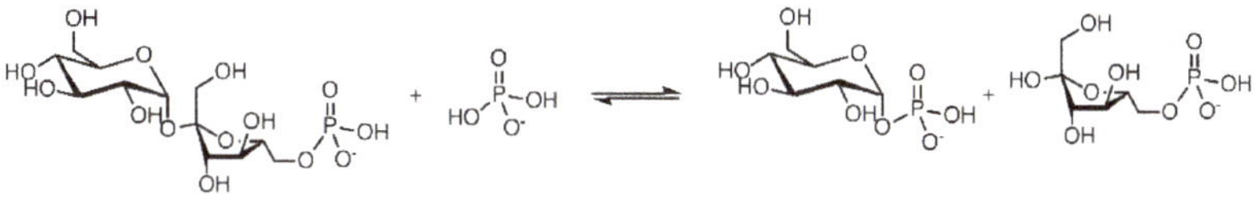

Figure 2. Reaction catalyzed by sucrose 6^F-phosphate phosphorylase.

The biochemical properties of IcSPP were investigated. The optimal pH in the synthesis direction of the reactions was 6.5, and more than 50% of the maximum activity was retained within the pH range of 5.5 to 8 (Figure S2A). In the phosphorolytic direction, the optimum was reached at pH 6. Optimal activity was achieved at a temperature of 35 °C (Figure S2B), which is in line with the mesophilic nature of the *Ilumatobacter* strain that the protein originates from [24]. The enzyme followed Michaelis–Menten kinetics at the tested substrate concentrations (Table 1). Notably, the affinity for fructose 6-phosphate appears to be considerably higher in IcSPP (K_M 2.0 ± 0.3 mM) than in TtSPP (K_M 15.1 ± 2.3 mM).

Table 1. Apparent kinetic parameters of *I. coccineus* sucrose 6^F-phosphate phosphorylase at 35 °C and 50 mM 3-(N-morpholino)propanesulfonic acid (MOPS) pH 7.0 (phosphorolysis) or 50 mM MES pH 6.5 (synthesis).

Reaction	Substrate	K_M (mM)	k_{cat} (s^{-1})	k_{cat}/K_M ($mM^{-1}s^{-1}$)
Phosphorolysis	sucrose 6^F-phosphate	11.3 ± 1.8	126 ± 15	11.2
	phosphate	7.8 ± 1.2	110 ± 9	14.1
Synthesis	α-D-glucose 1-phosphate	18.8 ± 3.6	45 ± 3	2.4
	fructose 6-phosphate	2.0 ± 0.3	40 ± 4	20

2.3. Structural Comparison of IcSPP and TtSPP

The crystal structures of the SPPs from *I. coccineus* and *T. thermosaccharolyticum* were determined to 2.05 Å and 1.83 Å resolution, respectively (Table S2; PDB codes 6S9U and 6S9V). The IcSPP crystals contained one protein molecule in the asymmetric unit, whereas the TtSPP crystals contained two. Overall, the monomers showed high structural similarity to the *Bifidobacterium adolescentis* sucrose phosphorylase (BaSP; PDB codes 2GDU, 2GDV, 1R7A), the only other GH13_18 enzyme of which a three-dimensional structure is currently known [13,25,26], despite the limited sequence identity of TtSPP and IcSPP to BaSP (35.1% and 26.3%, respectively). Four distinct domains can be discerned in these three enzymes. Domain A (residues 1–93, 198–328, and 389–470 in IcSPP) is the largest and forms the $(\beta/\alpha)_8$-barrel that is characteristic to the GH13 family. The domain at the C-terminal end (residues 471–523) is made up of a five-stranded antiparallel β-sheet. However, the backbone conformation of the two remaining domains B (residues 94–197) and B′ (residues 329–388) appears to be more variable (Figure 3). Domain B in BaSP and TtSPP contains two antiparallel β-sheets formed by two strands each, flanked by two short α-helices. In IcSPP, the inner sheet immediately flanking the barrel is considerably larger and formed by three strands instead of two. IcSPP also has an additional set of two strands constituting a third, outermost β-sheet. Domain B′ in BaSP is a coil region with two α-helices. In both SPPs, however, part of the coil region is replaced by a β-sheet made out of antiparallel strands. Furthermore, those strands are larger in IcSPP than they are in TtSPP.

IcSPP seems to exist as a monomer, whereas TtSPP in the crystal structure forms dimers. BaSP was also shown to form dimers, but the arrangements of the TtSPP and BaSP dimers are quite different (Figure S3), and the TtSPP dimer interface is somewhat smaller compared to that of BaSP (780 $Å^2$ versus 960 $Å^2$). Based on size exclusion chromatography and light scattering analysis, both IcSPP and TtSPP, unlike BaSPP, are present as monomers in solution at pH 7. Thus, the dimers observed in the TtSPP crystals probably do not represent a functional state of the protein.

A closer look was taken at the sequence and structure of the active site (Figure 4, Figure 5, Figure S4). It seems that the architecture of subsite −1 is essentially identical in BaSP and the SPPs and most likely across all GH13_18 enzymes. Subsite −1 is dedicated to the recognition of the glucosyl moiety of the donor substrate. Residues that form hydrogen bonds to the glucosyl group are highly conserved, both in sequence and in structure (Asp49, His87, Arg195, His295, Arg399 in TtSPP) (Figure 5a). The same is evidently true for the catalytic nucleophile and general acid/base at the tips of β-sheets 4 and 5 of the $(\beta/\alpha)_8$-barrel, and the aspartate that crucially stabilizes the transition state. These catalytic residues are, respectively, found at Asp197, Glu238, and Asp296 in TtSPP. In IcSPP, they are found at Asp223, Glu264, and Asp327. Only the arginine that interacts with OH3 and OH4 is an exception to the strict conservation in subsite −1 as it is substituted by lysine in IcSPP.

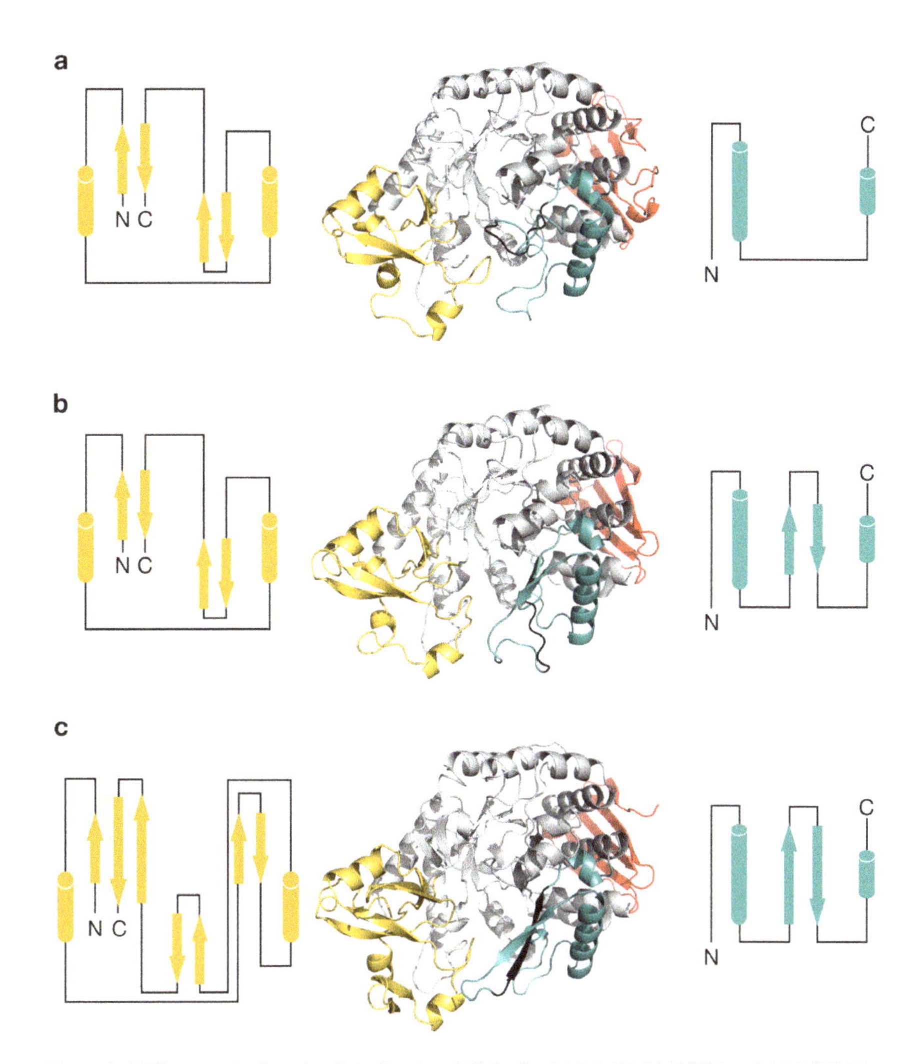

Figure 3. Differences in domains B (yellow) and B′ (teal) of (**a**) BaSP, (**b**) TtSPP, and (**c**) IcSPP. The C-terminal domain is shown in red, loop A within domain B′ is shown in black. Schematic secondary structure topology diagrams for loop B and B′ are shown left and right of the structure, respectively.

BaSP	133 | RPRPGLP	189 | IRLDAVGY	231 | IEVH	289 | HDGIGV
TtSPP	133 | LRR.TLP	194 | VRLDAVGY	237 | LEVH	295 | HDGIPV
IcSPP	151 | MRKPGLP	220 | VRLDAFAY	263 | PEIH	326 | HDGIPV

BaSP	329 | ..E	340 | NLDLYQVN	399 | RDINR
TtSPP	327 | LSL	340 | GFDVHQIN	399 | REINR
IcSPP	363 | VKN	373 | KVSYYQVN	434 | KEINR

Figure 4. Partial multiple sequence alignment of BaSP, TtSPP, and IcSPP. Catalytic residues are indicated by a purple sphere. Residues suggested to be involved in binding of fructose 6-phosphate, according to automated docking, are highlighted in blue.

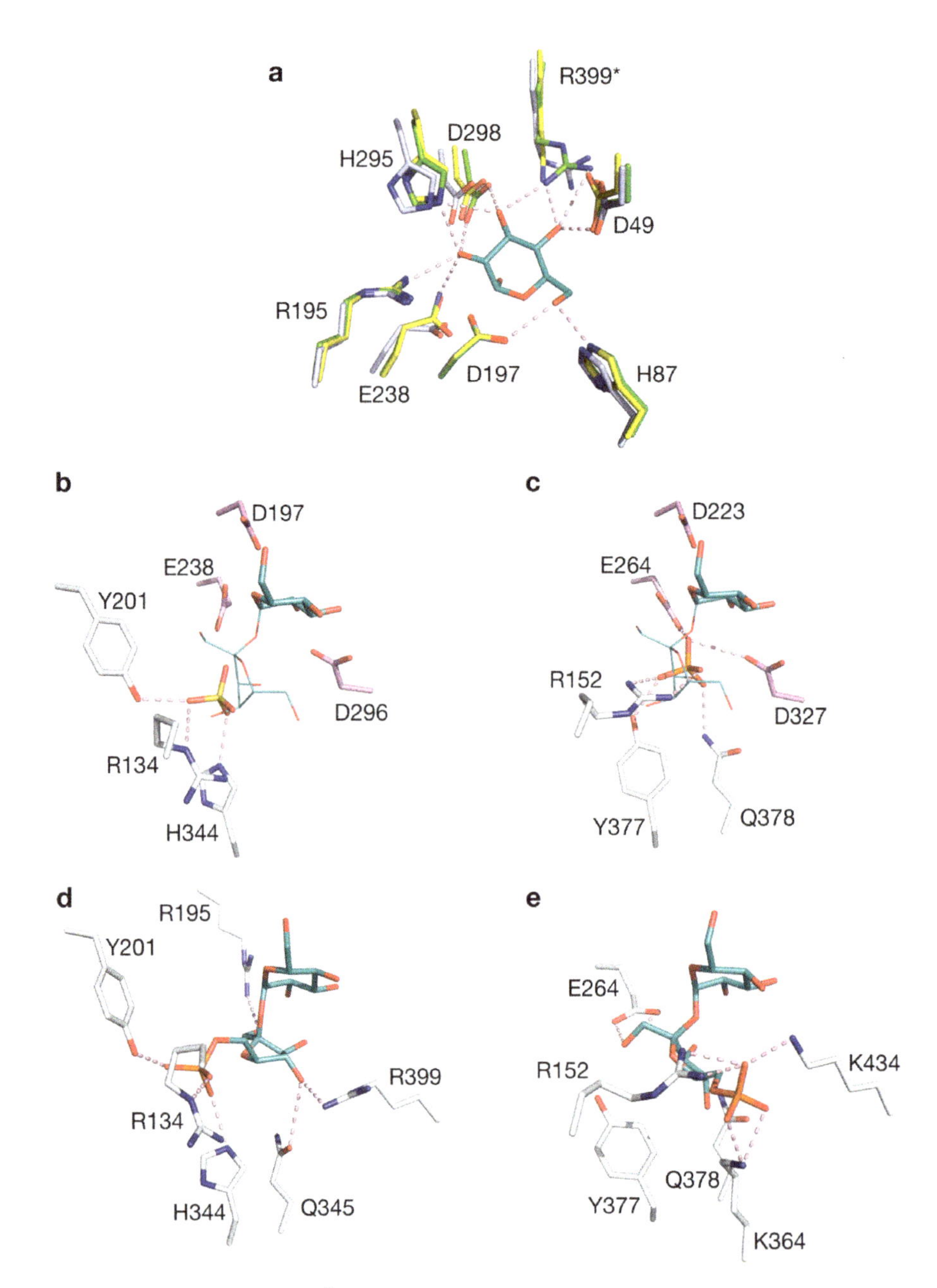

Figure 5. Active site of sucrose 6^F-phosphate phosphorylases (SPPs). (**a**) Overlay of the −1 subsite of BaSP (green), TtSPP (blue), and IcSPP (yellow), with the glucosyl moiety of sucrose from the crystal structure of BaSP (PDB code 2GDU). Residue numbering according to TtSPP. Arg399 is substituted by Lys in IcSPP (asterisk). (**b**) Close-up view of the active site in TtSPP with a bound sulfate ion. (**c**) Close-up view of the active site in IcSPP with a bound phosphate ion. The catalytic residues are shown in purple. An overlay of sucrose from the structure of BaSP is shown for reference. Automated docking of sucrose 6^F-phosphate in (**d**) TtSPP and (**e**) IcSPP. All figures of TtSPP were made using molecule B from the asymmetric unit of the crystal structure.

One tetrahedral sulfate ion is present in subsite +1 of TtSPP. It is held by hydrogen bonds with Tyr201, Arg134, and His344 (Figure 5b). Mutagenesis previously confirmed that the latter two are involved in binding the phosphate moiety of sucrose 6^F-phosphate [18]; hence, the sulfate group in the crystal structure is probably situated at that particular binding site. In the structure of IcSPP, a phosphate ion is found interacting with Arg152, Glu264, Asp327, Tyr377, and Gln378 in the active site. One of the oxygen atoms of phosphate is in close proximity to the likely position of the anomeric carbon of the glucosyl donor. This observation suggests that the complexed molecule takes on the binding mode of the phosphate that attacks the covalent glucosyl-enzyme intermediate during the enzyme's double displacement reaction (Figure 5c) [27,28].

Automated docking was performed to predict the binding mode of sucrose 6^F-phosphate in the SPP structures. For both enzymes, a cluster could be obtained where the glucosyl moiety of the substrate was located in subsite −1, in full agreement with the structure of BaSP in complex with sucrose. The anomeric carbon atom, as well as the oxygen atom of the glycosidic linkage, were at an appropriate distance and angle from the catalytic residues. Although the possibility of conformational rearrangements upon binding must be considered, several residues that participate in hydrogen bonding with the fructose 6-phosphate moiety could be identified in subsite +1, and these residues are fully conserved among sequences from the same phylogenetic branch. Arg134, Arg195, Tyr201, His344, Gln345, and Arg399 fulfill this role in TtSPP (Figure 5d). Furthermore, the position of the substrate's phosphate group indeed matches the position of the sulfate molecule that was seen in the crystal structure. In IcSPP, the predicted binding partners are Arg152, Glu264, Lys364, Gln378, and Lys434 (Figure 5e). Clearly, the set of amino acids responsible for substrate binding in subsite +1 is different between the two enzymes. For instance, Lys364 of IcSPP is considered to be properly situated to interact with fructose 6-phosphate, while no obvious function is predicted for the corresponding serine residue in TtSPP. As another example, IcSPP houses a tyrosine at position 377, just like sucrose phosphorylases do at the equivalent position. In TtSPP, on the other hand, a histidine residue is present instead. Mutating this histidine into tyrosine lowers the activity of TtSPP on fructose 6-phosphate almost fivefold [18]. Finally, it is worth noting that one of the strictly conserved residues in loop A of IcSPP-like enzymes, Lys373 (Figure 1b), is not among the predicted substrate-binding residues. Its sidechain is pointed away from the active site in the crystal structure (Figure S5), just like the conserved Asn residue that is present at the corresponding position in SP. The reason for their conservation remains unknown for now, although it is conceivable that they play a role in the backbone rearrangements that loop A undergoes during catalysis [26].

2.4. Mutagenesis

Despite the structural differences, an obvious reason for the strict specificity of IcSPP in contrast to the promiscuous behavior of TtSPP is not readily apparent from their active site layout alone. Therefore, the possibility of introducing alternative activities in IcSPP by means of mutagenesis was explored. We hypothesized that by disrupting the interaction network with the phosphate group from sucrose 6^F-phosphate, the acceptor site might recognize compounds lacking that phosphate group and take on a more relaxed shape in general. The residues participating in hydrogen bonds with the phosphate, according to the docking analysis, are Arg152, Lys364, and Lys434. Tyr377 was also targeted because of its close proximity, although its role in substrate binding is more ambiguous. All residues were substituted by the amino acid that is found at the equivalent position in TtSPP. However, position 152 is occupied by arginine in both SPPs and was therefore mutated into alanine to remove the side chain while retaining the structural integrity of the backbone [29]. An additional alanine mutant was also created at position 434 due to the similar properties of the lysine and arginine sidechains found in IcSPP and TtSPP, respectively.

All mutants were evaluated in the synthesis direction, using α-D-glucose 1-phosphate as the glucosyl donor (Table 2). Determination of the kinetic parameters for glucosyl acceptor fructose 6-phosphate pointed out that all mutants suffer a pronounced drop in activity and/or affinity. Mutation

K434R seemed to be the least detrimental in terms of activity, as was expected. However, the most drastic effect on turnover rate and affinity was observed when the same residue was mutated into alanine. The affinity of mutant Y377H for fructose 6-phosphate decreased only slightly, supporting the results of the docking experiment.

Table 2. (Left) Apparent kinetic parameters of *I. coccineus* sucrose 6^F-phosphate phosphorylase mutants using 50 mM glucose 1-phosphate at 35 °C and pH 6.5. (Right) Transglucosylation activity using 50 mM glucose 1-phosphate and 30 mM (fructose 6-phosphate; Fru6P), 50 mM (D-glycerate), or 200 mM (others) acceptor at 30 °C and pH 6.5, expressed with hydrolysis as reference.

Mutant	Kinetics Fru6P		$v_{acceptor}/v_{water}$			
	K_M (mM)	k_{cat} (s^{-1})	Fru6P	Fructose	Glycerol	D-Glycerate
Wild-type	2.0 ± 0.3	40.0 ± 3.8	1360 ± 120	-	-	-
R152A	3.3 ± 0.7	1.42 ± 0.11	98 ± 5	6.9 ± 0.3	2.4 ± 0.3	2.5 ± 0.2
K364S	6.3 ± 1.1	0.61 ± 0.04	18 ± 2	1.3 ± 0.1	1.3 ± 0.1	1.3 ± 0.1
Y377H	2.8 ± 0.3	0.80 ± 0.08	21 ± 2	-	-	-
K434R	10.6 ± 1.9	6.84 ± 0.15	680 ± 45	8.1 ± 0.4	-	4.9 ± 0.4
K434A	33.2 ± 2.9	0.10 ± 0.01	3.5 ± 0.5	2.8 ± 0.4	1.4 ± 0.2	-

(-): No significant transglucosylation activity could be detected.

To assess the ability of the mutants to employ alternative substrates, their activity was measured using fructose, glycerol, and D-glycerate as the glucosyl acceptor (Table 2). Those reactions represent the other known wild-type activities that have been discovered in subfamily GH13_18 thus far. Only variant Y377H, mutated at a position that was not predicted to interact with the phosphate group of fructose 6-phosphate, was unable to perform any of the alternate transglucosylation reactions. All others could effectively use fructose to synthesize sucrose. Mutants R152A and K364S also exhibited activity on glycerol and D-glycerate. However, the substitutions at position 434 had a dissimilar effect in that regard. Mutant K434R could use the negatively charged D-glycerate but not glycerol, whereas the opposite was true for K434A. Considering the properties of these variants, it can be concluded that the breadth of acceptor substrates tolerated by IcSPP can be extended rather easily. Although the measured activities may sound modest, being merely up to a few times higher than the hydrolytic side-activity that is inherent to the enzyme's double displacement mechanism, they are not insignificant. For reference, the ratio of the transglucosylation activity with glycerol over the hydrolytic activity ranges between 1.2 and 1.7 with *Leuconostoc mesenteroides* SP [6], yet this transglucosylation process was eventually optimized for the production of glucosylglycerol on a commercial scale [30,31].

3. Conclusions

The structures of the highly specific sucrose 6^F-phosphate phosphorylase from *Ilumatobacter coccineus* and its exceptionally promiscuous homologue from *Thermoanaerobacterium thermosaccharolyticum* provided insight into their similarities and differences. The architecture of subsite +1 showed the most prominent disparities, where a different set of residues appears to be responsible for substrate recognition. Mutational analysis revealed that the specificity of the *I. coccineus* phosphorylase can be loosened up by targeting positions around its predicted phosphate group binding site. More specifically, single point substitutions at positions Arg152, Lys364, or Lys434 were sufficient to introduce activity on fructose, glycerol, or D-glycerate, the natural acceptor substrates of related GH13_18 enzymes. This study finally expanded our structural knowledge of GH13_18, as the three-dimensional structure of only one enzyme from this subfamily has been described so far [25]. The findings also open up perspectives for further engineering of SPPs and other GH13_18 phosphorylases. Indeed, it has been established that engineering efforts can be more successful when starting from a collection of different scaffolds [32,33]. TtSPP was already proven to be a favorable starting point for mutagenesis [19], and the availability of a crystal structure should promote further exploration of its capabilities. Moreover, considering how the strict behavior of IcSPP can be loosened up, the same might be true for the related glucosylglycerol and glucosylglycerate phosphorylases. These relaxed enzymes may then turn out to be appealing templates for designing catalysts for the synthesis of various valuable glucosides and sugars as well.

4. Materials and Methods

4.1. Materials

All chemicals were obtained from Sigma-Aldrich, Merck or Carbosynth unless noted otherwise and were of the highest purity. α-D-Glucose 1-phosphate was produced in house using sucrose phosphorylase [34]. The acid glucose 1-phosphatase negative strain *Escherichia coli* CGSC 8974 was obtained from the Coli Genetic Stock Center. The *E. coli* BL21(DE3) strain was obtained from New England Biolabs.

4.2. Sequence Analysis

All 2405 full-length protein sequences in family GH13, subfamily 18, were extracted from the CAZy database (http://www.cazy.org) [17] and subsequently aligned with ClustalO using default settings [35]. A script was written in Python to remove duplicate sequences, retaining only 1254 unique sequences. A maximum-likelihood phylogenetic tree was constructed using PhyML 3.1 with default parameters [36] and visualized using iTOL v4 [37]. Sequence logos were generated with WebLogo [38]. Multiple sequence alignments were visualized using ESPript 3.0 [39].

4.3. Gene Cloning and Transformation

The *T. thermosaccharolyticum* sucrose 6^F-phosphate phosphorylase (UniProt code D9TT09) was expressed as described in earlier work [18]. The amino acid sequence for the *I. coccineus* phosphorylase (UniProt code M5A566) was codon-optimized for *E. coli*, synthesized, and subcloned into a pET21a vector by restriction digestion with *Nhe*I and *Xho*I and ligation by Life Technologies (Merelbeke, Belgium). The plasmid was transformed in *E. coli* CGSC 8974 electrocompetent cells. Sequences are listed in Table S3.

4.4. Protein Expression and Purification

TtSPP was expressed and purified as described in earlier work [18]. To express IcSPP, 2% of an overnight culture was inoculated in 500 mL LB medium containing 100 μg/mL ampicillin in a 2 L Erlenmeyer flask and grown at 37 °C with continuous shaking at 200 rpm. The culture was incubated until OD_{600} reached ~0.6. Then the temperature was lowered to 18 °C, and expression was induced by adding 0.1 mM isopropyl β-D-1-thio-galactopyranoside. Protein expression took place for 18 h. The culture was spun down, and the pellet was frozen at −20 °C for at least 4 h.

Cell pellets of a 250 mL culture were thawed at 4 °C and dissolved in 8 mL lysis buffer that consists of 0.1 mM phenylmethylsulfonyl fluoride, 1 mg/mL lysozyme, 10 mM imidazole, and 50 mM phosphate buffered saline. After incubation on ice for 30 min, the lysate was sonicated 3 times for 3 min (Branson Sonifier 250, level 3, 50% duty cycle) and the resulting extract was centrifuged (9000 rpm, 1 h, 4 °C). The supernatant was further purified by means of nickel-nitrilotriacetic acid chromatography as described by the supplier (Thermo Scientific, Waltham, MA, USA), and the buffer was exchanged to 2-morpholinoethanesulfonic acid (MES) buffer (pH 6.5) in a 30 kDa Amicon Ultra centrifugal filter (Merck Millipore, Darmstadt, Germany). A NanoDrop ND-1000 (Thermo Scientific, Waltham, MA, USA) was applied to measure the protein concentration in triplicate, using the extinction coefficients calculated with the ProtParam tool on the ExPASy server (http://web.expasy.org/protparam/). Molecular weight and purity were verified by sodium dodecyl sulfate-polyacrylamide gel electrophoresis (10% gel).

4.5. Site-Directed Mutagenesis

Site-directed mutations were introduced with a megaprimer-based whole-plasmid PCR method described elsewhere [40]. The forward primer contained the desired mutation, whereas the reverse primer (pET21a_Rv_seq1) was kept constant (Table 3). Template DNA was digested by *Dpn*I treatment (Westburg, Leusden, Netherlands) for at least 3 h at 37 °C. After transformation in *E. coli* BL21(DE3) electrocompetent cells, the obtained plasmid was subjected to nucleotide sequencing to verify the presence of the mutation (Macrogen, Amsterdam, Netherlands).

Table 3. Primers used in this study. Mutations are underlined.

Primer	DNA sequence (5′-3′)
IcSPP_Fw_R152A	CGTCTGTTTATGGCTAAACCGGGTCTGC
IcSPP_Fw_K364S	GGTGGTCGTGTGAGTAATCTGTATGGTG
IcSPP_Fw_Y377H	GGTACAAAAGTGAGCTATCATCAGGTTAACGCC
IcSPP_Fw_K434R	GGTGCGGATGGTCATCGTGAAATCAATCG
IcSPP_Fw_K434A	GGTGCGGATGGTCATGCAGAAATCAATCG
pET21a_Rv_seq1	TCCGCGCACATTTCC

4.6. Colorimetric Assays

The activity of IcSPP and TtSPP was measured in the phosphorolysis direction of the reversible reaction by measuring the reduction of NAD^+ in the presence of phosphoglucomutase and glucose 6-phosphate dehydrogenase (glucose 1-phosphate assay) [41]. In the synthesis direction, the release of inorganic phosphate from glucose 1-phosphate could be quantified with the phosphomolybdate assay [42]. To determine the hydrolytic side-activity, an enzymatic coupled assay with glucose oxidase and peroxidase was carried out (GOD-POD assay) [43]. The true transglucosylation activity could be calculated by subtracting the hydrolytic activity from the total activity in synthesis direction, as determined by the phosphomolybdate assay. Samples were inactivated by the acidic conditions of the assay solution (phosphomolybdate assay) or by heating for 5 min at 95 °C (other assays).

4.7. Characterization of IcSPP

To assess which acceptor substrates IcSPP is active on, ~50 µg/mL enzyme was incubated with 50 mM α-D-glucose 1-phosphate (Glc1P) and 100 mM acceptor in 50 mM MES buffer (pH 6.5) for 30 min at 30 °C in a total reaction volume of 1 mL. A sample was taken and analyzed with the phosphomolybdate and GOD-POD assays. To assess the donor substrates, ~50 µg/mL enzyme was incubated with 50 mM donor and 50 mM phosphate under the same conditions, and a sample was analyzed with the Glc1P assay. To precisely quantify the specific activity of IcSPP towards its wild-type substrates, 5 µg/mL enzyme was added. The reactions were carried out at 37 °C for 8 min (reaction volume of 1 mL), and a sample was taken and analyzed every min.

The influence of pH on activity was checked in the synthesis direction in 50 mM acetate (pH 4.5), MES (pH 5.0–6.5), or 3-(N-morpholino)propanesulfonic acid (pH 7.0–8.0) at 37 °C, and the optimal temperature was determined in 50 mM MES, pH 6.For each reaction, 5 µg/mL enzyme was incubated with 50 mM Glc1P and 50 mM fructose 6-phosphate (reaction volume of 1 mL). Samples of 50 µL were taken every 30 s for 4 min and analyzed with the phosphomolybdate and GOD-POD assays.

The apparent kinetic parameters of IcSPP for its natural substrates were determined at the optimal temperature and pH in 50 mM MES buffer (reaction volume of 1 mL), with samples taken every minute for 8 min. The enzyme concentration was 4 µg/mL, and the fixed cosubstrate was either 100 mM sucrose 6^F-phosphate or 100 mM phosphate in the degradation direction, or either 100 mM fructose 1-phosphate or 100 mM Glc1P in the synthesis direction. When evaluating the IcSPP mutants, the enzyme concentration was increased to 30 µg/mL (K434R) or 400 µg/mL (others) due to their lower activity. Parameters were calculated by non-linear regression of the Michaelis–Menten equation using SigmaPlot 11. The molecular weight of 55.9 kDa was used to calculate the turnover number k_{cat}.

The activity of IcSPP mutants on the alternative acceptor substrates was measured at 30 °C and in 50 mM MES buffer (pH 6.5), with 500 μg/mL enzyme, 50 mM Glc1P, and either 30 mM fructose 6-phosphate, 200 mM fructose, 200 mM glycerol, or 50 mM D-glycerate (reaction volume of 1 mL). Samples were taken every 2 min for 16 min and analyzed with the phosphomolybdate assay.

4.8. Crystallography

Crystallization screening was preceded by an extra protein purification step using a Superdex 200 10/300 GL column (GE Healthcare, Buckinghamshire, UK) preequilibrated with 100 mM Tris buffer (pH 7.5) for IcSPP or 20 mM Bis-Tris (pH 7.5) and 150 mM NaCl for TtSPP. Both proteins eluted as monomers with an apparent molecular weight of about 60 kDa, as confirmed by dynamic light scattering analysis. Purified IcSPP and TtSPP were concentrated to 9 mg/mL in 10 mM Tris (pH 7.5) and 6 mg/mL in 10 mM Bis-Tris (pH 7.5) with 75 mM NaCl, respectively. Screening for crystallization conditions was performed at room temperature in 96-well sitting-drop crystallization plates, using various commercial crystallization screens and the help of a Mosquito (TTP LabTech) pipetting robot. Drops were prepared by mixing 200 nL protein solution with 200 nL reservoir solution. Crystals for IcSPP were obtained in the presence of 20 mM fructose 6-phosphate with a reservoir solution containing 20% PEG 3350, 200 mM Na/K phosphate, and 100 mM Bis-Tris propane buffer (pH 7.5), while crystals for TtSPP grew in the presence of 40 mM fructose 6^F-phosphate with a reservoir solution containing 26% PEG 3350, 100 mM $(NH_3)_2SO_4$, and 100 mM Bis-Tris propane buffer (pH 5.5). Although fructose 6^F-phosphate was present in the crystallization trials, no sugar could be detected in the final structures. Diffraction data for IcSPP and TtSPP were recorded at the synchrotron (ESRF, Grenoble). Prior to X-ray data collection, the crystals were flash-cooled in liquid nitrogen using 20% PEG400 as a cryo-protectant. Processing of the X-ray diffraction data was performed with the program XDS [44] and with the AIMLESS routines of the CCP4 software suite [45]. The IcSPP crystals belong to space group $C222_1$ and contain a single polypeptide chain in the asymmetric unit (solvent content of 44%), whereas the TtSPP crystals belong to space group $P2_12_12_1$ and contain two polypeptide chains in the asymmetric unit (solvent content of 42%). Initial phases and electron maps were obtained by molecular replacement with the program Phaser [46], using the structure of BaSP (PDB code 1R7A) as a search model. The resulting models were improved by automatic model building using ARP/wARP [47], followed by several rounds of model building and refinement, using the program Coot [48] and routines from the Phenix software suite [49]. Validation of the final structures was performed with the Molprobity server [50].

4.9. Automated Docking and Protein Figures

The binding of sucrose 6^F-phosphate in IcSPP and TtSPP was simulated by ligand docking using the AutoDock VINA module implemented in YASARA [51]. The default settings were applied, except for the number of runs which was increased to The correct cluster was selected by comparing the binding mode of the ligand's glucosyl moiety in subsite −1 to the binding mode observed in BaSP (PDB code 2GDU).

All structure manipulations, such as superpositions, were carried out in YASARA. Figures were rendered in PyMOL [52].

Supplementary Materials: Supplementary materials can be found at http://www.mdpi.com/1422-0067/20/16/3906/s1.

Author Contributions: Conceptualization, J.F., N.C., T.D. and A.-M.W.H.T.; Investigation, J.F. and N.C.; Supervision, T.D. and A.-M.W.H.T.; Visualization, J.F.; Writing—original draft, J.F.; Writing—review & editing, T.D. and A.-M.W.H.T.

Funding: This research was funded by Research Foundation Flanders (FWO; PhD grant for J.F.).

Acknowledgments: The authors would like to thank Natan Van Welden for the help with experiments, and the staff of the ESRF and of EMBL-Grenoble for assistance and support in using beamline ID30A-3.

Conflicts of Interest: The authors declare no conflict of interest.

Abbreviations

BaSP	*Bifidobacterium adolescentis* sucrose phosphorylase
Fru6P	Fructose 6-phosphate
GGaP	Glucosylglycerate phosphorylase
GGoP	Glucosylglycerol phosphorylase
GH13_18	Subfamily 18 of glycoside hydrolase family 13
Glc1P	α-D-glucose 1-phosphate
GOD-POD	Glucose oxidase—peroxidase
GP	Glycoside phosphorylase
IcSPP	*Ilumatobacter coccineus* sucrose 6^F-phosphate phosphorylase
MES	2-morpholinoethanesulfonic acid
MOPS	3-(N-morpholino)propanesulfonic acid
SP	Sucrose phosphorylase
SPP	Sucrose 6^F-phosphate phosphorylase
TtSPP	*Thermoanaerobacterium thermosaccharolyticum* sucrose 6^F-phosphate phosphorylase

References

1. Desmet, T.; Soetaert, W. Enzymatic glycosyl transfer: Mechanisms and applications. *Biocatal. Biotransform.* **2011**, *29*, 1–18. [CrossRef]
2. Puchart, V. Glycoside phosphorylases: Structure, catalytic properties and biotechnological potential. *Biotechnol. Adv.* **2015**, *33*, 261–276. [CrossRef] [PubMed]
3. Kitaoka, M. Diversity of phosphorylases in glycoside hydrolase families. *Appl. Microbiol. Biotechnol.* **2015**, *99*, 8377–8390. [CrossRef] [PubMed]
4. Kok, D.S.; Yilmaz, D.; Suir, E.; Pronk, J.T.; Daran, J.M.; Van Maris, A.J.A. Increasing free-energy (ATP) conservation in maltose-grown *Saccharomyces cerevisiae* by expression of a heterologous maltose phosphorylase. *Metab. Eng.* **2011**, *13*, 518–526. [CrossRef]
5. Goedl, C.; Sawangwan, T.; Wildberger, P.; Nidetzky, B. Sucrose phosphorylase: A powerful transglucosylation catalyst for synthesis of α-D-glucosides as industrial fine chemicals. *Biocatal. Biotransform.* **2010**, *28*, 10–21. [CrossRef]
6. Aerts, D.; Verhaeghe, T.F.; Roman, B.I.; Stevens, C.V.; Desmet, T.; Soetaert, W. Transglucosylation potential of six sucrose phosphorylases toward different classes of acceptors. *Carbohydr. Res.* **2011**, *346*, 1860–1867. [CrossRef]
7. Gudiminchi, R.K.; Nidetzky, B. Walking a fine line with sucrose phosphorylase: Efficient single-step biocatalytic production of l-ascorbic acid 2-glucoside from sucrose. *ChemBioChem* **2017**, *18*, 1387–1390. [CrossRef]
8. De Winter, K.; Desmet, T. Biphasic catalysis with disaccharide phosphorylases: Chemoenzymatic synthesis of α-d-glucosides using sucrose phosphorylase. *Org. Process Res. Dev.* **2014**, *18*, 781–787. [CrossRef]
9. Luley-Goedl, C.; Sawangwan, T.; Brecker, L.; Wildberger, P.; Nidetzky, B. Regioselective O-glucosylation by sucrose phosphorylase: A promising route for functional diversification of a range of 1,2-propanediols. *Carbohydr. Res.* **2010**, *345*, 1736–1740. [CrossRef]
10. Sawangwan, T.; Goedl, C.; Nidetzky, B. Single-step enzymatic synthesis of (*R*)-2-*O*-α-D-glucopyranosyl glycerate, a compatible solute from micro-organisms that functions as a protein stabiliser. *Org. Biomol. Chem.* **2009**, *7*, 4267–4270. [CrossRef]
11. Cerdobbel, A.; De Winter, K.; Aerts, D.; Kuipers, R.; Joosten, H.J.; Soetaert, W.; Desmet, T. Increasing the thermostability of sucrose phosphorylase by a combination of sequence- and structure-based mutagenesis. *Protein Eng. Des. Sel.* **2011**, *24*, 829–834. [CrossRef]
12. Verhaeghe, T.; De Winter, K.; Berland, M.; De Vreese, R.; D'hooghe, M.; Offmann, B.; Desmet, T. Converting bulk sugars into prebiotics: Semi-rational design of a transglucosylase with controlled selectivity. *Chem. Commun.* **2016**, *52*, 3687–3689. [CrossRef]

13. Kraus, M.; Grimm, C.; Seibel, J. Redesign of the active site of sucrose phosphorylase through a clash-induced cascade of loop shifts. *ChemBioChem* **2016**, *17*, 33–36. [CrossRef]
14. Kraus, M.; Grimm, C.; Seibel, J. Switching enzyme specificity from phosphate to resveratrol glucosylation. *Chem. Commun.* **2017**, *53*, 12181–12184. [CrossRef]
15. Franceus, J.; Dhaene, S.; Decadt, H.; Vandepitte, J.; Caroen, J.; Van der Eycken, J.; Beerens, K.; Desmet, T. Rational design of an improved transglucosylase for production of the rare sugar nigerose. *Chem. Commun.* **2019**, *55*, 4531–4533. [CrossRef]
16. Stam, M.R.; Danchin, E.G.J.; Rancurel, C.; Coutinho, P.M.; Henrissat, B. Dividing the large glycoside hydrolase family 13 into subfamilies: Towards improved functional annotations of α-amylase-related proteins. *Protein Eng. Des. Sel.* **2006**, *19*, 555–562. [CrossRef]
17. Lombard, V.; Golaconda Ramulu, H.; Drula, E.; Coutinho, P.M.; Henrissat, B. The carbohydrate-active enzymes database (CAZy) in 2013. *Nucleic Acids Res.* **2014**, *42*, 490–495. [CrossRef]
18. Verhaeghe, T.; Aerts, D.; Diricks, M.; Soetaert, W.; Desmet, T. The quest for a thermostable sucrose phosphorylase reveals sucrose 6′-phosphate phosphorylase as a novel specificity. *Appl. Microbiol. Biotechnol.* **2014**, *98*, 7027–7037. [CrossRef]
19. Dirks-Hofmeister, M.E.; Verhaeghe, T.; De Winter, K.; Desmet, T. Creating space for large acceptors: Rational biocatalyst design for resveratrol glycosylation in an aqueous system. *Angew. Chem.* **2015**, *127*, 9421–9424. [CrossRef]
20. Decuyper, L.; Franceus, J.; Dhaene, S.; Debruyne, M.; Vandoorne, K.; Piens, N.; Dewitte, G.; Desmet, T.; D'hooghe, M. Chemoenzymatic approach toward the Synthesis of 3-*O*-(α/β)-Glucosylated 3-Hydroxy-β-lactams. *ACS Omega* **2018**, *3*, 15235–15245. [CrossRef]
21. Franceus, J.; Pinel, D.; Desmet, T. Glucosylglycerate phosphorylase, an enzyme with novel specificity involved in compatible solute metabolism. *Appl. Environ. Microbiol.* **2017**, *83*, e01434-17. [CrossRef]
22. Franceus, J.; Decuyper, L.; D'hooghe, M.; Desmet, T. Exploring the sequence diversity in glycoside hydrolase family 13_18 reveals a novel glucosylglycerol phosphorylase. *Appl. Microbiol. Biotechnol.* **2018**, *102*, 3183–3191. [CrossRef]
23. Tauzin, A.S.; Bruel, L.; Laville, E.; Nicoletti, C.; Navarro, D.; Henrissat, B.; Perrier, J.; Potocki-Veronese, G.; Giardina, T.; Lafond, M. Sucrose 6F-phosphate phosphorylase: A novel insight in the human gut microbiome. *Microb. Genom.* **2019**, 1–14. [CrossRef]
24. Fujinami, S.; Takarada, H.; Kasai, H.; Sekine, M.; Omata, S.; Fukai, R.; Hosoyama, A.; Horikawa, H.; Kato, Y.; Fujita, N.; et al. Complete genome sequence of Ilumatobacter coccineum YM16–304. *Stand. Genom. Sci.* **2013**, *8*, 430–440. [CrossRef]
25. Sprogoe, D.; van den Broek, L.A.M.; Mirza, O.; Kastrup, J.S.; Voragen, A.G.J.; Gajhede, M.; Skov, L.K. Crystal structure of sucrose phosphorylase from *Bifidobacterium adolescentis*. *Biochemistry* **2004**, *43*, 1156–1162. [CrossRef]
26. Mirza, O.; Skov, L.K.; Sprogøe, D.; van den Broek, L.M.; Beldman, G.; Kastrup, J.S.; Gajhede, M. Structural rearrangements of sucrose phosphorylase from *Bifidobacterium adolescentis* during sucrose conversion. *J. Biol. Chem.* **2006**, *281*, 35576–35584. [CrossRef]
27. Schwarz, A.; Nidetzky, B. Asp-196—Ala mutant of *Leuconostoc mesenteroides* sucrose phosphorylase exhibits altered stereochemical course and kinetic mechanism of glucosyl transfer to and from phosphate. *FEBS Lett.* **2006**, *580*, 3905–3910. [CrossRef]
28. Schwarz, A.; Brecker, L.; Nidetzky, B. Acid–base catalysis in *Leuconostoc mesenteroides* sucrose phosphorylase probed by site-directed mutagenesis and detailed kinetic comparison of wild-type and Glu237→Gln mutant enzymes. *Biochem. J.* **2007**, *403*, 441–449. [CrossRef]
29. Verhaeghe, T.; Diricks, M.; Aerts, D.; Soetaert, W.; Desmet, T. Mapping the acceptor site of sucrose phosphorylase from *Bifidobacterium adolescentis* by alanine scanning. *J. Mol. Catal. B Enzym.* **2013**, *96*, 81–88. [CrossRef]
30. Goedl, C.; Sawangwan, T.; Mueller, M.; Schwarz, A.; Nidetzky, B. A high-yielding biocatalytic process for the production of 2-*O*-(α-D-glucopyranosyl)-sn-glycerol, a natural osmolyte and useful moisturizing ingredient. *Angew. Chem. Int. Ed. Engl.* **2008**, *47*, 10086–10089. [CrossRef]
31. Bolivar, J.M.; Luley-Goedl, C.; Leitner, E.; Sawangwan, T.; Nidetzky, B. Production of glucosyl glycerol by immobilized sucrose phosphorylase: Options for enzyme fixation on a solid support and application in microscale flow format. *J. Biotechnol.* **2017**, *257*, 131–138. [CrossRef]

32. Khanal, A.; McLoughlin, S.Y.; Kershner, J.P.; Copley, S.D. Differential effects of a mutation on the normal and promiscuous activities of orthologs: Implications for natural and directed evolution. *Mol. Biol. Evol.* **2015**, *32*, 100–108. [CrossRef]
33. Newton, M.S.; Arcus, V.L.; Gerth, M.L.; Patrick, W.M. Enzyme evolution: Innovation is easy, optimization is complicated. *Curr. Opin. Struct. Biol.* **2018**, *48*, 110–116. [CrossRef]
34. De Winter, K.; Cerdobbel, A.; Soetaert, W.; Desmet, T. Operational stability of immobilized sucrose phosphorylase: Continuous production of α-glucose-1-phosphate at elevated temperatures. *Process Biochem.* **2011**, *46*, 1074–1078. [CrossRef]
35. Madeira, F.; Park, Y.M.; Lee, J.; Buso, N.; Gur, T.; Madhusoodanan, N.; Basutkar, P.; Tivey, A.R.N.; Potter, S.C.; Finn, R.D.; et al. The EMBL-EBI search and sequence analysis tools APIs in 2019. *Nucleic Acids Res.* **2019**, *47*, W636–W641. [CrossRef]
36. Guindon, S.; Dufayard, J.-F.; Lefort, V.; Anisimova, M.; Hordijk, W.; Gascuel, O. New algorithms and methods to estimate maximum-likelihood phylogenies: Assessing the performance of PhyML 3.0. *Syst. Biol.* **2010**, *59*, 307–321. [CrossRef]
37. Letunic, I.; Bork, P. Interactive Tree Of Life (iTOL) v4: Recent updates and new developments. *Nucleic Acids Res.* **2019**, 2–5. [CrossRef]
38. Crooks, G.; Hon, G.; Chandonia, J.; Brenner, S. WebLogo: A sequence logo generator. *Genome Res.* **2004**, *14*, 1188–1190. [CrossRef]
39. Robert, X.; Gouet, P. Deciphering key features in protein structures with the new ENDscript server. *Nucleic Acids Res.* **2014**, *42*, 320–324. [CrossRef]
40. Reetz, M.; Kahakeaw, D.; Sanchis, J. Shedding light on the efficacy of laboratory evolution based on iterative saturation mutagenesis. *Mol. Biosyst.* **2009**, *5*, 115–122. [CrossRef]
41. Silverstein, R.; Voet, J.; Reed, D.; Abeles, R. Purification and mechanism of action of sucrose phosphorylase. *J. Biol. Chem.* **1967**, *242*, 1338–1346.
42. Gawronski, J.D.; Benson, D.R. Microtiter assay for glutamine synthetase biosynthetic activity using inorganic phosphate detection. *Anal. Biochem.* **2004**, *327*, 114–118. [CrossRef]
43. Blecher, M.; Glassman, A.B. Determination of glucose in the presence of sucrose using glucose oxidase; effect of pH on absorption spectrum of oxidized o-dianisidine. *Anal. Biochem.* **1962**, *3*, 343–352. [CrossRef]
44. Kabsch, W. Integration, scaling, space-group assignment and post-refinement. *Acta Crystallogr. Sect. D Biol. Crystallogr.* **2010**, *66*, 133–144. [CrossRef]
45. CCP4 The CCP4 suite: Programs for protein crystallography. *Acta Crystallogr. Sect. D Biol. Crystallogr.* **1994**, *50*, 760–763. [CrossRef]
46. McCoy, A.J. Solving structures of protein complexes by molecular replacement with Phaser. *Acta Crystallogr. Sect. D Biol. Crystallogr.* **2006**, *63*, 32–41. [CrossRef]
47. Langer, G.; Cohen, S.X.; Lamzin, V.S.; Perrakis, A. Automated macromolecular model building for X-ray crystallography using ARP/wARP version 7. *Nat. Protoc.* **2008**, *3*, 1171–1179. [CrossRef]
48. Emsley, P.; Cowtan, K. Coot: Model-building tools for molecular graphics. *Acta Crystallogr. Sect. D Biol. Crystallogr.* **2004**, *60*, 2126–2132. [CrossRef]
49. Adams, P.D.; Afonine, P.V.; Bunkóczi, G.; Chen, V.B.; Davis, I.W.; Echols, N.; Headd, J.J.; Hung, L.W.; Kapral, G.J.; Grosse-Kunstleve, R.W.; et al. PHENIX: A comprehensive Python-based system for macromolecular structure solution. *Acta Crystallogr. Sect. D Biol. Crystallogr.* **2010**, *66*, 213–221. [CrossRef]
50. Chen, V.B.; Arendall, W.B.; Headd, J.J.; Keedy, D.A.; Immormino, R.M.; Kapral, G.J.; Murray, L.W.; Richardson, J.S.; Richardson, D.C. MolProbity: All-atom structure validation for macromolecular crystallography. *Acta Crystallogr. Sect. D Biol. Crystallogr.* **2010**, *66*, 12–21. [CrossRef]
51. Krieger, E.; Vriend, G. YASARA View—Molecular graphics for all devices—From smartphones to workstations. *Bioinformatics* **2014**, *30*, 2981–2982. [CrossRef]
52. Schrödinger LLC. The PyMOL Molecular Graphics System, v 2.0. Available online: https://sourceforge.net/p/pymol/mailman/message/36047137/ (accessed on 20 September 2017).

International Journal of *Molecular Sciences*

Article

A Novel PL9 Pectate Lyase from *Paenibacillus polymyxa* KF-1: Cloning, Expression, and Its Application in Pectin Degradation

Ye Yuan [1], Xin-Yu Zhang [2], Yan Zhao [2], Han Zhang [1], Yi-Fa Zhou [1] and Juan Gao [1,2,*]

1 School of Life Sciences, Northeast Normal University, Changchun 130024, China; yuany268@nenu.edu.cn (Y.Y.); zhangh800@nenu.edu.cn (H.Z.); zhouyf383@nenu.edu.cn (Y.-F.Z.)
2 School of Biological Science and Technology, University of Jinan, Jinan 250022, China; zhangxinyu950512@163.com (X.-Y.Z.); zhaoyan_1994@126.com (Y.Z.)
* Correspondence: bio_gaoj@ujn.edu.cn; Tel.: +86-531-89736825

Received: 10 June 2019; Accepted: 21 June 2019; Published: 22 June 2019

Abstract: Pectate lyases play an important role in pectin degradation, and therefore are highly useful in the food and textile industries. Here, we report on the cloning of an alkaline pectate lyase gene (*pppel9a*) from *Paenibacillus polymyxa* KF-1. The full-length gene (1350 bp) encodes for a 449-residue protein that belongs to the polysaccharide lyase family 9 (PL9). Recombinant PpPel9a produced in *Escherichia coli* was purified to electrophoretic homogeneity in a single step using Ni^{2+}-NTA affinity chromatography. The enzyme activity of PpPel9a (apparent molecular weight of 45.3 kDa) was found to be optimal at pH 10.0 and 40 °C, with substrate preference for homogalacturonan type (HG) pectins vis-à-vis rhamnogalacturonan-I (RG-I) type pectins. Using HG-type pectins as substrate, PpPel9a showed greater activity with de-esterified HGs. In addition, PpPel9a was active against water-soluble pectins isolated from different plants. Using this lyase, we degraded citrus pectin, purified fractions using Diethylaminoethyl (DEAE)-sepharose column chromatography, and characterized the main fraction MCP-0.3. High-performance gel permeation chromatography (HPGPC) analysis showed that the molecular mass of citrus pectin (~230.2 kDa) was reduced to ~24 kDa upon degradation. Ultra-performance liquid chromatography - tandem mass spectrometer (UPLC-MS) and monosaccharide composition analyses demonstrated that PpPel9a worked as an endo-pectate lyase, which acted primarily on the HG domain of citrus pectin. In vitro testing showed that the degradation product MCP-0.3 significantly promotes the growth of *Lactobacillus plantarum* and *L. rhamnosus*. In this regard, the enzyme has potential in the preparation of pharmacologically active pectin products.

Keywords: pectate lyase; *Paenibacillus polymyxa*; pectins; degradation; *Lactobacillus*

1. Introduction

Found widely in plants, pectins are composed of D-galacturonic acid (GalA), L-rhamnose (L-Rha), D-galactose (D-Gal), and L-arabinose (L-Ara) [1,2]. They consist of three structural domains: 1) homogalacturonan (HG) that is mainly composed of α-1,4-linked D-GalA and thought of as "smooth regions" in pectin, 2) rhamnogalacturonan I (RGI) domains, and 3) rhamnogalacturonan II (RGII) domains that are both rich in neutral saccharides and thought as the "hairy regions" [3]. The acidic residues in pectin are usually esterified or acetylated, which makes pectin a diverse family of anionic structural hetero-polysaccharides. Pectin shows great potential in the food, pharmaceutical and polymer industries[1,4,5]. Recent research has shown that pectin has many biological activities, such as anti-tumor and antioxidation activities, as well as immune regulation activity [4,6]. However, the molecular weight of natural pectin is too large to be absorbed and utilized by the body. Modified pectin with lower molecular weight fragments and lower degrees of esterification would be more favorable for absorption and utilization, and potentially exhibit greater biological activity. Previous studies reported that low-molecular-weight modified citrus pectin (MCP) had greater activity against colon, breast, and gastrointestinal cancer, likely due to decreased expression of galectin-3 [4,7,8]. Low-molecular-weight MCP (i.e., 3 kDa to 30 kDa) exhibited greater anti-cancer activity by inhibiting the migration, aggregation, and proliferation of cancer cells compared to parent MCP with molecular weights greater than 30 kDa [9]. Low-molecular-weight MCP also promoted the growth of *Bifidobacterium longum*, suggesting that it has potential as a prebiotic agent [10].

Pectin degradation can be achieved by temperature, pH, or enzymatic modification [11]. Among these, enzymatic modification has a significant advantage [12,13]. Pectinolytic enzymes, known as pectinases, are a group of enzymes that include hydrolases, lyases, and oxidases, which play important roles in the degradation and modification of pectin [12,14]. Pectate lyase (EC 4.2.2.2), a member of the pectinase family, catalyzes the hydrolysis of α-1,4-glycosidic bonds to produce oligosaccharides with 4-deoxy-α-D-mann-4-enuronosyl groups at their non-reducing end [15,16]. Pectate lyase can function in either an endo- or exo-mode of action. Unmethylated pectins or pectins with a low degree of methylation are the preferred substrates for pectate lyases [17]. Depending on their amino acid sequence, pectate lyases (PL) are divided into five families: PL1, 2, 3, 9, and 10 [18].

These enzymes are widely distributed in microorganisms and plants. Previously, pectate lyases cloned from bacteria and fungi, such as *Bacillus*, *Erwinia*, *Aspergillus*, *Clostridium*, *Paenibacillus*, *Streptomyces*, and *Klebsiella* sp., have been reported [16,17,19]. Various industrial applications with these enzymes have been investigated, including fruit juice/wine clarification, plant fiber processing, and paper production [19–21]. Pectinases are good tools in pectin degradation and preparation of bioactive oligosaccharides [22]. However, there are few reports on the use of pectate lyase in pectin degradation or pectin structure analysis. In the present study, we cloned a pectate lyase gene belonging to PL9 from *Paenibacillus polymyxa* KF-1 and determined the substrate specificity of the recombinant pectate lyase PpPel9a. The degradation product of citrus pectin by PpPel9a was prepared, analyzed, and its prebiotic activity was evaluated.

2. Results and Discussion

2.1. Gene Cloning and Sequence Analysis of PL9 Pectate Lyase from P. polymyxa KF-1

Previously, four pectate lyases were identified from the fermentation broth of *P. polymyxa* KF-1 by liquid chromatograph-mass spectrometer/mass spectrometer (LC-MS/MS) analysis. These enzymes belong to PL families 1, 3, 9, and 10 [23]. In our previous work, a pectate lyase named PpPel10a was cloned from *P. polymyxa* KF-1 and was identified to be a member of PL10 family [23]. Here, the protein with the UniProt accession number E3E7F9 (NCBI protein ID MK809514) was encoded by an open reading frame (ORF) of 1350 bp. The protein of 449 amino acids was named as PpPel9a, and the N-terminal 34-amino acids was predicted to be the signal peptide sequence by signalP 5.0 server, which suggested the extracellular location of the protein. The *p*I value of PpPel9a without the signal peptide was predicted to be pH 5.07, similar to that reported for PL9 pectate lyase Pel-15H from *Bacillus* sp. strain KSM-P15 (4.60) [24], but lower than other PL9 enzymes, such as PelX from *Erwinia chrysanthemi* 3937 (7.7) [25], and Exo-PL from *E. chrysanthemi* EC16 (8.6) [26]. Analysis of the amino acid sequence by Uniprot confirmed that PpPel9a was a member of PL family 9 pectate lyase family, and the Pfam analysis indicated that PpPel9a had a β-helix superfamily domain. Multiple sequence alignment analysis showed that the amino acid sequence of PpPel9a has moderate similarity to other PL9 enzymes, such as pectate lyase Pel9A from *E. chrysanthemi* (1RU4) [27], rPel9A from *Clostridium stercorarium* F-9 (AB106865) [17], and rhamnogalacturonan lyase from *Bacteroides thetaiotaomicron* VPI-5482 (PDB accession no. 5OLQ) [28]. The sequence identities were 39.34%, 37.57%, and 42.53%, respectively. The low homology between PpPel9a and the reported PL9 enzymes indicates that PpPel9a may be a novel enzyme. So far, only several pectate lyases belonging to PL family 9 have been characterized, including Pel-15H from *Bacillus* sp. strain KSM-P15 (AB028878) [24], rhamnogalacturonan lyase BT4170 from *B. thetaiotaomicron* VPI-5482 (5OLQ) [28], Exo-PL from *E. chrysanthemi* EC16 (AAA24850) [26], PelL1 from *E. chrysanthemi* PY35 (AAF05308) [29], and PelX from *E. chrysanthemi* 3937 (CAB39324) [25]. The substrate specificity and biochemical properties of PL9 enzymes have yet to be clarified. In addition, the use of PL9 enzymes in the degradation of pectins had not been explored. Therefore, in the present study, we chose the gene encoding for the PL family 9 enzyme from *P. polymyxa* KF-1 for cloning, expression, and characterization.

So far, only two crystal structures have been reported for enzymes in the PL9 family (http://www.cazy.org/PL9_structure.html), namely rhamnogalacturonan lyase from *B. thetaiotaomicron* VPI-5482 (PDB accession no. 5OLQ) [28] and Pel9A from *E. chrysanthemi* (PDB accession no. 1RU4) [27]. Both lyases have a right-handed parallel β-helix fold with three short parallel β-sheets (PB1, PB2, and PB3) and ten turns, with two calcium ion binding sites (Ca-1 and Ca-2). The key components of the active site comprise Ca-1 on coils 5 and 6, and a lysine in coil 7 that functions as a catalytic base to abstract a proton from C5. The three-dimensional structure prediction by SWISS-MODEL indicates that PpPel9a may comprise a 10-coil parallel β-helix domain (Figure S1), similar to Pel9A from *E. chrysanthemi* (1RU4) [27]. ClustalW alignment of PpPel9a from *P. polymyxa* with the two characterized PL9 pectate lyase amino acid sequences highlighted three highly conserved calcium-binding sites: Asp202, Asp226, and Asp230 (Figure 1). Similar to the reported PL9 enzymes, the catalytic base in PpPel9a was predicted to be Lys271, which is different from that in PL1 pectate lyases where the catalytic base is an arginine (Arg198 in PL1 enzyme BsPelA from *Bacillus* sp. N16-5 [30], Arg240 in EcPelE from *E. chrysanthemi* [31], and Arg300 in BsPel from *B. subtilis)* [32]. In addition, PpPel9a and PpPel10a have different structures. PpPel9a has a right-handed parallel β-helix fold, whereas PpPel10a is predominantly α-helical with short β-strands and irregular coils [23].

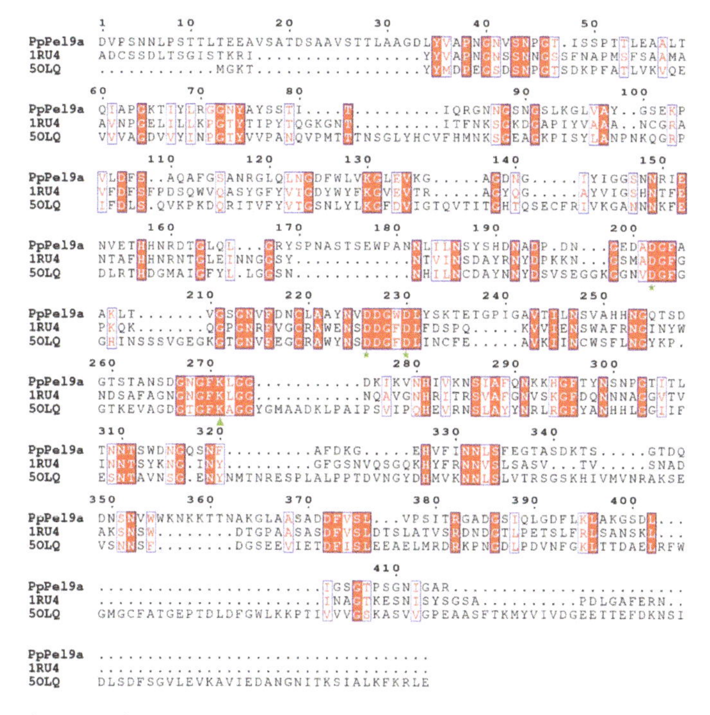

Figure 1. Alignment of amino acid sequence of PpPel9a with the PL9 family pectate lyases: Pel9A from *E. chrysanthemi* (1RU4) and rhamnogalacturonan lyase from *B. thetaiotaomicron* VPI-5482 (PDB accession no. 5OLQ). The conserved Asp were labeled by asterisk and the catalytic base was labeled by triangle.

2.2. Expression of Recombinant PpPel9a

After induction at 25 °C for 20 h, pectate lyase activity in the cytoplasm with polygalacturonic acid (PGA) as substrate, reached its maximum level (117.4 ± 4.82 U/mg). The recombinant enzyme was purified by Ni-NTA affinity chromatography, and the purified enzyme showed an approximate 4.5-fold increase in purity, and a recovery of 28.9% relative to the crude enzyme. Purification yielded 212 mg of enzyme/liter. This relatively high yield makes PpPel9a a good candidate for industrial applications.

Sodium dodecyl sulphate-polyacrylamide gel electrophoresis (SDS-PAGE) indicates that purified recombinant PpPel9a has a Mw of ~45.3 kDa, consistent with the predicted Mw (Figure 2). The Mw of PpPel9a was similar to that reported for PL9 enzyme PelL1 from *E. chrysanthemi* PY35 (43 kDa) [29], but lower than that reported for other PL9 pectate lyases, such as Pel-15H from *Bacillus* sp. strain KSM-P15 (69.55 kDa) [24], Exo-PL from *E. chrysanthemi* EC16 (76 kDa) [26], PelX from *E. chrysanthemi* 3937 (76.938 kDa) [25], and Pel9A from *C. stercorarium* F-9 (135.171 kDa) [17].

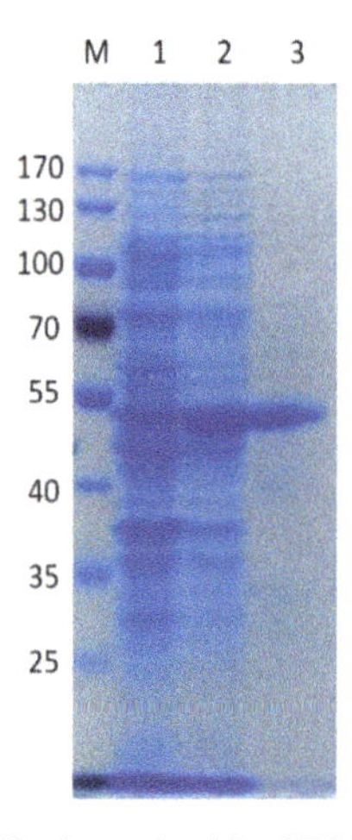

Figure 2. Molecular weight of PpPel9a determined by SDS-PAGE: (M) protein molecular weight markers (PageRuler Prestained Protein Ladder, Thermo Scientific) (1) supernatant of cell lysis from *E. coli* BL21 (DE3) cells; (2), supernatant of cell lysis from recombinant *E. coli* BL21 (DE3) cells harboring pET-28a-pppel9a plasmid; (3) recombinant enzyme PpPel9a purified from Ni-NTA agarose column.

The specific activity of purified PpPel9a with PGA as substrate, was 298.5 ± 3.6 U/mg, higher than that reported for PL9 enzymes rPel9A from *C. stercorarium* (58 U/mg) [17] and Pel-15H from *Bacillus* sp. strain KSM-P15 (10.6 U/mg) [24], but lower than the PL1 pectate lyases, such as BacPelA from *B. clausii* (675.5 U/mg)[19] and Apel from *B. subtilis* (1010.0 U/mg) [33]. PL9 enzymes show lower activity against PGA than PL1 enzymes. This is likely due to the Lys catalytic base in PL9 enzymes being less potent than Arg in PL1 enzymes. In addition, the third calcium is important in increasing the acidity of the C5 proton that may be absent in PL9 enzymes [27]. However, PpPel9a exhibited higher pectate lyase activity than that of PpPel10a when using PGA and citrus pectin (CP) as substrates [23]. The specific activities with PGA using PpPel9a and PpPel10a were 298.5 ± 3.6, and 289 ± 4.9 U/mg, respectively, while the specific activities on CP by PpPel9a and PpPel10a were 107 ± 0.2, and 98.0 ± 3.6 U/mg, respectively.

2.3. Biochemical Characterization of PpPel9a

The effect of pH on PpPel9a was studied using PGA as a substrate. PpPel9a showed high activity over the pH range of pH 6 to pH 11, with optimum activity at pH 10 (Figure 3a). The reported PL9 pectate lyases showed different pH optima, such as Exo-PL from *E. chrysanthemi* EC16 (pH 7.5–8.0) [26],

PelX from *E. chrysanthemi* 3937 (pH 8.5) [25], Pel-15H from Bacillus sp. strain KSM-P15 (pH 11.5) [24], and rPel9A from *C. stercorarium* (pH 7.0) [17]. The pH profile of PpPel9a was much broader than reported PL9 enzymes. So far, the pH stability of PL9 enzymes had not been reported. In this study, the pH stability profile of PpPel9a showed that the enzyme was stable over a wide pH range. After incubation between pH 5 and pH 11 for 24 h at 25 °C, residual activity remained greater than 75% (Figure 3b). The excellent alkali stability of PpPel9a makes the enzyme a good biocatalyst for pectin degradation.

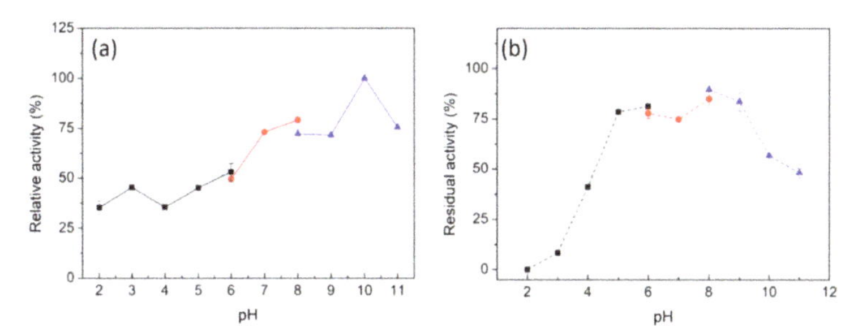

Figure 3. Effect of pH on activity (**a**) and stability (**b**) of PpPel9a. The enzyme activity was detected by using PGA as a substrate (0.2% *w*/*v*). The following buffers were used: 50 mM acetate buffer, pH 2.0–6.0; 50 mM phosphate buffer, pH 6.0–8.0; 50 mM glycine sodium buffer, pH 8.0–11.0.

The maximum activity of PpPel9a was observed at 40 °C (Figure 4a), which was lower than PL9 pectate lyases Pel-15H from *Bacillus* sp. strain KSM-P15 (55 °C) [24] and rPel9A from *C. stercorarium* F-9 (65 °C) [17]. A thermostability study showed that PpPel9a was stable below 50 °C, where more than 50% activity was maintained upon incubation for 60 min at 40 °C or 50 °C (Figure 4b). However, the activity decreased sharply when the incubation temperature reached 60 °C. Therefore, the thermostability of PpPel9a could be improved.

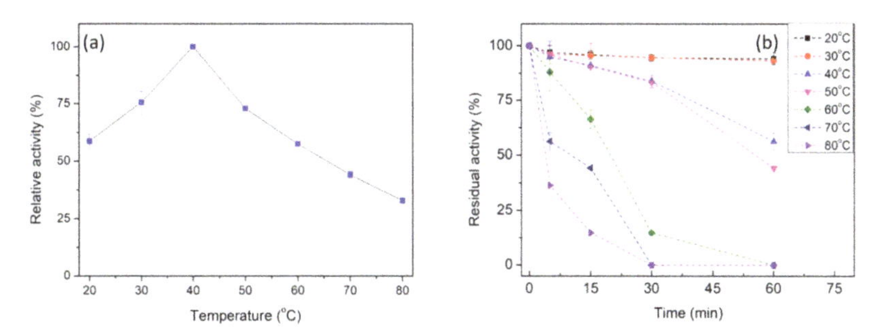

Figure 4. Effect of temperature on activity (**a**) and stability (**b**) of PpPel9a. The enzyme activity was detected by using 0.2% *w*/*v* PGA as a substrate.

The effect of metal ions and chemicals on the activity of PpPel9a was also evaluated, as shown in Table 1. 5 mM Tween-20 enhanced PpPel9a activity by 22.5%. Fe^{2+}, Co^{2+}, Tween-40, Tween-60, SDS, Tween-80, and Triton X-100 substantially decreased the activity of PpPel9a, and other metal ions examined did not significantly affect activity. Furthermore, different Tweens showed different effects on enzymatic activity.

Generally, Ca^{2+} plays an important role in the hydrolytic action of pectate lyase. This ion acts by acidifying the C5 proton of the galacturonate binding to the +1 subsite of pectate lyase [27]. The enzymatic activities of PL9 enzymes are generally inhibited by treatment with EDTA. For instance, PelX from *E. chrysanthemi* 3937 and rPel9A from *C. stercorarium* F-9 are completely inhibited by 1 mM or 0.2 mM EDTA, respectively [17,25]. The activity of the PL10 enzyme PpPel10a is independent of Ca^{2+}, although the enzymatic activity is significantly enhanced by its presence. Similar to PpPel10a, the activity of PpPel9a is not completely inhibited by EDTA, with over 76.5% of its enzymatic activity remaining. As shown in Figure 5, PpPel9a activity is enhanced by addition of Ca^{2+}. In the presence of 5 mM Ca^{2+}, PpPel19a activity is increased by 166.5%. Therefore, 5 mM Ca^{2+} was used in following experiments.

Table 1. Effect of various metal ions and chemicals on the activity of PpPel9a.

Metal Ions or Chemicals (5 mM)	Relative Activity (%) [1]
NaCl	86.7 ± 3.3
KCl	79.0 ± 0.6
$CaCl_2$	266.5 ± 7.0
$MgCl_2$	94.5 ± 4.6
$FeSO_4$	54.5 ± 0.7
EDTA	63.1 ± 0.5
$ZnCl_2$	88.0 ± 1.6
$AlCl_3$	106.1 ± 6.1
$CoCl_2$	51.4 ± 2.4
SDS	37.7 ± 0.2
Tween-20	122.5 ± 0.6
Tween-40	51.4 ± 2.1
Tween-60	40.0 ± 0.6
Tween-80	2.0 ± 0.1
TritonX-100	4.3 ± 0.3

[1] The activity was determined with 0.2% *w/v* PGA as a substrate.

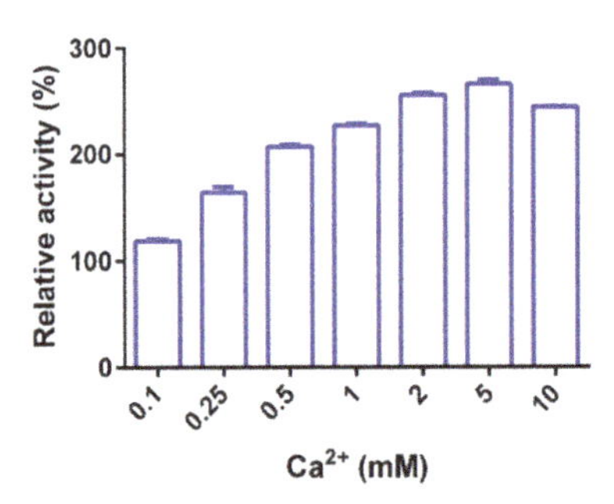

Figure 5. Effect of Ca^{2+} on the activity of PpPel9a. Values represent the mean ± SD (n = 3).

Michaelis–Menten parameters of PpPel9a on PGA and CP were determined. The K_m values of PpPel9a for PGA and CP were 0.18 and 0.32 g/L; V_{max} values were 298.5 and 107 μmol/min/mg, respectively, resulting in k_{cat}/K_m values of 225.4 and 80.8 s^{-1}, respectively.

2.4. *Substrate Specificity Analysis of PpPel9a*

Previously reported PL9 pectate lyases, such as Pel-15H from *Bacillus* sp. strain KSM-P15, PelX from *E. chrysanthemi* 3937, and rPel9A from *C. stercorarium* showed the highest catalytic activity for PGA, followed by pectins with low or medium methylation (<50%) [17,24,25]. Pectins with high degrees of methylation (≥50% methylation) were poor substrates for PL9 enzymes. However, the enzymatic activity of PL9 against RG pectins had yet to be reported. Here, the ability of PpPel9a to degrade various pectins, including RG-I type and HG-type pectins, were determined. As shown in Table 2, PpPel9a showed higher activity on HG-type pectins compared with RG-I type pectins. Using HG-type pectins as substrate, PpPel9a was more active on de-esterified HG-type pectins than on esterified HG-type pectins. The highest degradation activity was observed on de-esterified HG type pectin from citrus (CP-DeHG). Relative to the activity of PpPel9a on CP (100%), the activities of PpPel9a on CP-DeHG and CP-CeHG are 664% and 44%, respectively. Our results demonstrated that PL9 enzymes prefer low- or un-methylated HG pectins.

Table 2. Degradation of various pectins by PpPel9a.

Type of Pectin	Substrate [1]	Specific Activity (U/mg)
Pectin	PGP	48.1 ± 4.9
	RGP	59.1 ± 0.1
	CP	107.0 ± 0.2
RG-I	PGP-RGI	2.1 ± 0.9
	RGP-RGI	21.0 ± 0.3
HG	PGP-HG	106.8 ± 0.6
	RGP-HG	68.9 ± 0.5
Completely esterified HG	CP-CeHG	47.1 ± 0.1
De-esterified HG	CP-DeHG	710.7 ± 3.3

[1] CP: citrus pectin; PGP: pectin from *Panax ginseng*; RGP: pectin from red ginseng.

To explore the use of PpPel9a on the degradation of pectins, eight pectins from different plants were used as the substrates. The molecular weight distribution of the degradation products was detected by high-performance gel permeation chromatography (HPGPC). With the exception of ACP, PpPel9a degraded the pectins to varying degrees to produce low-molecular-weight fractions (Figure 6).

Previously, pectate lyases were reported to exhibit catalytic activity via an endo- or exo-acting mode by β-elimination [27]. An exo-acting enzyme can produce 4,5-unsaturated GalA (uG1), while endo-acting enzymes tend to produce only unsaturated dimers or higher-degree polymerization unsaturated products. Two PL9 enzymes, PelX from *E. chrysanthemi* 3937 and Exo-PL from *E. chrysanthemi* EC16, have been reported as exo-acting pectate lyases, whereas Pel9A from *E. chrysanthemi* (1RU4) was reported to act as an endo-pectate lyase [25–27]. To assess the catalytic mechanism of PpPel9a, the oligomers released from CP by PpPel9a were analyzed by ultra-performance liquid chromatography - tandem mass spectrometer (UPLC-MS). As shown in Figure 7, no uG1 was observed upon degradation, which demonstrated that PpPel9a functions as an endo-acting enzyme. Electrospray ionisation mass spectrometry (ESI-MS) confirmed the UPLC results. Negative ESI-MS gave a strong peak at m/z 350.92 (unsaturated bigalacturonide, uG2), followed by m/z 527.02 (unsaturated trigalacturonide, uG3). These results indicate that citrus pectin was degraded by PpPel9a to produce a mixture of 4, 5-unsaturated oligo-galacturonic acid, confirming the trans-elimination endo-reaction mechanism of PpPel9a. The catalytic mechanism of PpPel9a was different from that of PpPel10a, which degraded citrus pectin to produce uG1 and 4,5-unsaturated oligomers, including uG2, uG3, uG4, and uG5.

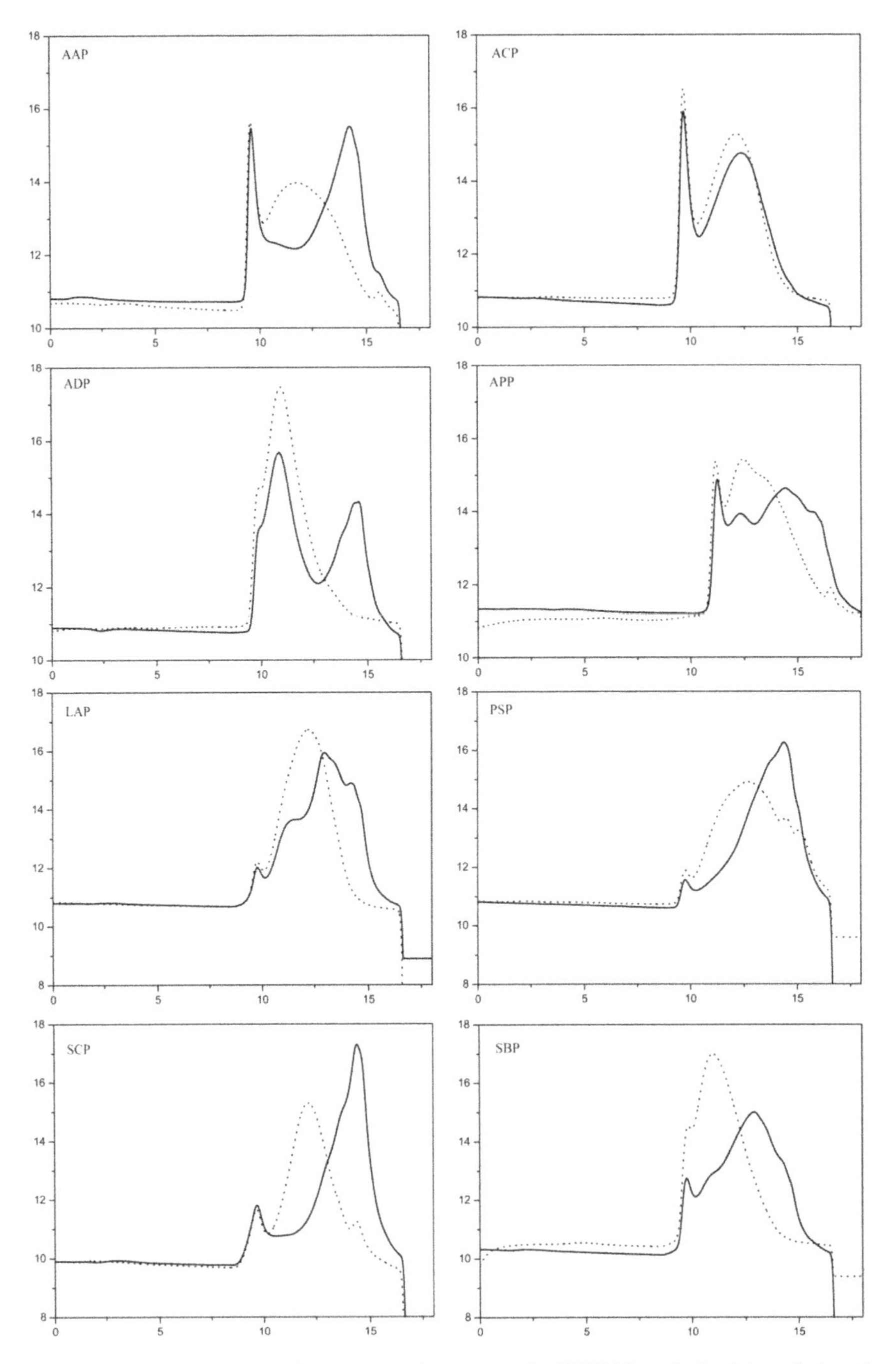

Figure 6. High-performance gel permeation chromatography (HPGPC) analysis of degradation of pectins from different plants by PpPel9a. Dash line represents the substrate and solid line represents the degradation product.

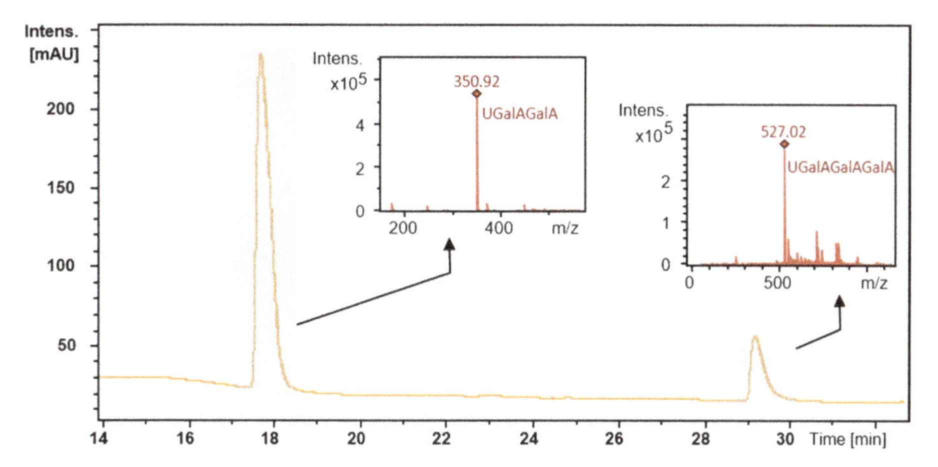

Figure 7. Analysis of the oligomers released from PGA by ultra-performance liquid chromatography-tandem mass spectrometer (UPLC-MS). The oligosaccharides separated by UPLC was further analyzed by ESI-MS analysis (small box).

2.5. Degradation of Citrus Pectin and Characterization of the Main Degradation Product

Previously, different methods including heat- and pH-modification were used to prepare modified citrus pectin. Jackson et al. showed that different treatment protocols of pectin can lead to different pectin activities, indicating that active molecules obtained from citrus pectin by different treatments are not the same [34]. Therefore, exploring new preparation techniques may be potentially useful for obtaining new active molecules. Here, citrus pectin was degraded by PpPel9a, and the degradation products were purified by using a DEAE sepharose fastflow column. As shown in Figure 8, a major peak was eluted by using 0.3 M NaCl and prolonging the degradation time or addition of more enzyme did not cause further degradation. This fraction is called MCP-0.3, and its molecular weight was determined to be ~24 kDa by HPGPC analysis (Figure S2). Monosaccharide composition analysis (Table 3) showed that after degradation, the content of GalA increased, while the content of Gal, Ara, and Glc decreased. The modified 2-Thiobarbituric acid (TBA) method indicated that the content of RGII domain increased in MCP-0.3 compared to that of CP (Figure S3). The Fourier Transform infrared spectroscopy (FT-IR) analysis showed that the degree of methyl esterification decreased from 65.5% in CP to 31.3% in MCP-0.3 (Figure S4), which indicated that the HG domain in CP was removed by PpPel9a degradation. Based on these results, we deduced that PpPel9a mainly cuts the HG domain in citrus pectin. Degradation experiments were repeated in triplicate, with similar results being obtained. The excellent reproducibility demonstrated that PpPel9a is suitable for citrus pectin degradation.

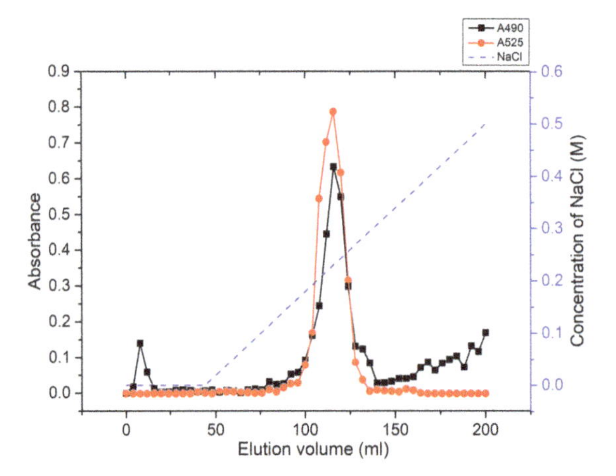

Figure 8. Elution profile of the degradation product of citrus pectin (CP) by DEAE sepharose fastflow column. The sugar content and uronic acid content were determined at 490 nm and 525 nm, respectively.

Table 3. Monosaccharide composition of CP and its degradation product MCP-0.3.

	Monosaccharide (%)								
	Man	GlcA	Rha	GalA	Glc	Gal	Xyl	Ara	Fuc
CP	1.20	0.57	3.62	62.18	4.12	13.73	- [1]	14.42	0.16
MCP-0.3	1.12	1.66	3.68	86.97	1.02	2.06	-	2.49	1.00

[1] Not detected.

2.6. *Growth Effect of MCP-0.3 on Lactobacillus Strains*

Pectin fragments have been reported to exhibit good gastroprotective effects. For example, arabinogalactans from soybean significantly inhibit ethanol-induced gastric lesions in rats [35]. Two pectins ALR-a and ALR-b from rhizomes of *Atractylodes lancea* DC showed intestinal immune system modulating activity [36]. Oligosaccharides from pectin (POS) have been suggested as a new class of prebiotics, with anti-obesity, anti-toxic, anti-infection, anti-bacterial, and antioxidant properties [37]. Therefore, pectin may be a good source of prebiotic molecules.

The growth effect of freeze-dried MCP-0.3 was assayed in vitro with *Lactobacillus plantarum* strains CH4, P3-18, S52, C88, K25 and *L. rhamnosus* CG strains JAAS8 and ITF-1. As shown in Figure 9, compared with CP, MCP-0.3 at a concentration of 5 mg/mL significantly promotes the growth of *L. plantarum* and *L. rhamnosus* CG strains. The highest promotion effect was observed on *L. plantarum* S52 strain. Previously, the *B. subtilis* pectate lyase rePelB was reported to hydrolyze citrus pectin in a 50 L reactor. The low-molecular-weight citrus pectin (LCP) obtained significantly promoted the growth of probiotics *Bifidobacterium longum* [10]. In addition, the pectin oligosaccharide (POS) fractions obtained by controlled chemical degradation of citrus peel pectin were reported to promote the selective growth of probiotics *L. paracasei* LPC-37 and *B. bifidum* ATCC 29521 [38]. Here, we reported that the degradation product of citrus pectin by PpPel9a showed better prebiotic effects than citrus pectin, indicating that MCP-0.3 may be a good prebiotic agent. PpPel9a had the advantage of reproducible production of a low-molecular-weight fraction from citrus pectin, which was not the case with PpPel10a. Based on this, PpPel9a may be useful in preparing pectin fractions with prebiotic activity from pectin-rich agricultural wastes, such as citrus peel and apple pomace.

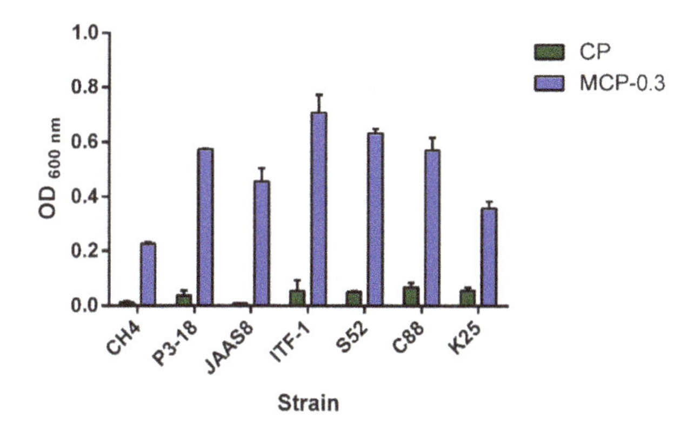

Figure 9. Effects of CP and MCP-0.3 on the growth of *L. plantarum* strains CH4, P3-18, S52, C88, K25, and *L. rhamnosus* CG strains JAAS8 and ITF-1.

3. Materials and Methods

3.1. Strains and Reagents

P. polymyxa KF-1 was isolated and identified by our lab and preserved in China Center for Type Culture Collection (accession number CCTCC AB 2018146). It was used for supplying the PL9 pectate lyase gene. *E. coli* DH5α cells and pGM-Simple-T vector (TIANGEN, Beijing, China) were used for gene cloning. *E. coli* BL21 (DE3) cells and pET-28a (+) vector (Novagen, Temecula, California, USA) were selected for protein expression. *L. plantarum* strains CH4, P3-18, S52, C88, K25, and *L. rhamnosus* CG strains JAAS8 and ITF-1 were kindly provided by Dr Shengyu Li of the Institute of Agrofood Technology, Jilin Academy of Agricultural Sciences, China. Restriction enzymes and DNA polymerase were from TaKaRa Biotechnology Co., Ltd. (Dalian, China). DEAE-sepharose fastflow gel were purchased from GE Healthcare (USA).

3.2. Substrates

Polygalacturonic acid (PGA) and citrus pectin (CP) were from Sigma-Aldrich (St. Louis, MO, USA). Pectins from *Panax ginseng* (PGP), red ginseng (RGP), *Anemarrhena asphodeloides* (AAP), *Asparagus cochinchinensis* (ACP), *Angelica dahurica* (ADP), *Angelica pubescens* (APP), *Leonurus Artemisia* (LAP), *Polygonatum sibiricum* (PSP), *Semen cassia* (SCP), and *Scutellaria baicalensis* (SBP) were isolated by hot water extraction followed by DEAE-cellulose fractionation according to the method previously reported by our lab [3,39]. The RG-I type pectins PGP-RGI from *P. ginseng*, and RGP-RGI from red ginseng, the HG type pectins PGP-HG from *P. ginseng*, RGP-HG from red ginseng, were prepared according to the method reported in Sun et al. [40]. The de-esterified HG type pectin CP-DeHG and the completely esterified HG type pectin CP-CeHG were prepared from citrus pectin according to the method reported in Sun et al. [40].

3.3. Analytical Methods

The unsaturated oligomers released from CP by PpPel9a were analyzed by UPLC-MS. The UPLC-MS was performed using a Waters Acquity UPLC system with tandem UV (232 nm) and ESI-MS detector. The oligosaccharides were first separated by Acquity UPLC BEH Amide column (1.7 μm, 2.1 mm × 150 mm) eluted at a flow rate of 300 μL/min and a column oven temperature of 35 °C. The mobile phase consisted of three mobile phase lines, (A) 20:80 (*v*/*v*) ACN/water, (B) 80:20 (*v*/*v*) ACN/water, and (C) 200 mM ammonium formate/50 mM formic acid buffer (pH 3.0). The elution program was as follows: 0–30 min, 95%–75% B; 30–31 min, 75%–60% B; 31–40 min, isocratic with 60% B; 40–41 min, 60%–95% B; 41–50 min, isocratic with 95% B; 5% buffer (C) was constantly added throughout the elution. ESI-MS analysis was performed with a Bruker Daltonics Amazon mass spectrometer equipped with an electrospray source and an ion trap mass analyzer (Bremen, Germany). The spectrometer was operated in the negative ion mode (capillary voltage, 4 kV; end plate off set: −300 V; temperature, 200 °C; nebulizer gas: 2 bar and dry gas, 6 mL/min). Monosaccharide composition was analyzed by high performance anion exchange chromatography (HPAEC) method using a CarboPac PA20 column (3 × 150 mm) assembled to a Dionex ICS-5000 Plus ion chromatographic system [41]. Homogeneity and molecular weight of polysaccharides were estimated by HPGPC with a TSK-gel G-3000PWXL column (7.8 × 300 mm, TOSOH, Japan) coupled to a Shimadzu high performance liquid chromatography (HPLC) system. The column was pre-calibrated using standard dextrans (1, 5, 12, 25, and 50 kDa) [42]. The degree of esterification was determined by FT-IR. The degree of methyl esterification = the area of methyl-esterified carboxyl group/the area of total carboxyl group × 100% [43]. Kdo and Dha were colorimetrically determined using the modified TBA method to check the presence of RG-II domain in polysaccharides [44].

3.4. Analysis of Pectate Lyase Gene

The ORF of the pectate lyase was analyzed by NCBI ORF finder [45]. The conserved domain was explored by Pfam [46]. The molecular weight (Mw) and isoelectric point (pI) were calculated by Compute pI/Mw too [47]. The signal peptide was predicted by SignalP 5.0 Server [48]. Multiple amino acid sequences alignments were performed by Clustal Omega and exhibited by ESPript 3.0 web server [49,50]. The structure of pectate lyase PpPel9a was predicted by SWISS-MODEL with the pectate lyase from *E. chrysanthemi* (PDB accession number: 1RU4) as the template [27]. The model was exhibited by Pymol software [51].

3.5. Cloning of PL9 Pectate Lyase and Its Expression

The mature pectate lyase encoding gene without the signal peptide sequence was obtained from the genomic DNA of *P. polymyxa* KF-1 by PCR using following primers: 5′-CGCATATGGATGTTCCATCTAACAACCTC-3′ and 5′-CGGGATCCTTACCGTGCCCCTATATTTCCGC-3′. The underlines indicate the *Nde*I and *Bam*HI restriction sites, respectively. The PCR amplification was performed by PrimeSTAR® HS DNA polymerase with the following program: Pre-denaturation at 98 °C for 10 s; 30 cycles of 98 °C for 10 s, 55 °C for 15 s, and 72 °C for 1.5 min; final extension at 72 °C for 10 min. The PCR product was ligased with pGM-Simple-T vector, transformed into *E. coli* DH5α competent cells, and plated onto Luria–Bertani (LB) agar plates supplemented with 100 μg/mL ampicillin according to the manufacturer's instructions. Recombinant plasmid was sequenced with T7 (5′-ACATCCACTTTGCCTTTCTC-3′) and SP6 (5′-ATTTAGGTGACACTATAG-3′) as sequencing primers. The recombinant plasmid with correct insert DNA was double digested by *Nde*I and *Bam*HI, and ligased into pET-28a (+) vector digested with the same restriction enzymes. The recombinant plasmid was transformed into *E. coli* BL21 (DE3) cells and plated onto LB agar plates supplemented with 30 μg/mL kanamycin.

The positive colony was incubated in 500 mL of LB broth at 37 °C for 3 h with shaking at 200 rpm. Then, the isopropyl β-D-thiogalactoside (IPTG) was added to a final concentration of 0.5 mM. After induced at 25 °C for 20 h, the cells were centrifuged at 5000 rpm and 4 °C for 10 min. Then, the cells were sonicated on ice and centrifuged at 12,000× *g* and 4 °C for 10 min to remove cell debris. The cell lysate was loaded onto a Ni^{2+}–NTA agarose chromatography (column volume 5 mL), which was pre-equilibrated with equilibration buffer (20 mM Tris–HCl, 100 mM NaCl, 20 mM imidazole, pH 8.0). Binding proteins were eluted with a linear gradient of imidazole (20–200 mM) in 20 mM Tris–HCl, 100 mM NaCl, pH 8.0. Fractions with enzymatic activity were pooled and dialyzed against 50 mM Tris–HCl buffer (pH 9.0). The cell lysate and purified enzyme were analyzed by SDS-PAGE using 10% separating gel and 3.9% stacking gel [52]. The protein concentration of the purified enzyme was determined using BCA Protein Assay Kit (Beyotime, Beijing, China).

3.6. Pectate Lyase Activity Assay and Substrate Specificity Analysis of PpPel9a

During expression and purification, the enzymatic activity of PpPel9a was measured by determining the formed unsaturated bonds at 235 nm with an UV-5100H spectrophotometer (METASH, Shanghai, China) with soluble PGA as the substrate [19]. The reaction mixture of 200 μL consists of 0.2% (*w*/*v*) PGA, 5 mM $CaCl_2$, and 2 μg of purified enzyme was incubated in 50 mM Tris-HCl buffer (pH 9.0). The reaction mixture was incubated at 40 °C for 30 min and the unsaturated bonds produced by enzymatic degradation were determined at 235 nm. One unit of pectate lyase activity was defined as the enzyme needed to produce 1 μmol of unsaturated bonds per min. The molar extinction coefficient was 4600 $M^{-1}cm^{-1}$.

The substrate specificity of recombinant PpPel9a was determined using different types of pectins, including HG, RG-I, de-esterified HG, and completely-esterified HG, under standard reaction conditions (pH 10.0, 40 °C, 5 mM $CaCl_2$, 0.2% *w*/*v* substrate). The unsaturated bonds released from the substrates were recorded at 235 nm. The molecular weights of the degradation products were analyzed by HPGPC method.

The kinetic parameters of PpPel9a on PGA and CP were determined by the A235 method. Purified recombinant PpPel9a was incubated with PGA or CP at different concentrations (0.1–5 mg/mL) at 40 °C in 50 mM glycine–NaOH buffer (pH 10.0) for 10 min. Then, the unsaturated bonds formed were measured at 235 nm. The K_m and k_{cat} values were calculated by GraphPad Prism5.0 software using non-linear regression. All data are expressed as the means of triplicate measurements.

3.7. Effect of pH, Temperature, and Metal Ions on the Activity and Stability of PpPel9a

The optimal pH of PpPel9a was determined at 40 °C for 30 min in 50 mM NaAC buffer (pH 2.0–6.0), 50 mM phosphate buffer (pH 6.0–8.0), and 50 mM glycine–NaOH buffer (pH 8.0–11.0). The temperature optimum of PpPel9a was measured at 20–80 °C for 30 min in 50 mM glycine–NaOH (pH 10.0). The pH stability of PpPel9a was assayed by measuring its residual activity after incubating the purified enzyme in 50 mM NaAC buffer (pH 2.0–6.0), 50 mM phosphate buffer (pH 6.0–8.0), and 50 mM glycine–NaOH buffer (pH 8.0–11.0) at 25 °C for 24 h. The thermal stability of PpPel9a was determined by measuring the residual enzyme activity after incubating the purified enzyme at 20–80 °C in 50 mM glycine–NaOH buffer (pH 10.0) for up to 1 h. In enzymatic characterization, soluble PGA was used as the substrate with the concentration of 0.2% (*w*/*v*), 5 mM $CaCl_2$ was added during enzymatic reaction. Formed unsaturated products were recorded at 235 nm. All data are expressed as the averages of triplicate measurements (± SD).

The effects of metal ions and chemical reagents on PpPel9a activity were detected by incubating the purified enzyme in 5 mM of metal ions and reagents at 25 °C for 24 h, then the residual activity was measured under the standard assay condition with 0.2% (*w*/*v*) PGA as the substrate. The enzymatic activity without adding any metal ions or chemical reagents was marked as 100%. The effect of Ca^{2+} at different concentrations (0.1–10 mM) on enzyme activity was determined in 50 mM glycine–NaOH buffers (pH 10.0) with 0.2% PGA (*w*/*v*) as the substrate. Formed unsaturated products were recorded at 235 nm. The enzymatic activity without adding $CaCl_2$ was marked as 100%.

3.8. Degradation of CP by PpPel9a and the Preparation of Degradation Product

First, 2.5 g of CP was incubated with 50 mg of purified recombinant PpPel9a in 50 mM Tris-HCl buffer (pH 8.0). The enzyamtic degradation lasted for 48 h at 40 °C. Then, the reaction mixture was boiled for 10 min and centrifuged at 10,000× *g* for 10 min to remove the protein. Then, the degradation product was freeze-dried and purified by DEAE-sepharose fastflow column. The column (bed volume 25 mL) was eluted by 2 column volume of water, followed by 6 column volume of linear gradient of 0–0.5 M NaCl. The eluate was collected at 4 mL/tube and detected for the distribution of total sugar by the phenol-sulphuric acid method, and uronic acid by the m-hydroxydiphenyl method [53,54]. The main fraction peak eluted by 0.3 M NaCl was combined, dialyzed against distilled water, and freeze-dried. The fraction was named MCP-0.3, and its monosaccharide composition, molecular weight, RG-II content, and degree of esterification were analyzed by HPAEC, HPGPC, modified TBA method, and FT-IR, respectively.

3.9. Prebiotic Effect of CP and Its Degradation Product MCP-0.3 on Lactobacillus Strains

The prebiotic effects of CP and MCP-0.3 on Lactobacillus strains were studied as described in Zheng et al. [41]. Briefly, Lactobacillus strains, including *L. rhamnosus* strains JAAS8 and ITF-1, and *L. plantarum* strains CH4, P3-18, S52, C88, and K25, were pre-cultured in De Man Rogosa Sharpe (MRS) liquid at 37 °C for 16 h. Then, the cells were inoculated with an inoculum size of 2% in 100 μL modified MRS media containing 5 mg/mL CP or MCP-0.3 as the sole carbon source. Then, the cells were collected by centrifugation after cultured at 37 °C for 16 h and washed twice with PBS buffer. The cell pellet was re-suspended in 200 μL PBS buffer, and the $OD_{600 nm}$ was detected by the Tecan Infinite F50 microplate reader. All data are shown as the averages of triplicate measurements (± SD).

4. Conclusions

In the present study, the PL9 pectate lyase PpPel9a was cloned from *P. polymyxa* KF-1 and was used to degrade pectins from different plants. Recombinant PpPel9a was used to degrade citrus pectin, reproducibly yielding the fraction MCP-0.3 that exhibited good prebiotic effects in vitro. This enzyme may be a valuable candidate for preparation of low-molecular-weight pectin fractions.

Supplementary Materials: Supplementary materials can be found at http://www.mdpi.com/1422-0067/20/12/3060/s1.

Author Contributions: Conceptualization, Y.Y. and J.G.; methodology, Y.Y., X.-Y.Z. and H.Z.; software, Y.Z.; writing—original draft preparation, J.G.; writing—review and editing, J.G.; supervision, Y.-F.Z.; project administration, J.G.; funding acquisition, Y.-F.Z.

Funding: This research was supported by the Fundamental Research Funds for the Central Universities (2412019QD011), Natural Science Foundation of Jilin Province (20190201258JC).

Conflicts of Interest: The authors declare no conflict of interest.

References

1. Christiaens, S.; Van Buggenhout, S.; Houben, K.; Jamsazzadeh Kermani, Z.; Moelants, K.R.N.; Ngouemazong, E.D.; Van Loey, A.; Hendrickx, M.E.G. Process-Structure-Function Relations of Pectin in Food. *Crit. Rev. Food Sci. Nutr.* **2016**, *56*, 1021–1042. [CrossRef] [PubMed]
2. Chan, S.Y.; Choo, W.S.; Young, D.J.; Loh, X.J. Pectin as a rheology modifier: Origin, structure, commercial production and rheology. *Carbohydr. Polym.* **2017**, *161*, 118–139. [CrossRef] [PubMed]
3. Zhang, X.; Li, S.; Sun, L.; Ji, L.; Zhu, J.; Fan, Y.; Tai, G.; Zhou, Y. Further analysis of the structure and immunological activity of an RG-I type pectin from *Panax ginseng*. *Carbohydr. Polym.* **2012**, *89*, 519–525. [CrossRef] [PubMed]
4. Minzanova, S.T.; Mironov, V.F.; Arkhipova, D.M.; Khabibullina, A. V.; Mironova, L.G.; Zakirova, Y.M.; Milyukov, V.A. Biological Activity and Pharmacological Application of Pectic Polysaccharides: A Review. *Polymers (Basel).* **2018**, *10*, 1407. [CrossRef] [PubMed]
5. Liu, Y.; Dong, M.; Yang, Z.; Pan, S. Anti-diabetic effect of citrus pectin in diabetic rats and potential mechanism via PI3K/Akt signaling pathway. *Int. J. Biol. Macromol.* **2016**, *89*, 484–488. [CrossRef] [PubMed]
6. Chen, J.; Ye, F.; Zhou, Y.; Zhao, G. Thiolated citrus low-methoxyl pectin: Synthesis, characterization and rheological and oxidation-responsive gelling properties. *Carbohydr. Polym.* **2018**, *181*, 964–973. [CrossRef] [PubMed]
7. Ruvolo, P.P. Galectin 3 as a guardian of the tumor microenvironment. *Biochim. Biophys. Acta Mol. Cell Res.* **2016**, *1863*, 427–437. [CrossRef] [PubMed]
8. Wang, S.; Li, P.; Lu, S.-M.; Ling, Z.-Q. Chemoprevention of Low-Molecular-Weight Citrus Pectin (LCP) in Gastrointestinal Cancer Cells. *Int. J. Biol. Sci.* **2016**, *12*, 746–756. [CrossRef]

9. do Prado, S.B.R.; Shiga, T.M.; Harazono, Y.; Hogan, V.A.; Raz, A.; Carpita, N.C.; Fabi, J.P. Migration and proliferation of cancer cells in culture are differentially affected by molecular size of modified citrus pectin. *Carbohydr. Polym.* **2019**, *211*, 141–151. [CrossRef]
10. Liu, M.; Huo, W.; Dai, X.; Dang, Y. Preparation of low-molecular-weight citrus pectin by recombinant *Bacillus subtilis* pectate lyase and promotion of growth of *Bifidobacterium longum*. *Catal. Commun.* **2018**, *107*, 39–42. [CrossRef]
11. Chen, J.; Liu, W.; Liu, C.-M.; Li, T.; Liang, R.-H.; Luo, S.-J. Pectin modifications: A review. *Crit. Rev. Food Sci. Nutr.* **2015**, *55*, 1684–1698. [CrossRef] [PubMed]
12. Patidar, M.K.; Nighojkar, S.; Kumar, A.; Nighojkar, A. Pectinolytic enzymes-solid state fermentation, assay methods and applications in fruit juice industries: A review. *3 Biotech* **2018**, *8*, 199. [CrossRef] [PubMed]
13. Jolie, R.P.; Duvetter, T.; Van Loey, A.M.; Hendrickx, M.E. Pectin methylesterase and its proteinaceous inhibitor: A review. *Carbohydr. Res.* **2010**, *345*, 2583–2595. [CrossRef] [PubMed]
14. Herron, S.R.; Benen, J.A.; Scavetta, R.D.; Visser, J.; Jurnak, F. Structure and function of pectic enzymes: Virulence factors of plant pathogens. *Proc. Natl. Acad. Sci. USA* **2000**, *97*, 8762–8769. [CrossRef] [PubMed]
15. Marin-Rodriguez, M.C.; Orchard, J.; Seymour, G.B. Pectate lyases, cell wall degradation and fruit softening. *J. Exp. Bot.* **2002**, *53*, 2115–2119. [CrossRef]
16. Hugouvieux-Cotte-Pattat, N.; Condemine, G.; Shevchik, V.E. Bacterial pectate lyases, structural and functional diversity. *Environ. Microbiol. Rep.* **2014**, *6*, 427–440. [CrossRef] [PubMed]
17. Hla, S.S.; Kurokawa, J.; Suryani; Kimura, T.; Ohmiya, K.; Sakka, K. A novel thermophilic pectate lyase containing two catalytic modules of *Clostridium stercorarium*. *Biosci. Biotechnol. Biochem.* **2005**, *69*, 2138–2145. [CrossRef]
18. Lombard, V.; Golaconda Ramulu, H.; Drula, E.; Coutinho, P.M.; Henrissat, B. The carbohydrate-active enzymes database (CAZy) in 2013. *Nucleic Acids Res.* **2014**, *42*, D490–D495. [CrossRef]
19. Zhou, C.; Xue, Y.; Ma, Y. Cloning, evaluation, and high-level expression of a thermo-alkaline pectate lyase from alkaliphilic *Bacillus clausii* with potential in ramie degumming. *Appl. Microbiol. Biotechnol.* **2017**, *101*, 3663–3676. [CrossRef]
20. Zhang, C.; Yao, J.; Zhou, C.; Mao, L.; Zhang, G.; Ma, Y. The alkaline pectate lyase PEL168 of *Bacillus subtilis* heterologously expressed in *Pichia pastoris* is more stable and efficient for degumming ramie fiber. *BMC Biotechnol.* **2013**, *13*, 26. [CrossRef]
21. van Rensburg, P.; Strauss, M.L.A.; Lambrechts, M.G.; Cordero Otero, R.R.; Pretorius, I.S. The heterologous expression of polysaccharidase-encoding genes with oenological relevance in *Saccharomyces cerevisiae*. *J. Appl. Microbiol.* **2007**, *103*, 2248–2257. [CrossRef] [PubMed]
22. Willems, J.L.; Low, N.H. Oligosaccharide formation during commercial pear juice processing. *Food Chem.* **2016**, *204*, 84–93. [CrossRef] [PubMed]
23. Zhao, Y.; Yuan, Y.; Zhang, X.; Li, Y.; Li, Q.; Zhou, Y.; Gao, J. Screening of a Novel Polysaccharide Lyase Family 10 Pectate Lyase from *Paenibacillus polymyxa* KF-1: Cloning, Expression and Characterization. *Molecules* **2018**, *23*, 2774. [CrossRef] [PubMed]
24. Ogawa, A.; Sawada, K.; Saito, K.; Hakamada, Y.; Sumitomo, N.; Hatada, Y.; Kobayashi, T.; Ito, S. A new high-alkaline and high-molecular-weight pectate lyase from a *Bacillus* isolate: Enzymatic properties and cloning of the gene for the enzyme. *Biosci. Biotechnol. Biochem.* **2000**, *64*, 1133–1141. [CrossRef] [PubMed]
25. Shevchik, V.E.; Kester, H.C.; Benen, J.A.; Visser, J.; Robert-Baudouy, J.; Hugouvieux-Cotte-Pattat, N. Characterization of the exopolygalacturonate lyase PelX of *Erwinia chrysanthemi* 3937. *J. Bacteriol.* **1999**, *181*, 1652–1663. [PubMed]
26. Brooks, A.D.; He, S.Y.; Gold, S.; Keen, N.T.; Collmer, A.; Hutcheson, S.W. Molecular cloning of the structural gene for exopolygalacturonate lyase from *Erwinia chrysanthemi* EC16 and characterization of the enzyme product. *J. Bacteriol.* **1990**, *172*, 6950–6958. [CrossRef] [PubMed]
27. Jenkins, J.; Shevchik, V.E.; Hugouvieux-Cotte-Pattat, N.; Pickersgill, R.W. The crystal structure of pectate lyase Pel9A from *Erwinia chrysanthemi*. *J. Biol. Chem.* **2004**, *279*, 9139–9145. [CrossRef]
28. Luis, A.S.; Briggs, J.; Zhang, X.; Farnell, B.; Ndeh, D.; Labourel, A.; Basle, A.; Cartmell, A.; Terrapon, N.; Stott, K.; et al. Dietary pectic glycans are degraded by coordinated enzyme pathways in human colonic Bacteroides. *Nat. Microbiol.* **2018**, *3*, 210–219. [CrossRef]

29. Park, S.R.; Kim, M.K.; Kim, J.O.; Bae, D.W.; Cho, S.J.; Cho, Y.U.; Yun, H.D. Characterization of *Erwinia chrysanthemi* PY35 cel and pel gene existing in tandem and rapid identification of their gene products. *Biochem. Biophys. Res. Commun.* **2000**, *268*, 420–425. [CrossRef]
30. Zhou, C.; Ye, J.; Xue, Y.; Ma, Y. Directed Evolution and Structural Analysis of Alkaline Pectate Lyase from the Alkaliphilic Bacterium *Bacillus* sp. Strain N16-5 To Improve Its Thermostability for Efficient Ramie Degumming. *Appl. Environ. Microbiol.* **2015**, *81*, 5714–5723. [CrossRef]
31. Hao, M.; Yuan, X.; Cheng, H.; Xue, H.; Zhang, T.; Zhou, Y.; Tai, G. Comparative studies on the anti-tumor activities of high temperature- and pH-modified citrus pectins. *Food Funct.* **2013**, *4*, 960–971. [CrossRef]
32. Pickersgill, R.; Jenkins, J.; Harris, G.; Nasser, W.; Robert-Baudouy, J. The structure of *Bacillus subtilis* pectate lyase in complex with calcium. *Nat. Struct. Biol.* **1994**, *1*, 717–723. [CrossRef] [PubMed]
33. Liu, Y.; Chen, G.; Wang, J.; Hao, Y.; Li, M.; Li, Y.; Hu, B.; Lu, F. Efficient expression of an alkaline pectate lyase gene from *Bacillus subtilis* and the characterization of the recombinant protein. *Biotechnol. Lett.* **2012**, *34*, 109–115. [CrossRef] [PubMed]
34. Jackson, C.L.; Dreaden, T.M.; Theobald, L.K.; Tran, N.M.; Beal, T.L.; Eid, M.; Gao, M.Y.; Shirley, R.B.; Stoffel, M.T.; Kumar, M.V.; et al. Pectin induces apoptosis in human prostate cancer cells: Correlation of apoptotic function with pectin structure. *Glycobiology* **2007**, *17*, 805–819. [CrossRef] [PubMed]
35. Cipriani, T.R.; Mellinger, C.G.; Bertolini, M.L.C.; Baggio, C.H.; Freitas, C.S.; Marques, M.C.A.; Gorin, P.A.J.; Sassaki, G.L.; Iacomini, M. Gastroprotective effect of a type I arabinogalactan from soybean meal. *Food Chem.* **2009**, *115*, 687–690. [CrossRef]
36. Yu, K.-W.; Kiyohara, H.; Matsumoto, T.; Yang, H.-C.; Yamada, H. Characterization of pectic polysaccharides having intestinal immune system modulating activity from rhizomes of *Atractylodes lancea* DC. *Carbohydr. Polym.* **2001**, *46*, 125–134. [CrossRef]
37. Di, R.; Vakkalanka, M.S.; Onumpai, C.; Chau, H.K.; White, A.; Rastall, R.A.; Yam, K.; Hotchkiss, A.T.J. Pectic oligosaccharide structure-function relationships: Prebiotics, inhibitors of *Escherichia coli* O157:H7 adhesion and reduction of Shiga toxin cytotoxicity in HT29 cells. *Food Chem.* **2017**, *227*, 245–254. [CrossRef]
38. Zhang, S.; Hu, H.; Wang, L.; Liu, F.; Pan, S. Preparation and prebiotic potential of pectin oligosaccharides obtained from citrus peel pectin. *Food Chem.* **2018**, *244*, 232–237. [CrossRef]
39. Zhang, X.; Yu, L.; Bi, H.; Li, X.; Ni, W.; Han, H.; Li, N.; Wang, B.; Zhou, Y.; Tai, G. Total fractionation and characterization of the water-soluble polysaccharides isolated from *Panax ginseng* C. A. Meyer. *Carbohydr. Polym.* **2009**, *77*, 544–552. [CrossRef]
40. Sun, L.; Ropartz, D.; Cui, L.; Shi, H.; Ralet, M.-C.; Zhou, Y. Structural characterization of rhamnogalacturonan domains from *Panax ginseng* C. A. Meyer. *Carbohydr. Polym.* **2019**, *203*, 119–127. [CrossRef]
41. Zheng, Y.; Li, L.; Feng, Z.; Wang, H.; Mayo, K.H.; Zhou, Y.; Tai, G. Preparation of individual galactan oligomers, their prebiotic effects, and use in estimating galactan chain length in pectin-derived polysaccharides. *Carbohydr. Polym.* **2018**, *199*, 526–533. [CrossRef] [PubMed]
42. Wu, D.; Cui, L.; Yang, G.; Ning, X.; Sun, L.; Zhou, Y. Preparing rhamnogalacturonan II domains from seven plant pectins using *Penicillium oxalicum* degradation and their structural comparison. *Carbohydr. Polym.* **2018**, *180*, 209–215. [CrossRef] [PubMed]
43. Peng, X.; Yang, G.; Fan, X.; Bai, Y.; Ren, X.; Zhou, Y. Controlled methyl-esterification of pectin catalyzed by cation exchange resin. *Carbohydr. Polym.* **2016**, *137*, 650–656. [CrossRef] [PubMed]
44. York, W.S.; Darvill, A.G.; McNeil, M.; Albersheim, P. 3-deoxy-d-manno-2-octulosonic acid (KDO) is a component of rhamnogalacturonan II, a pectic polysaccharide in the primary cell walls of plants. *Carbohydr. Res.* **1985**, *138*, 109–126. [CrossRef]
45. Sayers, E.W.; Barrett, T.; Benson, D.A.; Bolton, E.; Bryant, S.H.; Canese, K.; Chetvernin, V.; Church, D.M.; DiCuccio, M.; Federhen, S.; et al. Database resources of the National Center for Biotechnology Information. *Nucleic Acids Res.* **2011**, *39*, D38–D51. [CrossRef] [PubMed]
46. Finn, R.D.; Bateman, A.; Clements, J.; Coggill, P.; Eberhardt, R.Y.; Eddy, S.R.; Heger, A.; Hetherington, K.; Holm, L.; Mistry, J.; et al. Pfam: The protein families database. *Nucleic Acids Res.* **2014**, *42*, D222–D230. [CrossRef] [PubMed]
47. Wilkins, M.R.; Gasteiger, E.; Bairoch, A.; Sanchez, J.C.; Williams, K.L.; Appel, R.D.; Hochstrasser, D.F. Protein identification and analysis tools in the ExPASy server. *Methods Mol. Biol.* **1999**, *112*, 531–552. [PubMed]

48. Almagro Armenteros, J.J.; Tsirigos, K.D.; Sonderby, C.K.; Petersen, T.N.; Winther, O.; Brunak, S.; von Heijne, G.; Nielsen, H. SignalP 5.0 improves signal peptide predictions using deep neural networks. *Nat. Biotechnol.* **2019**, *37*, 420–423. [CrossRef]
49. Sievers, F.; Higgins, D.G. Clustal omega. *Curr. Protoc. Bioinforma.* **2014**, *48*, 3–13.
50. Gouet, P.; Robert, X.; Courcelle, E. ESPript/ENDscript: Extracting and rendering sequence and 3D information from atomic structures of proteins. *Nucleic Acids Res.* **2003**, *31*, 3320–3323. [CrossRef]
51. Mooers, B.H.M. Simplifying and enhancing the use of PyMOL with horizontal scripts. *Protein Sci.* **2016**, *25*, 1873–1882. [CrossRef] [PubMed]
52. Schagger, H. Tricine-SDS-PAGE. *Nat. Protoc.* **2006**, *1*, 16–22. [CrossRef] [PubMed]
53. DuBois, M.; Gilles, K.A.; Hamilton, J.K.; Rebers, P.A.; Smith, F. Colorimetric Method for Determination of Sugars and Related Substances. *Anal. Chem.* **1956**, *28*, 350–356. [CrossRef]
54. Blumenkrantz, N.; Asboe-Hansen, G. New method for quantitative determination of uronic acids. *Anal. Biochem.* **1973**, *54*, 484–489. [CrossRef]

International Journal of *Molecular Sciences*

Article

Characterizing a Halo-Tolerant GH10 Xylanase from *Roseithermus sacchariphilus* Strain RA and Its CBM-Truncated Variant

Seng Chong Teo [1], Kok Jun Liew [1], Mohd Shahir Shamsir [1], Chun Shiong Chong [1], Neil C. Bruce [2], Kok-Gan Chan [3,4,]* and Kian Mau Goh [1,]*

1 Faculty of Science, Universiti Teknologi Malaysia, Skudai 81310, Johor, Malaysia; scteo4@gmail.com (S.C.T.); kokjunliew@gmail.com (K.J.L.); shahir@utm.my (M.S.S.); cschong@utm.my (C.S.C.)
2 Centre for Novel Agricultural Products, Department of Biology, University of York, Wentworth Way, York YO10 5DD, UK; neil.bruce@york.ac.uk
3 Division of Genetics and Molecular Biology, Institute of Biological Science, Faculty of Science, University of Malaya, Kuala Lumpur 50603, Malaysia
4 International Genome Centre, Jiangsu University, Zhenjiang 212013, China
* Correspondence: kokgan@um.edu.my (K.-G.C.); gohkianmau@utm.my (K.M.G.); Tel.: +603-79677748 (K.-G.C.); +607-5557556 (K.M.G.)

Received: 14 March 2019; Accepted: 7 May 2019; Published: 9 May 2019

Abstract: A halo-thermophilic bacterium, *Roseithermus sacchariphilus* strain RA (previously known as *Rhodothermaceae* bacterium RA), was isolated from a hot spring in Langkawi, Malaysia. A complete genome analysis showed that the bacterium harbors 57 glycoside hydrolases (GHs), including a multi-domain xylanase (XynRA2). The full-length XynRA2 of 813 amino acids comprises a family 4_9 carbohydrate-binding module (CBM4_9), a family 10 glycoside hydrolase catalytic domain (GH10), and a C-terminal domain (CTD) for type IX secretion system (T9SS). This study aims to describe the biochemical properties of XynRA2 and the effects of CBM truncation on this xylanase. XynRA2 and its CBM-truncated variant (XynRA2ΔCBM) was expressed, purified, and characterized. The purified XynRA2 and XynRA2ΔCBM had an identical optimum temperature at 70 °C, but different optimum pHs of 8.5 and 6.0 respectively. Furthermore, XynRA2 retained 94% and 71% of activity at 4.0 M and 5.0 M NaCl respectively, whereas XynRA2ΔCBM showed a lower activity (79% and 54%). XynRA2 exhibited a turnover rate (k_{cat}) of 24.8 s^{-1}, but this was reduced by 40% for XynRA2ΔCBM. Both the xylanases hydrolyzed beechwood xylan predominantly into xylobiose, and oat-spelt xylan into a mixture of xylo-oligosaccharides (XOs). Collectively, this work suggested CBM4_9 of XynRA2 has a role in enzyme performance.

Keywords: glycoside hydrolase; xylanase; carbohydrate-binding module; CBM truncation; halo-tolerant; xylan hydrolysis

1. Introduction

Xylan is one of the most abundant polymers in plant biomass. The polymer consists of a β-1,3/1,4-linked xylopyranose backbone with side branches such as *O*-acetyl, α-4-*O*-glucuronic acid, α-L-arabinofuranose, *p*-coumaric acid, or ferulic acid at C-2 or C-3 positions [1]. Due to such complexity, a complete hydrolysis of xylan requires synergism of various xylanolytic enzymes including endo-β-1,4-D-xylanase, β-D-xylosidase, α-D-glucuronidase, α-L-arabinofuranosidase, and acetylesterase [2]. Among these enzymes, endo-β-1,4-xylanase (E.C. 3.2.1.8) plays a crucial role in hydrolyzing the β-1,4-glycosidic bonds of the xylan to form xylo-oligosaccharides (XOs) and xylose. According to the Carbohydrate-Active Enzyme (CAZy) database (http://www.cazy.org) [3], endo-β-1,4-xylanases are currently grouped in glycosyl hydrolase (GH) 5, 8, 10, 11, 30, 43, 51, 98, and

141 families. The majority of endo-β-1,4-xylanases belong to GH10 and GH11 families, which are distinctive of their respective origin, molecular properties, and protein structure [4].

Xylanases are produced by a diverse range of organisms, which include fungi, bacteria, yeast, algae, protozoa, crustaceans, and insects. Fungal and bacterial xylanases are important due to their superior properties, which could potentially be applied in industrial processes [5]. As summarized in a review article [6], xylanases are utilized for the delignification of paper pulp, modification of cereal food, improvement of digestibility of animal feedstock and production of xylo-oligosaccharides for pharmaceutical industries. Selection of the types of xylanases for these applications is based on suitability. For instance, thermostable alkaline xylanases are applicable for efficient biobleaching of pulp and paper [7], while thermostable acidic xylanases are applicable for animal feed processes [8]. Xylanases which are active and stable in low or high pH values are suitable for hemicellulosic biomass saccharification [5]. Xylanases with an optimum activity at low temperature and alkaline pH are applicable in detergent formulation additives [9]. Earlier reports suggested that xylanases obtained from psychrophilic species could improve the quality of bread and fruit juices [10,11]. Xylanases from halophilic microorganisms often exhibit salt-tolerance, which can be used for wastewater treatment and marine/saline food preparation [12].

The protein architecture of endo-β-xylanases comprises a glycoside hydrolase catalytic domain that is sometimes associated with one or more carbohydrate-binding modules (CBMs) [13]. Endo-β-xylanases without a CBM have also been reported [14–17]. CBMs do not contribute directly to catalytic mechanisms. However, they play a role in carbohydrate recognition and binding. The presence of CBM binding allows carbohydrate-active enzymes to concentrate on the polysaccharide surface and improve the overall catalytic efficiency [18]. There are currently 84 families of CBMs. These CBMs display considerable variations in substrate specificity against crystalline cellulose, non-crystalline cellulose, chitin, β-1,3/1,4-glucans, starch, glycogen, xylan, mannan, galactan, and inulin [19]. In the CAZy database, various families of CBMs were appended with GH10 xylanases, predominantly from GH 1, 2, 3, 4, 6, 9, 13, and 22, as well as from GH 10, 15, 35, and 37.

A rare halo-thermophilic bacterium, initially designated as *Rhodothermaceae* bacterium RA (NCBI taxonomy ID: 1779382), was isolated from a hot spring in Langkawi (6°25′22.31″ N, 99°48′48.97″ E), Malaysia [20]. The bacterium exhibited a low identity of 16S rRNA (89.3%) and ANI value (79.3) to the closest strain *Rhodothermus marinus* DSM 4252^T. This information indicates that *Rhodothermaceae* bacterium RA might represent a new genus in the family *Rhodothermaceae* [20,21]. In 2019, Park et al. reported a strain MEBiC09517^T isolated from a port in South Korea [22]. MEBiC09517^T was proposed as the first member of the new genus and the authors suggested that the strain be classified as *Roseithermus sacchariphilus* gen. nov., sp. nov. Due to high similarity (ANI value of 96.2%) between *Rhodothermaceae* bacterium RA and strain MEBiC09517^T, we propose that our strain is a subspecies of *Roseithermus sacchariphilus*. To differentiate both strains, we renamed our bacterium *Roseithermus sacchariphilus* RA. In this study, a xylanase gene (*xynRA2*) was cloned from this bacterium. The study aims to describe the biochemical properties of this enzyme and to understand the effects of CBM truncation on the xylanase.

2. Results and Discussion

2.1. Bioinformatic Analysis

Numerous xylanases have been discovered from extreme habitats such as soda lakes, marine sediments, and hot springs [23–25]. We previously isolated *R. sacchariphilus* RA from a saline hot spring (45 °C, pH 7.1, 13,000 mg/L for Cl^- ion, and 7900 mg/L for Na^+) [26]. A complete genome sequencing elucidated that strain RA harbors 57 GHs affiliated to 30 families [21]. Two non-homologous xylanases were identified and designated as XynRA1 and XynRA2 respectively. The xylanase XynRA1 (Genbank: ARA95075.1) has 379 amino acids and lacks a CBM, while the xylanase XynRA2 (Genbank: ARA92359.1) consists of 813 amino acids and a CBM. XynRA2 was chosen for further study, as we were interested in the function of the CBM attached to this xylanase.

XynRA2 and the putative xylanase annotated in the genome of *R. sacchariphilus* strain MEBiC09517^T are homologs with the identity of 98.6%. The protein sequence of XynRA2 has the identity of 50–65% with xylanases from *Rubrivirga marina*, *Verrucomicrobiae* bacterium DG1235, *Lewinella nigricans*, *Catalinimonas alkaloidigena*, *Fibrisoma* sp. HYT19, *Rhodohalobacter* sp. SW132, *Cellvibrio* sp. PSBB006, and *Ignavibacteria* bacterium GWC2_36_12. The xylanase XynRA2 shares less than 50% identity with xylanases from *Rhodothermaeota* bacterium MED-G12, *Hymenobacter chitinivorans*, and *Siccationidurans arizonensis*. The xylanases mentioned earlier including that of *R. sacchariphilus* MEBiC09517^T were deduced from genome annotation and have not been heterologously expressed and characterized. A phylogenetic tree utilizing a neighbor-joining algorithm was built to show a relationship of XynRA2 with selected counterparts (Figure 1a). In comparison to the well-characterized xylanases, XynRA2 is 73.1% in identity to that of xylanase Xyn10A produced by *R. marinus* DSM 4252 and 52.3% to a xylanase Xyl2091 from *Melioribacter roseus* P3M-2 [27,28]. The sequence identity is low between XynRA2 to other well-studied enzymes, including xylanases that are from *Fusarium graminearum* (38%) [29], *Trichoderma reesei* QM6α (28%) [30], *Bacillus stearothermophilus* T-6 (26%) [31], and *Thermoanaerobacterium saccharolyticum* B6A-RI (25%) [32].

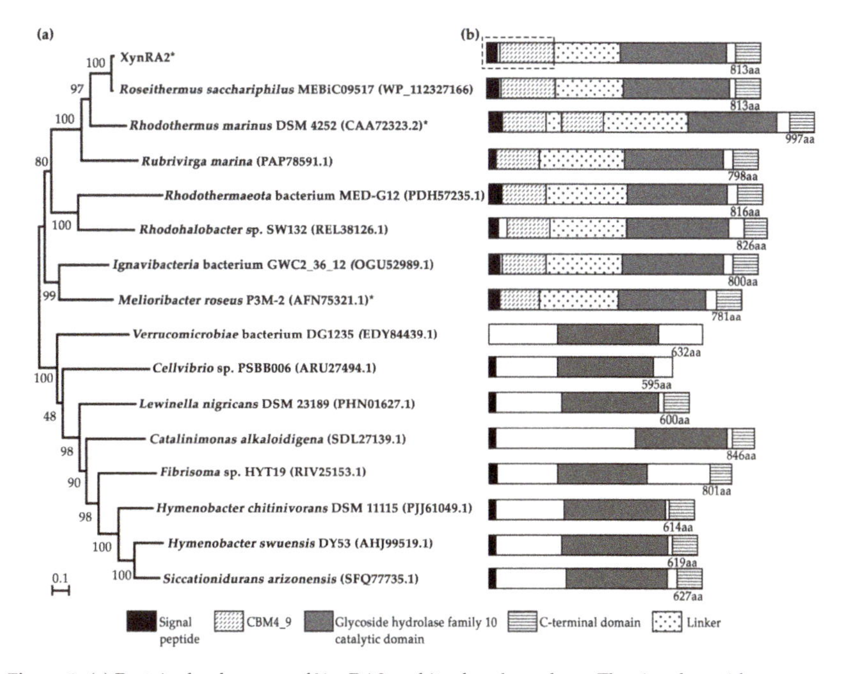

Figure 1. (**a**) Protein dendrogram of XynRA2 and its close homologs. The signal peptide sequences of the proteins were not included in the phylogenetic analysis. Asterisks (*) denote xylanases which have been characterized, otherwise represent genome annotated xylanases. (**b**) Schematic domains arrangement of the respective proteins identified by InterPro. Dotted-line box represents the truncated region in XynRA2ΔCBM.

A mature XynRA2 protein sequence comprises a CBM (Gln^{36}–Asn^{198}), a catalytic domain (Glu^{396}–Tyr^{712}), and a CTD (Trp^{735}–Val^{810}) (Figure 1b). The homology model of CBM clearly denoted the β-sandwich structure formed by eleven anti-parallel β-strands, while that of the catalytic domain is a typical TIM-barrel consisting of eight alternating β-strands and β-helices (Figure 2a,b). Similar to xylanases from *R. marinus* (Xyn10A) and *M. roseus* (Xyl2091), these enzymes possess a GH10 catalytic domain. From the multiple sequence alignment with six GH10 xylanases with crystal structures, the putative catalytic residues for XynRA2 were identified as Glu^{520} and Glu^{635}. The linker region connecting the GH10 domain and CBM comprises 198 residues.

It is likely that in the wild-type *R. sacchariphilus* RA, XynRA2 is exported across the cytoplasmic membrane by the Sec pathway due to the presence of a signal peptide (Met^{1} to Ala^{33}). In addition, XynRA2 has a CTD that enables the protein to be secreted across the outer membrane by T9SS. The T9SS protein secretion pathway is also known as Por secretion system (PorSS) [33,34] which was discovered in *Porphyromonas gingivalis* for secreting a potent protease gingipains [34]. Besides being identified in *R. sacchariphilus* RA, we noticed that in another genome sequencing project, some other annotated cellulases and hemicellulases incorporated a CTD; however, there has been little research describing the actual function of T9SS to GH enzymes. The two closest homologs of XynRA2, the Xyn10A from *R. marinus* and Xyl2091 from *M. roseus* also possessed a similar CTD [28,35]. The CTD possesses five short motifs, in which Motif B, Motif D and Motif E are important for the extensive modification by T9SS [36,37]. By aligning the CTD region of XynRA2 with other xylanases, the well-conserved Gly residues were identified in Motif B and Motif D, whereas Arg substituted the almost-conserved Lys in Motif E [37] (Figure 2c). Proteins that possessed the CTD were found to be cell-anchoring or rely on CTD for secretion, such as Xyn10A from *R. marinus* DSM 4252 as well as SprB, RemA, and ChiA from *Flavobacterium johnsoniae* [35,36]. Collectively, this suggests that XynRA2 could be a cell-anchoring enzyme. However, further experimental validation is required.

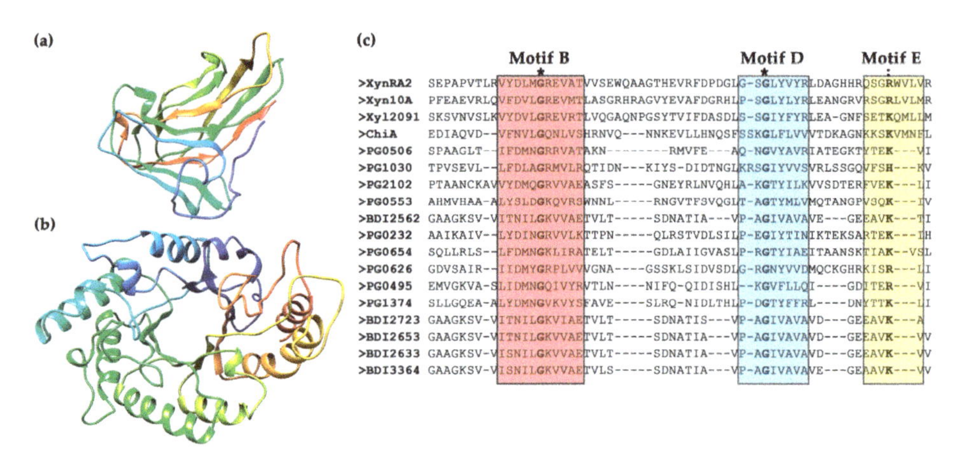

Figure 2. Putative structure of (**a**) CBM4_9 and (**b**) GH10 catalytic domain of XynRA2. The models were colored with the rainbow scheme (blue N-terminus, follows by green, yellow, and red C-terminus); (**c**) multiple sequence alignment of XynRA2 CTD with the counterpart of Xyn10A from *R. marinus*, Xyl2019 from *M. roseus*, ChiA from *F. johnsoniae*, as well as CTD proteins from *P. gingivalis* and *Parabacteroides distasonis*. Amino acid stretch for Motif B, D, and E are indicated by red, blue, and yellow boxes, respectively. Asterisks (*) indicate fully conserved amino acids while colon (:) indicates amino acid groups of similar properties.

Based on an InterPro analysis, the CBM of XynRA2 was annotated as CBM4_9. The closest biochemically characterized xylanase (Xyn10A) from *R. marinus* DSM 4252 has two dissimilar CBM4_9s arranged in tandem (Figure 1b), which were denoted as "CBM4-1" and "CBM4-2" in the original article [38]. Another close homolog, a characterized xylanase (Xyl2091) from *M. roseus*, also possessed a CBM4_9 [28]. Interestingly, the amino acid stretch of the CBM4_9 from *R. sacchariphilus* RA is only 70% and 51% identical to *R. marinus* and *M. roseus* counterparts respectively, suggesting that the affinity of the three enzymes against hemicellulose might be different. Different families of CBMs such as CBM6_36 for XynG1-1 [39], CBM13 for XynAS27 [40], and dual CBM9-CBM22 for XynSL3 [24] were often reported in GH10 xylanases. According to the CAZy, other CBMs associated with xylanases are from families 1, 2, 3, 10, 15, 35, and 37. The CBM4 family from xylanases usually binds to xylan β-glucan, and/or amorphous cellulose [41,42]. We anticipated the substrate specificity of CBM4_9 in XynRA2 to be similar. Several reports have shown that the removal of the CBMs affected the biochemical properties of their partnering xylanases [39,40,43]. Therefore, we constructed a mutant enzyme (designated as XynRA2ΔCBM) by deleting the CBM4_9 but retaining the linker connecting the CBM to the catalytic domain to evaluate the effect of its truncation on the xylanase.

2.2. Expression of Recombinant XynRA2 and XynRA2ΔCBM

The gene fragments encoding for mature XynRA2 (2349 bp) and XynRA2ΔCBM (1857 bp) were cloned in pET28a(+), expressed in *E. coli* BL21 (DE3) and purified using Ni-NTA columns. The purified enzymes migrated as two distinct bands around 90 kDa and 70 kDa on SDS-PAGE, which were consistent with the theoretical molecular weight of XynRA2 (89.5 kDa) and XynRA2ΔCBM (68.5 kDa) respectively (Figure 3a).

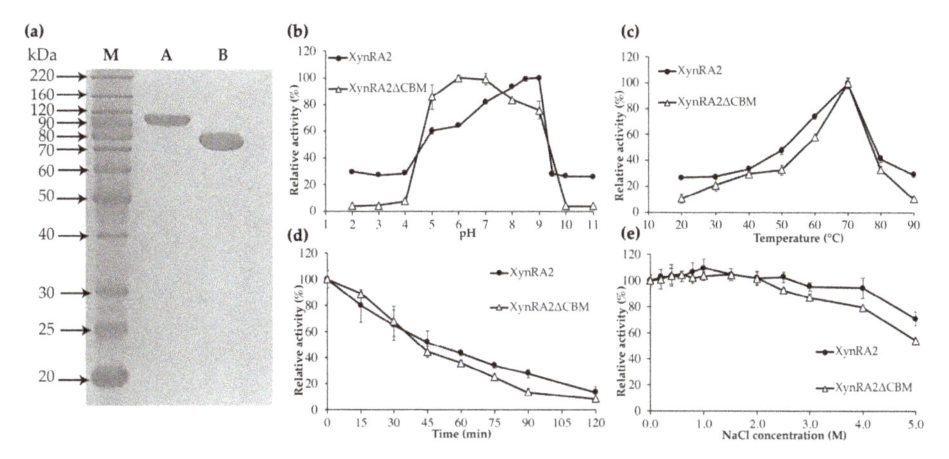

Figure 3. Molecular properties of purified XynRA2 and XynRA2ΔCBM. (**a**) SDS-PAGE analysis of purified XynRA2 and XynRA2ΔCBM. **M**, BenchMark™ Protein Ladder; **A**, purified XynRA2; **B**, purified XynRA2ΔCBM; (**b**) effect of pH in the range of 2–11; (**c**) effect of temperature from 20–90 °C; (**d**) thermostability at 70 °C across 120 min; (**e**) effect of NaCl from 0–5.0 M concentration.

2.3. Biochemical Characterization of XynRA2 and XynRA2ΔCBM

2.3.1. Effect of pH and Temperature

Using beechwood xylan as the substrate, the purified XynRA2 had maximum activity at pH 8.5 and retained a relatively high activity between pH 7–9. Truncation of the CBM broadened the pH profile (pH 5–9) with the optimum pH shifted to 6.0 (Figure 3b). Similarly, CBM4_9 truncation changed the optimum pH from 7.5 to 7.0 for a xylanase PX3 from *Paenibacillus terrae* HPL-003 [44]. The working pH for the mutant PX3 also narrowed to pH 5–10, while the native PX3 had an active pH ranging from 3–12. The optimum pHs of xylanase Xyn10A from *R. marinus* and Xyl2091 from *M. roseus* were 7.5 and 6.5 respectively [28,45], while that of truncated counterparts was not reported.

The optimum temperature for the activity of native XynRA2 and XynRA2ΔCBM was 70 °C. Overall, the temperature profiles for both enzymes were identical (Figure 3c). To evaluate the thermostability, XynRA2 and XynRA2ΔCBM were incubated at 70 °C without substrate for a specific interval prior to measuring residual activity. The half-life of both XynRA2 and XynRA2ΔCBM at 70 °C was approximately 45 min; however, XynRA2ΔCBM was more sensitive to prolonged temperature treatment (Figure 3d). The optimum temperatures of Xyn10A and Xyl2091 were 80 °C and 65 °C respectively and their half-lives were about 90 min (80 °C) and 160 min (60 °C), respectively. Truncation of the CBM in Xyn10A from *R. marinus* also resulted in a decrease in thermostability, indicating that the CBM with this xylanase also contributed to enzyme stability [46]. Truncation of the CBM from xylanases from *Streptomyce rochei* L10904 (Srxyn10) [43], *Paenibacillus campinasensis* G1-1 (XynG1-1) [39], and *Streptomyces* sp. S27 (XynAS27) [40] showed that removal of the CBM did not affect the optimum temperature of xylanases. However, the truncated versions of XynG1-1 and XynAS27 displayed a significant decrease in thermostability [39,40]. In contrast, the CBM-truncated variant of Srxyn10 from *S. rochei* L10904 exhibited a substantial increase in thermostability at 60–70 °C despite sharing similar optimum temperature with its native counterpart [43].

2.3.2. NaCl Tolerance

The *R. sacchariphilus* RA was capable of growing in media containing a high concentration of NaCl [20]. Since XynRA2 is probably expressed as an extracellular cell-bound enzyme, we decided to investigate the effect of NaCl on xylanase activity. Multiple xylanases are known to exhibited moderate halo-tolerance, but only limited reports have demonstrated extreme halo-tolerance ability as displayed by XynRA2 (Table 1). The relative activity of XynRA2 and XynRA2ΔCBM was slightly enhanced when the catalytic reactions were supplemented with 1.0 M NaCl. Notably, XynRA2 retained 94% of initial activity at 4.0 M, and 71% at 5.0 M NaCl. Although the mutant XynRA2ΔCBM was more salt-sensitive, the enzyme retained the relative activity of 79% at 4.0 M and 54% at 5.0 M (Figure 3e). The reason for the lower halo-tolerance is unknown. In addition, there is a lack of literature elucidating the relationship between CBM and halo-xylanase activity.

A homology model of XynRA2 catalytic domain demonstrated a high distribution of acidic amino acids on the protein surface resulting in an overall negative electrostatic potential (Figure 4), which might explain the excellent protein stability in high NaCl concentration. Theoretically, halo-tolerant enzymes contain more acidic residues (Asp and Glu) than non-polar residues (Val, Ile, Leu, Met, and Phe). Halo-tolerant enzymes are also enriched with small residues (Ala, Val, Ser, and Thr) but lack Lys residue [47]. It has been proposed that excess acidic residues could facilitate the weakening of hydrophobicity or strengthening of hydrophilic forces on the enzyme surface, which increases water-binding capacity and prevent proteins aggregation at high salt concentration [48,49].

Table 1. The reported halo-tolerant xylanases and their activity in high concentration of NaCl. Enzymatic reactions carried out at 0 M of NaCl was treated as 100%.

Xylanases	Activity in NaCl (%)			Strains	Reference
	3.0 M	4.0 M	5.0 M		
XynRA2	95	94	71	*R. sacchariphilus RA* strain RA	This study
M11	60	50	47	*Streptomyces viridochromogenes*	[12]
Xyn512	60	47	32	*Flammeovirga pacifica* WPAGA1	[23]
XynSL3	60	40	ND	*Alkalibacterium* sp. SL3	[24]
XynA	78	53	ND	*Glaciecola mesophila* KMM 241	[50]
XynA	180	140	100	*Zunongwangia profunda*	[49]
XynSL4	59	41	ND	*Planococcus* sp. SL4	[51]
XynRBM26	96	93	87	*Massilia* sp. RBM26	[52]
XynAHJ3	40	26	ND	*Lechevalieria* sp. HJ3	[53]
Xylanase	90	87	84	*Gracillibacillus* sp. TSCPVG	[54]

ND: not determined.

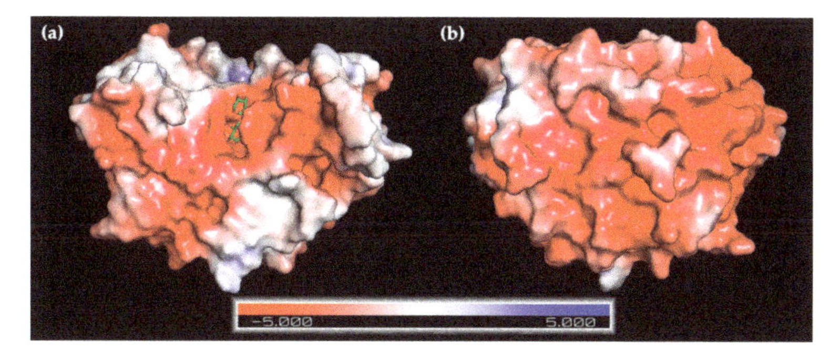

Figure 4. Predicted electrostatic potentials on the surface of XynRA2 catalytic domain from (**a**) top and (**b**) bottom views. The ligand (xylobiose) is bound at the catalytic binding pocket of XynRA2 as indicated in (**a**). Red and blue indicate negative and positive electrostatic potentials respectively.

2.3.3. Enzyme Kinetics

The specific activities and the turnover rate (k_{cat}) of the purified XynRA2 and XynRA2ΔCBM were determined by reacting the enzymes with soluble beechwood xylan. The specific activities of XynRA2 and XynRA2ΔCBM were 300 U/mg and 160 U/mg respectively. The k_{cat} of the native and mutant enzymes were 24.8 s^{-1} and 15.7 s^{-1}, respectively. We found that the truncation of CBM significantly affected the performance of the enzymes. This finding was in consistent with XynG1-1 from *P. campinasensis* that CBM truncation reduced the k_{cat} by 20% [39]. Removal of CBM alone did not affect the k_{cat} of XynAS27 from *Streptomyces* sp. S27. However, truncating CBM together with the linker reduced k_{cat} value by 25% [40]. On the other hand, the xylanase variant of Srxyn10 with a CBM truncation had a three-fold higher specific activity on beechwood xylan than its native counterpart [43].

2.3.4. Substrate and Product Specificities

The purified XynRA2 and XynRA2ΔCBM were incubated with various substrates before analyzing them using HPLC. Generally, XynRA2 and XynRA2ΔCBM showed similar substrate specificities. Both enzymes were active on beechwood xylan, oat-spelt xylan, and xylo-oligosaccharides (XOs) such as X_6, X_5, X_4 but not on X_3 and X_2. Except for xylose-based carbohydrates, the enzymes were unable to hydrolyze glucose-, maltose-, and arabinose-derived polymers such as carboxymethylcellulose (CMC), Avicel™, starch, pullulan, D-cellobiose, and arabinan. The results indicated that the enzymes did not possess either a cellulase or arabinase activity, suggesting that XynRA2 is a specific GH10 xylanase. This is in agreement with a recent statistical study that showed most of the characterized GH10 xylanases were mono-specific (96.8%, n = 350) towards xylanosic substrates [4].

We compared the product formation pattern of XynRA2 and XynRA2ΔCBM by reacting the purified enzymes with beechwood xylan and oat-spelt xylan (Figure 5). Upon reacting XynRA2 with beechwood xylan, the products constituted a mixture of XOs ranging X_6, X_5, X_4, X_3, and X_2 at the beginning of the reaction (15 min). After a prolonged hydrolysis (24 h), xylobiose (X_2) was accumulated as the primary product together with detectable X_3 and X_1 (Figure 5a), and the product formation pattern for XynRA2ΔCBM against beechwood xylan was shown in Figure 5b. For reactions of 15 min and 24 h, the product profile for XynRA2ΔCBM was similar to that of XynRA2. Yet, the ratio of X_3 and X_2 was slightly different in the 1 h and 3 h reactions. Previous reports on xylanase rXTMA from *Thermotoga maritima* and xylanase A from *Caldibacillus cellulovorans* also showed a variation in the profiles of XOs produced by native and CBM-depletion xylanases [55,56].

Although the same reaction conditions were used with oat-spelt xylan, we obtained lower sugar yields, probably due to the physical structure of oat-spelt xylan which consisted of both insoluble and soluble fractions. Furthermore, the product profiles were also different for beechwood xylan and oat-spelt xylan. After prolonged reaction, X_4, X_3, and X_2 were accumulated as the major products

(Figure 5c). Interestingly, oat-spelt xylan was a poor substrate for XynRA2ΔCBM (Figure 5d), as reported for other xylanases [39,40,57,58]. A lower activity against oat-spelt xylan might be due to the inefficient binding onto the substrate, as a result of CBM truncation in XynRA2ΔCBM. It has also been recurrently reported that the truncation of CBMs affects catalytic efficiency of GHs towards other insoluble substrates but not the soluble counterparts [18,19].

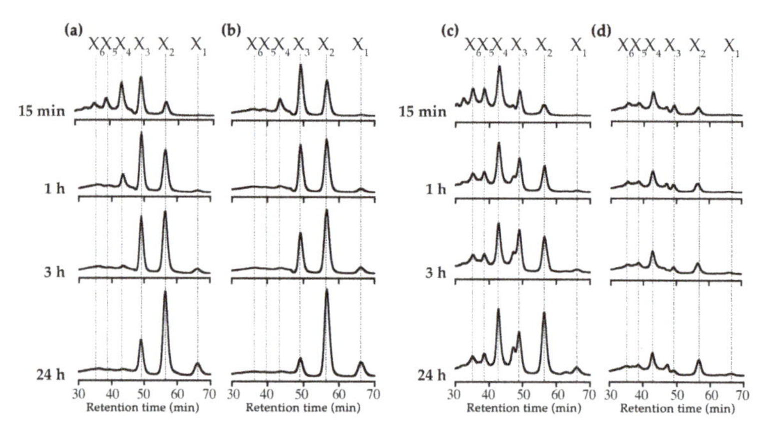

Figure 5. Product analysis of xylan degradation. (**a**,**b**) products formed from hydrolysis of beechwood xylan by XynRA2 and XynRA2ΔCBM, respectively; (**c**,**d**) products formed from hydrolysis of oat-spelt xylan by XynRA2 and XynRA2ΔCBM, respectively. The product peaks shown in this figure were normalized.

3. Materials and Methods

3.1. Sequence Analysis

The gene sequence of putative xylanase (GenBank: ARA92359.1) was extracted from the complete genome of *R. sacchariphilus* RA (GenBank: CP020382.1) [20,21]. The mature xylanase gene was designated as *xynRA2*. The protein sequence similarity was assessed using NCBI BLASTp program (https://blast.ncbi.nlm.nih.gov/Blast.cgi). Multiple sequence alignments were performed using Clustal Omega (https://www.ebi.ac.uk/Tools/msa/clustalo/). A phylogenetic tree of XynRA2 with its closest homologs was constructed with the neighbor-joining algorithm using MEGA v7.0 software [59] with a bootstrap value of 1000. The signal peptide sequence was predicted using SignalP v4.1 (http://www.cbs.dtu.dk/services/SignalP/). Conservative domains were identified using InterPro. The homology model of XynRA2 was performed using SWISS-MODEL (https://swissmodel.expasy.org/) with an evolved CBM of Xyn10A from *R. marinus* (PDB: 3JXS) [60] and GH10 xylanase (XynB) from *Xanthomonas axonopodis* pv *citri* (PDB: 4PN2) complexed with xylotriose as a template [61]. The predicted model and its surface electrostatic potential were assessed using APBS plugin in PyMOL (v2.2.3, Schrödinger Inc., New York, NY, USA).

3.2. Cloning of Xylanases

Genomic DNA of *R. sacchariphilus* RA was extracted using DNeasy Blood & Tissue Kit (Qiagen, Hilden, Germany). The gene sequence of *xynRA2* was amplified from the genomic DNA using a forward primer GH10F (5′-AGC<u>CATATG</u>CGTGCGCAGAGCAACACCA-3′) and a reverse primer GH10R (5′-CGATGGGTACTGGTCCGC<u>CTCGAG</u>CACC-3′). The underlined sequences represent *Nde*I and *Xho*I restriction sites respectively. N-terminal signal peptide was not included in the recombinant enzymes. The truncated gene *xynRA2ΔCBM* was amplified using primer GH10F-LC (5′-AGC<u>CATATG</u>CCCCTGGCGGGAGC-3′) and GH10R.

Both the gene fragments were amplified using Q5® High-Fidelity PCR kit (NEB, Ipswich, MA, USA). The PCR products were digested with *Nde*I and *Xho*I followed by ligation into pET28a(+) (Novagen, Madison, WI, USA) at the corresponding sites. The recombinant plasmids (pET28a_*xynRA2 and* pET28a_*xynRA2ΔCBM*) were separately transformed into *E. coli* BL21 (DE3) competent cells using the heat shock method. The transformants were grown in Luria-Bertani (LB) medium containing 50 μg/mL kanamycin at 37 °C for 18 h. Transformants harboring the recombinant plasmid were identified by restriction digestion and DNA sequencing.

3.3. Expression and Purification of Xylanases

The transformed cells were grown in LB medium containing 50 μg/mL kanamycin at 37 °C to an $A_{600\ nm}$ of 0.6. Protein expression was induced by addition of isopropyl-β-D-thiogalactopyranoside (IPTG) at a final concentration of 0.4 mM at 25 °C for 18 h. The cells were harvested by centrifugation (6000× *g*, 4 °C, 10 min) and lysed using B-PER™ Direct Bacterial Protein Extraction Kit (Thermo Scientific, Waltham, MA, USA) The crude enzyme was collected (12,000× *g*, 4 °C, 10 min) and dialyzed against 20 mM sodium phosphate buffer (pH 7.4) at 4 °C overnight in a SnakeSkin™ Dialysis Tubing 10k MWCO (Thermo Scientific, Waltham, MA, USA). To purify the His-tagged proteins, the crude enzyme was loaded onto a Ni-NTA Superflow column (Qiagen, Hilden, Germany) equilibrated with 20 mM sodium phosphate buffer (pH 7.4) and 50 mM imidazole. The enzymes were eluted with a linear gradient of 50–500 mM imidazole in 20 mM phosphate buffer (pH 7.4) containing 500 mM NaCl. Upon elution, fractions containing the XynRA2 and XynRA2ΔCBM were respectively pooled and dialyzed against 20 mM sodium phosphate buffer (pH 7.4) at 4 °C overnight to remove the remaining salts. The purity and apparent molecular mass of XynRA2 and XynRA2ΔCBM were validated by SDS-PAGE. The activity of the purified enzymes was assayed as described below.

3.4. Xylanase Assay

The xylanase activity of XynRA2 and XynRA2ΔCBM was calculated by measuring the reducing sugars released from substrates using 3,5-dinitrosalicylic acid (DNS) method. The reaction mixtures contained 50 μL of appropriately diluted enzymes and 500 μL of 1% (*w*/*v*) beechwood xylan (Megazyme, Bray, County Wicklow, Ireland) in 0.1 M Tris-HCl buffer (pH 8.5). The enzymatic reaction was carried out at 70 °C for 15 min, stopped with 500 μL DNS reagent and boiled for 5 min. The absorbance at 540 nm was measured when the reaction mixture is cooled to room temperature. The amount of sugar released was estimated using a standard curve of D-xylose (Sigma-Aldrich, St. Louis, MO, USA). One unit (U) of xylanase activity was defined as 1 μmol of reducing sugars released from substrate per minute per mL of enzyme under the assay condition. The enzyme activity was calculated by this standard procedure unless otherwise noted. All reactions were performed in at least triplicate.

3.5. Biochemical Characterization of XynRA2 and XynRA2ΔCBM

The optimum pH of XynRA2 was determined in a range of 2–11 at 50 °C. The buffers used were 0.1 M of glycine HCl (pH 2–3), sodium acetate (pH 4–6), Tris-HCl (pH 7–9), and glycine-NaOH (pH 10–11) containing 1% (*w*/*v*) purified beechwood xylan (Megazyme, Bray, County Wicklow, Ireland). The optimum temperature of the enzyme was determined over a range of temperature from 20 to 90 °C in Tris-HCl buffer (pH 8.5). The optimum pH of XynRA2ΔCBM was determined at 70 °C and the optimum temperature was determined in acetate buffer (pH 6.0). Thermostability of XynRA2 and XynRA2ΔCBM were determined by measuring the residual activity of the enzyme after pre-incubation in 0.1 M Tris-HCl buffer (pH 8.5) and 0.1 M acetate buffer (pH 6.0), respectively, at 70 °C without substrate for 2 h. The initial activity of enzymes without pre-incubation was set as 100%.

The effect of NaCl on the activity of XynRA2 and XynRA2ΔCBM was determined at 70 °C in 0.1 M Tris-HCl buffer (pH 8.5) and 0.1 M sodium acetate buffer (pH 6.0), respectively, in the presence of up to 5.0 M NaCl.

To determine the specific activities of purified XynRA2 and XynRA2ΔCBM, the enzyme activities were determined using a xylanase assay as described above, and the protein concentration was determined by Pierce™ BCA Protein Assay kit (Thermo Scientific, Waltham, MA, USA) using BSA as a standard. To determine the turnover rate (k_{cat}) of XynRA2 and XynRA2ΔCBM, the respective enzymatic reaction was carried out at 70 °C in 0.1 M Tris-HCl buffer (pH 8.5) and 0.1 M sodium acetate buffer (pH 6.0) containing 0.1–1.5% (*w*/*v*) of purified beechwood xylan. The k_{cat} of the enzymes were determined based on non-linear regression using PRISM 7 software (GraphPad Software Inc., San Diego, CA, USA).

3.6. Analysis of Substrate Specificity and Hydrolysis Products

The substrate specificity of XynRA2 and XynRA2ΔCBM was determined in 0.1 M Tris-HCl (pH 8.5) and 0.1 M acetate buffer (pH 6.0) containing 1% (*w*/*v*) of beechwood xylan, arabinan (Megazyme, Bray, County Wicklow, Ireland), oat-spelt xylan, CMC, Avicel™, D-cellobiose (Sigma, St. Louis, MO, USA), pullulan (TCI chemicals, Tokyo, Japan), starch (QReC, Auckland, New Zealand) or 5 mM of X_2–X_6 xylo-oligosaccharides (Bz Oligo Biotech, Qingdao, China). The reaction mixture contained 50 μL of purified enzymes and 500 μL of the substrate solutions. The substrate hydrolysis was detected as described below.

To analyze the hydrolysis products of XynRA2 and XynRA2ΔCBM, the reaction mixtures with 3 U of purified enzymes and 1% (*w*/*v*) beechwood xylan and oat-spelt xylan were incubated at 70 °C in Tris-HCl buffer (pH 8.5) for 24 h. The hydrolysis products were eluted using Rezex™ RSO-oligosaccharides Ag^+ (4%) column (Phenomenex, Torrance, CA, USA) at a flow rate of 1.0 mL/min at 80 °C for 80 min and detected using 1260 Infinity ELSD (Agilent Technologies, Santa Clara, CA, USA). Xylo- oligosaccharides (X_2–X_6) and xylose were used as its product standards.

4. Conclusions

Roseithermus is a newly proposed genus in family *Rhodothermaceae* affiliated to order *Rhodothermales*. Currently, the whole taxonomic order is comprised of only 14 type strains. So far, xylanase from *Rhodothermus marinus* is the only enzyme that was well characterized. This study described for the first time the biochemical properties of a xylanase from *Roseithermus*. The native XynRA2 was active at alkaline pH and elevated temperature (pH 8.5 and 70 °C) while retaining excellent activity even at 5.0 M NaCl. Such properties make XynRA2 a potential candidate for applications involving an alkaline environment, elevated temperature, and high salinity. In a separate part of the study, the CBM4_9 domain was removed. The data elucidated that CBM truncation affected enzyme specific activity, turnover rate, pH optimum, and NaCl tolerance, with an additional marginal effect on thermostability.

Author Contributions: Formal analysis, S.C.T.; Funding acquisition, M.S.S., C.S.C., N.C.B. and K.-G.C.; Methodology, K.J.L.; Supervision, K.M.G.; Writing—original draft, S.C.T. and K.M.G.; Writing—review & editing, K.J.L., M.S.S., C.S.C., N.C.B. and K.-G.C.

Funding: This work was financially sponsored by the Ministry of Education Malaysia and Biotechnology and Biological Sciences Research Council (BBSRC) United Kingdom under the program of United Kingdom-Southeast Asia Newton Ungku Omar Fund (UK-SEA-NUOF) with project number 4B297 and BB/P027717/1. This project was co-financially supported by Universiti Teknologi Malaysia RU grant (Grant number: 16H89). K-G Chan thanked University of Malaya for financial support (PPP grants: PG136-2016A, PG133-2016A, HIR grant: H50001-A-000027).

Acknowledgments: S.C. Teo appreciates UTM Zamalah scholarship for providing for financial support.

Conflicts of Interest: The authors declare no conflict of interest.

Abbreviations

ANI	average nucleotide identity
BCA	bicinchoninic acid
BSA	bovine serum albumin
BLAST	Basic Local Alignment Search Tool

CAZy	carbohydrate active enzymes database
CBM	carbohydrate-binding module
CMC	carboxymethylcellulose
CTD	C-terminal domain
DNS	dinitrosalicylic acid
ELSD	evaporative light scattering detector
GH	glycoside hydrolase
HPLC	high-performance liquid chromatography
IPTG	isopropyl-β-D-1-thiogalactopyranoside
MEGA 7	Molecular Evolutionary Genetics Analysis version 7
MWCO	molecular weight cut-off
NCBI	National Center for Biotechnology Information
Ni-NTA	nickel-nitrilotriacetic acid
PorSS	Por secretion system
SDS-PAGE	sodium dodecyl sulfate-polyacrylamide gel electrophoresis
TIM	triosephosphate isomerase
T9SS	type IX secretion system
XOs	xylo-oligosaccharides
X_1	xylose
X_2	xylobiose
X_3	xylotriose
X_4	xylotetraose
X_5	xylopentaose
X_6	xylohexaose

References

1. Collins, T.; Gerday, C.; Feller, G. Xylanases, xylanase families and extremophilic xylanases. *FEMS Microbiol. Rev.* **2005**, *29*, 3–23. [CrossRef]
2. Moreira, L.R.S. Insights into the mechanism of enzymatic hydrolysis of xylan. *Appl. Microbiol. Biotechnol.* **2016**, *100*, 5205–5214. [CrossRef] [PubMed]
3. Terrapon, N.; Lombard, V.; Drula, E.; Coutinho, P.M.; Henrissat, B. The CAZy Database/the Carbohydrate-Active Enzyme (CAZy) Database: Principles and Usage Guidelines. In *A Practical Guide to Using Glycomics Databases*; Aoki-Kinoshita, K.F., Ed.; Springer: Tokyo, Japan, 2017; pp. 117–131.
4. Nguyen, S.T.; Freund, H.L.; Kasanjian, J.; Berlemont, R. Function, distribution, and annotation of characterized cellulases, xylanases, and chitinases from CAZy. *Appl. Microbiol. Biotechnol.* **2018**, *102*, 1629–1637. [CrossRef]
5. Chakdar, H.; Kumar, M.; Pandiyan, K.; Singh, A.; Nanjappan, K.; Kashyap, P.L.; Srivastava, A.K. Bacterial xylanases: Biology to biotechnology. *3 Biotech* **2016**, *6*, 150. [CrossRef] [PubMed]
6. Juturu, V.; Wu, J.C. Microbial xylanases: Engineering, production and industrial applications. *Biotechnol. Adv.* **2012**, *30*, 1219–1227. [CrossRef]
7. Kumar, V.; Marin-Navarro, J.; Shukla, P. Thermostable microbial xylanases for pulp and paper industries: Trends, applications and further perspectives. *World J. Microb. Biotechnol.* **2016**, *33*, 1870–1874. [CrossRef]
8. Luo, H.; Wang, Y.; Li, J.; Yang, J.; Yang, Y.; Huang, H.; Fan, Y.; Yao, B. Cloning, expression and characterization of a novel acidic xylanase, XYL11B, from the acidophilic fungus *Bispora* sp. MEY-1. *Enzyme Microb. Technol.* **2009**, *45*, 126–133. [CrossRef]
9. Kumar, B.K.; Balakrishnan, H.; Rele, M. Compatibility of alkaline xylanases from an alkaliphilic *Bacillus* NCL (87-6-10) with commercial detergents and proteases. *J. Ind. Microbiol. Biot.* **2004**, *31*, 83–87. [CrossRef] [PubMed]
10. Dornez, E.; Verjans, P.; Arnaut, F.; Delcour, J.A.; Courtin, C.M. Use of psychrophilic xylanases provides insight into the xylanase functionality in bread making. *J. Agr. Food Chem.* **2011**, *59*, 9553–9562. [CrossRef]
11. Nagar, S.; Mittal, A.; Gupta, V.K. Enzymatic clarification of fruit juices (apple, pineapple, and tomato) using purified *Bacillus pumilus* SV-85S xylanase. *Biotechnol. Bioproc. Eng.* **2012**, *17*, 1165–1175. [CrossRef]

12. Liu, Z.; Zhao, X.; Bai, F. Production of xylanase by an alkaline-tolerant marine-derived *Streptomyces viridochromogenes* strain and improvement by ribosome engineering. *Appl. Microbiol. Biotechnol.* **2013**, *97*, 4361–4368. [CrossRef]
13. Talamantes, D.; Biabini, N.; Dang, H.; Abdoun, K.; Berlemont, R. Natural diversity of cellulases, xylanases, and chitinases in bacteria. *Biotechnol. Biofuels* **2016**, *9*, 133. [CrossRef]
14. Chawachart, N.; Anbarasan, S.; Turunen, S.; Li, H.; Khanongnuch, C.; Hummel, M.; Sixta, H.; Granström, T.; Lumyong, S.; Turunen, O. Thermal behaviour and tolerance to ionic liquid [emim]OAc in GH10 xylanase from *Thermoascus aurantiacus* SL16W. *Extremophiles* **2014**, *18*, 1023–1034. [CrossRef]
15. Evangelista, D.E.; Kadowaki, M.A.S.; Mello, B.L.; Polikarpov, I. Biochemical and biophysical characterization of novel GH10 xylanase prospected from a sugar cane bagasse compost-derived microbial consortia. *Int. J. Biol. Macromol.* **2018**, *109*, 560–568. [CrossRef]
16. Niderhaus, C.; Garrido, M.; Insani, M.; Campos, E.; Wirth, S. Heterologous production and characterization of a thermostable GH10 family endo-xylanase from *Pycnoporus sanguineus* BAFC 2126. *Process Biochem.* **2018**, *67*, 92–98. [CrossRef]
17. Sharma, K.; Antunes, I.L.; Rajulapati, V.; Goyal, A. Molecular characterization of a first endo-acting β-1,4-xylanase of family 10 glycoside hydrolase (PsGH10A) from *Pseudopedobacter saltans* comb. nov. *Process Biochem.* **2018**, *70*, 79–89. [CrossRef]
18. Guillén, D.; Sánchez, S.; Rodríguez-Sanoja, R. Carbohydrate-binding domains: Multiplicity of biological roles. *Appl. Microbiol. Biotechnol.* **2010**, *85*, 1241–1249. [CrossRef]
19. Varnai, A.; Mäkelä, M.R.; Djajadi, D.T.; Rahikainen, J.; Hatakka, A.; Viikari, L. Carbohydrate-binding modules of fungal cellulases: Occurrence in nature, function, and relevance in industrial biomass conversion. In *Advances in Applied Microbiology*; Elsevier: Amsterdam, The Netherlands, 2014; Volume 88, pp. 103–165.
20. Goh, K.M.; Chan, K.-G.; Lim, S.W.; Liew, K.J.; Chan, C.S.; Shamsir, M.S.; Ee, R.; Adrian, T.-G.-S. Genome analysis of a new *Rhodothermaceae* strain isolated from a hot spring. *Front. Microbiol.* **2016**, *7*, 1109. [CrossRef]
21. Liew, K.J.; Teo, S.C.; Shamsir, M.S.; Sani, R.K.; Chong, C.S.; Chan, K.-G.; Goh, K.M. Complete genome sequence of *Rhodothermaceae* bacterium RA with cellulolytic and xylanolytic activities. *3 Biotech* **2018**, *8*, 376. [CrossRef]
22. Park, M.-J.; Oh, J.H.; Yang, S.-H.; Kwon, K.K. *Roseithermus sacchariphilus* gen. nov., sp. nov. and proposal of *Salisaetaceae* fam. nov., representing new family in the order *Rhodothermales*. *Int. J. Syst. Evol. Microbiol.* **2019**, *69*, 1213–1219. [CrossRef]
23. Cai, Z.-W.; Ge, H.-H.; Yi, Z.-W.; Zeng, R.-Y.; Zhang, G.-Y. Characterization of a novel psychrophilic and halophilic β-1,3-xylanase from deep-sea bacterium, *Flammeovirga pacifica* strain WPAGA1. *Int. J. Biol. Macromol.* **2018**, *118*, 2176–2184. [CrossRef] [PubMed]
24. Wang, G.; Wu, J.; Yan, R.; Lin, J.; Ye, X. A novel multi-domain high molecular, salt-stable alkaline xylanase from *Alkalibacterium* sp. SL3. *Front. Microbiol.* **2017**, *7*, 2120. [CrossRef]
25. Yadav, P.; Maharjan, J.; Korpole, S.; Prasad, G.S.; Sahni, G.; Bhattarai, T.; Sreerama, L. Production, purification, and characterization of thermostable alkaline xylanase from *Anoxybacillus kamchatkensis* NASTPD13. *Front. Bioeng. Biotechnol.* **2018**, *6*, 65. [CrossRef] [PubMed]
26. Chan, C.S.; Chan, K.-G.; Ee, R.; Hong, K.-W.; Urbieta, M.S.; Donati, E.R.; Shamsir, M.S.; Goh, K.M. Effects of physiochemical factors on prokaryotic biodiversity in Malaysian circumneutral hot springs. *Front. Microbiol.* **2017**, *8*, 1252. [CrossRef] [PubMed]
27. Karlsson, E.N.; Bartonek-Roxå, E.; Holst, O. Cloning and sequence of a thermostable multidomain xylanase from the bacterium *Rhodothermus marinus*. *BBA - Gene Struct. Expr.* **1997**, *1353*, 118–124. [CrossRef]
28. Rakitin, A.L.; Ermakova, A.Y.; Ravin, N.V. Novel endoxylanases of the moderately thermophilic polysaccharide-degrading bacterium *Melioribacter roseus*. *J. Microbiol. Biotechnol.* **2015**, *25*, 1476–1484. [CrossRef]
29. Beliën, T.; Van Campenhout, S.; Van Acker, M.; Volckaert, G. Cloning and characterization of two endoxylanases from the cereal phytopathogen *Fusarium graminearum* and their inhibition profile against endoxylanase inhibitors from wheat. *Biochem. Biophys. Res. Commun.* **2005**, *327*, 407–414. [CrossRef]
30. La Grange, D.C.; Pretorius, I.S.; Van Zyl, W.H. Expression of a *Trichoderma reesei* beta-xylanase gene (XYN2) in *Saccharomyces cerevisiae*. *Appl. Environ. Microbiol.* **1996**, *62*, 1036–1044.
31. Khasin, A.; Alchanati, I.; Shoham, Y. Purification and characterization of a thermostable xylanase from *Bacillus stearothermophilus* T-6. *Appl. Environ. Microbiol.* **1993**, *59*, 1725–1730.

32. Lee, Y.-E.; Lowe, S.; Henrissat, B.; Zeikus, J.G. Characterization of the active site and thermostability regions of endoxylanase from *Thermoanaerobacterium saccharolyticum* B6A-RI. *J. Bacteriol.* **1993**, *175*, 5890–5898. [CrossRef]
33. Lasica, A.M.; Ksiazek, M.; Madej, M.; Potempa, J. The type IX secretion system (T9SS): Highlights and recent insights into its structure and function. *Front. Cell. Infect. Microbiol.* **2017**, *7*, 215. [CrossRef]
34. Sato, K.; Yukitake, H.; Narita, Y.; Shoji, M.; Naito, M.; Nakayama, K. Identification of *Porphyromonas gingivalis* proteins secreted by the Por secretion system. *FEMS Microbiol. Lett.* **2013**, *338*, 68–76. [CrossRef]
35. Karlsson, E.N.; Hachem, M.A.; Ramchuran, S.; Costa, H.; Holst, O.; Svenningsen, Å.F.; Hreggvidsson, G.O. The modular xylanase Xyn10A from *Rhodothermus marinus* is cell-attached, and its C-terminal domain has several putative homologues among cell-attached proteins within the phylum *Bacteroidetes*. *FEMS Microbiol. Lett.* **2004**, *241*, 233–242. [CrossRef]
36. Kharade, S.S.; McBride, M.J. *Flavobacterium johnsoniae* chitinase ChiA is required for chitin utilization and is secreted by the type IX secretion system. *J. Bacteriol.* **2014**, *196*, 961–970. [CrossRef]
37. Veith, P.D.; Nor Muhammad, N.A.; Dashper, S.G.; Likić, V.A.; Gorasia, D.G.; Chen, D.; Byrne, S.J.; Catmull, D.V.; Reynolds, E.C. Protein substrates of a novel secretion system are numerous in the *Bacteroidetes* phylum and have in common a cleavable C-terminal secretion signal, extensive post-translational modification, and cell-surface attachment. *J. Proteome Res.* **2013**, *12*, 4449–4461. [CrossRef]
38. Hachem, M.A.; Karlsson, E.N.; Bartonek-Roxå, E.; Raghothama, S.; Simpson, P.J.; Gilbert, H.J.; Williamson, M.P.; Holst, O. Carbohydrate-binding modules from a thermostable *Rhodothermus marinus* xylanase: Cloning, expression and binding studies. *Biochem. J.* **2000**, *345*, 53–60. [CrossRef]
39. Liu, Y.; Huang, L.; Li, W.; Guo, W.; Zheng, H.; Wang, J.; Lu, F. Studies on properties of the xylan-binding domain and linker sequence of xylanase XynG1-1 from *Paenibacillus campinasensis* G1-1. *J. Ind. Microbiol. Biot.* **2015**, *42*, 1591–1599. [CrossRef]
40. Li, N.; Shi, P.; Yang, P.; Wang, Y.; Luo, H.; Bai, Y.; Zhou, Z.; Yao, B. A xylanase with high pH stability from *Streptomyces* sp. S27 and its carbohydrate-binding module with/without linker-region-truncated versions. *Appl. Microbiol. Biotechnol.* **2009**, *83*, 99–107. [CrossRef]
41. Zhang, M.; Chekan, J.R.; Dodd, D.; Hong, P.-Y.; Radlinski, L.; Revindran, V.; Nair, S.K.; Mackie, R.I.; Cann, I. Xylan utilization in human gut commensal bacteria is orchestrated by unique modular organization of polysaccharide-degrading enzymes. *Proc. Natl. Acad. Sci. USA* **2014**, *111*, E3708–E3717. [CrossRef]
42. Karlsson, E.N.; Bartonek-Roxå, E.; Holst, O. Evidence for substrate binding of a recombinant thermostable xylanase originating from *Rhodothermus marinus*. *FEMS Microbiol. Lett.* **1998**, *168*, 1–7. [CrossRef]
43. Li, Q.; Sun, B.; Li, X.; Xiong, K.; Xu, Y.; Yang, R.; Hou, J.; Teng, C. Improvement of the catalytic characteristics of a salt-tolerant GH10 xylanase from *Streptomyce rochei* L10904. *Int. J. Biol. Macromol.* **2018**, *107*, 1447–1455. [CrossRef]
44. Lim, H.K.; Lee, K.I.; Hwang, I.T. Identification of a novel cellulose-binding domain within the endo-β-1,4-xylanase KRICT PX-3 from *Paenibacillus terrae* HPL-003. *Enzyme Microb. Technol.* **2016**, *93*, 166–173.
45. Karlsson, E.N.; Dahlberg, L.; Torto, N.; Gorton, L.; Holst, O. Enzymatic specificity and hydrolysis pattern of the catalytic domain of the xylanase Xyn1 from *Rhodothermus marinus*. *J. Biotechnol.* **1998**, *60*, 23–35. [CrossRef]
46. Hachem, M.A.; Olsson, F.; Nordberg Karlsson, E. Probing the stability of the modular family 10 xylanase from *Rhodothermus marinus*. *Extremophiles* **2003**, *7*, 483–491. [CrossRef]
47. Graziano, G.; Merlino, A. Molecular bases of protein halotolerance. *BBA - Proteins Proteom.* **2014**, *1844*, 850–858. [CrossRef]
48. Fukuchi, S.; Yoshimune, K.; Wakayama, M.; Moriguchi, M.; Nishikawa, K. Unique amino acid composition of proteins in halophilic bacteria. *J. Mol. Biol.* **2003**, *327*, 347–357. [CrossRef]
49. Liu, X.; Huang, Z.; Zhang, X.; Shao, Z.; Liu, Z. Cloning, expression and characterization of a novel cold-active and halophilic xylanase from *Zunongwangia profunda*. *Extremophiles* **2014**, *18*, 441–450. [CrossRef]
50. Guo, B.; Chen, X.-L.; Sun, C.-Y.; Zhou, B.-C.; Zhang, Y.-Z. Gene cloning, expression and characterization of a new cold-active and salt-tolerant endo-β-1,4-xylanase from marine *Glaciecola mesophila* KMM 241. *Appl. Microbiol. Biotechnol.* **2009**, *84*, 1107–1115. [CrossRef]

51. Huang, X.; Lin, J.; Ye, X.; Wang, G. Molecular characterization of a thermophilic and salt-and alkaline-tolerant xylanase from *Planococcus* sp. SL4, a strain isolated from the sediment of a soda lake. *J. Microbiol. Biotechnol.* **2015**, *25*, 662–671. [CrossRef]
52. Xu, B.; Dai, L.; Li, J.; Deng, M.; Miao, H.; Zhou, J.; Mu, Y.; Wu, Q.; Tang, X.; Yang, Y. Molecular and biochemical characterization of a novel xylanase from *Massilia* sp. RBM26 isolated from the feces of *Rhinopithecus bieti*. *J. Microbiol. Biotechnol.* **2015**, *26*, 9–19. [CrossRef]
53. Zhou, J.; Gao, Y.; Dong, Y.; Tang, X.; Li, J.; Xu, B.; Mu, Y.; Wu, Q.; Huang, Z. A novel xylanase with tolerance to ethanol, salt, protease, SDS, heat, and alkali from actinomycete *Lechevalieria* sp. HJ3. *J. Ind. Microbiol. Biot.* **2012**, *39*, 965–975. [CrossRef]
54. Poosarla, V.G.; Chandra, T. Purification and characterization of novel halo-acid-alkali-thermo-stable xylanase from *Gracilibacillus* sp. TSCPVG. *Appl. Biochem. Biotech.* **2014**, *173*, 1375–1390. [CrossRef]
55. Verjans, P.; Dornez, E.; Segers, M.; Van Campenhout, S.; Bernaerts, K.; Beliën, T.; Delcour, J.A.; Courtin, C.M. Truncated derivatives of a multidomain thermophilic glycosyl hydrolase family 10 xylanase from *Thermotoga maritima* reveal structure related activity profiles and substrate hydrolysis patterns. *J. Biotechnol.* **2010**, *145*, 160–167. [CrossRef]
56. Sunna, A.; Gibbs, M.D.; Bergquist, P.L. A novel thermostable multidomain 1,4-β-xylanase from '*Caldibacillus cellulovorans*' and effect of its xylan-binding domain on enzyme activity. *Microbiol.* **2000**, *146*, 2947–2955. [CrossRef]
57. Ali, M.K.; Hayashi, H.; Karita, S.; Goto, M.; Kimura, T.; Sakka, K.; Ohmiya, K. Importance of the carbohydrate-binding module of *Clostridium stercorarium* Xyn10B to xylan hydrolysis. *Biosci. Biotechnol. Biochem.* **2001**, *65*, 41–47. [CrossRef]
58. Bai, W.; Xue, Y.; Zhou, C.; Ma, Y. Cloning, expression, and characterization of a novel alkali-tolerant xylanase from alkaliphilic *Bacillus* sp. SN5. *Biotechnol. Appl. Biochem.* **2015**, *62*, 208–217. [CrossRef]
59. Kumar, S.; Stecher, G.; Tamura, K. MEGA7: Molecular evolutionary genetics analysis version 7.0 for bigger datasets. *Mol. Biol. Evol.* **2016**, *33*, 1870–1874. [CrossRef]
60. Gullfot, F.; Tan, T.-C.; von Schantz, L.; Karlsson, E.N.; Ohlin, M.; Brumer, H.; Divne, C. The crystal structure of XG-34, an evolved xyloglucan-specific carbohydrate-binding module. *Proteins* **2010**, *78*, 785–789. [CrossRef]
61. Santos, C.R.; Hoffmam, Z.B.; de Matos Martins, V.P.; Zanphorlin, L.M.; de Paula Assis, L.H.; Honorato, R.V.; de Oliveira, P.S.L.; Ruller, R.; Murakami, M.T. Molecular mechanisms associated with xylan degradation by *Xanthomonas* plant pathogens. *J. Biol. Chem.* **2014**, *289*, 32186–32200. [CrossRef]

International Journal of
Molecular Sciences

Review

β-Xylosidases: Structural Diversity, Catalytic Mechanism, and Inhibition by Monosaccharides

Ali Rohman [1,2], Bauke W. Dijkstra [3] and Ni Nyoman Tri Puspaningsih [1,2,*]

1 Department of Chemistry, Faculty of Science and Technology, Universitas Airlangga, Surabaya 60115, Indonesia; alirohman@fst.unair.ac.id
2 Laboratory of Proteomics, Research Center for Bio-Molecule Engineering (BIOME), Universitas Airlangga, Surabaya 60115, Indonesia
3 Laboratory of Biophysical Chemistry, University of Groningen, 9747 AG Groningen, The Netherlands; b.w.dijkstra@rug.nl
* Correspondence: ni-nyoman-t-p@fst.unair.ac.id; Tel.: +62-31-5922427

Received: 15 October 2019; Accepted: 4 November 2019; Published: 6 November 2019

Abstract: Xylan, a prominent component of cellulosic biomass, has a high potential for degradation into reducing sugars, and subsequent conversion into bioethanol. This process requires a range of xylanolytic enzymes. Among them, β-xylosidases are crucial, because they hydrolyze more glycosidic bonds than any of the other xylanolytic enzymes. They also enhance the efficiency of the process by degrading xylooligosaccharides, which are potent inhibitors of other hemicellulose-/xylan-converting enzymes. On the other hand, the β-xylosidase itself is also inhibited by monosaccharides that may be generated in high concentrations during the saccharification process. Structurally, β-xylosidases are diverse enzymes with different substrate specificities and enzyme mechanisms. Here, we review the structural diversity and catalytic mechanisms of β-xylosidases, and discuss their inhibition by monosaccharides.

Keywords: biomass; hemicellulose; bioethanol; xylanolytic enzyme; hemicellulase; glycoside hydrolase

---

1. Introduction

Xylan is a prominent component of cellulosic biomass, a heterogeneous complex of carbohydrate polymers (cellulose and hemicellulose) and lignin, a complex polymer of phenylpropane units [1–3]. Hemicellulose, including xylan, makes up approximately one-third of the carbohydrate content of common agricultural and forestry waste [1–3]. It is among the most inexpensive non-food biomass that is sustainably available in nature in large quantities, and that can be converted into biofuel or other value-added products, such as low-calorie sweeteners, prebiotics, surfactants and various specialty chemicals [1,4–6]. Structurally, xylan is a complex heteropolysaccharide with a glycosidic β-(1,4)-linked D-xylose backbone that is frequently substituted with side chains of arabinose, glucuronic acid and other groups. In turn, these side chains may be further esterified with acetic, ferulic, and *p*-coumaric acids (Figure 1). The type and frequency of the side chains and their substituents vary with the source of xylans [2,7–11].

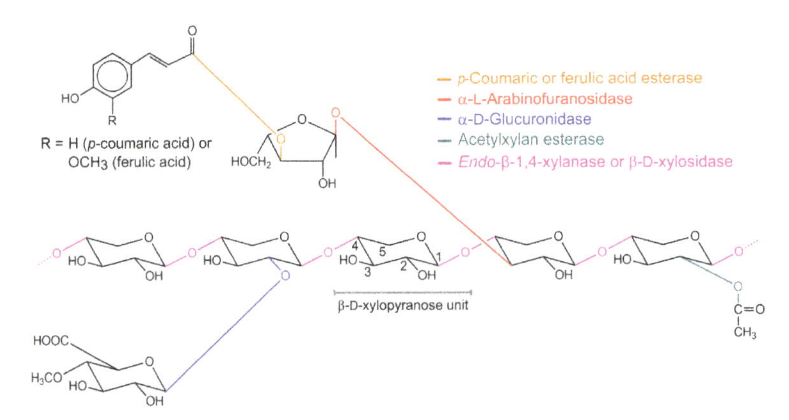

Figure 1. Example of the structure of a plant xylan with the cleavage sites of various xylanolytic enzymes indicated. A β-D-xylopyranose unit with numbered carbon atoms is shown in the middle. Glycosidic bonds and xylanolytic enzymes that hydrolyze them are depicted in the same color [7,9].

As a complex heteropolysaccharide, full degradation of xylan into its monosaccharide constituents requires the concerted action of various hydrolytic xylan-degrading enzymes with different specificities (Figure 1). These enzymes include α-L-arabinofuranosidase (EC 3.2.1.55), α-D-glucuronidase (EC 3.2.1.139), acetylxylan esterase (EC 3.1.1.72), and *p*-coumaric acid and ferulic acid esterases (EC 3.1.1.73), which release the side chain substituents from the xylan backbone, and *endo*-β-1,4-xylanase (EC 3.2.1.8), which works synergistically with β-xylosidase (EC 3.2.1.37) to break down the xylan backbone. *Endo*-β-1,4-xylanase hydrolyses the internal β-(1,4) linkages of the xylan backbone producing short xylooligosaccharides, while β-xylosidase removes xylose units from the non-reducing termini of these xylooligosaccharides [2,7,8,10]. In nature, xylanolytic enzymes are mainly found in numerous saprophytic microorganisms, such as fungi, actinomycetes and other bacteria, as well as in the rumen biota of higher animals. The microorganisms secrete the enzymes, for example, as a strategy for expanding their versatility to use primary carbon sources [7,11–14].

Xylan-degrading enzymes have found application as environmentally friendly agents in a wide range of industrial processes, such as bleaching of paper pulp, deinking of recycled paper, enhancing the digestibility and nutritional properties of animal feed, degumming of plant fiber sources, manufacturing of beer and wine, clarification of fruit juices and maceration of fruits and vegetables, preparation of high-fiber baked goods, and the extraction of coffee [2,7,15,16]. Furthermore, the enzymes are applied during the saccharification of pretreated agricultural and forestry cellulosic biomass into fermentable sugars [2,15], e.g., for producing biofuel.

In xylan saccharification, β-xylosidase is a crucial enzyme since, of all the xylanolytic enzymes, it cleaves the greatest number of glycosidic bonds. [17–19]. In addition, because xylooligosaccharides are potent inhibitors of *endo*-β-1,4-xylanases and cellulases, the activity of β-xylosidase can improve the efficiency of the saccharification process by degrading the xylooligosaccharides and thus alleviating inhibition of those enzymes [7,11,20–23]. However, most of the characterized β-xylosidases are, to some extent, also inhibited themselves by xylose, arabinose, glucose, and/or other monosaccharides [2,11,24–27]. This is an important problem, since in industrial cellulosic biomass saccharification, the monosaccharides may accumulate to high enough concentrations to significantly reduce the activity of β-xylosidase, even in simultaneous saccharification and fermentation processes, where monosaccharides are directly consumed by the fermenting organisms [26,27]. This adverse property may severely reduce the efficiency of β-xylosidases in the saccharification process.

In this review, the structural diversity, catalytic mechanisms and inhibition by monosaccharides of β-xylosidases are discussed.

2. Structural Diversity of β-xylosidases

β-Xylosidases are a group of structurally diverse enzymes with varying specificities, in line with the diversity of the organisms that produce them and the heterogeneity of their substrates [28]. However, as commonly observed in other glycoside hydrolases (GHs), they hydrolyze the glycosidic bond via one of two routes, either with overall retention or with overall inversion of the anomeric carbon configuration [29].

GHs are classified in the Carbohydrate-Active Enzymes database (CAZy; http://www.cazy.org/), which groups the enzymes into families based on their amino acid sequence similarities [30,31]. As there is a direct relationship between amino acid sequence similarity and similarity of folding, the classification also represents the structural features and commonality of the catalytic mechanism of the enzymes. Thus, enzymes in a particular family display highly similar three-dimensional structures and catalytic mechanisms [29,32]. At present, 161 GH families (GH1 to GH161) are represented on the CAZy server. Nevertheless, despite divergent amino acid sequences, several different GH families show significantly similar protein folding and active site architecture. Such GH families are considered to have a common ancestor and, therefore, have been grouped together into a clan [33]. To date, 18 GH clans (GH-A to GH-R) have been assigned in the database.

A search using the enzyme classification number for β-xylosidase (EC 3.2.1.37) in the CAZy database [31] revealed that enzymes with this number are presently found in 11 different GH families (Table 1). Nevertheless, a further literature examination suggests that 3 families, i.e., GH1, GH54 and GH116, may not contain enzymes with β-xylosidase activity on natural substrates. The enzymes from *Reticulitermes flavipes* (RfBGluc-1; GenPept accession No. ADK12988) [34] and *R. santonensis* De Feytaud (GenPept ADT62000) [35] in GH1, and *Trichoderma koningii* G-39 (TkAbf; GenPept AAA81024) [36] in GH54 were classified as β-xylosidases because they hydrolyze artificial nitrophenyl-β-D-xylopyranoside derivatives. However, to our knowledge there is no evidence that these enzymes are able to release xylose from natural substrates. Similarly, a bifunctional aryl β-glucosidase/β-xylosidase from the hyperthermophilic archaeon *Saccharolobus solfataricus* P2 (formerly *Sulfolobus solfataricus*; SSO1353; GenPept AAK41589) in GH116 is called so based on its activity on aryl β-glucosides and β-xylosides, but the enzyme likely does not hydrolyze xylooligosaccharides [37]. All in all, this suggests that enzymes with β-xylosidase activity on natural substrates currently occur in only 8 GH families in the CAZy database, i.e., in GH families 3, 5, 30, 39, 43, 51, 52 and 120.

Table 1. Distribution of the current β-xylosidases in the CAZy database, their catalytic domain fold, their type of catalytic mechanism, and their catalytic residues.

Family (GH)	Total Number of β-xylosidase Sequences	Clan	Overall Fold of the Catalytic Domain	Catalytic Mechanism †	Nucleophile	General Acid/Base
‡ 1	2	A	$(\beta/\alpha)_8$ TIM-barrel	Retention	Glu	Glu
3	103	n.a. #	$(\beta/\alpha)_8$ TIM-barrel	Retention	Asp	Glu
5	1	A	$(\beta/\alpha)_8$ TIM-barrel	Retention	Glu	Glu
30	4	A	$(\beta/\alpha)_8$ TIM-barrel	Retention	Glu	Glu
39	24	A	$(\beta/\alpha)_8$ TIM-barrel	Retention	Glu	Glu
43	96	F	5-bladed β-propeller	Inversion	Asp §	Glu
51	2	A	$(\beta/\alpha)_8$ TIM-barrel	Retention	Glu	Glu
52	11	O	$(\alpha/\alpha)_6$-barrel	Retention	Glu	Asp
‡ 54	2	n.a. #	β-sandwich %	Retention	Glu %	Asp %
‡ 116	1	O	$(\alpha/\alpha)_6$-barrel	Retention	Glu	Asp
120	2	n.a. #	right-handed parallel β-helix	Retention	Asp	Glu

†: Catalysis by GHs commonly proceeds with either retention or inversion of the substrate's anomeric carbon configuration. See main text for further information. ‡: It is unknown whether the enzymes from GH1, GH54, and GH116 have β-xylosidase activity on natural substrates. #: Not part of a clan. §: General base; General acid. %: Not assigned in the CAZy database. Data are from the crystal structure of the α-L-arabinofuranosidase from *Aspergillus kawachii* IFO4308 [38].

2.1. Glycoside Hydrolase Clan A (GH-A)

GH families 1, 5, 30, 39 and 51 are part of clan GH-A, the largest clan in the CAZy database with currently 23 GH families. Enzymes in this clan all have a $(\beta/\alpha)_8$ catalytic domain, also known as triose-phosphate isomerase (TIM) barrel domain [39].

Of clan GH-A, structural data for β-xylosidases are currently only available for GH39, i.e., β-xylosidases from *Thermoanaerobacterium saccharolyticum* B6A-RI (TsXynB; Protein Data Bank code 1px8; Figure 2e) [40], *Geobacillus stearothermophilus* T-6 (GsXynB1; PDB 2BS9) [41] and *Caulobacter crescentus* NA1000 (CcXynB2; PDB 4EKJ) [42]. These enzymes fold into a three-domain structure, consisting of an N-terminal $(\beta/\alpha)_8$-barrel catalytic domain, sequentially followed by a β-sandwich and an α-helical accessory domain. Their structures are very similar. Superposition of the structures of isolated proteins gave an overall root mean squared deviation (RMSD) of 1.6 Å for 462 amino acid residues. However, while CcXynB2 exists as a monomer in solution [42], TsXynB and GsXynB1 are present as tetramers [40,41]. The absence of a short amino acid sequence at the C-terminus of CcXynB2, compared to the other two enzymes, has been suggested to prevent the formation of a stable tetramer [42]. Additionally, it has been proposed that subtle structural differences in the accessory domains of these β-xylosidases slightly alter their overall structure and the accessibility of their catalytic region [42].

In the absence of structural data for β-xylosidases from families GH1, GH5, GH30, and GH51 we generated homology models to compare the 3D structures of β-xylosidases from the different families of the GH-A clan. Models were built of the β-glucosidase/β-xylosidase RfBGluc-1 (GH1; Figure 2a) [34], a β-xylosidase from *Phanerochaete chrysosporium* BKM-F-1767 (PcXyl5; GenPept AHL69750; GH5; Figure 2c) [43], a β-glucosidase/β-xylosidase from *Phytophthora infestans* (PiBGX1; GenPept AAK19754; GH30; Figure 2d) [44], and an α-L-arabinofuranosidase/β-xylosidase from *Arabidopsis thaliana* (AtAraf; GenPept AAF19575; GH51; Figure 2h) [45], using 3D structures of their nearest homologs as templates. All resulting models display a $(\beta/\alpha)_8$-barrel catalytic domain that is highly similar to the catalytic domain of GH39 β-xylosidases (e.g., Figure 2e) and that shows that the catalytic residues of GH39 are present at the equivalent positions in the GH1, GH5, GH30, and GH51 β-xylosidase families. A multiple structural alignment of the catalytic domains of these models and GH39 β-xylosidases gave an overall RMSD of 3.4 Å for 168 amino acid residues, with PcXyl being the most divergent from the other structures. In contrast, the structures of their accessory domains varied with the family. The accessory domains of GH39 β-xylosidases are absent in RfBGluc-1 and PcXyl5, but they are retained at a comparable position in PiBGX1 and AtAraf albeit with some modifications. The major differences are observed for the third domain, in which the GH39 α-helical domain is replaced by a β-sheet and a loop structure in PiBGX1 and AtAraf, respectively.

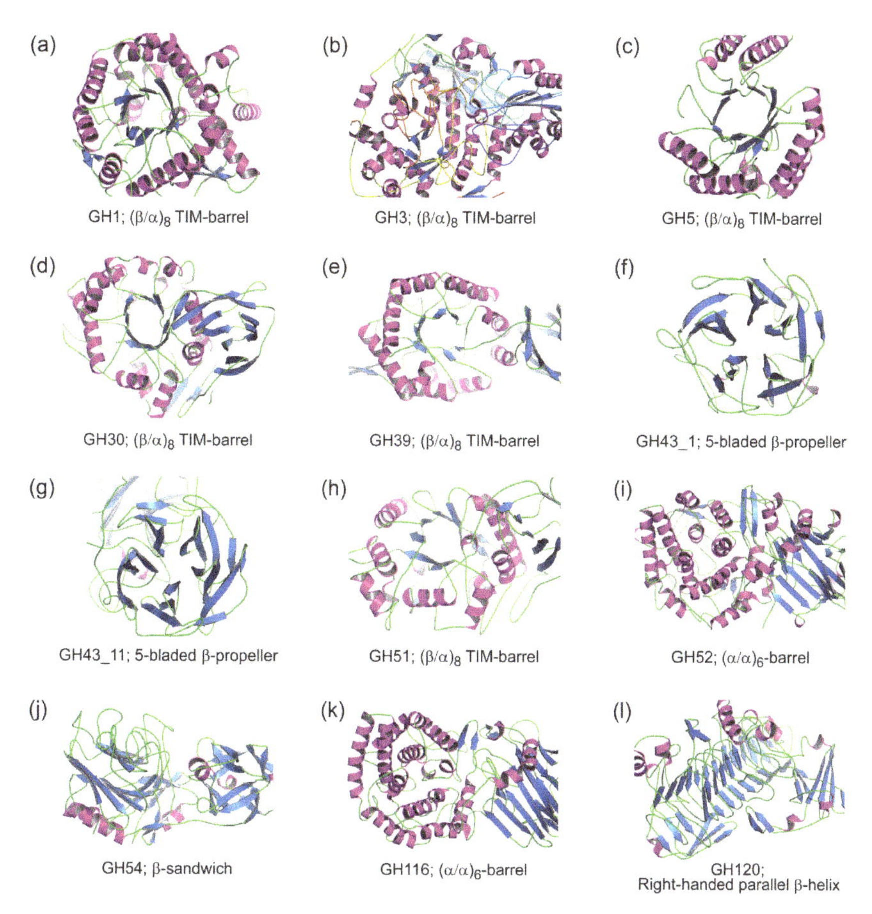

Figure 2. Three-dimensional (3D) structures of β-xylosidases from various GH families. Helix, strand, and loop structures are colored in magenta, blue, and green, respectively. GH family numbers and fold type of their catalytic domains are shown. The structures represented are (**a**) RfBGluc-1 from *Reticulitermes flavipes* (GenPept ADK12988); (**b**) GlyA1 from metagenomic cow rumen fluid (PDB 5K6L); (**c**) PcXyl5 from *Phanerochaete chrysosporium* BKM-F-1767 (GenPept AHL69750) (**d**) PiBGX1 from *Phytophthora infestans* (GenPept AAK19754); (**e**) TsXynB from *Thermoanaerobacterium saccharolyticum* B6A-RI (PDB 1PX8); (**f**) RS223-BX from an uncultured organism (PDB 4MLG); (**g**) GsXynB3 from *Geobacillus stearothermophilus* T-6 (PDB 2EXH); (**h**) AtAraf from *Arabidopsis thaliana* (GenPept AAF19575); (**i**) GT2_24_00240 from *Geobacillus thermoglucosidasius* TM242 (PDB 4C1O); (**j**) TkAbf from *Trichoderma koningii* G-39 (GenPept AAA81024); (**k**) SSO1353 from *Saccharolobus solfataricus* P2 (GenPept AAK41589); and (**l**) TsXylC from *Thermoanaerobacterium saccharolyticum* JW/SL-YS485 (PDB 3VST). The structures of (a), (c), (d), (h), (j), and (k) were modeled using PDB entries 3VIK, 1EQP, 2XWE, 2C8N, 1WD3, and 5BVU, respectively, which belong to the same GH family but do not have β-xylosidase activity. Structure modeling was performed using the Swiss-Model server [46]. Figure 2, Figure 3, and Figure 6 were produced using the program PyMol (The PyMOL Molecular Graphics System, v. 0.99, Schrödinger, LLC, http://www.pymol.org).

2.2. Glycoside Hydrolase Family 3 (GH3)

While GH families 1, 30, 39 and 51 are part of clan GH-A, other β-xylosidases belong to other families that are not part of this clan. GH3 is one of the largest and most diverse GH families in the CAZy database [28,47]. It contains more than 23400 entries with various enzyme activities, including β-xylosidase, β-glucosidase, β-glucosylceramidase, β-N-acetylhexosaminidase, and α-L-arabinofuranosidase activities. A number of GH3 enzymes are reported to be bi/multifunctional, particularly toward synthetic substrates.

Enzymes in family GH3 vary considerably in the lengths of their peptide chains [48,49] and, consequently, in the number of tertiary structure domains [48,50]. The basic structure of GH3 members is a single $(\beta/\alpha)_8$ TIM-barrel domain [48,50], similar to the domain that is observed in clan GH-A. In most members, the domain is followed by an (α/β)-sandwich domain that varies in size [48], e.g., $(\alpha/\beta)_6$ in *Kluyveromyces marxianus* NBRC1777 β-glucosidase [51], $(\alpha/\beta)_5$ in *Thermotoga neapolitana* β-glucosidase [49], or even only an αβα motif in *Bacillus subtilis* 168 β-N-acetylglucosaminidases [52]. Sometimes the order of the domains in the primary structure is reversed [48]. Although these two domains are generally sufficient to organize the active site of GH3 enzymes, frequently GH3 members are extended with a fibronectin type III (FnIII) domain of unknown function at the C-terminus of the (α/β)-sandwich domain [48,49]. Moreover, in some GH3 members, the (α/β)-sandwich domain is interrupted by a PA14 domain. This domain appeared to be important for the substrate specificity of the *Kluyveromyces marxianus* NBRC1777 β-glucosidase [51].

A total of 103 enzymes with β-xylosidase annotation are currently found in GH3, making it the largest β-xylosidase-containing GH family. A protein domain search using the program InterProScan 5 [53] revealed that the majority of the GH3 β-xylosidases are composed of three domains (TIM-barrel, (α/β)-sandwich and FnIII). However, a bifunctional β-xylosidase/β-glucosidase from *Erwinia chrysanthemi* D1 (EcBgxA; GenPept AAA80156) [54] has two domains (TIM-barrel and (α/β)-sandwich) and a β-xylosidase from an environmental sample (G06-24; GenPept ACY24766) [55] has four domains (TIM-barrel, (α/β)-sandwich, FnIII, and PA14). A phylogenetic analysis clustered these two enzymes divergently from the other GH3 β-xylosidases [56].

3D Structures of GH3 β-xylosidases are available for a β-xylosidase from the fungus *Trichoderma reesei* RutC-30 (TrBxl1; PDB 5A7M; GenPept CAA93248) [57] and a β-glucosidase/β-xylosidase from metagenomic cow rumen fluid (GlyA1; PDB 5K6L; Figure 2b) [58]. Both structures have a $(\beta/\alpha)_8$ TIM-barrel, a $(\alpha/\beta)_6$-sandwich, and a FnIII domain, but at different positions in the primary structure. As observed for the majority of GH3 structures [48], TrBxl1 has its TIM-barrel domain at the N-terminus, followed sequentially by the (α/β)-sandwich and FnIII domains. This order is reversed in GlyA1, where the (α/β)-sandwich domain is at the N-terminus, followed by the FnIII and TIM-barrel domains. In addition, GlyA1 has an additional domain with unknown structure at its C-terminus [58]. Despite this, the 3D structures of TrBxl1 and GlyA1 are conserved, with the TIM-barrel and (α/β)-sandwich domains, as well as the catalytic residues superimposing reasonably well when the domains are structurally aligned.

2.3. Glycoside Hydrolase Family 43 (GH43)

GH43 is the second largest β-xylosidase-containing GH family with currently 96 members annotated as β-xylosidase. In addition to β-xylosidases, this family also contains enzymes with (putative) α-L-arabinofuranosidase, arabinanase, xylanase, galactan 1,3-β-galactosidase, α-1,2-L-arabinofuranosidase, *exo*-α-1,5-L-arabinofuranosidase, *exo*-α-1,5-L-arabinanase, or β-1,3-xylosidase activities. As observed for the GH3 members, several enzymes in this family are bi/multifunctional.

Together with GH62, GH43 is grouped into clan GH-F in the CAZy database with a structural characteristic of a 5-bladed β-propeller catalytic domain [59]. Some of its members contain only this single catalytic domain, and, based on their domain architecture, were classified as type I [60]. In other members, the catalytic domain is extended with a family 6 carbohydrate-binding module (type II), or a unique β-sandwich domain that is designated as X19 [61] (type III), or contain an even more complex domain composition and organization (type IV). The extensions are commonly fused at the C-terminus of the catalytic domain [60,62], although in a β-xylosidase from *G. thermoleovorans* IT-08 (GbtXyl43B), for example, the extension is at the N-terminus [63]. Thus, GH43 contains enzymes that vary both in the lengths of their primary structure and in their number of structure domains.

For detailed characterization, enzymes in GH43 have been divided into 37 subfamilies, GH41_1 to GH43_37 [61]. In this classification, β-xylosidases are currently found in 16 different subfamilies, with the majority belonging to subfamilies GH43_1 and GH43_11. Two GH43_1 β-xylosidase crystal structures are currently present in the PDB database, i.e., from an uncultured organism (RS223-BX; PDB 4MLG; Figure 2f) [64] and from a compost metagenome (CoXyl43; PDB 5GLK) [65]. The most structurally characterized β-xylosidases are from GH43_11, with crystal structures available of seven different β-xylosidases, i.e., the β-xylosidases from *B. subtilis* (PDB 1YIF) (Patskovsky et al., unpublished work), *Clostridium acetobutylicum* ATCC 824 (CaXyl43_11; PDB 1YI7) (Teplyakov et al., unpublished work), *B. halodurans* (PDB 1YRZ) (Fedorov et al., unpublished work), *G. stearothermophilus* T-6 (PDB 2EXH; Figure 2g) [66], *Selenomonas ruminantium* GA192 (PDB 3C2U) [67], *B. pumilus* IPO (PDB 5ZQJ) [68], and *Bacillus* sp. HJ14 (PDB 6IFE) [69]. Additionally, GH43 β-xylosidase 3D structures are also found in subfamilies GH43_12 and GH43_26, i.e., the β-xylosidases from *G. thermoleovorans* IT-08 (PDB 5Z5D) [70] and *C. acetobutylicum* ATCC 824 (CaXyl43_26; PDB 3K1U) (Osipiuk et al., unpublished work), respectively.

While β-xylosidases from GH43_1 and GH43_26 have only a single 5-bladed β-propeller catalytic domain [64,65] and belong to type I GH43 [60], those from GH43_11 and GH43_12 possess an additional X19 domain at the C-terminus [66–68,70] and belong to type III GH43 [60]. Although the architecture of the 5-bladed β-propeller is highly conserved among the GH43 β-xylosidases [64], structural superposition of the type I and type III catalytic domains gave a high RMSD. This is because the catalytic domains of the type I enzymes have several significantly longer loops than those of type III. The single catalytic domain of the type I GH43 β-xylosidases is sufficient for activity, but the enzymes are strongly activated by divalent metal ions, particularly calcium. Indeed, those metal-containing enzymes contain a metal-binding site close to the enzymes' active site [64,65]. In contrast, the type III GH43 β-xylosidases have no such metal-binding site [66–68,70]. The X19 domain, which is only found in a subset of GH43 subfamilies [61], appeared to be crucial for catalytic activity of the type III GH43 β-xylosidases, since removing this domain abolished the activity of the GH43_11 β-xylosidases from *Thermobifida fusca* YX [71] and *Enterobacter* sp. [72]. In fact, a loop from the X19 domain contributes a Phe residue to the active site of the type III β-xylosidases [66–68,70], which is spatially conserved among all GH43 β-xylosidase structures. Only in CaXyl43_26 this Phe is missing. Unfortunately, no biochemical evidence is available on the enzyme's substrate preferences and catalytic activity, but given that all other enzymes in GH43_26 are α-L-arabinofuranosidases [61], some doubt that CaXyl43_26 is a genuine β-xylosidase seems justified. These observations suggest that although GH43 β-xylosidases adopt different overall folds, the enzymes have a common active site organization and use a conserved Phe to interact with the substrate in subsite −1 (see below; Figure 4b).

2.4. Glycoside Hydrolase Family 52 (GH52)

Currently, GH52 contains 112 entries, of which 11 enzymes are annotated as β-xylosidases. These β-xylosidases have comparable amino acid sequence lengths of about 700 amino acid residues, with the exception of a β-xylosidase from *G. stearothermophilus* 236 (GsXylA, GenPept AAA50863), which is composed of only 618 amino acid residues. The GH52 β-xylosidases are very similar to each other with amino acid sequence identities of around 41%–90%. In this family, crystal structures are available for β-xylosidases from *Parageobacillus thermoglucosidasius* TM242 (GT2_24_00240; PDB 4C1O; Figure 2i) [73] and *G. stearothermophilus* T-6 (Xyn52B2; PDB 4RHH) (Dann et al., unpublished work). With a sequence identity of 86%, the two proteins fold into almost the same structures; they display two distinct domains, an N-terminal β-sandwich domain and a C-terminal $(\alpha/\alpha)_6$-barrel domain. The catalytic residues of the GH52 enzymes, which are Glu-357 and Asp-517 in GT2_24_00240 [73], are located in the $(\alpha/\alpha)_6$-barrel domain. Protein homology modeling based on the structure of GT2_24_00240 suggested that the domains are conserved among the GH52 β-xylosidases, except for the C-terminal domain of GsXylA. Compared to other GH52 β-xylosidases, this latter domain lacks five α-helices of the C-terminal domain, such that it displays an open half-barrel structure.

2.5. Glycoside Hydrolase Family 54 (GH54)

Most of the characterized enzymes in GH54 are annotated as α-L-arabinofuranosidases. However, two sequences in this family are annotated as β-xylosidase. TkAbf from *T. koningii* G-39 was characterized as a bifunctional α-L-arabinofuranosidase/β-xylosidase due to its activity on synthetic nitrophenyl derivatives of α-L-arabinofuranoside and β-D-xylopyranoside with comparable k_{cat}/K_m values [36]. A three-dimensional structure of a GH54 enzyme is currently only available for an α-L-arabinofuranosidase from *A. kawachii* IFO4308 (AkAbfB; PDB 1WD3) [38]. The primary structures of TkAbf (500 residues) and AkAbfB (499 residues) are very similar with an amino acid sequence identity of 73%. Therefore, the three-dimensional structure of TkAbf was predicted by homology modeling using the crystal structure of AkAbfB as a template. The predicted model (Figure 2j) consists of two domains that correspond to the N-terminal catalytic domain and the C-terminal arabinose-binding domain of the AkAbfB structure [38]. The catalytic domain folds into a β-sandwich similar to that of clan GH-B enzymes, while the arabinose-binding domain has a β-trefoil structure that belongs to the family 42 carbohydrate-binding module [28,38].

2.6. Glycoside Hydrolase Family 116 (GH116)

In GH116, SSO1353 is the only enzyme that exhibits β-xylosidase activity. As mentioned above, this enzyme does not hydrolyze xylooligosaccharides, but it is active on artificial substrates such as *p*-nitrophenyl- and methylumbelliferyl-linked β-D-xylopyranosides [37]. Currently, a three-dimensional structure of a GH116 member is only available for the β-glucosidase from the thermophilic bacterium *T. xylanolyticum* LX-11 (TxGH116; PDB 5BVU). This protein folds into an N-terminal β-sandwich domain and a C-terminal $(\alpha/\alpha)_6$ solenoid catalytic domain [74]. The primary structure similarity of SSO1353 and TxGH116 is rather low with an amino acid sequence identity of only ~20%. However, homology modeling of SSO1353 based on the structure of TxGH116 using the Swiss-Model server [46] produced a relatively good quality model with a Global Model Quality Estimation (GMQE) value of 0.63 (on a scale of 0–1). As expected, the model displays a two-domain fold, i.e., an N-terminal β-sandwich domain and a C-terminal $(\alpha/\alpha)_6$-barrel domain (Figure 2k), very much like the domain organization of the GH52 proteins (see above). Importantly, the modeling placed the catalytic nucleophile and acid/base residues of SSO1353 (Glu-335 and Asp-462, respectively) [37] at about the same positions as those of the GH52 β-xylosidase GT2_24_00240 (Glu-357 and Asp-517, respectively) [73]. In view of this structural similarity and the conservation of the catalytic residues, GH families 52 and 116 were recently grouped into clan GH-O [74].

2.7. Glycoside Hydrolase Family 120 (GH120)

Of the 176 sequences that are currently available in the CAZy database for the GH120 family, two enzymes were characterized and identified as β-xylosidases, i.e., enzymes from *Thermoanaerobacterium* saccharolyticum JW/SL-YS485 (TsXylC; GenPept ABM68042) [75] and *Bifidobacterium adolescentis* LMG10502 (BaXylB; GenPept BAF39080) [56]. While TsXylC was shown to be active on xylobiose and xylotriose [75], BaXylB prefers xylotriose or longer xylooligosaccharides as its substrate [56]. The three-dimensional structure of TsXylC has been reported to fold into a core domain of a right-handed parallel β-helix, a common fold observed in several GHs, polysaccharide lyases, and carbohydrate esterases. This core domain is intervened by an Ig-like β-sandwich domain (PDB 3VST, Figure 2l). Both domains are important to organize the active site of the enzyme [25]. BaXylB shares 47% amino acid sequence identity with TsXylC. A homology model of BaXylB based on the structure of TsXylC suggested that the active site residues and their positions are conserved in the enzymes, except for Trp-362 in BaXylB, which is a histidine (His-352) in TsXylC.

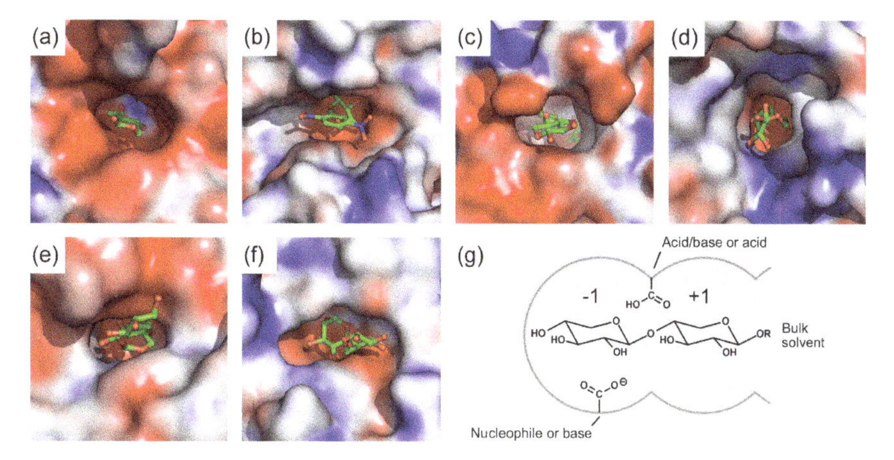

Figure 3. β-Xylosidase active site. Molecular surface drawing of active sites of β-xylosidases colored according to their electrostatic potential (negative, red; neutral, white; positive, blue). Complexed ligands are depicted in ball and stick representation with carbon atoms in green. The active sites are of (**a**) GlyA1 from metagenomic cow rumen fluid in complex with xylose (PDB 5K6N; GH3); (**b**) GsXynB1 from *Geobacillus stearothermophilus* T-6 in complex with 2,5-dinitrophenyl-β-D-xyloside (PDB 2BFG; GH39); (**c**) CoXyl43 from a compost metagenome in complex with xylose and xylobiose (PDB 5GLN; GH43_1); (**d**) GsXynB3 from *G. stearothermophilus* T-6 in complex with xylobiose (PDB 2EXJ; GH43_11); (**e**) GT2_24_00240 from *Parageobacillus thermoglucosidasius* TM242 in complex with xylobiose (PDB 4C1P; GH52); and (**f**) TsXylC from *Thermoanaerobacterium saccharolyticum* JW/SL-YS485 in complex with xylobiose (PDB 3VSU; GH120). The electrostatic potential was calculated using the APBS (Adaptive Poisson–Boltzmann Solver) implemented in the program PyMol [76]. (**g**) Generalized schematic diagram of a β-xylosidase active site with a ligand bound at subsites −1 and +1. Catalytic residues (see below) are represented by carboxylate groups and their catalytic roles are indicated. The exact positions of the catalytic residues vary with enzymes (see Figure 6).

3. Active Site of β-Xylosidases

Despite the diversity of their three-dimensional folds, all structurally characterized β-xylosidases display a typical pocket-shaped active site (Figure 3) that is very suitable for *exo*-acting enzymes [29]. The pocket is negatively charged due to the presence of several acidic residues, but contains also hydrophobic patches of aromatic residues (Figure 3a–f). It has only a single route for substrates to enter and products to exit. The active site pocket can be virtually divided into two subsites with each of them able to accommodate a monosaccharide residue (Figure 3g). One subsite is buried, and, in several enzyme-xylobiose complexes (e.g., PDBs 2EXJ, 4C1P and 3VSU), interacts with the −1 non-reducing-end xylose (subsite −1), while the other is more open and binds the +1 xylose (subsite +1). Substrates with more than two xylose residues must have the additional residues beyond +1 exposed in the bulk solvent [25,67]. The active site architecture seems to be both necessary and sufficient for β-xylosidase activity of the enzymes.

Furthermore, comparison of active site structures of several β-xylosidase-ligand complexes suggests that there are similar interactions between the enzymes and their ligands (Figure 4). The ligand in subsite −1 is strongly bound to the enzyme by a large number of hydrogen bonds and a few hydrophobic stacking interactions. In contrast, the ligand substrate in subsite +1 interacts less strongly with the enzyme with less hydrogen bonds but more hydrophobic stacking interactions.

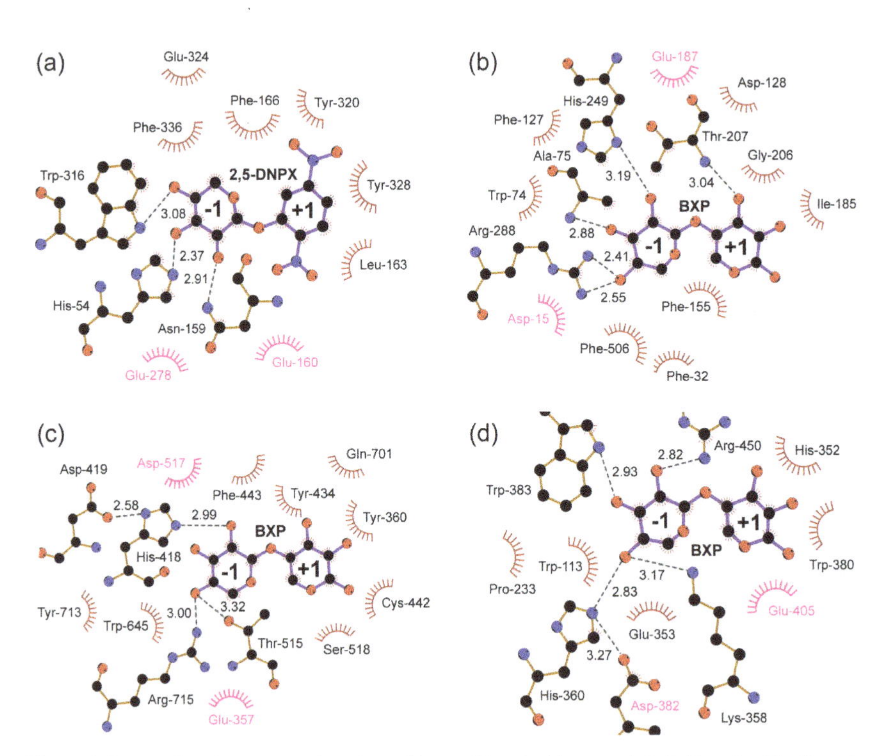

Figure 4. Interactions between active site residues of β-xylosidases and their ligands. The ligands 2,5-DNPX (2,5-dinitrophenyl-β-D-xyloside) and BXP (β-D-xylobiopyranose) are represented with purple bonds and their binding subsites -1 and +1 are indicated. Catalytic residues are labeled in magenta. Hydrogen bonds are shown as dashed lines and their distances are indicated in Å, while hydrophobic interactions are rendered with arcs. The active sites are of (**a**) GsXynB1 (PDB 2bfg); (**b**) GsXynB3 (PDB 2exj); (**c**) GT2_24_00240 (PDB 4c1p); and (**d**) TsXylC (PDB 3vsu), which represent β-xylosidases from GH families 39, 43, 52, and 120, respectively (see caption of Figure 3 for further details of the enzymes). Interaction analysis and figure preparation were performed using LigPlot$^+$ [77].

4. Catalytic Mechanism of β-Xylosidases

With respect to their catalytic mechanism most GHs can generally be classified into retaining and inverting enzymes [29,78]. The retaining GHs hydrolyze their substrates with overall retention of the stereochemistry of the anomeric carbon atom of the hydrolyzed glycosidic bond, while the inverting GHs yield a product with an inverted stereochemistry of the anomeric carbon atom [29,78]. In both mechanisms, the enzymes rely on two catalytic carboxylate groups that function as a nucleophile and a general acid/base in the retaining enzymes, or as a general base and a general acid in the inverting enzymes, respectively (Figure 5).

The retaining enzymes use a two-step double-displacement mechanism, in which enzyme glycosylation is followed by deglycosylation. In the glycosylation step, the nucleophile attacks the anomeric carbon to form a glycosyl–enzyme intermediate with the inverted configuration at the anomeric carbon. Concomitantly, the (protonated) acid/base residue transfers its proton to the glycosidic oxygen atom, to cleave the scissile glycosidic bond. Departure of the aglycone creates space allowing a catalytic water molecule to come closer to the anomeric center. In the deglycosylation step, the incoming catalytic water molecule, which is activated by the now negatively charged acid/base, attacks the anomeric carbon to release the glycone product from the intermediate. The attack re-inverts the inverted configuration of the anomeric carbon and hence the released glycone has the same stereochemistry as it had in the substrate. In contrast, the inverting enzymes follow a single-displacement mechanism to hydrolyze the glycosidic bond. A catalytic water molecule, which is deprotonated by the general base, does a nucleophilic attack on the anomeric carbon in concert with the general acid protonating the glycosidic oxygen. This cleaves the scissile glycosidic bond and frees the glycone with the inverted stereochemistry of its anomeric carbon.

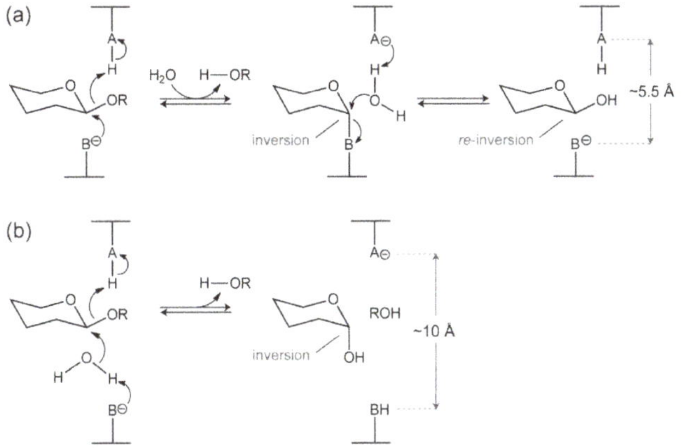

Figure 5. The two common types of catalysis by glycoside hydrolases as adapted from Davies and Henrissat [29]. (**a**) The retaining mechanism. The nucleophile and the general acid/base are represented as B^- and AH, respectively. (**b**) The inverting mechanism. The general base and the general acid are represented as B^- and AH, respectively. The typical distances of the catalytic residues in both mechanisms are indicated in Å. In most GHs, A and B are either Asp or Glu. See main text for further details.

Except for the enzymes from family GH43, all β-xylosidases in the CAZy database are predicted to have a retaining mechanism (Table 1). Among these retaining β-xylosidases, structural data with bound ligand are available for enzymes from GH3 [58], GH39 [40–42], GH52 [73], and GH120 [25]. A structural alignment of these β-xylosidases on the basis of their bound ligand revealed that the carboxylate group of their catalytic nucleophiles, which are Glu in GH39 and GH52, and Asp in GH3 and GH120, are spatially conserved relative to the bound ligand (Figure 6a). They are within good distance (~3.1 Å) and right position for reaction with the anomeric carbon of the scissile glycosidic bond. On the other hand, the carboxylate group of their catalytic acid/base, which is Glu in GH3, GH39, and GH120, and Asp in GH52, are spatially less conserved, although they are at productive hydrogen-bonding positions (~3.2 Å on average) to the corresponding glycosidic oxygen atom.

β-Xylosidases from GH43 are inverting enzymes [79]. They use Asp and Glu as the general base and general acid, respectively [70,80,81]. Similar to the catalytic acid/base of the retaining β-xylosidases, their catalytic acid is within hydrogen-bonding distance (~2.7 Å) to the glycosidic oxygen atom of the scissile bond (Figure 6b). However, compared to the catalytic nucleophile of the retaining β-xylosidases, their catalytic base is located further away from the anomeric carbon atom of the scissile glycosidic bond with a distance of ~5.2 Å [65,66,70]. This distance provides sufficient space for accommodating a catalytic water molecule that can be activated by the catalytic base to attack the anomeric carbon [66]. It has been observed generally for GHs that the distance between the carboxylate groups of the catalytic base and acid of retaining enzymes is shorter (~5 Å) than the distance between the carboxylates of the catalytic nucleophile and acid/base of inverting enzymes (~8–10 Å) [29,82]. This is also the case for the GH43 β-xylosidases. Indeed, in the inverting GH43 β-xylosidase from *G. stearothermophilus* T-6, for example, a distance of ~7.9 Å between the carboxylate groups of its catalytic residues has been observed [66].

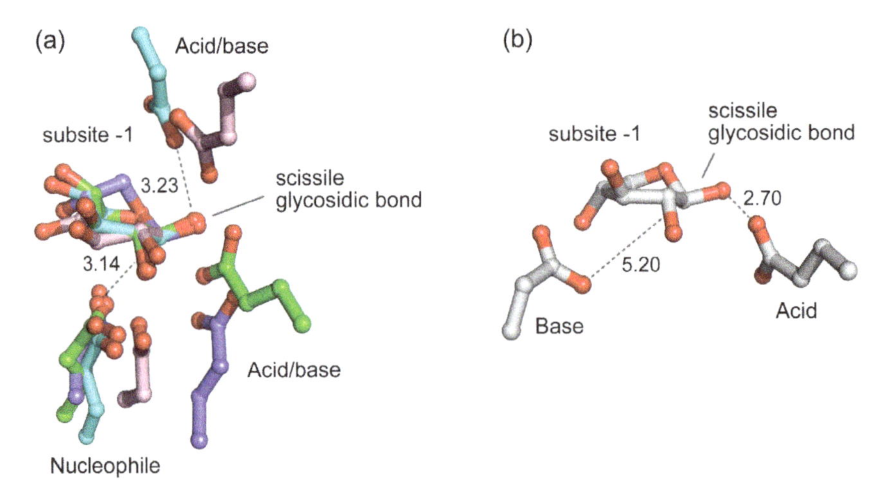

Figure 6. Positions of the catalytic residues relative to the xylosyl moiety bound in subsite -1 of the active sites of (**a**) retaining and (**b**) inverting β-xylosidases. The structures are of GlyA1 (PDB 5K6N; GH3; carbon atoms in pink), GsXynB1 (PDB 2BFG; GH39; green), GT2_24_00240 (PDB 4C1P; GH52; cyan), and TsXylC (PDB 3VSU; GH120; blue), which are retaining β-xylosidases, and GsXynB3 (PDB 2EXJ; GH43; white), which is an inverting β-xylosidase (see caption of Figure 3 for further details of the enzymes). Important distances (in Å) are shown next to dashed lines.

5. Inhibition of β-Xylosidases by Monosaccharides

5.1. Inhibition by D-xylose

Many β-xylosidases are inhibited to varying degrees by their main product D-xylose (Table 2). For example, the β-xylosidase from the fungus *Trichoderma harzianum* C is very sensitive to D-xylose inhibition; its activity is completely inhibited by the presence of only 2 mM D-xylose [83]. In contrast, the GH39 β-xylosidase from the extreme thermophilic bacterium *Dictyoglomus thermophilum* DSM 3960 is very resistant to D-xylose, with only 40% inhibition in the presence of 3 M of the sugar [84].

Table 2. Examples of microbial β-xylosidase inhibition by D-xylose.

Organism	GH Family	D-xylose Concentration (mM)	Inhibition (%)	Reference
Bacteria:				
Bacillus halodurans C-125	GH39	200	0	[85]
Bacillus subtilis M015	GH43_11	20	45	[86]
Corynebacterium alkanolyticum ATCC 21511	GH3	200	70	[87]
Dictyoglomus thermophilum DSM 3960	GH39	3000	40	[84]
Geobacillus sp. WSUCF1	GH39	300	50	[88]
Geobacillus thermodenitrificans NG80-2	GH39	400	50	[89]
Geobacillus thermodenitrificans NG80-2	GH43	300	50	[89]
Geobacillus thermodenitrificans NG80-2	GH52	600	50	[89]
Lactobacillus brevis ATCC 14869	GH43_11	100	20	[90]
Lactobacillus brevis ATCC 14869	GH43_12	100	66	[90]
Massilia sp. RBM26	GH43_11	500	50	[91]
Paenibacillus woosongensis KCTC 3953	GH43_35	100	25	[92]
Selenomonas ruminantium GA192	GH43_11	40	57	[93]
Sphingobacterium sp. HP455	GH43_1	247	50	[94]
Thermoanaerobacterium saccharolyticum JW/SL-YS485	GH120	200	30	[75]

Table 2. *Cont.*

Organism	GH Family	D-xylose Concentration (mM)	Inhibition (%)	Reference
Thermotoga petrophila DSM 13995	GH3	150	50	[95]
Thermotoga thermarum DSM 5069	GH3	1000	50	[96]
Fungi:				
Aspergillus nidulans CECT2544	n.a. #	25	44	[97]
Aspergillus niger 11	n.a. #	10	50	[98]
Aspergillus niger ADH-11	GH3	12	50	[99]
Aureobasidium pullulans CBS 58475	n.a. #	6,6	42	[100]
Candida utilis IFO 0639	n.a. #	300	0	[101]
Humicola grisea var. thermoidea	GH43_1	603	50	[102]
Humicola insolens Y1	GH43_1	79	50	[103]
Humicola insolens Y1	GH43_11	292	50	[103]
Paecilomyces thermophila J18	n.a. #	139	50	[104]
Phanerochaete chrysosporium BKM-F-1767	GH43_14	50	70	[43]
Pseudozyma hubeiensis NCIM 3574	n.a. #	75	50	[105]
Rhizophlyctis rosea Fischer NBRC 105426	GH43_1	100	49	[106]
Scytalidium thermophilum 77.7.8	n.a. #	200	0	[107]
Trichoderma harzianum C	n.a. #	2	100	[83]
Trichoderma reesei QM 9414	GH3	53	80	[108]
Metagenomes:				
Synthetic metagenome	GH43_1	20	44	[109]
Uncultured rumen metagenome	GH3	5	27	[110]
Yak rumen metagenome (RuBg3A §)	GH3	5	18	[111]
Yak rumen metagenome (RuBg3B §)	GH3	5	3	[111]

#: GH family is not assigned in the CAZy database; §: Protein symbol.

Most characterized β-xylosidases have significant affinity for D-xylose, with reported inhibition constants (K_i) of less than 10 mM (Table 3). To our knowledge, the β-xylosidase from *Talaromyces emersonii* has the highest affinity for D-xylose with a K_i value as low as 1.3 mM [112], suggesting that the enzyme is very sensitive to inhibition by this monosaccharide. Other β-xylosidases are much less prone to D-xylose inhibition, such as those from *B. pumilus* 12 [113], *G. thermoleovorans* IT-08 [114], and uncultured bacterium [115] with K_i's of 26.2, 76, and 145 mM, respectively. So far, the β-xylosidase from the bacterium *Cellulomonas uda* [116] is the β-xylosidase with the highest reported K_i value for D-xylose, i.e., 650 mM, with the caveat that this K_i was determined using crude enzyme. Thus, in general, and as summarized in Table 3, β-xylosidases have relatively high affinity (low K_i) for D-xylose and, therefore, they are susceptible to product inhibition by this sugar.

Table 3. Examples of inhibition constants for D-xylose of β-xylosidases.

Organism	GH Family	Inhibition Constant (K_i, mM)	Reference
Bacteria:			
Alkaliphilus metalliredigens QYMF	GH43_11	16.2	[18]
Anoxybacillus sp. 3M	GH52	21.3	[117]
Bacillus halodurans C-125	GH43_11	62.3	[118]
Bacillus pumilus 12	n.a. #	26.2	[113]
Bacillus pumilus IPO	GH43_11	70	[18]
Bacillus subtilis subsp. subtilis str. 168	GH43_11	15.6	[18]
Bacteroides ovatus V975	GH43_1	6.6	[119]
Caldocellum saccharolyticum Tp8T6.3.3.1	n.a. #	40.0	[120]
Cellulomonas uda	n.a. #	650.0	[116]
Enterobacter sp.	GH43_11	79.9	[72]
Geobacillus thermoleovorans IT-08	GH43_12	76.0	[114]
Lactobacillus brevis ATCC 367	GH43_11	30.1	[18]

Table 3. *Cont.*

Organism	GH Family	Inhibition Constant (K_i, mM)	Reference
Selenomonas ruminantium GA192	GH43_11	6.24	[121]
Streptomyces sp. CH7	GH3	40.0	[122]
Thermoanaerobacterium saccharolyticum B6A-RI	GH39	20	[123]
Thermobifida fusca TM51	GH43_11	67.0	[124]
Thermobifida halotolerans YIM 90462^T	GH43_11	43.8	[125]
Thermomonospora	n.a. #	35-100	[126]
Thermomonospora fusca BD21	n.a. #	19	[127]
Fungi:			
Arxula adeninivorans SBUG 724	n.a. #	5.8	[128]
Aspergillus awamori X-100	GH3	7.7	[129]
Aspergillus carbonarius KLU-93	n.a. #	1.9	[130]
Aspergillus fumigatus	n.a. #	4.5	[131]
Aspergillus japonicus	GH3	2.9	[132]
Aspergillus niger 15	n.a. #	2.9	[133]
Aspergillus niger 90196	GH3	8.3	[134]
Aspergillus niger ATCC 10864	GH3	3.3	[135]
Aspergillus niger NW147 (xlnD I §)	GH3	9.8	[136]
Aspergillus niger NW147 (xlnD II §)	GH3	13.2	[136]
Aspergillus niger van Tieghem (DSM 22593)	GH3	7.5	[137]
Aspergillus oryzae KBN616	GH3	2.7	[138]
Aspergillus terreus IJIRA 6.2	n.a. #	10.5	[139]
Aspergillus versicolor (xylose-induced)	n.a. #	5.3	[140]
Aspergillus versicolor (xylan-induced)	n.a. #	2.0	[140]
Aureobasidium pullulans CBS 135684	n.a. #	18.0	[141]
Colletotrichum graminicola	GH3	3.3	[142]
Fusarium proliferatum NRRL 26517	n.a. #	5.0	[143]
Fusarium verticillioides NRRL 26518	n.a. #	6.0	[144]
Humicola insolens Y1	GH3	29.0	[145]
Neocallimastix frontalis RK 21	n.a. #	4.0	[146]
Neurospora crassa ST A (74 A)	GH3	1.7	[147]
Penicillium janczewskii CRM 1348	n.a. #	6	[148]
Penicillium oxalicum 114-2	GH43	28.1	[149]
Penicillium sclerotiorum	n.a. #	28.7	[150]
Talaromyces amestolkiae	GH3	1.7	[151]
Talaromyces emersonii	GH3	1.3	[112]
Thermomyces lanuginosus CAU44	GH43_1	63.0	[152]
Trichoderma koningii G-39	n.a. #	5.0	[153]
Trichoderma reesei (βXTR §)	GH3	2.4	[112]
Trichoderma reesei	GH3	1.4	[132]
Trichoderma reesei QM 9414	n.a. #	11.0	[154]
Trichoderma reesei RUT C30	n.a. #	2.3	[155]
Trichoderma reesei RUT C30	n.a. #	2.4	[24]
Plant:			
Saccharum officinarum L.	n.a. #	8.0	[156]
Metagenomes:			
Compost starter	GH43	145.0	[115]
Mixed microorganism (RS223-BX §)	GH43_1	3.4	[19]
Uncultured rumen bacterium	GH30_2	10.6	[157]
Uncultured rumen bacterium	GH43_1	76.0	[157]

#: GH family is not assigned in the CAZy database; §: Protein symbol.

From Tables 2 and 3 it appears that no direct relationship exists between the inhibition of β-xylosidases by D-xylose and their organismal origin or GH family. Many β-xylosidases from different bacteria and fungi suffer from such product inhibition. Likewise, product inhibition is observed for β-xylosidases belonging to different GH families. This is reasonable because all β-xylosidases bind the same substrate (D-xylose oligomers), necessitating affinity for xylosyl residues and some commonality in their active site structure (see above).

Interestingly, the activity of several β-xylosidases is stimulated by D-xylose, particularly at low concentration. The β-xylosidase from *Thermotoga thermarum* DSM 5069 was stimulated by D-xylose concentrations of up to 500 mM; its maximum activity was observed in the presence of 200 mM D-xylose, which was ~20% higher than in the absence of the sugar [96]. A similar stimulatory effect has been reported for the bifunctional β-xylosidase/α-L-arabinofuranosidase from *Phanerochaete chrysosporium* BKM-F-1767 [43] and the β-xylosidase from *Dictyoglomus thermophilum* DSM 3960 [84,158]. However, the mechanism of such stimulation is currently not known.

5.2. Inhibition by L-arabinose

L-arabinose is one of the monosaccharides produced during enzymatic saccharification of cellulosic biomass [7,8,28]. It is liberated by the action of α-L-arabinofuranosidases, and, during the saccharification process, its concentration may sufficiently increase to inhibit the activity of hemicellulolytic enzymes [115]. Indeed, L-arabinose has been identified as an inhibitor of various β-xylosidases (Table 4). For instance, 50 mM L-arabinose reduces the activity of the aryl β-xylosidase from *Caldocellum saccharolyticum* Tp8T6.3.3.1 by 15% [120], the β-xylosidase from *B. pumilus* 12 by 21% [113], and the bifunctional β-xylosidase/α-L-arabinofuranosidase from *P. chrysosporium* BKM-F-1767 by ~70% [43]. This is not surprising, since the stereochemistry of L-arabinose near the glycosidic bond is similar to that of D-xylose [61], explaining its binding in the active site of a β-xylosidase.

Table 4. Examples of microbial β-xylosidase inhibition by L-arabinose.

Organism	GH Family	L-arabinose Concentration (mM)	Inhibition (%)	Reference
Bacteria:				
Bacillus pumilus 12	n.a. #	50	21	[113]
Caldocellum saccharolyticum Tp8T6.3.3.1	n.a. #	50	15	[120]
Corynebacterium alkanolyticum ATCC 21511	GH3	200	40	[87]
Lactobacillus brevis ATCC 14869	GH43_11	100	39	[90]
Lactobacillus brevis ATCC 14869	GH43_12	100	38	[90]
Paenibacillus woosongensis KCTC 3953	GH43_35	100	40	[92]
Selenomonas ruminantium GA192	GH43_11	80	61	[93]
Fungi:				
Aspergillus niger 11	n.a. #	25	10	[98]
Aspergillus niger van Tieghem (DSM 22593)	GH3	200	30	[137]
Colletotrichum graminicola	GH3	50	15	[142]
Penicillium oxalicum 114-2	GH43	20	11	[149]
Phanerochaete chrysosporium BKM-F-1767	GH43_14	50	70	[43]

#: GH family is not assigned in the CAZy database.

In general, L-arabinose is a weaker inhibitor of β-xylosidase activity than D-xylose. For example, the β-xylosidase from the fungus *Trichoderma reesei* RUT C30 is strongly inhibited by D-xylose with a K_i of 2.4 mM, but it is not inhibited by L-arabinose, even at a concentration of 500 mM [24]. Several other β-xylosidases can also withstand high concentrations of L-arabinose [95,122,141]. Finally, the β-xylosidase from *Enterobacter* sp. is competitively inhibited by L-arabinose, but with a quite high K_i value of 102 mM [72], indicating that the enzyme has only low affinity for L-arabinose.

Intriguingly, at low concentration (~5 mM) L-arabinose stimulates rather than inhibits the β-xylosidase activity of the bifunctional β-xylosidase/α-L-arabinofuranosidase from *P. chrysosporium* BKM-F-1767, as also noticed for D-glucose [43]. Activation by L-arabinose has also been observed for a furan aldehyde-tolerant β-xylosidase/α-L-arabinofuranosidase procured from a metagenomic sample, which showed 65% higher β-xylosidase activity compared with the control without L-arabinose [109].

Although there are many data describing the effects of L-arabinose on inhibition/activation of β-xylosidase activity, the molecular basis of the effects on activity still needs further investigation.

5.3. Inhibition by Other Monosaccharides

Apart from D-xylose and L-arabinose, D-glucose is another monosaccharide that has been frequently reported to affect β-xylosidase activity. D-glucose inhibits the β-xylosidase activity of the β-xylosidases from *S. ruminantium* GA192 (K_i 44 mM) [27], *B. pumilus* 12 (9% inhibition at 50 mM) [113], *T. harzianum* (3% inhibition at 5 mM) [83], and the bifunctional β-glucosidase/β-xylosidases RuBG3A and RuBG3B from the metagenome of yak rumen microorganisms (97.5% and 45.6% inhibition, respectively, at 5 mM) [111]. On the other hand, the sugar did not inhibit β-xylosidases from *Thermomonospora fusca* [159], *A. niger* 90196 [134], *A. oryzae* [138], and *N. crassa* ST A [147] at concentrations of up to 90, 20, 20, and 10 mM, respectively. The Mg^{2+}-activated β-xylosidase RS223-BX could even withstand much higher D-glucose concentrations displaying a *Ki* value of 1270 mM on the substrate *p*-nitrophenyl-α-L-arabinofuranoside (*p*NPA) [19]. Apparently, inhibition by D-glucose varies considerably among β-xylosidases.

Finally, besides D-xylose, L-arabinose, and D-glucose, also other monosaccharides have been reported to inhibit β-xylosidases, including D-arabinose [26,113], D-erythrose [26], D-fructose [111,138], D-galactose [26,113,120], D-ribose, and L-xylose [26,113]. Again, the molecular details of the interactions of these sugars with the enzymes are not known.

5.4. Inhibition Kinetics

The inhibition of β-xylosidases by monosaccharides follows competitive, non-competitive, or un-competitive inhibition kinetics. For most β-xylosidases, D-xylose acts as a competitive inhibitor when using *p*-nitrophenyl-β-D-xyloside (*p*NPX) as substrate [113,114,120,134]. However, the β-xylosidase from *N. crassa* ST A showed non-competitive inhibition by this sugar [147]. Furthermore, D-xylose inhibition of the β-xylosidase from *S. ruminantium* GA192, which also displays α-L-arabinofuranosidase activity, followed non-competitive kinetics for its β-xylosidase activity on *p*NPX as substrate, but competitive kinetics for its α-L-arabinofuranosidase activity on *p*NPA as substrate [26]. This differs slightly from *A. carbonarius* KLU-93 β-xylosidase, for which D-xylose was a competitive inhibitor for the conversion of both substrates [130]. Similarly, L-arabinose acts as a competitive inhibitor for the hydrolysis of *p*NPA by the RS223-BX β-xylosidase [19], but it is a non-competitive inhibitor of *S. ruminantium* GA192 β-xylosidase hydrolyzing *p*NPX or *p*NPA [26]. With respect to these substrates, D-arabinose, D-glucose, and D-ribose are competitive inhibitors of *S. ruminantium* GA192 β-xylosidase, whereas D-erythrose and L-xylose are non-competitive [26]. As also observed for *S. ruminantium* GA192 β-xylosidase, D-glucose inhibition of RS223-BX was competitive when using *p*NPA as substrate [19]. Uniquely, un-competitive inhibition was displayed by D-fructose for the activity of *A. oryzae* β-xylosidase on *p*NPX [138]. Thus, commonly, competitive inhibition by D-xylose is observed. Other monosaccharides can display both competitive and non-competitive inhibition, and in one case, D-fructose, un-competitive inhibition takes place. The exact mechanism and the structural details of non-competitive and un-competitive inhibition remain unknown.

5.5. Structural Details of Inhibitor Binding in the Active Site of β-Xylosidases

As discussed above, the active site pockets of β-xylosidases contain two substrate-binding subsites, subsites −1 and +1, on either side of the scissile bond. In the active site of *S. ruminantium* GA192 β-xylosidase, the monosaccharides D-arabinose, L-arabinose, D-erythrose, and D-ribose can bind in both subsites −1 and +1, but D-xylose and L-xylose bind only in subsite −1 [26]. Similarly, D-glucose binds in one subsite only. Its binding position was speculated to be in subsite −1, with partial occupancy of subsite +1, because glucose is too large to fit in subsite −1 only. Alternatively, it could bind in such a way that it excludes binding of a second sugar [26].

In contrast, *G. thermoleovorans* IT-08 β-xylosidase shows rather dissimilar binding properties for L-arabinose and D-xylose compared to *S. ruminantium* GA192 β-xylosidase. Crystal structures of *G. thermoleovorans* IT-08 β-xylosidase revealed that L-arabinose binds exclusively in subsite −1, while D-xylose prefers subsite +1 [70]. Thus, depending on the enzyme, the −1 and +1 subsites differ in preference for different monosaccharides, which could also contribute to the differences in inhibition kinetics observed for the different enzymes.

5.6. Engineering to Reduce β-Xylosidase Inhibition by Monosaccharides

During saccharification of cellulosic biomass, monosaccharides such as D-xylose, L-arabinose, and D-glucose, may reach concentrations that are high enough to inhibit β-xylosidase activity [26,27]. Therefore, β-xylosidases that are not affected by high monosaccharide concentrations are highly desirable for the efficiency of the saccharification process. To develop such β-xylosidase variants, the W145G mutation was introduced into *S. ruminantium* GA192 β-xylosidase, resulting in a variant with a 3-fold lower affinity for D-xylose and a 2-fold lower affinity for D-glucose [121]. Subjecting this variant to saturation mutagenesis of residue 145 yielded variants with even lower affinity for monosaccharides and higher catalytic activity than wild-type enzyme [27]. Mutation of Trp-145 alters the affinity of subsite +1 for D-xylose, but not that of subsite -1, where catalysis occurs, suggesting a strategy for reducing inhibition by monosaccharides by mutating residues of subsite +1 [27]. In the structure of *G. thermoleovorans* IT-08 β-xylosidase, D-xylose binds in subsite +1 interacting, among others, with Asp-198, which is not present in most other β-xylosidases [70]. Therefore, this residue may be a good target for mutation to obtain *G. thermoleovorans* IT-08 β-xylosidase variants with lower affinity for D-xylose.

6. Concluding Remarks

β-Xylosidases are highly diverse in their amino acid sequence. Currently, enzymes with β-xylosidase activity can be found in 11 different glycoside hydrolase families in the CAZy database, i.e., in GH families 1, 3, 5, 30, 39, 43, 51, 52, 54, 116, and 120. They fold into several distinct three-dimensional structures. While the enzymes from GH families 1, 3, 5, 30, 39, and 51 all show a $(\beta/\alpha)_8$ TIM-barrel structure, those from families 43, 52 and 116, 54, and 120 have as their main structural feature a 5-bladed β-propeller, a $(\alpha/\alpha)_6$-barrel, a β-sandwich, and a right-handed parallel β-helix, respectively. Likewise, the catalytic mechanism of β-xylosidases is also varied. Although generally β-xylosidases hydrolyze their substrates with retention of the substrate's anomeric carbon configuration, the enzymes from GH43 invert the anomeric configuration. However, despite their diversity in overall folds, all structurally characterized β-xylosidases have a typical pocket-shaped active site with two carbohydrate-binding subsites, which bind a xylobiosyl moiety of the xylooligosaccharide substrates. Unfortunately, the active sites of many β-xylosidases also possess a relatively high affinity for monosaccharides, such as xylose, arabinose, and erythrose, that competitively inhibit the enzymes' activity. Moreover, some β-xylosidases are also non-competitively or un-competitively inhibited by monosaccharides. Such product inhibition limits the application of β-xylosidases in xylan saccharification, since high monosaccharide concentrations may easily be generated during the process. Therefore, β-xylosidases with low monosaccharide affinity are highly desirable for various applications. A random mutagenesis approach has already shown success in reducing the affinity of a β-xylosidase for D-xylose. Furthermore, with 3D structures available for a variety of β-xylosidases, rational site-directed mutagenesis may also be a good approach to render the enzymes less prone to product inhibition.

Author Contributions: Conceptualization, A.R., B.W.D., and N.N.T.P.; software, A.R.; validation, A.R., B.W.D., and N.N.T.P.; data curation, A.R.; writing—original draft preparation, A.R.; writing—review and editing, B.W.D., and N.N.T.P.; visualization, A.R.; supervision, B.W.D., and N.N.T.P.; funding acquisition, A.R. and N.N.T.P.

Funding: This work was funded by Universitas Airlangga to A.R. and N.N.T.P. (Hibah Riset Mandat No. 624/UN3.14/LT/2017). N.N.T.P is a receiver of a Tahir Professorship of Universitas Airlangga (SK Rektor Unair No. 1149/UN3/2018).

Conflicts of Interest: The authors declare no conflict of interest. The funders had no role in the design of the study; in the collection, analyses, or interpretation of data; in the writing of the manuscript, or in the decision to publish the results.

References

1. Gray, K.A.; Zhao, L.; Emptage, M. Bioethanol. *Curr. Opin. Chem. Biol.* **2006**, *10*, 141–146. [CrossRef] [PubMed]
2. Saha, B.C. Hemicellulose bioconversion. *J. Ind. Microbiol. Biotechnol* **2003**, *30*, 279–291. [CrossRef] [PubMed]
3. Scheller, H.V.; Ulvskov, P. Hemicelluloses. *Annu. Rev. Plant. Biol.* **2010**, *61*, 263–289. [CrossRef] [PubMed]
4. Deutschmann, R.; Dekker, R.F.H. From plant biomass to bio-based chemicals: Latest developments in xylan research. *Biotechnol. Adv.* **2012**, *30*, 1627–1640. [CrossRef] [PubMed]
5. Hahn-Hagerdal, B.; Galbe, M.; Gorwa-Grauslund, M.F.; Lidén, G.; Zacchi, G. Bio-ethanol—The fuel of tomorrow from the residues of today. *Trends Biotechnol.* **2006**, *24*, 549–556. [CrossRef]
6. Menon, V.; Rao, M. Trends in bioconversion of lignocellulose: Biofuels, platform chemicals & biorefinery concept. *Prog. Energy Combust. Sci.* **2012**, *38*, 522–550. [CrossRef]
7. Beg, Q.K.; Kapoor, M.; Mahajan, L.; Hoondal, G.S. Microbial xylanases and their industrial applications: A review. *Appl. Microbiol. Biotechnol.* **2001**, *56*, 326–338. [CrossRef]
8. Biely, P. Microbial xylanolytic systems. *Trends Biotechnol.* **1985**, *3*, 286–290. [CrossRef]
9. Collins, T.; Gerday, C.; Feller, G. Xylanases, xylanase families and extremophilic xylanases. *FEMS Microbiol. Rev.* **2005**, *29*, 3–23. [CrossRef]
10. Polizeli, M.L.T.M.; Rizzatti, A.C.S.; Monti, R.; Terenzi, H.F.; Jorge, J.A.; Amorim, D.S. Xylanases from fungi: Properties and industrial applications. *Appl. Microbiol. Biotechnol.* **2005**, *67*, 577–591. [CrossRef]
11. Sunna, A.; Antranikian, G. Xylanolytic enzymes from fungi and bacteria. *Crit. Rev. Biotechnol.* **1997**, *17*, 39–67. [CrossRef] [PubMed]
12. Chávez, R.; Bull, P.; Eyzaguirre, J. The xylanolytic enzyme system from the genus *Penicillium*. *J. Biotechnol.* **2006**, *123*, 413–433. [CrossRef] [PubMed]
13. Kulkarni, N.; Shendye, A.; Rao, M. Molecular and biotechnological aspects of xylanases. *FEMS Microbiol. Rev.* **1999**, *23*, 411–456. [CrossRef] [PubMed]
14. Prade, R.A. Xylanases: From biology to biotechnology. *Biotechnol. Genet. Eng. Rev.* **1996**, *13*, 101–132. [CrossRef]
15. Subramaniyan, S.; Prema, P. Biotechnology of microbial xylanases: Enzymology, molecular biology, and application. *Crit. Rev. Biotechnol.* **2002**, *22*, 33–64. [CrossRef]
16. Viikari, L.; Kantelinen, A.; Sundquist, J.; Linko, M. Xylanases in bleaching: From an idea to the industry. *FEMS Microbiol. Rev.* **1994**, *13*, 335–350. [CrossRef]
17. Jordan, D.B.; Wagschal, K. Properties and applications of microbial β-D-xylosidases featuring the catalytically efficient enzyme from *Selenomonas ruminantium*. *Appl. Microbiol. Biotechnol.* **2010**, *86*, 1647–1658. [CrossRef]
18. Jordan, D.B.; Wagschal, K.; Grigorescu, A.A.; Braker, J.D. Highly active β-xylosidases of glycoside hydrolase family 43 operating on natural and artificial substrates. *Appl. Microbiol. Biotechnol.* **2013**, *97*, 4415–4428. [CrossRef]
19. Lee, C.C.; Braker, J.D.; Grigorescu, A.A.; Wagschal, K.; Jordan, D.B. Divalent metal activation of a GH43 β-xylosidase. *Enzym. Microb. Technol.* **2013**, *52*, 84–90. [CrossRef]
20. Qing, Q.; Wyman, C.E. Supplementation with xylanase and β-xylosidase to reduce xylo-oligomer and xylan inhibition of enzymatic hydrolysis of cellulose and pretreated corn stover. *Biotechnol. Biofuels* **2011**, *4*, 18. [CrossRef]
21. Qing, Q.; Yang, B.; Wyman, C.E. Xylooligomers are strong inhibitors of cellulose hydrolysis by enzymes. *Bioresour. Technol.* **2010**, *101*, 9624–9630. [CrossRef] [PubMed]
22. Royer, J.C.; Nakas, J.P. Purification and characterization of two xylanases from *Trichoderma longibrachiatum*. *Eur. J. Biochem.* **1991**, *202*, 521–529. [CrossRef] [PubMed]
23. Williams, S.J.; Hoos, R.; Withers, S.G. Nanomolar versus millimolar inhibition by xylobiose-derived azasugars: Significant differences between two structurally distinct xylanases. *J. Am. Chem. Soc.* **2000**, *122*, 2223–2235. [CrossRef]

24. Herrmann, M.C.; Vrsanska, M.; Jurickova, M.; Hirsch, J.; Biely, P.; Kubicek, C.P. The β-D-xylosidase of *Trichoderma reesei* is a multifunctional β-D-xylan xylohydrolase. *Biochem. J.* **1997**, *321*, 375–381. [CrossRef] [PubMed]
25. Huang, C.H.; Sun, Y.; Ko, T.P.; Chen, C.C.; Zheng, Y.; Chan, H.C.; Pang, X.; Wiegel, J.; Shao, W.; Guo, R.T. The substrate/product-binding modes of a novel GH120 β-xylosidase (XylC) from *Thermoanaerobacterium saccharolyticum* JW/SL-YS485. *Biochem. J.* **2012**, *448*, 401–407. [CrossRef]
26. Jordan, D.B.; Braker, J.D. Inhibition of the two-subsite β-D-xylosidase from *Selenomonas ruminantium* by sugars: Competitive, noncompetitive, double binding, and slow binding modes. *Arch. Biochem. Biophys.* **2007**, *465*, 231–246. [CrossRef]
27. Jordan, D.B.; Wagschal, K.; Fan, Z.; Yuan, L.; Braker, J.D.; Heng, C. Engineering lower inhibitor affinities in β-D-xylosidase of *Selenomonas ruminantium* by site-directed mutagenesis of Trp145. *J. Ind. Microbiol. Biotechnol.* **2011**, *38*, 1821–1835. [CrossRef]
28. Lagaert, S.; Pollet, A.; Courtin, C.M.; Volckaert, G. β-Xylosidases and α-L-arabinofuranosidases: Accessory enzymes for arabinoxylan degradation. *Biotechnol. Adv.* **2014**, *32*, 316–332. [CrossRef]
29. Davies, G.; Henrissat, B. Structures and mechanisms of glycosyl hydrolases. *Structure* **1995**, *3*, 853–859. [CrossRef]
30. Henrissat, B. A classification of glycosyl hydrolases based on amino acid sequence similarities. *Biochem. J.* **1991**, *280*, 309–316. [CrossRef]
31. Lombard, V.; Golaconda Ramulu, H.; Drula, E.; Coutinho, P.M.; Henrissat, B. The carbohydrate-active enzymes database (CAZy) in 2013. *Nucleic Acids Res.* **2014**, *42*, D490–D495. [CrossRef] [PubMed]
32. Claeyssens, M.; Henrissat, B. Specificity mapping of cellulolytic enzymes: Classification into families of structurally related proteins confirmed by biochemical analysis. *Protein Sci.* **1992**, *1*, 1293–1297. [CrossRef] [PubMed]
33. Henrissat, B.; Bairoch, A. Updating the sequence-based classification of glycosyl hydrolases. *Biochem. J.* **1996**, *316*, 695–696. [CrossRef] [PubMed]
34. Scharf, M.E.; Kovaleva, E.S.; Jadhao, S.; Campbell, J.H.; Buchman, G.W.; Boucias, D.G. Functional and translational analyses of a β-glucosidase gene (glycosyl hydrolase family 1) isolated from the gut of the lower termite *Reticulitermes flavipes*. *Insect Biochem. Mol. Biol.* **2010**, *40*, 611–620. [CrossRef] [PubMed]
35. Mattéotti, C.; Haubruge, E.; Thonart, P.; Francis, F.; De Pauw, E.; Portetelle, D.; Vandenbol, M. Characterization of a new β-glucosidase/β-xylosidase from the gut microbiota of the termite (*Reticulitermes santonensis*). *FEMS Microbiol. Lett.* **2011**, *314*, 147–157. [CrossRef]
36. Wan, C.-F.; Chen, C.-T.; Huang, L.; Li, Y.-K. Expression, purification and characterization of a bifunctional α-L-arabinofuranosidase/β-D-xylosidase from *Trichoderma koningii* G-39. *J. Chin. Chem. Soc.* **2007**, *54*, 109–116. [CrossRef]
37. Cobucci-Ponzano, B.; Aurilia, V.; Riccio, G.; Henrissat, B.; Coutinho, P.M.; Strazzulli, A.; Padula, A.; Corsaro, M.M.; Pieretti, G.; Pocsfalvi, G.; et al. A new archaeal β-glycosidase from *Sulfolobus solfataricus*: Seeding a novel retaining β-glycan-specific glycoside hydrolase family along with the human non-lysosomal glucosylceramidase GBA2. *J. Biol. Chem.* **2010**, *285*, 20691–20703. [CrossRef]
38. Miyanaga, A.; Koseki, T.; Matsuzawa, H.; Wakagi, T.; Shoun, H.; Fushinobu, S. Crystal structure of a family 54 α-L-arabinofuranosidase reveals a novel carbohydrate-binding module that can bind arabinose. *J. Biol. Chem.* **2004**, *279*, 44907–44914. [CrossRef]
39. Naumoff, D.G. Hierarchical classification of glycoside hydrolases. *Biochemistry* **2011**, *76*, 764–780. [CrossRef]
40. Yang, J.K.; Yoon, H.J.; Ahn, H.J.; Lee, B.I.; Pedelacq, J.-D.; Liong, E.C.; Berendzen, J.; Laivenieks, M.; Vieille, C.; Zeikus, G.J.; et al. Crystal structure of β-D-xylosidase from *Thermoanaerobacterium saccharolyticum*, a family 39 glycoside hydrolase. *J. Mol. Biol.* **2004**, *335*, 155–165. [CrossRef]
41. Czjzek, M.; Ben-David, A.; Bravman, T.; Shoham, G.; Henrissat, B.; Shoham, Y. Enzyme-substrate complex structures of a GH39 β-xylosidase from *Geobacillus stearothermophilus*. *J. Mol. Biol.* **2005**, *353*, 838–846. [CrossRef] [PubMed]
42. Santos, C.R.; Polo, C.C.; Corrêa, J.M.; Simão, R.C.G.; Seixas, F.A.V.; Murakami, M.T. The accessory domain changes the accessibility and molecular topography of the catalytic interface in monomeric GH39 β-xylosidases. *Acta Crystallogr. D Biol. Crystallogr.* **2012**, *68*, 1339–1345. [CrossRef] [PubMed]

43. Huy, N.D.; Thayumanavan, P.; Kwon, T.-H.; Park, S.-M. Characterization of a recombinant bifunctional xylosidase/arabinofuranosidase from *Phanerochaete chrysosporium*. *J. Biosci. Bioeng.* **2013**, *116*, 152–159. [CrossRef] [PubMed]
44. Brunner, F.; Wirtz, W.; Rose, J.K.C.; Darvill, A.G.; Govers, F.; Scheel, D.; Nürnberger, T. A β-glucosidase/ xylosidase from the phytopathogenic oomycete, *Phytophthora infestans*. *Phytochemistry* **2002**, *59*, 689–696. [CrossRef]
45. Minic, Z.; Rihouey, C.; Do, C.T.; Lerouge, P.; Jouanin, L. Purification and characterization of enzymes exhibiting β-D-xylosidase activities in stem tissues of Arabidopsis. *Plant. Physiol.* **2004**, *135*, 867–878. [CrossRef]
46. Biasini, M.; Bienert, S.; Waterhouse, A.; Arnold, K.; Studer, G.; Schmidt, T.; Kiefer, F.; Cassarino, T.G.; Bertoni, M.; Bordoli, L.; et al. SWISS-MODEL: Modelling protein tertiary and quaternary structure using evolutionary information. *Nucleic Acids Res.* **2014**, *42*, W252–W258. [CrossRef]
47. Suzuki, K.; Sumitani, J.-i.; Nam, Y.-W.; Nishimaki, T.; Tani, S.; Wakagi, T.; Kawaguchi, T.; Fushinobu, S. Crystal structures of glycoside hydrolase family 3 β-glucosidase 1 from *Aspergillus aculeatus*. *Biochem. J.* **2013**, *452*, 211–221. [CrossRef]
48. Harvey, A.J.; Hrmova, M.; De Gori, R.; Varghese, J.N.; Fincher, G.B. Comparative modeling of the three-dimensional structures of family 3 glycoside hydrolases. *Proteins* **2000**, *41*, 257–269. [CrossRef]
49. Pozzo, T.; Pasten, J.L.; Karlsson, E.N.; Logan, D.T. Structural and functional analyses of β-glucosidase 3B from *Thermotoga neapolitana*: A thermostable three-domain representative of glycoside hydrolase 3. *J. Mol. Biol.* **2010**, *397*, 724–739. [CrossRef]
50. Karkehabadi, S.; Helmich, K.E.; Kaper, T.; Hansson, H.; Mikkelsen, N.-E.; Gudmundsson, M.; Piens, K.; Fujdala, M.; Banerjee, G.; Scott-Craig, J.S.; et al. Biochemical characterization and crystal structures of a fungal family 3 β-glucosidase, Cel3A from *Hypocrea jecorina*. *J. Biol. Chem.* **2014**, *289*, 31624–31637. [CrossRef]
51. Yoshida, E.; Hidaka, M.; Fushinobu, S.; Koyanagi, T.; Minami, H.; Tamaki, H.; Kitaoka, M.; Katayama, T.; Kumagai, H. Role of a PA14 domain in determining substrate specificity of a glycoside hydrolase family 3 β-glucosidase from *Kluyveromyces marxianus*. *Biochem. J.* **2010**, *431*, 39–49. [CrossRef] [PubMed]
52. Litzinger, S.; Fischer, S.; Polzer, P.; Diederichs, K.; Welte, W.; Mayer, C. Structural and kinetic analysis of *Bacillus subtilis* N-acetylglucosaminidase reveals a unique Asp-His dyad mechanism. *J. Biol. Chem.* **2010**, *285*, 35675–35684. [CrossRef] [PubMed]
53. Jones, P.; Binns, D.; Chang, H.Y.; Fraser, M.; Li, W.; McAnulla, C.; McWilliam, H.; Maslen, J.; Mitchell, A.; Nuka, G.; et al. InterProScan 5: Genome-scale protein function classification. *Bioinformatics* **2014**, *30*, 1236–1240. [CrossRef] [PubMed]
54. Vroemen, S.; Heldens, J.; Boyd, C.; Henrissat, B.; Keen, N.T. Cloning and characterization of the *bgxA* gene from *Erwinia chrysanthemi* D1 which encodes a β-glucosidase/xylosidase enzyme. *Mol. Gen. Genet.* **1995**, *246*, 465–477. [CrossRef] [PubMed]
55. Beloqui, A.; Nechitaylo, T.Y.; López-Cortés, N.; Ghazi, A.; Guazzaroni, M.-E.; Polaina, J.; Strittmatter, A.W.; Reva, O.; Waliczek, A.; Yakimov, M.M.; et al. Diversity of glycosyl hydrolases from cellulose-depleting communities enriched from casts of two earthworm species. *Appl. Environ. Microbiol.* **2010**, *76*, 5934–5946. [CrossRef]
56. Lagaert, S.; Pollet, A.; Delcour, J.A.; Lavigne, R.; Courtin, C.M.; Volckaert, G. Characterization of two β-xylosidases from *Bifidobacterium adolescentis* and their contribution to the hydrolysis of prebiotic xylooligosaccharides. *Appl. Microbiol. Biotechnol.* **2011**, *92*, 1179–1185. [CrossRef]
57. Margolles-Clark, E.; Tenkanen, M.; Nakari-Setälä, T.; Penttilä, M. Cloning of genes encoding α-L-arabinofuranosidase and β-xylosidase from *Trichoderma reesei* by expression in *Saccharomyces cerevisiae*. *Appl. Environ. Microbiol.* **1996**, *62*, 3840–3846.
58. Ramírez-Escudero, M.; del Pozo, M.V.; Marín-Navarro, J.; González, B.; Golyshin, P.N.; Polaina, J.; Ferrer, M.; Sanz-Aparicio, J. Structural and functional characterization of a ruminal β-glycosidase defines a novel subfamily of glycoside hydrolase family 3 with permuted domain topology. *J. Biol. Chem.* **2016**, *291*, 24200–24214. [CrossRef]
59. Nurizzo, D.; Turkenburg, J.P.; Charnock, S.J.; Roberts, S.M.; Dodson, E.J.; McKie, V.A.; Taylor, E.J.; Gilbert, H.J.; Davies, G.J. *Cellvibrio japonicus* α-L-arabinanase 43A has a novel five-blade β-propeller fold. *Nat. Struct. Biol.* **2002**, *9*, 665–668. [CrossRef]

60. Yoshida, S.; Hespen, C.W.; Beverly, R.L.; Mackie, R.I.; Cann, I.K.O. Domain analysis of a modular α-L-arabinofuranosidase with a unique carbohydrate binding strategy from the fiber-degrading bacterium *Fibrobacter succinogenes* S85. *J. Bacteriol.* **2010**, *192*, 5424–5436. [CrossRef]
61. Mewis, K.; Lenfant, N.; Lombard, V.; Henrissat, B. Dividing the large glycoside hydrolase family 43 into subfamilies: A Motivation for detailed enzyme characterization. *Appl. Environ. Microbiol.* **2016**, *82*, 1686–1692. [CrossRef] [PubMed]
62. Ferrer, M.; Ghazi, A.; Beloqui, A.; Vieites, J.M.; López-Cortés, N.; Marín-Navarro, J.; Nechitaylo, T.Y.; Guazzaroni, M.-E.; Polaina, J.; Waliczek, A.; et al. Functional metagenomics unveils a multifunctional glycosyl hydrolase from the family 43 catalysing the breakdown of plant polymers in the calf rumen. *PLoS ONE* **2012**, *7*, e38134. [CrossRef] [PubMed]
63. Ratnadewi, A.A.I.; Fanani, M.; Kurniasih, S.D.; Sakka, M.; Wasito, E.B.; Sakka, K.; Nurachman, Z.; Puspaningsih, N.N.T. β-D-Xylosidase from *Geobacillus thermoleovorans* IT-08: Biochemical characterization and bioinformatics of the enzyme. *Appl. Biochem. Biotechnol.* **2013**, *170*, 1950–1964. [CrossRef] [PubMed]
64. Jordan, D.B.; Braker, J.D.; Wagschal, K.; Lee, C.C.; Chan, V.J.; Dubrovska, I.; Anderson, S.; Wawrzak, Z. X-ray crystal structure of divalent metal-activated β-xylosidase, RS223BX. *Appl. Biochem. Biotechnol.* **2015**, *177*, 637–648. [CrossRef] [PubMed]
65. Matsuzawa, T.; Kaneko, S.; Kishine, N.; Fujimoto, Z.; Yaoi, K. Crystal structure of metagenomic β-xylosidase/α-L-arabinofuranosidase activated by calcium. *J. Biochem.* **2017**, *162*, 173–181. [CrossRef] [PubMed]
66. Brüx, C.; Ben-David, A.; Shallom-Shezifi, D.; Leon, M.; Niefind, K.; Shoham, G.; Shoham, Y.; Schomburg, D. The structure of an inverting GH43 β-xylosidase from *Geobacillus stearothermophilus* with its substrate reveals the role of the three catalytic residues. *J. Mol. Biol.* **2006**, *359*, 97–109. [CrossRef]
67. Brunzelle, J.S.; Jordan, D.B.; McCaslin, D.R.; Olczak, A.; Wawrzak, Z. Structure of the two-subsite β-D-xylosidase from *Selenomonas ruminantium* in complex with 1,3-bis[tris(hydroxymethyl) methylamino]propane. *Arch. Biochem. Biophys.* **2008**, *474*, 157–166. [CrossRef]
68. Hong, S.; Kyung, M.; Jo, I.; Kim, Y.-R.; Ha, N.-C. Structure-based protein engineering of bacterial β-xylosidase to increase the production yield of xylobiose from xylose. *Biochem. Biophys. Res. Commun.* **2018**, *501*, 703–710. [CrossRef]
69. Zhang, R.; Li, N.; Liu, Y.; Han, X.; Tu, T.; Shen, J.; Xu, S.; Wu, Q.; Zhou, J.; Huang, Z. Biochemical and structural properties of a low-temperature-active glycoside hydrolase family 43 β-xylosidase: Activity and instability at high neutral salt concentrations. *Food Chem.* **2019**, *301*, 125266. [CrossRef]
70. Rohman, A.; van Oosterwijk, N.; Puspaningsih, N.N.T.; Dijkstra, B.W. Structural basis of product inhibition by arabinose and xylose of the thermostable GH43 β-1,4-xylosidase from *Geobacillus thermoleovorans* IT-08. *PLoS ONE* **2018**, *13*, e0196358. [CrossRef]
71. Morais, S.; Salama-Alber, O.; Barak, Y.; Hadar, Y.; Wilson, D.B.; Lamed, R.; Shoham, Y.; Bayer, E.A. Functional association of catalytic and ancillary modules dictates enzymatic activity in glycoside hydrolase family 43 β-xylosidase. *J. Biol. Chem.* **2012**, *287*, 9213–9221. [CrossRef] [PubMed]
72. Ontañon, O.M.; Ghio, S.; Marrero Díaz de Villegas, R.; Piccinni, F.E.; Talia, P.M.; Cerutti, M.L.; Campos, E. EcXyl43 β-xylosidase: Molecular modeling, activity on natural and artificial substrates, and synergism with endoxylanases for lignocellulose deconstruction. *Appl. Microbiol. Biotechnol.* **2018**, *102*, 6959–6971. [CrossRef] [PubMed]
73. Espina, G.; Eley, K.; Pompidor, G.; Schneider, T.R.; Crennell, S.J.; Danson, M.J. A novel β-xylosidase structure from *Geobacillus thermoglucosidasius*: The first crystal structure of a glycoside hydrolase family GH52 enzyme reveals unpredicted similarity to other glycoside hydrolase folds. *Acta Crystallogr. D Biol. Crystallogr.* **2014**, *70*, 1366–1374. [CrossRef] [PubMed]
74. Charoenwattanasatien, R.; Pengthaisong, S.; Breen, I.; Mutoh, R.; Sansenya, S.; Hua, Y.; Tankrathok, A.; Wu, L.; Songsiririthigul, C.; Tanaka, H.; et al. Bacterial β-glucosidase reveals the structural and functional basis of genetic defects in human glucocerebrosidase 2 (GBA2). *ACS Chem. Biol.* **2016**, *11*, 1891–1900. [CrossRef] [PubMed]
75. Shao, W.; Xue, Y.; Wu, A.; Kataeva, I.; Pei, J.; Wu, H.; Wiegel, J. Characterization of a novel β-xylosidase, XylC, from *Thermoanaerobacterium saccharolyticum* JW/SL-YS485. *Appl. Environ. Microbiol.* **2011**, *77*, 719–726. [CrossRef] [PubMed]

76. Baker, N.A.; Sept, D.; Joseph, S.; Holst, M.J.; McCammon, J.A. Electrostatics of nanosystems: Application to microtubules and the ribosome. *Proc. Natl. Acad. Sci. USA* **2001**, *98*, 10037–10041. [CrossRef]
77. Laskowski, R.A.; Swindells, M.B. LigPlot{+}: Multiple ligand-protein interaction diagrams for drug discovery. *J. Chem. Inf. Model.* **2011**, *51*, 2778–2786. [CrossRef]
78. McCarter, J.D.; Withers, S.G. Mechanisms of enzymatic glycoside hydrolysis. *Curr. Opin. Struct. Biol.* **1994**, *4*, 885–892. [CrossRef]
79. Braun, C.; Meinke, A.; Ziser, L.; Withers, S.G. Simultaneous high-performance liquid chromatographic determination of both the cleavage pattern and the stereochemical outcome of the hydrolysis reactions catalyzed by various glycosidases. *Anal. Biochem.* **1993**, *212*, 259–262. [CrossRef]
80. Jordan, D.B.; Li, X.-L.; Dunlap, C.A.; Whitehead, T.R.; Cotta, M.A. Structure-function relationships of a catalytically efficient β-D-xylosidase. *Appl. Biochem. Biotechnol.* **2007**, *141*, 51–76. [CrossRef]
81. Shallom, D.; Leon, M.; Bravman, T.; Ben-David, A.; Zaide, G.; Belakhov, V.; Shoham, G.; Schomburg, D.; Baasov, T.; Shoham, Y. Biochemical characterization and identification of the catalytic residues of a family 43 β-D-xylosidase from *Geobacillus stearothermophilus* T-6. *Biochemistry* **2005**, *44*, 387–397. [CrossRef] [PubMed]
82. Rye, C.S.; Withers, S.G. Glycosidase mechanisms. *Curr. Opin. Chem. Biol.* **2000**, *4*, 573–580. [CrossRef]
83. Ximenes, F.D.A.; de Paula Silveira, F.Q.; FFilho, E.X. Production of β-xylosidase activity by *Trichoderma harzianum* strains. *Curr. Microbiol.* **1996**, *33*, 71–77. [CrossRef]
84. Li, Q.; Wu, T.; Qi, Z.; Zhao, L.; Pei, J.; Tang, F. Characterization of a novel thermostable and xylose-tolerant GH 39 β-xylosidase from *Dictyoglomus thermophilum*. *BMC Biotechnol.* **2018**, *18*, 29. [CrossRef] [PubMed]
85. Wagschal, K.; Franqui-Espiet, D.; Lee, C.C.; Robertson, G.H.; Wong, D.W.S. Cloning, expression and characterization of a glycoside hydrolase family 39 xylosidase from *Bacillus halodurans* C-125. *Appl. Biochem. Biotechnol.* **2008**, *146*, 69–78. [CrossRef] [PubMed]
86. Banka, A.L.; Guralp, S.A.; Gulari, E. Secretory expression and characterization of two hemicellulases, xylanase, and β-xylosidase, isolated from *Bacillus subtilis* M015. *Appl. Biochem. Biotechnol.* **2014**, *174*, 2702–2710. [CrossRef] [PubMed]
87. Watanabe, A.; Hiraga, K.; Suda, M.; Yukawa, H.; Inui, M. Functional characterization of *Corynebacterium alkanolyticum* β-xylosidase and xyloside ABC transporter in *Corynebacterium glutamicum*. *Appl. Environ. Microbiol.* **2015**, *81*, 4173–4183. [CrossRef]
88. Bhalla, A.; Bischoff, K.M.; Sani, R.K. Highly thermostable GH39 β-xylosidase from a *Geobacillus* sp. strain WSUCF1. *BMC Biotechnol.* **2014**, *14*, 963. [CrossRef]
89. Huang, D.; Liu, J.; Qi, Y.; Yang, K.; Xu, Y.; Feng, L. Synergistic hydrolysis of xylan using novel xylanases, β-xylosidases, and an α-L-arabinofuranosidase from *Geobacillus thermodenitrificans* NG80-2. *Appl. Microbiol. Biotechnol.* **2017**, *101*, 6023–6037. [CrossRef]
90. Michlmayr, H.; Hell, J.; Lorenz, C.; Böhmdorfer, S.; Rosenau, T.; Kneifel, W. Arabinoxylan oligosaccharide hydrolysis by family 43 and 51 glycosidases from *Lactobacillus brevis* DSM 20054. *Appl. Environ. Microbiol.* **2013**, *79*, 6747–6754. [CrossRef]
91. Xu, B.; Dai, L.; Zhang, W.; Yang, Y.; Wu, Q.; Li, J.; Tang, X.; Zhou, J.; Ding, J.; Han, N.; et al. Characterization of a novel salt-, xylose- and alkali-tolerant GH43 bifunctional β-xylosidase/α-L-arabinofuranosidase from the gut bacterial genome. *J. Biosci. Bioeng.* **2019**, *128*, 429–437. [CrossRef] [PubMed]
92. Kim, Y.A.; Yoon, K.-H. Characterization of a *Paenibacillus woosongensis* β-xylosidase/α-arabinofuranosidase produced by recombinant *Escherichia coli*. *J. Microbiol. Biotechnol.* **2010**, *20*, 1711–1716. [PubMed]
93. Whitehead, T.R.; Cotta, M.A. Identification of a broad-specificity xylosidase/arabinosidase important for xylooligosaccharide fermentation by the ruminal anaerobe *Selenomonas ruminantium* GA192. *Curr. Microbiol.* **2001**, *43*, 293–298. [CrossRef] [PubMed]
94. Sheng, P.; Xu, J.; Saccone, G.; Li, K.; Zhang, H. Discovery and characterization of endo-xylanase and β-xylosidase from a highly xylanolytic bacterium in the hindgut of *Holotrichia parallela* larvae. *J. Mol. Catal. B Enzym.* **2014**, *105*, 33–40. [CrossRef]
95. Zhang, S.; Xie, J.; Zhao, L.; Pei, J.; Su, E.; Xiao, W.; Wang, Z. Cloning, overexpression and characterization of a thermostable β-xylosidase from *Thermotoga petrophila* and cooperated transformation of ginsenoside extract to ginsenoside 20(S)-Rg3 with a β-glucosidase. *Bioorg. Chem.* **2019**, *85*, 159–167. [CrossRef] [PubMed]
96. Shi, H.; Li, X.; Gu, H.; Zhang, Y.; Huang, Y.; Wang, L.; Wang, F. Biochemical properties of a novel thermostable and highly xylose-tolerant β-xylosidase/α-arabinosidase from *Thermotoga. thermarum*. *Biotechnol. Biofuels* **2013**, *6*, 27. [CrossRef]

97. Kumar, S.; Ramón, D. Purification and regulation of the synthesis of a β-xylosidase from *Aspergillus nidulans*. *FEMS Microbiol. Lett.* **1996**, *135*, 287–293. [CrossRef]
98. John, M.; Schmidt, B.; Schmidt, J. Purification and some properties of five endo-1,4-β-D-xylanases and a β-D-xylosidase produced by a strain of *Aspergillus niger*. *Can. J. Biochem.* **1979**, *57*, 125–134. [CrossRef]
99. Patel, H.; Kumar, A.K.; Shah, A. Purification and characterization of novel bi-functional GH3 family β-xylosidase/β-glucosidase from *Aspergillus niger* ADH-11. *Int. J. Biol. Macromol.* **2018**, *109*, 1260–1269. [CrossRef]
100. Dobberstein, J.; Emeis, C.C. Purification and characterization of β-xylosidase from *Aureobasidium pullulans*. *Appl. Microbiol. Biotechnol.* **1991**, *35*, 210–215. [CrossRef]
101. Yanai, T.; Sato, M. Purification and characterization of an β-D-xylosidase from *Candida utilis* IFO 0639. *Biosci. Biotechnol. Biochem.* **2001**, *65*, 527–533. [CrossRef] [PubMed]
102. Cintra, L.C.; Fernandes, A.G.; Oliveira, I.C.M.d.; Siqueira, S.J.L.; Costa, I.G.O.; Colussi, F.; Jesuíno, R.S.A.; Ulhoa, C.J.; Faria, F.P.d. Characterization of a recombinant xylose tolerant β-xylosidase from *Humicola grisea* var. thermoidea and its use in sugarcane bagasse hydrolysis. *Int. J. Biol. Macromol.* **2017**, *105*, 262–271. [CrossRef] [PubMed]
103. Yang, X.; Shi, P.; Huang, H.; Luo, H.; Wang, Y.; Zhang, W.; Yao, B. Two xylose-tolerant GH43 bifunctional β-xylosidase/α-arabinosidases and one GH11 xylanase from *Humicola insolens* and their synergy in the degradation of xylan. *Food Chem.* **2014**, *148*, 381–387. [CrossRef]
104. Yan, Q.J.; Wang, L.; Jiang, Z.Q.; Yang, S.Q.; Zhu, H.F.; Li, L.T. A xylose-tolerant β-xylosidase from *Paecilomyces thermophila*: Characterization and its co-action with the endogenous xylanase. *Bioresour. Technol.* **2008**, *99*, 5402–5410. [CrossRef] [PubMed]
105. Mhetras, N.; Liddell, S.; Gokhale, D. Purification and characterization of an extracellular β-xylosidase from *Pseudozyma hubeiensis* NCIM 3574 (PhXyl), an unexplored yeast. *AMB Express* **2016**, *6*, 73. [CrossRef]
106. Huang, Y.; Zheng, X.; Pilgaard, B.; Holck, J.; Muschiol, J.; Li, S.; Lange, L. Identification and characterization of GH11 xylanase and GH43 xylosidase from the chytridiomycetous fungus, *Rhizophlyctis rosea*. *Appl. Microbiol. Biotechnol.* **2019**, *103*, 777–791. [CrossRef]
107. Zanoelo, F.F.; Polizeli Md, M.d.L.T.d.M.; Terenzi, H.F.; Jorge, J.A. Purification and biochemical properties of a thermostable xylose-tolerant β-D-xylosidase from *Scytalidium thermophilum*. *J. Ind. Microbiol. Biotechnol.* **2004**, *31*, 170–176. [CrossRef]
108. Fujii, T.; Yu, G.; Matsushika, A.; Kurita, A.; Yano, S.; Murakami, K.; Sawayama, S. Ethanol production from xylo-oligosaccharides by xylose-fermenting *Saccharomyces cerevisiae* expressing β-xylosidase. *Biosci. Biotechnol. Biochem.* **2011**, *75*, 1140–1146. [CrossRef]
109. Maruthamuthu, M.; Jiménez, D.J.; van Elsas, J.D. Characterization of a furan aldehyde-tolerant β-xylosidase/α-arabinosidase obtained through a synthetic metagenomics approach. *J. Appl. Microbiol.* **2017**, *123*, 145–158. [CrossRef]
110. Gruninger, R.J.; Gong, X.; Forster, R.J.; McAllister, T.A. Biochemical and kinetic characterization of the multifunctional β-glucosidase/β-xylosidase/α-arabinosidase, Bgxa1. *Appl. Microbiol. Biotechnol.* **2014**, *98*, 3003–3012. [CrossRef]
111. Bao, L.; Huang, Q.; Chang, L.; Sun, Q.; Zhou, J.; Lu, H. Cloning and characterization of two β-glucosidase/xylosidase enzymes from yak rumen metagenome. *Appl. Biochem. Biotechnol.* **2012**, *166*, 72–86. [CrossRef] [PubMed]
112. Rasmussen, L.E.; Sorensen, H.R.; Vind, J.; Viksø-Nielsen, A. Mode of action and properties of the β-xylosidases from *Talaromyces emersonii* and *Trichoderma reesei*. *Biotechnol. Bioeng.* **2006**, *94*, 869–876. [CrossRef] [PubMed]
113. Kersters-Hilderson, H.; Loontiens, F.G.; Claeyssens, M.; De Bruyne, C.K. Partial purification and properties of an induced β-D-xylosidase of *Bacillus pumilus* 12. *Eur. J. Biochem.* **1969**, *7*, 434–441. [CrossRef] [PubMed]
114. Wagschal, K.; Heng, C.; Lee, C.C.; Robertson, G.H.; Orts, W.J.; Wong, D.W.S. Purification and characterization of a glycoside hydrolase family 43 β-xylosidase from *Geobacillus thermoleovorans* IT-08. *Appl. Biochem. Biotechnol.* **2009**, *155*, 304–313. [CrossRef]
115. Wagschal, K.; Heng, C.; Lee, C.C.; Wong, D.W.S. Biochemical characterization of a novel dual-function arabinofuranosidase/xylosidase isolated from a compost starter mixture. *Appl. Microbiol. Biotechnol.* **2009**, *81*, 855–863. [CrossRef]
116. Rapp, P.; Wagner, F. Production and properties of xylan-degrading enzymes from *Cellulomonas uda*. *Appl. Environ. Microbiol.* **1986**, *51*, 746–752.

117. Marcolongo, L.; La Cara, F.; Del Monaco, G.; Paixão, S.M.; Alves, L.; Marques, I.P.; Ionata, E. A novel β-xylosidase from *Anoxybacillus* sp. 3M towards an improved agro-industrial residues saccharification. *Int. J. Biol. Macromol.* **2019**, *122*, 1224–1234. [CrossRef]
118. Wagschal, K.; Jordan, D.B.; Braker, J.D. Catalytic properties of β-D-xylosidase XylBH43 from *Bacillus halodurans* C-125 and mutant XylBH43-W147G. *Process. Biochem.* **2012**, *47*, 366–372. [CrossRef]
119. Jordan, D.B.; Stoller, J.R.; Lee, C.C.; Chan, V.J.; Wagschal, K. Biochemical characterization of a GH43 β-Xylosidase from *Bacteroides ovatus*. *Appl. Biochem. Biotechnol.* **2017**, *182*, 250–260. [CrossRef]
120. Hudson, R.C.; Schofield, L.R.; Coolbear, T.; Daniel, R.M.; Morgan, H.W. Purification and properties of an aryl β-xylosidase from a cellulolytic extreme thermophile expressed in *Escherichia coli*. *Biochem. J.* **1991**, *273*, 645–650. [CrossRef]
121. Fan, Z.; Yuan, L.; Jordan, D.B.; Wagschal, K.; Heng, C.; Braker, J.D. Engineering lower inhibitor affinities in β-D-xylosidase. *Appl. Microbiol. Biotechnol.* **2010**, *86*, 1099–1113. [CrossRef] [PubMed]
122. Pinphanichakarn, P.; Tangsakul, T.; Thongnumwon, T.; Talawanich, Y.; Thamchaipenet, A. Purification and characterization of β-xylosidase from *Streptomyces* sp. CH7 and its gene sequence analysis. *World J. Microbiol. Biotechnol.* **2004**, *20*, 727–733. [CrossRef]
123. Vocadlo, D.J.; Wicki, J.; Rupitz, K.; Withers, S.G. A case for reverse protonation: Identification of Glu160 as an acid/base catalyst in *Thermoanaerobacterium saccharolyticum* β-xylosidase and detailed kinetic analysis of a site-directed mutant. *Biochemistry* **2002**, *41*, 9736–9746. [CrossRef] [PubMed]
124. Fekete, C.A.; Kiss, L. Purification and characterization of a recombinant β-D-xylosidase from *Thermobifida fusca* TM51. *Protein J.* **2012**, *31*, 641–650. [CrossRef]
125. Yin, Y.-R.; Xian, W.-D.; Han, M.-X.; Zhou, E.-M.; Liu, L.; Alkhalifah, D.H.M.; Hozzein, W.N.; Xiao, M.; Li, W.-J. Expression and characterisation of a pH and salt tolerant, thermostable and xylose tolerant recombinant GH43 β-xylosidase from *Thermobifida halotolerans* YIM 90462T for promoting hemicellulose degradation. *Antonie Leeuwenhoek* **2019**, *112*, 339–350. [CrossRef] [PubMed]
126. Ristroph, D.L.; Humphrey, A.E. The β-xylosidase of *Thermomonospora*. *Biotechnol. Bioeng.* **1985**, *27*, 909–913. [CrossRef] [PubMed]
127. Bachmann, S.L.; McCarthy, A.J. Purification and characterization of a thermostable β-xylosidase from *Thermomonospora fusca*. *J. Gen. Microbiol.* **1989**, *135*, 293–299. [CrossRef]
128. Büttner, R.; Bode, R. Purification and characterization of β-xylosidase activities from the yeast *Arxula adeninivorans*. *J. Basic Microbiol.* **1992**, *32*, 159–166. [CrossRef]
129. Eneyskaya, E.V.; Ivanen, D.R.; Bobrov, K.S.; Isaeva-Ivanova, L.S.; Shabalin, K.A.; Savel'ev, A.N.; Golubev, A.M.; Kulminskaya, A.A. Biochemical and kinetic analysis of the GH3 family β-xylosidase from *Aspergillus awamori* X-100. *Arch. Biochem. Biophys.* **2007**, *457*, 225–234. [CrossRef]
130. Kiss, T.; Kiss, L. Purification and characterization of an extracellular β-D-xylosidase from *Aspergillus carbonarius*. *World J. Microbiol. Biotechnol.* **2000**, *16*, 465–470. [CrossRef]
131. Kitpreechavanich, V.; Hayashi, M.; Nagai, S. Purification and characterization of extracellular β-xylosidase and β-glucosidase from *Aspergillus fumigatus*. *Agric. Biol. Chem.* **1986**, *50*, 1703–1711. [CrossRef]
132. Semenova, M.V.; Drachevskaya, M.I.; Sinitsyna, O.A.; Gusakov, A.V.; Sinitsyn, A.P. Isolation and properties of extracellular β-xylosidases from fungi *Aspergillus japonicus* and *Trichoderma reesei*. *Biochemistry* **2009**, *74*, 1002–1008. [CrossRef] [PubMed]
133. Rodionova, N.A.; Tavobilov, I.M.; Bezborodov, A.M. β-Xylosidase from *Aspergillus niger* 15: Purification and properties. *J. Appl. Biochem.* **1983**, *5*, 300–312. [PubMed]
134. La Grange, D.C.; Pretorius, I.S.; Claeyssens, M.; Van Zyl, W.H. Degradation of xylan to D-xylose by recombinant *Saccharomyces cerevisiae* coexpressing the *Aspergillus niger* β-xylosidase (*xlnD*) and the *Trichoderma reesei xylanase* II (*xyn2*) genes. *Appl. Environ. Microbiol.* **2001**, *67*, 5512–5519. [CrossRef] [PubMed]
135. Selig, M.J.; Knoshaug, E.P.; Decker, S.R.; Baker, J.O.; Himmel, M.E.; Adney, W.S. Heterologous expression of *Aspergillus niger* β-D-xylosidase (XlnD): Characterization on lignocellulosic substrates. *Appl. Biochem. Biotechnol.* **2008**, *146*, 57–68. [CrossRef]
136. van Peij, N.N.; Brinkmann, J.; Vrsanska, M.; Visser, J.; de Graaff, L.H. β-Xylosidase activity, encoded by *xlnD*, is essential for complete hydrolysis of xylan by *Aspergillus niger* but not for induction of the xylanolytic enzyme spectrum. *Eur. J. Biochem.* **1997**, *245*, 164–173. [CrossRef]

137. Boyce, A.; Walsh, G. Purification and characterisation of a thermostable β-xylosidase from *Aspergillus niger* van Tieghem of potential application in lignocellulosic bioethanol production. *Appl. Biochem. Biotechnol.* **2018**, *186*, 712–730. [CrossRef]
138. Kirikyali, N.; Wood, J.; Connerton, I.F. Characterisation of a recombinant β-xylosidase (xylA) from *Aspergillus oryzae* expressed in *Pichia pastoris*. *AMB Express* **2014**, *4*, 68. [CrossRef]
139. Chakrabarti, S.K.; Ranu, R.S. Characterization of a β-xylosidase from *Aspergillus terreus* (IJIRA 6.2). *J. Plant Biochem. Biotechnol.* **1995**, *4*, 117–120. [CrossRef]
140. Andrade, S.d.V.; Polizeli, M.d.L.T.d.M.; Terenzi, H.F.; Jorge, J.A.l. Effect of carbon source on the biochemical properties of β-xylosidases produced by *Aspergillus versicolor*. *Process Biochem.* **2004**, *39*, 1931–1938. [CrossRef]
141. Bankeeree, W.; Akada, R.; Lotrakul, P.; Punnapayak, H.; Prasongsuk, S. Enzymatic hydrolysis of black liquor xylan by a novel xylose-tolerant, thermostable β-xylosidase from a tropical strain of *Aureobasidium pullulans* CBS 135684. *Appl. Biochem. Biotechnol.* **2018**, *184*, 919–934. [CrossRef] [PubMed]
142. Carvalho, D.R.d.; Carli, S.; Meleiro, L.P.; Rosa, J.C.; Oliveira, A.H.C.d.; Jorge, J.A.; Furriel, R.P.M. A halotolerant bifunctional β-xylosidase/α-L-arabinofuranosidase from *Colletotrichum graminicola*: Purification and biochemical characterization. *Int. J. Biol. Macromol.* **2018**, *114*, 741–750. [CrossRef] [PubMed]
143. Saha, B.C. Purification and properties of an extracellular β-xylosidase from a newly isolated *Fusarium proliferatum*. *Bioresour. Technol.* **2003**, *90*, 33–38. [CrossRef]
144. Saha, B.C. Purification and characterization of an extracellular β-xylosidase from a newly isolated *Fusarium verticillioides*. *J. Ind. Microbiol. Biotechnol.* **2001**, *27*, 241–245. [CrossRef] [PubMed]
145. Xia, W.; Shi, P.; Xu, X.; Qian, L.; Cui, Y.; Xia, M.; Yao, B. High level expression of a novel family 3 neutral β-xylosidase from *Humicola insolens* Y1 with high tolerance to D-xylose. *PLoS ONE* **2015**, *10*, e0117578. [CrossRef] [PubMed]
146. Garcia-Campayo, V.; Wood, T.M. Purification and characterisation of a β-D-xylosidase from the anaerobic rumen fungus *Neocallimastix frontalis*. *Carbohydr. Res.* **1993**, *242*, 229–245. [CrossRef]
147. Kirikyali, N.; Connerton, I.F. Heterologous expression and kinetic characterisation of *Neurospora crassa* β-xylosidase in *Pichia pastoris*. *Enzym. Microb. Technol.* **2014**, *57*, 63–68. [CrossRef]
148. Terrasan, C.R.F.; Guisan, J.M.; Carmona, E.C. Xylanase and β-xylosidase from *Penicillium janczewskii*: Purification, characterization and hydrolysis of substrates. *Electron. J. Biotechnol.* **2016**, *23*, 54–62. [CrossRef]
149. Ye, Y.; Li, X.; Zhao, J. Production and characteristics of a novel xylose- and alkali-tolerant GH 43 β-xylosidase from *Penicillium oxalicum* for promoting hemicellulose degradation. *Sci. Rep.* **2017**, *7*, 11600. [CrossRef]
150. Knob, A.; Carmona, E.C. Cell-associated acid β-xylosidase production by *Penicillium sclerotiorum*. *N. Biotechnol.* **2009**, *26*, 60–67. [CrossRef]
151. Nieto-Domínguez, M.; de Eugenio, L.I.; Barriuso, J.; Prieto, A.; Fernández de Toro, B.; Canales-Mayordomo, Á.; Martínez, M.J. Novel pH-stable glycoside hydrolase family 3 β-xylosidase from *Talaromyces amestolkiae*: An Enzyme displaying regioselective transxylosylation. *Appl. Environ. Microbiol.* **2015**, *81*, 6380–6392. [CrossRef] [PubMed]
152. Chen, Z.; Jia, H.; Yang, Y.; Yan, Q.; Jiang, Z.; Teng, C. Secretory expression of a β-xylosidase gene from *Thermomyces lanuginosus* in *Escherichia coli* and characterization of its recombinant enzyme. *Lett. Appl. Microbiol.* **2012**, *55*, 330–337. [CrossRef] [PubMed]
153. Li, Y.K.; Yao, H.J.; Cho, Y.t. Effective induction, purification and characterization of *Trichoderma koningii* G-39 β-xylosidase with high transferase activity. *Biotechnol. Appl. Biochem.* **2000**, *31*, 119–125. [CrossRef] [PubMed]
154. Dekker, R.F. Bioconversion of hemicellulose: Aspects of hemicellulase production by *Trichoderma reesei* QM 9414 and enzymic saccharification of hemicellulose. *Biotechnol. Bioeng.* **1983**, *25*, 1127–1146. [CrossRef] [PubMed]
155. Poutanen, K.; Puls, J. Characteristics of *Trichoderma reesei* β-xylosidase and its use in the hydrolysis of solubilized xylans. *Appl. Microbiol. Biotechnol.* **1988**, *28*, 425–432. [CrossRef]
156. Chinen, I.; Oouchi, K.; Tamaki, H.; Fukuda, N. Purification and properties of thermostable β-xylosidase from immature stalks of *Saccharum officinarum* L. (sugar cane). *J. Biochem.* **1982**, *92*, 1873–1881. [CrossRef]
157. Zhou, J.; Bao, L.; Chang, L.; Zhou, Y.; Lu, H. Biochemical and kinetic characterization of GH43 β-D-xylosidase/α-L-arabinofuranosidase and GH30 α-L-arabinofuranosidase/β-D-xylosidase from rumen metagenome. *J. Ind. Microbiol. Biotechnol.* **2012**, *39*, 143–152. [CrossRef]

158. Li, Q.; Wu, T.; Zhao, L.; Pei, J.; Wang, Z.; Xiao, W. Highly efficient biotransformation of Astragaloside IV to Cycloastragenol by sugar-stimulated β-glucosidase and β-xylosidase from *Dictyoglomus thermophilum*. *J. Microbiol. Biotechnol.* **2018**. [CrossRef]
159. Bachmann, S.L.; McCarthy, A.J. Purification and cooperative activity of enzymes constituting the xylan-degrading system of *Thermomonospora fusca*. *Appl. Environ. Microbiol.* **1991**, *57*, 2121–2130.

International Journal of
Molecular Sciences

Review

Leloir Glycosyltransferases in Applied Biocatalysis: A Multidisciplinary Approach

Luuk Mestrom [1], Marta Przypis [2,3], Daria Kowalczykiewicz [2,3], André Pollender [4], Antje Kumpf [4,5], Stefan R. Marsden [1], Isabel Bento [6], Andrzej B. Jarzębski [7], Katarzyna Szymańska [8], Arkadiusz Chruściel [9], Dirk Tischler [4,5], Rob Schoevaart [10], Ulf Hanefeld [1] and Peter-Leon Hagedoorn [1,*]

1. Department of Biotechnology, Delft University of Technology, Section Biocatalysis, Van der Maasweg 9, 2629 HZ Delft, The Netherlands; l.mestrom@tudelft.nl (L.M.); s.r.marsden@tudelft.nl (S.R.M.); u.hanefeld@tudelft.nl (U.H.)
2. Department of Organic Chemistry, Bioorganic Chemistry and Biotechnology, Silesian University of Technology, B. Krzywoustego 4, 44-100 Gliwice, Poland; marta.przypis@polsl.pl (M.P.); daria.kowalczykiewicz@polsl.pl (D.K.)
3. Biotechnology Center, Silesian University of Technology, B. Krzywoustego 8, 44-100 Gliwice, Poland
4. Environmental Microbiology, Institute of Biosciences, TU Bergakademie Freiberg, Leipziger Str. 29, 09599 Freiberg, Germany; andre.pollender@ioez.tu-freiberg.de (A.P.); antje.kumpf@ruhr-uni-bochum.de (A.K.); dirk.tischler@rub.de (D.T.)
5. Microbial Biotechnology, Faculty of Biology & Biotechnology, Ruhr-Universität Bochum, Universitätsstr. 150, 44780 Bochum, Germany
6. EMBL Hamburg, Notkestraβe 85, 22607 Hamburg, Germany; ibento@embl-hamburg.de
7. Institute of Chemical Engineering, Polish Academy of Sciences, Bałtycka 5, 44-100 Gliwice, Poland; andrzej.jarzebski@polsl.pl
8. Department of Chemical and Process Engineering, Silesian University of Technology, Ks. M. Strzody 7, 44-100 Gliwice Poland.; katarzyna.szymanska@polsl.pl
9. MEXEO Wiesław Hreczuch, ul. Energetyków 9, 47-225 Kędzierzyn-Koźle, Poland; arkach@mexeo.pl
10. ChiralVision, J.H. Oortweg 21, 2333 CH Leiden, The Netherlands; schoevaart@chiralvision.com

* Correspondence: p.l.hagedoorn@tudelft.nl; Tel.: +31-15-278-2334

Received: 2 October 2019; Accepted: 18 October 2019; Published: 23 October 2019

Abstract: Enzymes are nature's catalyst of choice for the highly selective and efficient coupling of carbohydrates. Enzymatic sugar coupling is a competitive technology for industrial glycosylation reactions, since chemical synthetic routes require extensive use of laborious protection group manipulations and often lack regio- and stereoselectivity. The application of Leloir glycosyltransferases has received considerable attention in recent years and offers excellent control over the reactivity and selectivity of glycosylation reactions with unprotected carbohydrates, paving the way for previously inaccessible synthetic routes. The development of nucleotide recycling cascades has allowed for the efficient production and reuse of nucleotide sugar donors in robust one-pot multi-enzyme glycosylation cascades. In this way, large glycans and glycoconjugates with complex stereochemistry can be constructed. With recent advances, LeLoir glycosyltransferases are close to being applied industrially in multi-enzyme, programmable cascade glycosylations.

Keywords: glycosyltransferase; applied biocatalysis; enzyme cascades; chemoenzymatic synthesis; sugar chemistry; carbohydrate; Leloir; nucleotide

---

1. Introduction

Enzymes were already used for the conversion of glycosides even before all stereochemical details of the known carbohydrates were assigned [1,2]. In 1837, a crude formulation of almonds containing hydroxynitrile lyases catalyzed the enzymatic hydrolysis of the glycoside amygdalin [3]. Moving almost two centuries forward, the largest volumetric biocatalytic industrial process is the application of glucose isomerase for the production of high fructose syrup for food and drink applications, producing fructose from glucose at 10^7 tons per year [4]. The secret of the success of enzymes in the production or treatment of carbohydrates and glycosides is their exquisite stereo- and regioselectivity. The excellent selectivity of enzymes is required due to the diversity of structural features of carbohydrates [5], comprising D- and L-epimers, ring size, anomeric configuration, linkages, branching, and oxidation state(s). Since drug targets often exhibit specificity for all of these structural features, the production process should not contain any side-products to prevent undesired side-effects [6].

The challenge in the synthesis of carbohydrates is their wide variety of functionalities and stereochemistry (Figure 1). (Poly)hydroxyaldehydes containing a terminal aldehyde are referred to as aldoses and (poly)hydroxyketones are defined as ketoses. In aqueous solutions, monosaccharides form equilibrium mixtures of linear open-chain and ring-closed 5- or 6-membered furanoses or pyranoses, respectively. For aldoses, the asymmetric ring forms at C-1. For ketoses, it closes at C-2 as an axial (α) or equatorial (β) hemiacetal or hemiketal, respectively (commonly defined as the anomeric center). A glycosidic linkage is a covalent *O*-, *S*-, *N*-, or *C*-bond connecting a monosaccharide to another residue resulting in a glycoside, while glucoside is specific for a glucose moiety. The equatorial or axial position of the glycosidic bond is referred to as α- (axial) or β-linkage (equatorial). The number of carbohydrates linked via glycosidic bonds can be subdivided into oligosaccharides with two to ten linked carbohydrates, while polysaccharides (glycans) contain more than ten glycosidic bonds. A glycan either contains multiple different monosaccharides or more than ten glycosidic bonds. A glycoconjugate contains at least one or more monosaccharides or oligosaccharides covalently attached to a non-carbohydrate moiety (aglycon). If an oligosaccharide contains an aldose or ketose that is in equilibrium with its open-chain form, the aldehyde or ketone can be oxidized with chemical reagents (e.g., with the Benedict reagent). This is referred to as the reducing end in oligosaccharides. If there is no possibility for the sugar to form the open chain-form, then this is called a non-reducing end. Non-reducing sugars are found in glycoconjugates (i.e. nucleotides) and oligosaccharides (i.e., raffinose).

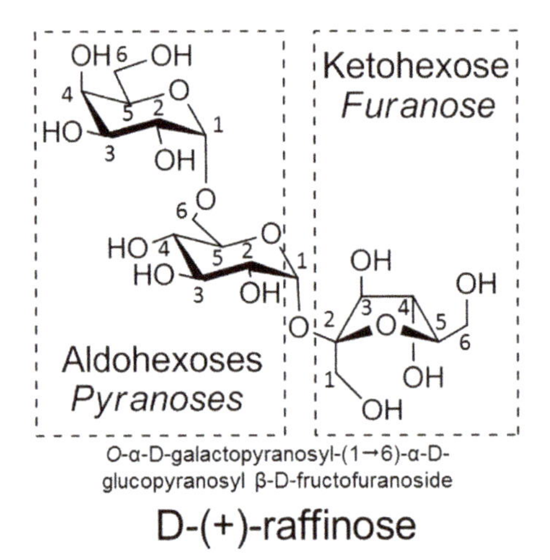

Figure 1. The nomenclature of glycosides and oligosaccharides.

Glycosyltransferases (GTs) catalyze the transfer of a carbohydrate acceptor from an activated sugar nucleotide donor with high selectivity and yield, enabling the stereo- and regioselective extension and branching of large glycans and glycoconjugates (Scheme 1). Upon formation of the glycosidic bond, the stereochemistry can either be retained or inverted by GTs with high selectivity for the α- or β-anomer. Leloir glycosyltransferases utilize carbohydrates linked to a nucleotide diphosphate (NDP) with an α-linked glycosidic bond, where non-Leloir glycosyl transferase utilize a phosphorylated sugar donor. For both types of glycosyltransferase, the main driving force for the reaction to go to completion is the exergonic release of either P_i or NDP from their respective sugar donors. The choice of nucleotide acceptor determines the (stereo) chemical outcome of the type of *O-*, *NH-*, *S-*, *C*-glycosidic bonds.

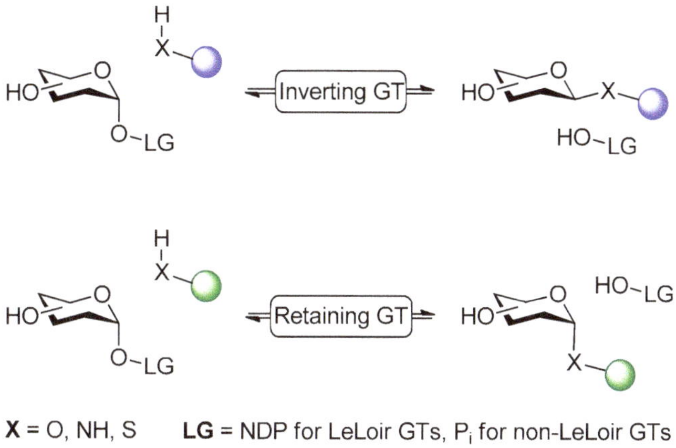

X = O, NH, S **LG** = NDP for LeLoir GTs, P_i for non-LeLoir GTs

Scheme 1. The overall scheme of an enzymatic glycosylation reaction for the biocatalytic synthesis of glycosides by retaining or inverting glycosyltransferases (GT) using NDP or P_i activated sugar donors for Leloir and non-Leloir GTs, respectively.

The enzymatic treatment of glycosides is mainly applied in the food industry using non-LeLoir GTs, enhancing flavors and functionality in complex food formulations, such as debittering [7–9], sweetening [10,11], or clarification [12–16]. The high costs of nucleotides, enzymes, and (enzymatic) regeneration systems for the treatment or production of low-value carbohydrate-containing products limited the application of nucleotide-dependent LeLoir GTs within the industry in the past. However, recent advancements in glycobiology have sparked interest in the (chemo) enzymatic production of high-value glycosides and glycoconjugates with high yield and selectivity for pharmaceutical applications [17–22]. As more LeLoir GTs are being reported with high protein expression, wide substrate scope, and high selectivity, industrial enzymatic glycosylation for the production of glycosides and glycoconjugates in vitro is becoming economically feasible. For instance, the expression of a large part of the human glycosyltransferases is a new hallmark for the production of human glycans or glycoconjugates [23], simplifying their chemoenzymatic synthesis. Besides these developments, the reaction methodologies are currently being further optimized. Multi-step enzymatic coupling with glycosyltransferases using non-natural sugar acceptors and nucleotide sugar donors have been performed with automated synthesizers and are under development, as was recently reviewed [24].

The enzymatic synthesis of glycosides has received increasing attention in organic synthesis. However, the application of Leloir glycosyltransferases in a multi-enzymatic sugar coupling process is challenging from a process design point of view. The high costs, low stability, and difficult or limited availability of nucleotide sugar donors, in addition to the challenging protein production of Leloir glycosyltransferases hamper the development of enzymatic glycosylation. As a compromise, separate nucleotide sugar regeneration cascades and optimization of the protein production of industrial biocatalysts has been pursued [25]. Although there is a large body of scientific literature reporting on the biochemical properties and the reactions that glycosyl transferring enzymes catalyze, the performance of these biocatalytic processes has only sparingly been described. Due to their inherent complexity, kinetic and thermodynamic parameters have often not been analyzed in detail for the production of larger oligosaccharides using Leloir glycosyltransferases. In this review, the possibilities and limitations for industrial applications of Leloir glycosyltransferases are highlighted from the intersection of biochemical, chemical, thermodynamic, and reaction engineering perspectives, giving an overview of the requirements of industrial processes involving glycosyltransferases.

2. Glycosyltransferases in Nature

Glycosyltransferases catalyze the formation of a glycosidic bond between an unactivated acceptor monosaccharide or aglycon and an activated sugar donor [26] to a di-, oligo-, polysaccharide [27], lipo(poly)saccharide [28] or peptidoglycan [29]. More than 484,620 glycosyltransferases in over 106 families have been identified according to the carbohydrate active enzymes (CAZy) database under the Enzyme Commission number E.C.2.4.x.y. (CAZy database, last updated 01/15/18 [30]), representing an enormous number of metabolic pathways [31]. Glycosyltransferases can be sub-classified based on four different criteria: (i) the class of substrates [32]; (ii) the protein structure [26]; (iii) the preference in stereochemistry [27,33]; (iv) the dependency on metals for catalytic activity [26]. Non-Leloir glycosyltransferases use phosphorylated donors (i.e., lipid polyprenol [34,35], sugar 1-phosphates [32]) and can be described as phosphorylases. The second class are transglycosidases accepting non-activated di- or polysaccharides as carbohydrate donors. The largest class of glycosyltransferases are the nucleotide-dependent Leloir glycosyltransferases [26,32,36], named in honor of Luis Federico Leloir, who received a Nobel prize for the discovery of nucleotide sugar donors in 1970 (Figure 2).

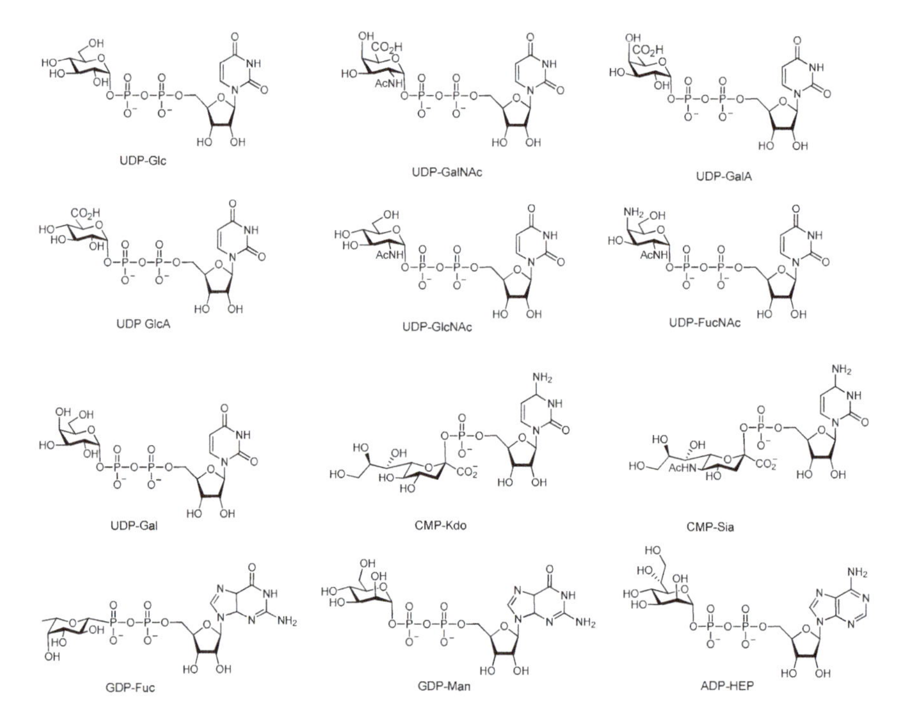

Figure 2. Common sugar nucleotides found in all kingdoms of life. Abbreviations: UDP-Glc, UDP-glucose; UDP-GalNAc, UDP-*N*-acetyl-2-deoxy-D-galactosamine; UDP-GalA, UDP-D-galacturonic acid; UDP-GlcA, UDP-D-glucuronic acid; UDP-GlcNAc, UDP-*N*-acetyl-2-deoxy-D-glucosamine; UDP-FucNAc, UDP-*N*-acetyl-L-fucosamine; UDP-Gal, UDP-D-galactose; CMP-Kdo, CMP-3-deoxy-D-manno-octulosonate; CMP-Sia, CMP-*N*-acetylneuraminic acid; GDP-Fuc, GDP-L-fucose; GDP-Man, GDP-D-mannose; ADP-HEP, ADP-L-*glycero*-D-*manno*-heptose.

The protein sequence and crystallographic data demonstrate that glycosyltransferases are mainly comprised of five different protein folds (Figure 3) [26,27,37]. Glycosyltransferases having a GT-A or GT-B fold consist of two β/α/β-Rossmann-like domains, abutting each other in case of the GT-A fold or facing each other for GT-B folds [26,27,38]. Both folds contain separate donor and acceptor binding sites [26]. Gloster et al. reported that glycosyltransferases with a GT-A fold belong to the divalent metal ion dependent class of these enzymes, whereas GT-B folds are often metal ion independent [26,27]. Interestingly, glycosyltransferases having a GT-C fold are non-Leloir glycosyltransferases, utilizing membrane integrated or membrane linked proteins with lipid phosphate sugar donors, also known as non-Leloir donors [27,32,37]. The Leloir glycosyltransferases containing a GT-D fold catalyze the transfer of glucose to hexasaccharide *O*-linked to serine-rich repeats of bacterial adhesins [39]. The most recent addition, is the *N*-acetyl-D-mannose transferase utilizing non-Leloir undecaprenyl-linked glycosyl diphosphates with a unique GT-E fold [40].

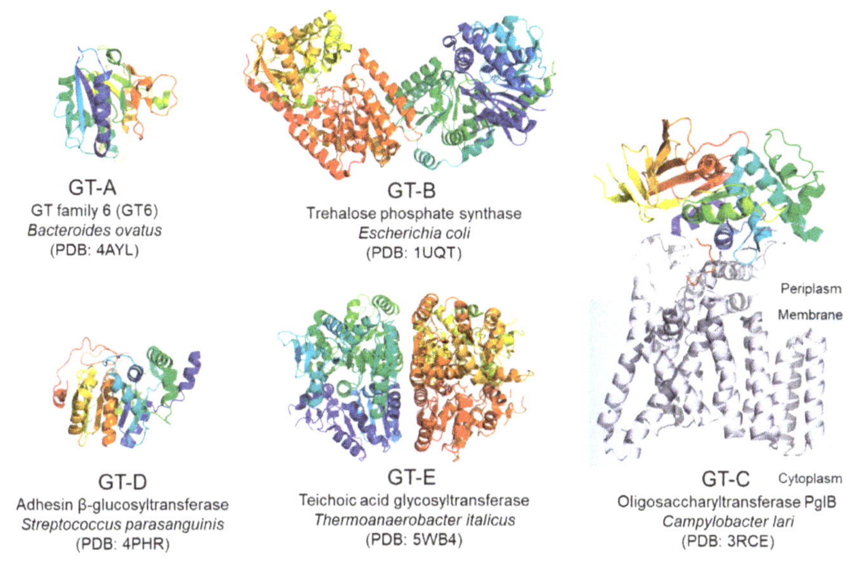

Figure 3. Protein folds of Leloir glycosyltransferases (GT-A, GT-B, GT-D) and non-Leloir glycosyltransferases (GT-C, GT-E).

For Leloir glycosyltransferases, the binding of the sugar donor nucleotide and acceptor follows a sequential ordered bi-bi catalytic mechanism via non-covalent interactions of the sugar donor nucleotide. The binding of the sugar or aglycone acceptor results in an enzyme-substrate ternary complex [41]. Hydrolysis of the sugar nucleotide donor is prevented by the tight binding in an unproductive state, where the high affinity of the enzyme for the sugar nucleotide donor is an indicator for product inhibition (K_i) by the released nucleotide [42]. For Leloir glycosyltransferases, a lower affinity or promiscuity towards the nucleotide donor results often in less product inhibition [43]. Upon binding of the sugar nucleotide donor, the enzyme undergoes a conformational change stabilizing the transition state, resulting in the formation of a glycosidic bond and the release of the nucleotide donor. Different reaction mechanisms of glycosyltransferases have been described and reviewed [26,27,44–47]. The *inverting* occurs via a S_N2 mechanism, while a *retaining* transfer can proceed via a concerted or ion-pair intermediate mechanism through a double displacement via a S_N2 mechanism. Also, a *transient covalent* intermediate via a S_Ni-type mechanism has been described for LeLoir GTs (Figure 4). *Inverting* glycosyltransferases use general base catalysis (i.e., aspartate or glutamate) [31,48,49] to form an oxocarbenium ion-like transition state. They show a catalytic rate enhancement by utilizing divalent metals (i.e. Mn (II) or Mg (II)), which are often coordinated by the amino acid motif Asp-X-Asp.

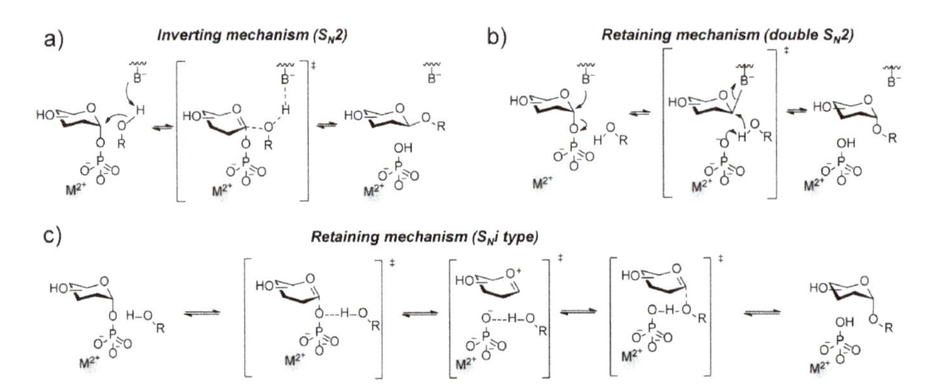

Figure 4. Reaction mechanism of glycosyltransferases upon inversion (**a**) or retention (**b**,**c**) of the anomeric glycosidic bond. The divalent metal (M^{2+}) is not necessarily a requirement for catalytic activity for GTs.

2.1. Distinguishing Glycosyl Transferases from Glycoside Hydrolases

Two main groups of enzymes can catalyze the regio-, stereo-, and enantioselective coupling of carbohydrates. *Glycoside hydrolases* and *glycosyltransferases* are often combined in biocatalytic retro-synthetic strategies for linear elongation and branching of oligosaccharides. Glycoside hydrolases are enzymes that condense a sugar donor with an aglycone acceptor. The broad substrate scope of glycoside hydrolases has resulted in numerous synthetic applications such as synthesis [50–53] or hydrolysis [54–57] of glycosidic bonds, and desymmetrization [58]. As a drawback, their broad substrate scope also leads to the formation of side-products. Glycosylations with glycoside hydrolases are under kinetic (transglycosylation) or thermodynamic control (direct glycosylation) using activated and non-activated sugars respectively (Figure 5). With *transglycosylation*, relatively high yields can be obtained in comparison with *direct glycosylation* due to a thermodynamically unfavorable reaction equilibrium (K_{eq}) in water. As a rule of thumb, transglycosylation should be faster than glycoside hydrolysis, as otherwise the activated sugars would hydrolyze before the glycosylation reaction is completed. Also, the rate of hydrolysis of the product should be slower than the activated glycosyl donor or the product yield decreases. As this is often not the case, an excess of the activated sugar donor is required under kinetic control. Similar to the coupling of protected glycosyl donors, the donors for transglycosylation, such as fluoro [59–62], -azido [63], *p*-nitrophenyl- [64] or *p*-nitropyridyl- [65], vinyl- [66], and allyl-glycosides [67] require their separate synthesis. The direct glycosylation is challenging due to the poor K_{eq} under aqueous reaction conditions, limiting the degree of conversion. The product yields with direct glycosylation can be improved by adding one substrate in excess, lowering the water activity [68], and in situ product removal [69].

Leloir glycosyltransferases couple NDP sugar donors with a wide range of sugar acceptors resulting in the formation of a glycosidic bond. The exclusion of hydrolysis activity of the nucleotide sugar donor separates glycoside hydrolases from glycosyltransferases. Nevertheless, hydrolysis of the nucleotide sugar donor in the absence of a sugar acceptor has been reported and is referred to as "error hydrolysis" [70–73]. Hence, the competition between water or a sugar acceptor as nucleophile is important for the efficiency of glycosylation. Only a handful of studies investigated the nature of the hydrolysis activity of Leloir glycosyltransferases with sugar nucleotide donors. For instance, the bacterial sialyltransferase from *Pasteurella dagmatis* hydrolyzed the rather hydrolysis-prone CMP-Neu5Ac in the absence of another substrate [74]. Directed evolution has been shown to be an effective tool to diminish the degree of hydrolysis of NDP sialyl donor [75,76]. In comparison to hemiketals, hemiacetals are more stable sugar donor nucleotides (i.e., GDP-L-fucose). Here, the Leloir glycosyltransferases catalyze hydrolysis to a lesser degree [77]. Interestingly, the affinity of water to the active site for the hydrolysis of sugar nucleotide donors has not been determined for Leloir glycosyltransferases.

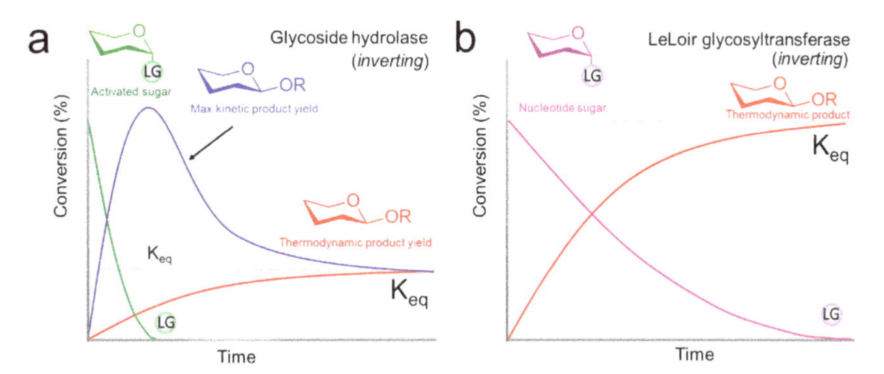

Figure 5. Exemplary enzymatic glycosylation of an activated sugar donor (green) and acceptor (R-group) to afford a maximum transient kinetic (blue) product yield catalyzed by a glycoside hydrolase, followed by reverse hydrolysis towards the thermodynamic product concentration. Direct esterification leads to the thermodynamic product yield K_{eq} without the requirement for an activated sugar (red) in (**a**). LeLoir GTs only catalyze the direct esterification of a nucleotide sugar donor (purple) to thermodynamic product (red) in (**b**).

2.2. Recombinant Expression of Glycosyl Transferases

Although protein structures and the reaction mechanism of Leloir glycosyltransferases are widely investigated, production of the enzyme is often challenging. Heterologous bacterial hosts such as *E. coli* often lead to poor expression or formation of inclusion bodies (IBs) in certain cases with retention of catalytic activity [78,79]. Besides the difficulties in recombinant protein production and isolation, the half-life of this class of enzymes is often less than a couple of hours [80–82]. Thermostable glycosyltransferases from thermophilic archaea show higher overall stability [43]. Leloir glycosyltransferases are often aggregation-prone in vitro [83,84]. As a solution to their aggregation, a large number of solubility tags have been successfully applied to increase the solubility of Leloir glycosyltransferases [23,79,85,86]. The recent advance of using the fluorescent proteins mCherry [79] or GFP [23] as tags allowed for both an increase in solubility as well as rapid protein quantification. For example, the fusion of GFP allowed for a modular expression approach of all human glycoenzymes in HEK293 cells enabling multi-milligram isolation from the culture media in 65% of all cases [23].

The optimization of protein expression, the number of enzymes discovered, and the characterization of a wide range of Leloir GTs has led to fundamental insights into their protein structures, reaction mechanism, and substrate spectrum. The result of this extensive biochemical knowledge is leading to the adoption of Leloir glycosyltransferases within the field of carbohydrate chemistry. Next, we will discuss how these biochemical insights have been developing alongside their application in chemoenzymatic glycosylations of glycoconjugates and oligosaccharides.

3. Application of Glycosyl Transferases in Organic Synthesis

The production of glycosides and glycans requires the use of highly selective catalysts to prevent the formation of side-products. The development of automated chemical methods such as the solid-phase production of oligosaccharides using the Seeberger method [87], the Demchenko synthesizer using HPLC-based platforms for automation [88], and the Yoshida procedure employing an electrochemical oxidation step [89], improved glycochemistry significantly. The basic principle of elongating a sugar on a solid particle by performing a coupling-wash-deprotection-wash cycle under computer control allows for the rapid production of a wide variety of carbohydrates [90–97]. The mechanism of action is the assembly of an oligosaccharide using protection group manipulation of either an activated glycosyl acceptor or donor [98–105]. The purification of the intermediates produced in sequential reactions remains the largest hurdle for chemical synthesis of an oligosaccharide or glycan. In particular, the low orthogonality of activated glycosyl donors and acceptors limits multiple glycosylation reactions in one-pot reactions. Also, the inherently low chemical reactivity of certain glycosidic bond forming reactions, such as α-sialylation [106–108] and β-mannosylation [109], restrict different types of linkages. Enzymes which catalyze one-pot glycosylation reactions with unprotected sugars can produce different types of glycosidic linkages and have expanded the synthetic toolbox of glycochemistry considerably.

Leloir glycosyltransferases (GTs) transfer a nucleotide sugar donor to an aglycon acceptor, forming *O*-, *N*- [110–124] or the rare *C*- [125–136] and *S*-glycosidic bonds [137–140] under thermodynamic control. In comparison to chemical methods, the enzymatic coupling of carbohydrates occurs without the use of protecting groups in a highly selective manner, allowing for orthogonal one-pot multi-enzymatic (OPME) reactions. With a few robust GTs, complete libraries of glycans can be constructed [141–143], which is particularly interesting since most of the human GTs are accessible in heterologous expression systems [23]. The advantages of employing GTs are their mild reaction conditions, short reaction times, pH tolerance, high specific activity, and high yields allowing for the (poly)glycosylation of a wide array of glycans.

3.1. Catalytic Reversibility of Glycosyltransferases

One of the notable discoveries on glycosyltransferases was the recognition that glycosyltransferases do not catalyze unidirectional reactions [144]. Alternatively, synthetic sugar donors and/or (chemo) enzymatic regeneration systems either alter the overall K_{eq} or regenerate the nucleotide in situ [145–149]. Such regeneration systems are not always a requirement; the glycosylation with nucleotide sugar donors allows for repeated glycosylation on a single aglycon (i.e., flavonol-*O*-diglycoside [150]) or elongation of a (poly)saccharide, such as glycogen with a molecular weight of up to 10^7 kDa [151]. The high glycosylation efficiency with Leloir GTs arises from a favorable thermodynamic equilibrium K_{eq} in these examples, determined by the sugar nucleotide donor and carbohydrate or aglycone acceptor, pH, and ionic strength. As mentioned earlier, in a few examples the hydrolysis of the NDP-sugar donor has been reported for Leloir GTs [70–77]. In these particular cases, it is important to emphasize that the glycosylation with Leloir GTs is under kinetic control, and the sugar acceptor and water are competing nucleophiles throughout the entire course of reaction [70–77].

A large impact on the field of glycobiology is the improved group estimation method [152,153] for the determination of the change in Gibbs free energy of formation of glycosylation reactions with increased accuracy, named eQuilibrator 2.0 [154,155]. In comparison to empiric thermodynamic data (i.e., Thermodynamics of Enzyme-Catalyzed reactions Database [156]), prediction tools allow for a much higher coverage of Gibbs free energies of formation for different compounds. As a drawback, such prediction methods can lead to contradictory observations due to either experimental uncertainties [157] or incorrect analysis of given data [158]. Using Equilibrator 2.0, the synthesis of naturally occurring glycosides with nucleotide diphosphates (NDPs) were shown to be thermodynamically favorable, as is known for the glycosylation of phenolic [159–170], amino [171,172], or alcoholic [173] aglycones (Figure 6), and has been reviewed recently [44]. Interestingly, the importance of the pH has been reported for the glycosylation of acids [167,174,175] resulting in a low $K_{eq} < 1$ at a neutral pH. The K_{eq} depends on the pK_a of the aglycone- or saccharide acceptor, as well as the terminal phosphate of the sugar nucleotide donor.

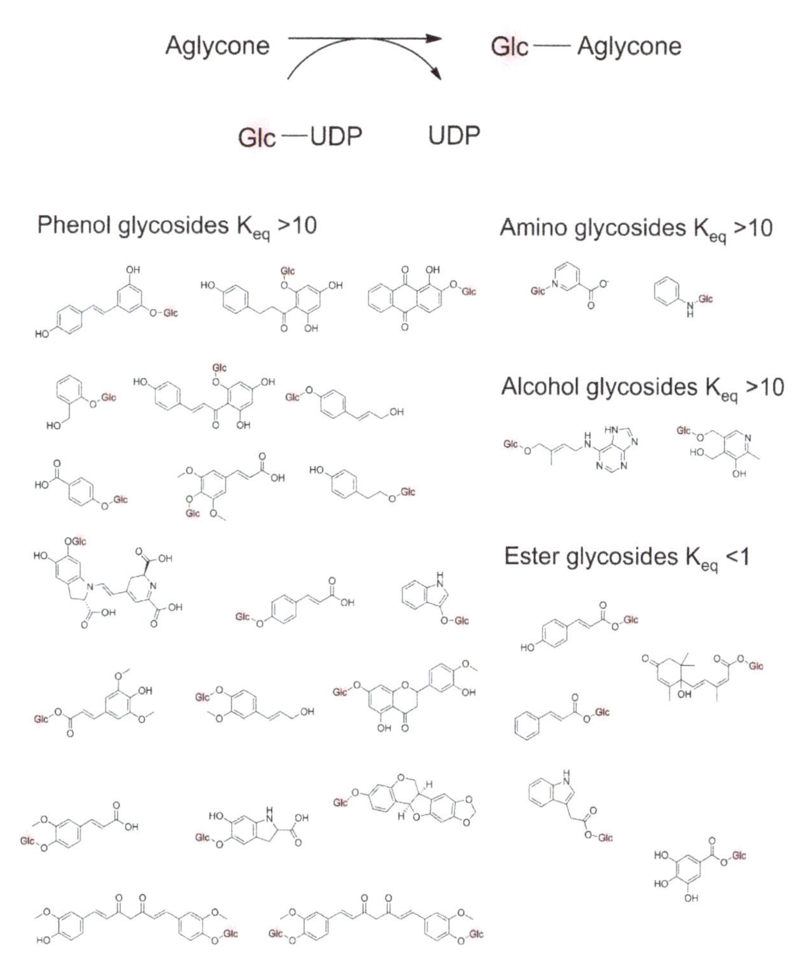

Figure 6. Glycosylation of aglycones producing phenolic glycosides, amino glycosides, alcohol glycosides, ester glycosides, and disaccharides with their estimated K_{eq}. The K_{eq} was calculated from the Gibbs free energy $\Delta G_r'^{\circ}$ using the eQuilibrator web interface (http://equilibrator.weizmann.ac.il) [155] assuming the following conditions: ionic strength 0.1 M, pH 7.0, aglycon (1 mM), UDP (1 mM), UDP-D-glucose (1 mM), glycosylated product (1 mM), and 298 K.

3.2. Sugar Donors and Acceptors and Their Glycosylation Efficiency

The thermodynamic constraints of enzymatic glycosylations of sugar acceptors with nucleotide donors for the synthesis of di-, oligo-, or polysaccharides has been explored to a lesser extent. Sucrose synthase has been employed for the regeneration of nucleotide sugars [25,42,80,176,177]. The equilibrium constant (K_{eq}) of the reaction of sucrose with UDP to afford the sugar donor UDP-glucose was determined [178]. The pH influences the K_{eq} for the synthesis of UDP-glucose with *Acidithiobacillus caldus* sucrose synthase (*Ac*SuSy) due to the (de)protonation of the phosphate group of the NDP: going from a pH of 5.0 to 7.0 lowered the K_{eq} of 1.14 to less than 0.1 [178]. Enzymatic regeneration of NDP-glucose can also be achieved using trehalose as substrate [179]. For the regeneration of nucleotide sugars, sucrose has been described as a more attractive D-glucopyranosyl donor than α, α-D-trehalose due to the lower free energy of the glycosidic bond, resulting in a more favorable thermodynamic equilibrium [44,180].

While the type of carbohydrate donor and acceptor determines the glycosylation product, the respective choice of nucleotide used for activation of the donor glycoside is important from a thermodynamic point of view. Similar enzyme activities and affinities were observed for the coupling of UDP-, GDP-, and ADP-glucose with α-D-glucose by trehalose transferase (TreT) from *Pyrococcous horikoshii*. However, different K_{eq} were observed for the enzymatic production of D-trehalose. [43] In line with these observations, a trehalose transferase from *Thermoproteus uzoniensis* fused to a mCherry solubility tag also reported different K_{eq} for ADP- and UDP-glucose for the production of D-trehalose [79]. Hence, the overall extent of conversion for the synthesis of disaccharides was determined by the thermodynamics of the nucleotide. Although a thorough examination of the Gibbs free energy of formation of NMP, NDP, or NTP salt or metals pairs in aqueous solution is beyond the scope of this review, it should be noted that the pKa of nucleotides differ affecting the Gibbs free energy of formation. Indeed, the ADP/ADP-glucose couple shows the largest Gibbs free energy change for a transfer of α-glucopyranosyl moiety to a nucleotide, followed by UDP, CDP, and dTDP according to Equilibrator 2.0 [155]. Nature might evolve enzymes to catalyze either the synthesis of nucleotide sugar donors or reactions based on the K_{eq} of nucleotides, as TreT of *Thermococcus litoralis* solely accepts ADP for the transfer of an α-glucopyranosyl moiety from trehalose to produce ADP-glucose [181]. Oppositely, TreT from *Thermoproteus tenax* utilizes UDP-glucose for the synthesis of trehalose since UDP favors synthesis [182]. Further work regarding this is required to elucidate the nature of the effect of nucleotides on the K_{eq} in a more comprehensive manner.

Under thermodynamic control, Leloir glycosyltransferases produce oligosaccharides if the overall glycosylation reaction is exergonic (Figure 7). A one-pot procedure using five enzymes allowed for the production of raffinose and stachyose from sucrose [183], using unpurified cell-free extract formulations and supplementation of UDP with a total-turnover number (TTN) of 337. Thermodynamic constraints were observed in the endergonic nucleotide sugar donor production, while coupling of the galactinol, raffinose, and stachyose were exergonic, thereby driving the overall reaction toward oligosaccharide synthesis. The estimation of the Gibbs free energy of individual components gives insights into energetic constraints of one-pot multi-enzyme Leloir glycosyltransferase catalyzed glycosylation reactions. An understanding of these limitations is essential for the optimization of industrial process conditions and reactor design (i.e., product removal) of a biocatalytic process.

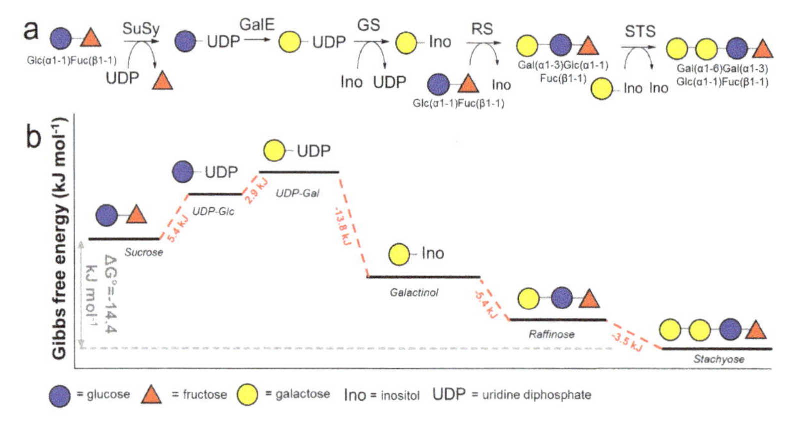

Figure 7. Enzymatic cascade for the production of stachyose from sucrose with glycosyltransferases. (**a**). The standard Gibbs free energy changes of the individual reactions (ΔG°s, red) and the total reaction (ΔG°, grey) shown in (**b**) [183]. The $\Delta_r G'^{\circ}$ represents the change of Gibbs free energy and was calculated using the eQuilibrator web interface (http://equilibrator.weizmann.ac.il) [155] using the following conditions: ionic strength 0.1 M, pH 7.0, 1 mM of component, 298 K. Abbreviations: UDP-D-glc, UDP-D-glucose; UDP-D-gal, UDP-D-galactose, SuSy, Sucrose synthase; GalE, UDP-D-glucose-4-epimerase; GS, galactinol synthase; RS, raffinose synthase; STS, stachyose synthase.

3.3. NTP Regeneration for NDP-Sugar Donor Production

Recycling of the nucleotide sugar donor is considered essential for the application of Leloir glycosyltransferases in large scale applications by preventing product inhibition from the released nucleotide and reducing costs of expensive nucleotides. The use of purified enzymes in comparison to whole-cell systems is often preferred, due to undesired side-reactions of endogenous enzymes of the recombinant hosts during glycosylation of complex oligosaccharides. Most of the glycosyltransferases and enzymes involved in the regeneration of nucleotides operate under neutral conditions and often require the presence of divalent metals, such as Mg^{2+} or Mn^{2+}. As Leloir glycosyltransferases use the elimination of the nucleotide as a driving force for the glycosylation reaction, the high energy gain poses a problem during the regeneration of NDP-sugar donors. For the production of NTP, the driving force then has to be derived from even more energy-rich donors.

Four of the most widely applied enzymatic methods for the regeneration of the nucleotide triphosphates (NTPs) are (see Figure 8): (1) pyruvate kinase using phospho(enol)pyruvate (PEP), (2) acetate kinase using acetyl phosphate, (3) creatine kinase using creatine phosphate, and (4) polyphosphate kinase using polyphosphate. The reaction equilibrium for PEP is highly favorable and the phosphate donor is stable in solution [184]. However, commercial phosphoenolpyruvate is expensive. Creatine phosphate is an alternative donor which is more affordable, but has considerably lower energetic advantages than PEP. A cheap energy-rich phosphate donor is acetyl phosphate, which can be synthesized directly from acetic anhydride and phosphate in excellent yields [185,186]. The disadvantage of using acetyl phosphate is the rapid spontaneous hydrolysis in water, requiring either continuous supplementation or an excess of acetyl phosphate. The inexpensive (poly)phosphate is a linear polymer that contains from ten to hundreds of energy-rich phosphate linkages [187]. (Poly)phosphate can drive the glycosylation reaction towards completion by the exergonic cleavage of the phosphoanhydride bond ($\Delta G° = -30 - -32$ kJ·mol^{-1} [158]) upon phosphorylation of nucleosides with polyphosphate kinase (PPK). Mono- or diphosphorylation with PPK have been reported for ATP [184,188–193], UTP [189,194,195], CTP [196,197], tTMP [198], often showing broad promiscuity towards different nucleotides [187].

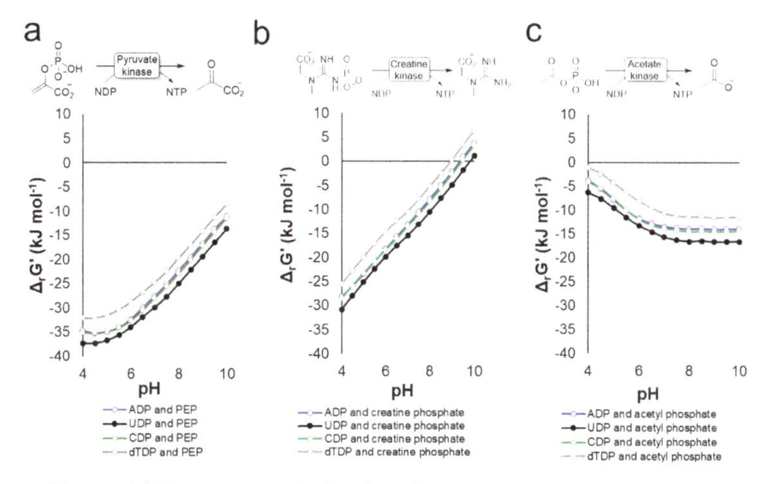

Figure 8. The use of different energy-rich phosphate donors to regenerate NTP using either pyruvate- (**a**), creatine- (**b**), or acetate kinase (**c**). The $\Delta_rG'°$ represents the standard change of Gibbs free energy and was calculated using the eQuilibrator web interface (http://equilibrator.weizmann.ac.il) [155] using the following conditions: ionic strength 0.1 M, pH 7.0, 1 mM of component, 298 K. Abbreviations: NDP, nucleotide diphosphate; NTP, nucleotide triphosphate; ADP, adenosine diphosphate; UDP, uridine diphosphate; CDP, cytidine diphosphate; dTDP; deoxythymidine diphosphate; PEP, phosphoenolpyruvate.

Different (re)generation schemes for the in-situ production of nucleotide sugars for the transfer of a galactosylpyranoside moiety with either stoichiometric amounts of NTP [199], PEP [200], poly(phosphate) [189], or acetyl phosphate [201] are shown in Figure 9. The main driving force for the glycosylation reaction is the exergonic hydrolysis of pyrophosphate to phosphate by pyrophosphatases or alkaline phosphatases. Although it has been suggested that the sacrificial hydrolysis of NTPs with alkaline phosphatases is beneficial due to the removal of the nucleotide mono-, di-, or triphosphate inhibitors [44,202], experimental evidence separating thermodynamics (additional hydrolysis of pyrophosphate) from kinetics (product inhibition) is often not investigated in detail. It is evident that under thermodynamic control nucleotide regeneration and enzymatic glycosylation can only occur with highly exergonic sacrificial substrates (i.e., hydrolysis of pyrophosphate), as was proposed by Hirschbein et al. who compared the energy of hydrolysis of the sacrificial donors as a rationale for glycosylation efficiency [203]. Besides, for the common nucleotide glycosylation donors UDP-Glc, UDP-GlcNAc [204], UDP-GlcA [205], UDP-Gal [204], UDP-GalA, UDP-Xyl, GDP-Man, GDP-Fuc [204], CMP-Neu5Ac [200,204] the (re)generation systems for the production have been employed for rare or synthetic nucleotide sugar donors, such as CMP-MAnNGc [200], CMP-Man [200], CMP-ManNac5OMe [200], CMP-Kdo [200], ADP-Hep [206], and dTDP-Rha [207].

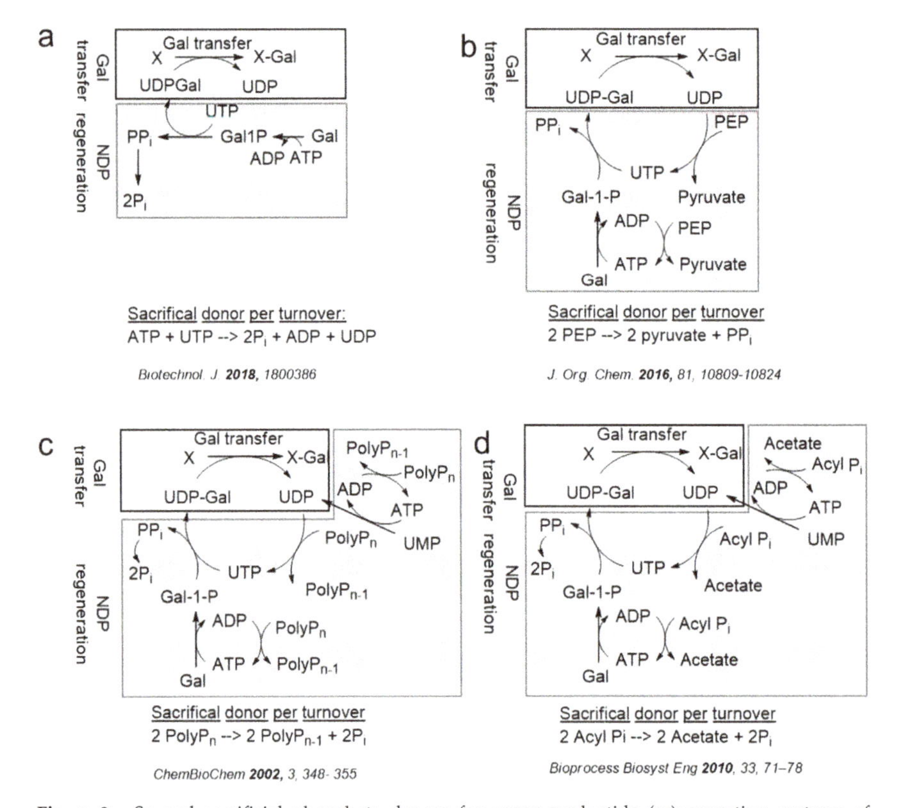

Figure 9. Several sacrificial phosphate donors for sugar nucleotide (re)generation systems of galactosyltransferases using a stoichiometric amount of NTPs (**a**) [199], PEP (**b**) [200], $PolyP_n$ (**c**) [189], and acyl P_i (**d**) [201]. Abbreviations: PP_i, pyrophosphate; P_i, orthophosphate; UMP, uridine monophosphate; UDP, uridine diphosphate; UTP, uridine triphosphate; UDP-Gal, UDP-D-galactose; ADP, adenosine diphosphate; ATP, adenosine triphosphate; Gal, D-galactose; Gal1P, D-galactose-1-phosphate; PEP, (phospho)enol pyruvate; $PolyP_n$, (poly)phosphate; acyl P_i, acetyl phosphate.

Often, one-pot multienzyme (OPME) cascade reactions do not go to completion without the thermodynamic driving force from in-situ regeneration systems of nucleotide sugar donors. For instance, the gram-scale OPME cascade of the glycoconjugate *N*-acetyl-D-lactosamine resulted in 85% isolated yields and TTN of 80 for UTP (Figure 10a) [208]. The PEP/UDP-regeneration system produces UTP at the expense of PEP while the hydrolysis of pyrophosphate provides the thermodynamic driving force to complete the glycosylation cycle (Figure 10b). Upon replacement of the PEP/UDP-regeneration system with (poly)phosphate/UDP for the enzymatic production of *N*-acetyl-D-lactosamine, no additional pyrophosphatases are required [195]. Here, the required driving force is generated by the hydrolysis of the energy rich phosphoanhydride bond in (poly) phosphate instead of pyrophosphate hydrolysis.

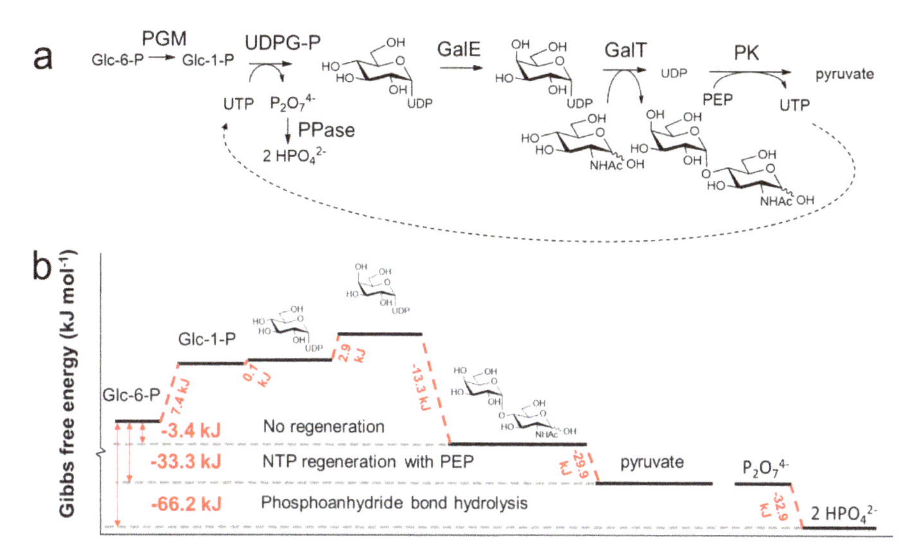

Figure 10. Enzymatic glycosylation for the production of *N*-acetyl-D-lactosamine from glucose-6-phosphate and *N*-acetyl-D-glucosamine (**a**) [208]. The $\Delta_r G'^\circ$ represents the standard change in Gibbs free energy (**b**) was calculated using the eQuilibrator web interface (http://equilibrator.weizmann.ac.il) [155] using the following conditions: ionic strength 0.1 M, pH 7.0, 1 mM of component, 298 K. Abbreviations: Glc-6-P, D-glucose-6-phosphate; Glc-1-P, D-glucose-1-phosphate; $P_2O_7^{4-}$, pyrophosphate; HPO_4^{2-}, orthophosphate; UDP, uridine diphosphate; PGM, Phosphoglucomutase; UDPG-P, UDP-glucose pyrophosphorylase; GalE, UDP-galactose epimerase; GalT, galactosyltransferase; PK, pyruvate kinase.

Besides enzymatic regeneration, the enzymatic NTP synthesis can be performed directly from nucleosides in the presence of an excess of a phosphate donor reducing the overall costs of reagents (i.e., less than US$10 per gram UTP). A mutant of uridine kinase from *Thermus thermophilus* phosphorylates a broad range of nucleosides [209]. The addition of an excess of acetyl phosphate allows for the phosphorylation of nucleosides to NMPs with lysates from recombinant *E. coli* containing the overexpressed and promiscuous uridine kinase [207]. Advantageously, cell-free extracts from *E. coli* contain naturally occurring kinases that catalyze sequential phosphorylations to NTP in high yields. Recently, a recombinant *E. coli* strain containing an enzymatic cascade of eight enzymatic steps showed promising production titers of 1.4 g UTP per liter within 2.5 h starting from uracil [210]. Such regeneration systems have been extended to non-natural nucleosides [207,211,212].

3.4. Chemoenzymatic NTP Regeneration Cascades

The chemoenzymatic synthesis of NDP-sugar donors has been investigated with activated glucose donors under kinetic control. Although the realization that glycosyltransferases catalyze the reverse reaction dates back to 1957 [213], the first application for the glycosylation of nucleotides with an activated glycosyl fluoride was reported much later in 1999 [214]. The use of β-glucosyl fluoride for the production of UDP-α-glucose using a flavonoid *O*- and *C*-β-glycosyltransferases has been successful for the production of 3′-β-*C*-glucosylated phloretin under kinetic control [215]. Disadvantageously, fast hydrolysis of β-glucosyl fluoride in water limits its practical application. The pioneering work using nitrophenol glycosides demonstrated the broad adaptability of activated sugar acceptors for the glycosylation of nucleotide donors by altering the thermodynamics of the reaction [216]. The engineered inverting macrolide-inactivating glycosyltransferase (OleD) from *Streptomyces antibioticus* accepts a wide range of conveniently synthesized aromatic β-D-glucopyranoside donors for the production of UDP-α-D-glucose in the presence of UDP [145]. The directionality of the reaction is dependent on the nitrophenol β-D-glucopyranoside donor, ranging from exergonic favoring UDP-sugar formation to endergonic favoring the production of the aromatic sugar donor [145]. Alternatively, by coupling the 2-chloro-4-nitrophenol glycosides to catalytic amounts of nucleotide diphosphate, the glycosylation of a wide variety of substrates has been demonstrated [217–219]. However, the undesired hydrolysis of 2-chloro-4-nitrophenol glycosides by Leloir glycosyltransferases was observed as well [149]. Hence, separating glycosyltransferase from glycoside hydrolase activity is not always evident in Leloir glycosyltransferases.

3.5. One-Pot Multi Enzyme Cascades

The use of a wide variety of NDP sugar donor regeneration systems coupled to exergonic sacrificial P_i donors inspired the extension of OPME systems towards oligosaccharides and glycans. Key to the success of Leloir glycosyltransferases is the selection of glycosyltransferases with high selectivities towards their substrates, and avoiding the formation of side-products. The human cancer antigen Globo H is a neutral hexasaccharide glycosphingolipid, which has been synthesized chemically by a linear sequence of 11 synthesis steps with predesigned building blocks resulting in a 2.6% overall yield [171]. The optimization of chemical glycosylation using the OptiMer program with custom-synthesized carbohydrate building blocks constructed Globo H in three consecutive steps, with an isolated yield of 41% [220]. The OptiMer program was improved to 83% isolated yield by a one-pot approach using a complex carbohydrate building block containing a Galα1-4Gal bond from a multi-step synthetic route [221], as is shown in Figure 11. A one-pot biocatalytic coupling of readily available nucleotide sugar donors UDP-galactose, UDP-*N*-acetyl-D-glucosamine, and GDP-fucose with three glycosyltransferases resulted in 54% isolated yield without any nucleotide sugar donor regeneration cycles [222]. Additional regeneration of the nucleotide sugar donor improved the overall yield to 94% at large-scale (i.e. 4.5 g allyl Globo-H) [129]. The efficiency of enzymatic glycosylation, the availability of the nucleotide sugar donors, the simplicity of the one-pot reaction, and the mild reaction conditions demonstrate the effectiveness of Leloir glycosyltransferases as catalysts for the production of complex saccharides.

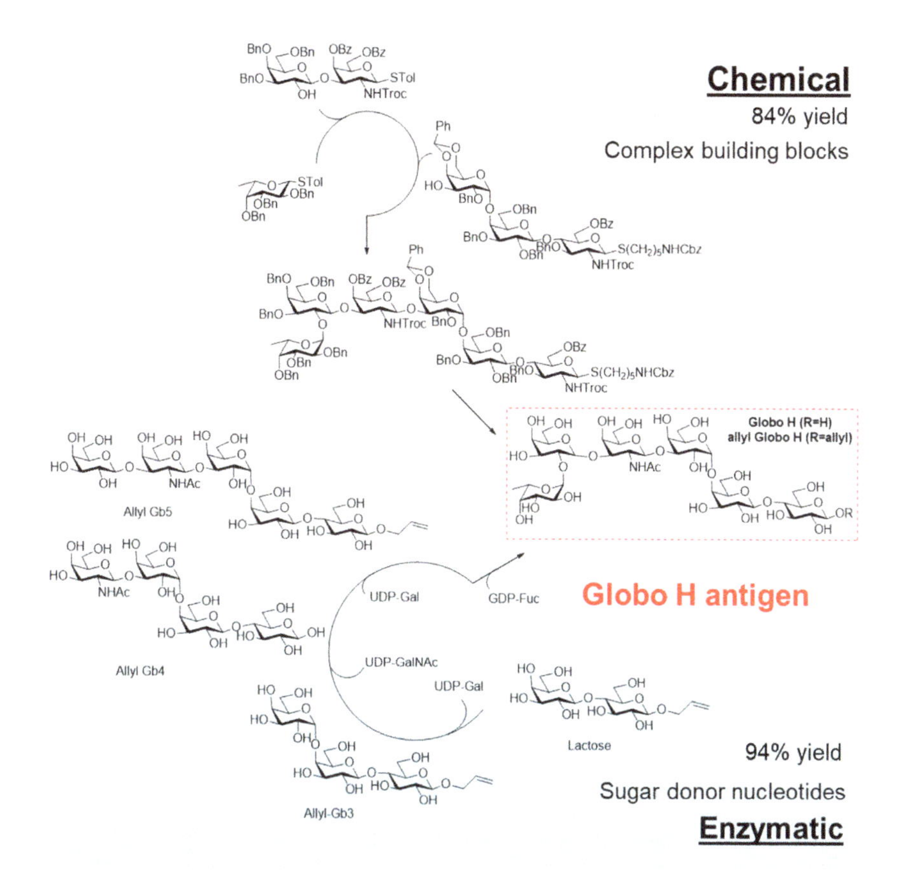

Figure 11. Reaction scheme of both the multistep chemical glycosylation utilizing previously synthesized carbohydrate building blocks [221] and enzymatic glycosylation utilizing nucleotide sugar donors for linear saccharide elongation capped with a terminal vinyl group [129]. The top half of the figure was adapted from [221], copyright 2006, National Academy of Sciences.

Sequential OPME synthesis allows for the coupling of glycosidic linkages which are synthetically challenging, such as sialylation. The production of disialoganglioside cancer antigens GD1b and its derivatives by two sequential α-sialylation reactions has been performed using an engineered Leloir glycosyltransferase [200] with limited nucleotide donor hydrolysis activity (Figure 12) [223]. Lactose was converted to the trisaccharide GM3 using α2-3 sialyltransferase 1 (M144D) from *Pasteuralla multocida*, followed by a second α3,8-sialylation in 85% yield to GD3 fusing α2-3/8-sialyltransferase from *Campylobacter jejuni*. Subsequently, the quantitative enzymatic β1-4-GalNAc coupling to GD2 with β1-4-GalNAc transferase from *Campylobacter jejuni* followed by β1-3-Gal transfer with β1-3-galactosyltransferase resulted in GD1b with an overall isolated yield of 73% [223].

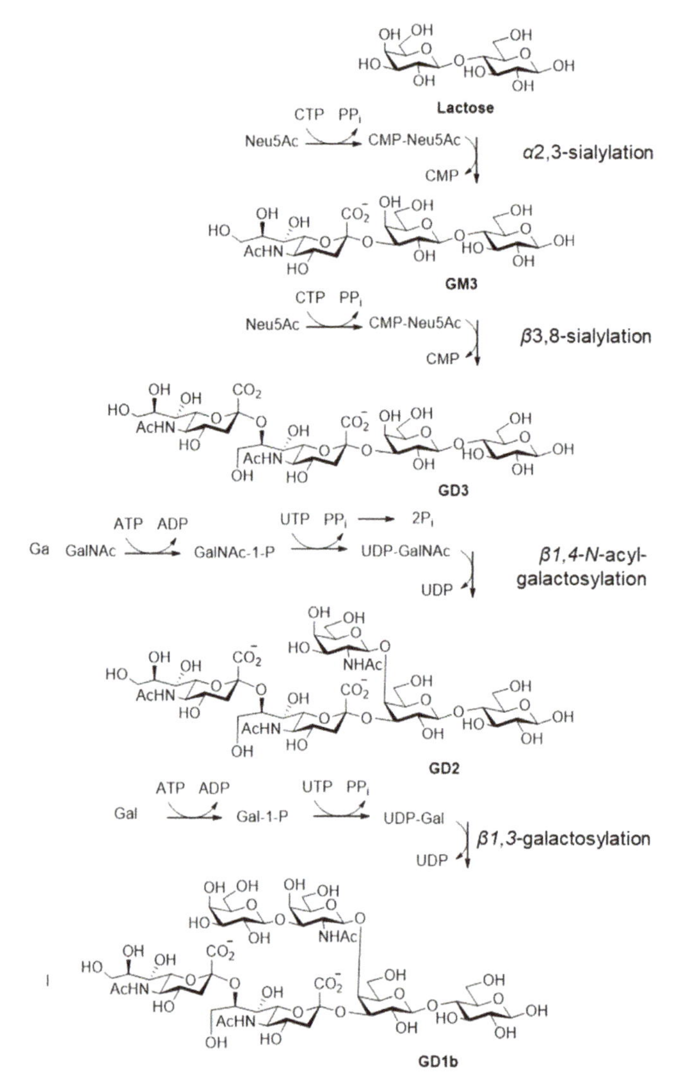

Figure 12. The enzymatic synthesis of GD1b glycan using OPME α2,3-sialylation, α3,8-sialylation, β1,4-*N*-acyl-galactosylation, and β1,3-galactosylation with a sacrificial (re)generation system for *N*-acetylneuraminic acid (Neu5Ac), *N*-acetylgalactosamine (GalNAc), and D-galactose (Gal).

4. Reactor Engineering for (Non)-LeLoir Glycosyltransferases

One of the advancements in integrated biocatalytic processes using glycosyltransferases is the development of automated enzymatic synthesis, using either immobilized substrates or enzymes. Immobilized substrates allow for the spatiotemporal control of the produced glycoconjugate or oligosaccharide in a reactor. Two prominent approaches using immobilized substrates exist: (i) enzymatic *solution-phase synthesis* with tagged products allowing for rapid purification and (ii) enzymatic *solid-phase synthesis* on the surface of an insoluble carrier with soluble substrates and products.

Solution-phase synthesis with a substrate bound to water-soluble [224] or thermo-responsive polymers [225], fluorous- [217,226–236], ion exchange [237], or lipid-like tags [238] has attracted much interest since it can bypass compatibility issues between enzymes and solid carriers. The main disadvantage of solution-phase assembly of oligosaccharides catalyzed by glycosyltransferases is the low catalytic efficiency and affinity for the substrates due to different steric and stereoelectronic properties induced by the substrate-bound tag. Automated enzymatic synthesis of oligosaccharides with Leloir glycosyltransferases has emerged as a promising approach with the thermoresponsive polymer poly(*N*-isopropylacrylamide) (PNIPAM) as a soluble or insoluble support of the sugars, allowing for the synthesis of the antigen of blood groups A, B, and O, as well as the production of the ganglioside GM1 in microchannel reactors (Figure 13) [239]. The Wong group reported a variety of water-soluble polymers of PNIPAM attached to carbohydrates with different linkers to minimize deleterious effects of the presence of the support on the activity of enzymes [24]. As a disadvantage, the covalent attachment of oligosaccharides to PNIPAM requires cleavage of the oligosaccharide with hydrogen peroxide (1M, pH 10), conditions which are incompatible with oxidative labile carbohydrates (i.e., thiosugars).

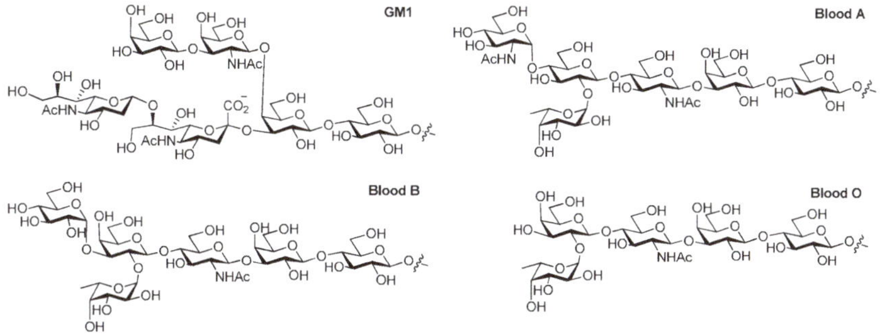

Figure 13. Examples of oligosaccharides synthesized with automated enzymatic synthesis where GM1 is a well-known ganglioside, and the antigens of blood types A, B, and O.

Alternatively, substrate bound *solid-phase synthesis* strategies also received attention [240,241]. The sugar is immobilized on the solid carrier, while the enzyme and sugar nucleotide donor are dissolved in the mobile phase [242–245]. Requirements for the full automatization of immobilized substrates are dependent on the development of (i) efficient enzymes; (ii) availability of glycosylation donors; (iii) the use of carrier and support material and (iv) linkers, spacers, and tags [24]. Continuous biocatalytic processes using *immobilized enzymes* are from an engineering perspective highly attractive due to the ease of reuse of the Leloir glycosyltransferase [189,246–255]. The immobilization of Leloir GTs enables a simplification of the reactor's structure and allows for precise control of the enzymatic glycosylation process [255–257]. The immobilization of glycosyltransferases has been achieved by attachment onto *solid supports* [78,128,258–260], *entrapment* inside a porous carrier [252,261], or *cross-linking* in larger aggregates (CLEA) [262], as is shown in Figure 14. Furthermore, after immobilization the reusability, thermal, pH, and operational stability of the enzymes was often increased [128,262,263]. In particular cases, enzyme immobilization even created a more favorable micro-environment for enzyme activity [264] and selectivity [265,266].

On the other hand, the reaction conditions in a continuous biocatalytic process with immobilized enzymes can be harsh from an engineering point of view, due to vigorous mixing, high pressures, and flow rates. High enzyme stability of Leloir glycosyltransferases is required to tolerate shear stress [267,268]. Also, due to the lack of an universal enzyme immobilization technique many factors must be considered [256,268,269], including mode of *interactions* (i.e., enzyme-substrate/product, enzyme-carrier, substrate/product-carrier), *compatibility* of the carrier to reaction conditions (solvent, temperature, pH, etc.), and the *type of reactor or process* (i.e., batch reactor, packed-bed reactor, basket reactor, microfluidic reactor).

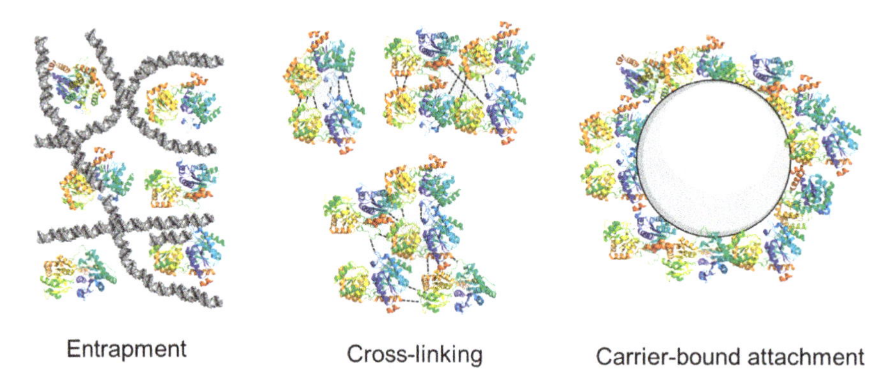

Figure 14. Different modes of immobilization of Leloir glycosyltransferases with entrapment, cross-linking, or carrier-bound attachment.

Reactor Design for Glycosyltransferases

For industrial applications, besides the appropriate immobilization of the enzymes, the chemical character of the carrier's surface is of high importance as are the particle size and pore structure which need to match the type of reactor: i.e. tank vs. tube/column and the mode of operation: periodic/batch vs. continuous flow. Conventional stirred tank reactors (STR) [259,266,270,271] are still frequently applied, but basket and in particular rotating bed reactors [272] are increasingly used in batch operations. In continuous flow applications, packed-bed (micro)reactors [273] have been most popular, but structural (micro)reactors [274,275], lab-on-a-chip [276], and capillary microreactors are on the rise [277]. The STRs are predominantly used in biotechnology owing to cost efficiency and versatility. However, vigorous mixing results in frequent collisions of fine biocatalyst particles resulting in tensile and shearing forces which enhance abrasion or disintegration of the enzyme or its carrier [272]. Moreover, the mixing on a medium and large scale can be insufficient to prevent "hot spot" formation, resulting in enzyme denaturation. But even more importantly, as mass transfer of reactants to fine particles of the biocatalysts or freely suspended enzymes is low, the mass transfer does not keep up the pace of intrinsic activity of the highly active enzymes expressed by turnover frequency values of about or over 10^4 s^{-1}. In effect, it is the mass transfer that strongly impacts, or even fully controls the apparent rate of the enzyme catalyzed reactions carried out in STRs, and that has unfortunately often been overlooked [277,278]. The recovery of the freely suspended biocatalyst particles can be challenging, requiring filtration or centrifugation [279]. The biocatalyst recovery can be simplified using a basket-rotating bed reactor (RBR), which enables simultaneous mixing and effective percolation of liquid through the bed of the catalyst packed in a cylindrical basket, thus avoiding the catalysts destruction, enhancing mass transport, and facilitating separation of the catalyst [280–282]. However, the size of the applied catalyst particles/enzyme carriers needs to be larger than 0.1–0.2 mm [280,281]. The scale-up of RBRs is challenging due to the large size of the rotor as well as the power required for rotating, although a 750-L scale has been successfully demonstrated (Chiralvision. Low-volume continuous flow reactors with flow through channels typically in the range of diameters 0.1–0.5 mm (capillary microreactors) are better scalable [277,283–285]. The problems related to scaling up are resolved by numbering up the (micro) reactors in a parallel process, designated as 'scaling out'. In particular, if thermal effects are not too strong, as is often the case, for the same type of packing the scalability is not a problem. Average linear velocity of flowing reactants and a mean residence time need to be identical on different scales to obtain the same conversion. In addition, owing to narrow channels the time needed for substrate transfer from the center of the liquid channel to the wall-attached catalysts is significantly reduced and the transfer of heat and mass increased, with a positive effect on the apparent reaction rate. Also, the ratio of the active surface of the reactor to its volume may be increased by a factor of 50 or even much more, (e.g., from about 10^3 m^{-1} in industrial reactor to ~5×10^4 m^{-1} for 0.1 mm capillaries) which results in an increase of volumetric productivity (space-time-yield) and hence more profit and lower investment costs [277,278,285]. Moreover, the low reaction volume favors the reduction of potential hazards, particularly important in the case of highly exothermic reactions or when hazardous substances are involved [278,284]. A simplified approach may be used to evaluate process boundaries for the capillary microreactors with enzymes attached to the wall surface [277]. But more in-depth analyses of design and modelling issues for various continuous flow microreactors have also been reported [285].

Different examples of reactors that have been used with glycosyltransferases are shown in Figure 15. *Stirred tank reactors* are flexible in design and operation conditions, but often require high operation costs and vary in the product quality per batch [128,259,266,270,271]. *Microchannel reactors* feature flow-through channels of micrometric sizes that contain the enzyme immobilized on their wall surface [277]. *Packed-bed (micro) reactors* contain fine particles with immobilized enzymes in a flow-through channel, allowing for a higher volumetric activity than microchannel reactors. The heterogenous biocatalyst should not be able to compact to avoid high pressure drops, while mass transport between liquid reactants and catalyst surface is enhanced owing to a more chaotic flow which facilitates mixing. The large pressure drop, even at low flow rates may, however, be a problem if fine catalyst particles are applied. Structured microreactors contain a reactor core made of a porous monolithic structure with open, usually curved pores/channels connected with each other offering excellent mixing and mechanical stability. Moreover, the pressure drop can be significantly reduced and flow rate increased, compared to the packed-bed reactors, and this boosts productivity. The enzymes are immobilized either on the external surface of the monolith or in its pores [273].

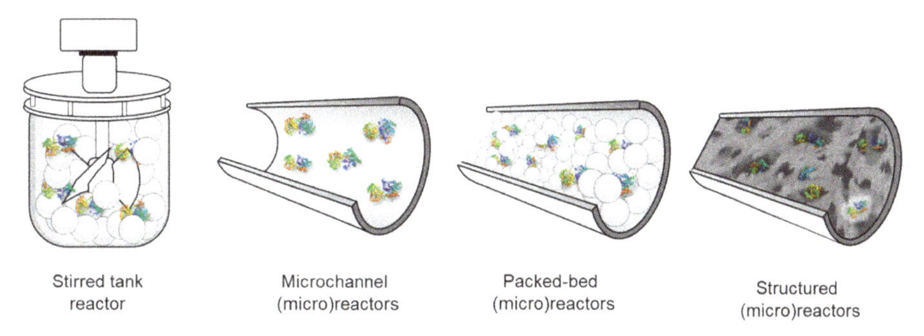

Figure 15. Different reactor types for the immobilization of Leloir glycosyltransferases.

A packed-bed reactor is a commonly used system for continuous production with a heterogeneous biocatalyst, especially because one can immobilize simultaneously different enzymes. Schöffer et al. used glutaraldehyde-activated chitosan spheres [259] and amino- or thiol- functionalized silica particles [260] as a support for cyclodextrin glucosyltransferase immobilization. The silica-based biocatalyst was successfully applied in a packed-bed reactor for continuous cyclodextrin production and maintained 100% of its initial activity after 200 h, whereas activity of the chitosan-based catalyst decreased to 78% of its initial value already after 50 h. This was ascribed to the super packing of spheres, resulting in the reduction of bed height by 45%, and thus in a decrease in the residence/reaction time. However, after washing and re-packing, the spheres recovered 100% of their initial activity.

An interesting effect was observed by Cho, et al. [271], who compared the performance of batch and continuous packed-bed reactors using Eupergit C250L as an enzyme support. The batch reaction was performed for trehalose production from maltose using trehalose synthase for more than 20 h. They found that the product composition was almost the same after 10 h and a maximum trehalose production yield of 25% was established. Trehalose production was improved using a packed-bed bioreactor, wherein the yield reached 42% with a retention time of 100 min. The authors claimed that continuous feeding of fresh substrate into the packed-bed reactor might have eliminated and removed inhibitory compounds from the solution such as by-products formed and accumulated during the batch reaction and thus increased the production yield. A combination of continuous-flow stirred tank reactor (CSTR) with a packed-bed reactor (PBR) was also studied [270]. A highly concentrated starch solution was first partially converted to β-cyclodextrins in a CSTR, which resulted in a decrease of starch viscosity. After that, the reaction mixture was pumped through the PBR. The integrated reactor offered much higher concentration of the final product than each of the reactors separately.

Integration of two or more reactors attracts increasing attention, especially if two or more enzymes are applied. For continuous flow nucleoside synthesis Cattaneo, et al. [273] combined a PBR, filled with purine nucleoside phosphorylase immobilized on silica particles, with uridine phosphorylase immobilized on a silica monolith. In the first approach, co-immobilization of both enzymes on a slightly longer silica-filled PBR was tested, and a high immobilization yield was obtained. However, a very high backpressure of the system (>10 MPa) was registered even at a low flow rate value of 0.1 mL/min, thus hampering the full characterization of the reactor system and resulting in a dramatic drop in conversion. Nonetheless, the application of a monolithic reactor, which exhibited only 6 MPa of pressure drop at a flow rate of 0.5 mL/min, combined with a shorter PBR, showed good activity and stability [273]. The additional advantage of this set-up would be the availability of a single bioreactor that could be used independently, either for "one-enzyme" synthesis or in a different sequence.

Recently Nidetzky, et al. [277], presented an elegant exemplary glycosylation process with sucrose phosphorylase immobilized on the internal surface of a microchannel (Figure 16). Its mathematical model clearly demonstrated that microreactors with the lower hydrolytic channel diameter (d_h) exhibit enhanced performance in terms of conversion and space-time-yield (STY). As the enzyme was attached on the microchannel's wall only the external mass transfer had to be considered [285], and the enzymatic transformations appeared to experience a shift from diffusion to reaction control with miniaturization of d_h (second Damköhler number—$Da_{II} < 1$). Thus, the microreactors, in consequence of their small d_h, emerge as an effective means of gaining full control of the reaction [277]. However, the practical limits to the decrease in d_h, due to high pressure drop, and increased tendency of microchannel clogging have to be kept in mind [274,277]. Therefore, to boost both STY and microreactor performance, a combination of d_h decrease and enzyme activity increase appears to be the rational solution [277].

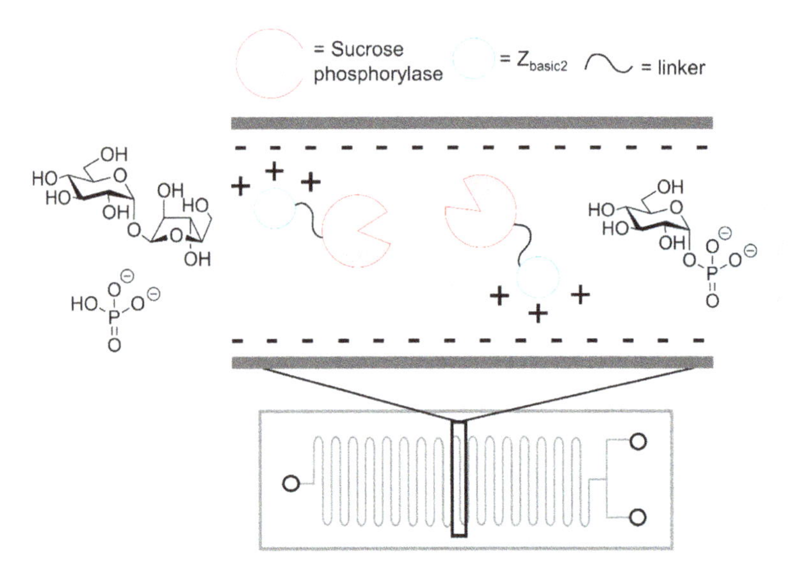

Figure 16. Glass microchannel reactor with immobilized enzyme sucrose phosphorylase attached to the siliceous wall [277]. Attachment occurs via the highly positively charged Z_{basic2} binding module, which binds to the negatively charged silica surface.

To summarize, the application of miniaturized synthetic systems and flow microreactors with immobilized enzymes in particular, attracts attention for the synthesis of more complex carbohydrates. The reported studies clearly demonstrate important advantages of microreactor-based synthetic processes: good stability and high activity that allow for very effective/highly productive syntheses of targeted carbohydrates. While the application of capillary or packed-bed microreactors has been best characterized, structured microreactors are emerging as a class of miniaturized devices that offer additional advantages.

5. Summary and Outlook

The discovery, characterization, and engineering of Leloir glycosyltransferases has expanded the synthetic toolbox to couple, elongate, or branch glycoconjugates, oligosaccharides, and glycans with high regio- and stereoselectivity. Efficient regeneration systems and large-scale production of nucleotide sugar donors under either kinetic or thermodynamic control increased the efficiency of enzymatic glycosylation, reducing overall process costs and the use of stoichiometric amounts of nucleotide phosphates. In this review, the importance of the thermodynamics of glycosylation reactions has been given attention, separating kinetics from thermodynamics for the coupling of a wide variety of NDP sugar donors with acceptors, including sacrificial phosphate donors for NDP (re)generation.

The wide range of different Leloir glycosyltransferases has allowed for the implementation of enzymes in glycochemistry, and consequently, industrially applicable glycosylation methodologies are now in progress. The protein production improved significantly by using (fluorescent) solubility tags, allowing for the production of the biocatalyst in high(er) titers. Despite of the production of enzymes becoming a routine, no industrial application of Leloir glycosyltransferases for the glycosylation of large oligosaccharides has yet been scaled to large volumes, in contrast to non-Leloir glycosyltransferase (i.e., cyclodextrin glycosyltransferases). Until now, one of the main limitations has been the cost-efficient production of NDP sugar donors, which has been an important topic of interest in the last decade. Indeed, due to the advance of many chemical and biocatalytic NDP-sugar production processes, their commercial cost price has been rapidly declining over the last few years.

As the demand for high-value antigens is increasing, the number of biocatalytic glycosylation processes applied for the synthesis of these complex oligosaccharides can be anticipated to rise. One trend is the embracement of automated enzymatic synthesizers for the computer-controlled synthesis of large oligosaccharides using Leloir GTs. In comparison to non-enzymatic coupling strategies, which mostly rely on protection group chemistry, Leloir GTs have now been shown to couple a wide spectrum of unprotected sugar acceptors with excellent regio- and enantioselectivity. Due to the availability of the nucleotide sugar donors and well-established NDP (re)generation systems, Leloir GTs matured as a competitive glycosylation strategy for the enzymatic synthesis of oligosaccharides.

Future developments for industrial enzymatic glycosylation are expected to mostly be focused on optimizing the overall glycosylation process conditions. The main drivers for selecting the most optimal process can be attributed to economic (i.e., revenue of products), development (i.e., time), and process (i.e., performance) parameters. Multiple aspects influence these important aspects, such as the selection of the most optimal reactor design, separating batch versus continuous process operation, the choice of either immobilized substrates to immobilized enzymes, or the use of NDP-regeneration system or stoichiometric use of NDPs. Different enzymatic solution-phase and solid-phase glycosylation strategies have been developed for the automated enzymatic synthesis of carbohydrates. The enzymatic synthesis of an immobilized substrate allows for the purification to become more straight-forward, but requires the stoichiometric use of enzymes. On the other hand, immobilized Leloir GTs in continuous operations have been described sparingly in cascade glycosylation reactions. Immobilized enzymes, ensuring process flexibility and the purification of the produced oligosaccharide are engineering design challenges which have yet to be met.

Author Contributions: Conceptualization, L.M., M.P., D.K., A.P., A.K., I.B., A.B.J., K.S., A.C., D.T., R.S., U.H. and P.-L.H.; writing—original draft preparation, L.M., M.P., D.K., A.P., A.K.; writing—review and editing, L.M., M.P., D.K., A.P., A.K., S.R.M., I.B., A.B.J., K.S., A.C., D.T., R.S., U.H. and P.-L.H.

Funding: The authors are grateful for the generous sponsoring via the ERA-IB scheme, grant ERA-IB-15-110.

Conflicts of Interest: The authors declare no conflict of interest. The funders had no role in the design of the study; in the collection, analyses, or interpretation of data; in the writing of the manuscript, or in the decision to publish the results.

References

1. Fischer, E. Ueber die configuration des traubenzuckers und seiner isomeren. *Ber. Dtsch. Chem. Ges.* **1891**, *24*, 1836–1845. [CrossRef]
2. Fischer, E. Ueber die configuration des traubenzuckers und seiner isomeren. II. *Ber. Dtsch. Chem. Ges.* **1891**, *24*, 2683–2687. [CrossRef]
3. Wöhler, F.; Liebig, J. Ueber die bildung des bittermandelöls. *Ann. Pharm.* **1837**, *22*, 1–24.
4. DiCosimo, R.; McAuliffe, J.; Poulose, A.J.; Bohlmann, G. Industrial use of immobilized enzymes. *Chem. Soc. Rev.* **2013**, *42*, 6437–6474. [CrossRef] [PubMed]
5. Laine, R.A. A calculation of all possible oligosaccharide isomers both branched and linear yields 1.05×10^{12} structures for a reducing hexasaccharide: The isomer barrier to development of single-method saccharide sequencing or synthesis systems. *Glycobiology* **1994**, *4*, 759–767. [CrossRef] [PubMed]
6. Varki, A.; Cummings, R.D.; Aebi, M.; Packer, N.H.; Seeberger, P.H.; Esko, J.D.; Stanley, P.; Hart, G.; Darvill, A.; Kinoshita, T.; et al. Symbol nomenclature for graphical representations of glycans. *Glycobiology* **2015**, *25*, 1323–1324. [CrossRef]
7. Puri, M.; Kaur, A.; Singh, R.S.; Kanwar, J.R. *Immobilized Enzyme Technology for Debittering Citrus Fruit Juices*; Transworld Research Network: Kerala, India, 2008; pp. 91–103.
8. Vila Real, H.J.; Alfaia, A.J.; Calado, A.R.T.; Ribeiro, M.H.L. High pressure-temperature effects on enzymatic activity: Naringin bioconversion. *Food Chem.* **2007**, *102*, 565–570. [CrossRef]
9. Puri, M.; Banerjee, U.C. Production, purification, and characterization of the debittering enzyme naringinase. *Biotechnol. Adv.* **2000**, *18*, 207–217. [CrossRef]
10. Kubota, M. New function and property of trehalose. *New Food Ind.* **2005**, *47*, 17–29.
11. Walmagh, M.; Zhao, R.; Desmet, T. Trehalose analogues: Latest insights in properties and biocatalytic production. *Int. J. Mol. Sci.* **2015**, *16*, 13729. [CrossRef]
12. Szaniawski, A.R.; Spencer, H.G. Effects of immobilized pectinase on the microfiltration of dilute pectin solutions by macroporous titania membranes: Resistance model interpretation. *J. Membr. Sci.* **1997**, *127*, 69–76. [CrossRef]
13. Alkorta, I.; Garbisu, C.; Llama, M.J.; Serra, J.L. Industrial applications of pectic enzymes: A review. *Process Biochem.* **1998**, *33*, 21–28. [CrossRef]
14. Lozano, P.; Manjón, A.; Iborra, J.; Cánovas, M.; Romojaro, F. Kinetic and operational study of a cross-flow reactor with immobilized pectolytic enzymes. *Enzyme Microb. Technol.* **1990**, *12*, 499–505. [CrossRef]
15. Lozano, P.; Manjón, A.; Romojaro, F.; Canovas, M.; Iborra, J.L. A cross-flow reactor with immobilized pectolytic enzymes for juice clarification. *Biotechnol. Lett.* **1987**, *9*, 875–880. [CrossRef] [PubMed]
16. Alkorta, I.; Garbisu, C.; Llama, M.J.; Serra, J.L. Viscosity decrease of pectin and fruit juices catalyzed by pectin lyase from *Penicillium italicum* in batch and continuous-flow membrane reactors. *Biotechnol. Tech.* **1995**, *9*, 95–100. [CrossRef]
17. Lisboa, M.P.; Khan, N.; Martin, C.; Xu, F.-F.; Reppe, K.; Geissner, A.; Govindan, S.; Witzenrath, M.; Pereira, C.L.; Seeberger, P.H. Semisynthetic glycoconjugate vaccine candidate against *Streptococcus pneumoniae* serotype 5. *Proc. Natl. Acad. Sci. USA* **2017**, *114*, 11063–11068. [CrossRef]
18. Emmadi, M.; Khan, N.; Lykke, L.; Reppe, K.G.; Parameswarappa, S.; Lisboa, M.P.; Wienhold, S.-M.; Witzenrath, M.; Pereira, C.L.; Seeberger, P.H. A *Streptococcus pneumoniae* type 2 oligosaccharide glycoconjugate elicits opsonic antibodies and is protective in an animal model of invasive pneumococcal disease. *J. Am. Chem. Soc.* **2017**, *139*, 14783–14791. [CrossRef]

19. Cavallari, M.; Stallforth, P.; Kalinichenko, A.; Rathwell, D.C.K.; Gronewold, T.M.A.; Adibekian, A.; Mori, L.; Landmann, R.; Seeberger, P.H.; De Libero, G. A semisynthetic carbohydrate-lipid vaccine that protects against *S. pneumoniae* in mice. *Nat. Chem. Biol.* **2014**, *10*, 950. [CrossRef]
20. Oldrini, D.; Fiebig, T.; Romano, M.R.; Proietti, D.; Berger, M.; Tontini, M.; De Ricco, R.; Santini, L.; Morelli, L.; Lay, L.; et al. Combined chemical synthesis and tailored enzymatic elongation provide fully synthetic and conjugation-ready *Neisseria meningitidis* serogroup x vaccine antigens. *ACS Chem. Biol.* **2018**, *13*, 984–994. [CrossRef]
21. Marciani, D.J.; Press, J.B.; Reynolds, R.C.; Pathak, A.K.; Pathak, V.; Gundy, L.E.; Farmer, J.T.; Koratich, M.S.; May, R.D. Development of semisynthetic triterpenoid saponin derivatives with immune stimulating activity. *Vaccine* **2000**, *18*, 3141–3151. [CrossRef]
22. Fiebig, T.; Romano, M.R.; Oldrini, D.; Adamo, R.; Tontini, M.; Brogioni, B.; Santini, L.; Berger, M.; Costantino, P.; Berti, F.; et al. An efficient cell free enzyme-based total synthesis of a meningococcal vaccine candidate. *NPJ Vaccines* **2016**, *1*, 16017. [CrossRef] [PubMed]
23. Moremen, K.W.; Ramiah, A.; Stuart, M.; Steel, J.; Meng, L.; Forouhar, F.; Moniz, H.A.; Gahlay, G.; Gao, Z.; Chapla, D.; et al. Expression system for structural and functional studies of human glycosylation enzymes. *Nat. Chem. Biol.* **2017**, *14*, 156. [CrossRef] [PubMed]
24. Wen, L.; Edmunds, G.; Gibbons, C.; Zhang, J.; Gadi, M.R.; Zhu, H.; Fang, J.; Liu, X.; Kong, Y.; Wang, P.G. Toward automated enzymatic synthesis of oligosaccharides. *Chem. Rev.* **2018**, *118*, 8151–8187. [CrossRef] [PubMed]
25. De Bruyn, F.; Maertens, J.; Beauprez, J.; Soetaert, W.; De Mey, M. Biotechnological advances in UDP-sugar based glycosylation of small molecules. *Biotechnol. Adv.* **2015**, *33*, 288–302. [CrossRef] [PubMed]
26. Lairson, L.L.; Henrissat, B.; Davies, G.J.; Withers, S.G. Glycosyltransferases: Structures, functions, and mechanisms. *Ann. Rev. Biochem.* **2008**, *77*, 521–555. [CrossRef] [PubMed]
27. Gloster, T.M. Advances in understanding glycosyltransferases from a structural perspective. *Curr. Opin. Struct. Biol.* **2014**, *28*, 131–141. [CrossRef]
28. Whitfield, C.; Trent, M.S. Biosynthesis and export of bacterial lipopolysaccharides. *Ann. Rev. Biochem.* **2014**, *83*, 99–128. [CrossRef]
29. Typas, A.; Banzhaf, M.; Gross, C.A.; Vollmer, W. From the regulation of peptidoglycan synthesis to bacterial growth and morphology. *Nat. Rev. Microbiol.* **2011**, *10*, 123–136. [CrossRef]
30. Cantarel, B.L.; Coutinho, P.M.; Rancurel, C.; Bernard, T.; Lombard, V.; Henrissat, B. The carbohydrate-active enzymes database (CAZy): An expert resource for glycogenomics. *Nucleic Acids Res.* **2009**, *37*, D233–D238. [CrossRef]
31. Campbell, J.A.; Davies, G.J.; Bulone, V.; Henrissat, B. A classification of nucleotide-diphospho-sugar glycosyltransferases based on amino acid sequence similarities. *Biochem. J.* **1997**, *326 Pt 3*, 929–939. [CrossRef]
32. Sinnott, M. *Carbohydrate Chemistry and Biochemistry: Structure and Mechanism*; Royal Society of Chemistry: Cambridge, UK, 2013; p. 1, online resource.
33. Tvaroška, I. Atomistic insight into the catalytic mechanism of glycosyltransferases by combined quantum mechanics/molecular mechanics (QM/MM) methods. *Carbohydr. Res.* **2015**, *403*, 38–47. [CrossRef] [PubMed]
34. Welzel, P. Syntheses around the transglycosylation step in peptidoglycan biosynthesis. *Chem. Rev.* **2005**, *105*, 4610–4660. [CrossRef]
35. Lovering, A.L.; de Castro, L.H.; Lim, D.; Strynadka, N.C.J. Structural insight into the transglycosylation step of bacterial cell-wall biosynthesis. *Science* **2007**, *315*, 1402–1405. [CrossRef] [PubMed]
36. Timm, M.; Görl, J.; Kraus, M.; Kralj, S.; Hellmuth, H.; Beine, R.; Buchholz, K.; Dijkhuizen, L.; Seibel, J. An unconventional glycosyl transfer reaction: Glucansucrase GTFA functions as an allosyltransferase enzyme. *ChemBioChem* **2013**, *14*, 2423–2426. [CrossRef] [PubMed]
37. Bi, Y.; Hubbard, C.; Purushotham, P.; Zimmer, J. Insights into the structure and function of membrane-integrated processive glycosyltransferases. *Curr. Opin. Struct. Biol.* **2015**, *34*, 78–86. [CrossRef] [PubMed]
38. Breton, C.; Snajdrová, L.; Jeanneau, C.; Koca, J.; Imberty, A. Structures and mechanisms of glycosyltransferases. *Glycobiology* **2006**, *16*, 29R–37R. [CrossRef]
39. Zhang, H.; Zhu, F.; Yang, T.; Ding, L.; Zhou, M.; Li, J.; Haslam, S.M.; Dell, A.; Erlandsen, H.; Wu, H. The highly conserved domain of unknown function 1792 has a distinct glycosyltransferase fold. *Nat. Commun.* **2014**, *5*, 4339. [CrossRef]

40. Kattke, M.D.; Gosschalk, J.E.; Martinez, O.E.; Kumar, G.; Gale, R.T.; Cascio, D.; Sawaya, M.R.; Philips, M.; Brown, E.D.; Clubb, R.T. Structure and mechanism of TagA, a novel membrane-associated glycosyltransferase that produces wall teichoic acids in pathogenic bacteria. *PLoS Pathog.* **2019**, *15*, e1007723. [CrossRef] [PubMed]
41. Ardèvol, A.; Rovira, C. Reaction mechanisms in carbohydrate-active enzymes: Glycoside hydrolases and glycosyltransferases. Insights from ab initio quantum mechanics/molecular mechanics dynamic simulations. *J. Am. Chem. Soc.* **2015**, *137*, 7528–7547. [CrossRef]
42. Gutmann, A.; Lepak, A.; Diricks, M.; Desmet, T.; Nidetzky, B. Glycosyltransferase cascades for natural product glycosylation: Use of plant instead of bacterial sucrose synthases improves the UDP-glucose recycling from sucrose and UDP. *Biotechnol. J.* **2017**, *12*, 1600557. [CrossRef]
43. Ryu, S.-I.; Kim, J.-E.; Kim, E.-J.; Chung, S.-K.; Lee, S.-B. Catalytic reversibility of *Pyrococcus horikoshii* trehalose synthase: Efficient synthesis of several nucleoside diphosphate glucoses with enzyme recycling. *Proc. Biochem.* **2011**, *46*, 128–134. [CrossRef]
44. Nidetzky, B.; Gutmann, A.; Zhong, C. Leloir glycosyltransferases as biocatalysts for chemical production. *ACS Catal.* **2018**, *8*, 6283–6300. [CrossRef]
45. Chang, A.; Singh, S.; Phillips, G.N.; Thorson, J.S. Glycosyltransferase structural biology and its role in the design of catalysts for glycosylation. *Curr. Opin. Biotechnol.* **2011**, *22*, 800–808. [CrossRef] [PubMed]
46. Breton, C.; Fournel-Gigleux, S.; Palcic, M.M. Recent structures, evolution and mechanisms of glycosyltransferases. *Curr. Opin. Struct. Biol.* **2012**, *22*, 540–549. [CrossRef] [PubMed]
47. Blanco Capurro, J.I.; Hopkins, C.W.; Pierdominici Sottile, G.; González Lebrero, M.C.; Roitberg, A.E.; Marti, M.A. Theoretical insights into the reaction and inhibition mechanism of metal-independent retaining glycosyltransferase responsible for mycothiol biosynthesis. *J. Phys. Chem. B* **2017**, *121*, 471–478. [CrossRef] [PubMed]
48. Albesa-Jové, D.; Sainz-Polo, M.Á.; Marina, A.; Guerin, M.E. Structural snapshots of α-1,3-galactosyltransferase with native substrates: Insight into the catalytic mechanism of retaining glycosyltransferases. *Angew. Chem. Int. Ed.* **2017**, *129*, 15049–15053. [CrossRef]
49. Charnock, S.J.; Davies, G.J. Structure of the nucleotide-diphospho-sugar transferase, spsa from *Bacillus subtilis*, in native and nucleotide-complexed forms. *Biochemistry* **1999**, *38*, 6380–6385. [CrossRef]
50. Huber, R.E.; Gaunt, M.T.; Hurlburt, K.L. Binding and reactivity at the "glucose" site of galactosyl-β-galactosidase (*Escherichia coli*). *Arch. Biochem. Biophys.* **1984**, *234*, 151–160. [CrossRef]
51. Ooi, Y.; Mitsuo, N.; Satoh, T. Enzymic synthesis of glycosides of racemic alcohols using beta-galactosidase and separation of the diastereomers by high-performance liquid chromatography using a conventional column. *Chem. Pharm. Bull.* **1985**, *33*, 5547–5550. [CrossRef]
52. Okuyama, M.; Mori, H.; Watanabe, K.; Kimura, A.; Chiba, S. A-glucosidase mutant catalyzes "α-glycosynthase"- type reaction. *Biosci. Biotechnol. Biochem.* **2002**, *66*, 928–933. [CrossRef]
53. Jahn, M.; Marles, J.; Warren, R.A.J.; Withers, S.G. Thioglycoligases: Mutant glycosidases for thioglycoside synthesis. *Angew. Chem. Int. Ed.* **2003**, *42*, 352–354. [CrossRef] [PubMed]
54. Aguirre, A.; Peiru, S.; Eberhardt, F.; Vetcher, L.; Cabrera, R.; Menzella, H.G. Enzymatic hydrolysis of steryl glucosides, major contaminants of vegetable oil-derived biodiesel. *Appl. Microbiol. Biotechnol.* **2014**, *98*, 4033–4040. [CrossRef] [PubMed]
55. Peng, X.; Su, H.; Mi, S.; Han, Y. A multifunctional thermophilic glycoside hydrolase from *Caldicellulosiruptor owensensis* with potential applications in production of biofuels and biochemicals. *Biotechnol. Biofuels* **2016**, *9*, 98. [CrossRef] [PubMed]
56. Ezeilo, U.R.; Zakaria, I.I.; Huyop, F.; Wahab, R.A. Enzymatic breakdown of lignocellulosic biomass: The role of glycosyl hydrolases and lytic polysaccharide monooxygenases. *Biotechnol. Biotechnol. Equip.* **2017**, *31*, 647–662. [CrossRef]
57. Michlmayr, H.; Varga, E.; Malachova, A.; Nguyen, N.T.; Lorenz, C.; Haltrich, D.; Berthiller, F.; Adam, G. A versatile family 3 glycoside hydrolase from *Bifidobacterium adolescentis* hydrolyzes β-glucosides of the *Fusarium* mycotoxins deoxynivalenol, nivalenol, and HT-2 toxin in cereal matrices. *Appl. Environ. Microbiol.* **2015**, *81*, 4885–4893. [CrossRef]
58. Mitsuo, N.; Takeichi, H.; Satoh, T. Synthesis of beta-alkyl glucosides by enzymic transglucosylation. *Chem. Pharm. Bull.* **1984**, *32*, 1183–1187. [CrossRef]

59. Drueckhammer, D.G.; Wong, C.H. Chemoenzymic syntheses of fluoro sugar phosphates and analogs. *J. Org. Chem.* **1985**, *50*, 5912–5913. [CrossRef]
60. Gold, A.M.; Osber, M.P. A-d-glucopyranosyl fluoride: A substrate of sucrose phosphorylase. *Biochem. Biophys. Res. Commun.* **1971**, *42*, 469–474. [CrossRef]
61. Williams, S.J.; Withers, S.G. Glycosyl fluorides in enzymatic reactions. *Carbohydr. Res.* **2000**, *327*, 27–46. [CrossRef]
62. Tolborg, J.F.; Petersen, L.; Jensen, K.J.; Mayer, C.; Jakeman, D.L.; Warren, R.A.J.; Withers, S.G. Solid-phase oligosaccharide and glycopeptide synthesis using glycosynthases. *J. Org. Chem.* **2002**, *67*, 4143–4149. [CrossRef]
63. Fialová, P.; Carmona, A.T.; Robina, I.; Ettrich, R.; Sedmera, P.; Přikrylová, V.; Petrásková-Hušáková, L.; Křen, V. Glycosyl azide—A novel substrate for enzymatic transglycosylations. *Tetrahedron Lett.* **2005**, *46*, 8715–8718. [CrossRef]
64. Strahsburger, E.; de Lacey, A.M.L.; Marotti, I.; DiGioia, D.; Biavati, B.; Dinelli, G. In vivo assay to identify bacteria with β-glucosidase activity. *Electr. J. Biotechnol.* **2017**, *30*, 83–87. [CrossRef]
65. Yasukochi, T.; Fukase, K.; Kusumoto, S. 3-nitro-2-pyridyl glycoside as donor for chemical glycosylation and its application to chemoenzymatic synthesis of oligosaccharide. *Tetrahedron Lett.* **1999**, *40*, 6591–6593. [CrossRef]
66. Chiffoleau-Giraud, V.; Spangenberg, P.; Rabiller, C. Β-galactosidase transferase activity in ice and use of vinyl-β-D-galactoside as donor. *Tetrahedron Asymmetry* **1997**, *8*, 2017–2023. [CrossRef]
67. Vic, G.; Crout, D.H.G. Synthesis of allyl and benzyl β-D-glucopyranosides, and allyl β-D-galactopyranoside from D-glucose or D-galactose and the corresponding alcohol using almond β-D-glucosidase. *Carbohydr. Res.* **1995**, *279*, 315–319. [CrossRef]
68. Hansson, T.; Andersson, M.; Wehtje, E.; Adlercreutz, P. Influence of water activity on the competition between β-glycosidase-catalysed transglycosylation and hydrolysis in aqueous hexanol. *Enzyme Microb. Technol.* **2001**, *29*, 527–534. [CrossRef]
69. Gavlighi, H.A.; Meyer, A.S.; Mikkelsen, J.D. Enhanced enzymatic cellulose degradation by cellobiohydrolases via product removal. *Biotechnol. Lett.* **2013**, *35*, 205–212. [CrossRef]
70. Cote, J.M.; Taylor, E.A. The glycosyltransferases of lps core: A review of four heptosyltransferase enzymes in context. *Int. J. Mol. Sci.* **2017**, *18*, 2256. [CrossRef]
71. Moréra, S.; Imberty, A.; Aschke-Sonnenborn, U.; Rüger, W.; Freemont, P.S. T4 phage β-glucosyltransferase: Substrate binding and proposed catalytic mechanism. *J. Mol. Biol.* **1999**, *292*, 717–730. [CrossRef]
72. Czabany, T.; Schmölzer, K.; Luley-Goedl, C.; Ribitsch, D.; Nidetzky, B. All-in-one assay for β-D-galactoside sialyltransferases: Quantification of productive turnover, error hydrolysis, and site selectivity. *Anal. Biochem.* **2015**, *483*, 47–53. [CrossRef]
73. Ding, L.; Zhao, C.; Qu, J.; Li, Y.; Sugiarto, G.; Yu, H.; Wang, J.; Chen, X. A *Photobacterium* sp. α2–6-sialyltransferase (Psp2,6ST) mutant with an increased expression level and improved activities in sialylating Tn antigens. *Carbohydr. Res.* **2015**, *408*, 127–133. [CrossRef]
74. Schmölzer, K.; Luley-Goedl, C.; Czabany, T.; Ribitsch, D.; Schwab, H.; Weber, H.; Nidetzky, B. Mechanistic study of CMP-Neu5Ac hydrolysis by α2,3-sialyltransferase from *Pasteurella dagmatis*. *FEBS Lett.* **2014**, *588*, 2978–2984. [CrossRef] [PubMed]
75. Sugiarto, G.; Lau, K.; Qu, J.; Li, Y.; Lim, S.; Mu, S.; Ames, J.B.; Fisher, A.J.; Chen, X. A sialyltransferase mutant with decreased donor hydrolysis and reduced sialidase activities for directly sialylating Lewis^x. *ACS Chem. Biol.* **2012**, *7*, 1232–1240. [CrossRef] [PubMed]
76. Schmölzer, K.; Eibinger, M.; Nidetzky, B. Active-site his85 of *Pasteurella dagmatis* sialyltransferase facilitates productive sialyl transfer and so prevents futile hydrolysis of CMP-Neu5Ac. *ChemBioChem* **2017**, *18*, 1544–1550. [CrossRef] [PubMed]
77. Zhao, C.; Wu, Y.; Yu, H.; Shah, I.M.; Li, Y.; Zeng, J.; Liu, B.; Mills, D.A.; Chen, X. One-pot multienzyme (OPME) synthesis of human blood group H antigens and a human milk oligosaccharide (HMOS) with highly active *Thermosynechococcus elongatus* α1–2-fucosyltransferase. *Chem. Commun.* **2016**, *52*, 3899–3902. [CrossRef]
78. Koszagova, R.; Krajcovic, T.; Palencarova-Talafova, K.; Patoprsty, V.; Vikartovska, A.; Pospiskova, K.; Safarik, I.; Nahalka, J. Magnetization of active inclusion bodies: Comparison with centrifugation in repetitive biotransformations. *Microb. Cell Factories* **2018**, *17*, 139. [CrossRef]

79. Mestrom, L.; Marsden, S.R.; Dieters, M.; Achterberg, P.; Stolk, L.; Bento, I.; Hanefeld, U.; Hagedoorn, P.-L. Artificial fusion of mCherry enhanced solubility and stability of trehalose transferase. *Appl. Environ. Microbiol.* **2019**, *85*, e03084-18.
80. Schmölzer, K.; Gutmann, A.; Diricks, M.; Desmet, T.; Nidetzky, B. Sucrose synthase: A unique glycosyltransferase for biocatalytic glycosylation process development. *Biotechnol. Adv.* **2016**, *34*, 88–111. [CrossRef]
81. Bungaruang, L.; Gutmann, A.; Nidetzky, B. Leloir glycosyltransferases and natural product glycosylation: Biocatalytic synthesis of the *C*-glucoside nothofagin, a major antioxidant of redbush herbal tea. *Adv. Synth. Catal.* **2013**, *355*, 2757–2763. [CrossRef]
82. Schmölzer, K.; Lemmerer, M.; Nidetzky, B. Glycosyltransferase cascades made fit for chemical production: Integrated biocatalytic process for the natural polyphenol *C*-glucoside nothofagin. *Biotechnol. Bioeng.* **2018**, *115*, 545–556. [CrossRef]
83. Wang, Q.M.; Peery, R.B.; Johnson, R.B.; Alborn, W.E.; Yeh, W.-K.; Skatrud, P.L. Identification and characterization of a monofunctional glycosyltransferase from *Staphylococcus aureus*. *J. Bacteriol.* **2001**, *183*, 4779–4785. [CrossRef]
84. Chen, L.; Sun, P.; Li, Y.; Yan, M.; Xu, L.; Chen, K.; Ouyang, P. A fusion protein strategy for soluble expression of stevia glycosyltransferase UGT76G1 in *Escherichia coli*. *3 Biotech* **2017**, *7*, 356. [CrossRef]
85. Welner, D.H.; Shin, D.; Tomaleri, G.P.; DeGiovanni, A.M.; Tsai, A.Y.-L.; Tran, H.M.; Hansen, S.F.; Green, D.T.; Scheller, H.V.; Adams, P.D. Plant cell wall glycosyltransferases: High-throughput recombinant expression screening and general requirements for these challenging enzymes. *PLoS ONE* **2017**, *12*, e0177591. [CrossRef]
86. Ortiz-Soto, M.E.; Seibel, J. Expression of functional human sialyltransferases ST3Gal1 and ST6Gal1 in *Escherichia coli*. *PLoS ONE* **2016**, *11*, e0155410. [CrossRef] [PubMed]
87. Plante, O.J.; Palmacci, E.R.; Seeberger, P.H. Automated solid-phase synthesis of oligosaccharides. *Science* **2001**, *291*, 1523–1527. [CrossRef] [PubMed]
88. Pistorio, S.G.; Nigudkar, S.S.; Stine, K.J.; Demchenko, A.V. HPLC-assisted automated oligosaccharide synthesis: Implementation of the autosampler as a mode of the reagent delivery. *J. Org. Chem.* **2016**, *81*, 8796–8805. [CrossRef]
89. Nokami, T.; Hayashi, R.; Saigusa, Y.; Shimizu, A.; Liu, C.-Y.; Mong, K.-K.T.; Yoshida, J.-I. Automated solution-phase synthesis of oligosaccharides via iterative electrochemical assembly of thioglycosides. *Org. Lett.* **2013**, *15*, 4520–4523. [CrossRef]
90. Huang, T.-Y.; Zulueta, M.M.L.; Hung, S.-C. Regioselective one-pot protection, protection–glycosylation and protection–glycosylation–glycosylation of carbohydrates: A case study with d-glucose. *Org. Biomol. Chem.* **2014**, *12*, 376–382. [CrossRef]
91. Wang, C.-C.; Lee, J.-C.; Luo, S.-Y.; Kulkarni, S.S.; Huang, Y.-W.; Lee, C.-C.; Chang, K.-L.; Hung, S.-C. Regioselective one-pot protection of carbohydrates. *Nature* **2007**, *446*, 896. [CrossRef]
92. Walvoort, M.T.C.; Volbeda, A.G.; Reintjens, N.R.M.; van den Elst, H.; Plante, O.J.; Overkleeft, H.S.; van der Marel, G.A.; Codée, J.D.C. Automated solid-phase synthesis of hyaluronan oligosaccharides. *Org. Lett.* **2012**, *14*, 3776–3779. [CrossRef]
93. Geert Volbeda, A.; van Mechelen, J.; Meeuwenoord, N.; Overkleeft, H.S.; van der Marel, G.A.; Codée, J.D.C. Cyanopivaloyl ester in the automated solid-phase synthesis of oligorhamnans. *J. Org. Chem.* **2017**, *82*, 12992–13002. [CrossRef]
94. Zhou, J.; Lv, S.; Zhang, D.; Xia, F.; Hu, W. Deactivating influence of 3-*O*-glycosyl substituent on anomeric reactivity of thiomannoside observed in oligomannoside synthesis. *J. Org. Chem.* **2017**, *82*, 2599–2621. [CrossRef]
95. Kaeothip, S.; Demchenko, A.V. On orthogonal and selective activation of glycosyl thioimidates and thioglycosides: Application to oligosaccharide assembly. *J. Org. Chem.* **2011**, *76*, 7388–7398. [CrossRef]
96. Kanie, O.; Ito, Y.; Ogawa, T. Orthogonal glycosylation strategy in oligosaccharide synthesis. *J. Am. Chem. Soc.* **1994**, *116*, 12073–12074. [CrossRef]
97. Tang, S.-L.; Linz, L.B.; Bonning, B.C.; Pohl, N.L.B. Automated solution-phase synthesis of insect glycans to probe the binding affinity of pea enation mosaic virus. *J. Org. Chem.* **2015**, *80*, 10482–10489. [CrossRef]
98. Senf, D.; Ruprecht, C.; de Kruijff, G.H.M.; Simonetti, S.O.; Schuhmacher, F.; Seeberger, P.H.; Pfrengle, F. Active site mapping of xylan-deconstructing enzymes with arabinoxylan oligosaccharides produced by automated glycan assembly. *Chem. Eur. J.* **2017**, *23*, 3197–3205. [CrossRef]

99. Naresh, K.; Schumacher, F.; Hahm, H.S.; Seeberger, P.H. Pushing the limits of automated glycan assembly: Synthesis of a 50mer polymannoside. *Chem. Commun.* **2017**, *53*, 9085–9088. [CrossRef]
100. Weishaupt, M.W.; Hahm, H.S.; Geissner, A.; Seeberger, P.H. Automated glycan assembly of branched β-(1,3)-glucans to identify antibody epitopes. *Chem. Commun.* **2017**, *53*, 3591–3594. [CrossRef]
101. Hahm, H.S.; Hurevich, M.; Seeberger, P.H. Automated assembly of oligosaccharides containing multiple cis-glycosidic linkages. *Nat. Commun.* **2016**, *7*, 12482. [CrossRef]
102. Hurevich, M.; Seeberger, P.H. Automated glycopeptide assembly by combined solid-phase peptide and oligosaccharide synthesis. *Chem. Commun.* **2014**, *50*, 1851–1853. [CrossRef]
103. Kandasamy, J.; Hurevich, M.; Seeberger, P.H. Automated solid phase synthesis of oligoarabinofuranosides. *Chem. Commun.* **2013**, *49*, 4453–4455. [CrossRef]
104. Eller, S.; Collot, M.; Yin, J.; Hahm, H.S.; Seeberger, P.H. Automated solid-phase synthesis of chondroitin sulfate glycosaminoglycans. *Angew. Chem. Int. Ed.* **2013**, *52*, 5858–5861. [CrossRef] [PubMed]
105. Kröck, L.; Esposito, D.; Castagner, B.; Wang, C.-C.; Bindschädler, P.; Seeberger, P.H. Streamlined access to conjugation-ready glycans by automated synthesis. *Chem. Sci.* **2012**, *3*, 1617–1622. [CrossRef]
106. Tanaka, H.; Nishiura, Y.; Takahashi, T. Stereoselective synthesis of oligo-α-(2,8)-sialic acids. *J. Am. Chem. Soc.* **2006**, *128*, 7124–7125. [CrossRef] [PubMed]
107. Crich, D.; Li, W. *O*-sialylation with *N*-acetyl-5-*N*,4-*O*-carbonyl-protected thiosialoside donors in dichloromethane: Facile and selective cleavage of the oxazolidinone ring. *J. Org. Chem.* **2007**, *72*, 2387–2391. [CrossRef] [PubMed]
108. Crich, D.; Wu, B. Stereoselective iterative one-pot synthesis of *N*-glycolylneuraminic acid-containing oligosaccharides. *Org. Lett.* **2008**, *10*, 4033–4035. [CrossRef] [PubMed]
109. Hashimoto, Y.; Tanikawa, S.; Saito, R.; Sasaki, K. β-stereoselective mannosylation using 2,6-lactones. *J. Am. Chem. Soc.* **2016**, *138*, 14840–14843. [CrossRef] [PubMed]
110. Chiu, H.-T.; Lin, Y.-C.; Lee, M.-N.; Chen, Y.-L.; Wang, M.-S.; Lai, C.-C. Biochemical characterization and substrate specificity of the gene cluster for biosyntheses of K-252a and its analogs by in vitro heterologous expression system of *Escherichia coli*. *Mol. BioSyst.* **2009**, *5*, 1192–1203. [CrossRef]
111. Gao, Q.; Zhang, C.; Blanchard, S.; Thorson, J.S. Deciphering indolocarbazole and enediyne aminodideoxypentose biosynthesis through comparative genomics: Insights from the AT2433 biosynthetic locus. *Chem. Biol.* **2006**, *13*, 733–743. [CrossRef]
112. Sánchez, C.; Méndez, C.; Salas, J.A. Indolocarbazole natural products: Occurrence, biosynthesis, and biological activity. *Nat. Prod. Rep.* **2006**, *23*, 1007–1045. [CrossRef]
113. Zhang, C.; Albermann, C.; Fu, X.; Peters, N.R.; Chisholm, J.D.; Zhang, G.; Gilbert, E.J.; Wang, P.G.; Van Vranken, D.L.; Thorson, J.S. RebG- and RebM-catalyzed indolocarbazole diversification. *ChemBioChem* **2006**, *7*, 795–804. [CrossRef]
114. Salas, A.P.; Zhu, L.; Sánchez, C.; Braña, A.F.; Rohr, J.; Méndez, C.; Salas, J.A. Deciphering the late steps in the biosynthesis of the anti-tumour indolocarbazole staurosporine: Sugar donor substrate flexibility of the StaG glycosyltransferase. *Mol. Microbiol.* **2005**, *58*, 17–27. [CrossRef]
115. Hyun, C.-G.; Bililign, T.; Liao, J.; Thorson, J.S. The biosynthesis of indolocarbazoles in a heterologous *E. coli* host. *ChemBioChem* **2003**, *4*, 114–117. [CrossRef]
116. Sánchez, C.; Butovich, I.A.; Braña, A.F.; Rohr, J.; Méndez, C.; Salas, J.A. The biosynthetic gene cluster for the antitumor rebeccamycin: Characterization and generation of indolocarbazole derivatives. *Chem. Biol.* **2002**, *9*, 519–531. [CrossRef]
117. Ohuchi, T.; Ikeda-Araki, A.; Watanabe-Sakamoto, A.; Kojiri, K.; Nagashima, M.; Okanishi, M.; Suda, H. Cloning and expression of a gene encoding *N*-glycosyltransferase (ngt) from *Saccharothrix aerocolonigenes* ATCC39243. *J. Antibiot.* **2000**, *53*, 393–403. [CrossRef]
118. Guo, Z.; Li, J.; Qin, H.; Wang, M.; Lv, X.; Li, X.; Chen, Y. Biosynthesis of the carbamoylated D-gulosamine moiety of streptothricins: Involvement of a guanidino-*N*-glycosyltransferase and an *N*-acetyl-D-gulosamine deacetylase. *Angew. Chem. Int. Ed.* **2015**, *54*, 5175–5178. [CrossRef]
119. Gawthorne, J.A.; Tan, N.Y.; Bailey, U.-M.; Davis, M.R.; Wong, L.W.; Naidu, R.; Fox, K.L.; Jennings, M.P.; Schulz, B.L. Selection against glycosylation sites in potential target proteins of the general HMWC *N*-glycosyltransferase in *Haemophilus influenzae*. *Biochem. Biophys. Res. Commun.* **2014**, *445*, 633–638. [CrossRef]

120. Naegeli, A.; Lin, C.-W.; Aebi, M.; Michaud, G.; Darbre, T.; Reymond, J.-L.; Schubert, M.; Lizak, C. Substrate specificity of cytoplasmic *N*-glycosyltransferase. *J. Biol. Chem.* **2014**, *289*, 24521–24532. [CrossRef]
121. Choi, K.-J.; Grass, S.; Paek, S.; St. Geme, J.W., III; Yeo, H.-J. The *Actinobacillus pleuropneumoniae* HMW1C-like glycosyltransferase mediates *N*-linked glycosylation of the *Haemophilus influenzae* HMW1 adhesin. *PLoS ONE* **2011**, *5*, e15888. [CrossRef]
122. Grass, S.; Lichti, C.F.; Townsend, R.R.; Gross, J.; St. Geme, J.W., III. The *Haemophilus influenzae* HMW1C protein is a glycosyltransferase that transfers hexose residues to asparagine sites in the HMW1 adhesin. *PLoS Pathog.* **2010**, *6*, e1000919. [CrossRef]
123. Zhao, P.; Bai, L.; Ma, J.; Zeng, Y.; Li, L.; Zhang, Y.; Lu, C.; Dai, H.; Wu, Z.; Li, Y.; et al. Amide *N*-glycosylation by Asm25, an *N*-glycosyltransferase of ansamitocins. *Chem. Biol.* **2008**, *15*, 863–874. [CrossRef] [PubMed]
124. Magarvey, N.A.; Haltli, B.; He, M.; Greenstein, M.; Hucul, J.A. Biosynthetic pathway for mannopeptimycins, lipoglycopeptide antibiotics active against drug-resistant gram-positive pathogens. *Antimicrob. Agents Chemother.* **2006**, *50*, 2167–2177. [CrossRef]
125. Chen, D.; Chen, R.; Wang, R.; Li, J.; Xie, K.; Bian, C.; Sun, L.; Zhang, X.; Liu, J.; Yang, L.; et al. Probing the catalytic promiscuity of a regio- and stereospecific *C*-glycosyltransferase from *Mangifera indica*. *Angew. Chem. Int. Ed.* **2015**, *54*, 12678–12682. [CrossRef]
126. Foshag, D.; Campbell, C.; Pawelek, P.D. The *C*-glycosyltransferase IroB from pathogenic *Escherichia coli*: Identification of residues required for efficient catalysis. *Biochim. Biophys. Acta (BBA) Proteins Proteom.* **2014**, *1844*, 1619–1630. [CrossRef]
127. Gutmann, A.; Krump, C.; Bungaruang, L.; Nidetzky, B. A two-step *O*- to *C*-glycosidic bond rearrangement using complementary glycosyltransferase activities. *Chem. Commun.* **2014**, *50*, 5465–5468. [CrossRef]
128. Panek, A.; Pietrow, O.; Synowiecki, J.; Filipkowski, P. Immobilization on magnetic nanoparticles of the recombinant trehalose synthase from *Deinococcus geothermalis*. *Food Bioprod. Process.* **2013**, *91*, 632–637. [CrossRef]
129. Tsai, T.-I.; Lee, H.-Y.; Chang, S.-H.; Wang, C.-H.; Tu, Y.-C.; Lin, Y.-C.; Hwang, D.-R.; Wu, C.-Y.; Wong, C.-H. Effective sugar nucleotide regeneration for the large-scale enzymatic synthesis of Globo H and SSEA4. *J. Am. Chem. Soc.* **2013**, *135*, 14831–14839. [CrossRef]
130. Gutmann, A.; Nidetzky, B. Switching between *O*- and *C*-glycosyltransferase through exchange of active-site motifs. *Angew. Chem. Int. Ed.* **2012**, *51*, 12879–12883. [CrossRef]
131. Härle, J.; Günther, S.; Lauinger, B.; Weber, M.; Kammerer, B.; Zechel, D.L.; Luzhetskyy, A.; Bechthold, A. Rational design of an aryl-*C*-glycoside catalyst from a natural product *O*-glycosyltransferase. *Chem. Biol.* **2011**, *18*, 520–530.
132. Mittler, M.; Bechthold, A.; Schulz, G.E. Structure and action of the C-C bond-forming glycosyltransferase UrdGT2 involved in the biosynthesis of the antibiotic urdamycin. *J. Mol. Biol.* **2007**, *372*, 67–76. [CrossRef]
133. Baig, I.; Kharel, M.; Kobylyanskyy, A.; Zhu, L.; Rebets, Y.; Ostash, B.; Luzhetskyy, A.; Bechthold, A.; Fedorenko, V.A.; Rohr, J. On the acceptor substrate of *C*-glycosyltransferase UrdGT2: Three prejadomycin *C*-glycosides from an engineered mutant of *Streptomyces globisporus* 1912 ΔlndE(UrdGT2). *Angew. Chem. Int. Ed.* **2006**, *45*, 7842–7846. [CrossRef]
134. Liu, T.; Kharel, M.K.; Fischer, C.; McCormick, A.; Rohr, J. Inactivation of gilGT, encoding a *C*-glycosyltransferase, and gilOIII, encoding a P450 enzyme, allows the details of the late biosynthetic pathway to gilvocarcin V to be delineated. *ChemBioChem* **2006**, *7*, 1070–1077. [CrossRef]
135. Fischbach, M.A.; Lin, H.; Liu, D.R.; Walsh, C.T. In vitro characterization of IroB, a pathogen-associated *C*-glycosyltransferase. *Proc. Natl. Acad. Sci. USA* **2005**, *102*, 571–576. [CrossRef]
136. Bililign, T.; Hyun, C.-G.; Williams, J.S.; Czisny, A.M.; Thorson, J.S. The hedamycin locus implicates a novel aromatic PKS priming mechanism. *Chem. Biol.* **2004**, *11*, 959–969. [CrossRef]
137. Kopycki, J.; Wieduwild, E.; Kohlschmidt, J.; Brandt, W.; Stepanova, A.N.; Alonso, J.M.; Pedras, M.; Soledade, C.; Abel, S.; Grubb, C.D.; et al. Kinetic analysis of *Arabidopsis* glucosyltransferase UGT74B1 illustrates a general mechanism by which enzymes can escape product inhibition. *Biochem. J.* **2013**, *450*, 37–46. [CrossRef]
138. Wang, H.; van der Donk, W.A. Substrate selectivity of the sublancin *S*-glycosyltransferase. *J. Am. Chem. Soc.* **2011**, *133*, 16394–16397. [CrossRef]

139. Almendros, M.; Danalev, D.; François-Heude, M.; Loyer, P.; Legentil, L.; Nugier-Chauvin, C.; Daniellou, R.; Ferrières, V. Exploring the synthetic potency of the first furanothioglycoligase through original remote activation. *Org. Biomol. Chem.* **2011**, *9*, 8371–8378. [CrossRef]
140. Douglas Grubb, C.; Zipp, B.J.; Ludwig-Müller, J.; Masuno, M.N.; Molinski, T.F.; Abel, S. *Arabidopsis* glucosyltransferase UGT74B1 functions in glucosinolate biosynthesis and auxin homeostasis. *Plant J.* **2004**, *40*, 893–908. [CrossRef]
141. Calderon, A.D.; Zhou, J.; Guan, W.; Wu, Z.; Guo, Y.; Bai, J.; Li, Q.; Wang, P.G.; Fang, J.; Li, L. An enzymatic strategy to asymmetrically branched *N*-glycans. *Org. Biomol. Chem.* **2017**, *15*, 7258–7262. [CrossRef]
142. Gagarinov, I.A.; Li, T.; Toraño, J.S.; Caval, T.; Srivastava, A.D.; Kruijtzer, J.A.W.; Heck, A.J.R.; Boons, G.-J. Chemoenzymatic approach for the preparation of asymmetric bi-, tri-, and tetra-antennary *N*-glycans from a common precursor. *J. Am. Chem. Soc.* **2017**, *139*, 1011–1018. [CrossRef]
143. Xiao, Z.; Guo, Y.; Liu, Y.; Li, L.; Zhang, Q.; Wen, L.; Wang, X.; Kondengaden, S.M.; Wu, Z.; Zhou, J.; et al. Chemoenzymatic synthesis of a library of human milk oligosaccharides. *J. Org. Chem.* **2016**, *81*, 5851–5865. [CrossRef]
144. Zhang, C.; Griffith, B.R.; Fu, Q.; Albermann, C.; Fu, X.; Lee, I.-K.; Li, L.; Thorson, J.S. Exploiting the reversibility of natural product glycosyltransferase-catalyzed reactions. *Science* **2006**, *313*, 1291–1294. [CrossRef]
145. Gantt, R.W.; Peltier-Pain, P.; Cournoyer, W.J.; Thorson, J.S. Using simple donors to drive the equilibria of glycosyltransferase-catalyzed reactions. *Nat. Chem. Biol.* **2011**, *7*, 685. [CrossRef]
146. Gantt, R.W.; Thorson, J.S. Chapter seventeen—High-throughput colorimetric assays for nucleotide sugar formation and glycosyl transfer. In *Methods in Enzymology*; Hopwood, D.A., Ed.; Academic Press: Cambridge, MA, USA, 2012; Volume 516, pp. 345–360.
147. Peltier-Pain, P.; Marchillo, K.; Zhou, M.; Andes, D.R.; Thorson, J.S. Natural product disaccharide engineering through tandem glycosyltransferase catalysis reversibility and neoglycosylation. *Org. Lett.* **2012**, *14*, 5086–5089. [CrossRef]
148. Zhou, M.; Hamza, A.; Zhan, C.-G.; Thorson, J.S. Assessing the regioselectivity of OleD-catalyzed glycosylation with a diverse set of acceptors. *J. Nat. Prod.* **2013**, *76*, 279–286. [CrossRef]
149. Gantt, R.W.; Peltier-Pain, P.; Singh, S.; Zhou, M.; Thorson, J.S. Broadening the scope of glycosyltransferase-catalyzed sugar nucleotide synthesis. *Proc. Natl. Acad. Sci. USA* **2013**, *110*, 7648–7653. [CrossRef]
150. Jourdan, P.S.; Mansell, R.L. Isolation and partial characterization of three glucosyl transferases involved in the biosynthesis of flavonol triglucosides in *Pisum sativum* L. *Archives of Biochemistry and Biophysics* **1982**, *213*, 434–443. [CrossRef]
151. Shearer, J.; Graham, T.E. Novel aspects of skeletal muscle glycogen and its regulation during rest and exercise. *Exerc. Sport Sci. Rev.* **2004**, *32*, 120–126. [CrossRef]
152. Mavrovouniotis, M.L. Estimation of standard Gibbs energy changes of biotransformations. *J. Biol. Chem.* **1991**, *266*, 14440–14445.
153. Jankowski, M.D.; Henry, C.S.; Broadbelt, L.J.; Hatzimanikatis, V. Group contribution method for thermodynamic analysis of complex metabolic networks. *Biophys. J.* **2008**, *95*, 1487–1499. [CrossRef]
154. Bar-Even, A.; Flamholz, A.; Davidi, D.; Noor, E.; Milo, R.; Lubling, Y. An integrated open framework for thermodynamics of reactions that combines accuracy and coverage. *Bioinformatics* **2012**, *28*, 2037–2044.
155. Noor, E.; Haraldsdóttir, H.S.; Milo, R.; Fleming, R.M.T. Consistent estimation of Gibbs energy using component contributions. *PLoS Comput. Biol.* **2013**, *9*, e1003098. [CrossRef]
156. Goldberg, R.N.; Bhat, T.N.; Tewari, Y.B. Thermodynamics of enzyme-catalyzed reactions—A database for quantitative biochemistry. *Bioinformatics* **2004**, *20*, 2874–2877. [CrossRef]
157. Minakami, S.; Yoshikawa, H. Thermodynamic considerations on erythrocyte glycolysis. *Biochem. Biophys. Res. Commun.* **1965**, *18*, 345–349. [CrossRef]
158. Held, C.; Sadowski, G. Thermodynamics of bioreactions. *Annual Rev. Chem. Biomol. Eng.* **2016**, *7*, 395–414. [CrossRef]
159. Ozaki, S.-I.; Imai, H.; Iwakiri, T.; Sato, T.; Shimoda, K.; Nakayama, T.; Hamada, H. Regioselective glucosidation of trans-resveratrol in *Escherichia coli* expressing glucosyltransferase from *Phytolacca americana*. *Biotechnol. Lett.* **2012**, *34*, 475–481. [CrossRef]

160. Yahyaa, M.; Davidovich-Rikanati, R.; Eyal, Y.; Sheachter, A.; Marzouk, S.; Lewinsohn, E.; Ibdah, M. Identification and characterization of UDP-glucose: Phloretin 4′-*O*-glycosyltransferase from *Malus* x *domestica* Borkh. *Phytochemistry* **2016**, *130*, 47–55. [CrossRef]
161. Zhang, T.; Liang, J.; Wang, P.; Xu, Y.; Wang, Y.; Wei, X.; Fan, M. Purification and characterization of a novel phloretin-2′-*O*-glycosyltransferase favoring phloridzin biosynthesis. *Sci. Rep.* **2016**, *6*, 35274. [CrossRef]
162. Lim, E.-K.; Doucet, C.J.; Li, Y.; Elias, L.; Worrall, D.; Spencer, S.P.; Ross, J.; Bowles, D.J. The activity of *Arabidopsis* glycosyltransferases toward salicylic acid, 4-hydroxybenzoic acid, and other benzoates. *J. Biol. Chem.* **2002**, *277*, 586–592. [CrossRef]
163. Tadera, K.; Yagi, F.; Kobayashi, A. Specificity of a particulate glycosyltransferase in seedlings of *Pisum sativum* L. Which catalyzes the formation of 5′-*O*-(β-D-glucopyranosyl)pyridoxine. *J. Nutr. Sci. Vitaminol.* **1982**, *28*, 359–366. [CrossRef]
164. Lin, J.-S.; Huang, X.-X.; Li, Q.; Cao, Y.; Bao, Y.; Meng, X.-F.; Li, Y.-J.; Fu, C.; Hou, B.-K. UDP-glycosyltransferase 72B1 catalyzes the glucose conjugation of monolignols and is essential for the normal cell wall lignification in *Arabidopsis thaliana*. *Plant J.* **2016**, *88*, 26–42. [CrossRef]
165. Ibrahim, R.K.; Grisebach, H. Purification and properties of UDP-glucose: Coniferyl alcohol glucosyltransferase from suspension cultures of Paul's scarlet rose. *Arch. Biochem. Biophys.* **1976**, *176*, 700–708. [CrossRef]
166. Hyung Ko, J.; Gyu Kim, B.; Joong-Hoon, A. Glycosylation of flavonoids with a glycosyltransferase from *Bacillus cereus*. *FEMS Microbiol. Lett.* **2006**, *258*, 263–268. [CrossRef]
167. Ostrowski, M.; Hetmann, A.; Jakubowska, A. Indole-3-acetic acid UDP-glucosyltransferase from immature seeds of pea is involved in modification of glycoproteins. *Phytochemistry* **2015**, *117*, 25–33. [CrossRef]
168. Tahara, K.; Nishiguchi, M.; Frolov, A.; Mittasch, J.; Milkowski, C. Identification of UDP glucosyltransferases from the aluminum-resistant tree *Eucalyptus camaldulensis* forming β-glucogallin, the precursor of hydrolyzable tannins. *Phytochemistry* **2018**, *152*, 154–161. [CrossRef]
169. Wang, Q.; Xu, Y.; Xu, J.; Wang, X.; Shen, C.; Zhang, Y.; Liu, X.; Yu, B.; Zhang, J. Molecular cloning and expression of a glycosyltransferase from *Bacillus subtilis* for modification of morin and related polyphenols. *Biotechnol. Lett.* **2017**, *39*, 1229–1235. [CrossRef]
170. Marcinek, H.; Weyler, W.; Deus-Neumann, B.; Zenk, M.H. Indoxyl-UDPG-glucosyltransferase from *Baphicacanthus cusia*. *Phytochemistry* **2000**, *53*, 201–207. [CrossRef]
171. Mandal, S.S.; Liao, G.; Guo, Z. Chemical synthesis of the tumor-associated Globo H antigen. *RSC Adv.* **2015**, *5*, 23311–23319. [CrossRef]
172. Frear, D.S. Herbicide metabolism in plants—I: Purification and properties of UDP-glucose: arylamine *N*-glucosyl-transferase from soybean. *Phytochemistry* **1968**, *7*, 381–390. [CrossRef]
173. Martin, R.C.; Mok, M.C.; Mok, D.W.S. Isolation of a cytokinin gene, ZOG1, encoding zeatin *O*-glucosyltransferase from *Phaseolus lunatus*. *Proc. Natl. Acad. Sci. USA* **1999**, *96*, 284–289. [CrossRef]
174. Landmann, C.; Fink, B.; Schwab, W. FaGT2: A multifunctional enzyme from strawberry (*Fragaria* x *ananassa*) fruits involved in the metabolism of natural and xenobiotic compounds. *Planta* **2007**, *226*, 417–428. [CrossRef]
175. Lunkenbein, S.; Bellido, M.; Aharoni, A.; Salentijn, E.M.J.; Kaldenhoff, R.; Coiner, H.A.; Muñoz-Blanco, J.; Schwab, W. Cinnamate metabolism in ripening fruit. Characterization of a UDP-glucose: cinnamate glucosyltransferase from strawberry. *Plant Physiol.* **2006**, *140*, 1047–1058. [CrossRef]
176. Trobo-Maseda, L.; Orrego, A.H.; Moreno-Pérez, S.; Fernández-Lorente, G.; Guisan, J.M.; Rocha-Martin, J. Stabilization of multimeric sucrose synthase from *Acidithiobacillus caldus* via immobilization and post-immobilization techniques for synthesis of UDP-glucose. *Appl. Microbiol. Biotechnol.* **2018**, *102*, 773–787. [CrossRef]
177. Dewitte, G.; Walmagh, M.; Diricks, M.; Lepak, A.; Gutmann, A.; Nidetzky, B.; Desmet, T. Screening of recombinant glycosyltransferases reveals the broad acceptor specificity of stevia UGT-76G1. *J. Biotechnol.* **2016**, *233*, 49–55. [CrossRef]
178. Gutmann, A.; Nidetzky, B. Unlocking the potential of leloir glycosyltransferases for applied biocatalysis: Efficient synthesis of uridine 5′-diphosphate-glucose by sucrose synthase. *Adv. Synth. Catal.* **2016**, *358*, 3600–3609. [CrossRef]
179. Ryu, S.-I.; Park, C.-S.; Cha, J.; Woo, E.-J.; Lee, S.-B. A novel trehalose-synthesizing glycosyltransferase from *Pyrococcus horikoshii*: Molecular cloning and characterization. *Biochem. Biophys. Res. Commun.* **2005**, *329*, 429–436. [CrossRef]

180. Tewari, Y.B.; Goldberg, R.N. Thermodynamics of hydrolysis of disaccharides: Lactulose, α-D-melibiose, palatinose, D-trehalose, D-turanose and 3-*O*-β-D-galactopyranosyl-D-arabinose. *Biophys. Chem.* **1991**, *40*, 59–67. [CrossRef]
181. Qu, Q.; Lee, S.-J.; Boos, W. TreT, a novel trehalose glycosyltransferring synthase of the hyperthermophilic archaeon *Thermococcus litoralis*. *J. Biol. Chemistry* **2004**, *279*, 47890–47897. [CrossRef]
182. Kouril, T.; Zaparty, M.; Marrero, J.; Brinkmann, H.; Siebers, B. A novel trehalose synthesizing pathway in the hyperthermophilic crenarchaeon *Thermoproteus tenax*: The unidirectional tret pathway. *Arch. Microbiol.* **2008**, *190*, 355. [CrossRef]
183. Tian, C.; Yang, J.; Zeng, Y.; Zhang, T.; Zhou, Y.; Men, Y.; You, C.; Zhu, Y.; Sun, Y. Biosynthesis of raffinose and stachyose from sucrose via an in vitro multienzyme system. *Appl. Environ. Microbiol.* **2019**, *85*, e02306–e02318. [CrossRef]
184. Resnick, S.M.; Zehnder, A.J.B. In vitro atp regeneration from polyphosphate and AMP by polyphosphate: AMP phosphotransferase and adenylate kinase from *Acinetobacter johnsonii* 210A. *Appl. Environ. Microbiol.* **2000**, *66*, 2045–2051. [CrossRef]
185. Crans, D.C.; Whitesides, G.M. A convenient synthesis of disodium acetyl phosphate for use in in situ ATP cofactor regeneration. *J. Org. Chem.* **1983**, *48*, 3130–3132. [CrossRef]
186. Tasnádi, G.; Jud, W.; Hall, M.; Baldenius, K.; Ditrich, K.; Faber, K. Evaluation of natural and synthetic phosphate donors for the improved enzymatic synthesis of phosphate monoesters. *Adv. Synth. Catal.* **2018**, *360*, 2394–2401. [CrossRef]
187. Kulmer, S.T.; Gutmann, A.; Lemmerer, M.; Nidetzky, B. Biocatalytic cascade of polyphosphate kinase and sucrose synthase for synthesis of nucleotide-activated derivatives of glucose. *Adv. Synth. Catal.* **2017**, *359*, 292–301. [CrossRef]
188. Andexer, J.N.; Richter, M. Emerging enzymes for atp regeneration in biocatalytic processes. *ChemBioChem* **2015**, *16*, 380–386. [CrossRef]
189. Liu, Z.; Zhang, J.; Chen, X.; Wang, P.G. Combined biosynthetic pathway for de novo production of udp-galactose: Catalysis with multiple enzymes immobilized on agarose beads. *ChemBioChem* **2002**, *3*, 348–355. [CrossRef]
190. Murata, K.; Uchida, T.; Kato, J.; Chibata, I. Polyphosphate kinase: Distribution, some properties and its application as an atp regeneration system. *Agric. Biol. Chem.* **1988**, *52*, 1471–1477.
191. Restiawaty, E.; Iwasa, Y.; Maya, S.; Honda, K.; Omasa, T.; Hirota, R.; Kuroda, A.; Ohtake, H. Feasibility of thermophilic adenosine triphosphate-regeneration system using *Thermus thermophilus* polyphosphate kinase. *Process. Biochem.* **2011**, *46*, 1747–1752. [CrossRef]
192. Sato, M.; Masuda, Y.; Kirimura, K.; Kino, K. Thermostable ATP regeneration system using polyphosphate kinase from *thermosynechococcus elongatus* BP-1 for D-amino acid dipeptide synthesis. *J. Biosci. Bioeng.* **2007**, *103*, 179–184. [CrossRef]
193. Iwamoto, S.; Motomura, K.; Shinoda, Y.; Urata, M.; Kato, J.; Takiguchi, N.; Ohtake, H.; Hirota, R.; Kuroda, A. Use of an *Escherichia coli* recombinant producing thermostable polyphosphate kinase as an ATP regenerator to produce fructose 1,6-diphosphate. *Appl. Environ. Microbiol.* **2007**, *73*, 5676–5678. [CrossRef]
194. 1Shiba, T.; Tsutsumi, K.; Ishige, K.; Noguchi, T. Inorganic polyphosphate and polyphosphate kinase: Their novel biological functions and applications. *Biochemistry (Moscow)* **2000**, *65*, 315–323.
195. Noguchi, T.; Shiba, T. Use of *Escherichia coli* polyphosphate kinase for oligosaccharide synthesis. *Biosci. Biotechnol. Biochem.* **1998**, *62*, 1594–1596. [CrossRef]
196. Nahálka, J.; Pätoprstý, V. Enzymatic synthesis of sialylation substrates powered by a novel polyphosphate kinase (PPK3). *Org. Biomol. Chem.* **2009**, *7*, 1778–1780. [CrossRef]
197. Lee, S.-G.; Lee, J.-O.; Yi, J.-K.; Kim, B.-G. Production of cytidine 5′-monophosphate *N*-acetylneuraminic acid using recombinant *Escherichia coli* as a biocatalyst. *Biotechnol. Bioeng.* **2002**, *80*, 516–524. [CrossRef]
198. Rupprath, C.; Kopp, M.; Hirtz, D.; Müller, R.; Elling, L. An enzyme module system for in situ regeneration of deoxythymidine 5′-diphosphate (dTDP)-activated deoxy sugars. *Adv. Synt. Catal.* **2007**, *349*, 1489–1496. [CrossRef]
199. Fischöder, T.; Wahl, C.; Zerhusen, C.; Elling, L. Repetitive batch mode facilitates enzymatic synthesis of the nucleotide sugars UDP-Gal, UDP-GlcNAc, and UDP-GalNAc on a multi-gram scale. *Biotechnol. J.* **2019**, *14*. [CrossRef]

200. Wen, L.; Zheng, Y.; Li, T.; Wang, P.G. Enzymatic synthesis of 3-deoxy-D-manno-octulosonic acid (KDO) and its application for LPS assembly. *Bioorg. Med. Chem. Lett.* **2016**, *26*, 2825–2828. [CrossRef]
201. Lee, J.-H.; Chung, S.-W.; Lee, H.-J.; Jang, K.-S.; Lee, S.-G.; Kim, B.-G. Optimization of the enzymatic one pot reaction for the synthesis of uridine 5′-diphosphogalactose. *Bioprocess. Biosyst. Eng.* **2009**, *33*, 71. [CrossRef]
202. Unverzagt, C.; Kunz, H.; Paulson, J.C. High-efficiency synthesis of sialyloligosaccharides and sialoglycopeptides. *J. Am. Chem. Soc.* **1990**, *112*, 9308–9309. [CrossRef]
203. Hirschbein, B.L.; Mazenod, F.P.; Whitesides, G.M. Synthesis of phosphoenolypyruvate and its use in ATP cofactor regeneration. *J. Org. Chem.* **1982**, *47*, 3765–3766. [CrossRef]
204. Yu, H.; Santra, A.; Li, Y.; McArthur, J.B.; Ghosh, T.; Yang, X.; Wang, P.G.; Chen, X. Streamlined chemoenzymatic total synthesis of prioritized ganglioside cancer antigens. *Org. Biomol. Chem.* **2018**, *16*, 4076–4080. [CrossRef] [PubMed]
205. Muthana, M.M.; Qu, J.; Xue, M.; Klyuchnik, T.; Siu, A.; Li, Y.; Zhang, L.; Yu, H.; Li, L.; Wang, P.G.; et al. Improved one-pot multienzyme (OPME) systems for synthesizing UDP-uronic acids and glucuronides. *Chem. Commun.* **2015**, *51*, 4595–4598. [CrossRef] [PubMed]
206. Read, J.A.; Ahmed, R.A.; Tanner, M.E. Efficient chemoenzymatic synthesis of ADP-D-glycero-β-D-manno-heptose and a mechanistic study of ADP-L-glycero-D-manno-heptose 6-epimerase. *Org. Lett.* **2005**, *7*, 2457–2460. [CrossRef] [PubMed]
207. Alissandratos, A.; Caron, K.; Loan, T.D.; Hennessy, J.E.; Easton, C.J. ATP recycling with cell lysate for enzyme-catalyzed chemical synthesis, protein expression and PCR. *ACS Chem. Biol.* **2016**, *11*, 3289–3293. [CrossRef]
208. Wong, C.H.; Haynie, S.L.; Whitesides, G.M. Enzyme-catalyzed synthesis of *N*-acetyllactosamine with in situ regeneration of uridine 5′-diphosphate glucose and uridine 5′-diphosphate galactose. *J. Org. Chem.* **1982**, *47*, 5416–5418. [CrossRef]
209. Tomoike, F.; Nakagawa, N.; Kuramitsu, S.; Masui, R. A single amino acid limits the substrate specificity of *Thermus thermophilus* uridine-cytidine kinase to cytidine. *Biochemistry* **2011**, *50*, 4597–4607. [CrossRef]
210. Loan, T.D.; Easton, C.J.; Alissandratos, A. Recombinant cell-lysate-catalysed synthesis of uridine-5′-triphosphate from nucleobase and ribose, and without addition of ATP. *New Biotechnol.* **2019**, *49*, 104–111. [CrossRef]
211. Scism, R.A.; Bachmann, B.O. Five-component cascade synthesis of nucleotide analogues in an engineered self-immobilized enzyme aggregate. *ChemBioChem* **2010**, *11*, 67–70. [CrossRef]
212. Fernández-Lucas, J. Multienzymatic synthesis of nucleic acid derivatives: A general perspective. *Appl. Microbiol. Biotechnol.* **2015**, *99*, 4615–4627. [CrossRef]
213. Glaser, L.; Brown, D.H. The synthesis of chitin in cell-free extracts of *Neurospora crassa*. *J. Biol. Chem.* **1957**, *228*, 729–742.
214. Lougheed, B.; Ly, H.D.; Wakarchuk, W.W.; Withers, S.G. Glycosyl fluorides can function as substrates for nucleotide phosphosugar-dependent glycosyltransferases. *J. Biol. Chem.* **1999**, *274*, 37717–37722. [CrossRef] [PubMed]
215. Lepak, A.; Gutmann, A.; Nidetzky, B. β-glucosyl fluoride as reverse reaction donor substrate and mechanistic probe of inverting sugar nucleotide-dependent glycosyltransferases. *ACS Catal.* **2018**, *8*, 9148–9153. [CrossRef]
216. Lairson, L.L.; Wakarchuk, W.W.; Withers, S.G. Alternative donor substrates for inverting and retaining glycosyltransferases. *Chem. Commun.* **2007**, *4*, 365–367. [CrossRef]
217. Li, C.; Zhang, Z.; Duan, Q.; Li, X. Glycopeptide synthesis on an ionic liquid support. *Org. Lett.* **2014**, *16*, 3008-3011. [CrossRef] [PubMed]
218. Hughes, R.R.; Shaaban, K.A.; Zhang, J.; Cao, H.; Phillips, G.N.; Thorson, J.S. OleD Loki as a catalyst for tertiary amine and hydroxamate glycosylation. *ChemBioChem* **2017**, *18*, 363–367. [CrossRef]
219. Lee, A.A.; Chen, Y.C.S.; Ekalestari, E.; Ho, S.Y.; Hsu, N.S.; Kuo, T.F.; Wang, T.S.A. Facile and versatile chemoenzymatic synthesis of enterobactin analogues and applications in bacterial detection. *Angew. Chem. Int. Ed.* **2016**, *55*, 12338–12342. [CrossRef]
220. Burkhart, F.; Zhang, Z.; Wacowich-Sgarbi, S.; Wong, C.H. Synthesis of the Globo H hexasaccharide using the programmable reactivity-based one-pot strategy. *Angew. Chem. Int. Ed.* **2001**, *40*, 1274–1277. [CrossRef]

221. Huang, C.-Y.; Thayer, D.A.; Chang, A.Y.; Best, M.D.; Hoffmann, J.; Head, S.; Wong, C.-H. Carbohydrate microarray for profiling the antibodies interacting with Globo H tumor antigen. *Proc. Natl. Acad. Sci. USA* **2006**, *103*, 15–20. [CrossRef]
222. Su, D.M.; Eguchi, H.; Yi, W.; Li, L.; Wang, P.G.; Xia, C. Enzymatic synthesis of tumor-associated carbohydrate antigen Globo-H hexasaccharide. *Org. Lett.* **2008**, *10*, 1009–1012. [CrossRef]
223. Yu, H.; Lau, K.; Li, Y.; Sugiarto, G.; Chen, X. One-pot multienzyme synthesis of Lewis x and sialyl Lewis x antigens. *Curr. Protoc. Chem. Biol.* **2012**, *4*, 233–247.
224. Nishimura, S.-I.; Yamada, K. Transfer of ganglioside GM3 oligosaccharide from a water soluble polymer to ceramide by ceramide glycanase. A novel approach for the chemical-enzymatic synthesis of glycosphingolipids. *J. Am. Chem. Soc.* **1997**, *119*, 10555–10556. [CrossRef]
225. Huang, X.; Witte, K.L.; Bergbreiter, D.E.; Wong, C.-H. Homogenous enzymatic synthesis using a thermo-responsive water-soluble polymer support. *Adv. Synt. Catal.* **2001**, *343*, 675–681. [CrossRef]
226. Jaipuri, F.A.; Pohl, N.L. Toward solution-phase automated iterative synthesis: Fluorous-tag assisted solution-phase synthesis of linear and branched mannose oligomers. *Org. Biomol. Chem.* **2008**, *6*, 2686–2691. [CrossRef] [PubMed]
227. Yang, B.; Jing, Y.; Huang, X. Fluorous-assisted one-pot oligosaccharide synthesis. *Eur. J. Org. Chem.* **2010**, *2010*, 1290–1298. [CrossRef] [PubMed]
228. Liu, L.; Pohl, N.L.B. A fluorous phosphate protecting group with applications to carbohydrate synthesis. *Org. Lett.* **2011**, *13*, 1824–1827. [CrossRef]
229. Carrel, F.R.; Geyer, K.; Codée, J.D.C.; Seeberger, P.H. Oligosaccharide synthesis in microreactors. *Org. Lett.* **2007**, *9*, 2285–2288. [CrossRef]
230. Carrel, F.R.; Seeberger, P.H. Cap-and-tag solid phase oligosaccharide synthesis. *J. Org. Chem.* **2008**, *73*, 2058–2065. [CrossRef]
231. Zhang, F.; Zhang, W.; Zhang, Y.; Curran, D.P.; Liu, G. Synthesis and applications of a light-fluorous glycosyl donor. *J. Org. Chem.* **2009**, *74*, 2594–2597. [CrossRef]
232. Huang, W.; Gao, Q.; Boons, G.-J. Assembly of a complex branched oligosaccharide by combining fluorous-supported synthesis and stereoselective glycosylations using anomeric sulfonium ions. *Chem. Eur. J.* **2015**, *21*, 12920–12926. [CrossRef]
233. Macchione, G.; de Paz, J.L.; Nieto, P.M. Synthesis of hyaluronic acid oligosaccharides and exploration of a fluorous-assisted approach. *Carbohydr. Res.* **2014**, *394*, 17–25. [CrossRef]
234. Chai, Y.-H.; Feng, Y.-L.; Wu, J.-J.; Deng, C.-Q.; Liu, A.-Y.; Zhang, Q. Recyclable benzyl-type fluorous tags: Preparation and application in oligosaccharide synthesis. *Chin. Chem. Lett.* **2017**, *28*, 1693–1700. [CrossRef]
235. Hwang, J.; Yu, H.; Malekan, H.; Sugiarto, G.; Li, Y.; Qu, J.; Nguyen, V.; Wu, D.; Chen, X. Highly efficient one-pot multienzyme (OPME) synthesis of glycans with fluorous-tag assisted purification. *Chem. Commun.* **2014**, *50*, 3159–3162. [CrossRef] [PubMed]
236. Tanaka, H.; Tanimoto, Y.; Kawai, T.; Takahashi, T. A fluorous-assisted synthesis of oligosaccharides using a phenyl ether linker as a safety-catch linker. *Tetrahedron* **2011**, *67*, 10011–10016. [CrossRef]
237. Zhu, H.; Wu, Z.; Gadi, M.R.; Wang, S.; Guo, Y.; Edmunds, G.; Guan, W.; Fang, J. Cation exchange assisted binding-elution strategy for enzymatic synthesis of human milk oligosaccharides (HMOS). *Bioorg. Med. Chem. Lett.* **2017**, *27*, 4285–4287. [CrossRef]
238. Santra, A.; Li, Y.; Yu, H.; Slack, T.J.; Wang, P.G.; Chen, X. Highly efficient chemoenzymatic synthesis and facile purification of α-Gal pentasaccharyl ceramide Galα3nLc$_4$βCer. *Chem. Commun.* **2017**, *53*, 8280–8283. [CrossRef]
239. Zhang, J.; Chen, C.; Gadi, M.R.; Gibbons, C.; Guo, Y.; Cao, X.; Edmunds, G.; Wang, S.; Liu, D.; Yu, J.; et al. Machine-driven enzymatic oligosaccharide synthesis by using a peptide synthesizer. *Angew. Chem. Int. Ed.* **2018**, *57*, 16638–16642. [CrossRef]
240. Schuster, M.; Wang, P.; Paulson, J.C.; Wong, C.-H. Solid-phase chemical-enzymic synthesis of glycopeptides and oligosaccharides. *J. Am. Chem. Soc.* **1994**, *116*, 1135–1136. [CrossRef]
241. Halcomb, R.L.; Huang, H.; Wong, C.-H. Solution- and solid-phase synthesis of inhibitors of *H. pylori* attachment and E-selectin-mediated leukocyte adhesion. *J. Am. Chem. Soc.* **1994**, *116*, 11315–11322. [CrossRef]
242. Blixt, O.; Norberg, T. Enzymatic glycosylation of reducing oligosaccharides linked to a solid phase or a lipid via a cleavable squarate linker. *Carbohydr. Res.* **1999**, *319*, 80–91. [CrossRef]

243. Houseman, B.T.; Mrksich, M. The role of ligand density in the enzymatic glycosylation of carbohydrates presented on self-assembled monolayers of alkanethiolates on gold. *Angew. Chem. Int. Ed.* **1999**, *38*, 782–785. [CrossRef]
244. Yamada, K.; Fujita, E.; Nishimura, S.-I. High performance polymer supports for enzyme-assisted synthesis of glycoconjugates. *Carbohydr. Res.* **1997**, *305*, 443–461. [CrossRef]
245. Yan, F.; Wakarchuk, W.W.; Gilbert, M.; Richards, J.C.; Whitfield, D.M. Polymer-supported and chemoenzymatic synthesis of the *Neisseria meningitidis* pentasaccharide: A methodological comparison. *Carbohydr. Res.* **2000**, *328*, 3–16. [CrossRef]
246. Sears, P.; Wong, C.-H. Toward automated synthesis of oligosaccharides and glycoproteins. *Science* **2001**, *291*, 2344–2350. [CrossRef] [PubMed]
247. Ivannikova, T.; Bintein, F.; Malleron, A.; Juliant, S.; Cerutti, M.; Harduin-Lepers, A.; Delannoy, P.; Augé, C.; Lubineau, A. Recombinant (2→3)-α-sialyltransferase immobilized on nickel-agarose for preparative synthesis of sialyl Lewis^x and Lewis^a precursor oligosaccharides. *Carbohydr. Res.* **2003**, *338*, 1153–1161. [CrossRef]
248. Augé, C.; Fernandez-Fernandez, R.; Gautheron, C. The use of immobilised glycosyltransferases in the synthesis of sialyloligosaccharides. *Carbohydr. Res.* **1990**, *200*, 257–268. [CrossRef]
249. Augé, C.; Gautheron, C. An efficient synthesis of cytidine monophospho-sialic acids with four immobilized enzymes. *Tetrahedron Lett.* **1988**, *29*, 789–790. [CrossRef]
250. David, S.; Auge, C. Immobilized enzymes in preparative carbohydrate chemistry. *Pure Appl. Chem.* **1987**, *59*, 1501–1508. [CrossRef]
251. Augé, C.; David, S.; Mathieu, C.; Gautheron, C. Synthesis with immobilized enzymes of two trisaccharides, one of them active as the determinant of a stage antigen. *Tetrahedron Lett.* **1984**, *25*, 1467–1470. [CrossRef]
252. Chen, X.; Fang, J.; Zhang, J.; Liu, Z.; Shao, J.; Kowal, P.; Andreana, P.; Wang, P.G. Sugar nucleotide regeneration beads (superbeads): A versatile tool for the practical synthesis of oligosaccharides. *J. Am. Chem. Soc.* **2001**, *123*, 2081–2082. [CrossRef]
253. Yu, C.-C.; Kuo, Y.-Y.; Liang, C.-F.; Chien, W.-T.; Wu, H.-T.; Chang, T.-C.; Jan, F.-D.; Lin, C.-C. Site-specific immobilization of enzymes on magnetic nanoparticles and their use in organic synthesis. *Bioconjugate Chem.* **2012**, *23*, 714–724. [CrossRef]
254. Nahalka, J.; Liu, Z.; Chen, X.; Wang, P.G. Superbeads: Immobilization in "sweet" chemistry. *Chem. Eur. J.* **2003**, *9*, 372–377. [CrossRef] [PubMed]
255. Wang, P.-Y.; Tsai, S.-W.; Chen, T.-L. Improvements of enzyme activity and enantioselectivity via combined substrate engineering and covalent immobilization. *Biotechnol. Bioeng.* **2008**, *101*, 460–469. [CrossRef] [PubMed]
256. Zhang, Y.; Ge, J.; Liu, Z. Enhanced activity of immobilized or chemically modified enzymes. *ACS Catal.* **2015**, *5*, 4503–4513. [CrossRef]
257. Chibata, I.; Tosa, T.; Sato, T.; Mori, T. Production of L-amino acids by aminoacylase adsorbed on DEAE-sephadex. In *Methods in Enzymology*; Academic Press: Cambridge, MA, USA, 1976; Volume 44, pp. 746–759.
258. Petronijević, Ž.; Ristić, S.; Pešić, D.; Šmelcerović, A. Immobilization of dextransucrase on regenerated benzoyl cellulose carriers. *Enzyme Microb. Technol.* **2007**, *40*, 763–768. [CrossRef]
259. Schöffer, J.d.N.; Klein, M.P.; Rodrigues, R.C.; Hertz, P.F. Continuous production of β-cyclodextrin from starch by highly stable cyclodextrin glycosyltransferase immobilized on chitosan. *Carbohydr. Polym.* **2013**, *98*, 1311–1316. [CrossRef]
260. Schöffer, J.d.N.; Matte, C.R.; Charqueiro, D.S.; de Menezes, E.W.; Costa, T.M.H.; Benvenutti, E.V.; Rodrigues, R.C.; Hertz, P.F. Directed immobilization of CGTase: The effect of the enzyme orientation on the enzyme activity and its use in packed-bed reactor for continuous production of cyclodextrins. *Process. Biochem.* **2017**, *58*, 120–127. [CrossRef]
261. Rakmai, J.; Cheirsilp, B.; Prasertsan, P. Enhanced thermal stability of cyclodextrin glycosyltransferase in alginate–gelatin mixed gel beads and the application for β-cyclodextrin production. *Biocatal. Agric. Biotechnol.* **2015**, *4*, 717–726. [CrossRef]
262. Jung, D.-H.; Jung, J.-H.; Seo, D.-H.; Ha, S.-J.; Kweon, D.-K.; Park, C.-S. One-pot bioconversion of sucrose to trehalose using enzymatic sequential reactions in combined cross-linked enzyme aggregates. *Bioresour. Technol.* **2013**, *130*, 801–804. [CrossRef]

263. Orrego, A.H.; Trobo-Maseda, L.; Rocha-Martin, J.; Guisan, J.M. Immobilization-stabilization of a complex multimeric sucrose synthase from *Nitrosomonas europaea*. Synthesis of UDP-glucose. *Enzyme Microb. Technol.* **2017**, *105*, 51–58. [CrossRef]
264. De Winter, K.; Soetaert, W.; Desmet, T. An imprinted cross-linked enzyme aggregate (iCLEA) of sucrose phosphorylase: Combining improved stability with altered specificity. *Int. J. Mol. Sci.* **2012**, *13*, 11333. [CrossRef]
265. Kaulpiboon, J.; Pongsawasdi, P.; Zimmermann, W. Molecular imprinting of cyclodextrin glycosyltransferases from *Paenibacillus* sp. A11 and *Bacillus macerans* with γ-cyclodextrin. *FEBS J.* **2007**, *274*, 1001–1010. [CrossRef] [PubMed]
266. Sun, J.; Wang, S.; Li, W.; Li, R.; Chen, S.; Ri, H.I.; Kim, T.M.; Kang, M.S.; Sun, L.; Sun, X.; et al. Improvement of trehalose production by immobilized trehalose synthase from *Thermus thermophilus* HB27. *Molecules* **2018**, *23*, 1087. [CrossRef]
267. Sheldon, R.A.; van Pelt, S. Enzyme immobilisation in biocatalysis: Why, what and how. *Chem. Soc. Rev.* **2013**, *42*, 6223–6235. [CrossRef]
268. Liese, A.; Hilterhaus, L. Evaluation of immobilized enzymes for industrial applications. *Chem. Soc. Rev.* **2013**, *42*, 6236–6249. [CrossRef]
269. Hanefeld, U.; Gardossi, L.; Magner, E. Understanding enzyme immobilisation. *Chem. Soc. Rev.* **2009**, *38*, 453–468. [CrossRef]
270. Rakmai, J.; Cheirsilp, B. Continuous production of β-cyclodextrin by cyclodextrin glycosyltransferase immobilized in mixed gel beads: Comparative study in continuous stirred tank reactor and packed bed reactor. *Biochem. Eng. J.* **2016**, *105*, 107–113. [CrossRef]
271. Cho, Y.-J.; Park, O.-J.; Shin, H.-J. Immobilization of thermostable trehalose synthase for the production of trehalose. *Enzyme Microb. Technol.* **2006**, *39*, 108–113. [CrossRef]
272. Szymańska, K.; Odrozek, K.; Zniszczoł, A.; Pudło, W.; Jarzębski, A.B. A novel hierarchically structured siliceous packing to boost the performance of rotating bed enzymatic reactors. *Chem. Eng. J.* **2017**, *315*, 18–24. [CrossRef]
273. Cattaneo, G.; Rabuffetti, M.; Speranza, G.; Kupfer, T.; Peters, B.; Massolini, G.; Ubiali, D.; Calleri, E. Synthesis of adenine nucleosides by transglycosylation using two sequential nucleoside phosphorylase-based bioreactors with on-line reaction monitoring by using HPLC. *ChemCatChem* **2017**, *9*, 4614–4620. [CrossRef]
274. Szymańska, K.; Odrozek, K.; Zniszczoł, A.; Torrelo, G.; Resch, V.; Hanefeld, U.; Jarzębski, A.B. MsAcT in siliceous monolithic microreactors enables quantitative ester synthesis in water. *Catal. Sci. Technol.* **2016**, *6*, 4882–4888. [CrossRef]
275. Szymańska, K.; Pudło, W.; Mrowiec-Białoń, J.; Czardybon, A.; Kocurek, J.; Jarzębski, A.B. Immobilization of invertase on silica monoliths with hierarchical pore structure to obtain continuous flow enzymatic microreactors of high performance. *Microporous Mesoporous Mater.* **2013**, *170*, 75–82. [CrossRef]
276. Lawrence, J.; O'Sullivan, B.; Lye, G.J.; Wohlgemuth, R.; Szita, N. Microfluidic multi-input reactor for biocatalytic synthesis using transketolase. *J. Mol. Catal. B Enzym.* **2013**, *95*, 111–117. [CrossRef] [PubMed]
277. Valikhani, D.; Bolivar, J.M.; Pfeiffer, M.; Nidetzky, B. Multivalency effects on the immobilization of sucrose phosphorylase in flow microchannels and their use in the development of a high-performance biocatalytic microreactor. *ChemCatChem* **2017**, *9*, 161–166. [CrossRef]
278. Strub, D.J.; Szymańska, K.; Hrydziuszko, Z.; Bryjak, J.; Jarzębski, A.B. Continuous flow kinetic resolution of a non-equimolar mixture of diastereoisomeric alcohol using a structured monolithic enzymatic microreactor. *React. Chem. Eng.* **2019**, *4*, 587–594. [CrossRef]
279. Sheldon, R.A.; Woodley, J.M. Role of biocatalysis in sustainable chemistry. *Chem. Rev.* **2018**, *118*, 801–838. [CrossRef]
280. Aurell, C.-J.; Karlsson, S.; Pontén, F.; Andersen, S.M. Lipase catalyzed regioselective lactamization as a key step in the synthesis of *N*-BOC-(2*R*)-1,4-oxazepane-2-carboxylic acid. *Org. Process. Res. Develop.* **2014**, *18*, 1116–1119. [CrossRef]
281. Mallin, H.; Muschiol, J.; Byström, E.; Bornscheuer, U.T. Efficient biocatalysis with immobilized enzymes or encapsulated whole cell microorganism by using the spinchem reactor system. *ChemCatChem* **2013**, *5*, 3529–3532. [CrossRef]
282. Carberry, J.J. Designing laboratory catalytic reactors. *Ind. Eng. Chem.* **1964**, *56*, 39–46. [CrossRef]

283. Woodley, J.M.; Titchener-Hooker, N.J. The use of windows of operation as a bioprocess design tool. *Bioprocess. Eng.* **1996**, *14*, 263–268. [CrossRef]
284. Wohlgemuth, R.; Plazl, I.; Žnidaršič-Plazl, P.; Gernaey, K.V.; Woodley, J.M. Microscale technology and biocatalytic processes: Opportunities and challenges for synthesis. *Trends Biotechnol.* **2015**, *33*, 302–314. [CrossRef]
285. Rossetti, I. Continuous flow (micro-)reactors for heterogeneously catalyzed reactions: Main design and modelling issues. *Catal. Today* **2018**, *308*, 20–31. [CrossRef]

International Journal of
Molecular Sciences

Review

Pyruvate Substitutions on Glycoconjugates

Fiona F. Hager [1], Leander Sützl [2], Cordula Stefanović [1], Markus Blaukopf [3] and Christina Schäffer [1,*]

1 Department of NanoBiotechnology, NanoGlycobiology unit, Universität für Bodenkultur Wien, Muthgasse 11, A-1190 Vienna, Austria; fiona.hager@boku.ac.at (F.F.H.); cordula.brinskele@boku.ac.at (C.S.); christina.schaeffer@boku.ac.at (C.Sch.)
2 Department of Food Science and Technology, Food Biotechnology Laboratory, Muthgasse 11, Universität für Bodenkultur Wien, A-1190 Vienna, Austria; leander.suetzl@boku.ac.at
3 Department of Chemistry, Division of Organic Chemistry, Universität für Bodenkultur Wien, Muthgasse 18, A-1190 Vienna, Austria; markus.blaukopf@boku.ac.at
* Correspondence: christina.schaeffer@boku.ac.at; Tel.: +43-1-47654 (ext. 80203)

Received: 13 August 2019; Accepted: 27 September 2019; Published: 5 October 2019

Abstract: Glycoconjugates are the most diverse biomolecules of life. Mostly located at the cell surface, they translate into cell-specific "barcodes" and offer a vast repertoire of functions, including support of cellular physiology, lifestyle, and pathogenicity. Functions can be fine-tuned by non-carbohydrate modifications on the constituting monosaccharides. Among these modifications is pyruvylation, which is present either in enol or ketal form. The most commonly best-understood example of pyruvylation is enol-pyruvylation of *N*-acetylglucosamine, which occurs at an early stage in the biosynthesis of the bacterial cell wall component peptidoglycan. Ketal-pyruvylation, in contrast, is present in diverse classes of glycoconjugates, from bacteria to algae to yeast—but not in humans. Mild purification strategies preventing the loss of the acid-labile ketal-pyruvyl group have led to a collection of elucidated pyruvylated glycan structures. However, knowledge of involved pyruvyltransferases creating a ring structure on various monosaccharides is scarce, mainly due to the lack of knowledge of fingerprint motifs of these enzymes and the unavailability of genome sequences of the organisms undergoing pyruvylation. This review compiles the current information on the widespread but under-investigated ketal-pyruvylation of monosaccharides, starting with different classes of pyruvylated glycoconjugates and associated functions, leading to pyruvyltransferases, their specificity and sequence space, and insight into pyruvate analytics.

Keywords: pyruvylation; pyruvyltransferase; exopolysaccharides; capsular polysaccharides; cell wall glycopolymers; *N*-glycans; lipopolysaccharides; biosynthesis; sequence space; pyruvate analytics

---

1. Introduction

Pyruvylation is a widespread non-carbohydrate modification of monosaccharides found in various classes of glycoconjugates. In most cases, the modification is present as a pyruvate (Pyr) ketal (cyclic acetal/ketal) bridging two hydroxyl groups of a monosaccharide residue and forming a ring structure [1], where pyruvate is most frequently placed across the 2,3-, 4,6-, or 3,4-positions (Figure 1I–III). The best-known example of pyruvylation, however, occurs as enol pyruvate (Figure 1IV), which is elaborated during the biosynthesis of the bacterial cell wall component peptidoglycan [2]. In both modes of pyruvylation, a dedicated pyruvyltransferase catalyses the transfer of the pyruvate moiety to the monosaccharide target.

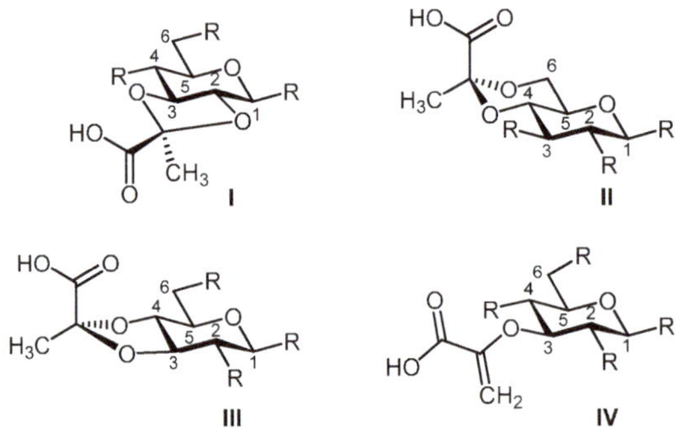

R = OH or glycosidic linkage

Figure 1. Overview on the most common modes of monosaccharide pyruvylation. Shown is pyruvate ketal at bridging positions 2,3 (**I**), 4,6 (**II**), and 3,4 (**III**), and enol pyruvate (**IV**). Arabic numbers indicate ring positions.

Pyruvate-ketal-modified (henceforth abbreviated as "pyruvylated") glycoconjugates are found in various phylogenetic orders of life, including bacteria, yeast, and algae, but not in humans. Pyruvylated glycoconjugates are typically present in the cell envelope to which they impart a net negative charge; this is necessary for vital biological functions, such as regulation of the cell influx/efflux processes and cell–cell interactions including cell aggregation and pathogenic adhesion. Notably, besides pyruvylation, nature offers various alternate strategies to create anionic cell surfaces, including a wide range of acidic saccharides (e.g., muramic acid, hexuronic acids, sialic acids) and saccharide modifications (e.g., succinate, lactate, phosphate) [3]. These compounds further lead to an increased capability of cells to electrostatically bind cations at the surface, which, in turn, may foster the packing density of the saccharide portion of glycoconjugates [4].

The repertoire of monosaccharide targets of pyruvylation is quite diverse. The most abundant pyruvylated monosaccharide with 59 hits in the Carbohydrate Structure Database (CSDB, http://csdb.glycoscience.ru/) [5,6] is galactose (Gal). Examples of pyruvylated galactose include, among others, the capsular polysaccharide of *Bacteroides fragilis* [7] and *Streptococcus pneumoniae* [8], *N*-glycans of the fission yeast *Schizosaccharomyces pombe* [9], as well as carragenans [10], and galactans from algae [11–13]. Recently, pyruvylated *N*-acetylmannosamine (ManNAc) has emerged as an important epitope on bacterial "non-classical" secondary cell wall glycopolymers, serving as a cell wall ligand for cell surface (S-) layer proteins, such as those of the pathogen *Bacillus anthracis* and the honeybee saprophyte *Paenibacillus alvei* [14,15]. There are also examples of pyruvylated monosaccharides on capsular polysaccharides serving as an immunostimulatory effector [7,16], or contributor to virulence as in the case of the secondary cell wall polymer of *Bacillus cereus* [17] or the exopolysaccharide xanthan of *Xanthomonas* spp., where pyruvylation is essential for successful colonization and pathogenesis in planta [18]. Pyruvyl groups on terminal glucose (Glc) and *N*-acetylgalactosamine (GalNAc) residues in the lipooligosaccharide of *Pseudomonas stutzeri* OX1, in contrast, are assumed to have biosynthetic implications [4]. All these examples are, among others, discussed in detail below.

Several studies dealing with pyruvylated glycans and their structural elucidation are available in the literature. However, knowledge of the enzymatic machinery governing pyruvylation is scarce. This is mainly due to missing sequencing data of the organisms, which produce pyruvylated glycoconjugates of known structure. Thus, despite their predictably widespread occurrence, pyruvyltransferases are an under-investigated class of enzymes.

This review summarizes the current state of knowledge about pyruvylated glycoconjugates in nature—focusing on bacterial sources—with an emphasis on the pyruvyltransferases involved in their biosynthesis.

2. Enol-Pyruvylation in Peptidoglycan Biosynthesis

The UDP-*N*-acetylglucosamine-3-*O*-enol-pyruvyltransferase MurA (EC 2.5.1.7) targets UDP-*N*-acetylglucosamine (UDP-GlcNAc) as an acceptor substrate for enol-pyruvyl transfer from a phosphoenolpyruvate (PEP) substrate, thereby releasing free phosphate and yielding the UDP-activated form of the essential bacterial cell wall compound *N*-acetylmuramic acid (enolpyruvyl-UDP-*N*-acetylglucosamine; MurNAc). This first committed step of peptidoglycan biosynthesis is inhibited by the epoxide antibiotic fosfomycin [19]. As a PEP analogue, fosfomycin binds covalently to the key cysteine residue at position 115/116 (position depending on the source of enzyme) in the active site of MurA, preventing the formation of UDP-MurNAc [19–21]. The co-crystal structure of a Cys (cysteine)-to-Ser (serine) mutant of *Enterobacter cloacae*, MurA, together with its substrates, revealed that the Cys residue is essential for product release and not directly involved in the chemical reaction of enol-pyruvyl transfer. The comparison of the product state with the intermediate state and an unliganded state of MurA indicated that the dissociation of the products is an ordered event, with inorganic phosphate leaving first, followed by conformational changes that lead to the opening of the two-domain structure of MurA and the final release of UDP-MurNAc [20]. A recent study on MurA of the opportunistic pathogen *Acinetobacter baumannii* revealed that the enzyme exists as a monomer in solution and has a pH optimum of 7.5 at 37 °C. The Km for UDP-GlcNAc is 1.062 ± 0.09 mM and 1.806 ± 0.23 mM for PEP [21]. The relative enzymatic activity is inhibited approximately threefold in the presence of 50 mM fosfomycin. Superimposition of a model for the *A. baumannii* enzyme with MurA of *Escherichia coli (E. coli)* confirmed the structural similarity in the fosfomycin binding site. Because of the worldwide spread of antimicrobial resistance and the paucity of novel drugs in the development pipeline, there has been a renewed interest in fosfomycin as an alternative option for the treatment of infections caused by multidrug-resistant Gram-negative bacteria [22]. However, it has to be considered that natural MurA mutants exist that render the respective organisms fosfomycin resistant. This includes *Mycobacteria* and *Chlamydia* species, where Cys-to-Asp mutants occur [20].

Interestingly, NikO, another enol-pyruvyltransferase that is structurally closely related to the common MurA enzymes and, consequently, inhibited by fosfomycin, plays an essential role in the biosynthesis of nikkomycins. Nikkomycins are peptide-nucleoside antibiotics, which strongly inhibit chitin synthesis and, therefore, are effective against fungi and insects. NikO was shown to transfer the enol-pyruvyl moiety from PEP to the 3′-hydroxyl group of UMP and to be inactivated by fosfomycin because of alkylation of Cys130. However, the degree of inactivation is not as pronounced as in the case of common MurA enzymes [23].

3. Ketal-Pyruvylated Glycoconjugates

In this section, an overview of the different classes of pyruvylated glycoconjugates is given, including glycan composition and structure, as well as functional and biosynthetic aspects, when known.

There is a recent interest in understanding the biosynthetic pathways of pyruvylated glycoconjugates from bacterial sources, as these pathways might unravel novel targets for therapeutic intervention. However, the current biosynthesis models for the different classes of glycoconjugates are in most cases only fragmentarily available and, frequently, they consider in silico predictions of involved components without experimental evidence.

3.1. Exopolysaccharides

Many organisms produce extracellular polysaccharides (exopolysaccharide, EPS) that are actively secreted during growth, including bacteria, yeasts, and microalgae [24]. EPSs are a diverse class of carbohydrate polymers that are composed of either linear or branched repeating units that are connected with varying stereochemistry. Monosaccharide constituents include pentoses (ribose and arabinose—especially in *Mycobacterium* spp.), hexoses (mannose (Man), glucose, fructose, galactose), deoxysugars (rhamnose (Rha), fucose (Fuc), uronic acids (glucuronic and galacturonic acids), and amino sugars (glucosamine, galactosamine, in several cases modified by *N*-acetylation) [24]. Depending on the monosaccharide composition, homo- or hetero-polymers are differentiated.

EPSs have a "jelly-like" appearance and are part of the glycocalyx—with which the "cellular sugar coat" is referred to [24]; as a common feature, they create a protective matrix around cells. The shielding effect against macromolecules that is conferred by EPS makes some bacteria 1000 times more resistant to antibiotics than their EPS-free counterparts [25].

Given the high application potential of microbial EPSs in medical fields, biomaterials, food applications, and in the replacement of petro-based chemicals [26], these glycoconjugates are currently of high interest.

3.1.1. Xanthan

Xanthan is the main EPS produced by *Xanthomonas campestris* and other phytopathogenic *Xanthomonas* spp. that cause various economically important diseases in mono- and di-cotyledonous crops. Xanthan enhances the attachment to plant surfaces through its effect on biofilm formation, promotes pathogenesis by Ca^{2+} chelation and, thereby, suppression of the plant defence responses in which Ca^{2+} acts as a signal [27]. In practical applications, xanthan is frequently used as a viscosifying agent [28,29].

The pentasaccharide-repeating unit of xanthan consists of two β-(1→4) linked D-Glc residues as backbone and a trisaccharide side chain, α-(1→3)-linked to every other glucose. The side chain is composed of α-D-Man, β-D-glucuronic acid (GlcA), and β-D-Man, which are β-(1→2)- and β-(1→4)-linked to another, respectively [30]. In its natural state, the α-D-Man residue is acetylated and the β-D-Man is either acetylated or pyruvylated. It was found that the 4,6-ketal-pyruvate (4,6Pyr) specifically and, to a lesser extent, the acetyl groups that decorate the mannose residues are involved in Ca^{2+} chelation [27] and affect bacterial adhesion and biofilm architecture and, hence, contribute to the bacterium's virulence [18]. Furthermore, the rheological properties of xanthan are influenced by its pyruvylation and acetylation pattern [28,31].

Xanthan biosynthesis is encoded in a so-called *gum*-cluster. Of the 13 encoded genes, *gumDMHK* are involved in the synthesis of the pentasaccharide repeat, and *gumBCEJ* in polymerization and xanthan export across the outer membrane in a flippase/polymerase (Wzx/Wzy)-dependent pathway. Regarding the modifications of xanthan, the predicted pyruvyltransferase GumL is hypothesised to catalyse pyruvylation of β-D-Man residues, while GumF and GumG are involved in β-D-Man acetylation [28,32]. It remains to be determined at which stage of xanthan biosynthesis the modifications are elaborated; it might be either at the cytoplasmic membrane or in the periplasmic space [32]. To this end, it was shown that GumK, a glucuronic acid transferase, is active on the lipid-linked trisaccharide precursor α-Man-(1→3)-β-Glc-(1→4)-β-Glc-P-P-polyisoprenyl, and shows reduced activity on the acetylated precursor substrate 6-*O*-acetyl-α-Man-(1→3)-β-Glc-(1→4)-β-Glc-PP-polyisoprenyl [33]. This suggests that mannose acetylation occurs after the completion of the trisaccharide side chain [32,33]; this might also hold true for pyruvylation. The xanthan biosynthetic enzymes seem to be highly conserved among different organisms, except for the mannose-transferase GumI and the pyruvyltransferase GumL, for which no homologues are found in other organisms [34].

Interestingly, *Bacillus* sp. strain GL1, which utilizes xanthan for its growth, produces an extra-cellular xanthan lyase, which catalyses the cleavage of the glycosidic bond between 4,6Pyr-β-D-Man and β-D-GlcA residues in xanthan side chains and, thus, contributes to depolymerisation of xanthan [35,36].

3.1.2. Succinoglycan

Succinoglycan is a pyruvylated EPS that is produced by *Agrobacterium* [37,38], *Alcaligenes*, *Pseudomonas* [39], and *Rhizobium* strains [40], and is of great importance in plant symbiosis.

It is a heteropolymer that is multiply decorated with pyruvate, succinate, and acetate substituents. While the extent of acetylation and succinylation depends on the strain and the cultivation conditions, pyruvate is always found in a stoichiometric manner at the terminal β-Glc residue [32]. The repeat unit structure of succinoglycan is composed of β-Glc and β-Gal in a molar ratio of 7:1. Nineteen genes are involved in the polymer's biosynthesis, which are referred to as *exo* genes and encoded in a 16 kb gene cluster. The biosynthesis starts with the production of the nucleotide-activated sugars UDP-Glc and UDP-Gal, where ExoC (phosphoglucomutase), ExoB (UDP-glucose-4-epimerase), and ExoN (UDP-pyrophosphorylase) are involved [32,41,42] (Figure 2). The initial step in the biosynthesis is executed by ExoY, a priming galactosyltransferase transferring a single, reducing-end Gal residue onto an undecaprenylphosphate (undp-P) carrier. *ExoA*, *exoL*, *exoM*, *exoO*, *exoU*, and *exoW* encode subsequent glycosyltransferases, which complete the octa-saccharide repeat in a step-wise manner, with each enzyme transferring a single monosaccharide, each, except for ExoW which transfers the subterminal and terminal glucoses. Prior to the export of the octasaccharide via a Wzx-dependent pathway, pyruvylation (at the terminal, non- reducing-end glucose), acetylation, and succinylation reactions catalysed by ExoV, ExoZ, and ExoH, handed over to ExoQ, which is responsible for polymerization of the fully modified repeats [43]. Studies on ExoV, the pyruvyltransferase of *Shinorhizobium* (previously *Rhizobium*) *meliloti*, suggest that pyruvylation is important for polymerization of repeating units and efficient succinoglycan export [44].

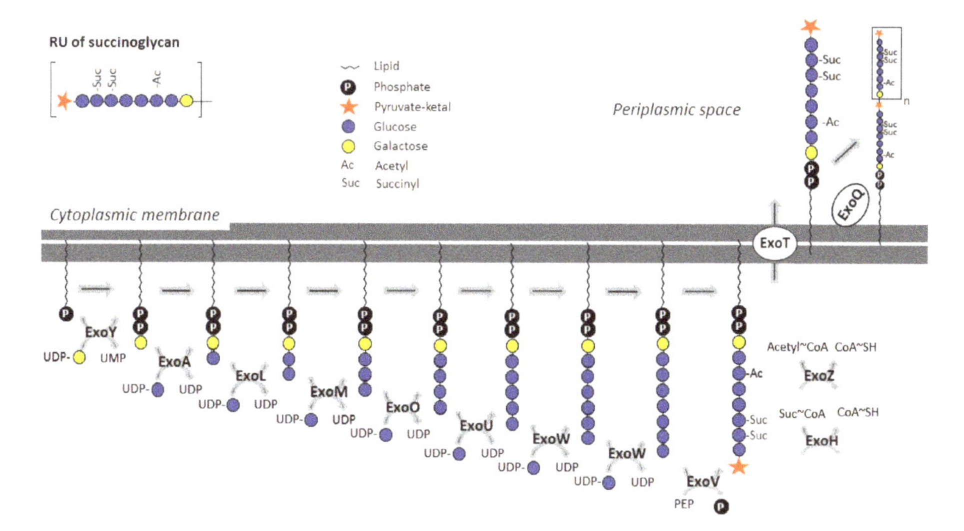

Figure 2. Scheme of succinoglycan biosynthesis in *Shinorhizobium meliloti* [43]. The pyruvylation step occurs in the cytoplasm at the stage of the undp-PP-linked RU prior to export and polymerization in the periplasmic space. Pyruvylation (ExoV) is indicated by a star. The order of pyruvylation, acetylation (ExoZ), and succinylation (Exo) is unknown. RU: repeating unit. Monosaccharide symbols are shown according to the Symbol Nomenclature for Glycans (SNFG) [45].

For *Rhizobium leguminosarum*, it was shown that missing pyruvylation on the terminal glucose residue of the succinoglycan impairs the formation of the nitrogen-fixing symbiosis with *Pisum sativum*, supportive of a signalling role of pyruvylation in this process [46]. PssK was identified as the pyruvyltransferase involved in succinoglycan modification of *R. leguminosarum* [46].

3.1.3. Salecan

The salt-tolerant soil bacterium *Agrobacterium* sp. ZX09 is the producer of salecan, a soluble, succinylated, and pyruvylated EPS with a β-(1→3) glucan structure that is of interest because of its multiple bioactivities and unusual rheological properties. Its basic repeating unit structure was initially elucidated as →3)-β-D-Glc*p*-(1→3)-[β-D-Glc*p*-(1→3)-β-D-Glc*p*-(1→3)]-α-D-Glc*p*-(1→3)-α-D-Glc*p*-(1→ [47]. On the basis of amino acid homology with the respective *exo* genes, it can be concluded that succinyl- and pyruvyl-groups are conferred to salecan upon catalysis of SleA (succinyl-transferase) and SleV (pyruvyltransferase), respectively, both of which are located in a 19.6-kb gene cluster [48]. The exact positions of the salecan modifications remain to be determined.

3.1.4. Colonic Acid

Colonic acid (CA) or M-antigen is another class of pyruvylated EPS mostly found in *Enterobacteriaceae*, including the majority of *Escherichia coli* strains. CA forms a loosely associated saccharide mesh that coats the bacteria, often within biofilms. CA is composed of hexasaccharide repeat units consisting of glucose, two fucoses, two galactoses, and glucuronic acid [49]. Additionally, acetylation is found on fucose or/and galactose, while pyruvylation is found on the terminal galactose only, with both modifications occurring non-stoichiometrically [32,50]. The overall structure of CA is 4,6Pyr-α-Gal*p*-(1→4)-β-Glc*p*A-(1→3)-*O*Ac-α-Gal*p*-(1→3)-α-Fuc*p*-(1→4)-*O*Ac-*O*Ac-α-Fuc*p*-(1→3)-β-Glc-(1→7)-α-Hep*p*-(1→6)-α-Glc*p*-(1→2)-α-Glc*p*-(1→3)-[α-Gal*p*-(1→6]-α-Glc*p*-(1→3)-[α-Hep*pP*-(1→7)] -α-Hep*pP*-(1→3)-[PEtN]α-Hep*p*-(1→5)-αKdo*p*-(1→, where Kdo is 3-deoxy-D-*manno*-oct-2-ulosonic acid.

The genetic determinants for CA biosynthesis reside in a 19-gene *wca* (*cps*) cluster and are tightly regulated by a complex signal transduction cascade [51]. The gene cluster encodes six glycosyl-transferases, named WcaJ, WcaI, WcaE, WcaC, WcaL, and WcaA. Furthermore, a putative pyruvyl-transferase (WcaK) is encoded next to two predicted acetyltransferases (WcaF and WcaB), although there are up to three acetylation positions described in CA [50]. Interestingly, WcaF seems to contribute to biofilm formation of the bacterium, since knocking out of this enzyme led to biofilm disruption under *in vitro* conditions [52].

3.1.5. Unclassified Pyruvylated EPS

Up to now, there are several unclassified types of pyruvylated EPSs. The repeating unit structure of the acidic EPS produced by a mucoid strain of *Burkholderia cepacia* isolated from a cystic fibrosis patient was established as →3)-β-D-Gal*p*-(1→3)-4,6Pyr-α-Gal*p*-(1→ [53].

The freshwater biofilm isolate *Pseudomonas* strain 1.15 produces considerable amounts of an acidic EPS that is composed of repeating units with the structure →4)-[4,6Pyr-α-D-Gal*p*-(1→4)-β-D-GlcA*p*-(1→3)-α-D-Gal*p*-(*O*→3]-α-L-Fuc*p*-(1→4)-α-L-Fuc*p*-(1→3)-β-D-Glc*p*-(1→. Furthermore, of the four different *O*-acetyl groups present in non-stoichiometric amounts, two were established to be on O-2 of the 3-linked galactose and on O-2 of the 4-linked fucose [54].

Enterobacter amnigenus, a bacterium isolated from sugar beets, produces an EPS that is rich in L-Fuc and has a terminal, pyruvylated α-D-Man residue [55].

The cystic fibrosis lung pathogen *Inquilinus limosus* produces two EPSs with unique structures—an α-(1→2)-linked mannan and a β-(1→3)-linked glucan—both fully substituted with 4,6-linked pyruvate ketals [56,57]. Cystic fibrosis is an autosomal recessive disorder and its mortality is due to chronic microbial colonisation of the major airways that leads to exacerbation of pulmonary infection [58]. While *Pseudomonas aeruginosa* is one of the most threatening microbes colonizing cystic fibrosis lungs, the EPS of *I. limosus* is suspected to play a role in the pathogenesis of the disease.

The bacterium *Azorhizobium caulinodans* produces a linear homopolysaccharide-type EPS composed of α-(1→3)-linked 4,6Pyr-D-Gal residues. The bacterium undergoes a symbiotic interaction with *Sesbania rostrata* as a legume host plant, which results in the development of root nodules, accompanied by a massive production of H_2O_2. In situ H_2O_2 localization demonstrated that increased EPS production during early stages of invasion prevents the incorporation of H_2O_2 inside the bacteria, suggesting a role for EPS in protecting the microsymbiont against H_2O_2 [59].

A special K-antigen-like EPS is found in the marine bacterium *Cobetia marina* DSMZ 4741, with its repeating unit composed of ribose and pyruvylated Kdo [60].

Within the EPS structure of the lactic acid bacterium *Pediococcus pentosaceus* LP28, a pyruvate modification was described to occur on one of the four constituting monosaccharides (Glc, Gal, Man, and GlcNAc) [61]. The EPS biosynthetic gene cluster consists of 12 ORFs containing a priming enzyme, five glycosyltransferases, and a putative polysaccharide: pyruvyltransferase [61].

EPSs produced by an *Erwinia* spp. in association with the bacterium *Coniothyrium zuluense* are linked to a fungal canker disease of *Eucalyptus* [62]. One of these EPSs is that of *Erwinia stewartii*; another is that of *Erwinia futululu*, whose structures are identical except for the replacement of one terminal Glc residue by 4,6Pyr-Gal*p* in the latter, yielding →3)-β-D-Gal*p*-(1→3)[4,6Pyr-α-D-Gal*p*-(1→4)-β-D-Glc*p*A-(1→4)][β-D-Glc*p*-(1→6)]-α-D-Gal*p*-(1→6)-β-D-Glc*p*-(1→ [63].

Agrobacterium radiobacter (ATCC 53271) produces an anionic EPS that gives aqueous dispersions, exhibiting high viscosity at low concentrations. The *A. radiobacter* EPS is composed of a complex heptadekasaccharide repeating unit, which exposes a subterminal 4,6Pyr-α-D-Glc residue on each of the two identical tetraglycosyl branches [64].

Methylobacterium sp. is a slime-forming bacterium isolated from a Finnish paper machine, which is a high EPS producer. Its EPS repeating unit has the structure →3)-4,6Pyr-α-D-Gal*p*-(1→3)-4,6Pyr-α-D-Gal*p*-(1→3)-α-D-Gal*p*-(1→ [65].

The marine bacterium *Alteromonas macleodii* subsp. *fijiensis* isolated from deep-sea hydrothermal vents displays a pyruvylated mannose in its EPS hexasaccharide repeating unit structure →4)-β-D-Glc*p*-(1→4)[4,6Pyr-β-D-Man*p*-(1→4)-β-D-Glc*p*A-(1→3)-α-D-Glc*p*A-(1→3)]α-D-Gal*p*A-(1→4)-α-D-Gal*p*-(1→ [66]. Aside from its use in the food industry, this marine polymer has been suggested to be used for the treatment of cardiovascular diseases and bone healing [67].

3.2. *Capsular Polysaccharides*

Capsular polysaccharides (CPSs) are also part of the glycocalyx but, in contrast to EPSs, are covalently connected to the bacterial cell surface via membrane phospholipids [24]. Because of their prominent cellular localization, CPSs are the first interaction zone of bacteria with the host immune system, and thus are important virulence factors of many bacteria. Very often, encapsulated bacteria are pathogenic, whereas capsule-deficient isolates are not [68]. Hence, CPSs are frequently used for the production of polysaccharide conjugate vaccines [69].

Bacterial capsules are formed primarily from long-chain polysaccharides with repeat-unit structures. A given bacterial species can produce a range of CPSs with different structures, and these aid in distinguishing isolates by serotyping [68]. The widespread occurrence and the high structural differences of CPSs are reflected by 84 capsular serotypes (K-antigens) found alone in *E. coli* strains. Essentially, there are four groups of capsules [70]. Group I- and IV-CPS, which are often found in organisms leading to gastrointestinal diseases, use the Wzx/Wzy-dependent export pathway and their biosynthesis proceeds on a polyprenol linker. Capsules from groups II and III use the ABC-transporter export pathway and are frequently present in mucosal pathogens such as *Neisseria meningitidis*. Interestingly, a CPS attached via a novel β-linked poly-3-deoxy-D-*manno*-oct-2-ulosonic acid linker to the phospholipid *lyso*-phosphatidylglycerol is present, which in earlier studies was described as a diacylglycerol because of hydrolysis experiments [68].

3.2.1. *Streptococcus pneumoniae* CPS

Streptococcus pneumoniae (pneumococcus) is a leading cause of bacterial-induced pneumonia, meningitis, and bacteraemia globally [71]. Prevnar 13, the most broadly protective pneumococcal conjugate vaccine, is composed of 13 protein-polysaccharide conjugates consisting of pneumococcal CPS serotypes 1, 3, 4, 5, 6A, 6B, 7F, 9V, 14, 18C, 19A, 19F, and 23F—each individually linked to the genetically inactivated diphtheria toxoid CRM_{197}. Currently, approaches on the basis of biocon-jugation and glycosylation engineering are being pursued as manufacturing alternatives to enable the production of vaccines with higher protection rates, especially in children [69,72,73].

S. pneumonia CPS serotype 4 (ST4) is a prevalent serotype in vaccine formulations, containing 2,3-pyruvate ketal on the Gal residue of its repeating units—with the structure

→3)-β-D-ManpNAc-β-(1→3)-α-L-FucpNAc-(1→3)-GalpNAc-α-(1→4)-2,3Pyr-α-D-Gal-(1→—which has been shown to be the key component of its specific immunogenic motif [74].

Thus, the pyruvate modification is essential for designing minimal synthetic carbohydrate vaccines for ST4, as vaccine formulations without pyruvylation would not recognize the natural CPS [8]. It is, therefore, highly recommended to include the pyruvate ketal epitope in glycoconjugate vaccines [16].

3.2.2. *Acinetobacter baumannii* CPS

A clinically relevant producer of CPS is the opportunistic pathogen *Acinetobacter baumannii*, which triggers infections in immunocompromised patients causing severe nosocomial, bloodstream, pneumonia, urinary tract infections, and septicaemia [75]. Its clinical importance is related to its low susceptibility towards most of the antibiotics commonly used [76].

There are seven different capsule loci—KL1, KL2, KL4 KL6, KL7, KL8, KL9—in *A. baumannii* genomes. Five of these were found in clonal group 2, whereas two were found in clonal group 1, indicating that isolates with developing antibiotic resistance have a lot of variations of these loci [77]. The K4 CPS of isolate D78, which is a multiple antibiotic resistant strain, contains the KL4 cluster. The KL4 CPS backbone repeating structure is composed of a trisaccharide of α-*N*-acetyl-D-quinovosamine (D-QuipNAc), α-*N*-acetyl-D-galactosamine uronic acid (α-D-GalpNAcA), and α-D-GalpNAc, which contains a branching 4,6-pyruvylated GalNAc residue. The trisaccharide structure was elucidated as →4)[4,6Pyr-α-D-GalpNAc-(1→6)]-α-D-GalpNAc-(1→4)-α-D-GalpNAcA-(1→3)-α-D-QuipNAc-(1→. The pyruvate ketal is predicted to be transferred by the putative pyruvyltransferase PtrA, however, without biochemical evidence [78].

3.2.3. *Klebsiella* CPSs

Pyrogenic liver abscess-causing *Klebsiella pneumoniae* produces a CPS, which is composed of trisaccharide repeating units with the structure →4)-β-D-Glc-(1→4)-2,3(*S*)Pyr-β-D-GlcA-(1→4)-β-L-Fuc-(1→, in which each glucuronic acid residue is pyruvylated and additional acetylation of the fucose residue occurs at the C2-OH or C3-OH [79]. The CPS induces secretion of tumour necrosis factor and interleukin-6 by macrophages through the Toll-like receptor 4 dependent pathway, which is abandoned when pyruvylation is missing in the trisaccharide. This finding indicates that pyruvylation on glycoconjugates may be relevant for immune system stimulation [80]. Previously, the recognition of pyruvylated CPS from *K. pneumoniae* by IgM antibodies has been described [81].

The structures of several other pyruvylated *Klebsiella* CPS structures have been elucidated, however, without any functional information.

The structure of the CPS from *Klebsiella* serotype K14 was the first report on the rare case of a *Klebsiella* polysaccharide to contain a Gal*f* residue. The repeating hexasaccharide structure was shown to terminate with a glucose residue carrying a 4,6Pyr modification—→4)-β-D-GlcpA-(1→3)-β-D-Gal*f*-(1→3)-β-D-Glc*p*-(1→4)[4,6Pyr-β-D-Glc*p*-(1→2)][α-L-Rha(1→3)]β-D-Man*p*-(1→ [82].

Also in the doubly pyruvylated CPS of *Klebsiella* K12, Gal*f* residues are found; its repeating unit has the structure 5,6Pyr-β-D-Gal*f*-(1→4)-β-D-GlcpA-(1→3)-β-D-Gal*f*-(1→6)-β-D-Glc*p*-(1→3)-α-L-Rha-(1→3)-α-D-Gal*p*-(1→2)[5,6Pyr-β-D-Gal*f*-(1→4)-β-D-GlcpA-(1→3)]-β-D-Gal*f*-(1→6)-β-D-Glc*p*-(1→3)-α-L-Rha-(1→3)-α-D-Gal*p* [83].

The structure of the CPS from *Klebsiella* serotype K70 is composed of linear hexasaccharide repeating units that contain a Pyr group attached to a (1→2)-linked α-L-Rha residue in every second repeating unit. The full structure of the *Klebsiella* K70 CPS is →4)-β-D-GlcpA-(1→4)-α-L-Rha*p*-(1→2)-α-L-Rha*p*-(1→2)-α-D-Glc*p*-(1→3)-β-D-Gal*p*-(1→2)-3,4Pyr-α-L-Rhap(1→ [84].

The *Klebsiella* serotype K64 CPS consists of hexasaccharide repeating units, composed of a →4)-α-D-GlcpA-(1→3)-α-D-Man*p*-(1→3)-β-D-Gal*p*-(1→4)-α-D-Man*p*-(1→ backbone with a 4,6Pyr-β-D-Gal*p* and a L-Rha residue attached to the (1→4)-linked α-D-Man*p* residue at O-2 and O-3, respectively; the repeating unit further contains one *O*-acetyl substituent [85].

The structure of the CPS from *Klebsiella* type K46 consists of a hexasaccharide repeating unit, which is unique in having a 4,6Pyr residue on a lateral, but non-terminal sugar residue—[β-D-Glc*p*-(1→3)-4,6Pyr-β-D-Man*p*-(1→4)]→3)-α-D-Glc*p*A-(1→3)-α-D-Man*p*-(1→3)-α-D-Gal*p*-(1→ [86].

The *Klebsiella* K33 CPS revealed to be a tetrasaccharide alditol with the structure β-D-Glc*p*-(1→4)[3,4Pyr-β-D-Gal*p*-(1→6)]-β-D-Man*p*-(1→2)-Ery-ol, where Ery-ol is erythritol [87].

Interestingly, there are two human monoclonal macroglobulins, IgM^{WEA} and IgM^{MAY} [88], which show specificity for *Klebsiella* polysaccharides containing 3,4-(K30, K33) and 4,6-(K21, K11) pyruvylated D-Gal in a pH-dependent manner and with differences in co-precipitation in dependence of the number of the CPS repeating units (i.e., CPS length) [81]. Of note, agar, which has an internal 4,6Pyr-Gal residue in its repeating unit, cross-reacts with IgM^{WEA} [81].

Klebsiella rhinoscleromatis is a heavily capsulated bacterium that possesses a K3-type capsule. The repeating unit of K3 is a pentasaccharide with the structure →2)-4,6-*S*-Pyr-α-D-Man-(1→4)-α-D-GalA-(1→3)-α-D-Man-(1→2)-α-D-Man-(1→3)-β-D-Gal-(1→ [89]. The *Klebsiella* K3 capsule has been shown to be one of the few *Klebsiella* K types that are able to bind to the eukaryotic mannose receptor [90].

3.2.4. *Bacteroides fragilis* CPS A

Bacteroides fragilis is an opportunistic anaerobe, most frequently isolated from intra-abdominal abscesses [7,91–93]. Its most prominent CPS—CPS A—is composed of tetrasaccharide repeating units with the structure →4)-α-D-2-*N*-acetylamido-4-amino-galactopyranose (AADGal*p*)-(1→3)-4,6PyrGal*p*-(1→3)-[β-D-Gal*f*-(1→3)]α-D-Gal*p*NAc-(1→) [94]. CPS A has been shown to have a tremendous effect on the immune system of a mammalian host and to be internalized by antigen-presenting cells [7]. Upon genetic deletion of CPS A, the abscess-inducing capability of the bacterium was drastically reduced [93]. CPS A from *B. fragilis* caught recent interest as a carbohydrate antigen to be used in vaccine formulations instead of conventional cationic proteins such as bovine serum albumin (BSA) and keyhole limpet hemocyanin (KHL) [92].

CPS A biosynthesis is encoded by a single ~10.7-kb gene locus on the *B. fragilis* genome [93], and predictably employs a Wzx/Wzy-dependent pathway, on the basis of genomic evidence (Figure 3).

The gene locus encodes four transferases (WcfN, WcfP, WcfQ, and WcfS), where WcfS is responsible for the transfer of the AADGal*p* residue from its nucleotide activator to an undp-P-lipid carrier as the first step in the synthesis of the CPS A repeating unit, and WcfR is responsible for the prior transfer of the amino group on the ADGal*p* residue to yield AADGal*p*, which is crucial for virulence. A recent *in vitro* study of the individual enzymatic steps involved in the repeating unit biosynthesis of CPS A yielded first insight into the sugar pyruvylation reaction, with phosphoenolpyruvate (PEP) serving as a donor substrate. There is evidence that pyruvylation occurs on the undp-PP-linked disaccharide repeat unit precursor prior to tetrasaccharide repeat completion, export, and polymerization by a Wzx/Wzy-dependent system [7]. The pyruvyltransferase WcfO from the CPS A biosynthesis of *B. fragilis* [7] is one of the few biochemically characterized enzyme ketal-pyruvyltransferases (for details, see Section 5.1.2).

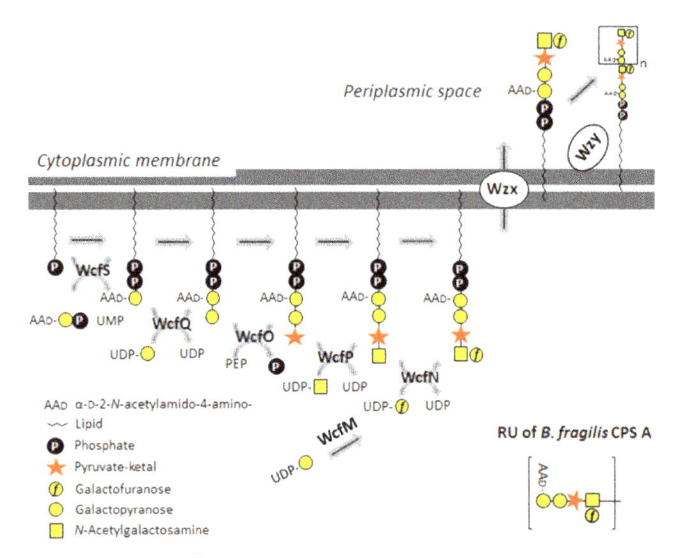

Figure 3. Scheme of capsular polysaccharide (CPS) A biosynthesis in *Bacteroides fragilis*. The pyruvylation step occurs in the cytoplasm at the stage of the lipid-PP-linked RU precursor. Pyruvylation (WcfO) is indicated by a star. Notably, in contrast to succinoglycan biosynthesis (Figure 2), pyruvylation of the internal Gal*p* of the RU needs to proceed prior to completion of the lipid-PP-linked tetrasaccharide repeat. RU: repeating unit. Monosaccharide symbols are shown according to the Symbol Nomenclature for Glycans (SNFG) [45].

3.2.5. *Rhodococcus equi* CPS

The bacterial horse pathogen *Rhodococcus equi* elaborates a serotype-specific CPS that functions as a potential virulence factor [94]. This CPS is a high-molecular-weight acidic polymer composed of D-Glc, D-Man, pyruvate, and 5-amino-3,5-dideoxynonulosonic acid (rhodaminic acid, Rho) in the molar ratio of 2:1:1:1. Structural analysis revealed that the CPS consists of linear pyruvylated tetra-saccharide repeats with the structure →3)-β-D-Man*p*-(1→4)-β-D-Glc*p*-(1→4)-α-D-Glc*p*-(1→4)-7,9Pyr-α-RhoAmAc-(2→ [95].

3.3. *"Non-Classical" Secondary Cell Wall Glycopolymers with Pyruvylated β-D-ManNAc*

Anionic secondary cell wall glycopolymers (SCWPs)—among which the wall teichoic acids (WTA) and the lipoteichoic acids (LTA) are best known—serve as a rich source for both validated and unexploited pathways that are essential for bacterial virulence and survival [96–98]. Pyruvylated SCPWs are a less-investigated class of peptidoglycan-attached SCWPs that arouse interest because they can be hijacked for a predictably widespread mechanism of protein cell surface display in Gram-positive bacteria [99]. These SCWPs are 5–20 kDa in size, composed of species–specific repeats, but lack repetitive alditol phosphates and phosphodiester bonds typical of WTAs and LTAs [100–102]—hence the terminology "non-classical" SCWPs. Importantly, they contain a 4,6-ketal pyruvylated β-D-ManNAc residue (4,6Pyr-β-D-ManNAc), imparting a negative charge and serving as a specific cell wall ligand for S-layer homology (SLH) domains usually present in triplicate at the termini of cell surface proteins [99,103,104]. Among such proteins are S-layer proteins, which self-assemble into 2D crystalline arrays on the bacterial cell surface [105,106]; they are important for many biological functions such as maintenance of cell integrity, enzyme display, protection to phagocytosis, and interactions with the host and its immune system [107]. Because of their unique 2D crystallization ability, S-layer proteins are of great interest for drug delivery, biomaterial engineering, and vaccine development [108].

Fifty-four thousand specific hits within the conserved protein domain family SLH (pfam00395), showing up most prevalent in the *Firmicutes*, *Cyanobacteria*, and *Actinobacteria* phyla of bacteria, emphasize the prevalence of this protein domain. Several bacteria synthesizing a suite of SLH proteins contain pyruvate in their cell wall and have a pyruvyltransferase CsaB ortholog [99,109], which indicates a functional coupling of SLH domains and SCWP pyruvylation.

The best-known pyruvylated SCWPs are those from *B. anthracis* and *B. cereus* strains [14,110, 111], and from *P. alvei* [15]. The most evident difference between these SCWPs is the presence of 4,6Pyr-β-D-ManNAc exclusively at the terminal repeat in the former SCWPs, while in that of *P. alvei*, the β-D-ManNAc of each repeat is pyruvylated. This might explain the essentiality of the pyruvyltransferase CsaB in the latter organism [109]. It is conceivable to assume that mono- versus poly-pyruvylation of β-D-ManNAc has implications with regard to the biosynthetic pathway of the respective SCWP, especially the mode of activity of the cognate pyruvyltransferase (compare with Section 5.1.1.).

The genetic determinants and assembly reactions of pyruvylated SCWPs are only beginning to be discovered. Pyruvylated SCWPs are peptidoglycan-linked polymers, as are WTAs [97,98], however, they lack experimental evidence of a comparable biosynthetic route.

3.3.1. *Bacillus anthracis* SCWP

The *B. anthracis* SCWP is composed of →4)-β-ManNAc-(1→4)-β-GlcNAc-(1→6)-α-GlcNAc-(1→ trisaccharide repeats [102] with strain-dependent galactosylation occurring at the GlcNAc residues, and are bound to peptidoglycan via a murein linkage unit [14,112]. Notably this SCWP contains a modified terminal repeat with the structure 4,6-Pyr-β-ManNAc-(1→4)-[3*O*Ac-β-GlcNAc-(1→6)]β-GalN-(1→4) that is pyruvylated and *O*-acetylated [14].

A scenario for the biosynthesis of the *B. anthracis* SCWP has been proposed [113] on the basis of the bioinformatic prediction of four contributing genomic gene clusters and their genetic manipulation, accompanied with analyses of mutant cells. The S-layer gene cluster [99,112–114] encodes, among others, components for pyruvylation (*B. anthracis* CsaB) and *O*-acetylation of the terminal SCWP trisaccharide, and a Wzy-like protein. The other gene clusters play predicted roles in the formation of lipid-linked precursors of the murein linkage unit and the trisaccharide repeat [112,114–116]. The SCWP biosynthesis model proposes the separate assembly of different undp-PP-linked building blocks in the cytoplasm—the murein linkage unit, trisaccharide repeat, and terminal modified trisaccharide [113]—followed by the individual translocation across the cytoplasmic membrane via Wzx, followed by SCWP polymerization at the outer face of the membrane involving Wzy [113]. This model does not explain how the different building block precursors converge and how pyruvylation of β-D-ManNAc is elaborated. A recent study identified PatB1 as *O*-acetyl-transferase in the terminal repeat biosynthesis of *B. cereus* SCWP, proposing an extracellular *O*-acetylation mechanism [117]. Previously, it was surmised that the modifications at the terminal trisaccharide emerge post-polymerization and ligation to peptidoglycan [14].

B. anthracis CsaB is not essential for survival, but it is important for the pathogenesis of infection; Δ*csaB* mutants lacking SCWP pyruvylation fail to retain SLH-domain containing proteins in the cell wall, leading to an atypical cell morphology [99,118]. Of note, in addition to SCWP [14], the *B. anthracis* cell wall contains a polyglycerol phosphate LTA [119] and a poly-γ-D-glutamic acid capsule [101], which could support the cell wall integrity in a strain devoid of pyruvylation providing the anionic character.

3.3.2. *Paenibacillus alvei* SCWP

The SCWP of *P. alvei* consists of →3)-β-D-ManNAc-(1→4)-β-D-GlcNAc-(1→ repeats, where β-D-Man*p*NAc of each disaccharide is modified with 4,6-linked pyruvate ketal [15,120]. Notably, an identical SCWP composition was found in the S-layer carrying *Lysinibacillus sphaericus* CCM 2177, where on average, every second β-D-ManNAc residue contains a 4,6Pyr-modification [121].

P. alvei possesses a polycistronic SCWP biosynthesis gene cluster comprised of *csaB*, *tagA*, *tagO*, two SLH-protein encoding genes—*slhA* and *spaA*—and two ORFs of unknown function [109,122]. For SCWP assembly in *P. alvei*, distinct enzymatic steps have been investigated [109]. The bacterium utilizes the subsequent TagO [123] and TagA [124,125] catalysed reactions—typically involved in WTA biosynthesis to produce the undp-PP-bound murein linkage unit [97,126,127] for the biosynthesis of the lipid-linked disaccharide substrate needed to generate the repeat unit backbone of its SCWP [15]. Pyruvylation with PEP as donor substrate was experimentally determined *in vitro* at the stage of the lipid-linked disaccharide repeat precursor [109] (see Section 5.1.1.).

In *P. alvei*, no viable deletion mutant could be obtained of either *tagO*, *tagA*, or *csaB* [109]—each of which is located on the *P. alvei* genome as a single copy—indicating essentiality of the pyruvylated SCWP for the bacterium. This might be explained by the presence of pyruvylated SCWP as exclusive anionic SCWP in *P. alvei* and may be supportive of the necessity of at least one anionic polymer in the Gram-positive cell wall [128,129].

3.3.3. Cell Wall Polysaccharide of *Paenibacillus polymyxa*

Paenibacillus (previously *Bacillus*) *polymyxa* AHU 1385 was among the first bacteria for which a pyruvylated ManNAc residue was described [130]. The pyruvylated epitope is contained in a →3)-4,6Pyr-ManNAc-(1→4)-GlcNAc-(1→-repeating unit of an SCWP that is presumably peptidoglycan-linked. However, its clear that assignment to a specific SCWP class has not yet been reported, nor have any functional implications such as protein binding. Notably, *P. polymyxa* does not possess an S-layer.

3.4. SCWPs with Other Pyruvylated Sugar Epitopes

Some SCWPs containing pyruvylated epitopes other than β-D-ManNAc have been reported from the phylum *Actinobacteria*. There, sugar pyruvylation mainly serves as a chemotaxonomic marker of distinct strains, without further knowledge of putative associated functions.

3.4.1. SCWPs from the Genus *Promicromonospora*

Two strains of the genus *Promicromonospora* are recently uncovered examples of bacteria, which possess non-phosphorylated anionic glycopolymers ("non-classical" SCWPs) with pyruvic acid acetals of *R*-configuration in their cell wall [131]. Members of this genus produce a mycelium that fragments into rod-shaped or coccoid elements and are characterised according to different genus-specific chemotaxonomic markers, including the peptidoglycan of the A4α type [132].

The type strain *Promicromonospora citrea* 665^T contains two "non-classical" SCWPs, namely a 2-keto-3-deoxy-D-*glycero*-D-*galacto*-nononic acid (Kdn)-teichulosonic acid containing polymer with the repeating unit structure →6)-α-D-Glc*p*(1→6)-α-D-Glc*p*3SO_3^--(1→4)-α-7,9Pyr-Kdn-(2→, where the Kdn residue is 7,9-pyruvylated, and a galactan with the repeating unit structure →3)-α-4,6Pyr-D-Gal*p*-2OAc-(1→, including 4,6-pyruvylation of Gal.

The cell wall of *Promicromonospora* sp. VKM Ac-1028 contains a teichuronic acid-like structure with the repeating unit →6)-α-D-Glc*p*-(1→4)-β-2,3Pyr-D-Glc*p*A-(1→, where glucuronic acid is 2,3-pyruvylated [131].

3.4.2. Teichoic Acids from the Genus *Nocardiopsis*

The first description of a pyruvate ketal modification on a classical teichoic acid (TA) was reported in *Nocardiopsis* strains, a widespread group among the *Actinobacteria* [133]. The genus *Nocardiopsis* is of pharmaceutical and biotechnological interest because of its ability to produce a variety of secondary metabolites—accounting for its wide range of biological activities—and, thus, holds promises as a source of novel bioactive compounds [134].

The major TA of *Nocardiopsis metallicus* VKM Ac-2522T is a 1,5-poly(ribitol phosphate) TA, with each ribitol unit carrying a pyruvate ketal group at positions 2 and 4. The major TA of *N. halotolerans* is a poly(glycerol phosphate-*N*-acetyl-β-galactosaminylglycerol phosphate) structure in which the GalNAc residue carries a 4,6-ketal pyruvate modification.

3.4.3. Teichoic Acids of *Brevibacterium iodinum*

Brevibacterium iodinum VKM Ac-2106 produces two distinct WTAs, namely a mannitol-WTA and a glycerol-TA, present in minor amounts. Mannitol-WTA is a 1,6-poly(mannitol phosphate) bearing β-D-Glc*p* residues at the C-2 of mannitol (Man-ol) and, optionally, 4,5-*S*-Pyr residues. Glycerol-WTA is a 1,3-poly(glycerol phosphate) substituted at the C-2 of glycerol by α-D-Gal*p*NAc residues bearing 4,6-*R*-Pyr [135].

3.5. *Lipopolysaccharides and Lipooligosaccharides*

Lipopolysaccharides (LPSs) of Gram-negative bacteria are a unique family of glycolipids based on a highly conserved lipid moiety known as lipid A. These molecules are produced by most Gram-negative bacteria, in which they play important roles in the integrity of the outer-membrane permeability barrier and participate extensively in the host–pathogen interplay [136,137]. Complete LPSs have a three-domain molecule architecture; the two-domain variants without an O-antigenic polysaccharide (O-PS) are termed lipooligosaccharides (LOSs) [138]. Lipid A is the hydrophobic anchor of LPSs; it is a unique phosphoglycolipid containing glucosamine (GlcN) residues, which are present as β-(1→6)-linked dimers. The disaccharide contains phosphoryl groups and (*R*)-3-hydroxy fatty acids in ester and amide linkages. Variations in the fine structure can arise from the type of hexosamine present, the degree of phosphorylation, the presence of phosphate substituents, and, importantly, in the nature, chain length, number, and position of the acyl groups. Lipid A is glycosylated with a core oligosaccharide (core OS)—typically containing Kdo, a signature molecule of LPS [139], and heptose residues, which may provide an attachment site for a long-chain O-PS of varying repeating unit composition. The O-PS provides a major cellular antigen (O-antigen) used for serological typing of clinical isolates of a given species. Notably, the O-antigen is expressed by most of the clinically relevant strains and is an important phage receptor; LOS, in contrast has been found to be expressed by a group of Gram-negatives that colonize genital and respiratory mucosal surfaces [140].

3.5.1. *Pseudomonas stutzeri*

Analysis of the LPS of *Pseudomonas stutzeri* OX1 revealed a novel type of highly negatively charged LOS, containing two 4,6-linked pyruvate ketals linked to *N*-acetylglucosamine and glucose, independently [4]. The overall LOS structure has been determined to be 4,6(*S*)Pyr-β-Glc-(1→3)[4,6(S)Pyr-β-GlcNAc](1→4)GalNAc(1→3)-Hep7Cm2*P*4*P*-(1→3)-Hep2*P*4*P*-(1→5)[Kdo-(2→4)]Kdo-(2→6)-β-GlcN4*P*-(1→6)-GlcN1*P*, where *P* represents a phosphate group and Cm is carbamoyl. *P. stutzeri* OX1 was isolated from the activated sludge of a wastewater treatment plant, where unusual metabolic capabilities for the degradation of aromatic hydrocarbons were found. Pyruvate residues might be used to block elongation of the LPS chain to yield an LOS. This would lead to a less hydrophilic cellular surface, indicating an adaptive response of *P. stutzeri* OX1 to a hydrocarbon-containing environment [4].

3.5.2. *Providencia alcalifaciens*

Bacteria of the genus *Providencia* are opportunistic human pathogens that cause intestinal and urinary tract infections. The O-antigen-based serological classification scheme of *Providencia alcalifaciens, Providencia rustigianii*, and *Providencia stuartii* includes 63 O-serogroups [141], most of which are acidic. Complex core structures have been elucidated in several *Providencia* O-serogroups [142,143].

P. alcalifaciens O19 differs in its O-PS from other *Providencia* strains. The O-PS repeat contains a 4,6-pyruvylated GlcNAc residue and has the complete structure →2)-β-Fuc3NAc4Ac-(1→3)-4,6(*S*)Pyr-α-GlcNAc-(1→4)-α-Gal-(1→4)-β-Gal-(1→3)-β-GlcNAc-(1→. In the NMR spectra of the oligosaccharide, signals of the methyl group of pyruvic acid have been observed and the presence of pyruvic acid in two thirds of the O-units in the polysaccharide has been proven [141].

3.5.3. *Shigella dysenteriae*

Shigella dysenteriae is an aetiological agent of various intestinal disorders, including shigellosis. The strains of this bacterium are serologically heterogeneous because of the diversity of the structures of their O-antigens [144].

The structure of the *S. dysenteriae* type 10 O-antigen has been revised by Perepelov et al. in order to account for the acid-labile pyruvate modification that has been lost in a previous investigation due to acidic treatment of the sample. The full O-PS structure has been elucidated to be →2)-4,6(*S*)Pyr-β-D-Man*p*-(1→3)-α-D-Man*p*NAc-(1→3)-β-L-Rha*p*-(1→4)-α-D-Glc*p*NAc-(1→ [145].

3.5.4. *Raoultella terrigena*

The enterobacterium *Raoultella terrigena* is another bacterium that carries a pyruvic acid modification on its O-PS β-Man residue, located at the O-4 and O-6 positions [146]. The structure of the repeating unit of the O-PS has been determined by means of chemical and spectroscopic methods and found to be a linear tetrasaccharide with the structure →2)-4,6(*S*)-Pyr-β-D-Man*p*-(1→3)-α-D-Man*p*NAc-(1→3)-β-L-Rha*p*-(1→4)-α-D-Glc*p*NAc-(1→ [146], which is identical to that of *S. dysenteriae* type 10 [145].

3.5.5. *Proteus mirabilis*

The O-PS repeating unit of *Proteus mirabilis* O16 has been established to be →3)-β-D-Glc*p*NAc-(1→3)-4,6-*R*-Pyr-α-D-Gal*p*NAc1→4)-α-D-Gal*p*A-(1→3)-α-L-Rha*p*2Ac-(1→ [147]. This structure is significantly different from the O-PS structures of other *Proteus* spp. from the serogroup O19, such as *Proteus vulgaris, Proteus hauseri*, and *Proteus penneri* strains, and thus was a key for the reclassification of various *Proteus* strains.

3.5.6. *Cobetia pacifica*

The O-PS form the LPS of *Cobetia pacifica* KMM 3878—an aquatic isolate form Japan—is composed of sulphated and pyruvylated trisaccharide repeats with the structure →4)-β-D-Gal-2,3-SO_3H-(1→6)-β-D-Gal-3,4-*S*-Pyr-(1→6)-β-D-Gal-(1→ [148].

3.6. *Mycobacterial Glycolipids with Pyruvate*

Mycobacteria contain a variety of glycolipids, including, among others, acylated glucose, acylated trehalose, sulfatides, mannophosphoinositides, and glycopeptidolipids [149]. A crude glycolipid fraction from *Mycobacterium smegmatis* ATCC 356 obtained by ethanolic extraction and silica gel chromatography revealed the presence of hitherto unknown anionic glycolipids [150]. The corresponding glycan moiety has the structure 4,6Pyr-3-*O*-Me-β-D-Glc*p*-(1→3)-4,6Pyr-β-D-Glc*p*-(1→4)-β-D-Glc*p*-(1→6)-β-D-Glc*p*-(1→1)-α-D-Glc.

Members of the *Mycobacterium avium–Mycobacterium intracellulare* (MAI) complex are typeable on the basis of their specific antigenic glycolipid. For instance, the dominant epitope of the MAI serovar 8-specific glycopeptidolipid is a terminal 4,6Pyr-*O*-Me-α-D-Glc*p* unit, whereas that of the MAI serovar 21 has the same terminal pyruvylated glucose devoid of the 3-methoxy group [151]. Healthy individuals of some populations are carriers of antibodies that are specific to these pyruvylated epitopes on the glycopeptidolipids. It is currently unclear, if the antibody reflects previous experience with one or both of these serovars or whether some other common cross-reacting pyruvylated environmental antigen is involved [151]. However, this finding might have protective implications against mycobacterioses and other infectious diseases.

3.7. *Pyruvylated Glycoconjugates in Eukaryotes*

3.7.1. Eukaryotic Glycolipids

Information on pyruvylated glycoconjugates in eukaryotes is scarce in comparison to their description in bacteria. It is currently not clear whether this reflects the natural distribution of pyruvylation or if pyruvylation on eukaryotic glycoconjugates has escaped detection. Notably, pyruvylation has so far not been detected in humans.

An "exotic" example of a pyruvylated eukaryotic glycoconjugate is the phosphonoglyco-sphingolipid containing pyruvylated galactose in the nerve fibres of the sea hare *Aplysia kurodai*. The glycan structure of this phosphonoglycosphingolipid is 3,4Pyr-β-Gal-(1→3)-α-GalNAc-(1→3)-α-Fuc-(1→)-2-aminoethylphosphonyl-Fuc-(1→6)-β-Gal-(1→4)-Glc-(l→ [152].

3.7.2. *N*-Linked Glycans in Yeast

Yeast species are known for the production of high- or oligo-mannosidic *N*-glycans that are displayed on various cell surface proteins [153]. In several yeast species (e.g., *Saccharomyces cerevisiae*, *Candida albicans*, *Pichia holstii*, and *Pichia pastoris*), phosphate groups or, to a lesser extent, sialic acids present on these extracellular glycans provide the necessary negative cell surface charge [153–155].

S. pombe is a notable example of a yeast whose net negative surface charge is neither conferred by phosphate nor by sialic acid. Instead, the *N*-linked galactomannans of *S. pombe* have pyruvylated β-Gal-(1→3)-(PvGal) caps on a portion of the α-Gal-(1→2)-residues in their outer *N*-glycan chains [156]. *S. pombe* lacks the ER Man_9-α-mannosidase function as known from, for example, *Saccharomyces cerevisiae*. Therefore, it adds further mannose and galactose residues to the common *N*-glycan core structures, yielding galactomannans [157–159]. At least five different genes are required to synthesize the PvGal epitope. It is assumed that 4,6Pyr-β-Gal-(1→3) synthesis is carried out by a coordinated enzymatic system in which the β-Gal-(1→3) residues are first added to the *S. pombe* galactomannans and subsequently pyruvylated by the pyruvyltransferase Pvg1p [3] (see Section 5.1.3.). However, the complete mechanism for PvGal biosynthesis is currently unknown [3].

4,6Pyr-β-Gal is predicted to be the only contributor to the net negative cell surface charge of yeast, as disruption of the *pvg1+* gene resulted in charge abolishment [160].

3.7.3. Pyruvylated Galactans of Algae

Pyruvylated galactan sulphates are often found in red algal polysaccharides, which generally contain 3-substituted 4,6Pyr-D-Gal*p* residues. Among these galactans is that of *Palisada flagellifera*, which represents a highly complex structure with at least 18 different types of derivatives that are found mostly pyruvylated, 2-sulfated, and 6-methylated [161]. Another galactan is that of *Solieria chordalis*, the structure of which remains unknown but was shown to have high immunostimulating potential [162]. Other examples include the carragenans from Australian red algae of the family *Solieriaceae* [10] and galactans of the red seaweed *Cryptonemia crenulata* [13].

Examples of green algae include the highly pyruvylated and sulfated galactans from tropical green seaweeds of the order *Bryopsidalesor*, which have anticoagulant activity [11], such as that of *Codium divaricatum* with the structure Gal*p*-($4SO_4$)-(1→3)-Gal*p*-(1→3)-Gal*p*-(1→3)-Gal*p* and 3,4Pyr-Gal*p*-($6SO_4$)-(1→3)-Gal*p* [12].

3.7.4. Pyruvylated Proteoglycan

The marine sponge *Microcionia prolifera* produces a pyruvylated adhesion proteoglycan with the structure 4,6Pyr-β-Gal-(l→4)-β-GlcNAc-(l→3)-Fuc-(1→ that is involved in species-specific cell re-aggregation [163].

4. Methods for Research of Pyruvylated Glycoconjugates

4.1. Isolation of Pyruvylated Bacterial Glycoconjugates

Several protocols for the isolation of glycoconjugates are in use; however, there is no specific general protocol for pyruvylated glycoconjugates. The procedures are strongly dependent on the source of the glycoconjugate—with a special emphasis on the cell wall architecture (i.e., Gram-positive versus Gram–negative bacteria)—and the class of glycoconjugate. Further, for each studied organism, the extraction protocol needs to be optimised. Because of the chemical nature of the acid-labile pyruvate entity, as the only commonality, for the isolation of pyruvylated glycoconjugates, acidic conditions should be avoided to prevent the loss of pyruvate [9,100,145].

For the extraction of EPS, for instance, the types of interactions by which the EPS matrix is created need to be taken into account, including variable extents of electrostatic interactions, van der Waal forces, hydrogen bonds, and hydrophobic interactions [164]. In most cases, physical forces are used to extract EPSs, such as centrifugation and filtration [28], stirring, pumping or shaking, heat treatment, or sonication [164]. Chemical steps include alkaline treatment with NaOH, addition of EDTA for removal of cations, addition of NaCl, use of ion exchange resins (e.g., Dowex), or enzymatic treatment [63,165]. If proteases are used for break-down of co-isolated proteins, an *O*-deacetylation step needs to be introduced to avoid the loss of putative acetyl groups on the EPS [55]. All mentioned chemical additives increase the solubility of the EPS in the aqueous phase; to solubilise EPS with hydrophobic portions, such as that from *Klebsiella pneumoniae*, detergents are necessary [166]. For the precipitation of EPS from the aqueous phase, ethanol is routinely used [65]. To enhance the EPS yield, often a combination of physical and chemical methods is applied [164].

For the extraction of CPS from Gram-negative bacteria, again, NaCl and EDTA are recommended [167]. Other protocols for the release of CPS are based on heat treatment of cells followed by precipitation of the CPS with acetone [80,90].

The isolation of SCWPs—classical and "non-classical" forms—is divided in two main steps: the purification of the peptidoglycan sacculus, which includes treatment of cells with heat, SDS, nuclease, and a protease such as trypsin, and extraction of SCWP by either ethanol precipitation for WTAs, or hydrofluoric acid treatment followed by ethanol precipitation for "non-classical" SCWPs [115,168,169]. The isolation of "non-classical" pyruvylated SCWP of *B. anthracis* was recently described in detail [169].

The extraction of LPS and other cell surface polysaccharides has been described previously [170, 171]. Prior to extraction of LPS, pelleted Gram-negative bacteria are usually depleted from CPS by aqueous washing [171]; most commonly, LPS is extracted [146,172,173] or, in the case of LOS, with phenol/chloroform/petrol ether (PCP) [174,175]. The crude extracts are subsequently de-*O*/*N*-acylated under mild acidic or basic conditions, with a preference for the latter. Further purification of the samples can be achieved by size exclusion and/or ion exchange chromatography [176].

4.2. Pyruvate Analytics

4.2.1. Lectin Approach

Serum amyloid P component (SAP)—a normal plasma glycoprotein—has a Ca^{2+}-dependent binding specificity for 4,6Pyr-*O*Me-β-D-Gal*p* (MOPDG) [177], and thus behaves like a lectin and may be a useful probe for this epitope as present in the cell walls of bacteria and other organisms [178]. SAP has been found to bind *in vitro* to *K. rhinoscleromatis* [89], the cell wall of which is known to contain this particular pyruvylated epitope. Binding was shown to be less pronounced to *X. campestris*, which contains a 4,6Pyr-Man*p* epitope [18], and no SAP bound to *E. coli*, which contains pyruvate 4,6-linked to glucose or to *S. pneumoniae* type 4, which contains pyruvate 2,3-linked to Gal*p* [74]. Binding of SAP to those organisms, which it did recognise, was completely inhibited or reversed by millimolar concentrations of free MOPDG.

4.2.2. Biochemical Pyruvate Assays

A specifically developed colorimetric/fluorometric assay for ketal-pyruvate detection via enzymatic oxidation has been incorporated in a recently introduced high throughput screening platform for the structural analysis of novel EPS structures [179], which underlines the importance of pyruvylated epitopes. The platform is based on ultra-high performance liquid chromatography coupled with ultra-violet and electrospray ionization ion trap detection following EPS isolation.

A similar procedure for detection of free pyruvate is used in clinics. Pyruvate serves as an important metabolite in the citric acid cycle for the screening of liver diseases and genetic disorders in humans, as these are reflected by high pyruvate levels [180]. The procedure is based on the oxidation of pyruvate by pyruvate oxidase in the presence of acetyl phosphate, which leads to the production of CO_2 and H_2O_2. The latter is detected via a fluorometric probe followed by a horseradish peroxidase reaction, which leads to the formation of resorufin. Colour development can be detected at 570 nm, and fluorescence at 530–540 nm for excitation and 585–595 nm for emission (Cayman pyruvate assay kit: https://www.caymanchem.com/pdfs/700470.pdf).

Other methods for pyruvate detection stem from food analytics, as pyruvate is involved in the degree of pungency of onions. Different methods are on the basis of the determination of total 2,4-dinitrophenylhydrazine-reacting carbonyls in a sample by photometric detection. Furthermore, oxidation of reduced diphosphopyridine nucleotide (DPNH) by pyruvate can be measured in a coupled reaction with lactic dehydrogenase. Decrease of the absorbance at 340 nm correlates with the oxidation of DPNH and, therefore, the concentration of pyruvate [181,182].

Assaying pyruvylation reactions of monosaccharides using HPLC-based approaches is dependent on the intended mode of detection. Frequently, specifically introduced saccharide modifications are used for detection purposes. One prominent example is the chemical attachment of *para*-nitrophenol (pNP) to the saccharides of interest for monitoring at 265 nm. To determine, for instance, the activity of the yeast pyruvyltransferase Pvg1p, the pyruvylated product species was separated from the unpyruvylated educt species using a COSMOSIL 5C18-P revered phase (RP) C18 column with 0.3% ammonium acetate, pH 7.4, containing 13% acetonitrile as a solvent. The pyruvylated product eluted from the column earlier than the educt, as monitored by recording the absorbance at 265 nm [160]. Another option is the use of a RP-C18 column in combination with a 1-propanol gradient in 88% 100 mM ammonium bicarbonate, accompanied by the detection of the nitrophenyl-modified sugar at an absorbance of 405 nm [7].

A more sophisticated fluorescent polyisoprenoid chemical probe—2-amideaniline-undP-PP-AAᴅGal-Gal, which equals an acceptor substrate mimic from the *B. fragilis* CPS A tetrasaccharide biosynthesis pathway—was established by Sharma et al. to monitor pyruvylation of the fluorescent lipid-linked substrate by the pyruvyltransferase WcfO directly by HPLC on a C18 column. An isocratic gradient of 35% 1-propanol with 65% 100 mM ammonium bicarbonate was used, and detection was done by fluorescence at excitation at 340 nm, and emission at 390 nm [7].

4.2.3. NMR Analysis of Pyruvylation

Nuclear magnetic resonance (NMR) is a versatile tool for the non-invasive structure elucidation of bacterial polysaccharides, including substitutions such as pyruvic acid [183], which can be frequently found as 4,6-*O*, 3,4-*O*, or 2,3-*O* acetals. Systematic investigations of defined pyruvylated monosaccharides revealed stereospecific repeating patterns from which the absolute configuration of pyruvic acid acetals can be inferred [184]. It was shown that the ^{13}C signal of an equatorial 4,6-pyruvate methyl group (Figure 4I,II) resonates at ~26 ppm, while the axial methyl group can be found at ~17 ppm. For 3,4 acetals, the ^{13}C shifts have been studied in detail, and the difference between axial and equatorial methyl groups was found to be much smaller in comparison to 4,6. The ^{1}H difference, however, is in this case more noticeable [185]. The ring form of the acetal being either 5- (for 3,4-O) or 6- (for 4,6-O) membered is reflected by ^{13}C shifts [186]. It has also been shown that for most 4,6 acetals, the configuration of the methyl group is equatorial, which results in an *S* configuration for the D-*gluco*- and D-*manno*-pyranosyls and an *R* configuration for the D-*galacto*-pyranosyls [187].

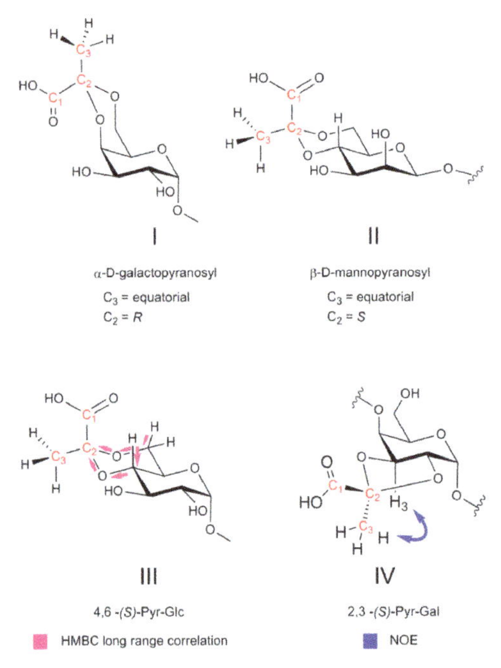

Figure 4. **I** and **II**, equatorial-oriented methyl groups of 4,6 *galacto*- and *manno*-pyranosyls. **III**, identification of the attachment site of pyruvylation via hetero multiple bond correlation (HMBC). **IV**, through-space correlation of the pyruvate methyl group to a ring proton—in this case H_3. Pink arrows indicate though-bond interactions of neighbouring protons (HMBC), while blue arrows indicate through-space interactions of neighbouring protons (NOE).

In ^{1}H NMR, the presence of pyruvic acid (4,6-, 3,4-, 2,3-) is usually indicated by a single prominent signal of the methyl group between 1.3–1.7 ppm, with a threefold higher relative intensity (peak area) in relation to another indicative signal such as the anomeric proton. For repeating units of polymers, the peak area of the methyl signal relative to another indicative signal reveals the degree of pyruvate substitution.

In ^{13}C NMR, the presence of pyruvate substitution is usually indicated by the presence of signals for the pyruvic methyl group around 17–30 ppm (Figure 4, C_3). The quaternary acetal carbon (Figure 4, C_2) resonates in the anomeric region around 100 ppm (for 4,6-*O*) or 110 ppm (for 3,4-*O* and 2,3-*O*). Additionally, the quaternary signal of the carboxylic acid (Figure 4, C_1) can be found between 170 and 180 ppm, with the 4,6-pyruvates present more towards 170 ppm and the 2,3 and 3,4 acetals found closer to 180 ppm [131].

The connectivity between the pyruvate and a saccharide is routinely determined by the employment of long-range ^{1}H-^{13}C correlation detection methods such as hetero multiple bond correlation (HMBC) experiments, which usually give correlation information over three and more bonds from the corresponding ring protons to the quaternary carbon of the acetal (Figure 4III). Therefore, 4,6-, 3,4-, or 2,3-pyruvic acetal identification is straightforward. The absolute configuration of the pyruvic acid acetal can be confirmed by through-space correlation experiments such as 1D or 2D NOESY (nuclear Overhauser and exchange spectroscopy), ROESY (rotating frame Overhauser enhancement spectroscopy), or GOESY (gradient nuclear Overhauser and exchange spectroscopy) (Figure 4, IV) [74,146,188,189].

4.2.4. MS analysis of Pyruvylation

Mass spectrometry in combination with NMR is a very powerful tool to determine the presence and position of pyruvylation in oligo- or polysaccharides. Common approaches are based on the break-up of polysaccharides by acid hydrolysis, methanolysis, and then either silylation [59,66] or acetylation [148], followed by gas chromatography (GC) or electrospray ionization-mass spectrometry (ESI-MS) analysis of the resulting monosaccharides [190]. Usually, characteristic patterns in the mass spectrum at the monosaccharide level are observed in the presence of pyruvates, such as a characteristic fragment ion at *m/z* 363 (M-COOMe) consistent with a molecular mass of 422, which would be indicative of a methyl *O*-(1-carboxyethylidene)hexopyranoside methyl ester di-O-trimethylsilyl-ether. Comparison of the methylation analysis on native and depyruvylated polysaccharides allows for the pinning down the initial position of the acetalic linkages. Methanolysis and reductive cleavage have been described for the analysis of pyruvate-containing polysaccharides [191]. General procedures for the MS analysis of oligosaccharides have been reviewed in detail elsewhere [192,193].

5. Ketal-Pyruvyltransferases

5.1. Substrate Specificity of Ketal-Pyruvyltransferases

The pyruvyltransferase CsaB from the SCWP biosynthesis pathways of *P. alvei* [109], the pyruvyltransferase WcfO from CPS A biosynthesis of *B. fragilis* [7], and the pyruvyltransferase Pvg1p from the *N*-glycan biosynthesis of *S. pombe* [161] are among the few studied enzyme orthologues.

5.1.1. CsaB from *P. alvei*

P. alvei CsaB catalyses the pyruvate modification on a β-D-ManNAc residue present in every SCWP repeat; the resulting 4,6-β-D-ManNAc is the essential epitope for the binding of the bacterium's SLH domain-containing S-layer protein SpaA, as revealed from the co-crystal structure of synthetic pyruvylated ligand with truncated $SpaA_{SLH}$ [104,194]. Supporting data comes from isothermal titration calorimetry, revealing binding between these modules only when the pyruvate entity was present [104].

Notably, a comparable mode of binding is elaborated between the S-layer protein Sap of *B. anthracis* and its pyruvylated SCWP [195]. However, poly-pyruvylation, as in the *P. alvei* SCWP, is not found in the *B. anthracis* SCWP, where only the β-D-ManNAc of the terminal repeat is modified [14].

For *B. anthracis*, a model for a Wzx/Wzy-dependent biosynthesis has been proposed, including cytoplasmic pyruvylation of the terminal repeating unit; however, the model is without any evidence of the nature of the acceptor substrate and biochemical proof of CsaB activity [113].

Concerning the pyruvyltransferase CsaB from *P. alvei*, in an *in vitro* enzyme assay, the necessity of a lipid-portion of the acceptor could be unambiguously demonstrated; neither free ManNAc nor a nucleoside-diphosphate-linked substrate was accepted as a substrate [109]. Using *P. alvei* TagA in combination with a synthetic 11-phenoxy-undecyl-PP-α-GlcNAc acceptor and UDP-ManNAc as substrate, it was demonstrated that TagA is an inverting UDP-α-D-ManNAc:GlcNAc-lipid carrier transferase of *P. alvei*. The produced 11-phenoxyundecyl-PP-α-D-GlcNAc-(1→4)-β-D-ManNAc compound was an acceptor substrate for 4,6-ketalpyruvyl transfer catalysed by recombinant *P. alvei* CsaB using PEP as a donor substrate [109], yielding a lipid-PP-linked pyruvylated disaccharide precursor (Figure 5).

Subsequent steps of SCWP biosynthesis in *P. alvei* remain elusive, as there is currently neither in silico nor experimental data available favouring a Wzx/Wzy- or ABC transporter-dependent pathway.

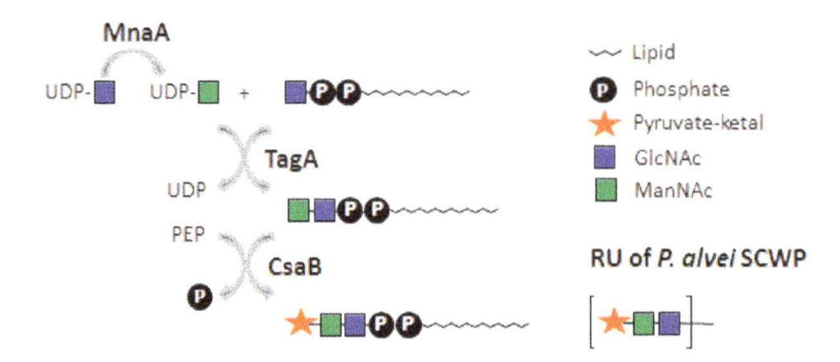

Figure 5. *In vitro* one-pot reaction involving the *Paenibacillus alvei* enzymes MnaA, TagA, and CsaB in combination with a synthetic lipid-PP-GlcNAc primer, demonstrating the preference of *P. alvei* CsaB for a lipid-PP-linked disaccharide substrate [109]. Pyruvylation (CsaB) is indicated by a star. RU: repeating unit. Monosaccharide symbols are shown according to the Symbol Nomenclature for Glycans (SNFG) [45].

5.1.2. WcfO from *B. fragilis*

The necessity of a lipid-PP-bound substrate for pyruvyltransferase activity is supported by studies on the pyruvyltransferase WcfO from the *B. fragilis* CPS A biosynthesis; CPS A is composed of tetrasaccharide repeats containing an internal 4,6Pyr-Gal residue. On the basis of the stepwise enzymatic processing of an undp-PP-AADGal*p* precursor *in vitro*, pyruvylation by WcfO was predicted to occur in the cytoplasm at the stage of the lipid-linked CPS A repeat unit precursor undp-PP-AADGal*p*-Gal before completion of the tetrasaccharide repeat and completion of the CPS A in the periplasm [7,93] (Figure 3). Importantly, WcfO was inactive on UDP-galactose or pNP-galactose, supporting the requirement of a lipid-P carrier for pyruvyltransferase activity of WcfO [7].

Enzymatic transfer of pyruvate onto lipid-bound sugar intermediates has also been previously described in CPS biosynthesis of *Rhizobium trifolii* [196,197] and in xanthan biosynthesis of *Xanthomonas campestris* [31].

5.1.3. Pvg1p from *S. pombe*

In contrast, the third functionally characterized 4,6-ketal-pyruvyltransferase, Pvg1p from *S. pombe*, was proven *in vitro* with both pNP-β-Gal and pNP-β-lactose serving as suitable acceptor substrates [160]. According to studies of the pyruvylation mechanism, Pvg1p resides in the membrane of the Golgi apparatus where it adds the pyruvate moiety to the Gal caps of its *N*-glycans. For this purpose, PEP is transported by two transporters, Pet1p and Pet2p, into the lumen of the Golgi apparatus where it serves as a donor substrate for the pyruvylation reaction [160].

A recent study has determined the crystal structure of the Pvg1p enzyme [160]. Pvg1p consists of 12 α-helices and 12 β-sheets, with 2 α/β/α domains at the N- and C-terminal half regions. Charged surface representation analysis revealed a positively charged cleft situated between the N- and C-terminal halves of Pvg1p, which suggests a possible mode of binding that may accommodate the negatively charged PEP donor substrate. Since neither PEP- nor pNP-β-Gal-co-crystal structures with the enzyme could be obtained, the empty substrate-binding cleft was used as a scaffold for computational substrate modelling using PEP [198]. In the proposed computational model, residues R217, R337, L338, and H339 form direct hydrogen bond contacts with PEP. Residues L338, H339, and D240 also appear to function in maintaining the shape of the PEP-binding pocket via a set of specific interactions. The crystallization study indicated that the pyruvylation process mimics sialyation; interestingly, Pvg1p shows resistance to sialidase digestion. Thus, a better characterization of the effects of pyruvylation might facilitate the development of pharmaceutical glycoproteins [198]. From the same research group, an enzyme was characterised as a 4,6Pyr-β-D-Gal-releasing enzyme (PyrGal-ase) with specificity for the (1→3) yeast linkage; mammalian (1→4)-linked PyrGal could not be hydrolysed. The physiological role of the PyrGal-ase in the *Bacillus* strain from where it was isolated is currently unknown [199].

Except for the three characterized pyruvyltransferases, no data on neither the activity nor the substrate specificity of pyruvyltransferases is available in the literature. This is surprising, considering that pyruvylation on glycoconjugates is widely distributed in nature. Future research on ketal-pyruvyltransferases should be directed towards mechanistic investigations of the enzyme's mode of catalysis, as well as inhibitor screening, similar to that of the enol-pyruvyltransferase MurA, which is a prominent target of antibiotics.

5.2. Challenges in Research of Ketal-Pyruvyltransferases

Currently, no definite classification of pyruvyltransferases is possible, although orthologous enzymes are predicted in various organisms. The Carbohydrate-Active enZYme (CAZy) database (http://www.cazy.org/) reveals, for instance, putative polysaccharide pyruvyltransferases from *Clostridium stercorarium* subsp. *stercorarium* DSM 8532 and *Clostridium thermosuccinogenes* DSM 5807 belonging to the glycosyltransferase 4 (GT4) family, with a classification as retaining GT type B fold-like glycosyltransferases.

The reasons for the limited number of characterised pyruvyltransferases are due to the challenges faced with the set-up of *in vitro* enzyme assays. While commercially available PEP has been proven to be a suitable donor substrate for the transfer of the pyruvyl moiety in distinct cases [7,109,195], the availability of suitable acceptor substrates is a limiting factor. Free saccharides have not been recognised as acceptor substrates by the pyruvyltransferases investigated so far [7,109]. According to our current knowledge, these enzymes instead require more elaborate intermediates from the pyruvylated glycoconjugate's biosynthesis pathway. Depending on the glycoconjugate structure and its mode of biosynthesis—which might be an *en bloc* (involving an ABC transporter) or sequential synthesis (involving a Wzx flippase and a Wzy polymerase) according to the terminology introduced for LPS biosynthesis routes [97,98,138,200], yielding pyruvylation as either a pre- or post-polymerization modification—di-, tri-, or even oligosaccharide repeating units might be required. Furthermore, most glycoconjugates are biosynthesized on a membrane-embedded lipid carrier, such as undp-P or diacylglycerol. Such lipid-linked glycan precursors usually cannot be purified from the natural source in sufficient quantity and purity because of the high turnover rates and efficient recycling pathways of these lipid carriers, which are shared between several cellular glycoconjugate biosynthesis pathways, including that of peptidoglycan [201].

Thus, complex saccharide acceptor substrates are required, which are not commercially available. These compounds need to be produced along sophisticated and laborious chemical synthesis schemes, which also need to account for a lipophilic portion, either in the form of the native lipid carrier or a simplified mimic thereof.

For identifying a suitable acceptor substrate for an *in vitro* pyruvyltransferase assay, a delicate balance between the best possible acceptor mimic and solubility needs to be found in order to enable subsequent analytical procedures. To overcome all these challenges, the development of novel chemical, enzymatic, or chemo-enzymatic synthesis strategies for acceptor substrate production is a current major focus in pyruvyltransferase research.

5.3. Sequence Space of Ketal-Pyruvyltransferases

This review aimed at exploring the currently known sequence variation (extant sequence space) of pyruvyltransferases and their taxonomic distribution on the basis of the three functionally characterized sequences—*P. alvei* CsaB (K4ZGN3), *S. pombe* Pvg1p (Q9UT27), and *B. fragilis* WcfO (Q5LFK7).

5.3.1. Methods

The best 50 sequence hits of BLAST searches with K4ZGN3_CsaB, Q9UT27_Pvg1p, and Q5LFK7_WcfO were aligned with MAFFT using the algorithm FFT-NS-2. The three alignments were then used as queries for hmmsearch [202] on the UniProtKB database, setting significant E-values for sequences $<9.0 \times 10^{-30}$ and for hits $<9.9 \times 10^{-30}$. Results were restricted to hits showing a pyruvyltransferase domain (PS_pyruv_trans domain; Pfam: PF04230). The resulting three sequence selections were filtered for incomplete sequences and annotated according to their taxonomy using the online tool SeqScrub [203]. The sequences were further submitted to the Enzyme Function Initiative-Enzyme Similarity Tool (EFI-EST) [204] with an initial BLAST E-value of 1×10^{-5} to calculate sequence similarity networks (SSNs). Sequences were restricted to a length between 250 and 600 amino acids, and the calculated networks were displayed at an alignment score cut-off of 1×10^{-50}.

5.3.2. Results

Three independent database searches based on the biochemically characterized pyruvyltransferases CsaB, Pvg1p, and WcfO resulted in three sequence selections of 2053, 1019, and 233 sequences, respectively. When comparing these selections, it was found that they did not share any protein sequences, implicating that the sequence space covered by these searches does not overlap. It is, therefore, conceivable to assume that more pyruvyltransferase sequences and organisms harbouring a pyruvyltransferase gene exist, which are not covered in this study. Additionally, the comparison shows that there are at least three different types of pyruvyltransferases that do not share a close sequence relationship. Judging from the number of sequences in the selections and the extent of their taxonomic distribution, pyruvyltransferases from the SSN of CsaB (CsaB-like), and pyruvyltransferases from the SSN of Pvg1P (Pvg1P-like) seem to be the most common types of pyruvyltransferases, while WcfO-like pyruvyltransferases might be more of a specialized type of pyruvyltransferase.

Looking at the taxonomic distribution of these types of pyruvyltransferases in the SSNs (Figure 6), it can be seen that CsaB-like pyruvyltransferases occurred mainly in the phyla of *Firmicutes* and *Cyanobacteria*; Pvg1p-like pyruvyltransferases occurred mainly in the phyla of *Proteobacteria* and *Firmicutes*; and WcfO-like pyruvyltransferases occurred almost exclusively in the phyla of *Bacteroidetes, Proteobacteria,* and *Firmicutes*. In most cases, these different phyla separated nicely into different clades. For the SSN of Pvg1p-like pyruvyltransferases, however, there were two clusters where *Proteobacteria* were heavily mixed with *Firmicutes* and *Bacteroidetes*, and in the SSN of WcfO-like pyruvyltransferases, *Proteobacteria* were found to be heavily mixed with *Bacteroidetes*. Such mixed sequence populations might occur because of the high rates of lateral gene transfer among *Proteobacteria* [173]. Looking across all three SSNs, there was typically only one major cluster for each phylum. The only exception was pyruvyltransferase sequences from *Firmicutes*, which showed multiple big clusters in all three SSNs, indicating that *Firmicutes* might carry multiple types of pyruvyltransferases.

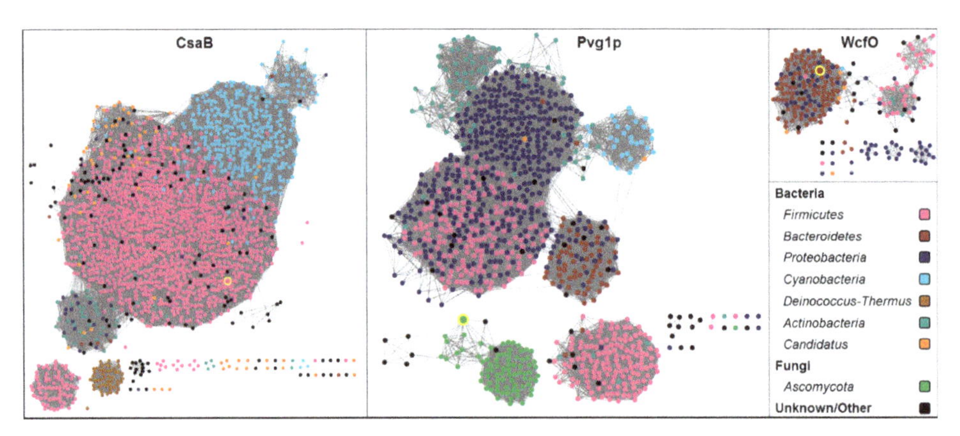

Figure 6. Sequence similarity networks illustrating the extant sequence space around *Paenibacillus alvei* CsaB (K4ZGN3), *Schizosaccharomyces pombe* Pvg1p (Q9UT27), and *Bacteroides fragilis* WcfO (Q5LFK7). The three characterized sequences—CsaB, Pvg1p, and WcfO—are highlighted by yellow circles.

Analysing the functionally characterized pyruvyltransferase CsaB in the context of its surrounding sequence space showed the enzyme to be a typical representative of the biggest cluster (*Firmicutes*) in the CsaB-like SSN. The same goes for WcfO, which was also found within the biggest cluster (*Bacteroidetes* and *Proteobacteria*) of the WcfO-like SSN. Pvg1p, on the other hand, was found at the border of a minor *Ascomycota* clade in the Pvg1p-like network and, therefore, cannot be considered a typical representative of this network. It is interesting to note, however, that Pvg1p is a pyruvyltransferase from the fungal phylum of *Ascomycota*, but the sequence search based on Pvg1p resulted mainly in bacterial sequences from *Proteobacteria* and *Firmicutes*, rather than other fungal sequences.

In addition to CsaB, Pvg1p, and WcfO, this review further discussed 48 putative pyruvyltransferases, and about half of their corresponding amino acid sequences were present within the calculated SSNs. Possible reasons for this incomplete recovery of sequences in the SSNs were the lack of genome sequencing data, missing or faulty taxonomic annotation of sequences, and the fragmentary coverage of the pyruvyltransferase sequence space in the performed SSN analysis.

Note that this study refrained from removing sequences showing 100% sequence identities (possible duplicates), meaning that the utilized datasets included all currently known pyruvyl-transferase entries found under the given search parameters on UniProtKB. It is inevitable that this leads to a possible bias in sequence counts towards organisms that are more heavily sequenced than others, but at the same time, it guarantees the representation of the full taxonomic distribution of pyruvyltransferases. From these datasets, the phyla *Firmicutes*, *Proteobacteria*, *Cyanobacteria*, and *Bacteroidetes* were found to be the phyla where pyruvyltransferases are most common.

From this study, it is evident that the pyruvyltransferase sequences available in public databases are extremely diverse, and without the availability of further biochemically characterized pyruvyltransferases, predictions of pyruvyltransferases based on amino acid sequences have to be interpreted with care.

6. Discussion

Pyruvyltransferases are a widespread but little investigated class of carbohydrate-active enzymes, which transfer a pyruvate moiety from a PEP donor to various monosaccharide targets (Table 1). This leads to a wealth of glycoconjugates carrying this modification. Pyruvylation can be found in almost all classes of glycoconjugates—including EPS, CPS, CA, LPS, LOS, SCWP, and *N*-glycans—occurring in bacteria, algae, and yeast, but not in humans. Importantly, pyruvylation imparts an anionic character to the glycoconjugates, which is pivotal to many biological functions. Described functions include the influence on the viscosity of the EPS, bacterial symbiosis with plants [18,28,46], immunostimulatory effects (mostly of CPSs [7,93]), employment of sialylation-like properties in human-type oligosaccharides [198], and cell wall anchoring relying on the Pyr-β-D-ManNAc epitope [14,99,104,195], to name a few. However, learning more about the biological significance of pyruvylated glycoconjugates and delineating a possible association between the position of pyruvylation and functionality are remaining challenges for future research.

Regrettably, for most of the described pyruvylated glycoconjugates, the genetic determinants of the modification are unknown because of missing genome sequencing data of the respective organisms. Given the widespread occurrence and the importance of sugar pyruvylation in nature, there is a high interest in the research community to identify pyruvyltransferases and gain insight into the mechanism of pyruvylation, especially with regard to the high potential to reveal novel functions and drug target points.

Table 1. Pyruvylated saccharides and positions of pyruvate found in the organisms described in this review.

Organism	Saccharide (Stereochemistry)	Position of Pyr	Biological Significance *		References
			Pyr-Saccharide	Pyr-Glycoconjugate	
Exopolysaccharides					
Xanthomonas spp.	Man (β)	4,6		Plant pathogenesis	[18,28,32]
Xanthomonas campestris	Man (β)	4,6	Virulence		[18,28,32]
Sinorhizobium meliloti	Glc (?)	?	Glycan polymerization and export		[41]
Rhizobium leguminosarum	Gal; Glc (β)	4,6	Signalling		[46]
Agrobacterium sp. ZX09	Glc (?)	?		Plant symbiosis	[47]
Escherichia coli	Gal (α)	4,6		Biofilm	[32,50]
Burkholderia cepacia	Gal (α)	4,6		Pathogenesis	[53]
Pseudomonas strain 1.15	Gal (α)	4,6			[54]
Enterobacter amnigenus	Man (α)	4,6			[55]
Inquilinus limosus	Man (α); Glc (β)	4,6		Pathogenesis	[56]
Azorhizobium caulinodans	Gal (α)	4,6		Plant symbiosis	[59]
Cobetia marina DSMZ 4741	Kdo (α)	7,8			[60]
Pediococcus pentosaceus	hexose (?)	?			[61]
Erwinia futululu	Gal (α)	4,6		Pathogenesis	[63]
Agrobacterium radiobacter ATCC 53271	Glc (α)	4,6			[64]
Methylobacterium sp.	Gal (α)	4,6			[65]
Alteromonas macleodii subsp. *fijiensis*	Man (?)	4,6			[66,67]
Capsular polysaccharides					
Streptococcus pneumoniae	Gal (α)	2,3	Immunostimulant		[74]
Acinetobacter baumannii	GalNAc (α)	4,6		Virulence	[78]
Klebsiella pneumoniae	GlcA (β)	2,3	Stimulation of immune system		[79]
Klebsiella serotype K14	Glc (β)	4,6			[82]
Klebsiella serotype K12	Gal*f* (α)	5,6			[83]
Klebsiella serotype K70	Rha (α)	3,4			[84]
Klebsiella serotype K64	Gal (β)	4,6			[85]
Klebsiella serotype K46	Man (β)	4,6			[86]
Klebsiella serotype K33	Gal (β)	3,4			[87]
Klebsiella serotype K3	Man (α)	4,6		Immunogenicity; binding to human mannose receptor	[89]
Bacteroides fragilis	Gal (β)	4,6		Virulence	[7]
Rhodococcus equi serotype 4	RhoAc (α)	7,9		Pathogenesis	[95]

Table 1. *Cont.*

Organism	Saccharide (Stereochemistry)	Position of Pyr	Biological Significance *		References
			Pyr-Saccharide	Pyr-Glycoconjugate	
Non-classical secondary cell wall polymers					
Bacillus anthracis	ManNAc (β)	4,6	Binding of SLH-proteins		[102]
Bacillus cereus	ManNAc (β)	4,6	Binding of SLH-proteins		[17]
Paenibacillus alvei	ManNAc (β)	4,6	Binding of SLH-proteins		[15]
Lysinibacillus sphaericus CCM 2177	ManNAc (β)	4,6	Binding of SLH-proteins		[121]
Paenibacillus (Bacillus) polymyxa	ManNAc (β)	4,6			[130]
Secondary cell wall glycopolymers					
Promicromonospora citrea 665^T	Kdn (α); Gal (α)	7,9; 4,6			[131]
Promicromonospora sp. VKM Ac-1028	GlcA (β)	2,3			[131]
Nocardiopsis metallicus VKM Ac-2522T	Rib-ol-P (?)	2,4			[133]
Nocardiopsis halotolerans	GalNAc (?)	4,6			[133]
Brevibacterium iodinum VKM Ac-2106	Man-ol; GalNAc (α)	4,5; 4,6			[135]
Lipopolysaccharides and lipooligosaccharides					
Pseudomonas stutzeri	Glc (β); GlcNAc (β)	4,6	Glycoconjugate biosynthesis		[4]
Providencia alcalifaciens	GlcNAc (α)	4,6		Pathogenesis	[141]
Shigella dysenteriae	Man (β)	4,6		Pathogenesis	[145]
Proteus mirabilis O16	GalNAc (α)	4,6			[147]
Raoultella terrigena	Man (β)	4,6			[146]
Cobetia pacifica KMM 3878	Gal (β)	3,4			[148]
Mycobacterial glycolipids					
Mycobacterium smegmatis	Me-Glc (α)	4,6			[150]
Mycobacterium avium-Mycobacterium intracellulare (MAI) serovar 8	Me-Glc (α)	4,6	Immunostimulant		[151]
MAI serovar 21	Glc (α)	4,6			[151]
Pyruvylated glycoconjugates in eukaryotes					
Aplysia kurodai	Gal (β)	3,4			[152]
Schizosaccharomyces pombe	Gal (β)	4,6	Charge effect (mimic of sialylation?)		[9]
Red algae	Gal (β)	4,6		Immunostimulant	[10,13,161,162]
Green algae	Gal (?)	3,4		Anticoagulant	[11,12]
Microciona prolifera	Gal (β)	4,6		Species-specific cell reaggregation	[163]

* The biological significance is given, when known. SLH: S-layer homology.

Up until now, three orthologous pyruvyltransferases have been biochemically investigated [7,111,161]. However, they do not show any close sequence relationship (compare with Figure 6). This finding might point towards a convergent evolution of pyruvyltransferases or a very high evolutionary rate that underlines the high sequence variability present in this enzyme class. The SSNs established within the frame of this review indicate that the described sequence space around the three hitherto characterized sequences was not sufficient to cover the whole extent of sequence variation of pyruvyltransferases. Based on the currently available sequence information, pyruvyltransferases mainly occur in bacterial phyla of *Firmicutes*, *Proteobacteria*, *Cyanobacteria*, and *Bacteroidetes*, and to a lesser extent in eukaryotic species.

Given the relentless spread of antibiotic-resistant organisms, new chemotherapeutic strategies to overcome infections could be based on intervening in the mechanisms of pyruvylation, an enzymatic modification detected in almost all classes of cell envelope glycoconjugates.

Author Contributions: F.F.H. and C.S. (Cordula Stefanović) wrote the initial manuscript; L.S. conducted and edited the sequence space part (SSNs) of the manuscript; M.B. contributed to the pyruvate analytics part of the manuscript; C.S. (Christina Schäffer) wrote, revised, and edited the manuscript. All authors approved the final version of the manuscript.

Funding: This research was funded by the Austrian Science Fund FWF, projects P27374-B22 and P31521-B22 (to C.Sch.); the Hochschuljubiläumsstiftung der Stadt Wien, project H-318348/2018 (to F.F.H.), and the PhD Program "Biomolecular Technology of Proteins", FWF project W1224.

Conflicts of Interest: The authors declare no conflict of interest. The funders had no role in the design of the study; in the collection, analyses, or interpretation of data; in the writing of the manuscript, or in the decision to publish the results.

Abbreviations

4,6Pyr	4,6-ketal-pyruvate
AADGal*p*	α-D-2-*N*-acetylamido-4-amino-galactopyranose
Ac	acetyl
Am	aecetamido
CA	colonic acid
Cm	carbamoyl
CPS	capsular polysaccharide
CSDB	Carbohydrate Structure Database
Cys	cysteine
D	aspartic acid
DPNH	reduced diphosphopyridine nucleotide
EPS	exopolysaccharide
ER	endoplasmic reticulum
Ery-ol	erythritol
ESI-MS	electrospray ionization mass spectrometry
f	furanosidic
Fuc	fucose
Gal	galactose
GalN	galactosamine
GalNAc	*N*-acetylgalactosamine
GalNAcA	α-*N*-acetyl-D-galactosamine uronic acid
GC	gas chromatography
Glc	glucose
GlcA	glucuronic acid
GlcNAc	*N*-acetylglucosamine
GOESY	gradient nuclear Overhauser and exchange spectroscopy
GT	glycosyltransferase
H	histidine
HMBC	hetero multiple bond correlation

HMM	Hidden Markow Model
HPLC	high-performance liquid chromatography
Kdn	2-keto-3-deoxy-D-*glycero*-D-*galacto*-nononic acid
Kdo	3-deoxy-D-*manno*-oct-2-ulosonic acid
L	leucine
LOS	lipooligosaccharide
LPS	lipopolysaccharide
LTA	lipoteichoic acid
MAI complex	Mycobacterium avium-Mycobacterium intracellulare complex
Man	mannose
Man-ol	mannitol
ManNAc	*N*-acetylmannosamine
Me	methyl
MOPDG	4,6Pyr-*O*Me-β-D-Gal*p*
MurNAc	*N*-acetylmuramic acid
NOESY	nuclear Overhauser and exchange spectroscopy
p	pyranosidic
P	phosphate
PEP	phosphoenolpyruvate
PEtN	phosphoetanolamine
pNP	*para*-nitrophenyl
Pyr	pyruvate
pyrGal-ase	pyruvylated galactose-releasing enzyme
QuiNAc	*N*-acetylquinovosamine
R	arginine
Rha	rhamnose
Rho	5-amino-3,5-dideoxynonulosonic (rhodaminic) acid
Rib-ol	ribitol
ROESY	rotating frame Overhauser enhancement spectroscopy
RP	reversed phase
RU	repeating unit
SAP	plasma glycoprotein serum amyloid P component
SCWP	secondary cell wall glycopolymer
Ser	serine
SLH	surface layer homology
sp.	species
spp.	species (*plural*)
TA	teichoic acid
undp-P	undecaprenylphosphate
WTA	wall teichoic acid

References

1. Bennet, L.G.; Bishop, C.T. The pyruvate ketal as a stereospecific immunodeterminant in the type XXVII *Streptococcus pneumoniae* (pneumococcal) capsular polysaccharide. *Immunochemistry* **1977**, *14*, 693–696. [CrossRef]
2. Lovering, A.L.; Safadi, S.S.; Strynadka, N.C. Structural perspective of peptidoglycan biosynthesis and assembly. *Annu. Rev. Biochem.* **2012**, *81*, 451–478. [CrossRef] [PubMed]
3. Andreishcheva, E.N.; Kunkel, J.P.; Gemmill, T.R.; Trimble, R.B. Five genes involved in biosynthesis of the pyruvylated Galβ1,3-epitope in *Schizosaccharomyces pombe N*-linked glycans. *J. Biol. Chem.* **2004**, *279*, 35644–35655. [CrossRef] [PubMed]
4. Leone, S.; Izzo, V.; Silipo, A.; Sturiale, L.; Garozzo, D.; Lanzetta, R.; Parrilli, M.; Molinaro, A.; Di Donato, A. A novel type of highly negatively charged lipooligosaccharide from *Pseudomonas stutzeri* OX1 possessing two 4,6-*O*-(1-carboxy)-ethylidene residues in the outer core region. *Eur. J. Biochem.* **2004**, *271*, 2691–2704. [CrossRef] [PubMed]

5. Toukach, P.V.; Egorova, K.S. Carbohydrate structure database merged from bacterial, archaeal, plant and fungal parts. *Nucleic Acids Res.* **2016**, *44*, D1229–D1236. [CrossRef]
6. Toukach, P.V.; Egorova, K.S. Bacterial, plant, and fungal carbohydrate structure databases: Daily usage. *Methods Mol. Biol.* **2015**, *1273*, 55–85. [PubMed]
7. Sharma, S.; Erickson, K.M.; Troutman, J.M. Complete tetrasaccharide repeat unit biosynthesis of the immunomodulatory *Bacteroides fragilis* capsular polysaccharide A. *ACS Chem. Biol.* **2017**, *12*, 92–101. [CrossRef]
8. Geissner, A.; Pereira, C.L.; Leddermann, M.; Anish, C.; Seeberger, P.H. Deciphering antigenic determinants of *Streptococcus pneumoniae* serotype 4 capsular polysaccharide using synthetic oligosaccharides. *ACS Chem. Biol.* **2016**, *11*, 335–344. [CrossRef]
9. Gemmill, T.R.; Trimble, R.B. *Schizosaccharomyces pombe* produces novel pyruvate-containing *N*-linked oligosaccharides. *J. Biol. Chem.* **1996**, *271*, 25945–25949. [CrossRef]
10. Chiovittia, A.; Bacica, A.; Craik, D.J.; Munro, S.L.A.; Kraft, G.T.; Liao, M.-L. Cell-wall polysaccharides from Australian red algae of the family *Solieriaceae* (*Gigartinales, Rhodophyta*): Novel, highly pyruvated carrageenans from the genus *Callophycus*. *Carbohydr. Res.* **1997**, *299*, 229–243. [CrossRef]
11. Arata, P.X.; Quintana, I.; Canelon, D.J.; Vera, B.E.; Compagnone, R.S.; Ciancia, M. Chemical structure and anticoagulant activity of highly pyruvylated sulfated galactans from tropical green seaweeds of the order *Bryopsidales*. *Carbohydr. Polym.* **2015**, *122*, 376–386. [CrossRef] [PubMed]
12. Li, N.; Mao, W.; Liu, X.; Wang, S.; Xia, Z.; Cao, S.; Li, L.; Zhang, Q.; Liu, S. Sequence analysis of the pyruvylated galactan sulfate-derived oligosaccharides by negative-ion electrospray tandem mass spectrometry. *Carbohydr. Res.* **2016**, *433*, 80–88. [CrossRef] [PubMed]
13. Zibetti, R.G.; Duarte, M.E.; Noseda, M.D.; Colodi, F.G.; Ducatti, D.R.; Ferreira, L.G.; Cardoso, M.A.; Cerezo, A.S. Galactans from *Cryptonemia* species. Part II: Studies on the system of galactans of *Cryptonemia seminervis* (*Halymeniales*) and on the structure of major fractions. *Carbohydr. Res.* **2009**, *344*, 2364–2374. [CrossRef] [PubMed]
14. Forsberg, L.S.; Abshire, T.G.; Friedlander, A.; Quinn, C.P.; Kannenberg, E.L.; Carlson, R.W. Localization and structural analysis of a conserved pyruvylated epitope in *Bacillus anthracis* secondary cell wall polysaccharides and characterization of the galactose-deficient wall polysaccharide from avirulent *B. anthracis* CDC 684. *Glycobiology* **2012**, *22*, 1103–1117. [CrossRef] [PubMed]
15. Schäffer, C.; Müller, N.; Mandal, P.K.; Christian, R.; Zayni, S.; Messner, P. A pyrophosphate bridge links the pyruvate-containing secondary cell wall polymer of *Paenibacillus alvei* CCM 2051 to muramic acid. *Glycoconj. J.* **2000**, *17*, 681–690. [CrossRef] [PubMed]
16. Khatun, F.; Stephenson, R.J.; Toth, I. An overview of structural features of antibacterial glycoconjugate vaccines that influence their immunogenicity. *Chemistry* **2017**, *23*, 4233–4254. [CrossRef] [PubMed]
17. Wang, Y.T.; Oh, S.Y.; Hendrickx, A.P.; Lunderberg, J.M.; Schneewind, O. *Bacillus cereus* G9241 S-layer assembly contributes to the pathogenesis of anthrax-like disease in mice. *J. Bacteriol.* **2013**, *195*, 596–605. [CrossRef] [PubMed]
18. Bianco, M.I.; Toum, L.; Yaryura, P.M.; Mielnichuk, N.; Gudesblat, G.E.; Roeschlin, R.; Marano, M.R.; Ielpi, L.; Vojnov, A.A. Xanthan pyruvilation is essential for the virulence of *Xanthomonas campestris* pv. *campestris*. *Mol. Plant. Microbe Interact.* **2016**, *29*, 688–699. [CrossRef]
19. Silver, L.L. Fosfomycin: Mechanism and resistance. *Cold Spring Harb Perspect. Med.* **2017**, *7*, a025262. [CrossRef]
20. Eschenburg, S.; Priestman, M.; Schönbrunn, E. Evidence that the fosfomycin target Cys115 in UDP-*N*-acetylglucosamine enolpyruvyl transferase (MurA) is essential for product release. *J. Biol. Chem.* **2005**, *280*, 3757–3763. [CrossRef]
21. Sonkar, A.; Shukla, H.; Shukla, R.; Kalita, J.; Pandey, T.; Tripathi, T. UDP-*N*-Acetylglucosamine enolpyruvyl transferase (MurA) of *Acinetobacter baumannii* (AbMurA): Structural and functional properties. *Int. J. Biol. Macromol.* **2017**, *97*, 106–114. [CrossRef] [PubMed]
22. Falagas, M.E.; Vouloumanou, E.K.; Samonis, G.; Vardakas, K.Z. Fosfomycin. *Clin. Microbiol. Rev.* **2016**, *29*, 321–347. [CrossRef] [PubMed]
23. Oberdorfer, G.; Binter, A.; Ginj, C.; Macheroux, P.; Gruber, K. Structural and functional characterization of NikO, an enolpyruvyl transferase essential in nikkomycin biosynthesis. *J. Biol. Chem.* **2012**, *287*, 31427–31436. [CrossRef] [PubMed]

24. Cescutti, P. Chapter 6—Bacterial capsular polysaccharides and exopolysaccharides. In *Microbial Glycobiology*; Holst, O., Brennan, P.J., Itzstein, M.V., Moran, A.P., Eds.; Academic Press: San Diego, CA, USA, 2010; pp. 93–108.
25. Stewart, P.S.; Costerton, J.W. Antibiotic resistance of bacteria in biofilms. *Lancet* **2001**, *358*, 135–138. [CrossRef]
26. Freitas, F.; Alves, V.D.; Reis, M.A. Advances in bacterial exopolysaccharides: From production to biotechnological applications. *Trends Biotechnol.* **2011**, *29*, 388–398. [CrossRef] [PubMed]
27. Aslam, S.N.; Newman, M.A.; Erbs, G.; Morrissey, K.L.; Chinchilla, D.; Boller, T.; Jensen, T.T.; De Castro, C.; Ierano, T.; Molinaro, A.; et al. Bacterial polysaccharides suppress induced innate immunity by calcium chelation. *Curr. Biol.* **2008**, *18*, 1078–1083. [CrossRef]
28. Gansbiller, M.; Schmid, J.; Sieber, V. In-depth rheological characterization of genetically modified xanthan-variants. *Carbohydr. Polym.* **2019**, *213*, 236–246. [CrossRef]
29. Becker, A.; Katzen, F.; Puhler, A.; Ielpi, L. Xanthan gum biosynthesis and application: A biochemical/genetic perspective. *Appl. Microbiol. Biotechnol.* **1998**, *50*, 145–152. [CrossRef]
30. Jansson, P.E.; Kenne, L.; Lindberg, B. Structure of extracellular polysaccharide from *Xanthomonas campestris*. *Carbohydr. Res.* **1975**, *45*, 275–282. [CrossRef]
31. Marzocca, M.P.; Harding, N.E.; Petroni, E.A.; Cleary, J.M.; Ielpi, L. Location and cloning of the ketal pyruvate transferase gene of *Xanthomonas campestris*. *J. Bacteriol.* **1991**, *173*, 7519–7524. [CrossRef]
32. Schmid, J.; Sieber, V.; Rehm, B. Bacterial exopolysaccharides: Biosynthesis pathways and engineering strategies. *Front. Microbiol.* **2015**, *6*, 496. [CrossRef] [PubMed]
33. Barreras, M.; Abdian, P.L.; Ielpi, L. Functional characterization of GumK, a membrane-associated beta-glucuronosyltransferase from *Xanthomonas campestris* required for xanthan polysaccharide synthesis. *Glycobiology* **2004**, *14*, 233–241. [CrossRef] [PubMed]
34. Vorhölter, F.J.; Schneiker, S.; Goesmann, A.; Krause, L.; Bekel, T.; Kaiser, O.; Linke, B.; Patschkowski, T.; Ruckert, C.; Schmid, J.; et al. The genome of *Xanthomonas campestris* pv. *campestris* B100 and its use for the reconstruction of metabolic pathways involved in xanthan biosynthesis. *J. Biotechnol.* **2008**, *134*, 33–45.
35. Nankai, H.; Hashimoto, W.; Miki, H.; Kawai, S.; Murata, K. Microbial system for polysaccharide depolymerization: Enzymatic route for xanthan depolymerization by *Bacillus* sp. strain GL1. *Appl. Environ. Microbiol.* **1999**, *65*, 2520–2526.
36. Hashimoto, W.; Miki, H.; Tsuchiya, N.; Nankai, H.; Murata, K. Xanthan lyase of *Bacillus* sp. strain GL1 liberates pyruvylated mannose from xanthan side chains. *Appl. Env. Microbiol.* **1998**, *64*, 3765–3768.
37. Thompson, M.A.; Onyeziri, M.C.; Fuqua, C. Function and regulation of *Agrobacterium tumefaciens* cell surface structures that promote attachment. *Curr. Top. Microbiol. Immunol.* **2018**, *418*, 143–184. [PubMed]
38. Matthysse, A.G. Exopolysaccharides of Agrobacterium tumefaciens. *Curr. Top. Microbiol. Immunol.* **2018**, *418*, 111–141.
39. Meade, M.J.; Tanenbaum, S.W.; Nakas, J.P. Production and rheological properties of a succinoglycan from *Pseudomonas* sp. 31260 grown on wood hydrolysates. *Can. J. Microbiol.* **1995**, *41*, 1147–1152. [CrossRef] [PubMed]
40. Andhare, P.; Delattre, C.; Pierre, G.; Michaud, P.; Pathak, H. Characterization and rheological behaviour analysis of the succinoglycan produced by *Rhizobium radiobacter* strain CAS from curd sample. *Food Hydrocoll.* **2017**, *64*, 1–8. [CrossRef]
41. Becker, A.; Rüberg, S.; Baumgarth, B.; Bertram-Drogatz, P.A.; Quester, I.; Pühler, A. Regulation of succinoglycan and galactoglucan biosynthesis in *Sinorhizobium meliloti*. *J. Mol. Microbiol. Biotechnol.* **2002**, *4*, 187–190.
42. Stredansky, M. Succinoglycan. In *Biopolymers Online*; Wiley: New York, NY, USA, 2005.
43. Becker, A. Challenges and perspectives in combinatorial assembly of novel exopolysaccharide biosynthesis pathways. *Front. Microbiol.* **2015**, *6*, 687. [CrossRef] [PubMed]
44. Glucksmann, M.A.; Reuber, T.L.; Walker, G.C. Genes needed for the modification, polymerization, export, and processing of succinoglycan by *Rhizobium meliloti*: A model for succinoglycan biosynthesis. *J. Bacteriol.* **1993**, *175*, 7045–7055. [CrossRef] [PubMed]
45. Varki, A.; Cummings, R.D.; Aebi, M.; Packer, N.H.; Seeberger, P.H.; Esko, J.D.; Stanley, P.; Hart, G.; Darvill, A.; Kinoshita, T.; et al. Symbol nomenclature for graphical representations of glycans. *Glycobiology* **2015**, *25*, 1323–1324. [CrossRef] [PubMed]

46. Ivashina, T.V.; Fedorova, E.E.; Ashina, N.P.; Kalinchuk, N.A.; Druzhinina, T.N.; Shashkov, A.S.; Shibaev, V.N.; Ksenzenko, V.N. Mutation in the *pssM* gene encoding ketal pyruvate transferase leads to disruption of *Rhizobium leguminosarum* bv. *viciae-Pisum sativum* symbiosis. *J. Appl. Microbiol.* **2010**, *109*, 731–742. [PubMed]
47. Xiu, A.; Kong, Y.; Zhou, M.; Zhu, B.; Wang, S.; Zhang, J. The chemical and digestive properties of a soluble glucan from *Agrobacterium* sp. ZX09. *Carbohydr. Polym.* **2010**, *82*, 623–628. [CrossRef]
48. Xu, L.; Cheng, R.; Li, J.; Wang, Y.; Zhu, B.; Ma, S.; Zhang, W.; Dong, W.; Wang, S.; Zhang, J. Identification of substituent groups and related genes involved in salecan biosynthesis in *Agrobacterium* sp. ZX09. *Appl. Microbiol. Biotechnol.* **2017**, *101*, 585–598. [CrossRef] [PubMed]
49. Meredith, T.C.; Mamat, U.; Kaczynski, Z.; Lindner, B.; Holst, O.; Woodard, R.W. Modification of lipopolysaccharide with colanic acid (M-antigen) repeats in *Escherichia coli*. *J. Biol. Chem.* **2007**, *282*, 7790–7798. [CrossRef] [PubMed]
50. Scott, P.M.; Erickson, K.M.; Troutman, J.M. Identification of the functional roles of six key proteins in the biosynthesis of *Enterobacteriaceae* colanic acid. *Biochemistry* **2019**, *58*, 1818–1830. [CrossRef]
51. Stevenson, G.; Andrianopoulos, K.; Hobbs, M.; Reeves, P.R. Organization of the *Escherichia coli* K-12 gene cluster responsible for production of the extracellular polysaccharide colanic acid. *J. Bacteriol.* **1996**, *178*, 4885–4893. [CrossRef]
52. Zhang, J.; Poh, C.L. Regulating exopolysaccharide gene *wcaF* allows control of *Escherichia coli* biofilm formation. *Sci. Rep.* **2018**, *8*, 13127. [CrossRef]
53. Cerantola, S.; Marty, N.; Montrozier, H. Structural studies of the acidic exopolysaccharide produced by a mucoid strain of *Burkholderia cepacia*, isolated from cystic fibrosis. *Carbohydr. Res.* **1996**, *285*, 59–67. [CrossRef]
54. Cescutti, P.; Toffanin, R.; Pollesello, P.; Sutherland, I.W. Structural determination of the acidic exopolysaccharide produced by a *Pseudomonas* sp. strain 1.15. *Carbohydr. Res.* **1999**, *315*, 159–168. [CrossRef]
55. Cescutti, P.; Kallioinen, A.; Impallomeni, G.; Toffanin, R.; Pollesello, P.; Leisola, M.; Eerikainen, T. Structure of the exopolysaccharide produced by *Enterobacter amnigenus*. *Carbohydr. Res.* **2005**, *340*, 439–447. [CrossRef] [PubMed]
56. Kuttel, M.; Ravenscroft, N.; Foschiatti, M.; Cescutti, P.; Rizzo, R. Conformational properties of two exopolysaccharides produced by *Inquilinus limosus*, a cystic fibrosis lung pathogen. *Carbohydr. Res.* **2012**, *350*, 40–48. [CrossRef]
57. Herasimenka, Y.; Cescutti, P.; Impallomeni, G.; Rizzo, R. Exopolysaccharides produced by Inquilinus limosus, a new pathogen of cystic fibrosis patients: Novel structures with usual components. *Carbohydr. Res.* **2007**, *342*, 2404–2415. [CrossRef] [PubMed]
58. Savant, A.P.; McColley, S.A. Cystic fibrosis year in review 2016. *Pediatr. Pulmonol.* **2017**, *52*, 1092–1102. [CrossRef]
59. D'Haeze, W.; Glushka, J.; De Rycke, R.; Holsters, M.; Carlson, R.W. Structural characterization of extracellular polysaccharides of *Azorhizobium caulinodans* and importance for nodule initiation on *Sesbania rostrata*. *Mol. Microbiol.* **2004**, *52*, 485–500. [CrossRef]
60. Lelchat, F.; Cerantola, S.; Brandily, C.; Colliec-Jouault, S.; Baudoux, A.C.; Ojima, T.; Boisset, C. The marine bacteria *Cobetia marina* DSMZ 4741 synthesizes an unexpected K-antigen-like exopolysaccharide. *Carbohydr. Polym.* **2015**, *124*, 347–356. [CrossRef]
61. Yasutake, T.; Kumagai, T.; Inoue, A.; Kobayashi, K.; Noda, M.; Orikawa, A.; Matoba, Y.; Sugiyama, M. Characterization of the LP28 strain-specific exopolysaccharide biosynthetic gene cluster found in the whole circular genome of *Pediococcus pentosaceus*. *Biochem. Biophys. Rep.* **2016**, *5*, 266–271. [CrossRef]
62. Van Zyl, L.M.; Coutinho, T.A.; Wingfield, M.J.; Pongpanich, K.; Wingfield, B.D. Morphological and molecular relatedness of geographically diverse isolates of *Coniothyrium zuluense* from South Africa and Thailand. *Mycol. Res.* **2002**, *106*, 51–59. [CrossRef]
63. Yang, B.Y.; Ding, Q.; Montgomery, R. Extracellular polysaccharides of *Erwinia futululu*, a bacterium associated with a fungal canker disease of *Eucalyptus* spp. *Carbohydr. Res.* **2002**, *337*, 2469–2480. [CrossRef]
64. O'Neill, M.A.; Robison, P.D.; Chou, K.J.; Darvill, A.G.; Albersheim, P. Evidence that the acidic polysaccharide secreted by *Agrobacterium radiobacter* (ATCC 53271) has a seventeen glycosyl-residue repeating unit. *Carbohydr. Res.* **1992**, *226*, 131–154. [CrossRef]

65. Verhoef, R.; de Waard, P.; Schols, H.A.; Siika-aho, M.; Voragen, A.G. *Methylobacterium* sp. isolated from a Finnish paper machine produces highly pyruvated galactan exopolysaccharide. *Carbohydr. Res.* **2003**, *338*, 1851–1859. [CrossRef]
66. Rougeaux, H.; Talaga, P.; Carlson, R.W.; Guezennec, J. Structural studies of an exopolysaccharide produced by *Alteromonas macleodii* subsp. *fijiensis* originating from a deep-sea hydrothermal vent. *Carbohydr. Res.* **1998**, *312*, 53–59. [CrossRef]
67. Poli, A.; Anzelmo, G.; Nicolaus, B. Bacterial exopolysaccharides from extreme marine habitats: Production, characterization and biological activities. *Mar. Drugs* **2010**, *8*, 1779–1802. [CrossRef] [PubMed]
68. Willis, L.M.; Whitfield, C. Structure, biosynthesis, and function of bacterial capsular polysaccharides synthesized by ABC transporter-dependent pathways. *Carbohydr Res.* **2013**, *378*, 35–44. [CrossRef] [PubMed]
69. Kay, E.; Cuccui, J.; Wren, B.W. Recent advances in the production of recombinant glycoconjugate vaccines. *npj Vaccines* **2019**, *4*, 16. [CrossRef] [PubMed]
70. Whitfield, C. Biosynthesis and assembly of capsular polysaccharides in *Escherichia coli*. *Annu. Rev. Biochem.* **2006**, *75*, 39–68. [CrossRef] [PubMed]
71. Hausdorff, W.P.; Hoet, B.; Adegbola, R.A. Predicting the impact of new pneumococcal conjugate vaccines: Serotype composition is not enough. *Expert Rev. Vaccines* **2015**, *14*, 413–428. [CrossRef] [PubMed]
72. Harding, C.M.; Nasr, M.A.; Scott, N.E.; Goyette-Desjardins, G.; Nothaft, H.; Mayer, A.E.; Chavez, S.M.; Huynh, J.P.; Kinsella, R.L.; Szymanski, C.M.; et al. A platform for glycoengineering a polyvalent pneumococcal bioconjugate vaccine using *E. coli* as a host. *Nat. Commun.* **2019**, *10*, 891. [CrossRef] [PubMed]
73. Kaplonek, P.; Khan, N.; Reppe, K.; Schumann, B.; Emmadi, M.; Lisboa, M.P.; Xu, F.F.; Calow, A.D.J.; Parameswarappa, S.G.; Witzenrath, M.; et al. Improving vaccines against *Streptococcus pneumoniae* using synthetic glycans. *Proc. Natl. Acad. Sci. USA* **2018**, *115*, 13353–13358. [CrossRef] [PubMed]
74. Pereira, C.L.; Geissner, A.; Anish, C.; Seeberger, P.H. Chemical synthesis elucidates the immunological importance of a pyruvate modification in the capsular polysaccharide of *Streptococcus pneumoniae* serotype 4. *Angew Chem. Int. Ed. Engl.* **2015**, *54*, 10016–10019. [CrossRef] [PubMed]
75. Falagas, M.E.; Karveli, E.A.; Siempos, I.I.; Vardakas, K.Z. *Acinetobacter infections*: A growing threat for critically ill patients. *Epidemiol. Infect.* **2008**, *136*, 1009–1019. [CrossRef] [PubMed]
76. Antunes, L.C.; Visca, P.; Towner, K.J. *Acinetobacter baumannii*: Evolution of a global pathogen. *Pathog. Dis.* **2014**, *71*, 292–301. [CrossRef] [PubMed]
77. Kenyon, J.J.; Hall, R.M. Variation in the complex carbohydrate biosynthesis loci of *Acinetobacter baumannii* genomes. *PLoS ONE* **2013**, *8*, e62160. [CrossRef] [PubMed]
78. Kenyon, J.J.; Speciale, I.; Hall, R.M.; De Castro, C. Structure of repeating unit of the capsular polysaccharide from *Acinetobacter baumannii* D78 and assignment of the K4 gene cluster. *Carbohydr. Res.* **2016**, *434*, 12–17. [CrossRef]
79. Yang, C.C.; Yen, C.H.; Ho, M.W.; Wang, J.H. Comparison of pyogenic liver abscess caused by non-*Klebsiella pneumoniae* and *Klebsiella pneumoniae*. *J. Microbiol. Immunol. Infect.* **2004**, *37*, 176–184.
80. Yang, F.L.; Yang, Y.L.; Liao, P.C.; Chou, J.C.; Tsai, K.C.; Yang, A.S.; Sheu, F.; Lin, T.L.; Hsieh, P.F.; Wang, J.T.; et al. Structure and immunological characterization of the capsular polysaccharide of a pyrogenic liver abscess caused by *Klebsiella pneumoniae: Activation* of macrophages through Toll-like receptor 4. *J. Biol. Chem.* **2011**, *286*, 21041–22151. [CrossRef]
81. Rao, A.S.; Liao, J.; Kabat, E.A.; Osserman, E.F.; Harboe, M.; Nimmich, W. Immunochemical studies on human monoclonal macroglobulins with specificities for 3,4-pyruvylated D-galactose and 4,6-pyruvylated D-glucose. *J. Biol. Chem.* **1984**, *259*, 1018–1026.
82. Dutton, G.G.; Parolis, H.; Parolis, L.A. A structural investigation of the capsular polysaccharide of *Klebsiella* K14. *Carbohydr. Res.* **1985**, *140*, 263–275. [CrossRef]
83. Beurret, M.; Joseleau, J.P.; Vignon, M.; Dutton, G.G.; Savage, A.V. Proof of the occurrence of 5,6-O-(1-carboxyethylidene)-D-galactofuranose units in the capsular polysaccharide of *Klebsiella* K12. *Carbohydr. Res.* **1989**, *189*, 247–260. [CrossRef]
84. Dutton, G.G.; Mackie, K.L. Structural investigation of *Klebsiella* serotype K70 polysaccharide. *Carbohydr. Res.* **1978**, *62*, 321–335. [CrossRef]
85. Merrifield, E.H.; Stephen, A.M. Structural studies on the capsular polysaccharide from *Klebsiella* serotype K64. *Carbohydr. Res.* **1979**, *74*, 241–257. [CrossRef]

86. Okutani, K.; Dutton, G.G. Structural investigation of *Klebsiella* serotype K46 polysaccharide. *Carbohydr. Res.* **1980**, *86*, 259–271. [CrossRef]
87. Rao, A.S.; Kabat, E.A.; Nilsson, B.; Zopf, D.A.; Nimmich, W. Isolation and characterization of 4-*O*-[3,4-*O*-(1-carboxyethylidene)-β-D-galactopyranosyl]erythritol from *Klebsiella* K33 polysaccharide. *Carbohydr. Res.* **1983**, *121*, 205–209. [CrossRef]
88. Kabat, E.A.; Liao, J.; Bretting, H.; Franklin, E.C.; Geltner, D.; Frangione, B.; Koshland, M.E.; Shyong, J.; Osserman, E.F. Human monoclonal macroglobulins with specificity for *Klebsiella* K polysaccharides that contain 3,4-pyruvylated-D-galactose and 4,6-pyruvylated-D-galactose. *J. Exp. Med.* **1980**, *152*, 979–995. [CrossRef] [PubMed]
89. Dutton, G.G.; Parolis, H.; Joseleau, J.P.; Marais, M.F. The use of bacteriophage depolymerization in the structural investigation of the capsular polysaccharide from *Klebsiella serotype* K3. *Carbohydr. Res.* **1986**, *149*, 411–423. [CrossRef]
90. Zamze, S.; Martinez-Pomares, L.; Jones, H.; Taylor, P.R.; Stillion, R.J.; Gordon, S.; Wong, S.Y. Recognition of bacterial capsular polysaccharides and lipopolysaccharides by the macrophage mannose receptor. *J. Biol. Chem.* **2002**, *277*, 41613–41623. [CrossRef] [PubMed]
91. De Silva, R.A.; Wang, Q.; Chidley, T.; Appulage, D.K.; Andreana, P.R. Immunological response from an entirely carbohydrate antigen: Design of synthetic vaccines based on Tn-PS A1 conjugates. *J. Am. Chem. Soc.* **2009**, *131*, 9622–9623. [CrossRef]
92. Trabbic, K.R.; De Silva, R.A.; Andreana, P.R. Elucidating structural features of an entirely carbohydrate cancer vaccine construct employing circular dichroism and fluorescent labeling. *Medchemcomm* **2014**, *5*, 1143–1149. [CrossRef]
93. Coyne, M.J.; Tzianabos, A.O.; Mallory, B.C.; Carey, V.J.; Kasper, D.L.; Comstock, L.E. Polysaccharide biosynthesis locus required for virulence of *Bacteroides fragilis*. *Infect. Immun.* **2001**, *69*, 4342–4350. [CrossRef] [PubMed]
94. Baumann, H.; Tzianabos, A.O.; Brisson, J.R.; Kasper, D.L.; Jennings, H.J. Structural elucidation of two capsular polysaccharides from one strain of *Bacteroides fragilis* using high-resolution NMR spectroscopy. *Biochemistry* **1992**, *31*, 4081–40899. [CrossRef] [PubMed]
95. Severn, W.B.; Richards, J.C. The structure of the specific capsular polysaccharide of *Rhodococcus equi* serotype 4. *Carbohydr. Res.* **1999**, *320*, 209–222. [CrossRef]
96. Percy, M.G.; Gründling, A. Lipoteichoic acid synthesis and function in gram-positive bacteria. *Annu. Rev. Microbiol.* **2014**, *68*, 81–100. [CrossRef] [PubMed]
97. Brown, S.; Santa Maria, J.P., Jr.; Walker, S. Wall teichoic acids of Gram-positive bacteria. *Annu. Rev. Microbiol.* **2013**, *67*, 313–336. [CrossRef]
98. Pasquina, L.W.; Santa Maria, J.P.; Walker, S. Teichoic acid biosynthesis as an antibiotic target. *Curr. Opin. Microbiol.* **2013**, *16*, 531–537. [CrossRef]
99. Mesnage, S.; Fontaine, T.; Mignot, T.; Delepierre, M.; Mock, M.; Fouet, A. Bacterial SLH domain proteins are non-covalently anchored to the cell surface via a conserved mechanism involving wall polysaccharide pyruvylation. *EMBO J.* **2000**, *19*, 4473–4484. [CrossRef]
100. Schäffer, C.; Messner, P. The structure of secondary cell wall polymers: How Gram-positive bacteria stick their cell walls together. *Microbiology* **2005**, *151*, 643–651. [CrossRef]
101. Fouet, A. The surface of Bacillus anthracis. *Mol. Asp. Med.* **2009**, *30*, 374–385. [CrossRef]
102. Choudhury, B.; Leoff, C.; Saile, E.; Wilkins, P.; Quinn, C.P.; Kannenberg, E.L.; Carlson, R.W. The structure of the major cell wall polysaccharide of *Bacillus anthracis* is species-specific. *J. Biol. Chem.* **2006**, *281*, 27932–27941. [CrossRef]
103. Sychantha, D.; Chapman, R.N.; Bamford, N.C.; Boons, G.J.; Howell, P.L.; Clarke, A.J. Molecular basis for the attachment of S-layer proteins to the cell wall of *Bacillus anthracis*. *Biochemistry* **2018**, *57*, 1949–1953. [CrossRef] [PubMed]
104. Blackler, R.J.; López-Guzmán, A.; Hager, F.F.; Janesch, B.; Martinz, G.; Gagnon, S.M.L.; Haji-Ghassemi, O.; Kosma, P.; Messner, P.; Schäffer, C.; et al. Structural basis of cell wall anchoring by SLH domains in *Paenibacillus alvei*. *Nat. Commun.* **2018**, *9*, 3120. [CrossRef] [PubMed]
105. Messner, P.; Schäffer, C.; Egelseer, E.M.; Sleytr, U.B. Occurrence, structure, chemistry, genetics, morphogenesis and functions of S-layers. In *Prokaryotic Cell Wall Compounds - Structure and Biochemistry*; König, H., Claus, H., Varma, A., Eds.; Springer: Berlin, Germany, 2010; pp. 53–109.

106. Chung, S.; Shin, S.H.; Bertozzi, C.R.; De Yoreo, J.J. Self-catalyzed growth of S layers via an amorphous-to-crystalline transition limited by folding kinetics. *Proc. Natl. Acad. Sci. USA* **2010**, *107*, 16536–16541. [CrossRef]
107. Messner, P.; Allmaier, G.; Schäffer, C.; Wugeditsch, T.; Lortal, S.; König, H.; Niemetz, R.; Dorner, M. Biochemistry of S-layers. *FEMS Microbiol. Rev.* **1997**, *20*, 25–46. [CrossRef] [PubMed]
108. Sleytr, U.B.; Schuster, B.; Egelseer, E.M.; Pum, D. S-layers: Principles and applications. *FEMS Microbiol. Rev.* **2014**, *38*, 823–864. [CrossRef] [PubMed]
109. Hager, F.F.; Lopez-Guzman, A.; Krauter, S.; Blaukopf, M.; Polter, M.; Brockhausen, I.; Kosma, P.; Schäffer, C. Functional characterization of enzymatic steps involved in pyruvylation of bacterial secondary cell wall polymer fragments. *Front. Microbiol.* **2018**, *9*, 1356. [CrossRef]
110. Forsberg, L.S.; Choudhury, B.; Leoff, C.; Marston, C.K.; Hoffmaster, A.R.; Saile, E.; Quinn, C.P.; Kannenberg, E.L.; Carlson, R.W. Secondary cell wall polysaccharides from *Bacillus cereus* strains G9241, 03BB87 and 03BB102 causing fatal pneumonia share similar glycosyl structures with the polysaccharides from *Bacillus anthracis*. *Glycobiology* **2011**, *21*, 934–948. [CrossRef]
111. Mesnage, S.; Tosi-Couture, E.; Mock, M.; Fouet, A. The S-layer homology domain as a means for anchoring heterologous proteins on the cell surface of *Bacillus anthracis*. *J. Appl. Microbiol.* **1999**, *87*, 256–260. [CrossRef]
112. Kern, J.; Ryan, C.; Faull, K.; Schneewind, O. *Bacillus anthracis* surface-layer proteins assemble by binding to the secondary cell wall polysaccharide in a manner that requires *csaB* and *tagO*. *J. Mol. Biol.* **2010**, *401*, 757–775. [CrossRef]
113. Missiakas, D.; Schneewind, O. Assembly and function of the *Bacillus anthracis* S-layer. *Annu. Rev. Microbiol.* **2017**, *71*, 79–98. [CrossRef]
114. Oh, S.Y.; Lunderberg, J.M.; Chateau, A.; Schneewind, O.; Missiakas, D. Genes required for *Bacillus anthracis* secondary cell wall polysaccharide synthesis. *J. Bacteriol.* **2017**, *199*, e00613–e00616. [CrossRef]
115. Chateau, A.; Lunderberg, J.M.; Oh, S.Y.; Abshire, T.; Friedlander, A.; Quinn, C.P.; Missiakas, D.M.; Schneewind, O. Galactosylation of the secondary cell wall polysaccharide of *Bacillus anthracis* and Its contribution to anthrax pathogenesis. *J. Bacteriol.* **2018**, *200*, e00562-17. [CrossRef] [PubMed]
116. Lunderberg, J.M.; Liszewski Zilla, M.; Missiakas, D.; Schneewind, O. *Bacillus anthracis tagO* is required for vegetative growth and secondary cell wall polysaccharide synthesis. *J. Bacteriol.* **2015**, *197*, 3511–3520. [CrossRef] [PubMed]
117. Sychantha, D.; Little, D.J.; Chapman, R.N.; Boons, G.J.; Robinson, H.; Howell, P.L.; Clarke, A.J. PatB1 is an *O*-acetyltransferase that decorates secondary cell wall polysaccharides. *Nat. Chem. Biol.* **2018**, *14*, 79–85. [CrossRef] [PubMed]
118. Schneewind, O.; Missiakas, D.M. Protein secretion and surface display in Gram-positive bacteria. *Philos Trans. R. Soc. Lond. B Biol. Sci.* **2012**, *367*, 1123–1139. [CrossRef] [PubMed]
119. Garufi, G.; Hendrickx, A.P.; Beeri, K.; Kern, J.W.; Sharma, A.; Richter, S.G.; Schneewind, O.; Missiakas, D. Synthesis of lipoteichoic acids in *Bacillus anthracis*. *J. Bacteriol.* **2012**, *194*, 4312–4321. [CrossRef] [PubMed]
120. Janesch, B.; Messner, P.; Schäffer, C. Are the surface layer homology domains essential for cell surface display and glycosylation of the S-layer protein from *Paenibacillus alvei* CCM 2051^{T}? *J. Bacteriol.* **2013**, *195*, 565–575. [CrossRef] [PubMed]
121. Ilk, N.; Kosma, P.; Puchberger, M.; Egelseer, E.M.; Mayer, H.F.; Sleytr, U.B.; Sára, M. Structural and functional analyses of the secondary cell wall polymer of *Bacillus sphaericus* CCM 2177 that serves as an S-layer-specific anchor. *J. Bacteriol.* **1999**, *181*, 7643–7646. [PubMed]
122. Zarschler, K.; Janesch, B.; Kainz, B.; Ristl, R.; Messner, P.; Schäffer, C. Cell surface display of chimeric glycoproteins via the S-layer of *Paenibacillus alvei*. *Carbohydr. Res.* **2010**, *345*, 1422–1431. [CrossRef] [PubMed]
123. D'Elia, M.A.; Pereira, M.P.; Chung, Y.S.; Zhao, W.; Chau, A.; Kenney, T.J.; Sulavik, M.C.; Black, T.A.; Brown, E.D. Lesions in teichoic acid biosynthesis in *Staphylococcus aureus* lead to a lethal gain of function in the otherwise dispensable pathway. *J. Bacteriol.* **2006**, *188*, 4183–4189. [CrossRef] [PubMed]
124. Zhang, Y.H.; Ginsberg, C.; Yuan, Y.; Walker, S. Acceptor substrate selectivity and kinetic mechanism of *Bacillus subtilis* TagA. *Biochemistry* **2006**, *45*, 10895–10904. [CrossRef] [PubMed]
125. D'Elia, M.A.; Henderson, J.A.; Beveridge, T.J.; Heinrichs, D.E.; Brown, E.D. The *N*-acetylmannosamine transferase catalyzes the first committed step of teichoic acid assembly in *Bacillus subtilis* and *Staphylococcus aureus*. *J. Bacteriol.* **2009**, *191*, 4030–4034. [CrossRef] [PubMed]

126. Ginsberg, C.; Zhang, Y.H.; Yuan, Y.; Walker, S. *In vitro* reconstitution of two essential steps in wall teichoic acid biosynthesis. *ACS Chem. Biol.* **2006**, *1*, 25–28. [CrossRef] [PubMed]
127. Swoboda, J.G.; Campbell, J.; Meredith, T.C.; Walker, S. Wall teichoic acid function, biosynthesis, and inhibition. *Chembiochem* **2010**, *11*, 35–45. [CrossRef] [PubMed]
128. Schirner, K.; Marles-Wright, J.; Lewis, R.J.; Errington, J. Distinct and essential morphogenic functions for wall- and lipo-teichoic acids in *Bacillus subtilis*. *EMBO J.* **2009**, *28*, 830–842. [CrossRef] [PubMed]
129. Chapot-Chartier, M.P.; Kulakauskas, S. Cell wall structure and function in lactic acid bacteria. *Microb Cell Fact.* **2014**, *13* (Suppl. 1), S9. [CrossRef]
130. Kojima, N.; Kaya, S.; Araki, Y.; Ito, E. Pyruvic-acid-containing polysaccharide in the cell wall of *Bacillus polymyxa* AHU 1385. *Eur. J. Biochem.* **1988**, *174*, 255–260. [CrossRef]
131. Dmitrenok, A.S.; Streshinskaya, G.M.; Tul'skaya, E.M.; Potekhina, N.V.; Senchenkova, S.N.; Shashkov, A.S.; Bilan, M.I.; Starodumova, I.P.; Bueva, O.V.; Evtushenko, L.I. Pyruvylated cell wall glycopolymers of *Promicromonospora citrea* VKM A succeeds, similar-665(T) and *Promicromonospora* sp. *VKM* A succeeds, similar-1028. *Carbohydr. Res.* **2017**, *449*, 134–142.
132. Schleifer, K.H.; Kandler, O. Peptidoglycan types of bacterial cell walls and their taxonomic implications. *Bacteriol. Rev.* **1972**, *36*, 407–477.
133. Tul'skaya, E.M.; Shashkov, A.S.; Streshinskaya, G.M.; Potekhina, N.V.; Evtushenko, L.I. New structures and composition of cell wall teichoic acids from *Nocardiopsis synnemataformans*, *Nocardiopsis halotolerans*, *Nocardiopsis composta* and *Nocardiopsis metallicus*: A chemotaxonomic value. *Antonie Van Leeuwenhoek* **2014**, *106*, 1105–1117. [CrossRef]
134. Ibrahim, A.H.; Desoukey, S.Y.; Fouad, M.A.; Kamel, M.S.; Gulder, T.A.M.; Abdelmohsen, U.R. Natural product potential of the genus Nocardiopsis. *Mar. Drugs* **2018**, *16*, 147. [CrossRef] [PubMed]
135. Potekhina, N.V.; Evtushenko, L.I.; Senchenkova, S.N.; Shashkov, A.S.; Naumova, I.B. Structures of cell wall teichoic acids of *Brevibacterium iodinum* VKM Ac-2106. *Biochem. (Mosc.)* **2004**, *69*, 1353–1359. [CrossRef] [PubMed]
136. Whitfield, C.; Trent, M.S. Biosynthesis and export of bacterial lipopolysaccharides. *Annu. Rev. Biochem.* **2014**, *83*, 99–128. [CrossRef] [PubMed]
137. Kanipes, M.I.; Guerry, P. Chapter 44—Role of microbial glycosylation in host cell invasion. In *Microbial Glycobiology*; Holst, O., Brennan, P.J., Itzstein, M.V., Moran, A.P., Eds.; Academic Press: San Diego, CA, USA, 2010; pp. 871–883.
138. Raetz, C.R.H.; Whitfield, C. Lipopolysaccharide endotoxins. *Annu. Rev. Biochem.* **2002**, *71*, 635–700. [CrossRef] [PubMed]
139. Zhang, G.; Meredith, T.C.; Kahne, D. On the essentiality of lipopolysaccharide to Gram-negative bacteria. *Curr. Opin. Microbiol.* **2013**, *16*, 779–785. [CrossRef] [PubMed]
140. Broeker, N.K.; Barbirz, S. Not a barrier but a key: How bacteriophages exploit host's O-antigen as an essential receptor to initiate infection. *Mol. Microbiol.* **2017**, *105*, 353–357. [CrossRef]
141. Kocharova, N.A.; Vinogradov, E.; Kondakova, A.N.; Shashkov, A.S.; Rozalski, A.; Knirel, Y.A. The full structure of the carbohydrate chain of the lipopolysaccharide of *Providencia alcalifaciens* O19. *J. Carbohydr. Chem.* **2008**, *27*, 320–331. [CrossRef]
142. Kondakova, A.N.; Vinogradov, E.; Lindner, B.; Kocharova, N.A.; Rozalski, A.; Knirel, Y.A. Elucidation of the lipopolysaccharide core Structures of bacteria of the genus *Providencia*. *J. Carbohydr. Chem.* **2006**, *25*, 499–520. [CrossRef]
143. Kocharova, N.A.; Kondakova, A.N.; Vinogradov, E.; Ovchinnikova, O.G.; Lindner, B.; Shashkov, A.S.; Rozalski, A.; Knirel, Y.A. Full structure of the carbohydrate chain of the lipopolysaccharide of *Providencia rustigianii* O34. *Chemistry* **2008**, *14*, 6184–6191. [CrossRef]
144. Perepelov, A.V.; Senchenkova, S.N.; Shashkov, A.S.; Lu, B.; Feng, L.; Vang, L.; Knirel, Y.A. Antigenic polysaccharides of bacteria: 40. The structures of O-specific polysaccharides from *Shigella dysenteriae* types 3 and 9 and *S. boydii* type 4 revised by NMR spectroscopy. *Russ. J. Bioorgan. Chem.* **2008**, *34*, 121–128. [CrossRef]
145. Perepelov, A.V.; Senchenkova, S.N.; Shashkov, A.S.; Knirel, Y.A.; Lu, B.; Feng, L.; Wang, L. A completed structure of the O-polysaccharide from *Shigella dysenteriae* type 10. *Rus. J. Bioorgan. Chem.* **2009**, *35*, 131–133. [CrossRef]

146. Leone, S.; Molinaro, A.; Dubery, I.; Lanzetta, R.; Parrilli, M. The *O*-specific polysaccharide structure from the lipopolysaccharide of the Gram-negative bacterium *Raoultella terrigena*. *Carbohydr. Res.* **2007**, *342*, 1514–1518. [CrossRef] [PubMed]
147. Perepelov, A.V.; Rozalski, A.; Bartodziejska, B.; Senchenkova, S.N.; Knirel, Y.A. Structure of the O-polysaccharide of *Proteus mirabilis* O19 and reclassification of certain *Proteus* strains that were formerly classified in serogroup O19. *Arch. Immunol. Ther. Exp. (Warsz.)* **2004**, *52*, 1881–1896.
148. Kokoulin, M.S.; Kalinovsky, A.I.; Komandrova, N.A.; Tomshich, S.V.; Romanenko, L.A.; Vaskovsky, V.E. The new sulfated O-specific polysaccharide from marine bacterium *Cobetia pacifica* KMM 3878, containing 3,4-*O*-[(S)-1-carboxyethylidene]-D-galactose and 2,3-*O*-disulfate-D-galactose. *Carbohydr. Res.* **2014**, *397*, 46–51. [CrossRef] [PubMed]
149. Barksdale, L.; Kim, K.S. Mycobacterium. *Bacteriol. Rev.* **1977**, *41*, 217–372. [PubMed]
150. Saadat, S.; Ballou, C.E. Pyruvylated glycolipids from *Mycobacterium smegmatis*. Structures of two oligosaccharide components. *J. Biol. Chem.* **1983**, *258*, 1813–1818. [PubMed]
151. Thayer, W.R., Jr.; Bozic, C.M.; Camphausen, R.T.; McNeil, M. Implications of antibodies to pyruvylated glucose in healthy populations for mycobacterioses and other infectious diseases. *J. Clin. Microbiol.* **1990**, *28*, 714–718. [PubMed]
152. Araki, S.; Abe, S.; Ando, S.; Kon, K.; Fujiwara, N.; Satake, M. Structure of phosphonoglycosphingolipid containing pyruvylated galactose in nerve fibers of *Aplysia kurodai*. *J. Biol. Chem.* **1989**, *264*, 19922–19927. [PubMed]
153. Gemmill, T.R.; Trimble, R.B. Overview of *N*- and *O*-linked oligosaccharide structures found in various yeast species. *Biochim. Biophys. Acta* **1999**, *1426*, 227–237. [CrossRef]
154. Cutler, J.E. *N*-glycosylation of yeast, with emphasis on *Candida albicans*. *Med. Mycol.* **2001**, *39* (Suppl. 1), 75–86. [CrossRef]
155. Parolis, L.A.; Duus, J.O.; Parolis, H.; Meldal, M.; Bock, K. The extracellular polysaccharide of *Pichia (Hansenula) holstii* NRRL Y-2448: The structure of the phosphomannan backbone. *Carbohydr. Res.* **1996**, *293*, 101–117. [CrossRef]
156. Gemmill, T.R.; Trimble, R.B. All pyruvylated galactose in *Schizosaccharomyces pombe N*-glycans is present in the terminal disaccharide, 4, 6-O-[(R)-(1-carboxyethylidine)]-Galβ1,3Galα1. *Glycobiology* **1998**, *8*, 1087–1095. [CrossRef] [PubMed]
157. Schüle, G.; Ziegler, T. Efficient block synthesis of a pyruvated tetrasaccharide 5-amino-pentyl glycoside related to *Streptococcus pneumoniae* type 27. *Tetrahedron* **1996**, *52*, 2925–2936. [CrossRef]
158. Ziegler, F.D.; Cavanagh, J.; Lubowski, C.; Trimble, R.B. Novel *Schizosaccharomyces pombe N*-linked GalMan9GlcNAc isomers: Role of the Golgi GMA12 galactosyltransferase in core glycan galactosylation. *Glycobiology* **1999**, *9*, 497–505. [CrossRef] [PubMed]
159. Ballou, C.E.; Ballou, L.; Ball, G. *Schizosaccharomyces pombe* glycosylation mutant with altered cell surface properties. *Proc. Natl. Acad. Sci. USA* **1994**, *91*, 9327–9331. [CrossRef]
160. Yoritsune, K.; Matsuzawa, T.; Ohashi, T.; Takegawa, K. The fission yeast Pvg1p has galactose-specific pyruvyltransferase activity. *FEBS Lett.* **2013**, *587*, 917–921. [CrossRef]
161. Ferreira, L.G.; Noseda, M.D.; Goncalves, A.G.; Ducatti, D.R.; Fujii, M.T.; Duarte, M.E. Chemical structure of the complex pyruvylated and sulfated agaran from the red seaweed *Palisada flagellifera* (*Ceramiales, Rhodophyta*). *Carbohydr. Res.* **2012**, *347*, 83–94. [CrossRef]
162. Bondu, S.; Deslandes, E.; Fbre, M.S.; Berthou, C.; Guangli, A. Carrageenan from *Solieria chordalis* (Gigartinales): Structural analysis and immunological activities of the low molecular weight fractions. *Carbohydr. Polym.* **2010**, *81*, 448–460.
163. Spillmann, D.; Thomas-Oates, J.E.; van Kuik, J.A.; Vliegenthart, J.F.; Misevic, G.; Burger, M.M.; Finne, J. Characterization of a novel sulfated carbohydrate unit implicated in the carbohydrate-carbohydrate-mediated cell aggregation of the marine sponge *Microciona prolifera*. *J. Biol. Chem.* **1995**, *270*, 5089–5097. [CrossRef]
164. Nielsen, P.H.; Jahn, A. Extraction of EPS. In *Microbial extracellular polymeric substances: Characterization, structure and function*; Wingender, J., Neu, T.R., Flemming, H.-C., Eds.; Springer: Berlin/Heidelberg, Germany, 1999; pp. 49–72.
165. Yang, B.Y.; Brand, J.; Montgomery, R. Pyruvated galactose and oligosaccharides from *Erwinia chrysanthemi* Ech6 extracellular polysaccharide. *Carbohydr. Res.* **2001**, *331*, 59–67. [CrossRef]

166. Domenico, P.; Diedrich, D.L.; Cunha, B.A. Quantitative extraction and purification of exopolysaccharides from *Klebsiella pneumoniae*. *J. Microbiol. Methods* **1989**, *9*, 211–219. [CrossRef]
167. Brimacombe, C.A.; Beatty, J.T. Surface polysaccharide extraction and quantification. *Bio-Protocol* **2013**, *3*, e934. [CrossRef]
168. Kho, K.; Meredith, T.C. Extraction and analysis of bacterial teichoic acids. *Bio-Protocol* **2018**, *8*, e3078. [CrossRef]
169. Chateau, A.; Schneewind, O.; Missiakas, D. Extraction and purification of wall-bound polymers of Gram-positive bacteria. *Methods Mol. Biol.* **2019**, *1954*, 47–57. [PubMed]
170. Whitfield, C.; Perry, M.B. 6.7 Isolation and purification of cell surface polysaccharides from Gram-negative bacteria. In *Methods in Microbiology*; Williams, P., Ketley, J., Salmond, G., Eds.; Academic Press: Cambridge, MA, USA, 1998; Volume 27, pp. 249–258.
171. Grice, I.D.; Wilson, J.C. Chapter 13 - Analytical approaches towards the structural characterization of microbial wall glycopolymers. In *Microbial Glycobiology*; Holst, O., Brennan, P.J., Itzstein, M.V., Moran, A.P., Eds.; Academic Press: San Diego, CA, USA, 2010; pp. 233–252.
172. Perepelov, A.V.; Weintraub, A.; Liu, B.; Senchenkova, S.N.; Shashkov, A.S.; Feng, L.; Wang, L.; Widmalm, G.; Knirel, Y.A. The O-polysaccharide of *Escherichia coli* O112ac has the same structure as that of *Shigella dysenteriae* type 2 but is devoid of *O*-acetylation: A revision of the *S. dysenteriae* type 2 *O*-polysaccharide structure. *Carbohydr. Resc.* **2008**, *343*, 977–981. [CrossRef] [PubMed]
173. Westphal, O.; Jann, K. Bacterial lipopolysaccharides extraction with phenol-water and further applications of the procedure. *Methods Carbohydr. Chem.* **1965**, *5*, 83–91.
174. Galanos, C.; Lüderitz, O.; Westphal, O. A new method for the extraction of R lipopolysaccharides. *Eur. J. Biochem.* **1969**, *9*, 245–249. [CrossRef] [PubMed]
175. Darveau, R.P.; Hancock, R.E. Procedure for isolation of bacterial lipopolysaccharides from both smooth and rough *Pseudomonas aeruginosa* and *Salmonella typhimurium* strains. *J. Bacteriol.* **1983**, *155*, 831–838.
176. Holst, O.; Broer, W.; Thomas-Oates, J.E.; Mamat, U.; Brade, H. Structural analysis of two oligosaccharide bisphosphates isolated from the lipopolysaccharide of a recombinant strain of *Escherichia coli* F515 (Re chemotype) expressing the genus-specific epitope of *Chlamydia* lipopolysaccharide. *Eur. J. Biochem.* **1993**, *214*, 703–710. [CrossRef]
177. Hind, C.R.; Collins, P.M.; Renn, D.; Cook, R.B.; Caspi, D.; Baltz, M.L.; Pepys, M.B. Binding specificity of serum amyloid P component for the pyruvate acetal of galactose. *J. Exp. Med.* **1984**, *159*, 1058–1069. [CrossRef]
178. Hind, C.R.; Collins, P.M.; Baltz, M.L.; Pepys, M.B. Human serum amyloid P component, a circulating lectin with specificity for the cyclic 4,6-pyruvate acetal of galactose. Interactions with various bacteria. *Biochem. J.* **1985**, *225*, 107–111. [CrossRef] [PubMed]
179. Rühmann, B.; Schmid, J.; Sieber, V. Automated modular high throughput exopolysaccharide screening platform coupled with highly sensitive carbohydrate fingerprint analysis. *J. Vis. Exp.* **2016**. [CrossRef] [PubMed]
180. Bjerring, P.N.; Hauerberg, J.; Frederiksen, H.J.; Jorgensen, L.; Hansen, B.A.; Tofteng, F.; Larsen, F.S. Cerebral glutamine concentration and lactate-pyruvate ratio in patients with acute liver failure. *Neurocrit. Care* **2008**, *9*, 3–7. [CrossRef] [PubMed]
181. Schwimmer, S.; Weston, W.J. Onion flavor and odor - Enzymatic development of pyruvic acid in onion as a measure of pungency. *J. Agric. Food Chem.* **1961**, *9*, 127–132. [CrossRef]
182. Anthon, G.E.; Barrett, D.M. Modified method for the determination of pyruvic acid with dinitrophenylhydrazine in the assessment of onion pungency. *J. Sci. Food Agric.* **2003**, *83*, 1210–1213. [CrossRef]
183. Ovodov, Y.S. Bacterial capsular antigens. Structural patterns of capsular antigens. *Biochem. (Mosc.)* **2006**, *71*, 937–954. [CrossRef] [PubMed]
184. Garegg, P.J.; Jansson, P.E.; Lindberg, B.; Lindh, F.; Lönngren, J.; Kvarnström, I.; Nimmich, W. Configuration of the acetal carbon-atom of pyruvic-acid acetals in some bacterial polysaccharides. *Carbohydr. Res.* **1980**, *78*, 127–132. [CrossRef]
185. Jansson, P.E.; Lindberg, J.; Widmalm, G. Syntheses and NMR-studies of pyruvic acid 4,6-acetals of some methyl hexopyranosides. *Acta Chem. Scand.* **1993**, *47*, 711–715. [CrossRef]

186. Gorin, P.A.J.; Mazurek, M.; Duarte, H.S.; Iacomini, M.; Duarte, J.H. Properties of ^{13}C-NMR spectra of O-(1-carboxyethylidene) derivatives of methyl-β-D-galactopyranoside - Models for determination of pyruvic-acid acetal structures in polysaccharides. *Carbohydr. Res.* **1982**, *100*, 1–15. [CrossRef]
187. Lindberg, B. Components of bacterial polysaccharides. In *Advances in Carbohydrate Chemistry and Biochemistry*; Tipson, R.S., Horton, D., Eds.; Academic Press: Washington, DC, USA, 1990; Volume 48, pp. 279–318.
188. Zhao, G.; Perepelov, A.V.; Senchenkova, S.N.; Shashkov, A.S.; Feng, L.; Li, X.M.; Knirel, Y.A.; Wang, L. Structural relation of the antigenic polysaccharides of *Escherichia coli* O40, *Shigella dysenteriae* type 9, and *E. coli* K47. *Carbohydr. Res.* **2007**, *342*, 1275–1279. [CrossRef]
189. Ravenscroft, N.; Parolis, L.A.S.; Parolis, H. Bacteriophage degradation of *Klebsiella* K30 capsular polysaccharide - an NMR investigation of the 3,4-pyruvated galactose-containing repeating oligosaccharide. *Carbohydr. Res.* **1994**, *254*, 333–340. [CrossRef]
190. Dudman, W.F.; Lacey, M.J. Identification of pyruvated monosaccharides in polysaccharides by gas-liquid chromatography mass spectrometry. *Carbohydr. Res.* **1986**, *145*, 175–191. [CrossRef]
191. Fontaine, T.; Talmont, F.; Dutton, G.G.; Fournet, B. Analysis of pyruvic acid acetal containing polysaccharides by methanolysis and reductive cleavage methods. *Anal. Biochem.* **1991**, *199*, 154–161. [CrossRef]
192. Kailemia, M.J.; Ruhaak, L.R.; Lebrilla, C.B.; Amster, I.J. Oligosaccharide analysis by mass spectrometry: A review of recent developments. *Anal. Chem.* **2014**, *86*, 196–212. [CrossRef] [PubMed]
193. Harvey, D.J. Matrix-assisted laser desorption/ionization mass spectrometry of carbohydrates. *Mass Spec. Rev.* **1999**, *18*, 349–451. [CrossRef]
194. Kern, J.; Wilton, R.; Zhang, R.; Binkowski, T.A.; Joachimiak, A.; Schneewind, O. Structure of surface layer homology (SLH) domains from *Bacillus anthracis* surface array protein. *J. Biol. Chem.* **2011**, *286*, 26042–26049. [CrossRef] [PubMed]
195. Chapman, R.N.; Liu, L.; Boons, G.J. 4,6-O-pyruvyl ketal modified *N*-acetyl mannosamine of the secondary cell polysaccharide of *Bacillus anthracis* is the anchoring residue for its surface layer proteins. *J. Am. Chem. Soc.* **2018**, *140*, 49. [CrossRef]
196. Gardiol, A.E.; Dazzo, F.B. Biosynthesis of *Rhizobium trifolii* capsular polysaccharide: Enzymatic transfer of pyruvate substitutions into lipid-bound saccharide intermediates. *J. Bacteriol.* **1986**, *168*, 1459–1462. [CrossRef] [PubMed]
197. Bossio, J.C.; de Iannino, N.I.; Dankert, M.A. *In vitro* synthesis of a lipid-linked acetylated and pyruvilated oligosaccharide in *Rhizobium trifolii*. *Biochem. Biophys. Res. Commun.* **1986**, *134*, 205–211. [CrossRef]
198. Higuchi, Y.; Yoshinaga, S.; Yoritsune, K.; Tateno, H.; Hirabayashi, J.; Nakakita, S.; Kanekiyo, M.; Kakuta, Y.; Takegawa, K. A rationally engineered yeast pyruvyltransferase Pvg1p introduces sialylation-like properties in neo-human-type complex oligosaccharide. *Sci. Rep.* **2016**, *6*, 26349. [CrossRef]
199. Higuchi, Y.; Matsufuji, H.; Tanuma, M.; Arakawa, T.; Mori, K.; Yamada, C.; Shofia, R.; Matsunaga, E.; Tashiro, K.; Fushinobu, S.; et al. Identification and characterization of a novel β-D-galactosidase that releases pyruvylated galactose. *Sci. Rep.* **2018**, *8*, 12013. [CrossRef] [PubMed]
200. Cuthbertson, L.; Kos, V.; Whitfield, C. ABC transporters involved in export of cell surface glycoconjugates. *Microbiol. Mol. Biol. Rev.* **2010**, *74*, 341–362. [CrossRef] [PubMed]
201. Workman, S.D.; Worrall, L.J.; Strynadka, N.C.J. Crystal structure of an intramembranal phosphatase central to bacterial cell-wall peptidoglycan biosynthesis and lipid recycling. *Nat. Commun.* **2018**, *9*, 1159. [CrossRef] [PubMed]
202. Potter, S.C.; Luciani, A.; Eddy, S.R.; Park, Y.; Lopez, R.; Finn, R.D. HMMER web server: 2018 update. *Nucleic Acids Res.* **2018**, *46*, W200–W204. [CrossRef] [PubMed]
203. Foley, G.; Sützl, L.; D'Cunha, S.A.; Gillam, E.M.; Boden, M. SeqScrub: A web tool for automatic cleaning and annotation of FASTA file headers for bioinformatic applications. *Biotechniques* **2019**, *67*, 50–54. [CrossRef]
204. Gerlt, J.A.; Bouvier, J.T.; Davidson, D.B.; Imker, H.J.; Sadkhin, B.; Slater, D.R.; Whalen, K.L. Enzyme Function Initiative-Enzyme Similarity Tool (EFI-EST): A web tool for generating protein sequence similarity networks. *Biochim. Biophys. Acta* **2015**, *1854*, 1019–1037. [CrossRef]

MDPI
St. Alban-Anlage 66
4052 Basel
Switzerland
Tel. +41 61 683 77 34
Fax +41 61 302 89 18
www.mdpi.com

International Journal of Molecular Sciences Editorial Office
E-mail: ijms@mdpi.com
www.mdpi.com/journal/ijms

Ingram Content Group UK Ltd.
Milton Keynes UK
UKHW050809070323
418120UK00004B/197